WILDFLOWERS
of Orange County and the Santa Ana Mountains

Robert L. Allen
Fred M. Roberts, Jr.

© 2013 Laguna Wilderness Press

Text, photographs, and illustrations ©Robert Lee Allen and Fred Marke Roberts, Jr., unless otherwise indicated

ISBN: 978-0-9840007-1-5

Library of Congress Control Number: 2013932890

1. Plants—Wildflowers–California—Orange County—Identification; 2. Plants—Wildflowers—California—Orange County—Pictorial works.

Cover: School bells, *Dichelostemma capitatum* ssp. *capitatum* (Themidaceae), in Crystal Cove State Park (photo Robert Lee Allen)

Design and Layout: Robert Lee Allen and Paul Paiement

Laguna Wilderness Press

P.O. Box 149, Laguna Beach, CA 92652

www.lagunawildernesspress.com

e-mail: orders@lagunawildernesspress.com

Tel: 951-827-1571

Printed in Singapore

Laguna Wilderness Press is a non-profit press dedicated to publishing books concerning the presence, preservation, and importance of wilderness environments.

For Charles Lee Allen
(13 September 1936 – 14 February 2011)
Co-Founder, National Scholastic Surfing Association
Founder, United States Amateur Snowboarding Association
Husband, Father, Grandfather, Organic Gardener, Inspiration
He read through an early draft of this book and looked forward to seeing it in print.

Contents

Acknowledgments	viii
Preface	x
Introduction	1
About Wildflowers	9
Watching Wildflowers	27
About the Entries	35
MAGNOLIIDS	37
EUDICOTS	39
MONOCOTS	413
Where to Go Wildflower-Watching	449
References	462
Glossary	466
General Index	475
Quick Index by Flower Color	495
About the Authors	498

Acknowledgments

Had we realized just how much work this book would be when we started this project in January of 2003, you might not be reading it right now. We are indebted to many professional and amateur botanists, naturalists, rangers, docents, hikers, friends, and family members who provided the encouragement that kept us moving toward our goal. Among them are the following:

Bob's wife, Linnell R. Allen, and son, Charles W. Allen, and Fred's wife, Carol Roberts, for love and patience.

Bob's parents, Charles L. and Chris Allen, and Lonnie and Dana J. Caldwell, and Fred's parents, Fred M. Roberts, Sr., and Mary Jane Roberts, for lifelong encouragement.

Our financial sponsors, the Orange County Chapter of the California Native Plant Society, the Sea and Sage Chapter of the National Audubon Society, the Laguna Canyon Foundation, the Riverside-San Bernardino Chapter of the California Native Plant Society, and Bob and Linnell Allen.

Our scientific technical reviewers, Sarah Jayne, C. Eugene Jones, Celia Kutcher, Marvin S. Sherrill, Dan Songster, and Ron Vanderhoff, who undertook the colossal job of reading every word and suggesting improvements.

Christopher M. Barnhill, Curator of Living Collections, Fullerton Arboretum, who was initially second author of this book and for two years spent numerous hours in the field with us, hunting down and photographing plants. His increased responsibilities at the Arboretum necessitated his withdrawal from the project, but the book is graced with some of his photographs and helpful information.

C. Eugene Jones, Professor of Botany (retired), Department of Biological Science, California State University, Fullerton, for inspiration, encouragement, and review of the entire book.

The late John D. Cooper, Professor of Geology, Department of Geological Science, California State University, Fullerton, for editing our geology section.

Leo C. Song, Jr., Greenhouse Manager (retired), Department of Biological Science, California State University, Fullerton, for field work, friendship, and tips for cultivation.

David J. Keil, Professor of Botany (retired), California Polytechnic State University, San Luis Obispo, for his review of the Asteraceae and for answering many questions over the years.

Steve Boyd, Herbarium Curator (retired), Rancho Santa Ana Botanic Garden, and Andy Sanders, Herbarium Curator, University of California, Riverside, for identification and field work.

Jon Rebman, Curator, San Diego Museum of Natural History, for his review of the Cactaceae. Judy Gibson, also of SDMNH, for specimen data.

Michael G. Simpson, San Diego State University, and Ron Kelley, Eastern Oregon University, for their review of the Boraginaceae.

Our friends in the Orange County Chapter of the California Native Plant Society (OC-CNPS), many of whom who attended our field trips/talks, reviewed portions of the book, and provided gentle encouragement, among them Sarah Jayne, Dan and Elizabeth Songster, Laura Camp, Brad Jenkins, Joan R. Hampton, Celia Kutcher, Rich Schilk, Ron Vanderhoff, Lois Taylor, and Allison Rudalevige.

Trisha Smith, former Director of the Nature Conservancy's Irvine Open Space Reserve, now Senior Ecologist, and her former staff members Joel Robinson and Mike Kahle, for access to the Irvine Ranch when the lands were under their management.

The Irvine Ranch Conservancy, which now oversees the historic Irvine Ranch lands, for kindly allowing us to explore many parts of the land, usually accompanying us, and often even doing the driving. Our hosts included Michael O'Connell, John Graves, David Olson, David Raetz, Megan Lulow, and Jutta Burger. Others who helped us include Evelyn Brown, Jared Considine, Melissa Fowler, Brian Hughes, Lindsay Kircher, and Kelley Reetz.

Laura Cohen and Geordie Shaw of the Richard and Donna O'Neill Conservancy, part of the Reserve at Rancho Mission Viejo, who allowed us to study and photograph on Conservancy lands.

Richard S. "Dick" Newell, Roger Reinke, Don Millar, Bob DeRuff, Len Gardner, Ron Vanderhoff, and Michael Hearst who found wildflowers that we needed and alerted us to their whereabouts.

Thea Gavin, Orange County naturalist and poet, for allowing our use of her poem "My Southern California Neighbors" and an excerpt of her poem "Foothill Fugitives."

Lawrence Harder, University of Calgary, for the use of Heather C. Proctor's illustration of a bumble bee on shooting star.

Theodore W. Palmer, Mt. Pisgah Arboretum in Eugene, Oregon, and his brother, Macdougall Palmer, for allowing us to reprint an excerpt from "The Botanist," a poem written by their father, Ernest Jesse Palmer, and originally published in *Gathered Leaves: Green Gold and Sere* (New York: William-Frederick Press, 1958), p. 43.

David Hollombe, Michael Charters, and J. Mark Sugars, who provided Latin translations and biographical information about the people for whom plants are named and the people who named them.

Karin Klein for reviewing the section, *Favorite Places on Public Lands*.

The Santa Ana Mountains Natural History Association, including Debra Clarke, Dave Taylor, Ed Hill, Tom Maloney, Gary Meredith, Lee Shoemaker, Larry Shaw, and Dick and Jan Sherman, for organizing and leading hikes in the Santa Ana Mountains and other local sites.

G. Victor "Vic" Leipzig and Trude Hurd for reviewing our information about birds.

Pete and Sandy DeSimone for providing access to Audubon Starr Ranch Sanctuary and notifying us of interesting things they found such as graceful tarplant and a new population of Allen's daisy.

Ronald A. Coleman and Robert Lauri for information on *Piperia* orchids, Nancy Morin for a review of the Campanulaceae, and Ronald A. Russo for information on plant galls.

Botanists Tony Bomkamp, David Bramlet, Dylan Hannon, Michelle Balk, Megan Enright, Sandy Leatherman, Diana Menuz, Allan Schoenherr, Tim Thibault, Ron Vanderhoff, and Justin Wood for localities and tips.

Entomologists the late Roy R. Snelling, Arthur V. Evans, James, N. Hogue, Rosser W. Garrison, Jeffrey A. Cole, Ken Osborne, and Greg Ballmer for identifying some of the insects from photographs that appear in this book. Olle Pellmyr, University of Idaho, Moscow, for assistance with yucca and greya moths.

Bart O'Brien, Susan Jett, Joan McGuire, Helen Smisko, and Michael Wall, of Rancho Santa Ana Botanic Garden, for providing live material from the garden, especially helpful during drought years.

The fine folks at Tree of Life Nursery, especially Mike Evans, Jeff Bohn, Laura Camp, Debbie Cressey, Gene Ratcliffe, Patty Roess, and Junior Rodriguez, who also provided live material for study.

From Laguna Wilderness Press, Ron Chilcote, Barbara Metzger, Paul Paiement, and Doug McCulloh, for believing in our work and taking on the project.

The local park rangers who allowed us access to their lands and often kept us informed of what was in flower: for Orange County Parks, John Gannaway, Donna Krucki, John Bovee, Candy Hubert, Vicki Malton, Larry Sweet, Barbara Norton, Tom Maloney, Ron Slimm, Bobbie Tumolo, Kathy Williams, Jennifer Naegele, Laura Cohen, and Chris Taylor; for the state parks, Winter Bonnin and Dave Pryor, and for the Trabuco District, Cleveland National Forest, USDA Forest Service, Debra Clarke, Mary Thomas, Lisa Young, Kirsten Winter, and Linh Davis.

The photographers, Daymond L. Allen, Wayne P. Armstrong, Greg Ballmer, Christopher M. Barnhill, Andrew Borcher, Pat Brennan, Brian V. Brown, Peter J. Bryant, Michael L. Charters, Michelle Cloud-Hughes, Joseph M. DiTomaso, Reggie I. Durant, John Green, Robb Hamilton, Joan R. Hampton, Trude Hurd, Donna Krucki, Robert Lauri, Alison Linden, Steve Matson, Gary S. Meredith, Jason Mintzer, Nancy Morin, Keir Morse, Bruce Perry, Olle Pellmyr, Jon Rebman, Aaron Schusteff, Michael G. Simpson, Trisha Smith, Daniel Songster, David Tharp, Ron Vanderhoff, Genevieve Walden, and Hartmut Wisch, for providing photographs (identified in adjacent credits) in addition to our own.

Ground Pink (*Linanthus dianthiflorus*) at Limestone Canyon, Spring 2008.

Preface

Wildflowers! The very name conjures up vivid scenes of fragile flowers in fierce wild lands, competing for soil, water, sunlight, and room to grow, then flowering, setting seed, and producing the next generation. And they are just that: wild, part of the landscape, the natural world, the bigger picture, *the wilderness.*

But what draws us toward vivid wildflowers and their associates such as nectar-sipping butterflies and watery-eyed deer? **Why do we seek wildlife?**

In his 1984 book, *Biophilia,* Edward O. Wilson says that "to explore and affiliate with life is a deep and complicated process in mental development." Following the thought to its natural conclusion, to deny yourself the thrill of exploring and bonding with other living things is a step toward mental disaster. For reasons still not completely understood, we have a deep-seated psychological *need* to travel down a trail or dirt road; we *have to* be exposed to wilderness and wildlife, and we *must* stop and experience wildflowers.

For some people just introducing themselves to wildlife, wildflowers are a safety net, something pretty to look at, smell, and touch. Getting to know them is a gentle first step toward learning about the natural world. Set yourself a goal of recognizing five of the most common types you find in your area. Attend nature walks and talks given by people who know wildflowers. Ask questions. Learn another five types, then another. Find out how they live and what lives with them. Soon you'll find that you know quite a bit and realize that other people are asking *you* questions.

As Gary Snyder said in *The Practice of the Wild* (1990), "The wild requires that we learn the terrain, nod to all the plants and animals and birds, ford the streams and cross the ridges, and tell a good story when we get back home."

We hope you will take this book along to use in the wild, in school, and at home. Nothing would make us prouder than to see your copy with curled pages and roughed edges, filled with your own notes. As your companion, the book will help you tell your own story of the wild.

Peace,

Bob Allen
Mission Viejo, California

Fred Roberts, Jr.
Oceanside, California

January 2013

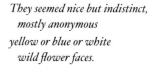

They seemed nice but indistinct,
 mostly anonymous
yellow or blue or white
 wild flower faces.

When I'd pass them on the trail
 they were easy to overlook;
since we hadn't been introduced
 I avoided eye contact.

Now that I've learned their names
 it's harder to just rush by
and pretend I don't see them
 or recognize them today.

How are you Mariposa Lily,
 Wild Hyacinth, Golden Yarrow?
Nice to see you Yerba Santa,
 Red Paintbrush, Fiesta Flower.

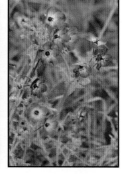

– Thea Gavin (1959-)
"My Southern California Neighbors"

Introduction

This book is a guide to the wildflowers of coastal southern California – specifically, the wildflowers found within the boundaries of Orange County and the Santa Ana Mountains, plus adjacent areas of the surrounding counties of Los Angeles, San Bernardino, Riverside, and San Diego. Because of their proximity to Orange County and their similar floras, we include wildflowers of the Whittier, Puente, and Chino Hills, Temescal Valley, Elsinore Basin, Santa Rosa Plateau, San Mateo Canyon Wilderness Area, and San Onofre State Beach. Most of the wildflowers discussed are found throughout southern California and northern Baja California Norte, so this book will also be useful in many areas outside of its title.

Southern California is geologically complex. What appears to be a single unit of solid earth is actually composed of several jumbled landmasses brought together through ongoing processes of tectonic plate movement, volcanism, uplifting, sinking, faulting, massive landslides, erosion, and sedimentation. As a result, California is a collection of very different landscapes and very different plant communities.

By their very presence, plants create habitat for all other living things. As our high school biology teacher Marvin S. Sherrill explained, "If you want to find insects, lizards, birds, and mammals, you must learn the plants." Plants provide food and shelter, cast shade, block wind, break rock, hold the ground in place, and change the physical nature of soil by adding organic matter to it. Where plants can live is determined by the physical environment: rocks, soils, topography, water, and climate. Consequently, in order to understand plant distributions and locate wildflowers, you need an understanding of the physical factors that control where they can grow.

Geology and Geography

Orange County occupies 800 square miles and forms the southern part of the Los Angeles Basin. It lies between Los Angeles County to the north-northwest and San Diego County to the southeast. The Pacific Ocean forms its southwest border. To the north it is bounded by the Chino Hills, to the east by the Santa Ana Mountains and Riverside County. Elevations range from sea level at the coast, through plains, valleys, foothills, up to 5,687 feet (1,733 meters) atop Santiago Peak in the Santa Ana Mountains.

The county's major ranges and hill systems are the Santa Ana Mountains, the Elsinore Mountains, the Peralta Hills, the El Modena Hills, the Whittier, Puente, and Chino Hills, the Coyote Hills, the Lomas de Santiago, and the San Joaquin Hills. The Los Angeles Basin forms a broad flat plain in the middle of the area. The largest canyons toward the coast include Santa Ana, Santiago, Silverado, Trabuco, Laguna, Aliso, San Juan, and San Mateo. Inland canyons are Temescal, Bedford, Indian, and Coldwater.

The exposed land mass of the county is geologically young; its oldest exposed rocks are only about 170 million years old, mere children when compared with the exposed igneous and metamorphic rocks of the San Bernardino Mountains at 1700 million (1.7 billion) and of some parts of Arizona at 1800 million (1.8 billion) years old. The major landforms of the county are the Los Angeles Basin, the San Joaquin Hills, the Chino-Puente-Whittier Hills, and the Santa Ana Mountains (see Roberts 2008 for details).

Los Angeles Basin. This is the relatively flat plain occupied by much of southern Los Angeles and Orange Counties. The portion from the Santa Ana River west to the base of the Santa Monica Mountains in Los Angeles County is known as the Downey Plain and the portion east of the Santa Ana River as the Tustin Plain. The rock layers here are primarily fossil-bearing marine sediments, uplifted and overlain with sediments from dry-land sources. Today this material is visible at the surface as sandy, gravelly soils, with occasional pockets of clay.

Much of the plant life of the Los Angeles Basin has been removed from its plains by human intrusion, including farming, land development, roadways, and house construction. Presumably the uplands were mostly coastal sage scrub and grasslands interspersed with oak woodlands. The lowlands were primarily marshes, other wetlands, wildflower fields, and vernal pools associated with the broad path of the Santa Ana River, which has since been channelized. Few of these lowland habitats exist today. Waterways were vegetated by riparian woodlands.

Los Angeles Basin. Orange in foreground, Palos Verdes Peninsula in distant background.

Chino-Puente-Whittier Hills. These hills extend from Yorba Linda northeast to Whittier and straddle the borders of Orange, Los Angeles, and San Bernardino Counties. Geologically they are a single unit and their boundaries political rather than geographical. For brevity we often simply call them the Chino Hills or hyphenate the portions of the hills that contain the plant being described (e.g., Whittier-Puente Hills). Major peaks include Gilman Peak (1685 feet, 514 meters), San Juan Hill (1781 feet, 543 meters), and Sculley Hill (about 893 feet, 272 meters). Major

Chino Hills from Diamond Bar. Modjeska and Santiago Peaks at distant center-right.

canyons include Turnbull, Brea, Tonner, Carbon, Soquel, Telegraph, and Aliso.

The soils of these hills include loose sandstones, cobbles, and clay. Prevalent plants are members of the coastal sage scrub, chaparral, oak woodland, and walnut woodland communities. Of particular importance here is southern California black walnut. These and the San Jose Hills are the largest remaining native stands of them and they are seriously imperiled by development. The walnut woodland community often includes coast live oak and blue elderberry with an understory of coastal sage scrub, chaparral, and oak woodland species. The plants and animals of these hills are understudied and largely undocumented.

San Joaquin Hills. This range parallels the coastline of Newport Bay south through Laguna Beach and Laguna Niguel and then grades into the foothills of Dana Point. Its highest peaks are Signal Peak (1164 feet, 355 meters), Temple Hill (1036 feet, 316 meters), and Niguel Hill (936 feet, 285 meters). Locally significant prominences are the Sycamore Hills (624 feet, 190 meters) between Laguna Canyon and El Toro Roads and the Sheep Hills (665 feet, 203 meters) between Wood and Aliso Canyons. Some of the canyons that drain the San Joaquin Hills are Coyote, Bommer, Shady, Los Trancos, Moro, Emerald, Bluebird, Hobo, and Wood. Major canyons that completely bisect these hills are Laguna, Aliso, and Sulphur Creek. These last three watercourses also drain the Santa Ana Mountains. As the San Joaquin Hills uplifted, they cut their channels through the hills.

Soils in the San Joaquin Hills are primarily unconsolidated sand with cobbles and larger rocks. Plant life in the hills includes coastal sage scrub, grasslands, chaparral, oak woodlands, and riparian woodlands. Some of their more spectacular plants are bushrue, bigpod lilac, summer holly, canyon sunflower, and Laguna Beach live-forever.

Lomas de Santiago. This range of hills lies between Santiago Canyon and Irvine. It is made up of many types of sandstones, mudstones, and conglomerates. Along its southwest-facing slope is a sharp ridge known as Loma Ridge, with elevations that range from 1000 to 1775 feet (305 to 541 meters). Its major canyons include Limestone, B-flat, Rattlesnake, Hicks, Bee, Round, Agua Chinon, Borrego, and Serrano. At the head of Agua Chinon Canyon lies stunning erosional features called the Sinks, nicknamed "Orange County's Grand Canyon."

Santa Ana Mountains. The county's prominent physical feature, this is a steep-sided, inland range of moderately high elevation that parallels the coast. It extends from Santa Ana Canyon southeast to El Cariso along the Ortega Highway and continues southeast to the San Mateo Wilderness Area and Alamos Canyon. From there, the Santa Margarita Mountains pick up and extend into the Marine Corps base, Camp Joseph H. Pendleton. The branch of the Santa Ana Range above Lake Elsinore, from the Ortega Highway to Wildomar and the Santa Rosa Plateau, is commonly known as the Elsinore Mountains. Foothills of the southeastern portion of the range extend south to Mission Viejo, Rancho Santa Margarita, and the area surrounding Caspers Wilderness Park, and then southeast toward San Mateo Canyon.

The lands of southern California were once part of a rocky sea floor that became overlain with sediments eroded from adjacent island landmasses and deposited on the sea floor. Heat and pressure generated by the accumulation of overlying sediments consolidated the loose material into solid sedimentary rock. Additional heat and pressure from even more overlying sediments changed the sedimentary rock physically and chemically. These rocks were later uplifted, often by fault activity. Today these lands include sedimentary, igneous, and metamorphic rocks, plus soils weathered from them, with great regional variation in their concentrations of sand, silt, and clay.

From northwest to southeast, the major peaks of the Santa Ana Mountains are Sierra (3045 feet, 928 meters), Pleasants (4007 feet, 1221 meters), Bedford (3800 feet, 1158 meters), Bald (3947 feet, 1203 meters), Modjeska (5496 feet, 1675 meters), Santiago (5687 feet, 1733 meters), Trabuco (4604 feet, 1403 meters), Los Piños (4514 feet, 1376 meters), Sugarloaf (3227 feet, 984 meters), and Sitton (3273 feet, 998 meters). In the Elsinore Mountain region, Elsinore Peak stands at (3575 feet, 1090 meters).

Major canyons that drain into Santa Ana Canyon are Coal, Gypsum, North Weir, Walnut, and Santiago. The Santa Ana River has meandered around the lowlands and at various times drained into Anaheim Bay in Seal Beach, through Huntington Beach, and into Newport Bay. In 1920 the Bitter Point Dam was built to prevent the river from entering Newport Bay, forever changing the character of the upper bay. It now empties into the Pacific Ocean through the Talbert Channel between Huntington Beach and Newport Beach.

San Joaquin Hills from Niguel Hill. Aliso Canyon in lower center. Temple Hill in middle-right.

Santiago Canyon. Santa Ana Mountains at far left. Lomas de Santiago at far right.

GEOLOGY • GEOGRAPHY • CLIMATE

The head of Santiago Canyon is at the intersection of the Joplin Trail with Main Divide Road, in the saddle between Santiago and Modjeska Peaks. Canyons that drain into Santiago Creek include Modjeska, Harding, Williams, Silverado, Ladd, Baker, Black Star, Fremont, Blind, and South Weir. Santiago Creek drains into the Santa Ana River just west of Main Place Mall in Santa Ana. Trabuco Creek begins high up in Trabuco Canyon, just below Trabuco Peak. Its tributaries come from Holy Jim, Falls, and Tijeras Canyons. In San Juan Capistrano, Oso and Horno Creeks drain into Trabuco Creek. San Juan Canyon begins high in the Santa Ana Mountains and drains Long, Decker, Bear, Lion, Hot Spring, Cold Spring, Lucas, Bell, Verdugo, Trampas, Chiquita, and Gobernadora Canyons. Once in San Juan Capistrano, Trabuco and San Juan Creeks merge and empty into the Pacific Ocean at Dana Point.

The inland north-facing slope of the Santa Ana Mountains is drained by major canyons such as Tin Mine, Hagador, Main Street, Eagle, Bedford, Coldwater, Mayhew, Indian, and Horsethief. Most of these currently or historically drained into Temescal Wash, a waterway in the Temescal Valley that is more or less paralleled by the Interstate 15 Freeway. Major canyons that drain directly into Lake Elsinore include Rice, McVicker, and Leach. To the southeast, other canyons such as Slaughterhouse drain into Murrieta Creek.

Plant life in the Santa Ana Mountains includes coastal sage scrub, grasslands, chaparral, oak woodlands, riparian woodlands, southern mixed evergreen forest, and coniferous forests. Large trees are primarily California bay laurel, coast live oak, canyon live oak, California sycamore, big-leaf maple, white alder, big-cone Douglas-fir, pines, and tecate cypress. Flowering shrubs include sumacs, goldenbushes, barberries, Pacific madrone, manzanitas, chaparral pea, currants, San Miguel savory, Parish's bluecurls, sages, bush poppies, monkeyflowers, beardtongues, Fish's milkwort, buckwheats, California lilacs, coffeeberries, redberries, chamise, toyon, southern California storax, desert wild grape, chaparral yucca, and chaparral beargrass. Wildflowers include rayless arnica, mountain dandelions, popcorn flowers, phacelias, mustards, campions, dodders, live-forevers, lupines, Parry's deer's ears, evening-primroses, California fuchsia, paintbrushes, broomrapes, poppies, snapdragons, spineflowers, buckwheats, larkspurs, violets, soap plants, onions, lilies, and orchids. Endemic or near-endemic species are Santiago Peak phacelia, Hammitt's clay-cress, goldenrod broomrape, heart-leaved pitcher-sage, intermediate thick-leaved monardella, Hall's large-flowered monardella, and Munz's onion.

Climate

Southern California enjoys a Mediterranean climate with hot dry summers and mild winters. Coastal areas such as San Clemente, Laguna Beach, and Huntington Beach often begin the day covered with a layer of low clouds and fog. The immediate coast also enjoys moderated temperatures, spared from the temperature extremes of inland areas. Most rainfall arrives in the winter months of January and February, but in good years it continues through early spring. December rain is quite variable, sometimes more than January-February, sometimes less. The month of June is known for a dense cover of low clouds and fog that often persists until noon. Summer monsoons in late July or August often spill over from Arizona and Mexico, bringing high humidity and stunning thunderheads above the mountains; sometimes they even provide a light dose of rain.

Freezing temperatures are uncommon in the county's cities, except when cold Arctic winds blast through. Snow, too, is rare here at all but the highest elevations. In most years, the higher peaks receive at least a fine layer of snow, especially on their north-facing summits. In some exceptional years, such as in late November 2004, snow blankets the Santa Ana Mountains from its peaks down to an elevation of about 3000 feet! In those storms, places such as Silverado Canyon, Acjachemen Meadow, Los Pinos Potrero, and El Cariso are covered in snow drifts that rival those of southern California's larger mountains.

Santa Ana Mountains. Pleasants Peak at distant far left, San Gabriel Mountains at very distant far right.

Temescal Valley and Santa Ana Mountains. Santiago Peak at left, Modjeska Peak at right.

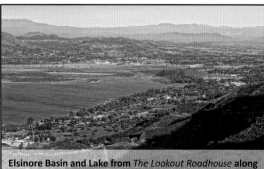

Elsinore Basin and Lake from *The Lookout Roadhouse* along Ortega Highway in the Santa Ana Mountains.

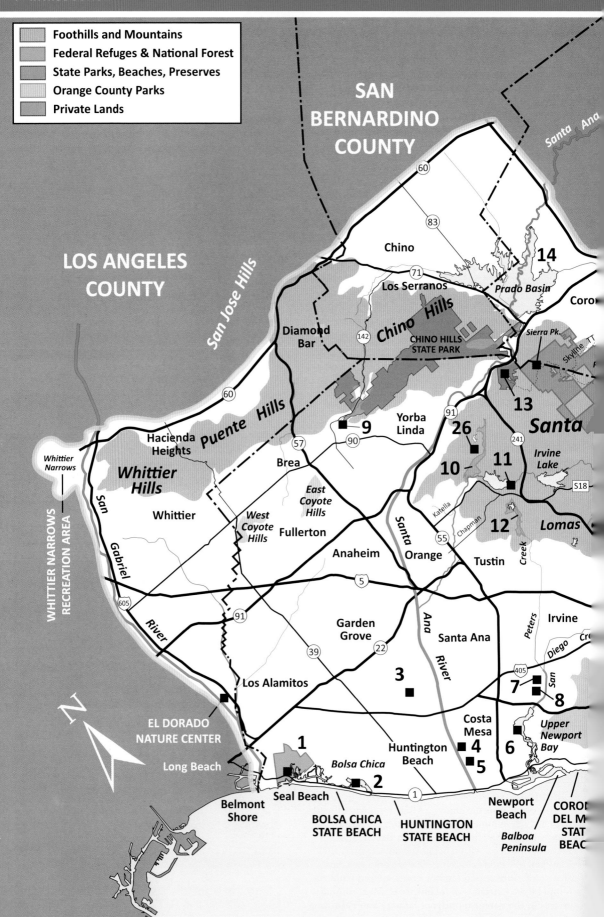

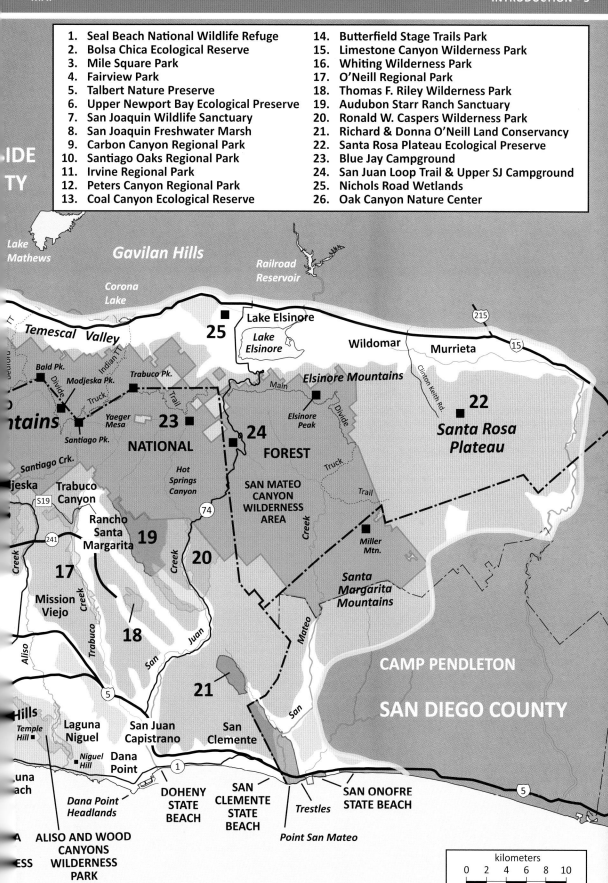

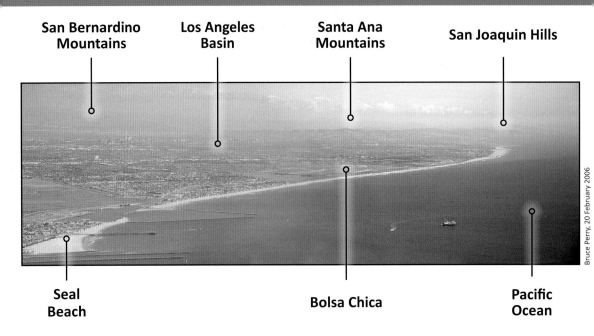

Above, coastal Orange County from off the coast of Belmont Shore, looking southeast. At lower left is the outfall of the San Gabriel River. Following the coastline, near upper right is the broad sandy beach of Huntington Beach State Park. In the distance at far left are the San Bernardino Mountains. Near center and much closer to the coast are the Santa Ana Mountains. At upper right are the San Joaquin Hills above Laguna Beach.

Below, southern-central Orange County from above Crystal Cove State Park, looking north-northeast. At far left is the newly constructed Seawatch road with new home construction evident. Across the Tustin Plain and out of view behind the Lomas de Santiago is Santiago Canyon. "Old Saddleback" is an outline made up of Modjeska and Santiago Peaks.

AERIAL PHOTOGRAPHS

Above, coastal Orange County off the coast of Monarch Beach, looking east. At lower left is Salt Creek Beach Park. Near center, behind the Dana Point Headlands, is the Dana Point marina.

Below, southeastern Orange County from above Caspers Wilderness Park, looking north. The Santa Ana Mountains dominate the view from left to right.

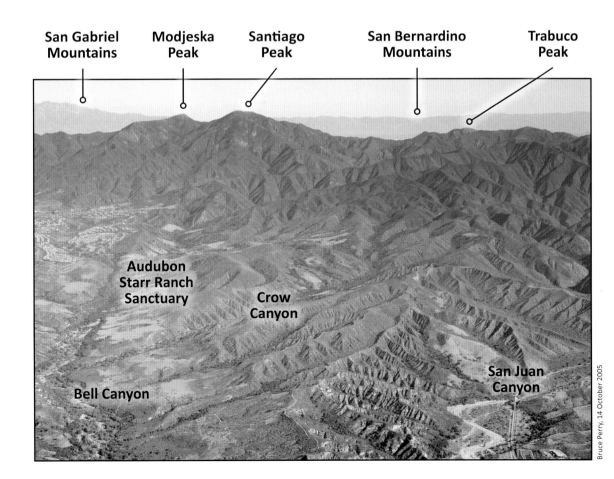

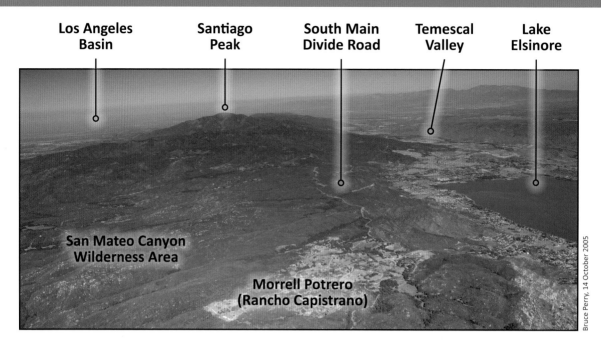

Above, the Santa Ana Mountains from above Elsinore Peak, looking northwest.

Below, the Santa Ana Mountains from above San Juan Canyon, looking south. The bulk of the view is San Mateo Canyon Wilderness Area; beyond it are Camp Pendleton and the Pacific Ocean. At lower right, the Ortega Highway (State Route 74) follows San Juan Canyon through the mountains.

About Wildflowers

The flower is the poetry of reproduction. It is an example of the eternal seductiveness of life.
— Jean Giraudoux (1882-1944)

In this book, we use the term "wildflower" to refer to any flowering plant. Flowers are more than pretty objects produced by plants; they are essential for plant reproduction. They have definite structures that work together toward this purpose.

Formal scientific classification of plants is based on the structure of flowers, fruits, seeds, and leaves, plus their growth form and genetics. The process of identification requires understanding those structures and growth forms and recognizing them on the plants you study. Because there are many parts to a flower and some look alike, it is easy to get confused. *When locating the parts, it is best to start your examination below the flower, working from base to tip and from outside to inside.* As each part is discussed, locate it on the diagram and trace it with your finger or shade it with a colored pencil before you continue reading about it. Better yet, draw your own flower and label it. This tactile approach will help you learn the parts.

Parts of a Flower

Most flowers are attached to the plant by some sort of **stalk**. When a flower has no stalk and is directly attached to a stem (e.g., as in cacti), we say the flower is **sessile**. The area at the top of the stalk, to which the other flower parts are attached, is the **receptacle**.

Next come two parts that protect the innermost parts and often attract pollinators. These leaflike structures, usually green and relatively small, are the **sepals**. They form a protective enclosure around the rest of the flower while it is in bud. When we refer to all of the sepals as a group, we call it a **calyx**. The calyx may be formed of completely separate sepals or the sepals may be fused together into some sort of a tube. If fused, the tips of the sepals often remain free, and we call each free tip a **calyx lobe**.

Moving inward are the **petals**, collectively called a **corolla**. Like the calyx, the corolla may be composed of separate or fused petals, and if fused, the free tips are called **corolla lobes**. Petals may be of single or multiple colors, and their shapes vary widely. In some cases, the calyx and corolla lobes are similar and you cannot easily determine their identity. In this situation, those lobes are called **tepals**. Collectively, the calyx and corolla (or tepals) form the **perianth** (Greek, peri = around; anthos = flower). Sometime after pollination, the perianth usually withers and falls off, exposing the developing **fruit**.

Reproductive organs are toward the center of the flower. First are the **male** parts, the **stamens**. Each stamen has a stalk, the **filament**, topped with an **anther** that holds and dispenses the grains of **pollen** each containing two male sex cells called **sperm**. The filament base is attached to the receptacle or sometimes to a petal. Anthers and their pollen are usually yellow, but may be of other colors such as white, red, orange, blue, or purple. Rarely, anthers are differently colored within a species, on the same individual, and even in the same flower. Sometimes anthers are empty or absent, leaving only the filament; this sterile type of stamen is called a **staminode**. The number of stamens and their attachment and color are often consistent among closely related species, similar within a plant family, and are *very* important to identification.

In the center of the flower is the **female** structure, called the **pistil**. At its base is the **ovary**, which is attached to the **receptacle**. Each chamber within an ovary is called a **locule**. Within the ovary are the **ovules**; each ovule contains an egg that, when fertilized, develops into a **seed**. A stalk called the **style** extends upward from the ovary. In most local flowers, there is only one style but in others there are more; it may even be divided into two or more **lobes**. The style is topped with a **stigma**, the structure that receives pollen. The stigma may be sticky-tipped or dry and is shaped like a thread, a cup, or a ball, sometimes divided into multiple parts.

All flowers are defined by their parts. By studying the parts of a flower, you can determine to what family the plant belongs. Note the position, form, and quantity of each part. How is this bush poppy similar to, yet different from, the illustration? Seeing those similarities and differences is a crucial step in your wildflower-watching experience.

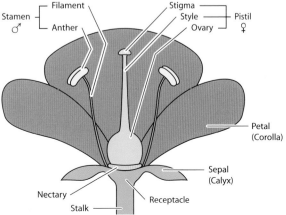

The position of the ovary is variable with respect to attachment of the perianth to the receptacle. If the ovary sits *above* the perianth attachment, it is called a **superior ovary**; if *below*, it is an **inferior ovary**. Plant families with superior ovaries include dogbanes and milkweeds (Apocynaceae), stonecrops (Crassulaceae), peas (Fabaceae), poppies (Papaveraceae), buckwheats (Polygonaceae), buttercups (Ranunculaceae), and violets (Violaceae). Those with inferior ovaries include sunflowers (Asteraceae), cactus (Cactaceae), honeysuckles (Caprifoliaceae), gourds (Cucurbitaceae), gooseberries (Grossulariaceae), evening-primroses (Onagraceae), and orchids (Orchidaceae).

Many flowers have a nectar-producing gland called a **nectary**. Most often, it resides near or under the ovary, but some species have it on their ovary, style, petals, sepals, stems, or leaves. Nectar contains water and sugars, often also vitamins and oils.

Flower, Ovary, and Fruit Formation

Flowering plants are characterized by a **flower**, a reproductive organ that has a pistil and/or stamens, often surrounded by perianth parts. Some flowers possess a pistil and lack stamens; they are all-female flowers. Others possess stamens but lack a pistil; they are all-male flowers. Flowers that possess at least one pistil and one stamen are said to be **perfect** or **bisexual**. Flowers of only one sex are said to be **imperfect** or **unisexual**.

In the ancestors of early flowering plants (Angiosperms), the ovules were attached to the edges of some leaves and modifications to those leaves resulted in their edges' curling and fusing into a cylinder that enclosed the ovules, forming the **pistil**. Further modifications to the pistil produced the **style** and **stigma** from its tip. A **simple pistil** is the product of a single modified leaf, evidenced by its single style. The ovary of a simple pistil is made up of a single unit called a **carpel**. The fruit produced by each flower consists of a single unit. Many plants produce simple pistils, such as the laurel and pea families. A pea pod is a good example of a fruit produced from a simple pistil.

The flowers of some plants have two or more simple pistils that remain distinct (not fused to each other). The fruit produced by each flower consists of multiple individual units. When the fruit is ripe, each unit falls off or releases seeds separately. This occurs in plants such as buttercups, larkspurs, and live-forevers.

When two or more simple pistils in the same flower fuse together, they form a **compound pistil**. The ovary of a compound pistil is made up of **multiple carpels**. The fruit produced by each flower consists of a single unit with one or more internal chambers. The vast majority of flowering plants, including all members of the cucumber, pink, and lily families, have this condition. In a compound pistil, the style tip is often lobed or branched. Externally, a compound pistil can be identified by features such as multiple styles, stigmas, and ovary lobes. The number of style lobes or branches often indicates the number of simple pistils that fused together to form the compound pistil. Internally, compound pistils generally have multiple chambers.

In a citrus fruit (such as an orange, tangerine, or grapefruit), each crescent-shaped inner section that you can remove from it is a carpel; its chambers are filled with seeds and succulent tissue. In the fruit of a bell pepper there are two carpels, each with a single locule, although the locules often form a single air-filled cavity. The number of locules does not necessarily equal the number of carpels.

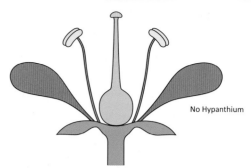

Superior Ovary: *Ovary is attached above the perianth (calyx and corolla)*

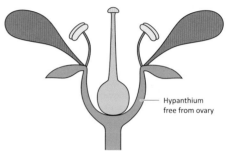

Superior Ovary: *Ovary is attached above the perianth (calyx and corolla), but the perianth itself is attached to an elongated receptacle, the hypanthium.*

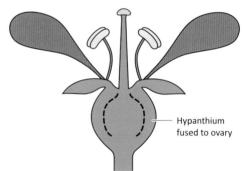

Inferior Ovary: *Ovary is attached below the perianth (calyx and corolla). Historically, there was a hypanthium but it has been fused to the ovary wall and is rarely detectable.*

Cross-section of a Simple Pistil: *one or more ovules enclosed by an ovary wall. Plants such as peas have only one simple pistil per flower. Plants such as buttercups, larkspurs, and live-forevers have a collection of simple pistils per flower.*

Cross-section of a Compound Pistil: *one or more simple pistils fused together, as found in most flowering plants.*

Arrangement of Flower Parts

Flowers are arranged on plants in recognizable patterns that are often useful for identification. Some flowers are borne individually on their own stalks or attached directly to the stem. The stalk of an individual flower is technically called a **peduncle**. Plants with individual flowers include southern bluecup, cacti, and violets.

Frequently, multiple flowers are arranged in a cluster called an **inflorescence**. The stalk of an inflorescence below the lowest flower is called a **peduncle**. The stalk of an individual flower within an inflorescence is called a **pedicel**. The leaf-like structure usually found just beneath each peduncle or pedicel is called a **bract**. There are numerous types of inflorescences, named for their overall outlines and branching patterns. Instead of using their technical names in this book, we refer to them with simple descriptive terms such as tight clusters (goldenrods), open clusters (live-forevers), drooping clusters (manzanitas), flat-topped clusters (elderberry and carrot families), and coiled (borage family). In the sunflower family (Asteraceae), individual flowers are arranged into complex groups called **composite heads**, often simply called **flowerheads**.

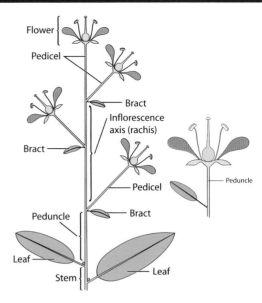

Left: Flowers in an inflorescence
Right: Individual flower not in an inflorescence

Leaves

Leaves are attached to the stem at a **leaf node** or simply a **node**. The stalk of a leaf is called a **petiole**. Leaves that have no petiole and are directly attached to the leaf node are said to be **sessile**. In Eudicots, an **axillary bud** is nestled in the upper angle of the petiole-node junction (**axil**). From it grows a new stem, a thorn, or an inflorescence. On the side(s) of the petiole-node junction there may be one or two appendages (**stipules**) that look like tiny leaves, scales, or spines.

The section of the leaf stalk that continues through the center of the leaf is the **midrib**. The expanded part of the leaf is called the **blade**. The area of the blade closest to the petiole is called the **base** and the area of the blade farthest from the petiole the **tip** or **apex**.

Leaves may be simple and undivided (**simple leaves**) or divided into smaller distinct sections called leaflets (**compound leaves**). Simple leaves may be smooth-edged, toothed, or with expanded pieces called **lobes** (**lobed leaves**). If the lobes arise from either side of the midrib, they are said to be **pinnately lobed**, as they are in many phacelias and warrior's plume. If the lobes are arranged like the palm of your hand, they are said to be **palmately lobed**, as they are in maples, sycamores, wild cucumber, castor bean, and larkspurs.

Compound leaves with leaflets that arise from either side of the midrib are said to be **pinnate**, **pinnately compound**, or **pinnately divided**. A pinnate leaf that has one set of (primary) leaflets is said to be 1-pinnate. A pinnate leaf where each primary leaflet is divided into secondary leaflets is said to be 2-pinnate. Blue elderberry, purple sanicle, California aralia, barberries, and locoweeds all have 1-pinnate leaves. Parish's tauschia, common yellow carpet, and tansy phacelia have 2-pinnate leaves. Plants with leaves that vary in their number of divisions are hyphenated, such as 1-2-pinnate (purple sanicle) and 1-4-pinnate (California poppy.)

If the leaflets are arranged like the palm of a hand, they are said to be **palmate** or **palmately compound**. Scarlet larkspur, California buttercup, and Shelton's violet have palmately compound leaves.

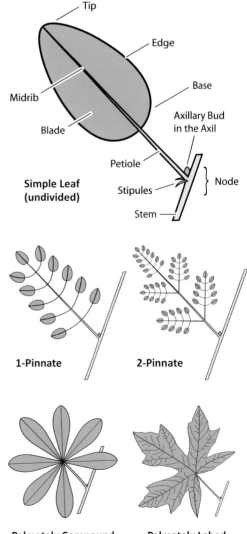

Sometimes it is difficult to tell whether a lobed leaf is simple or compound. The key is to look for the axillary bud, which appears at the base of a leaf and never at the base of an individual leaflet.

The arrangement of leaves on the stem is also important for identification. The condition in which only one leaf is attached at each leaf node is said to be **alternate**. When two leaves are attached at each leaf node, usually on opposite sides of the stem, they are **opposite**. When there are three or more leaves attached to a single leaf node, they are said to be **whorled**. Leaves that arise from the base of the plant are called **basal**. Basal leaves may grow on or near the ground, or a bit above it. When basal leaves are arranged in a very tight cluster, the cluster is said to be a **basal rosette**. Leaves along the stem are called **stem leaves** or **cauline leaves**. In some plants, the basal and stem leaves are identical in shape but smaller up the stem. In other plants, basal and stem leaves are very different in both size and shape.

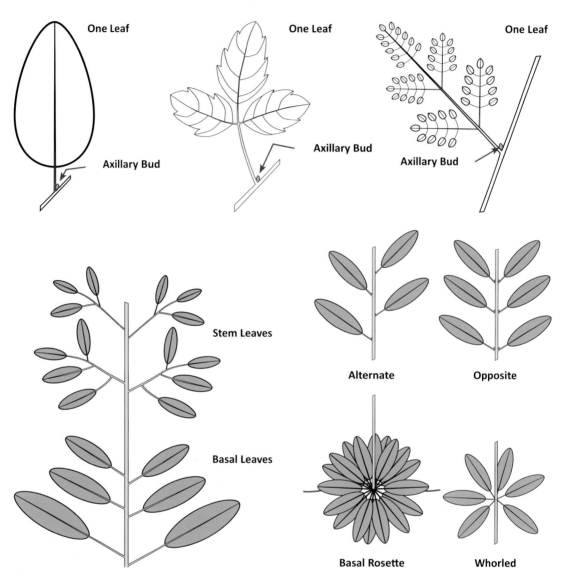

Tips for identification

- Take inventory: Begin at the bottom of the flower and work upward and inward
- Count sepals, petals, stamens, ovary, style, and stigma
- Note shape, structure, and color of all parts
- Calyx and corolla lobes: completely or partly fused or free?
- Ovary position: superior or inferior?
- Flowers: in an inflorescence or solitary?
- Leaves: simple, compound, lobed, divided, palmate?
- Leaf arrangement: alternate, opposite, whorled, basal?

Plant Origins and C[lassification]

The living things we ca[ll plant]s includes **green algae**, most spec[ies of which are] aquatic, and **land plants**, most of w[hich are on lan]d. The earliest land plants colonized th[e land during t]he Silurian Period, about 400 mill[ion years ago. I]mportant features that allowed t[hem to leave th]e water and invade the land includ[ed the creation] of an embryo and the addition of a [new phase] to their life cycle (called alternation of [generations]).

Various other chan[ges in struc]ture and development of earl[y plants le]d to the evolution of **Bryophytes**: liv[erworts, hornw]orts, and true mosses. These plant[s absor]b food and water through their bodie[s like a] kitchen sponge. Since the process o[f absorp]tion can only carry materials a relative[ly short distanc]e, they are limited in size and comple[xity.]

Other early land [plants evolv]ed to develop innovative means [of transport. Im]portant among them was the dev[elopment of a s]ystem to transport food and water t[hrough bo]dies similar to our circulatory syster[n. This system] involves two types of tissues, xylem an[d phloem, which] usually form tubular structures. Xyle[m carries water] from the roots upward to the rest of the plant. Phloem carries food, most often from the photosynthetic leaves, downward to the rest of the plant. Plants that possess these tissues are called **vascular plants**.

Like their ancestors, primitive vascular plants reproduce by means of spores. Among these are **Lycophytes** (quillworts and spike-mosses) and **ferns,** among them **Equisetophytes** (horsetails), **Leptosporangiate ferns** (chain, bracken, polypody, maidenhair, lip, coffee, and goldenback ferns), and **Ophioglossoid ferns** (adder's-tongue ferns).

Other vascular plants developed more complicated reproductive structures: **seeds**. A seed is an embryo surrounded by stored food (endosperm) and wrapped in a seed coat. Seeds develop from fertilized plant "eggs" (ovules), part of the female reproductive structure. An embryo within a seed is protected from physical damage and drying out, can be transported to new locations, allows the plant to remain dormant for a while (which is helpful when rainfall is less than desirable for plant growth), and provides food when the young plant escapes the seed coat and begins its initial growth (germination).

Some forms of early seed plants had simple leaves (often needle-shaped or awl-shaped) and their seeds held in **cones**, structures made of modified leaves and scales. Many of these plant forms still exist. Their male and female reproductive cells are formed in separate cones. Male cones produce and release pollen to the wind. Female cones produce and retain ovules. When pollen lands on an ovule-bearing scale of a female cone, it fertilizes an ovule and the female cone grows in size, often for a year or more. The seed develops between adjacent cone scales but is not encased in tissue. If you pry apart cone scales, you'll see that their seeds are somewhat exposed to the elements and easily fall out. Because of this, these are called naked-seeded plants or **Gymnosperms**. In our area, Gymnosperms are represented by cone-bearing plants such as cypresses, junipers, pines, and firs.

Later forms of seed plants developed more complex ovules and specialized tissues (carpels) that encase and protect the ovules. After fertilization of the ovule, a carpel(s) plus its developing seed(s) is called a fruit. They also developed **true flowers** with sepals, petals, and pollen in anthers, often mounted on a filament. Plants with protected seeds, true flowers, and fruit are called flowering plants or **Angiosperms**. Over 95% of extant land plants are Angiosperms.

The four major groups of Angiosperms found in our area are the **Magnoliids** (California bay laurel and yerba mansa), the **Ceratophyllales** (aquatic hornworts), the **Eudicots** (such as poppies, peas, snapdragons, roses, mistletoes, and grapes), and the **Monocots** (such as onions, lilies, orchids, and grasses). In the past, Magnoliids, Ceratophyllales, and Eudicots were combined into an artificial group called the Dicots; this grouping is invalid and no longer used. The Eudicots represent 75% of all Angiosperms while the Monocots make up 22%. The largest Eudicot family in the world is the Asteraceae (sunflower family) with about 23,000 species, while the largest Monocot family in the world is the Orchidaceae (orchid family), with about 25,000 species.

This book presents local Angiosperms in phylogenetic (evolutionary) order by major lineage: Magnoliids, Eudicots, and Monocots. Within those groups, plant families are listed alphabetically and species most often alphabetically.

Simplified cladogram of land plants. After Rebman and Simpson (2006), Simpson (2006), and Baldwin et al. (2012).

Common and Scientific Names

In this book, we present common and scientific names for every species we discuss. A **common name** is a non-technical name such as "chaparral currant" or "sapphire woolly-star." Most plants have multiple common names. Instead of listing every one of them, we selected the most commonly used and, admittedly, often used our favorite ones. A plant's common name may describe its appearance, its habitat, or even its behavior. Common names often have historical significance or suggest a human cultural relationship with the plant, such as its use in medicine or cooking. For example, "soap plant" is a common name historically used for three different plants: *Agave americana*, *Hesperoyucca whipplei*, and *Chlorogalum pomeridianum*, each of which has a root or a bulb that can be used as soap.

A pitfall of common names is that they may lead to incorrect or even dangerous conclusions. Poison oak, for example, is not an oak at all but a sumac. You certainly would not want to climb a poison oak tree, decorate your house with its foliage, or eat its fruits. To biologists, common names do not provide important information about plant relationships; we cannot tell from a common name to what other plants it is related. For example, black sage and purple sage are true sages, members of the mint family. But California sagebrush is a type of sunflower; the name "sage*brush*" is a good clue that is not a true *sage*.

Common names that include the word "weed," as in coastweed, tarweed, or deerweed, cause biologists particular frustration because they imply to non-scientists that the plants so named are garden weeds and must be eliminated. This is especially problematic when a rare plant whose common name includes "weed" is in need of legal protection, as with some of the rare species of tarweeds. It is difficult to convince public agencies, landowners, and the general public that something with "weed" in its name should be preserved. To be clear, "weed" is a subjective term; *a weed is a plant that grows in a place where a human does not want it to grow.* In this book, we have minimized the use of common names that include the word "weed."

A **scientific name** is the technical name for a life form, such as *Ribes malvaceum*. The first word of a scientific name is called the **genus** or generic name. Plants that share a genus name are closely related and are physically similar. In this way, scientific names convey information.

The second word of a scientific name is the **specific epithet** (sometimes shortened to "specific" or "epithet"). It usually describes the plant's physical attributes (color, shape, size) or where it lives, but sometimes it is named in honor of a person or a place.

Also part of a scientific name is the last name of the person who published it, called the **authority** or **author**. For people with particularly common names (such as Jones, Smith, Greene), we use their first initials as well (as in C.E. Jones, J.D. Smith, E.L. Greene). In the case of the Swedish botanist Carolus Linnaeus, he gave names to so many living things that we often use just the first letter of his last name: "L."

The **year** in which the name was published in a scientific paper is an important part of the scientific record but is usually not included as part of the scientific name.

The combination of genus and specific epithet (and, technically, the authority), is called a **species name**. In this book, we provide the common and scientific names, the name of the author, the year of publication, and the meaning behind the name, for example "Chaparral currant, *Ribes malvaceum* Smith. The epithet was given in 1815 by J.E. Smith, for the similiarity between its leaves and those of mallow (Latin, malva = mallow; -ceu = like)."

Scientific names are generally stable but may be changed for technical reasons. Let's suppose that two plant species were named and described. Later on, someone studied those plants and concluded that they were in fact the same species of plant. Since a living thing can have only one valid scientific name, we can only use the first name that was published and must relegate the other to synonymy (this is the Law of Priority). In 1842, Thomas Nuttall described California sycamore as *Platanus racemosa*, from specimens he collected in Santa Barbara. In 1844, George Bentham described *Platanus californica* from specimens collected in San Francisco. Since *Platanus racemosa* was published first, that is the correct name for the plant. *Platanus californica* is no longer used.

Sometimes a species is moved from one genus to another when a species is thought to be more closely related to members of that genus. In 1838, Thomas Nuttall named laurel sumac as *Rhus laurina*. In 1883, he moved it into the genus *Malosma,* thus making its name *Malosma laurina*. When a species is moved from one genus to another, the name of the original author is placed in parentheses and the name of the person who moved it is added. Sticky live-forever was named *Cotyledon viscida* in 1882 by Sereno Watson. In 1942, Reed Moran moved it into the genus *Dudleya,* which made its name *Dudleya viscida* (S. Watson) Moran.

If some individuals of a plant species are found to be consistently different from the typical form of that species, the different individuals are given a formal designation, that of **subspecies** (abbreviated **ssp.** or **subsp.**) or **variety** (abbreviated **var.**). Basically, if the different form does not live with the typical form, it is called a subspecies; if it does live with the typical form, it is called a variety. In 1840, Thomas Nuttall published the name *Pentachaeta aurea* for a golden daisy; its ray flowers are all golden yellow. A different form of the plant occurs only in Orange County; its ray flowers are golden yellow at their bases and white toward their tips. In 2006, Dr. David J. Keil published a name for the Orange County form, *Pentachaeta aurea* ssp. *allenii*. To make the names parallel, the golden daisy is now called *Pentachaeta aurea* ssp. *aurea* (the epithet of the typical form is used to make the subspecific epithet for it).

When a name change has occurred recently, in this book we place the older name in square brackets below the currently accepted name. Knowing the older name will help you find information about that plant in other publications.

Plant Communities

When you're out watching wildflowers, you may notice that plants seem to grow in somewhat distinct groups at consistent types of locations. Each of these groups of plants has a particular look because of the appearance of the individual plant species that live within them. For example, soft-leaved scented shrubs like California sagebrush (*Artemisia californica*) and black sage (*Salvia mellifera*) grow along the coast in valleys and foothills at low elevations, while tough-leaved manzanitas (*Arctostaphylos* spp.) and trees such as bigcone Douglas-fir (*Pseudotsuga macrocarpa*) and knobcone pines (*Pinus attenuata*) grow high up in the Santa Ana Mountains.

These plant groups, called **plant communities** or **vegetation types**, are groups of plant species that occur in a given area under specific conditions, including soil, slope, moisture, wind, temperature, sun exposure, and shade.

In some areas, a single plant species may stand out from the rest. This species may be the largest, the most common, or perhaps the only species in its community. This plant, called the **dominant species**, influences the other plants and thus determines the overall character of the community. Dominant species may change from one part of the community to the next as local environmental conditions change. For example, in the coastal sage scrub community *near the coast*, the dominant species is often California sagebrush. *Inland* where the climate is a bit hotter and drier, California sagebrush is no longer so common and is greatly outnumbered by white sage (*Salvia apiana*) which is the dominant species in this location.

Many wildflowers grow as members of one or two types of plant communities but not others. To help you locate wildflowers, we list the communities within which they usually occur. With practice, you'll learn to recognize these communities and improve your chances of finding wildflowers. A good way to get started is to attend field trips and nature walks led by experienced naturalists and docents who will show you the communities in person.

Several quite different systems for classifying plant communities have been proposed. None is perfect and none is accepted by all botanists. The arrangement we present here is a fairly straightforward system that you should find easy to follow and understand. Only the communities that contain wildflowers presented in this book are included here.

Below, Biotic Zonation in the Santa Ana Mountains Region. Adapted from Schoenherr, 2011.

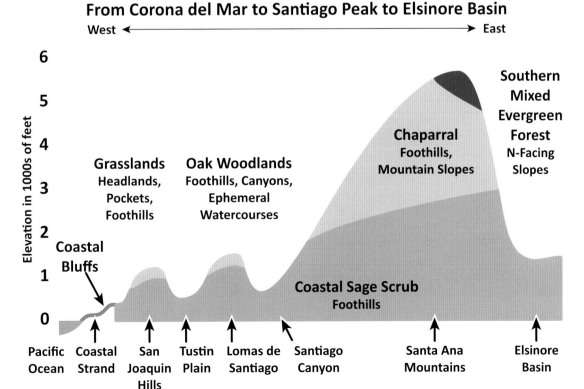

Coastal salt marsh, North Beach, upper Newport Bay. Visible in the foreground are pickleweeds, alkali heath, and California sea-lavender. In the background at slightly higher elevation are several individuals of southwestern spiny rush.

Freshwater marsh near Nichols Road wetlands, Lake Elsinore. Surrounding the open water are plants such as willows, cattails, sedges, rushes, spike-rushes, salt grass, and smartweeds.

Vernal pool, Fairview Park, Costa Mesa (27 February 2005). Full of water in winter and early spring. Along the edges are San Diego silverpuffs and grasses such as vernal spring barley.

Vernal pool, Fairview Park, Costa Mesa (21 May 2005). As the pool dries out in late spring, specialty annuals appear such as prostrate navarretia and woolly marbles.

Lowland riparian woodland in lower San Mateo Creek at Trestles. Cottonwoods, willows, blue elderberry, mule fat, and wildflowers such as yerba mansa, California mugwort, canyon dodder, monkeyflowers, hedge-nettles, and cattails. Poison oak is surprisingly abundant so near to the ocean.

Foothill riparian woodland in Trabuco Canyon through Rancho Santa Margarita. California sycamore, blue elderberry. Coast live oaks visible mostly on far bank. Wildflowers include California mugwort, canyon dodder, virgin's-bowers, monkeyflowers, hedge-nettles, and cattails.

Coastal salt marsh. This community occurs in the upper reaches of salty waters along the coast and receives only periodic inundation by seawater. Freshwater streams often flow through this community and serve to dilute the salinity of the seawater. Typical plants are seaside arrow-grass (*Triglochin concinna*), pickleweeds (*Arthrocnemum subterminale, Salicornia bigelovii, Salicornia pacifica*), fleshy jaumea (*Jaumea carnosa*), alkali heath (*Frankenia salina*), California sea-lavender (*Limonium californicum*), and salt grass (*Distichlis spicata*). Southwestern spiny rush (*Juncus acutus* ssp. *leopoldii*) sometimes grows at its upper edges. Good places to see the coastal salt marsh are Seal Beach National Wildlife Refuge, the Bolsa Chica State State Ecological Reserve, and upper Newport Bay.

Freshwater marsh. This community occurs mostly in coastal plains near permanent slow-moving ponded waters. Near the coast, freshwater marsh meets the upper reaches of the coastal salt marsh. Some of the plants spring up from the middle of ponds, lakes, or streams; others float upon deep water, but most thrive at the margins where the soil is more compact. Typical plants are cattails (*Typha* spp.), sedges (*Carex* spp.), rushes, (*Juncus* spp.) spike-rushes (*Eleocharis* spp.), duckweeds (*Lemna* spp.), and smartweeds (*Persicaria* spp.). Good examples exist at Laguna Lakes, U.C. Irvine's San Joaquin Freshwater Marsh Reserve, the San Joaquin Marsh Wildlife Sanctuary, and the Nichols Road wetlands in Lake Elsinore.

Vernal pools. This rare community is endemic to California, but most of it has been destroyed by human activities. Most plants that occur in vernal pools grow in no other community. The habitat begins as a soil depression, underlain by an impermeable layer of clay or rock. Winter rains fill the depression with water. As the water dries up in late spring, several annual plants on the drying margins bear flowers, set seed, and die off. Individual flowers are often small, but they are numerous. The effect of these multicolored wildflowers in bloom is spectacular.

Five vernal pool areas are known to have existed in Orange County, in Costa Mesa, Lake Forest (El Toro), Rancho Mission Viejo, Dana Point, and San Clemente. Only a few large pools remain, such as those in Costa Mesa's Fairview Park and on the ridges west of Cañada Chiquita in Rancho Mission Viejo. Smaller pools still exist in San Clemente State Park and in the San Joaquin Hills above Laguna Beach. The large vernal pool complex at Fairview Park was confirmed by local botanist Tony Bomkamp in the mid-1990s. It contains plants such as hairy pepperwort (*Marsilea vestita*), American pillwort (*Pilularia americana*), woolly marbles (*Psilocarphus brevissimus*), Virginia rock cress (*Planodes virginicum*), water pygmy-stonecrop (*Crassula aquatica*), truncate sack clover (*Trifolium depauperatum* var. *truncatum*), prostrate navarretia (*Navarretia prostrata*), little mouse-tail (*Myosurus minimus*), Mexican speedwell (*Veronica peregrina* ssp. *xalapensis*), bracted vervain (*Verbena bracteata*), spike-rushes (*Eleocharis* spp.), and vernal spring barley (*Hordeum intercedens*). Apparently, this large pool and some near Lake Elsinore were known to botanists in the early 1900s but "forgotten" through the years (see Tracy, 1936). Vernal pools also exist on the plateau above Surfer's Beach; they are host to adobe popcorn flower (*Plagiobothrys acanthocarpus*) and Pendleton button-celery (*Eryngium pendletonense*). Spectacular vernal pools still exist on the Santa Rosa Plateau Ecological Reserve. In addition to most of the plants listed above, they are home to two more very beautiful wildflowers, Hoover's calicoflower (*Downingia bella*) and toothed calicoflower (*Downingia cuspidata*).

Riparian woodland. This community occurs along permanent standing, running, and/or subterranean waters at the coastline, valleys, and foothills. It may be shaded by its tall trees or remain exposed with only a few shrubs and annuals present. Three general types exist locally.

In sandy soils at low elevations, lowland riparian woodlands are usually made up of medium-sized to large trees and may form dense forests. In such habitats are Fremont's cottonwood (*Populus fremontii*), arroyo willow (*Salix lasiolepis*), sandbar willow (*Salix exigua*), mule fat (*Baccharis salicifolia*), California buckwheat (*Eriogonum fasciculatum*), and poison oak (*Toxicodendron diversilobum*) with an understory of desert wild grape (*Vitis girdiana*), water cresses (*Nasturtium* and *Rorippa* spp.), green willow-herb (*Epilobium ciliatum ssp. ciliatum*), and seep monkeyflower (*Erythranthe guttata*). Such woodlands are well developed at Prado Basin, Featherly Park along the Santa Ana River, portions of Santiago Creek, Arroyo Trabuco, San Juan Creek, San Mateo Creek estuary, and many places along the Temescal Valley, and in the Elsinore Basin. In areas where surface water flows year-round or nearly so, white alder (*Alnus rhombifolia*) often grows as well.

In sandy-gravelly soils in foothill channels, foothill riparian woodlands develop as open woodlands dominated by California sycamore (*Platanus racemosa*), blue elderberry (*Sambucus nigra* ssp. *caerulea*), toyon (*Heteromeles arbutifolia*), mule fat, poison oak, and California mugwort (*Artemisia douglasiana*). In some areas, coast live oak (*Quercus agrifolia*), also occurs in this community. Look for this type in foothill canyons that drain the Santa Ana Mountains such as Santiago, Trabuco, San Juan, and Indian Canyons. Along San Juan Creek just below Upper San Juan Campground is an unusual form of this woodland composed of California sycamore, willows, mule fat, and black cottonwood (*Populus trichocarpa*). With surface water, white alder joins the community.

In shallow rocky soils of higher-elevation streams where water is plentiful, montane riparian woodlands appear, often as a dense shady forest with white alder, big-leaf maple (*Acer macrophyllum*), California bay laurel (*Umbellularia californica*), poison oak, and willows (*Salix* spp.). Understory plants include Durango root (*Datisca glomerata*), ocellated Humboldt lily (*Lilium humboldtii* ssp. *ocellatum*), California loosestrife (*Lythrum californicum*), scarlet monkeyflower (*Erythranthe cardinalis*), seep monkeyflower, California threadtorch (*Castilleja minor* ssp. *spiralis*), and stream orchid (*Epipactis gigantea*). It grows along upper Silverado, Trabuco, and San Juan Canyons.

Montane riparian woodland in upper Trabuco Canyon. Dominated by white alder, big-leaf maple, and California bay laurel.

Coastal strand, Newport Beach, between the Santa Ana River outfall and the northern portion of Balboa Peninsula. Dominated by beach bur-sage, beach evening-primrose, and beach sand-verbena.

Coastal sage scrub, coastal form dominated by California sagebrush and California buckwheat. Associates include bush monkeyflowers, coastal deer broom, laurel sumac, holly-leaved redberry, coast prickly-pear cactus, blue elderberry, and California dodder.

Coastal sage scrub, inland form dominated by California sagebrush and white sage. Other wildflowers include chaparral yucca, Parry's deer's ears, Weed's mariposa lily, shiny wing-fruit, and felt paintbrush. Nearby shrubs include laurel sumac, toyon, and holly-leaved redberry. Ron Vanderhoff and Doug Peltz were added for scale.

Coastal bluff scrub, Crystal Cove State Park, north end, right above the beach. Lemonade berry, California bush sunflower, California buckwheat, salt grass, California wishbone bush, coast prickly-pear cactus, and coast cholla.

Alluvial scrub, Caspers Wilderness Park, San Juan Creek. Dominated by scale broom, coast prickly-pear cactus, and California buckwheat. In spring, the area lights up with fields of yellow pincushion.

Coastal strand. This community is found on undisturbed sandy beaches and dunes above the high-tide level. Onshore winds deposit sand at the base of plants, forming mounds called dune hummocks. Plants there are low-growing deep-rooted perennials and annuals. Common plants are beach sand-verbena (*Abronia umbellata*), beach evening-primrose (*Camissoniopsis cheiranthifolia* ssp. *suffruticosa*), beach bur-sage (*Ambrosia chamissonis*), beach morning-glory (*Calystegia soldanella*), sea rocket (*Cakile maritima*), California croton (*Croton californicus*), and bluff buckwheat (*Eriogonum parvifolium*). It has been largely eliminated from our area by development, trampling, and invasive non-native plants but still exists in patches from the south end of Huntington State Beach to the tip of Balboa Peninsula, in Crystal Cove State Park, and from San Clemente to San Onofre State Beaches.

Coastal sage scrub. This widespread and abundant community begins just inland of the immediate coastline, covers much of the San Joaquin Hills, the Los Angeles Basin, portions of the Santa Ana Mountains (where it is largely replaced by chaparral). It becomes common again on the inland base of the Santa Ana Mountains and throughout the Temescal Valley and the Elsinore Basin. As an adaptation to summer drought, several of its plant species drop their leaves in summer and halt growth until winter. Many of its plants contain aromatic oils that give the community a characteristic scent. Shrubs within it are mostly 2-6 feet in height and shallow-rooted. Typical plants are California sagebrush (*Artemisia californica*), coyote brush (*Baccharis pilularis*), bush monkeyflowers (*Diplacus* spp.), lemonade berry (*Rhus integrifolia*), laurel sumac (*Malosma laurina*), blue elderberry, California scrub oak (*Quercus berberidifolia*), California buckwheat, California bush sunflower (*Encelia californica*), desert bush sunflower (*Encelia farinosa*), black sage (*Salvia mellifera*), white sage (*Salvia apiana*), toyon, coastal deer broom (*Acmispon glaber* var. *glaber*, primarily coastal), desert deer broom (*Acmispon glaber* var. *brevialatus*, primarily inland), wild cucumber (*Marah macrocarpus*), and western poison oak. When dominated principally by California sagebrush, as happens in areas of substantial fog, it is often called California sagebrush scrub. It is abundant in Crystal Cove State Park, all of our wilderness parks, on the Dana Point Headlands, in San Juan and Santiago Canyons, and in the inland foothills of the Santa Ana Mountains.

Coastal bluff scrub. Somewhat a subset of Coastal Sage Scrub, this community is found along sea bluffs that receive salt-laden moisture from rain, fog, and splash. It is often overtaken and crowded out by non-native vegetation planted to stabilize slopes. Dominants include lemonade berry, California buckwheat, bluff buckwheat, California bush sunflower, prostrate goldenbush (*Isocoma menziesii* var. *sedoides*), bladderpod (*Peritoma arborea*), and saltbushes (*Atriplex* spp.). Uncommon-to-rare associates include Parry's larkspur (*Delphinium parryi*), southern California virgin's-bower (*Clematis pauciflora*), sea-blite (*Aphanisma blitoides*), cliff spurge (*Euphorbia misera*), Orange County turkish rugging (*Chorizanthe staticoides* ssp. *chrysacantha*), and California box-thorn (*Lycium californicum*). Find it on bluffs along upper Newport Bay, Little Corona Beach, Crystal Cove State Park, Laguna Beach, Dana Point Headlands, San Clemente, and San Onofre State Beach.

Alluvial scrub. This is a community of short woody shrubs that occur in deep sandy-gravelly soils that have washed down from higher elevations. The plants are generally spaced far apart, with areas of open ground between them; it is within these open areas that displays of annual wildflowers appear. Its shrubs often drop leaves in summer-fall drought; others drop in fall-winter. In some areas, this scrub looks rather like desert scrub. The community often meets Foothill Riparian Woodlands, evidenced primarily by California sycamore. Typical plants include scabrid sweetbush (*Bebbia juncea* var. *aspera*), yellow pincushion (*Chaenactis glabriuscula*), scale-broom (*Lepidospartum squamatum*), valley cholla (*Cylindropuntia californica* var. *parkeri*), coast prickly-pear (*Opuntia littoralis*), California croton, chaparral sand-verbena (*Abronia villosa* var. *aurita*), Coulter's matilija poppy (*Romneya coulteri*), and California buckwheat. It is found in medium-to-lower elevations of major canyons such as Santa Ana, Santiago, Trabuco, San Juan (especially through Caspers Wilderness Park), Indian, and Temescal.

Southern California grassland. These variable communities are dominated by grasses and other herbaceous plants, usually with neither shrubs nor trees. As a result, these communities are composed of plants that are much less than one meter tall. They generally occur in soils of deep clay, sometimes interspersed with rocky regions. Commonly they occur on gentle slopes and playas, exactly the kind desired for human civilization. As a result many of our historic grasslands are long gone because of development.

In some areas, the end of domestic animal grazing and lack of human disturbance has allowed regeneration of the grasslands, but for this to occur, true native grasslands must exist nearby to provide seed. Our native grasslands are dominated by perennial bunchgrasses that generally do not totally dry up and turn brown in summer, some portion of each plant remains green all year. They also support a beautiful array of colorful spring, summer, and fall wildflowers. Native grassland plants include three-awned grass (*Aristida* spp.), foothill needle grass (*Stipa lepida* [formerly called *Nassella lepida*]), purple needle grass (*Stipa pulchra* [formerly called *Nassella pulchra*]), school bells (*Dichelostemma capitatum* ssp. *capitatum*), Johnny jump-ups (*Viola pedunculata*), padre's shooting stars (*Dodecatheon clevelandii*), western blue-eyed-grass (*Sisyrinchium bellum*), purple owl's-clover (*Castilleja exserta*), blue toadflax (*Nuttallanthus texanus*), lupines (*Lupinus* spp.), ground pink (*Linanthus dianthiflorus*), coastal checkerbloom (*Sidalcea malviflora*), Catalina mariposa lily (*Calochortus catalinae*), winecup clarkia (*Clarkia purpurea*), Parry's larkspur (*Delphinium parryi*), common fiddleneck (*Amsinckia intermedia*), and California poppy (*Eschscholzia californica*). Our remaining native grasslands exist primarily in narrow portions of Dove, Bell, San Juan, Cristianitos, and San Mateo Canyons in the Santa Ana Mountains.

PLANT COMMUNITIES

Southern California grassland, north-facing slope, Richard and Donna O'Neill Conservancy. Dominated by purple needle grass. Seasonal wildflowers include winter-spring plants such as padre's shooting stars, school bells, and rusty popcorn flower. Late-season wildflowers include mariposa lilies and tarplants.

Southern California grassland, south-facing flank of Miller Mountain in San Diego County, looking northwest toward Orange County. Fascicled tarplant, red-skinned onion, milkweeds, Brodiaeas, and native grasses such as purple needle grass, punctuated by California buckwheat. This grassland has been invaded by non-native grasses that die and turn brown in summer.

Chamise chaparral, Santa Ana Canyon, slopes above Upper San Juan Campground. Common chamise is clearly the dominant shrub in this region.

Chamise chaparral, Fremont Canyon. Common chamise, California scrub oaks, thick-leaved yerba santa, California poppies, white sage, and California rush-rose occur here.

Ceanothus chaparral. Bedford Canyon, north-facing slopes. Dominated by hairy lilac and California scrub oak, with wildflowers such as virgin's-bowers, royal beardtongue, and Coulter's matilija poppy.

Manzanita chaparral on a north-facing slope along the South Main Divide Road. Dominated by Eastwood manzanita and California scrub oak. Associates include golden yarrow, royal beardtongue, and thick-leaved lilac.

Chaparral. This is certainly the most characteristic vegetation type in California. It is dominated by woody shrubs with leaves that are usually leathery, small, and/or wax-covered; these are adaptations to prevent moisture loss. The plants grow quite near one another, making an impenetrable thicket. Many species are fire-adapted and can readily resprout or reseed after fire. Most of its woody plants are deep-rooted perennials that reach up to 10 feet in height. Some of its plants are summer-deciduous, though less so than those in coastal sage scrub. Associated plants include California lilacs (*Ceanothus* spp., especially *Ceanothus crassifolius*), California buckwheat, California scrub oak, holly-leaved redberry (*Rhamnus ilicifolia*), holly-leaved cherry (*Prunus ilicifolia* ssp. *ilicifolia*), California mountain-mahogany (*Cercocarpus betuloides*), southern honeysuckle (*Lonicera subspicata* var. *denudata*), sugar bush (*Rhus ovata*), yellow bush beardtongue (*Keckiella antirrhinoides* ssp. *antirrhinoides*), and chaparral yucca (*Hesperoyucca whipplei*). Three distinct chaparral subcommunities occur here, dominated by different species. Each contains the same associated species, but in different abundances.

Chamise chaparral is dominated by chamise (*Adenostoma fasciculatum*), a large woody member of the rose family. Associated plants include California scrub oak, California buckwheat, white sage, and chaparral yucca. It grows on low to middle elevations on very hot, dry sites, especially on south-and west-facing slopes. It exists throughout all of our hills and mountains.

Ceanothus chaparral is dominated by species of California lilac (*Ceanothus* spp.), large woody shrubs with fairly small leaves and dense sprays of tiny flowers. It usually grows on north-facing slopes that receive a fair amount of winter rain and in well-developed soils that retain moisture. On slopes that face east or west, it may intermix with chamise chaparral. When in bloom, the hillsides light up with colors that change as different species of California lilac come into flower. Typical species include southern deer brush (*Ceanothus integerrimus* var. *macrothyrsus*), chaparral whitethorn (*Ceanothus leucodermis*), bigpod lilac (*Ceanothus megacarpus*), hairy lilac (*Ceanothus oliganthus* var. *oliganthus*), Jim brush (*Ceanothus oliganthus* var. *sorediatus*), and woollyleaf lilac (*Ceanothus tomentosus*). It is known from many areas, such as on north-facing and north-west slopes of San Juan Canyon right above Ortega Highway and Santiago Peak. It is much more common on the inland side of the Santa Ana Mountains just over the ridgeline and in canyons such as Bedford, Coldwater, Mayhew and Indian.

Manzanita chaparral is dominated by manzanitas, attractive shrubs with smooth red bark. Locally, the dominant species is usually Eastwood manzanita (*Arctostaphylos glandulosa* ssp. *glandulosa*). It occupies higher elevations that receive less heat and more moisture than chamise chaparral. It receives its precipitation from winter snow, rain, and frequent fog drip. Generally it exists above chamise chaparral and below southern mixed evergreen forest. In some areas, interior live oak (*Quercus wislizenii*) and Coulter pine (*Pinus coulteri*) join the community. Extensive stands occur near Los Pinos and Trabuco Peaks, along the Trabuco Trail and West Horsethief Trail, and in many places along the North Main Divide Truck Trail.

Oak woodland. This variable community is found on gentle slopes and rugged canyon walls in our hills and mountains. It also grows in place of a riparian woodland in acutely steep canyons, especially at higher elevations. It is dominated by coast live oak and/or canyon live oak (*Quercus chrysolepis*), large stately trees that form broad canopies and cast plentiful shade. In some locations it is the only woody plant in the community; in others it is joined by a suite of shrubs and wildflowers. Associated plants include California scrub oak, poison oak, toyon, California coffeeberry (*Frangula californica* ssp. *californica*), southern California hillside gooseberry (*Ribes californica* var. *hesperium*), California rose (*Rosa californica*), California sweet-cicely (*Osmorhiza brachypoda*), Pacific sanicle (*Sanicula crassicaulis*), sticky cinquefoil (*Drymocallis glandulosa* ssp. *glandulosa*), horkelias (*Horkelia* spp.), and many types of ferns. We have two forms of oak woodland.

Southern oak woodland lives at low to middle elevations and in meadows and mesas at higher elevations. Good places to see it are Laguna Coast Wilderness Park, Trabuco Canyon, and Oak Canyon Nature Center.

Canyon live oak woodland is a higher-elevation form dominated by canyon live oak and varying numbers of coast live oak. It occupies the higher reaches of the Santa Ana Mountains, for example, upper Trabuco Canyon, and the moist canyons and north-facing slopes of the major peaks. The community often follows a watercourse.

Walnut woodland. This is a rare community in serious decline, occuring only in the Chino Hills, and is dominated by southern California black walnut (*Juglans californica*). Associated plants include coast live oak, blue elderberry, California ash (*Fraxinus dipetala*), and toyon. Some members of the adjacent coastal sage scrub community often form its understory, as do native and introduced grasses. Much of it has been lost to development and cattle grazing. It can be seen in Chino Hills State Park and Carbon Canyon Regional Park.

Southern mixed evergreen forest. This is a forest of broad-leaved trees and shrubs mixed with needle-leaved conifers and has the look of larger forests found in taller mountain ranges. Here it occurs at high elevations in the Santa Ana Mountains, most often on north-facing slopes and canyons that receive seasonal fog. No single plant species is consistently dominant throughout its range. Instead it is populated by a handful of tree species that vary in abundance. Its characteristic trees are big-leaf maple, California bay laurel, canyon live oak, and interior live oak (*Quercus wislizenii*), and conifers such as bigcone Douglas-fir (*Pseudotsuga macrocarpa*) and Coulter pine (*Pinus coulteri*). In a few locations, Pacific madrone (*Arbutus menziesii*) joins this community. Manzanita chaparral often forms its understory. Associated wildflowers include heart-leaved pitcher-sage (*Lepechinia cardiophylla*), southern tauschia (*Tauschia arguta*), and yellow violet (*Viola purpurea* ssp. *purpurea*). Perhaps the most impressive stand of southern mixed evergreen

Southern oak woodland, Weir Canyon. Dominated by coast live oak; those in flower appear lighter in color than those not in flower. Along drainages, coast live oak forms dense woodlands; away from drainages, the scattered individuals form an oak savannah. The understory is coastal sage scrub and grassland.

Canyon live oak woodland, north-facing slopes of Modjeska and Santiago Peaks. Dominated by canyon live oak. Patches of ceanothus chaparral and manzanita chaparral are interspersed among the oaks.

Walnut woodland, Chino Hills. Dominated by southern California black walnut. Common associates include coast live oak, toyon, and blue elderberry.

Southern mixed evergreen forest, upper Silverado Canyon. Dominant conifers are bigcone Douglas-fir (foreground) and Coulter pine (upper right); both of these conifers appear in the background.

Knobcone pine forest, Pleasants Peak. The diagonal band of conifers is made up of knobcone pines with an understory of wartleaf lilac and manzanita chaparral.

Tecate cypress forest, Coal Canyon. Frequent fires have seriously damaged the population of Tecate cypress. Today, most of the surviving plants are in pockets on hillsides.

forest is along the north-facing slope of upper Trabuco Canyon, easily viewed from the Trabuco Trail. Protection of this forest in Trabuco Canyon led to establishment of the Cleveland National Forest. Other dense stands occupy upper Silverado Canyon, the north-facing slope of Santiago Peak, and the upper portion of Indian Truck Trail.

Interior closed cone conifer forest. This community grows in rocky soils at high elevations where it is often surrounded by chamise chaparral. It is dominated by conifers that have closed cones. Their cones mature and remain on the trees, staying tightly closed until heated by fire; the parent tree is often killed outright. Later the seeds exit the cones, fall to the ground, and germinate in the ashen soil. The dominant conifers and many of their associates require fire for their reproduction, and therefore fire must be allowed to burn through this community occasionally. We have two types of interior closed cone conifer forest, named for their dominant conifers.

Knobcone pine forest is dominated by knobcone pine (*Pinus attenuata*), a relatively short-statured, often multiple-trunked tree. Its understory often consists of manzanita chaparral. The community grows in rocky serpentine soils on steep slopes and ridges where it receives precipitation in the form of snow, rain, and fog. Other plants in the community include wartleaf lilac (*Ceanothus papillosus*), Eastwood manzanita, common chamise, southern California currant (*Ribes malvaceum* var. *viridifolium*), and Santiago Peak phacelia (*Phacelia keckii*). It occurs only near Pleasants Peak, most abundantly on north-facing slopes. Many years ago, the U.S. Forest Service planted knobcone pines in other parts of the Santa Ana Mountains; you may occasionally encounter these stray trees.

Tecate cypress forest is dominated by Tecate cypress (*Hesperocyparis forbesii*), a large shrub to short-statured tree with multiple trunks. It grows in clays and sandstones on slopes and gets moisture from rain and fog. Associated plants include hairy lilac, common chamise, chaparral beargrass (*Nolina cismontana*), heart-leaved pitcher-sage, Fremont's death camas (*Toxicoscordion fremontii*), Braunton's milkvetch (*Astragalus brauntonii*), and San Diego reed grass (*Calamagrostis koelerioides*). It is restricted to the north- and west-facing slopes of Sierra Peak and Coal, Gypsum, and Fremont Canyons. A few stray Tecate cypress also occur in the nearby Santa Ana River channel.

On 11 February 2002, a 2,400-acre fire in Coal Canyon killed most of its Tecate cypress. On 7 May 2005, we examined the remains of the forest and found thousands of its seedlings flourishing. However, on 8 February 2006, a controlled burn recently conducted on Pleasants Peak flared up and burned Sierra Peak and Fremont Canyon, taking with it several Tecate cypress. As a result, there are now very few adult Tecate cypress in the Santa Ana Mountains.

Unusual habitats and vegetation types. Some assemblages of plants are adapted to grow under harsh conditions that other plants cannot tolerate. They are found in pockets within other plant communities and are often difficult to locate. The wildflowers they support are quite stunning and worth your while to seek out.

Sand lenses are pockets of ancient marine sand that are low in soil nutrients and organic compounds, too porous to retain water, and usually exposed to high temperatures. Plants that inhabit it include turkish rugging (*Chorizanthe staticoides* ssp. *staticoides*), sandmat (*Cardionema ramosissimum*), sand pygmy-stonecrop (*Crassula connata*), California plantain (*Plantago erecta* ssp. *erecta*), tarplants (*Deinandra* spp.), false rosinweed (*Osmadenia tenella*), and littleseed muhly grass (*Muhlenbergia microsperma*). Now largely extirpated, sand lenses occur from the Dana Point Headlands inland to Mission Viejo, Chiquita Ridge, and Caspers Wilderness Park.

Open rock and talus (rubble, decomposing rock) slopes are also poor at holding water, and they have shallow or no soils, offer few or no nutrients, and are often physically unstable. Additionally, they usually receive high wind, sun, and temperature extremes. Plants growing on open rock anchor their roots into cracks that develop with time. These are plants such as Bigelow's spike-moss (*Selaginella bigelovii*), coffee fern (*Pellaea andromedifolia*), California cotton fern (*Cheilanthes newberryi*), canyon live-forever (*Dudleya cymosa*), lance-leaved live-forever (*Dudleya lanceolata*), and granny's hairnet (*Pterostegia drymarioides*). Regions of open rock can be seen within most any plant community. Plants on talus slopes grow between the pieces of rubble, sometimes with deep roots anchored to soil beneath it. Quite a variety of plants live on them, including tuberous sanicle (*Sanicula tuberosa*), Parish's tauschia (*Tauschia parishii*), leafy daisy (*Erigeron foliosus*), golden yarrow (*Eriophyllum confertiflorum* var. *confertiflorum*), Douglas's stitchwort (*Minuartia douglasii*), San Francisco campion (*Silene verecunda*), Sonora morning-glory (*Calystegia occidentalis* ssp. *fulcrata*), Phacelias (*Phacelia* spp.), clustered broomrape (*Orobanche fasciculata*), rock buckwheat (*Eriogonum saxatile*), liver-leaved larkspur (*Delphinium patens* ssp. *hepaticoideum*), and San Bernardino Mountains onion (*Allium monticola*). Talus slopes are well-developed in the upper reaches of the Santa Ana Mountains.

Shady slopes, seeps, and waterfalls have soils that are deeper and able to retain water. In some shady areas, water visibly flows over bare rock. Shade is provided by hillsides, rocks, and trees, reducing evaporation and sun exposure. These are spots favored by delicate plants such as ferns and a suite of graceful wildflowers. Among them are liver-leaved larkspur, woodland star (*Lithophragma affine*), round-leaved boykinia (*Boykinia rotundifolia*), California saxifrage (*Micranthes californica*), common miner's lettuce (*Claytonia perfoliata*), purple Chinese houses (*Collinsia heterophylla*), sticky live-forever (*Dudleya viscida*), Rothrock's lobelia (*Lobelia dunnii* var. *serrata*), California loosestrife (*Lythrum californicum*), scarlet monkeyflower, and seep monkeyflower. Shady areas are found on north-facing slopes within most plant communities. Our better-known seeps, springs, and waterfalls are Dripping Springs, Ladd Canyon Spring, Bigcone Spring, Maple Spring, Bear Spring, San Juan Hot Springs, Hot Spring Falls, and Holy Jim Falls.

Talus slope on the north-facing western arm of Modjeska Peak. Wildflowers here include tuberous sanicle, mountain dandelion, San Francisco campion, rock buckwheat, liver-leaved larkspur, and San Bernardino Mountains onion.

Ruderal vegetation, Hicks Canyon, after the Santiago Fire of October 2007. The dominant yellow flower is black mustard, an aggressive invader from Europe. Non-natives often take hold after catastrophes and fire break construction that damage the soil.

Small fire along Ortega Highway, just above Lake Elsinore, September 2009. Man-made fires often start along roadways from discarded cigarettes, overheated/burning cars, or acts of arson.

Early stages of post-fire recovery, Fremont Canyon, after the Sierra Peak fire of Spring 2006. The area later recovered and native plants flourish again, in part because its soils were not damaged.

The Station Fire in the San Gabriel Mountains, as seen from the face of Modjeska Peak, in the Santa Ana Mountains, August 2009. The San Gabriel Mountains are visible just below the skyline in the far right, behind the haze. The fire's source is in the center of the photograph, just below the billowing plume of smoke. The hills in the foreground are above Silverado and Harding Canyons; they burned in the Santiago Fire of October 2007 and lost nearly all of their large trees. In the center-midground is Pleasants Peak, which did not burn in the Santiago Fire but did burn about 1980.

PLANT COMMUNITIES

Ruderal communities. This is something of a catch-all term for plants that grow in disturbed soils. Most have been introduced from other continents, although some native plants prefer this situation as well. In the wild, ruderal plants usually grow along trails, unpaved roads, firebreaks, and landslides. After wildfires, ruderal vegetation may appear and crowd out native plants, slowing or preventing the natural recovery process. Further, when these plants die and dry out, they create a severe fire hazard. This situation occurs frequently with non-native invasive species such as black mustard (*Brassica nigra*), which blankets the landscape with bright yellow flowers. Some people, totally unaware of the damage done by this aggressive weed, flock to these mustard fields to admire and photograph them, much to the chagrin of land managers and biologists. Typical natives in ruderal areas include deer broom, California croton, telegraph-weed (*Heterotheca grandiflora*), sessileflower golden aster (*Heterotheca sessiliflora* ssp. *echioides*), and tarplants. Introduced ruderal plants include common pineapple weed (*Matricaria discoidea*), black mustard, shortpod mustard (*Hirschfeldia incana*), filarees (*Erodium* spp.), sweetclovers (*Melilotus* spp.), Russian thistle (*Salsola tragus*), star-thistles (*Centaurea* spp.), and numerous grasses.

Fire-following Plants

Many local plants are adapted to fire. During the first few years following a fire, several wildflowers bloom in profusion and put on stunning displays of color. Some of these continue to grow in an area, but others will not reappear until the area is burned again. Here are some of our most abundant and showy wildflowers that come up after fire, arranged in the order in which they appear in this book. Observing an area while it recovers during the years after fire is a rewarding treat for the wildflower watcher and may be your only chance to see many of these plants.

Golden eardrops and scarlet larkspur, Long Canyon Road, 24 June 2011

Eudicots

Apiaceae
Rattlesnake plant (*Daucus pusillus*)

Boraginaceae
Cat's-eyes (*Cryptantha* spp.)
Popcorn flowers (*Plagiobothrys* spp.)
Whispering bells (*Emmenanthe penduliflora*)
Cluster flowers (*Phacelia* spp.)

Campanulaceae
Bluecups (*Githopsis* spp.)
Small Venus's looking-glass (*Triodanis biflora*)

Caryophyllaceae
Snapdragon catchfly (*Silene antirrhina*)

Cistaceae
California rush-rose (*Helianthemum scoparium*)

Fabaceae
Deer broom (*Acmispon glaber* ssp.)
Locoweeds (*Astragalus* spp.)
Lupines (*Lupinus* spp.)
Clovers (*Trifolium* spp.)

Lamiaceae
Chia (*Salvia columbariae*)
Danny's skullcap (*Scutellaria tuberosa*)
Mustang mint (*Monardella lanceolata*)

Loasaceae
Small-flowered stick-leaf (*Mentzelia macrantha*)

Malvaceae
Bush mallows (*Malacothamnus* spp.)

Onagraceae
Evening-primroses (*Camissonia* spp., *Cammisoniopsis* spp., *Eulobus californicus*)
Farewell-to-spring (*Clarkia* spp.)

Orobanchaceae
Warrior's plume (*Pedicularis densiflora*)

Papaveraceae
Bush poppy (*Dendromecon rigida*)
Golden eardrops (*Ehrendorferia chrysantha*)
White eardrops (*Ehrendorferia ochroleuca*)
California poppy (*Eschscholzia californica*)
Small-flowered fairy poppy (*Meconella denticulata*)
Fire poppy (*Papaver californicum*)
Wind poppy (*Papaver heterophylla*)
Cream cups (*Platystemon californicus*)

Phrymaceae
Annual monkeyflowers (*Diplacus* spp., *Erythranthe* spp., *Mimetanthe* spp.)

Plantaginaceae
Snapdragons (*Antirrhinum* spp.)
Purple Chinese houses (*Collinsia heterophylla*)

Polemoniaceae
Blue false-gilia (*Allophyllum glutinosum*)

Montiaceae
Common miner's lettuce (*Claytonia perfoliata*)

Ranunculaceae
Scarlet larkspur (*Delphinium cardinale*)

Rhamnaceae
California lilacs (*Ceanothus* spp.)

Monocots

Liliaceae
Mariposa lilies (*Calochortus* spp.)

Melanthiaceae
Fremont's death camas (*Toxicoscordion fremontii*)

Themidaceae
School bells (*Dichelostemma capitatum* ssp. *capitatum*)

Fire-following wildflowers, Limestone Canyon, after the Santiago Fire of October 2007. Seen here are California poppy, Parry's phacelia, Los Angeles gilia, school bells, and rusty popcorn flower.

Fire-following wildflowers, Limestone Canyon, after the Santiago Fire of October 2007. Seen here are California poppy, Parry's phacelia, Los Angeles gilia, school bells, rusty popcorn flower, stinging lupine, collar lupine, chia, San Diego jewelflower, hairy fringepod, and coast prickly-pear cactus.

Mosaic on the landscape, South Main Divide Road. Plant community types and the species within them change as the environmental conditions change. Thus, multiple communities can co-occur in an area. In the foreground, we see coastal sage scrub dominated by black sage. In the middle, southern oak woodland dominated by coast live oak and riparian woodland dominated by willows. In the background, chamise chaparral blankets the slopes.

Wildflowers abound among coastal sage scrub and chaparral, San Juan Trail. The pink flowers are fringed spineflower, yellow flowers are Weed's mariposa lily, the tall grasses are giant needle grass, the white pom-poms are California buckwheat.

Watching Wildflowers

When you take a flower in your hand and really look at it, it's your world for the moment. I want to give that world to someone else. Most people in the city rush around so, they have no time to look at a flower. I want them to see it whether they want to or not.
— Georgia O'Keeffe (1887-1986)

At first, you'll find wildflower-watching to be a simple diversion, enjoyed occasionally when time allows. Once you're hooked, it becomes a bit of an obsession. The need-to-know takes over and your mind will ask, "What was that flower I saw yesterday while hiking?"

Like any natural history hobby, watching wildflowers can be rewarding, relaxing, and challenging. Here are some tips for increasing your enjoyment, followed by suggestions for documenting what you see with illustration and photography.

You don't need special equipment to study wildflowers, but some form of magnification will help you see small flowers and their parts. A basic 10- or 20-power hand lens held from a lanyard or kept in a pocket can even be shared among fellow watchers. Binoculars are helpful to search for flowers that may be a long way off or on rough terrain. Some binoculars have close-focusing ability to allow you to see wildflowers and other wildlife up close and personal.

Take your time to study wildflowers. Learn their parts and try to see them on each flower in the field. Compare what you see in nature with the photographs in this book. This will take some getting used to, since photographs are two-dimensional and you see three dimensions in the field. We have tried to select photographs that accurately depict each species in one or two images. Natural variation may produce a flower that is deep purple in our photograph but looks pale lavender when you find it. Pay attention to more than just color. Look at structures and read the descriptions we provide, often with key characteristics in italics and/or bold type.

Browse through this book at home to become accustomed to the look and names of the wildflowers. Don't try to memorize everything, just try to get a feeling for the general forms of wildflowers in our area. You may find it helpful and enjoyable to select a group of wildflowers, the monkeyflowers for example, and read that section of the book to become somewhat of an expert on them. This is especially helpful when a group of fellow watchers does this and can share their knowledge during a group hike.

Keep a fieldbook or journal to jot down notes about the flowers you see. Some people are listmakers and keep a life-list of species they have seen along with when and where they have seen them. Such lists are helpful when trying to relocate the species next season in the same or different places.

Attend talks, walks, and hikes hosted by knowledgeable people such as park rangers, biologists, naturalists, and docents. Wildflower presentations are given locally by the California Native Plant Society, Southern California Botanists, Santa Ana Mountains Natural History Association, and the National Audubon Society. Short classes are taught at the Fullerton Arboretum, the Rancho Santa Ana Botanic Garden, and the Theodore Payne Foundation. Their contact information is listed in the Resources and References section of this book.

Documenting Wildflowers

In addition to watching wildflowers, you might want to record them in some way. Two popular methods are drawing and photographing them. Both force you to slow down and observe your subject, taking note of the flower's color, physical shape, structure, and arrangement on the plant. The results provide you with a permanent record of the plant and document your excursion into its habitat. You don't need to be a talented artist to succeed at either activity. Learn some basic techniques and practice at home with simple subjects before venturing into the wildlands. Approach this with the three P's: Positive attitude, Persistence, and Patience. Your rewards will be a greater appreciation for and knowledge of the natural world.

Quick Tips for Watching

- Get at least a 10-power hand lens
- Use close-focusing binoculars
- Bend, kneel, sit, or lie down
- Learn flower structures
- Browse through this book
- Become an expert on one group
- Attend talks, walks, and hikes
- Keep a fieldbook, journal, or life-list
- Draw and/or photograph what you see
- Relax and have fun with it!

Sarah Jayne. Photo by Joan R. Hampton.

Do NOT...

- Pick wildflowers at random. In many places, wildflowers are protected and picking them is unlawful unless you have a valid scientific collecting permit. Check with the administering agency or land owner for clarification.
- Trample wildflowers or other wildlife. When hiking, try to avoid harming flora and fauna. It is ironic that those of us seeking wildlife can inadvertently harm the very life forms we are there to enjoy.
- Enter wet meadows and seeps. Moisture-loving plants have sensitive roots and cannot tolerate trampling of any of their parts. Watch and photograph them from a distance.
- Touch or eat anything in the wild unless you are certain that it's not poisonous or irritating to the skin. Some of our local plants are quite poisonous.
- Dig up wildflowers or other plants from the wild. If you would like to have them for yourself, purchase them from a reputable source of native plants, such as those listed near the end of this book.

Drawing

If your goal is to draw what the wildflower really looks like (as opposed to artistic interpretation), then there is only one secret to drawing it: *know its parts!* A flower is a plant's reproductive structure. As such, it has well-defined parts that are genetically consistent. Observe the subject, find the flower parts, and record them as you see them.

To begin, all you need is a sketchbook with firm white pages, the proper drawing pencils, and a good art eraser. If you really get into this, you'll want to transform your pencil sketches into permanent ink illustrations. All of the gear you need is readily available at an art supply store.

Draw a simple outline, concentrating on relative sizes of the flower's parts and perspective. Find and sketch the main stem, branches, leaves, and flowers. Note the overall form of the plant: Are the leaves opposite or alternate? Are the flowers upright or pendulous? Can you find the sepals, petals, pistil, and stamens? Are there any accessory structures? Are you sketching a typical flower or one that is deformed or has been eaten by an insect? It's perfectly acceptable to draw a part as being perfect even if it isn't. Just look at other flowers or leaves as models and correct yours as needed. Once the basic sketch is complete, use your other pencils to darken lines, add depth, and create contrast.

Pencil sketches and ink drawings are useful for keeping in a notebook, adorning the walls of your home, and publishing in your own field guide.

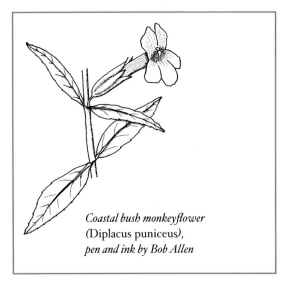

Coastal bush monkeyflower (Diplacus puniceus), *pen and ink by Bob Allen*

Photography

Another way to enjoy wildflowers is to photograph them. This is an extensive topic and one dear to the hearts of the authors of this book. We'll discuss the basics of wildflower photography but we strongly suggest a photography class if you are new to nature photography. Also spend time reading your camera's instruction manual to acquaint yourself with its controls and capabilities.

You'll need a camera body and a lens; for most close-up images, you'll also need an external flash unit. Today's single lens reflex (SLR) camera bodies (digital or film) are superb. Since most plants and their flowers are relatively small, a macro lens will let you get close to your subject and provide the sharpest images. Off-camera flash provides more light, enabling you to photograph at smaller apertures and greatest depth of field. When using flash on a close-up, Manual mode on your camera will produce the best results. Aperture Priority mode allows you to set the aperture (and thus the depth of field) for plant and habitat photos in daylight. Compact ("point-and-shoot") cameras have an attached lens and are lightweight, easy to carry, and simple to use. They are best for plant and habitat photos.

What types of images should you make? The answer depends on your predicted uses for your photographs. Photograph wide open spaces with fields of flowers, entire plants, colorful flower clusters, individual flowers, and fruits. Make images of the leaves and their various arrangements and patterns. Use close-up gear to get inside flowers. If you plan on teaching, be sure to dissect flowers and clearly photograph their parts. For a real treat, print your best images at large sizes and have them professionally matted and framed. Proudly displayed on the walls of your home, they serve as a constant reminder of your experiences with natural beauty and diversity.

What can you do with your wildflower photographs?

- Use them to give presentations about wildflowers.
- Make prints for yourself, friends, and family. Hang them on the wall at home, work, and school.
- Post them on a wildflower web site.
- License high-quality images for books, magazines, calendars, and postcards.
- Publish your work in books and guides, whether professional publications or self-devised.
- *Above all, have fun!*

Quick Tips for Wildflower Photography

Do Photograph...
- The plant in its habitat
- In vertical and horizontal orientation
- Complete leaves and their attachment
- Perfect flowers, not deformed or eaten ones
- Fully open flowers and their reproductive parts
- Fruits or seed pods (you may have to return later when they are ripe)
- Anything needed to identify the plant
- Diagnostic features of the species, genus, and family
- Suspected pollinators, visitors, or nectar thieves
- People enjoying nature, including other photographers

Photography is a great way to enjoy wildflowers. Right, Michelle Balk with Parry's deer's ears (Frasera parryi) *on San Juan Trail. Far right, Chris Barnhill with goldenrod broomrape* (Orobanche sp.) *in Whiting Ranch Wilderness Park.*

Gardening with Native Plants

Another way to enjoy wildflowers and other native plants is to grow them in a garden at your home, work, school, and community. You can enjoy them any time and repetition is a helpful way to learn their names, growth, and flowers.

Using natives has numerous benefits over using non-natives. When all gardens have the same plant species, the neighborhood looks monotonous and downright boring. Spice it up with native species that provide forms, fragrances, and attractive flowers to make your garden stand out. A suite of sages, golden yarrows, beardtongues, verbenas, buckwheats, California poppies, and California fuchsias will make a garden pop with color.

Not only can you wildflower-watch in a native garden, you can watch other creatures as well. Bring in the butterflies with sages and verbenas; hummingbirds with hummingbird sages, California fuchsias, gooseberries, and currents; and goldfinches with plants that produce copious amounts of seed such as black sage and California bush sunflower. Any plant that provides cover near the ground will bring in harmless fence lizards that will live beneath them and feed on all sorts of insects. You'll be mesmerized when flocks of tiny fast-moving birds called bushtits descend on your garden to pick small insects from your plants, chittering to each other all the while. Experiencing the interplay between native plants and animals helps you reconnect with the natural environment, a welcome de-stressor for our all-too-busy lives.

Be sure to select plant species that are appropriate for your local conditions (climate, soil, sun exposure). Species with similar soil and water requirements should be planted together; those with vastly different requirements should be planted away from each other. Do you want a low-maintenance garden? Use species that require little care, such as slow-growing manzanitas and drought-tolerant California fuchsias. Since 70-90% of a home's water use is for its lawn, consider removing the lawn and installing a native garden in its place. As water prices continue to rise, your decreased water bill will make that native garden a sound financial investment.

We both have native gardens at our homes. We often use our garden plants as indicators of what's going on with the same species in the wild. For example, when lemonade berry, woollyleaf lilac, and stream orchids are in bloom in the garden, they are almost certainly in bloom in the wild, so we know it's time to go out and photograph them.

To learn about gardening with California native plants, we suggest these books: *California Native Plants for the Garden* (Bornstein, Fross, and O'Brien, 2005), *Reimagining the California Lawn* (Bornstein, Fross, and O'Brien, 2011), *California Native Gardening: A Month-by-Month Guide* (Popper, 2012), *Growing California Natives* (Schmidt, Greenberg, and Merrick, 2012), and *Designing California Native Gardens* (Keator and Middlebrook, 2007).

Take a class or attend workshops about native plant gardening. Visit native plant gardens to see how the plants are selected, arranged, and maintained. In Orange County, we are fortunate to have Tree of Life Nursery in San Juan Capistrano. They have demonstration gardens, superb plant selections, and knowledgeable staff; plus they hold workshops nearly every weekend. You'll find more information about native gardens under Native Plant Gardens, Arboreta, and Local Retail Sources in the Where to Go chapter near the end of this book.

A garden of California native plants will make your garden and home truly Californian!

Safety First

Wildflower-watching can be fun and rewarding, but nothing can botch the experience faster than an injury. A little planning in advance may help you deal with situations encountered on the trail. If you are new to hiking and fieldwork, partner with more experienced people and learn the ropes. Trained naturalists and docents can make a field trip enjoyable by keeping you safe while showing you wildflowers and other wildlife. Contact the local parks and reserves, listed in the Resources section of this book, to get their schedule of organized hikes. Meanwhile, the following brief discussion will get you thinking in the right direction.

Be prepared and do not hike or bike alone! Kids especially must not hike alone, even for a minute. Many children want to run or walk rapidly on the trails, even though you want to stop and look at the wildlife. Agree to a common pace or vary the pace, but stay together; don't let kids get ahead or behind you! In both of the two unrelated mountain lion attacks on children here in March and October of 1986, parents and their kids were separated by only brief minutes while on the same trails. You certainly wouldn't let your young child wander alone in a crowded shopping mall; you should never do so in the wilderness either.

Attire and Field Gear

Temperature? Wear warm clothing on cool days and cooler clothing on hot days. Take a light jacket, even if you carry it in your pack. When on the trail very early or very late in the day, plan for cooler weather than expected.

Long or short pants? Long pants will protect you from scrapes and exposure to poison oak but shorts are more comfortable. The location should dictate your selection: long pants for brushy areas with narrow trails, shorts for open areas with soft low-growing plants. If in doubt, wear long pants.

Shoes or boots? Some hikers swear by rugged boots, while others prefer the comfort of running shoes. Rugged hiking sandals are good for wide trails and beaches but not for brushy places.

To sleeve or not to sleeve? A long-sleeved shirt is always better than a short-sleeved one. It protects you from sunburn, many insect bites, and scrapes. An earth-toned shirt (muted green or tan) helps you fade into the background a bit, giving you the opportunity to see deer, birds, and insects that may be alarmed by bright colors or tones.

Fred Roberts in proper field attire

Always, always, always wear a hat! This is needed to prevent sunburn, sunstroke, heatstroke, and dehydration. It's also good for fanning away those bothersome canyon flies. Bob's beloved shade cap from Sun Precautions is goofy-looking but quite protective and comfortable.

Wear a daypack and use it to carry supplies. Be sure it has room for this field guide, small notepad, water, and a light jacket. Consider also carrying toiletry supplies (toilet paper, facial tissues) to address the occasional call of nature or a runny nose.

Drinking water is very important to a comfortable hike. Generally, each person needs a bottle or canteen, possibly more for extended hikes. Do not drink from natural water sources!

Basic first aid supplies and a guidebook. A great one is *Mosby's Outdoor Emergency Medical Guide* by Manhoff and Vogel (1996). It includes a nice list of emergency supplies. There are other great guides as well. Even if you don't take along such a guide, read it at home and learn a few things.

Sunscreen. Greasy sunscreen lotions are often sticky and uncomfortable and help dust to adhere to your body. Use something like Bullfrog Gel. It absorbs into your skin and doesn't leave you greasy.

Polarized sunglasses. They help cut glare so you can see more detail in the landscape, minimize eye fatigue, and sunburn. Since they change the character of light and make wildflowers look odd, be sure to take them off while wildflower-watching.

Physical Hazards

Watch for cliffs, eroding trails, falling trees and limbs, roots, sharp plants, sharp rocks on the trail, and loose rocks above you. Be mindful of steep trails and the velocity you pick up while heading down them. As a child hiking alone, Bob broke his ankle on the way down a steep hill – another reason not to hike alone.

Biological Hazards

We have only a few living things here that can be a problem. *Mosby's Outdoor Emergency Medical Guide* (Manhoff and Vogel 1996) discusses what to do when faced with most of these situations.

Water. Be wary of dehydration. You lose water more than you expect to, so carry your own water and drink it as you hike. Do not drink from a stream! Consider all free-flowing water sources as contaminated. The days of drinking pure natural water in the outdoors are long gone. Parasites such as *Giardia*, *Entamoeba*, and *Cryptosporidium* can give you a nasty case of diarrhea that you'll never forget.

Ingestion. Do not eat any plant or part thereof unless you are absolutely positive that it is safe! Some poisonings occur when visitors come here and mistake our plants for edible ones found elsewhere.

Stinging hairs. Some of our plants have stiff hairs that may cause dermatitis in sensitive-skinned people. Other species, however, have stiff hairs that inject toxins and produce a more serious dermatitis. Such hairs are common among plants in the borage family (Boraginaceae) and nettle family (Urticaceae). Skim though the entries for our hairy plants and learn to recognize them. See Crosby (2004) for more information about plants toxic to skin.

Gooseberry — Prickles and spines

Cactus — Long spines and glochids

Prickles, thorns, and spines. Some local plants have prickly, thorny, or spiny branches, often hidden by leaves. Cacti have both long straight spines and very short barbed ones (glochids) that are difficult to see and remove. To remove long spines embedded in your skin but still attached to plant material, place a stiff comb next to your skin around the spines, then pull it *away* from your body; do not swat it with your hand. For removal of individual spines and slivers, use fine-tipped tweezers available from outdoor stores.

Western poison oak (*Toxicodendron diversilobum*). This plant grows along all of our local trails, from the coast, valleys, foothills, and mountains. It's very fast-growing and can overtake a trail in a few weeks. Physical contact with it can cause a very itchy dermatitis that is slow to heal. Learn to recognize and avoid it. See the sumac family (Anacardiaceae).

Western Poison Oak

Poison Oak Dermatitis

"Leaflets three, let it be."

Spiders. Many species are native to our area. Nearly all can use their fangs to bite into skin and inject venom. The only acutely dangerous spider here is the western black widow spider (*Latrodectus hesperus*). Its venom is dangerous and painful, but fortunately the dose is very small. Black widows are shy and generally not aggressive. When the web is touched, they run to a protected place to hide. If bitten, seek medical help right away.

Scorpions. About five species are native to our area. Their sharp-tipped tail can inject venom into your skin, but most rarely do so. All are mildly venomous and generally do not require medical assistance unless the victim is allergic to venom. Play it safe and do not touch scorpions! If stung, apply ice and seek medical assistance.

Western Black Widow Spider

Ticks. Small arachnids about the size of a pencil-tip eraser, often smaller, they insert their mouthparts into the skin of animals and suck their blood. In the process, they can transfer ("vector") to you the bacteria that cause disorders such as Rocky Mountain spotted fever (vectored locally by the western dog tick, *Dermacenter variabilis*) and Lyme disease (vectored locally by the western black-legged tick, *Ixodes pacificus*). During your hike, check for ticks often. They prefer to settle down in snug areas such as at the top of your socks, near your belt line, and at the base of your hairline. During and after outings, check yourself and companions for hitchhiking ticks. After every hike, go home and strip off all clothing and wash it, check your body in a mirror, then shower. As you shampoo, work your fingers through your scalp in search of them.

Burrowing Scorpion

If you find a tick embedded in your skin, don't panic. Use narrow tweezers or a tick-grabbing tool (available from outdoor suppliers). Grab the head area (reach all the way to the tip of the head; you might get a small piece of your skin, but that's fine), and pull it straight out. Don't twist, don't turn, just pull straight out. Wash with soap and water, dry, then apply a topical antibiotic to prevent bacterial infection. You might want to keep the tick and have it identified by a qualified entomologist such as at the Orange County Vector Control District in Garden Grove (714-971-2421) or Northwest Mosquito and Vector Control District in Corona (951-340-9792). If you are bitten by a species that is a vector for Lyme disease, your physician may give you a preventative antibiotic.

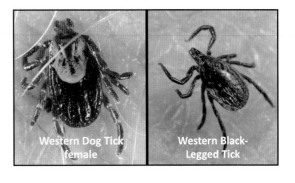

Western Dog Tick female

Western Black-Legged Tick

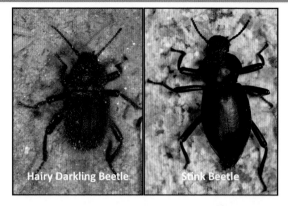

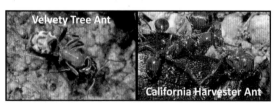

Beetles. A few of our beetles give off nasty defensive chemicals. **Darkling beetles** such as stink beetles (*Eleodes* spp.) are commonly found along trails throughout our area. When disturbed, some of them stand on their head and release benzoquinones from their abdomen. In addition to their awful smell, these chemicals irritate skin and the respiratory tract. As Bob learned in April 2005, a squirt in the eye from the **hairy darkling beetle** (*Eleodes osculans*) causes extremely painful ulceration of the cornea, requiring a trip to the emergency room and a week with an ophthalmologist. Keep your face away from them!

Flies. Female **mosquitoes** suck blood from vertebrate hosts and can transmit ("vector") bacteria, protozoans, and viruses that cause disease in the host, such as West Nile virus. **Deer flies** and **horse flies** painfully lacerate skin with their scissor-like mouthparts, then lap up blood. **Snipe flies** also inflict a painful bite.

Wasps, ants, and bees. These can give a nasty bite. Females of many species can also inflict a venomous sting (among insects, a stinger is a modified egg-laying apparatus; since only females lay eggs, only females can have a stinger.) Most of the time they will not bother you, but you may encounter and accidentally annoy them. Do not intentionally disturb them or their nests. Soldiers and workers may consider you to be an intruder and attack. People who are highly allergic to venom should always carry an adrenaline stick (or "epi pen") issued by their physician to treat incidental stings. **Yellow jacket wasps** (*Vespula pensylvanica*) build their paper-like nests in soil openings and rotten logs. **Velvet ants** (*Dasymutilla* spp.) are actually solitary wasps; their body is covered with long white, orange, or red hairs. Females are wingless (males have dark wings) and are often seen running about on the ground in search of the nests of ground-dwelling solitary bees in which to lay their eggs. They pack a powerful venom in their sting. **Tarantula hawk wasps** (*Pepsis chrysothemis*) are most often seen nectaring from flowers. They inflict what is reported to be the most painful sting in North America. **Velvety tree ants** (*Liometopum occidentale*) live under loose tree bark near watercourses. They cannot sting but will give a strong bite. **California harvester ants** (*Pogonomyrmex californicus*) are large orange-red ants that nest in dry sandy soils. They are quick to bite and sting. About five hundred bee species are native to our area. During the process of collecting nectar and/or pollen from flowers, they carry out vital pollination services without which most wildflower species could not reproduce. Most do not sting unless aggressively provoked. **Carpenter bees** (*Xylocopa* spp.) are very large. Most are all black, though the male of one species is golden. **Bumble bees** (*Bombus* spp.) are large and black with yellow stripes. They are the most likely native bee to sting. **Leafcutter bees** carry pollen on their bellies instead of in a pollen basket on their leg, as do most other bees. The non-native **European honey bee** (*Apis mellifera mellifera*) commonly stings people and pets when disturbed. **Africanized honey bees** (called killer bees by the media, a hybrid between the European honey bee and its African subspecies, *Apis mellifera scutellata*) and **red imported fire ants** (*Solenopsis invicta*) are found throughout Orange County, including some of the wilderness parks. These insects sting with little provocation; be sure to leave them alone.

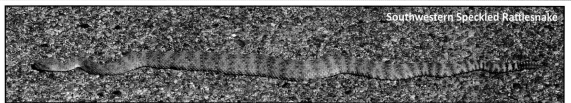
Southwestern Speckled Rattlesnake

Southern Pacific Rattlesnake

Snakes. About twenty species of snakes are native to our area, including three species of **rattlesnakes** that are the only dangerous poisonous snakes here. They have special organs that detect heat from your body, so they can sense you before you see them. As you approach, they often rattle their tail to warn you of their presence, which sounds like the shaking of a very fast baby rattle. A harmless insect, the rattlesnake grasshopper (*Chloealtis gracilis*), makes a similar though softer sound but stops "rattling" as you get closer to it. Rattlesnakes *increase* their rattling sound as you get closer. If you hear rattling, freeze and try to determine the snake's location. Only then should you try to back away. If you don't know where it is, you may accidentally go toward it.

Since snakes often rest under rocks, you should step *on* rocks, not over the edges of them. They also rest on rock outcrops and ledges, so when climbing, look where you are going to place your hands and feet *before* you place them there. Always "Think snake!"

When a rattlesnake bites, it injects a strong, painful toxin into the wound. Loss of a limb or death is possible if untreated. If someone is bitten, *do not* cut the wound open and suck out the poison as often shown in movies. *Seek medical assistance immediately.*

Mountain lion (also called cougar or puma, *Puma concolor*). Mountain lions live here. They were here first and we have intruded deep into their territory. You may encounter one when hiking, so be prepared. First, don't hike or bike alone. Keep children close to adults, with the tallest adults positioned toward the front and the back of your hiking group. Lions and their relatives often attack from behind, striking the base of the skull and breaking the neck. Do not intentionally put your back toward a lion.

If you see one or suspect that one is near, *do not run away!* Running may trigger the lion's instinct to attack and chase. Hold your ground, wave your hands, and shout at it. If the lion behaves aggressively, throw stones toward it to scare it away. Convince it that you are not food and that you might harm it. Do all you can to appear larger. Stand up straight and raise your arms above your head. Adults should pick up small children and carry them on their shoulders. Do not crouch down and try to hide from the lion. If you can see it, then it has already seen and smelled you. Report any mountain lion sightings in detail to a park ranger or other land manager.

Final Thoughts...

This review of hazards and safety is not meant to frighten you. Rather it is educate you on the reality of the wilderness and your responsibility to maintain your own safety while you enjoy it.

Red Diamond Rattlesnake

Above, local rattlesnakes. **Southwestern speckled rattlesnake** (*Crotalus mitchelli pyrrhus*) *is usually pinkish, speckled with black and white.* **Southern Pacific rattlesnake** (*Crotalus oreganus helleri*) *is grey to brown with dorsal "diamonds"; a white stripe runs from below the eye to the back corners of its mouth.* **Red diamond rattlesnake** (*Crotalus ruber*) *is orange to brown with dorsal "diamonds"; bold black and white bands go all the way around its tail.* **Below, mountain lion** (*Puma concolor*). *A large tawny cat with darker face and a very long tail. The young are spotted and resemble bobcats* (*Lynx rufus*) *which have very short tails.*

Mountain Lion

Allen's Daisy
(*Pentachaeta aurea* ssp. *allenii*)
CRPR 1B.1

The last word in ignorance is the man who says of an animal or plant: "What good is it?"
— Aldo Leopold (1887-1948)

Heart-leaved Pitcher-sage
(*Lepechinia cardiophylla*)
CRPR 1B.2

Much is missed if we have eyes only for the bright colors. Nature should be viewed without distinction... She makes no choice herself; everything that happens has equal significance.
— Eliot Porter (1901-1990)

Cleveland's Bush Monkeyflower
(*Diplacus clevelandii*)
CRPR 4.2

I think the environment should be put in the category of our national security. Defense of our resources is just as important as defense abroad. Otherwise what is there to defend?
— Robert Redford (1936-)

About the Entries

We include over 670 species of local wildflowers, selected from well over 1,000 that grow here naturally according to the western Riverside County and Orange County checklists (Roberts, et al. 2004, 2008).

Family names and the placement of species within them generally follows *The Jepson Manual*, second edition (Baldwin et al., 2012), Jepson Flora Project (2012), the Angiosperm Phylogeny Group (APG and APGII), Judd, et al. (2007), Simpson (2010), and current research. Descriptions are primarily based on information published in Baldwin et al. (2012), Hickman (1993), Abrams (1904, 1940, 1944, 1960), Munz (1973, 1974), and our own experiences.

Space prevented us from including every local wildflower in this book. Selecting which wildflowers to exclude was a difficult task. Within the Eudicots, we left out amaranths, most goosefoots, saltwort, white alder, waterworts, oaks, water-milfoil, walnut, myrsines, San Diego button bush, mignonette, and willows. Within the Monocots, we left out duckweeds, sedges, frog-bits, rushes, arrow-grasses, grasses, pondweeds, horned pondweeds, and eel-grasses.

Sequence

Wildflowers are presented in three groups: first the Magnoliids, then the Eudicots, then the Monocots. Next, each family within that group is organized alphabetically. In each family are presented the genera and the species within it, usually in alphabetical order. Especially important items are presented in italics.

Families. The entry for each family begins with its common and scientific names. General traits for each family are described. The family traits most helpful for recognizing our local members of the family are *italicized*.

Genera and Species. Genera are arranged alphabetically, as are most species within them. In some families, however, they are not in alphabetical order. Instead, we grouped similar-looking genera and species together to make it easier to compare and identify them. The entry for each wildflower species contains the items below.

Common Name. Many plants have several different common names, so we selected only the most commonly used and, in our opinion, best common name. We rarely provide more than one common name. Most are from the Orange and Riverside County checklists by Roberts, et al. (2004, 2008) If your favorite common name for a wildflower was not chosen, simply write it into your copy of the book.

We must confess to a few changes in common names. We dislike the use of a plant's scientific name *within* its common name, such as Grinnell's penstemon for *Penstemon grinnellii*. In these cases, we removed the misplaced scientific term and inserted the proper common name, which in this case is Grinnell's beardtongue. Also, we have tried to remove the word "weed" from the common names of native plants, such as *tarplants* (instead of *tarweeds*).

Quick Codes. To the right of a common name is a character that indicates if the species is sensitive, locally sensitive, non-native, and/or toxic. The characters and their designations are:

- ★ **Sensitive**. A native plant or animal that has federal and/or state legal status as Endangered or Threatened, and/or a native plant that has a California Rare Plant Rank (CRPR, explained on the following page).
- ✿ **Locally sensitive**. Few or dwindling in number, and/or in danger of extirpation within our area.
- ✖ **Non-native**. A species that is not native to our area. We include relatively few non-native plants in this guide. Most non-native plants are found only in urban/suburban areas; neither of those regions is within the scope of this book. However, we do present a few non-native species that are found in wild areas.
- ☠ **Toxic**. The toxin can be acquired through contact, inhalation, and/or ingestion. For many species, toxicity is unknown and individual reactions vary. Do not assume that a species is "safe" because it lacks a toxicity symbol.

Scientific Name. Generally, the most recently accepted scientific names are presented. There are some exceptions. We include some synonyms in square brackets, but only very common ones or those recently published. Most are from the Orange and Riverside County checklists by Roberts, et al. (2004, 2008) and from recently published scientific research.

Description. First is a description of the plant itself, seasonal growth (such as annual or perennial), its height (length if it's a vine or sprawling plant), leaf shape and size, inflorescence, and flowers. The flower description begins on the *outside* of the flower and works its way *inside*, from calyx to corolla, stamens to pistil, and sometimes fruits and seeds. The first paragraph always ends with its flowering period so you can find that information quickly. Keep in mind that *flowering times can change* with location, elevation, rainfall, temperature, and sun exposure, so use them only as guidelines.

Habitat and Location. Next is a note about the general type of habitat and plant communities in which it grows, followed by a list of local places where it has been found. Sometimes the locations are precise, sometimes quite general. Again, note that environmental conditions change over time, so a plant may not occur where or when this book directs you to look for it. If this happens, try similar habitats or slightly different times of year.

Natural History. Here we tell you how the plant lives, its abundance, vulnerability (if any, see Sensitive Species, below), scent, and pollinators when known. A plant's use in garden situations is added for especially worthy horticultural species.

We intentionally do *not* provide information about eating plants from the wild. Several local plants are poisonous and many look just like non-poisonous ones, especially to novice wildflower-watchers. Note that a longstanding urban legend says that if you see wild animals eat something, then it is safe for you to eat. *This is false! Do not graze in the wildlands!*

Origin and Meaning. We include names of people who gave each plant its name; sometimes we also give the name of its discoverer. We had no room to discuss these often-remarkable people, but such information is available online from Michael Charters's excellent website, *California Plant Names: Latin and Greek Meanings and Derivations. A Dictionary of Botanical and Biographical Etymology* (Charters 2011).

Next, we dissect the scientific name and provide its root word(s) and meaning(s). You'll find that many specific epithets often use the same root words. Knowing the meaning of a name can help you to understand why it was given and perhaps even to remember it.

Sensitive Species and Other Designations

In the second or a successive paragraph, we mention when a wildflower has been designated as a **sensitive species**. This indicates that the species is suffering from a threat to its existence, often from factors such as habitat lost to land development, cattle grazing, non-native plants, crushing by off-road vehicle use, trampling, pesticides, illicit horticultural collecting, drought, repeated fires, poor air quality, or loss of pollinators, which diminish its reproduction. The intent of designating a plant as sensitive is to make people aware of a plant's plight before it dwindles to a population that is too low to maintain itself and faces extinction.

When a native species is in serious trouble, it is proposed to be given formal legal protection under federal law (the Endangered Species Act, ESA) and/or state law (the California Endangered Species Act, CESA). The legal designations are **Federally Endangered**, **Federally Threatened**, **California Endangered**, and **California Threatened**.

In this state, the California Native Plant Society (CNPS) tracks native plants and if a species is found to be in trouble (usually dwindling in numbers and/or faces a threat that could significantly lower its populations), proposes to consider it a sensitive species. Each sensitive species is assigned a numerical designation (rank) from one of five categories (1A, 1B, 2, 3, 4). The ranks are determined by groups of interested parties such as CNPS, U.S. Fish and Wildlife Service (USFWS), and California Department of Fish and Game (CDFG). The ranking system is known as the **California Rare Plant Rank** (CRPR). Until early 2011, it was known as the CNPS List but was changed to CRPR to indicate that the ranks are not determined by CNPS alone.

In addition to CRPR, the California Native Plant Society assigns **Threat Rank Extensions** (.1, .2, .3) to the designation for each ranked species. These are a measure of how badly threatened the plant is in California. The extensions are expressed as a suffix added to the CRPR rank, such as 1B.2.

Native plant species that are rare, dwindling in number, and/or in danger of extirpation within the area defined by this book are given an informal rank called **Locally Rare** (also called Rare Local or Local Concern). It is determined by local botanists familiar with the species and its population trends in the area. It is given only to qualifying native plant species that have no other sensitive species designation but probably should have one.

Federal and State Designations

- Federally Endangered, as recognized by the U.S. Fish and Wildlife Service
- Federally Threatened, as recognized by the U.S. Fish and Wildlife Service
- California Endangered, as recognized by the California Department of Fish and Game
- California Threatened, as recognized by the California Department of Fish and Game

California Rare Plant Ranks (CRPR)

1A   Plants Presumed Extinct in California
1B   Plants Rare, Threatened, or Endangered in California and Elsewhere
2   Plants Rare, Threatened, or Endangered in California But More Common Elsewhere
3   Plants About Which We Need More Information — A Review List
4   Plants of Limited Distribution — A Watch List

CNPS Threat Rank Extensions

.1   Seriously threatened in California (high degree/immediacy of threat)
.2   Fairly threatened in California (moderate degree/immediacy of threat)
.3   Not very threatened in California (low degree/immediacy of threat or no current threats known)

Resources and References

After the wildflower section is a discussion of places to go for more wildflower viewing and learning. It includes parks, nature centers, herbaria, native plant gardens, and native plant nurseries. It ends with a list of natural history-type organizations that you might want to join. That is followed by a long list of most of the references consulted during the preparation of this book.

Indices

We provide three different types of index, each with a different purpose.

General Index is a traditional alphabetical list of plants, plant families, major plant groups, and associated animals mentioned in the book, with page numbers.

Quick Index by Flower Color lists major colors, the species of flowers that contain those colors, and page numbers to those species. This index was the brainchild of Linnell Allen, who painstakingly assembled it with Bob.

Quick Index is a simplified index to the major plant groups and some genera, printed on the inside back cover.

MAGNOLIIDS

Terrestrial flowering plants, generally trees, shrubs, or vines. Cotyledons 2 per seed; after seed germination, they usually serve as short-lived storage organs and/or photosynthetic leaves. *Leaf veins branched and net-like. Leaf base sheathed or not. Flower parts in multiples of 3 or flowers spirally arranged on a central stalk.* Many (e.g., bay laurel, cinnamon, and camphor) contain aromatic oils and have characteristic scents. Pollen aperture 1 per pollen grain.

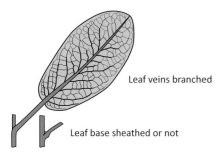
Leaf veins branched
Leaf base sheathed or not

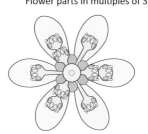
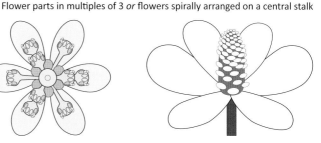
Flower parts in multiples of 3 *or* flowers spirally arranged on a central stalk

LAURACEAE • LAUREL FAMILY

Woody, evergreen trees, shrubs, and rarely leafless vines. *Leaves undivided,* rarely lobed; *dotted with tiny pits; alternate.* Stipules absent. Flowers yellow or greenish. Some species have all flowers on a plant of only one sex, other species have flowers of both sexes on the same plant. *Tepals sepal-like or petal-like, in 1-2(3) whorls. Stamens in 3-4 whorls of 3 each. Filaments of many species with a pair of nectar- or odor-producing appendages.* In some species, the innermost whorl of stamens has no anthers (they are "sterile staminodes") that may also have a pair of nectar- or odor-producing appendages. *Anthers with 4 chambers; pollen is shed when 2-4 flap-like valves curl upward and pull out the pollen. The pistil is simple (undivided) with a single style.* The *ovary is superior* with a single chamber and ovule; thus, their fruit has only 1 (unusually large) seed. Fruit a berry or drupe.

A family of 2,200 mostly tropical species, only 1 in California. Among its members are some familiar cultivated plants such as avocado, cinnamon, camphor, and sassafras. Rosewood is used in perfumes and cosmetics.

California Bay Laurel

California Bay Laurel
Fred M. Roberts, Jr.

California Bay Laurel
Umbellularia californica (Hook. & Arn.) Nutt.

A rounded shrub or tree up to 40 meters tall. Young bark green and smooth, older bark reddish-brown and rougher. *Leaves 3-10 cm long, oblong to lance-shaped, smooth and shiny. Flowers yellow-green, the 6 petal-like tepals in 2 whorls of 3, easily fall off.* Stamens 9, in 3 whorls of 3; the 3 innermost have orange-colored nectar-secreting appendages. Fruit 2-2.5 cm wide, spherical, green with tiny yellow dots, ripens to dark purple. In flower Dec-May.

A common plant, generally of north-facing slopes and moist canyons in the Santa Ana Mountains. Less common in lower-elevation foothills as in Caspers Wilderness Park. Easily seen in Silverado Canyon and along Holy Jim and Trabuco Trails. The scent of its cut or crushed leaves is quite strong, pleasant to some, not so to others. In some people, inhaling the vapors cures a headache, in others it causes a severe headache; in a few people it causes unconsciousness! Makes a fine evergreen tree in the garden where it casts a great deal of year-round shade.

The epithet was given in 1833 by W.J. Hooker and G.A.W. Arnott for the region of its discovery.

California Bay Laurel

SAURURACEAE • LIZARD'S-TAIL FAMILY

A small family of 7 species in North America and eastern Asia. They are all *perennials* that grow from *rhizomes or stolons, with undivided alternate leaves. Each inflorescence is a dense spike of individual flowers with a group of large petal-like bracts at its base. The flowers have neither petals nor sepals.* Stamens 6 or 8 in number, rarely 3. Ovary superior or inferior. Fruit is a fleshy capsule.

Yerba Mansa

Anemopsis californica (Nutt.) Hook. & Arn.

This unusual plant grows from a thick rhizome that creeps along just under or on top of the ground. Its main stem reaches 15-50 cm tall, is hollow, and hairy or hairless. It springs from a *basal tuft of large elliptic or oblong leaves,* 5-15 cm long, often with rounded lobes at their base. There are a few oval-shaped stem leaves as well. The *conical inflorescence* stands upright, 1.5-4 cm tall. *Below it is a ring of 5-8 bracts that look like large white petals, sometimes tinged or dotted with red. Each flower is tiny and has no petals but instead has a small white bract just below it.* Ovary inferior. In flower Mar-Sep.

Once a common wetland plant in our area, but much of its habitat has been drained, filled, and built upon. It is now found in only a few places here. Currently known from coastal places such as Bolsa Chica State Ecological Reserve and upper Newport Bay, in lowlands along Laguna Canyon Road in Laguna Beach, in Whiting Ranch Wilderness Park, and at Trestles where San Mateo Creek meets the ocean. Found inland at the Nichols Road wetlands of Lake Elsinore and on the Santa Rosa Plateau. In cultivation it generally requires partial shade and plentiful watering. It does especially well in and near garden ponds.

The genus was established in 1838 by W.J. Hooker in reference to the similarity of its inflorescence to that of wind flowers in the genus *Anemone* (Greek, -opsis = like). The epithet was given in 1838 by T. Nuttall for the place of its discovery.

Upper: At the edge of salt water at Bolsa Chica. Lower: Each inflorescence is subtended by 5-8 large white bracts. Individual flowers have no petals but each is subtended by its own tiny white bract.

Lower: Yerba mansa often turns red in the fall and when growing in highly alkaline soils or experiencing water stress. These plants were growing along San Mateo Creek near the ocean at Trestles.

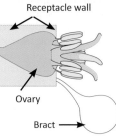

Individual flower of yerba mansa, with its associated spoon-shaped bract. The ovary is embedded in the receptacle wall, up to the base of the stamens. Style branches curl backward with age, easily seen in the photograph above.

EUDICOTS

Terrestrial or aquatic flowering plants with a variety of growth forms. Cotyledons 2 per seed; after seed germination, they usually serve as short-lived storage organs and/or photosynthetic leaves. Leaf veins branched and net-like. Leaf base usually not sheathed (sheathed in the carrot family, Apiaceae). *Flower parts generally in multiples of 4 or 5*. Pollen apertures 3 or more per pollen grain.

Leaf veins branched
Leaf base usually not sheathed

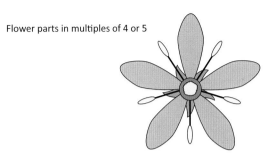
Flower parts in multiples of 4 or 5

ADOXACEAE • MUSKROOT FAMILY

Perennial herbs, shrubs, and trees with single or multiple trunks. *Leaves opposite* on the stem. In our area each leaf is *pinnate;* some species found elsewhere have undivided leaves. *Flowers arranged in flat-topped clusters at branch tips.* Calyx and corolla very small and cup-like, with 2-5 (calyx) and 4-5 (corolla) free lobes. Stamens 5. *Ovary inferior,* styles (0)3-5. Fruit is a berrylike drupe.

Elderberries were formerly placed within the honeysuckle family (Caprifoliaceae) but reinterpretation of their anatomy and DNA sequencing results suggest that they should be treated as separate families. Two genera live in California (*Sambucus* and *Viburnum*), but only *Sambucus* occurs in our area.

Blue Elderberry

Blue Elderberry
Sambucus nigra L. **ssp.** ***caerulea*** (Raf.) Bolli
[*S. caerulea* Raf.; *S. mexicana* C. Presl, misapplied]

A deciduous shrub or small tree, 2-8 meters tall, about as wide, with *multiple slender trunks and brittle branches that arch* back toward the ground. *Leaves often strongly arched; opposite;* 3-20 cm long, *odd-pinnate,* with 3-9 elliptical leaflets and toothed edges. The *tiny flowers* are *cream to yellow,* arranged *in dense flat-topped clusters,* 4-33 cm across. Calyx and corolla have 5 free lobes each. The 5 stamens alternate with the petals. Ovary inferior, styles 3-5. Fruits 3-5 seeded. In flower Mar-Sep.

The flowers do not produce nectar, thus insects visiting them (mostly bees, wasps, flies, and beetles) probably do so only to gather pollen. Its plentiful round fruits are 5-6 mm in diameter, blue or white, and are covered with a dense white powder. The blue fruits are actually blue-black but the powder makes them appear whitish. As the fruits ripen and swell, the cluster droops downward under the weight. After losing its leaves in fall (earlier in summer drought), the plant looks like a large bundle of arching sticks.

Blue elderberry is a common component of many plant communities such as coastal sage scrub, chaparral, riparian, oak woodland, and coniferous forests. It lives along the foggy coastline, in warm sunny hills and valleys, throughout the Santa Ana Mountains, the Puente-Chino Hills, and in the Temescal Valley.

Blue Elderberry

Elderberries are the source of folklore, music, and wine. *Nearly all parts contain cyanogenic glycosides and iridoides that are toxic to mammals.* Some birds eat the berries with impunity. The ripe berries are not overly toxic to humans but cause nausea if eaten raw; only cooked berries can be eaten, but even these will cause nausea if eaten in quantity. Elderberry wine and jams are made from *properly cooked* berries. The stems are mostly pith that can be removed and the stem hollowed out to make a flute, blowgun, or whistle, but human poisonings are known from this use. Thus, it is best to admire and not ingest elderberry.

Sambucus is the Latin term for the European elder tree and was used in 1753 by C. Linnaeus as the name of its genus. Earlier, the Greeks called a stringed musical instrument *sambuke* because it had parts made from European elderberry.

For many years, our elderberries were called Mexican elderberry (*Sambucus mexicana*). That was based on a misidentification many years ago, propagated by multiple authors. A recent study set this straight (Bolli 1994).

The epithet was given in 1838 by C.S. Rafinesque-Schmaltz for the blue color of its fruits (Latin, caeruleus = dark colored, dark blue).

Similar. **Ash trees** (*Fraxinus* spp., Oleaceae): leaves not strongly arched; opposite; petioles with a groove on their upper surface; flowers not in flat-topped clusters,; flowers have 2 petals each (*F. dipetala*) or no petals (*F. velutina*).
Walnuts (*Juglans* spp., Juglandaceae, not pictured): leaves alternate; petioles not grooved; when crushed, leaves have an oily odor; flowers have no petals, each flower is unisexual.

Blue Elderberry

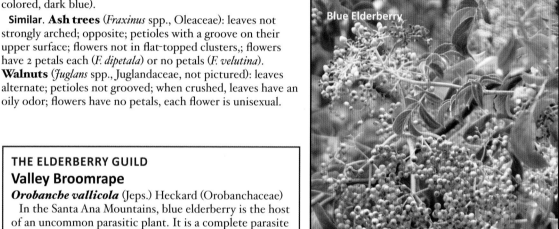
Blue Elderberry

THE ELDERBERRY GUILD
Valley Broomrape
Orobanche vallicola (Jeps.) Heckard (Orobanchaceae)

In the Santa Ana Mountains, blue elderberry is the host of an uncommon parasitic plant. It is a complete parasite (holoparasite) that cannot manufacture (photosynthesize) its own food and must steal nutrients and water from the host plant. Currently known here only from lower Holy Jim Canyon. It is pictured and discussed under Orobanchaeae.

California Elderberry Longhorn Beetle
Desmocerus californicus californicus Horn (Cerambycidae)

Adults up to 2 cm long, *rough-surfaced, black;* antennae as long as the body or nearly so. *Front wings (elytra) black, often with a blue-purple sheen, bordered with orange.* After mating on the tree, the female lays eggs on bark; larvae hatch and tunnel into the plant where they feed for up to 2 years. Before they pupate, larvae chew an oval exit hole about 1 cm wide, just above ground level. Adults emerge when the trees begin to flower in spring, generally Mar-June, and remain on the trees, where they feed on leaves and flowers. They are not abundant and are not considered a threat to the trees. With the loss of elderberry trees and their riparian habitat in California's central valley, the form of the beetle found there (*Desmocerus californicus dimorphus* Fisher) is listed as Federally Endangered. Though uncommon here, our local form of the beetle is not listed as endangered. Bob Allen has found the adult and exit holes in Whiting Ranch Wilderness Park, only exit holes in Crystal Cove State Park. Known in the 1960s from upper Newport Bay, but not found there recently.

AIZOACEAE • CARPET-WEED FAMILY

Hairless herbs or low shrubs that sprawl on the ground, though a few species stand upright. The *stems and leaves* of most are *fleshy and succulent.* Leaves alternate, opposite, or whorled. *Flowers cup-shaped,* those of some species outwardly similar to those of cactus (Cactaceae). Sepals 4-8, overlapping. Petals (0-)many, narrow and linear. Ovary superior or inferior, styles 0-20, stigmas 1-20; stamens variable in number. Fruit a capsule.

A small family of 2,500 species, primarily from southern Africa; 13 in California, 9 in our area. Most are escapes from cultivation, now found on our beaches and alkaline areas, where they crowd out and smother native vegetation.

Hottentot Fig ✖
Carpobrotus edulis (L.) N.E. Brown

A large mat-growing perennial. Leaves 6-10 cm long, *sharply triangular in cross-section, opposite, not covered with a fine whitish powder. Each leaf has teeth on the outer angle near the tip.* Flowers large, 8-10 cm in diameter, petals pink to yellow, age to pink. In flower Apr-Oct.

Originally planted for soil-stabilization, especially along sandy areas and beaches, but has escaped into the wild where it chokes out native plants and alters habitats. Now a common weed along the coast as at Bolsa Chica Ecological Reserve, Balboa Peninsula, Newport Bay, Crystal Cove State Park, San Clemente State Beach, etc. Native to South Africa. Named in 1759 by C. Linnaeus, who claimed it to be edible (Latin, edulis = edible).

Similar. Sea-fig (*C. chilensis,* not pictured): *leaves rounded-triangular in cross-section, not toothed;* flowers much smaller, 3-5 cm in diameter, rose-magenta; much less commonly found.

Croceum Iceplant ✖
Malephora crocea (Jacq.) Schwantes

A dense perennial with stout stems. *Leaves 2.5-6 cm long, round in cross-section, blue-gray in color, may be tinted red. Inner surface of the petals orange, the outer surface purple.* In flower much of the year.

A garden plant that often escapes into the wild to become a weed of coastal areas and bluffs such as at Newport Bay, cliffs of Dana Point Headlands, and San Clemente. Native to South Africa. Named in 1800 by N.J.B. von Jacquin for its orange-yellow "saffron" petals (Latin, croceus = saffron-colored).

Crystal Iceplant ✖
Mesembryanthemum crystallinum L.

A low-growing annual *covered with large crystal-like water-filled "warts." Leaves 2-20 cm long with a broad, flat, wavy-edged blade,* green to red in color. Flowers 7-10 mm across. Petals white, age to pink. In flower Mar-Oct.

An abundant weed of the sandy coastline, rocky bluffs, and disturbed soils nearby. Native to South Africa. Named in 1753 by C. Linnaeus for its crystal-like "warts" (Greek, krystallos = clear ice, glass).

Similar. Small-flowered iceplant (*M. nodiflorum*): water-warts small; leaves short, 1-2 cm, finger-like; flowers small, 4-5 mm across.

Small-flowered Iceplant ✖
Mesembryanthemum nodiflorum L.

A mat-forming annual, *covered with small crystal-like water-filled "warts." Leaves 1-2 cm long, round in cross-section, finger-like,* bright green to reddish. Flowers 4-5 mm across. Petals white, age to yellow. In flower Apr-Nov.

A common weed of our shoreline, especially common at Crystal Cove State Park. Native to South Africa. Named in 1753 by C. Linnaeus for its flowers, which appear at the ends of knot-like branches crowded with leaves (Latin, nodus = knotty, nobby; floris = flower).

Similar. Crystal iceplant (*M. crystallinum*): water-warts larger; leaves long, 2-20 cm long, flat; flowers large, 7-10 mm across.

Western Sea-purslane
Sesuvium verrucosum Raf.

A low-growing hairless perennial covered with minute "warts". *Leaves 0.5-4 cm long, linear to spoon-shaped,* the edges sometimes rolled under. Its short flowers arise from the nodes. *The calyx is 5-lobed and purple-pink.* Petals are absent. The ovary is green, the style short, immediately 3-branched. Stamens numerous, the filaments purple-pink. In flower Apr-Nov.

The only native member of this family in our area. It occurs along our coastline at Anaheim Bay, Bolsa Chica Ecological Reserve, Newport Bay, San Joaquin Marsh, and Laguna Canyon. It appears sparingly inland such as at Lambert Reservoir in the Lomas de Santiago and at Nichols Road wetlands in Lake Elsinore.

Named in 1838 by C.S. Rafinesque-Schmaltz in reference to the warts that cover the plant (Latin, verrucosus = full of warts).

Similar. Alkali heath (*Frankenia salina,* Frankeniaceae): shrubby; no water-warts; leaves elliptical to linear, the edges rolled under; *long tubular flowers;* sepals 4-7, fused; petals 4-7, fused below, free above; stamens 6, pale pink filaments, anthers red to purple; single slender style tipped with 3-branches; grow together at Bolsa Chica Ecological Reserve and upper Newport Bay.

Western Sea-purslane

New Zealand Spinach ✖
Tetragonia tetragonioides (Pallas) Kuntze

A spreading or upright annual that often grows in dense stands. It is covered with tiny water-filled "warts." *Leaf blades 2-5 cm long, triangular to ovate.* Its tiny flowers appear along the stem at the nodes. *Sepals are tiny and yellow-green; petals are absent.* The fruit has four flattish sides and 2-5 horns. In flower Apr-Sep.

An uncommon plant of the coast, mostly found at Bolsa Chica Ecological Reserve, Crystal Cove State Park, Laguna Beach, and Dana Point. Native to southeast Asia and Australian regions.

Named in 1781 by P.S. Pallas for its four-sided fruit (Latin, tetragonum = a quadrangle; -oiodes = like).

New Zealand Spinach

ANACARDIACEAE • SUMAC FAMILY

These are *shrubs, trees, or rarely vines.* Leaves are deciduous or evergreen, simple or divided into three lobes (more lobes in eastern species). Numerous flowers appear in their inflorescence, which is usually held upright or arching up and over. Though *flowers are small,* they occur *in dense clusters* that are easily noticed. *Sepals and petals 5, stamens 5-10. Ovary superior, the single style has 3 branches near its tip.* Most flowers possess both male and female parts, but some have only one sex. Lemonade berry and sugar bush, for example, have standard flowers with both sexual parts, plus some flowers that are either all-male or all-female. Such plants are termed *polygamous.*

The nectary is a flat doughnut-like disk below the ovary, visited most often by bees and flies. The fruit is a drupe; at its center is a single seed within a hard case. Most sumacs contain a sticky odorous sap; contact with the sap of some species produces contact dermatitis in humans. Well-known members of this family include poison oak, poison ivy, cashew, mango, and pistachio. Our 5 local sumacs are all common and easily found in local parks and preserves.

Lemonade Berry

Laurel Sumac
Malosma laurina (Nutt.) Abrams

A *large shrub or small tree, 2-6 meters tall*. Its *pliable evergreen leaves* are folded upward at the sides to direct dew and rain water downward to the roots. Leaves and stems are green with a suffusion of purple. *Petioles and twigs are often purple or burgundy.* Flowers are *cream-colored* and appear in pyramid-like clusters at the branch ends. In flower June-Aug, much later than its local relatives.

Widespread and common, it occurs in coastal canyons and foothills to low elevations in the Santa Ana Mountains. Found in nearly all of our wilderness parks. It is especially *sensitive to frost* and unusually cold winter nights, which kill off many of the upper leaves and branches, leaving dead areas. Most of these plants grow back from lower branches.

The genus and epithet were given in 1838 by T. Nuttall. The genus refers to the *apple-like smell of the cut leaves,* although some people liken it to the smell of bitter almonds (Latin, malum = apple; Greek, osme = smell or scent). The epithet refers to its similarity to a laurel tree (Latin, laurinus = of laurel).

Similar. Sugar bush (*Rhus ovata*): leaves very thick and waxy, petioles red; inflorescence much shorter, in flower much earlier.

Laurel Sumac

Laurel Sumac

Skunkbrush, Basketbrush
Rhus aromatica Aiton
  [*R. trilobata* Nutt. ex Torr. & A. Gray]

A *deciduous, arching, open shrub* from 0.5-2.5 m tall that resembles poison oak or currant (*Ribes* spp., Grossulariaceae). *Leaves are flat and divided into 3 (rarely more) leaflets with wavy edges, each often with square-edged lobes.* Its small yellow *flowers* generally *appear before the leaves.* In flower Mar-Apr.

Not toxic, though when bruised or cut it releases a strong odor considered unpleasant by some people. It is common but usually not abundant in oak woodlands often under partial shade. A few are found in Silverado Canyon and along San Juan Canyon but more common in higher elevations along San Juan Loop Trail, Upper San Juan Campground, the Ortega Candy Store, Blue Jay Campground, along South Main Divide Road, atop Santiago Peak, south into the San Mateo Canyon Wilderness Area.

The epithet was given in 1789 by W.T. Aiton in reference to its odor (Greek, aromatikos = fragrant).

Similar. Western poison oak (*Toxicodendron diversilobum*): leaflets generally not lobed; when lobed, usually round-edged; *petiolule long; flowers appear with or after the leaves;* herb, shrub, vine, or tree-like; abundant.

Fruits

Skunkbrush

Skunkbrush

Lemonade Berry
Rhus integrifolia (Nutt.) Brewer & S. Watson

A *large rounded shrub*, 1-8 meters tall. Its *stiff evergreen leaves* are *very thick, covered in wax,* and usually *flat* or wavy-edged, most always with small teeth at their edge. Petiole is green to pink. Flowers appear in small clusters near the branch ends. *Corolla white to pinkish.* In flower Feb-Mar.

The hard fruit is shaped like a fresh corn kernel, brown at first, then ripens to bright red in April-June. It is coated with a thick, white, pasty edible substance that tastes bitter, like unsweetened lemonade. If you elect to taste the substance, do not chew or eat the fruit, just spit it out. Occurs on coastal bluffs, in coastal canyons, and at lower elevations of the Santa Ana Mountains and Puente-Chino Hills. Quite common in our wilderness parks. A solid garden plant.

The genus was established in 1753 by C. Linnaeus (Latin, rhus = sumac). T. Nuttall provided the epithet in 1838, in reference to its entire (not divided) leaves (Latin, integ = entire; folium = leaf). Many sumac species, especially those in the eastern half of the U.S. where he lived, have divided leaves, so this sumac must have appeared unusual to him.

Similar. Sugar bush (*R. ovata*): *leaves folded up at the sides;* fruit without lemon-tasting paste; usually higher elevations.

Sugar Bush
Rhus ovata S. Watson

Grows to become a *large shrub*, 2-10 meters tall. *Its stiff evergreen oval leaves are waxy, folded up at the sides, and may be wavy-edged.* Petioles red or green. Flowers appear in small clusters near the branch ends. In winter, its bright red buds stand out against its contrasting green foliage. The *red sepals* open up (and *often fall off*) to reveal white to pinkish corollas. Fruits are red and short-haired like lemonade berry but are not coated with a white paste. In flower Mar-May.

Sugar bush lives in foothill canyons of the Santa Ana Mountains at higher elevations than lemonade berry. The two species hybridize in areas where both live. Common in Silverado Canyon, San Juan Canyon near the San Juan Loop Trail, Upper San Juan Campground, and North Main Divide Road not far from Ortega Highway.

The epithet was given in 1885 by S. Watson in reference to its oval leaves (Latin, ovatus = egg-shaped).

Similar. Lemonade berry (*R. integrifolia*): leaves round to oval, flat; fruits with lemonade paste; usually lower elevations. **Laurel sumac** (*Malosma laurina*): leaves not as thick and waxy, petioles purple; inflorescence much longer, in flower much later.

Western Poison Oak ☠

Toxicodendron diversilobum (Torr. & A. Grey) E. Greene

A lovely plant with variable forms. *It can grow as a low weak-stemmed herb-like shrub, as a woody shrub 0.5-4 meters tall, or as a vine up to 4 meters long that trails up and over other plants.* In wetter parts of the state, its stems reach up to 25 meters long; occasionally those individuals are treelike, growing alongside larger plants such as true oak trees and toyon. *Leaves shiny,* hairless above, *divided into 3 leaflets* (rarely 5) *that may be small or large, smooth-edged or lobed. Terminal leaflet 1-13 cm long; its petiolule is long and slender, not winged with leaf blade tissue.* Leaves are typically bronze when first out of bud, turn green with maturity, and change to red, orange, or yellow before they drop off in fall. *Flowers unisexual, generally appear with or after the leaves break bud. Stamens 5 in male flowers,* vestigial in female flowers (anthers, if present, are empty). Pistil 1 in female flowers, vestigial in male flowers. Styles 3 in female flowers, fused at base, with 3 terminal branches. *Petals 5(4), creamy-yellow to greenish.* Fruit a spherical white berry, 1.5-6 mm wide. In flower (mostly) Mar-Apr.

One of the most hazardous plants in California, according to Dieter H. Wilken in *The Jepson Manual* (Baldwin et al. 2012), western poison oak grows abundantly throughout our area along the coast, coastal canyons, waterways, foothills, woodlands, and mountains. It occurs in all of our wilderness parks and along most forest trails.

The toxin within its sap, *urushiol* (pronounced oo-RU-she-all), is present year-round within and on the surface of the plant. When you contact the leaves or stems, it oozes onto your skin and can cause allergic contact dermatitis (ACD), an intense itching sensation accompanied by fluid-filled blisters. Scratching the rash only makes it worse by tearing open the blisters and may lead to secondary bacterial infection. It generally takes 1-2 (to 10) days after contact for ACD to appear, and it worsens and persists for 2-3 weeks. According to Crosby (2004), "Although the liquid in the blisters is harmless, contact with the affected area can spread surface allergen to other parts of the body." It can also be transferred by contaminated clothing, shoes, tools, and pet fur. Local biologists and naturalists report that Tecnu® products by Tec Labs and Zanfel® by Zanfel Laboratories can be used to wash affected skin and contaminated clothing.

Western Poison Oak
Leaflets three, let it be!

Its attractive clusters of creamy yellow flowers are best watched through binoculars or a camera lens

female flowers

The extremely shiny lacquer that coats Japanese-made pottery and furniture is made from the sap of a related plant, the **Japanese lacquer tree** or **urushi** (*Toxicodendron vernicifluum* (Stokes) F. Barkley). The toxin urushiol was first isolated from the sap of this tree and named after it (Japanese, urushi = lacquer). Urushi lacquer can cause ACD when in liquid form but not after it dries and cures. Legend has it that people of Asian or Native American descent are less sensitive to urushiol than others. We do not know if this is true. The most prudent course of action, regardless of your ancestry, is to avoid touching this plant.

The genus was originally named by J.P. de Tournefort for its toxicity (Greek, toxicos = poison; dendron = tree) and formally published in 1735 by C. Linnaeus. The epithet was given in 1838 by J. Torrey and A. Gray for its lobed leaves (Latin, diversus = separated or turned; lobos = lobe).

Similar. Skunkbrush (*Rhus aromatica*): in open sun among chaparral; a shrub with arching branches; *flowers appear before the leaves; leaves divided into three or more wavy- or square-edged leaflets.* Usually grows in dry areas near coast live oak (*Quercus agrifolia,* Fagaceae). **California blackberry** (*Rubus ursinus,* Rosaceae): in moist soil near shady watercourses; always a vine, never a shrub; sharp *prickles along stems, petioles, and leaf undersides;* leaves 3-lobed or pinnately divided into 3-5 lobes, most always pointed and toothed; *large white petals,* 5-25 mm long; fruit an egg-shaped aggregate of tiny fleshy berries, each dark red to black. Especially common near watercourses. **Virgin's-bowers** (*Clematis* spp., Ranunculaceae): in moist or dry soil; always a vine; *leaves opposite, 1-2 pinnate, leaflets often lobed and toothed;* petioles grasp and twine around other plants and objects; flowers much larger and obvious; fruit a cluster of dry seeds, each with a fuzzy tail.

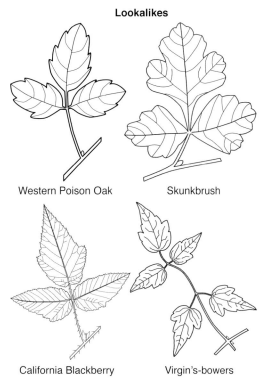

Lookalikes

Western Poison Oak | Skunkbrush

California Blackberry | Virgin's-bowers

APIACEAE • CARROT FAMILY

A family of *herbaceous annuals and perennials, geophytes, woody perennials, and creepers*. Most have a *stout taproot* like that of cultivated carrot. Their *leaves* are *alternate* (rarely opposite) and *pinnate*. *Leaf bases have sheaths that wrap around the stem for attachment.* Their *tiny flowers* are arranged *in* small rounded clusters called **umbels** (Latin, umbella = sunshade) in reference to the umbrella-like flower clusters. The stem that leads to each flower in an umbel is called a **ray** and may be of the same or different lengths within the same umbel. These umbels are, in turn, themselves arranged into a **primary umbel**, connected by long rays of different lengths, which may have a leafy bract at its base. Each flower has *5 sepals, 5 petals, and 5 stamens. The inferior ovary has 2 styles attached to a swollen base.* Petals are yellow or white, very few are purple or greenish. *Every flower produces 1 fruit that splits into 2 dry halves; each half contains only 1 seed.* For many species in this family, you need the fruit in order to identify it.

Many foreign species such as dill, caraway, garden parsley, and carrot are cultivated for food, spice, or ornamental garden use. Some species are highly toxic and even fatal if ingested. Because of this, one should never eat any member of the carrot family from the wild. A tea made from common poison hemlock, which was introduced here from Europe, is reported to have been used to kill the Greek scholar Socrates in 399 B.C.

Mock Parsley
Apiastrum angustifolium Nutt.

A *weak-stemmed, upright annual*, 5-50 cm tall. *Leaves* mostly opposite, bright green, 1-5 cm long, *finely dissected into thin leaflets, each leaflet often further dissected into 3 smaller leaflets*. Flowers are in a loose umbel. Sepals absent. Petals small, bright white. Fruits 1-1.5 mm long, shaped like an upside-down heart or oval, not flattened, covered with tiny bumps (not spines), and without obvious ribs. In flower Mar-Apr.

Very common in grasslands and along trailsides, from coastline to the mountains.

Named in 1840 by T. Nuttall in reference to its narrow leaflets (Latin, angustus = narrow, small; folium = leaf).

Similar. California hedge-parsley (*Yabea microcarpa*): *leaves alternate, 1-2-pinnate;* petals alternate between large and small sizes; fruit elongate, ribbed, and spiny.

California Hedge-parsley
Yabea microcarpa (Hook. & Arn.) Koso-Pol.

A *slender annual* that stands 3-40 cm tall. *Leaves 1-2-pinnate* with threadlike or linear leaflets. *Leaflets round near the end and come to a short point.* Its very tiny flowers have obvious sepals and oval, narrow-tipped, white petals. *Petals often alternate from large to small sizes. Fruits 3-7 mm long, elongate, oblong, flattened on one side, ribbed, winged, covered with stiff yellow spines that point upward.* In flower Apr-June.

Common in moist grasslands, chaparral, and woodlands from our coast to foothills and canyons. Often grows in a tight group.

The genus was established in 1915 by B.M. Koso-Poljansky to honor Y. Yabe. The epithet was given in 1838 by W.J. Hooker and G.A.W. Arnott in reference to its small fruits (Greek, mikros = small; karpos = fruit).

Similar. Mock parsley (*Apiastrum angustifolium*): *leaves opposite, divided into 3 leaflets,* often again into 3 smaller leaflets; petals all same-sized; fruit heart-shaped, bumpy, and spineless.

Common Celery ✖
Apium graveolens L.

An *upright perennial with a strong celery scent,* usually to 0.5 meter but known to reach 1.5 meter tall. Its *long wand-like stems* are *concave on the inner surface, convex and grooved on the outer.* Leaves 1-pinnate, leaflets narrow-based, lobed, angular. *Petals minute, white to yellow-green.* Fruits oval to round; flattened, spineless, hairless, and ribbed. In flower May-July.

Native to Eurasia, this is the store-bought celery plant that escaped from cultivation and now grows in creeks and other moist places such as upper Newport Bay, Laguna Beach, Aliso Viejo, Mission Viejo, and Whiting Ranch Wilderness Park. Since it often grows in dirty and contaminated waters, wild-collected plants should not be eaten.

Named in 1753 by C. Linnaeus in reference to its strong odor (Latin, graveolens = strong-scented).

Similar. Cutleaf water-parsnip ☠? (*Berula erecta*): no celery scent; leaves basal, upright; leaflets oval, broad-based; uncommon perennial native of watercourses; co-occurs in Peters Canyon. **Common poison hemlock** ☠✖ (*Conium maculatum*): taller; annual; purple spots on stems; leaves 2-pinnate into tiny leaflets; abundant weed.

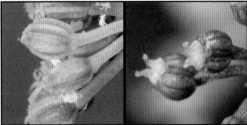

Right: The fruits of common celery are round and have prominent ribs. Far right: the fruits of cutleaf water-parsnip are round to heart-shaped and have low ribs.

Cutleaf Water-parsnip ☠?
Berula erecta (Hudson) Coville

An upright, hairless, *odorless perennial* from an underground stolon. *Non-flowering plants have clusters of mostly basal leaves that stand upright,* each with a long petiole; the blade to 15 cm long, *1-pinnate,* with 7-12 oval, broad-based leaflets. Inflorescence to 1.5 meters tall, of compound umbels. Petals white. Fruits 2 mm long, round or heart-shaped, slightly flattened, spineless, hairless, and slightly ribbed. In flower June-Sep.

An uncommon plant of ponds and active watercourses. Known from Moro Canyon in Crystal Cove State Park, Peters Canyon Creek in Peters Canyon Regional Park, and a pond at the junction of Arroyo Trabuco and Tijeras Canyon in O'Neill Regional Park. Suspected to be toxic and should not be handled.

The epithet was given in 1762 by W. Hudson for its upright growth (Latin, erectus = erect).

Similar. Common celery ✖ (*Apium graveolens*): celery odor; leaflets much broader, narrow-based, lobed; common perennial weed of watercourses; co-occurs in Peters Canyon. **Common poison hemlock** ☠✖ (*Conium maculatum*): taller; annual; purple spots on stems; leaves 2-pinnate into tiny leaflets; abundant weed.

American Bowlesia
Bowlesia incana Ruíz & Pavón

A *weak-stemmed, trailing annual or perennial* that grows to 60 cm long, *covered with star-shaped branched hairs. Leaves opposite, rounded, and often palmately lobed like the leaf of a maple tree.* The rays in its umbels are very short and form a tight cluster in the leaf axils. Sepals minute. Petals yellowish-green, often pink-tinged. Fruit oval, smooth or weakly ribbed, covered with star-shaped hairs. In flower Mar-Apr.

An uncommon to occasional understory herb in moist ground among coastal sage scrub and chaparral where it grows over the ground or atop other small plants. Known from Chino Hills above the Santa Ana River channel, UCI Ecological Reserve, Quail Hill, Laguna Beach, Caspers Wilderness Park, Lucas Canyon, and along San Juan Canyon.

The plant was named in 1802 by H. Ruíz Lopez and J.A. Pavón in reference to the hairs that cover the plant (Latin, incanus = hoary).

Similar. Granny's hairnet (*Pterostegia drymarioides*, Polygonaceae): same growth habit but its leaves are not palmate. **Wild cucumber** (*Marah macrocarpus*, Cucurbitaceae) has *much larger* palmate leaves arranged alternately on the stem; petals white to creamy yellow.

American Bowlesia

Common Poison Hemlock ☠ ✖
Conium maculatum L.

Very poisonous – do not touch! A *tall annual* 0.5-3 meters. *Stems green, marked with purple. Leaves 1-2 pinnate, leaflets finely dissected.* Petals white, arranged in large conspicuous umbels atop the plant. Fruit oval to spherical, slightly flattened, spineless, and hairless; ribs low and wavy. In flower Apr-Sep.

This native of Eurasia is *extremely toxic if ingested*. It is all too common, growing along many of our creeks and in moist disturbed places and roadways.

Named in 1753 by C. Linnaeus in reference to its purple spots (Latin, maculatus = spotted).

Similar. Douglas's water hemlock ☠ (*Cicuta douglasii* (DC.) C. & R., leaf pictured): perennial, 1.5-3 meters tall; leaves 1-3 pinnate, leaflets long, linear, narrow, toothed; upper Newport Bay (1932), San Juan Creek (1985), no recent records. **Sweet fennel** ✖ (*Foeniculum vulgare*): stems green; leaflets thread-like; petals yellow; licorice-scented; abundant weed.

Common Poison Hemlock / Purple Spots

Common Poison Hemlock / Douglas's Water Hemlock

Rattlesnake Plant, Yerba de la Vibora
Daucus pusillus Michx.

A few-branched upright annual, 30-90 cm tall. Leaves 2-3-pinnate, 3-10.5 cm long. *Flowers in a tight, cup-shaped terminal cluster at the top of a very long, nearly leafless stalk.* Leafy bracts just below the cluster have short bristly segments. Petals white. *Fruit elliptical, flattened on one side, armed with stiff, barbed bristles; often purple in age.* In flower Apr-June.

Occasionally found in coastal sage scrub and oak woodland, *more common after fire*. Known from upper Newport Bay, Dana Point Headlands, Santa Ana River Canyon, Turtle Rock, Sycamore Hills, Lomas de Santiago, Rancho Mission Viejo, and San Mateo Canyon Wilderness Area. It was reportedly used by indigenous peoples as an antidote for rattlesnake bite (thus its common names); to our knowledge its efficacy has not been substantiated.

Named in 1803 by A. Michaux in reference to its weak, slender stems (Latin, pusillus = very small and weak).

Rattlesnake Plant / Fruit

San Diego Button-celery ★
Eryngium aristulatum Jeps. var. ***parishii*** (J.M. Coult. & Rose) Mathias & Constance

A *painfully spine-tipped hairless annual or biennial, upright to spreading*, stems 10-90 cm long. *Basal leaf petiole 8-10 cm, blade 3-5 cm, pinnately lobed. Flowers subtended by stiff spiny bracts*. Sepals short; petals shorter, white. Fresh anthers blue-purple, pollen yellow. Fruit body oval, covered in small, flat, spine-tipped scales. In flower May-June.

A very rare vernal pool endemic found primarily along the coast and foothill regions of San Diego County from southern Camp Pendleton to the Mexican border and into northwestern Baja California. Known from Santa Rosa Plateau in Riverside County. Recently discovered by Justin Wood at Fairview Park in Costa Mesa. Federal and State listed as endangered, and CRPR 1B.1. Threatened by habitat loss and non-native plants.

The epithet was given in 1893 by W.L. Jepson for its bract spines (Latin, aristulatum = having a small bristle-like appendage). The subepithet was given in 1900 by J.M. Coulter and J.N. Rose in honor of S.B. Parish.

San Diego Button-celery

The inflorescence of San Diego button-celery (right) is often more spherical than that of Pendleton button-celery (far right). We know of no place where they co-occur.

Pendleton Button-celery ★
Eryngium pendletonense K.L. Marsden & M.G. Simpson
[*E. pendletonensis* K.L. Marsden & M.G. Simpson]

A *painfully spine-tipped hairless annual or biennial* with low-lying branches up to 20 cm long. *Leaves 8-25 cm long, 1-2 pinnate, from a basal rosette*. Stem leaves similar but smaller. *Flowers subtended by stiff, stout, spiny bracts*. Sepals short; petals shorter, white. Fresh anthers blue-purple, pollen yellow. Fruit body oval, covered in small, flat, spine-tipped scales. In flower Apr-June.

Very rare, quite limited, known only along sea bluff-tops from just above Surfer's Beach in San Onofre State Beach south through Camp Pendleton. Grows in clay soils among vernal pools, coastal sage scrub, and grasslands, sometimes even near coastal railroad tracks. CRPR 1B.1. Both of our button-celeries are threatened by non-native plants, trampling, weed abatement, development, etc.

The epithet was given in 1999 by K.L. Marsden & M.G. Simpson in reference to the place of its discovery, on Marine Corps Base Camp Pendleton (Latin, -ensis = from).

Pendleton Button-celery

Sweet Fennel, Wild Anise ✖
Foeniculum vulgare Mill. **var. *vulgare***

This *tall licorice-scented perennial* may reach 1-2 meters in height. It is *hairless* and *may be covered in a fine whitish powder.* Its *dark green leaves* can reach 30 cm long, each triangular or oval in outline, *finely dissected into thread-like leaflets* from 0.4-4 cm long. Sepals absent, petals yellow with a narrow tip. *Flowers* arranged *in large, broad, compound umbels.* Primary rays 1-4 cm long. In flower (Feb-)May-Sep.

Native to southern Europe. Widespread and common in disturbed ground and along roadways, mostly along the coast and lowlands such as Fullerton, Yorba Linda, Santiago Canyon Road, San Juan Capistrano, and many parks.

The common name "anise" for this plant is misleading. Anise (*Pimpinella anisum* L., Apiaceae), a native of Eurasia, is used as a licorice-flavored spice in cooking. Thus, it is best to use the common name "sweet fennel" for *Foeniculum vulgare*.

The epithet was named in 1768 by P. Miller in reference to its abundance (Latin, vulgaris = common).

Similar. Common poison hemlock ✖ (*Conium maculatum*): stems purple-dotted; leaflets broader; petals white, not licorice-scented.

Sweet Fennel

Sweet Fennel

THE CARROT GUILD
Western Carrot Swallowtail, Anise Swallowtail
Papilio zelicaon Lucas (Papilionidae)

Wingspan 7-9 cm, *wings with wide yellow patches surrounded by black.* Dorsal surface of each hindwing has a black-centered red or orange spot near the back edge. Larvae commonly feed on sweet fennel. Their natural foodplants are other native carrots such as those in the genera *Oenanthe*, *Lomatium*, and *Tauschia*. They occasionally feed on common celery (*Apium graveolens*), garden parsley (*Petroselinum crispum* (Mill.) Fuss), rue (*Ruta* spp., Rutaceae), and cultivated citrus (*Citrus* spp., Rutaceae). Young larvae resemble bird droppings, which discourages predators. Mature larvae are green with black and orange spots. The chrysalis can be either brown or green. Adults emerge nearly all year.

Mature Larva

Pupa

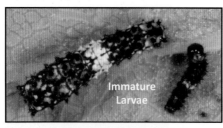
Immature Larvae

Adult

Field Hedge-parsley, Tall Sock-destroyer ✖

Torilis arvensis (Hudson) Link

Light gray to green upright annual, 30-100 cm tall, hairs held upward, flat against its central stem. *Leaf blade* 5-12 cm long, *triangular in outline,* lower leaves 2-3-pinnate, uppermost 1-pinnate. *Inflorescence longer than leaves, opposite a leaf.* Petals tiny, white, heart-shaped. *Fruit 3-5 mm long, oblong, covered with prickles that stand straight out away from the fruit body, curved only at the very tip; purple from tip to about halfway back to the fruit body.* In flower Apr-July.

Native to Europe, common *in moist soils along shady trails.*

The epithet was given in 1762 by W. Hudson for its occurrence in cultivated fields (Latin, arvensis = of or belonging to a field).

Field Hedge-parsley

Field Hedge-parsley

Field Hedge-parsley

Knotted Hedge-parsley

Knotted Hedge-parsley, Short Sock-destroyer ✖

Torilis nodosa (L.) Gaertner

Hairy annual, branches from the base, 10-50 cm long. Stems covered with tiny backward-pointing hairs. Leaf blade 3-9 cm long, oblong in outline, 2-pinnate. *Inflorescence stalk much shorter than leaves, opposite a leaf.* Petals tiny, white, heart shaped. *Fruits in short tight clusters, at the end of a stout stalk held out from the stem* (not at the top). In flower Apr-June.

Native to Eurasia, it lives in disturbed shaded areas below 2000 feet such as Foothill Ranch, on Rancho Mission Viejo, and in Trabuco, San Juan, Lucas, and Devil Canyons.

Named in 1753 by C. Linnaeus in reference to the small, short-stalked flower clusters that look like knots along the stem (Latin, nodosus = full of knots or knobs).

Notted Hedge-parsley

The following native perennials are members of 4 genera: *Osmorhiza*, *Lomatium*, *Sanicula*, and *Tauschia*. They are similar in appearance and can be difficult to identify. For most species, you will need the fruits to be certain of your identification.

	Sweet-cicely *Osmorhiza*	**Wing-fruits** *Lomatium*	**Sanicles, Snakeroots** *Sanicula*	**Tauschias** *Tauschia*
Central stem	Well-defined	Poorly defined	Well-defined	Poorly defined
Leaves	2-3 pinnate, leaflets toothed	Pinnate, leaflets variable	Undivided or palmate to pinnately lobed; leaf bases obviously sheathed	1-2 pinnate, leaflets broad and sharply toothed
Inflorescence	Small rounded secondary umbels, each with 2-6 obvious narrow bractlets just below	Round-topped secondary umbels, each with obvious leaflike bractlets just below	Small round-topped secondary umbels; primary rays long, of different lengths, each with a large leaf-like bractlet at its base	Round-topped secondary umbels, each with slender leaflike bractlets just below
Sepals	Absent	Absent (rarely present, small)	Present, obvious (larger than *Tauschia*)	Present, obvious (smaller than *Sanicula*)
Petals	Yellow-green, end in a narrow point, curl up-and-over toward the middle of the flower	Yellow, greenish-white, or purplish	Yellow, rarely purple	Yellow
Fruits	Long, club-like; narrow at base, fatter above; ribbed, bristly; straight or slightly curved like a banana	Oblong; flattened side-to-side; obviously winged (surrounded with thin tissue)	Oval, covered with curved prickles (1 with fat bumps, no prickles)	Oblong with prominent ribs; smooth-surfaced: no wings, bumps, or spines

California Sweet-cicely   California Sweet-cicely

California Sweet-cicely

Sweet-cicely: *Osmorhiza*
California Sweet-cicely
Osmorhiza brachypoda Torr.

Upright, 30-80 cm tall, covered with fine hairs. *Leaves* 10-20 cm long, *soft, and divided 2-3 times into toothed leaflets*. There are *2-6 obvious narrow bracts just below the inflorescence*. Sepals are absent. *Petals* are *yellow-green*, end in a narrow point, and curl up-and-over toward the middle of the flower. *Fruits* are 12-20 mm long, *club-like*, narrow at their base and fatter above, ribbed and bristly, straight or slightly curved like a banana. In flower Mar-May.

A widespread yet often overlooked plant of shady forest understory. It smells like licorice. Found in places such as Black Star Canyon, Silverado Canyon, upper Santiago Canyon, along San Juan Trail south of Blue Jay Campground, moist ground along the Main Divide Truck Trail and Bedford Road. Very common and easy to observe in Trabuco Canyon along lower sections of the Trabuco Trail, in the shade of big-leaf maples and alders.

Named in 1855 by J. Torrey, probably in reference to the short stalk at the base of each fruit (brachy = short; poda = foot).

Wing-fruits: *Lomatium*

Plants in this genus are small herbs with divided leaves, flowers in round-topped compound umbels, and *fruits that are oblong, flattened side-to-side, and obviously winged (surrounded with thin tissue)*. In our area, their petals are yellow, greenish-white, or purplish. They grow in meadows or among chaparral in rocky soils. We have 3 species.

The genus was established in 1819 by C.S. Rafinesque-Schmaltz in reference to the wing around the fruit (Greek, lomatos = fringe or border).

Woolly-fruited Wing-fruit
Lomatium dasycarpum (Torr. & A. Gray) J.M. Coulter & Rose **ssp.** ***dasycarpum***

Generally low-growing, 10-50 cm in height. The entire plant is *covered in soft white hairs. Leaves are finely divided into small, thin, narrow leaflets.* Its *petals* are greenish-white or purplish and are *very hairy. Fruit body very hairy* but may become hairless with age. In flower Mar-June.

Woolly-fruited Wing-fruit

Found commonly in clay soils among grasslands, oak woodlands, and chaparral, in the Lomas de Santiago and Santa Ana Mountains. Look for it in such places as Loma Ridge, Limestone Canyon, near Santiago Dam, Robinson Ranch, Preusker Peak on the Audubon Starr Ranch Sanctuary, Lucas Canyon near Caspers Wilderness Park, along South Main Divide Road, Elsinore Peak, the San Mateo Canyon Wilderness Area, and ridges near Sierra Peak and Black Star Canyon.

Named in 1840 by J. Torrey and A. Gray in reference to its hairy fruits (Greek, dasys = hairy or shaggy; karpos = fruit).

Shiny Wing-fruit
Lomatium lucidum (Nutt.) Jeps.

Generally low-growing or short-stemmed, 20-60 (-120) cm in height. Its stems and leaves are *completely hairless* and *shiny green to blue-green. Leaves divided into 3 coarse leaflets that are 3-lobed or may be divided again into another 3 leaflets.* These leaflets are the largest of our 3 wing-fruits; they are 1.5-5 cm long, oblong, and have teeth around their margins. The petals are yellow. In flower May-June.

Generally uncommon in our area but localized, it grows in grasslands and in clearings in chaparral, especially in the years following a fire. Near the coast, it is found in the San Joaquin Hills. More common inland along hillsides and trails in the Santa Ana Mountains. Easily seen along Holy Jim Canyon Trail, Main Divide Truck Trail near Blue Jay Campground, Los Pinos Saddle, and along San Juan Trail.

Shiny Wing-fruit

Named in 1840 by T. Nuttall for its shiny leaves (Latin, lucidus = clear, bright, shining).

Similar. Southern tauschia (*Tauschia arguta*): grows at higher elevations away from the coast; no wings on its fruit; light to medium green soft leaves, and toothed leaflets.

Foothill Wing-fruit
Lomatium utriculatum (Nutt.) J.M. Coulter & Rose

Generally low-growing, 10-50 cm in height. It may be *hairless or covered with short hairs. Its leaves are pinnate with very thin, linear leaflets and very large obvious open sheaths at the base of each leaf.* There are leafy bracts just below the inflorescence, as if to hold them up. Petals are yellow. In flower Mar-Apr.

Grows in *heavy clay soils* on slopes, *in meadows,* in oak woodlands in the Puente-Chino Hills, Santa Ana Mountains (South Main Divide Road, Elsinore Peak), and in the San Mateo Canyon Wilderness (Miller Mountain).

Foothill Wing-fruit

Named in 1840 by T. Nuttall in reference its wide-open bag-like leaf sheaths (Latin, utriculus = little bag).

Sanicles, Snakeroots: *Sanicula*

Snakeroots have divided leaves but not as finely divided as most *Lomatium* species. *Flowers are arranged in small rounded secondary umbels.* Rays of the primary umbel are long and of different lengths, with a leafy bract at the base of each ray. Petals of most species are yellow, but one is purple. *Fruits are oval and covered with bumps and/or prickles.* These are relatively common plants of seasonally or perennially moist meadows, hillside seeps, and among scrub.

The genus was named in 1753 by C. Linnaeus, perhaps in reference to species used in ancient medicine (Latin, sano = to heal).

Sharp-toothed Sanicle
Sanicula arguta E. Greene ex. J.M. Coulter & Rose

Upright or spreading, 10-50 cm tall. *Leaves shiny*, 3-10 cm long, *dissected into several broad leaflets with toothed edges.* Petals yellow. Fruits covered with stout curved prickles. In flower Mar-Apr.

Found on grassy slopes among grasslands, coastal sage scrub, and chaparral in our coastal foothills. Look for it on the hills surrounding Upper Newport Bay, UCI Ecological Reserve, San Joaquin Hills, Laguna Beach, Aliso and Wood Canyons Wilderness Park, and hills in Laguna Niguel. Occasional in the Puente-Chino Hills and Temescal Valley. Uncommon in the Santa Ana Mountains but found occasionally in the San Mateo Canyon Wilderness.

Named in 1900 by E.L. Greene for its shiny leaves (Latin, argutus = bright, clear, quick).

Poison Sanicle ☠?
Sanicula bipinnata Hook. & Arn.

An *upright* plant, 12-60 cm tall, with a somewhat swollen taproot. *Leaves light green*, 3-10 cm long, *divided 2-3 times, and far apart which gives the plant an open airy look.* Stems and petioles green. Petals bright yellow. Fruits oval or rounded, covered with fat bumps that bear short curved prickles. In flower Apr-May.

Scattered and uncommon, known from grassland, oak woodland, and juniper woodland; Verdugo Canyon, Temescal Valley, Puente Hills, and on the Santa Rosa Plateau, perhaps more widespread.

Named in 1838 by W.J. Hooker and G.A.W. Arnott for its pinnate leaves (Latin, bi = two; pinna, penna = a wing or feather).

Steve Matson

Purple Sanicle
Sanicula bipinnatifida Douglas ex Hooker

Spreading or upright, 12-60 cm tall. Its *green or purplish leaves* are 4-19 cm long and *dissected into very broad tooth-edged leaflets that are crowded together.* Stems and leaf petioles are green. *Petals purple or yellow. Stamens and styles very long, twice as long as the petals.* Fruits have stout, curved prickles that are swollen at their base. In flower Mar-May.

Found only among grasslands, chaparral, and pines mostly in the foothills and upper reaches of the Santa Ana Mountains. Look for it in the Puente Hills (Turnbull Canyon), Weir and Sierra Canyons, Caspers Wilderness Park, along South Main Divide Road, Temescal Valley, Santa Rosa Plateau, and on Miller Mountain.

Named in 1832 by D. Douglas for its divided leaves (Latin, bi = two; pinna, penna = a wing or feather; fidus = divided usually within outer third).

Pacific Sanicle

Sanicula crassicaulis Poepp. **var. *crassicaulis***

With a height of 24-120 cm, this is *California's tallest snakeroot.* It is *quite variable in growth form* but is *generally stiffly upright.* Its *leaves* are 3-12 cm long, *palmately 3-5 lobed* like a maple leaf, have toothed margins, and often droop down. Each umbel bears 2-5 secondary umbels. Its flowers are yellow and its fruits are covered with stout, curved prickles. In flower Mar-May.

Common on shaded hillsides in oak woodlands, chaparral, and mixed conifer forest, from the coast and foothills to peaks in the Santa Ana Mountains and San Mateo Canyon Wilderness. Look for it in lower Trabuco Canyon along Trabuco Creek Road, Holy Jim Trail, and upper Trabuco Canyon just below Los Pinos Saddle.

Named in 1830 by E.F. Poeppig for its thick stem (Latin, crassus = thick or heavy; caulis = stem).

Pacific Sanicle

Sierra Sanicle

Sanicula graveolens Poepp.

Low-growing to upright, 5-20(-45) cm tall, from a taproot or tuber. *Leaves shiny dark green, purplish,* 1.5-4 cm long, divided 2-3 times into broad leaflets. Its *stout stems and leaf petioles are red-purple. Bracts below the inflorescence linear to oval, 6-10 in number, fused together at their base, each about 1 mm long, not divided, and end in a narrow tip.* Petals pale yellow, short-lived. Fruits 3-5 mm long, red-purple, oval or rounded, swollen, covered with stout curved prickles. In flower Apr-June.

A plant of open or coniferous forests of higher mountains. In the Santa Ana Mountains, uncommonly found in rock crevices atop Santiago Peak (though mostly eliminated there) and its northern arm, also on talus slopes near Modjeska Peak where it grows with **tuberous sanicle** (*S. tuberosa*). Also found in Powder Canyon in the Puente Hills.

Named in 1830 by E.F. Poeppig for its strong scent (Latin, graveolens = strong-scented).

Similar. Parish's tauschia (*Tauschia parishii*): overall larger and more upright; central stem absent; stems, petioles, and leaf blades green; leaflets narrow, carrot-like; inflorescence bracts linear, few, longer (each 5-12 mm); fruits oblong, smooth-surfaced, prominently-ribbed (no wings, bumps, or spines); not known to grow together here.

Sierra Sanicle

Fruit: Steve Matson

Tuberous Sanicle

Sanicula tuberosa Torr.

Thin and dainty-stemmed, it stands upright 5-80 cm tall and *grows from a round tuber.* Leaves are *dark green to purplish,* 2-13 cm long, and divided 2-3 times into *very narrow leaflets.* Its stem and leaf petioles are red-purple. Bracts below the inflorescence are toothed or pinnate. Petals are bright yellow. *Fruits* are 1.5-2 mm long, oval or round, and covered with rounded bumps (*no prickles*). In flower Mar-June.

Grows in rocky ground on north- or west-facing slopes, sometimes in full or partial shade. *Widespread but not generally abundant.* Found sparingly in the Santa Ana Mountains and San Mateo Canyon Wilderness. Known from Upper Hot Springs Canyon, San Juan Canyon, along Sitton Peak Truck Trail, Trabuco Canyon, Leach Canyon, Los Pinos Trail, Elsinore Peak, scattered sites in the San Mateo Canyon Wilderness Area, and on talus slopes near Modjeska Peak where it grows with **sierra sanicle** (*S. graveolens*).

The epithet was given in 1857 by J. Torrey in reference to its tuberous root (Latin, tuberosus = swollen into a tuber).

Tuberous Sanicle

Tauschias: *Tauschia*

These plants lack a well-defined central stem. Their leaves are once- or twice-pinnate, the *leaflets broad and sharply toothed.* Tiny yellow flowers appear in small rounded secondary umbels, *the umbels are round-topped like Lomatium.* Fruits are oblong, smooth-surfaced, and have prominent ribs on them; they have neither wings, bumps, nor spines. We have two species, both found at higher elevations in the Santa Ana Mountains.

The genus was named in 1834 by D.F.L. von Schlechtendal in honor of I.F. Tausch.

Southern Tauschia
Tauschia arguta (Torr. & A. Gray) J.F. Macbr.

Grows 30-70 cm tall in flower; *its flower stalks are often quite long.* Leaves are shiny, *1-pinnate,* from 6-20 cm long, *with broad angled leaflets.* Petals yellow, with large, broad umbels. In flower Apr-June.

Commonly found in chaparral and woodlands in upper regions of the Santa Ana Mountains, such as Claymine Canyon, Sierra Canyon, Black Star Canyon, Harding Canyon, on roadcuts along Main Divide Truck Trail, and in scattered sites in the San Mateo Canyon Wilderness.

Named in 1840 by J. Torrey and A. Gray for its bright shiny leaves (Latin, argutus = bright, clear, quick).

Similar. Shiny wing-fruit (*Lomatium lucidum*): lower elevations; wide wings on its fruit; light gray-green stiff leaves and irregular spiny-edged leaflets; sometimes found growing near each other.

Southern Tauschia

Southern Tauschia

Parish's Tauschia
Tauschia parishii (Coulter & Rose) J.F. Macbr.

Low and spreading, 10-40 cm in flower, with flower stalks of variable lengths. Sometimes covered in a fine white powder (glaucous). *Leaves twice-pinnate (or more) and look like carrot leaves. Bracts below the inflorescence linear, few in number, each 5-12 mm long, not divided, and end in a narrow tip.* Petals yellow, arranged in large, broad umbels. In flower May-July.

Grows on dry slopes in chaparral, oak and pine woodlands on the highest peaks of the Santa Ana Mountains, usually among cracks in rocks. Far less common than southern tauschia. To date, we have not found our two species of *Tauschia* growing together.

Named in 1888 by J.M. Coulter and J.N. Rose in honor of S.B. Parish.

Similar. Sierra sanicle (*Sanicula graveolens*): overall smaller and lower; central stem present, stout; stems, petioles, and leaf blades red-purple; leaflets wider; inflorescence bracts linear to oval, 6-10 in number, fused together at their base, short (each about 1 mm long); fruits oval, covered with stout hooked prickles that are swollen at their base; not known to grow together here.

Parish's Tauschia

Parish's Tauschia

APOCYNACEAE • DOGBANE FAMILY

Shrubs or vines with *white milky sap* that may irritate your skin upon contact. Their *undivided leaves* are arranged in *opposite* pairs (rarely alternate) *or in whorls* of 3-4 leaves. The inflorescence emerges from the leaf bases or the top of the plant, the *flowers often* arranged *in a rounded umbel*. There are 5 *sepals and petals,* each fused at the base, free at the tip. The corolla may be funnel-like or tube-like (dogbanes, some milkweeds) or star-like with the 5 lobes held out or backward (most milkweeds). *Stamens 5, fused to the corolla;* hidden within the flower tube (dogbanes) or highly modified, obvious, and held above the petals (milkweeds). *Each flower has two superior ovaries.* In the dogbanes, both ovaries develop and produce "twin" fruits that are sometimes fused at their tips. In the milkweeds, 1 ovary generally shrivels, leaving the other to develop into a fruit. The fruits look like paired stringbeans (dogbanes) or a football (milkweeds). Their *flat seeds* each have a *long comb of silky white hairs* that enables them to float on the wind and disperse to new areas. Some species are *highly toxic;* a few are used in medicine.

For many years, the milkweeds were placed in their own family (Asclepiadaceae) but recent studies show that this group should be combined with the Apocynaceae. Dogbanes and milkweeds are common in our foothills and canyons within grasslands, coastal sage scrub, chaparral, and oak woodland communities. Ours exhibit two basic growth forms, **vine-like perennials** and **bushy perennials**, arranged here in those groups.

VINE-LIKE PERENNIALS

The flowers of these vine-like plants have five petals held straight out or slightly forward.

Climbing Milkweed ☠?

Funastrum cynanchoides (Decne.) Schltr. **ssp. *hartwegii*** (Vail) Krings
  [*Sarcostemma cyn.* Decne. ssp. *hartwegii* (Vail) R. Holm]

A slender climbing vine, its light to medium green leaves have spear-shaped or blunt leaf bases, arranged in opposite pairs on its long thin trailing stems that may reach 2 meters in length. Its *pink-purplish or whitish flowers* are borne in umbels at the leaf bases, have a shallow tube or no tube, and have 5 petals held outward or slightly upward. Its fruits are elliptical and held upward on the plant. In flower Apr-July.

An uncommon but widespread plant in our region. It is found sparingly in Laguna Coast Wilderness Park (right above Big Bend), Moro Canyon in Crystal Cove State Park, and Turtle Rock in Irvine. More common in foothills and upper elevations of the Santa Ana Mountains such as in Coal, Agua Chinon, Hot Springs, and Temescal Canyons. Also in scattered sites in the San Mateo Canyon Wilderness Area such as San Mateo and Los Alamos Canyons.

In 1897, A.M. Vail named this plant to honor K.T. Hartweg.

Periwinkle ✖

Vinca major L.

A *hairless, sprawling vine* that sprouts additional roots along its stems as it grows. *Leaves* about 7 cm long, *oval, dark green,* and mostly opposite. The tubular flowers arise singly from the leaf bases. *Petals* are *periwinkle blue* in color and have large rounded lobes. In flower Mar-July, sometimes year-round.

A commonly planted groundcover in rural canyon communities. Its aggressive root masses are reported to be so thick that it can prevent the growth and spread of poison oak and other plants into one's property. It often escapes into the wild and blankets shady creekbeds, trails, and roadsides, choking out native plants. Once established, it is quite difficult to eradicate. Look for it in Silverado Canyon, Trabuco Canyon, Holy Jim Canyon, Bell Canyon on the Audubon Starr Ranch Sanctuary, lower Hot Springs Canyon, and San Juan Canyon. Native to the Mediterranean region.

The epithet was given in 1753 by C. Linnaeus for its large flowers (Latin, majus = greater, larger).

BUSHY PERENNIALS

These are all perennial herbs, often from deep-seated roots. Their flowers are somewhat tubular, highly modified in the genus *Asclepias*.

Indian Hemp, Hemp Dogbane
Apocynum cannabinum L.

An odd-looking perennial with no central trunk. Instead several parallel stems emerge directly from underground rootstocks and stand straight up, sometimes forming thickets. Reaches 30-120 cm in height. Branching, if any, is toward the top of the plant. *Leaves* are *lance-shaped*, 5-8 cm long, *yellow-green, opposite on the stem, and face upward. Flowers* are arranged in clusters *hidden among the leaf axils.* The corolla is 2.5-5 mm long with greenish or white petals fused into a bell that has 5 petal tips spreading outward. *Fruits* 6-9 cm long, *very thin,* and *hang downward; two pods per fertilized flower.* In flower June-Aug.

Not common here, found in *seasonally moist sandy soil along watercourses* such as the Santa Ana River channel along the base of the Chino Hills, Agua Chinon near Limestone Canyon, along upper San Juan Canyon downstream from Lion Canyon, Temescal Valley, and San Mateo Canyon in the San Mateo Canyon Wilderness Area. Its fibers were used by Native Americans to make fishing nets, rope, and rough cloth.

The epithet was named in 1753 by C. Linnaeus in reference to its fibrous stems used as twine and rope (Greek, kannabis = hemp).

Milkweeds: *Asclepias*

Quite unusual flowers in structure and function. They have *5 petals held backward* and *10 structures formed from the 5 stamen filaments.* Five of these are *central hoods* (cup-like) that are held upward. *Each hood may contain a small horn that curves up and inward toward the flower center.* We have 3 species in our area.

The genus was named in 1753 by C. Linnaeus to honor Aesculapius, the Greek god of medicine and healing and/or Asklepios, the Roman god of medicine and healing, as a tribute to the usefulness of these plants in medicine.

Narrow-leaved Milkweed ☠
Asclepias fascicularis Decne in A. DC.

A thin, wiry, upright perennial herb to 80 cm tall. It is hairless or has a few scattered hairs. Its *leaves are thin and linear,* arranged *in whorls of 3-6 leaves per node, but the upper leaves may be in opposite pairs.* Flower umbels arise from the top of the plant and stand upright. *The flowers are small, white to greenish-white, sometimes with a pink tinge. The hoods are cup-like, the horns easily seen and usually longer than the hoods.* In flower June-Sep.

Very common in coastal canyons, Puente-Chino Hills, Lomas de Santiago (such as in Limestone Canyon), and up into the Santa Ana Mountains, even along roadsides. Often cultivated in gardens, usually as a larval foodplant and adult nectar source for **monarch butterflies** (*Danaus plexippus* L., Nymphalidae; see below). It commonly grows with **woolly-fruited milkweed** (*A. eriocarpa*).

The epithet was named in 1844 by J. Decaisne probably in reference to the bundles of leaves on its main stems (Latin, fasciculus = a bundle).

California Milkweed
Asclepias californica E. Greene

A semi-upright perennial herb to 15-50 cm tall, *with spreading stems that tilt over at a 45 degree angle. The entire plant is covered with soft white velvety hairs.* The oval-shaped leaves are 5-15 cm long, arranged *in opposite pairs* on the stem, and have no petiole or a very short one. *Flower umbels* arise from the top of the stems and *hang downward. Flowers are dark maroon, hoods are rounded like sacks, and the horn is lacking or hidden.* In flower Apr-July.

Uncommon in our area, found primarily *on dry slopes* in the Santa Ana Mountains generally at elevations higher than woolly milkweed, such as along the Holy Jim Canyon Trail, upper- and lowermost regions of the San Juan Trail, above Falcon Group Camp, above the lower trailhead for East Horsethief Trail, in the Chino Hills, and Temescal Valley. Its flowers are visited by small to medium-sized native bees.

The epithet was named in 1893 by E.L. Greene to honor of the state of its discovery.

Similar. Woolly-fruited milkweed (*A. eriocarpa*): upright stems; covered with long white hairs; *leaves very long, narrower, in whorls of 3-4 (rarely opposite); flowers creamy white* or tinted yellow or pink, hoods cup-like, horns barely peek out of the hood; more common.

California milkweed is covered with soft white hairs. Its stems are tilted; leaves oval, in opposite pairs; flowers maroon, often hang down.

California Milkweed

Woolly-fruited Milkweed, Kotolo
Asclepias eriocarpa Benth.

A stiff upright perennial herb that stands nearly 1 meter tall. Usually *covered with long white hairs,* less hairy in shade. *The oblong, often crinkled leaves have short petioles and are arranged in whorls of 3-4 leaves (rarely opposite pairs).* Flower umbels are at the top of the stems and stand mostly upright but topple over with age. Its *flowers* are *creamy white* and may be tinted yellow or pink. *Hoods are clearly cup-like, and the horns barely peek out of the hood though they are easy to see.* In flower June-Aug.

Common in our coastal valleys, foothills, Puente-Chino Hills, Santa Rosa Plateau, and the Santa Ana Mountains, especially so in Caspers and Riley Wilderness Parks and on the Richard and Donna O'Neill Conservancy. Widespread and occasionally found in the San Mateo Canyon Wilderness Area. Its flowers are visited by beetles, butterflies, flies, ants, wasps, and bees.

The epithet was named in 1849 by G. Bentham for its woolly fruits (Greek, erion = wool; karpos = fruit).

Similar. California milkweed (*A. californica*): spreading stems lean over; covered with white velvety-soft hairs; *leaves opposite; flowers dark maroon,* hoods rounded and sack-like, horn lacking or hidden; higher elevation, much less common.

Woolly-fruited milkweed is found mostly in foothills on the coastal side of the Santa Ana Mountains. California milkweed usually lives in dry rocky soils at higher elevations.

Woolly-fruited Milkweed

Woolly-fruited Milkweed

THE MILKWEED AND DOGBANE GUILD

Small Milkweed Bug

Lygaeus kalmii Stål (Lygaeidae)

A medium-sized true bug, 11-12 mm long when adult, red to salmon, marked with grey and black. *The head has a single orange dot, the thorax 2 round black dots.* Nymphs and adults feed mostly on milkweeds, occasionally on yarrows and ragweeds (Asteraceae), sometimes even on insects. Found in the same areas but usually less common than the large milkweed bug, with which it is often confused. *When its wings are closed, they appear to have a red letter "X" across them.*

The epithet was given in 1874 by C. Stål for its beauty (Greek, kalos = beautiful).

Large Milkweed Bug

Oncopeltus fasciatus (Dallas) (Lygaeidae)

A large true bug, 13-18 mm long when adult, orange to red, banded with black. *The head is mostly orange, the thorax black, edged with orange.* Nymphs and adults congregate together on dogbanes and milkweeds on which they feed, especially on the fruits in summer and fall. Common where foodplants are plentiful such as Agua Chinon and Caspers Wilderness Park.

The epithet was given in 1852 by W.S. Dallas for the the black band across the adult's wings (Latin, fasciatus = bundled, banded).

Leafhopper Assassin Bug

Zelus renardii Kolenati (Reduviidae)

A slender insect, 1-1.6 cm long, *with an unusually long narrow head.* The body is green or light gray-brown, marked with red, black, and white; the legs are usually green. Found on several plants, including milkweeds, where it feeds on small insects. Nymphs and adults gather plant sap or glandular excretions and adhere it to their bodies as an aid in capturing insects.

The epithet was given in 1856 by F.A. Kolenati probably in reference to the sly hunting technique of the insect, like a fox (French, renard = fox).

Oleander Aphid, Milkweed Aphid ✖

Aphis nerii Boyer de Fonscolombe (Aphididae)

A bright yellow-orange, pear-shaped insect 1.5-2.6 mm long, with black legs, long black antennae, and a pair of black hollow tubes (cornicles) near the hind end. A non-native pest species, it feeds on the sap of milkweeds and dogbanes, storing the plant's poisons. When threatened, it secretes a tiny amount of the poison from its cornicles. Predators, mostly spiders and birds, that contact this poison retreat and try to clean it from themselves.

Originally from the Mediterranean region, it was introduced around the world with **oleander** ☠ (*Nerium oleander* L., not shown), a commonly cultivated shrub and its primary host. There are no males; females give live birth to the young, all female, a process known as parthenogenesis.

Aphid populations are controlled by tiny parasitoid wasps. A female wasp stings the aphid and deposits an egg within its body. The wasp larva hatches and devours the aphid's insides. Parasitized aphids swell and turn brown or black as the wasp matures. It pupates within the aphid and emerges as an adult through an opening it makes near the back end of the aphid. The remaining aphid "mummy" is commonly found among live aphids, as seen in the left-center of the photograph.

The epithet was given in 1841 by E.L.J.H.B. de Fonscolombe in reference to the host plant (Greek, nerion = oleander).

Milkweed Borer Beetle

Tetraopes basalis LeConte (Cerambycidae)

A red-orange beetle, 1-1.3 cm long, with black legs, antennae, and spots. They are slow-moving and generally do nothing when disturbed, but may let out a squeeky sound. Their larvae tunnel through the roots of milkweed plants, pupate near them, and emerge as adults in spring and summer. Adults are commonly seen as they feed on leaves and flowers. Locally, they feed only on woolly-fruited milkweed.

The epithet was given in 1852 by J.L. LeConte for the black spot on either side of the base of the front wings (Latin, basalis = basal).

Milkweed Leaf Beetle
Chrysochus cobaltinus LeConte (Chrysomelidae)

A brilliantly-colored *steel-blue beetle* about 1 cm long. When disturbed they fold up their legs and drop from the plant, then play dead where they lie. Their larvae probably bore into the roots of the milkweed plants. Adults appear in late spring and summer, feeding on the leaves and stems of milkweeds and dogbanes.

The epithet was given in 1857 by J.L. LeConte for its cobalt blue color (cobalt; Latin, -inus = like).

Milkweed Leaf Beetle

Western Dogbane Moth
Saucrobotys futilalis inconcinnalis (Lederer), (Crambidae)

Adult moths have *pale yellow-brown forewings,* each wing roughly 1.2-1.4 cm long; hindwings are whitish. They are active only at night. *Larvae feed voraciously on dogbane,* devouring many of the leaves. Caterpillars are found abundantly on the plants, where they live close together and use their silk to fasten leaves together for protection. When young, they are green in color; older larvae are orangish with black spots. They generally feed at night and hide during the day. When disturbed, they regurgitate their meal and the odorous dogbane toxins within it to repel predators. They pupate on the plant in spring and emerge as adults in June-Aug. So far, known locally only from Agua Chinon wash.

The epithet was given in 1863 by J. Lederer for its abundance (Latin, futi = abundant; -alis = pertaining to). He named the western form of the moth at the same time, perhaps for the broken band of darker color on the forewing (Latin, in- = not; concinnus = neat, skillfully joined; -alis = pertaining to).

Western Dogbane Moth (larva)

Adult Moth: Brian V. Brown

The Monarch
Danaus plexippus L.
(Nymphalidae: Danainae)

A *large butterfly* with a wingspread of 8-12 cm. Above, the wings are burnt-orange-brown with black veins and edges, dotted with white; the hindwing is tan below. The hairless *larvae* are banded with yellow, black, and white and *have two fleshy filaments at each end*. Larvae feed on milkweeds, primarily narrow-leaved and woolly-fruited in our area; we have yet to see them feed on California milkweed though they are reported to feed on it elsewhere in California. Adult flight is year-round.

The epithet was given in 1758 by C. Linnaeus in honor of Plexippos, a son of Belus, the King of Egypt. Two of his brothers were a pair of twins named Aegyptus and Danaus.

The Striated Queen
Danaus gilippus thersippus (H. Bates)
(Nymphalidae: Danainae)

Overall similar to the monarch but *smaller* (wingspan 6.7-9.8 cm) with *darker burnt-orange tones. Larvae* also similar but with maroon banding and *a third pair of fleshy filaments about one-third back from the head.* Adult flight: Apr-Nov.

Fairly common in our deserts. In summer-fall, adults fly through inland and coastal areas of southern California. They often lay eggs on milkweed plants in our area; sometimes both monarch and queen larvae are on the same plants.

The epithet was given in 1776 by P. Cramer in honor of Gylippus, a Spartan (Greek) commander. The subepithet was given in 1863 by H.W. Bates to honor Thersippos, envoy of Alexander III the Great.

The Monarch

The Striated Queen

ARALIACEAE • GINSENG FAMILY

Herbs, vines, shrubs, or trees, common in the tropics and temperate regions. *Leaves alternate* (rarely opposite), entire or divided. *Flowers very small, in groups of rounded umbels. Sepals 5, small or absent. Petals usually 5, yellow-green or whitish.* Stamens equal the petals in number, arranged opposite to them. *Ovary inferior.* Fruit a berry or drupe. Very closely related to the carrot family (Apiaceae) and sometimes placed with them. Only 4 native species are known from California. Canary Islands ivy (*Hedera canariensis* Willd.) and English ivy (*Hedera helix* L.) are members of this family; both have escaped from cultivation and invaded wild areas such as Silverado and Hot Springs Canyons.

California Aralia, Elk Clover ✿
Aralia californica S. Watson

A *large herbaceous perennial*, 1-3 meters tall, with large roots and milky sap. *Leaves are very large,* up to 1-2 meters long, hairless, and *pinnately divided.* Each leaflet is 15-30 cm long, oval, with toothed margins. The inflorescence is 30-45 cm long, each tiny flower in a rounded umbel. Fruit is spherical, black, 5 mm across. In flower June-Aug.

Grows on moist, north-facing, shady slopes among chaparral and oak-conifer woodlands. Known locally only from upper Silverado Canyon near Bigcone Spring and at the head of Silverado Canyon near Main Divide Road.

The epithet was given in 1876 by S. Watson for the state of its discovery.

California Aralia

Pennyworts: *Hydrocotyle*

Aquatic perennials of standing water and saturated mud. Stems usually horizontal and creeping, rooting along the nodes. Leaves rounded in outline, entire or with rounded lobes. Inflorescence is a small simple umbel on a stalk attached to the main stem or rhizome. Sepals tiny or absent. Petals greenish or yellow-white to purplish. Fruits elliptic to round, flattened. Three species here, historically known from the Santa River Channel through Yorba Linda, but no recent records or sightings; probably extirpated from that area.

Floating Marsh Pennywort
Hydrocotyle ranunculoides L.f.

Leaves kidney-shaped in outline, deeply lobed. Inflorescence stalks (peduncles) short, only 1-5 cm long; umbels crowded with 5-10 flowers, each on a short stalk (pedicel). In flower Mar-Aug.

Currently known from a streambed at 23rd Street that flows into upper Newport Bay. Also in standing vernal pools on the Santa Rosa Plateau.

Named in 1781 by C. von Linne (son of C. Linnaeus) for its lobed leaves, which look like those of buttercup (*Ranunculus* spp.; -oides = resemblance).

Many-flowered Marsh Pennywort
Hydrocotyle umbellata L.

Leaves round and entire or with shallow lobes; petiole attached near the middle of the leaf, like an umbrella (peltate). Inflorescence stalks (peduncles) 1.5-35 cm long, umbels open with 10-60 flowers, each on an individual stalk (pedicel) of variable length. In flower Mar-July.

Currently known from San Mateo Canyon Wilderness Area, in San Mateo Canyon between Los Alamos and Tenaja Canyons. Common along the Santa Ana River channel upstream from our area.

Named in 1753 by C. Linnaeus (Latin, umbella = sunshade) for the obvious umbrella-like umbels of its flowers, unlike others of this genus, which have small hidden umbels.

Similar. Whorled marsh pennywort (*H. verticillata* Thurber): *leaves also peltate; inflorescence a spike, not an umbel; flowers whorled on short pedicels,* up to 15 flowers per whorl; Prado Basin; in flower Apr-Sep.

ASTERACEAE • SUNFLOWER FAMILY

A family of diverse attributes, its members range from annuals to biennials and perennials; delicate herbs to woody vines, shrubs, and trees. Leaves basal and/or on stems; alternate, opposite, or whorled; undivided to 2-pinnate. Their tiny flowers (**florets**) are densely packed into a cluster and attached to a common receptacle, a unit called a **composite flowerhead**. Though each flowerhead superficially resembles a single flower, it actually contains few-to-many florets (rarely only one). On the upper surface of the receptacle in some flowerheads, slender bracts called **paleae** individually subtend (occur below) the florets (paleae are also called pales, chaffy bracts, chaff scales, or receptacular bracts). A receptacle with paleae is described as **paleate**; a receptable without paleae is **epaleate**. The lower outer surface of the flowerhead is surrounded by leaf-like or sepal-like overlapping bracts called **phyllaries** (or involucral bracts). Phyllaries are arranged in one or more whorls or spirals, sometimes crowded, sometimes spread out. Together, the phyllaries form the **involucre**.

Each floret contains parts of both sexes (**bisexual**), of only one sex (**staminate** [male] or **pistillate** [female]), or none (**sterile**). The **calyx** is modified into a **pappus**, which consists of one or more rings of hairlike **bristles** (sometimes with feather-like side branches), narrow to broad flat membranous **scales**, or long stiff, smooth or barbed **awns** (sometimes a mixture of bristles, scales, and/or awns); in some taxa the pappus is totally absent. Pappus units are whitish or brownish, never green or sepal-like. The **corolla** is radial or bilateral (rarely absent); lobes (0)3-5. Stamens (3-)5, filament bases fused to the corolla, free above, *anthers fused into a cylinder that surrounds the style. Ovary inferior,* pistil 1, 1-chambered; ovule 1, attached in base of ovary. *Style 2-branched near tip; stigmas are borne on the basal facing (inner) sides of the style branches* and generally end below the tip. Fruit a dry 1-seeded **cypsela** (think of a sunflower seed still in its case, which is the ovary), sometimes called an achene. The pappus may remain attached to the fruit or may fall off. Fruit features are usually helpful for identification.

In bisexual florets, anthers release pollen *into* the cylinder they form around the style. Pollen adheres to the outside of the style and is carried upward as the style grows through the cylinder and elongates. Pollen adheres to insects as they move among the styles in search of food. When they visit other flowers, the pollen is inadvertently transferred to stigmas of mature florets. In unfertilized florets, the style branches continue to grow and curl back, forcing contact of the stigmas with the pollen-laden style, thus effecting self-pollination.

Sunflowers display quite a number of floret types. Those most commonly encountered in our area are **disk**, **ray**, and **ligulate** florets. Other types are **filiform**, **naked**, and **bilabiate** florets. **Disk floret** – corolla radially symmetric (rarely bilateral), slender and tubular with 5 (rarely 4) small equal-sized lobes at the tip; each includes both sexes (bisexual disk floret, such as in scabrid sweetbush, *Bebbia juncea* var. *aspera*); in a few genera the pistil is nonfunctional (staminate disk floret). **Ray floret** – corolla bilaterally symmetric, a short to long tube topped on one side with a strap-like blade that is generally 3-lobed at tip (occasionally unlobed or with a different number of teeth); ray florets lack stamens; each contains a fertile pistil (pistillate ray floret) or a small infertile pistil (sterile ray floret). Common tidy-tips (*Layia platyglossa*) has both disk and ray florets. **Ligulate floret** – corolla bilaterally symmetric; a short to long tube is topped on one side with a strap-like blade that is 5-lobed at tip; ligulate florets are always bisexual (such as in wreath-plants, *Stephanomeria* spp.). **Filiform floret** – resembles a disk floret in that the corolla is radially symmetric and tubular but differs in the corolla's being cylindric, very narrow, and blunt-tipped (rarely with minute lobes); it also lacks anthers. The female (pistillate) florets of baccharis (*Baccharis* spp.) are filiform. **Naked florets** – similar to filiform florets but have no corolla. The female (pistillate) florets of bur-sages (*Ambrosia* spp.) and cockleburs (*Xanthium* spp.) are naked. **Bilabiate florets** – corolla distinctly two-lipped; one lip has two slender lobes, the other has a single broad three-lobed ray. Our only aster with bilabiate florets is sacapellote (*Acourtia microcephala*).

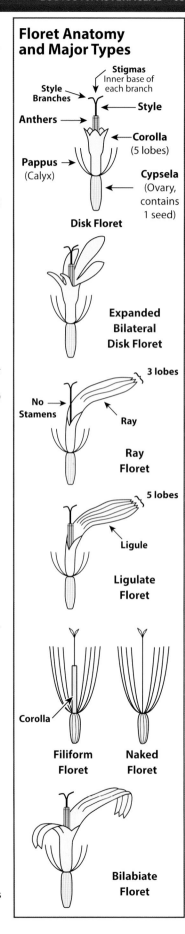

Floret Anatomy and Major Types

Disk Floret

Expanded Bilateral Disk Floret

Ray Floret

Ligulate Floret

Filiform Floret

Naked Floret

Bilabiate Floret

Flowerheads vary in form, floret composition, and sexuality. Some species have flowerheads with only disk, disk-like, or ligulate florets; others have both disk and ray florets in each flowerhead. Some species have male and female reproductive parts in each floret (**bisexual**), others have separate male and female florets on the same plant, still others have entire plants with only male or female florets. We present the asters in three groups based on their flowerhead type: **Discoid**, **Radiant**, and **Disciform** (all disk or disk-like florets), **Radiate** (disk and ray florets), and **Liguliflorous** (ligulate florets only).

Cultivated members include western sunflower, safflower, thistles, artichoke, tarragon, sages, European chamomile, and chicory. With 23,000 species worldwide and about 200 in our area, this is the largest family of eudicots and the second largest family of vascular plants (second only to the orchid family).

	Discoid	**Radiant**	**Disciform**	**Radiate**	**Liguliflorous**
Florets	Disk florets only	A type of discoid head; disk florets only	Resemble discoid heads; typically disk florets in center, surrounded by filiform florets; filiform florets only in pistillate *Baccharis* heads	Disk florets in center, surrounded by a ring of ray florets	Ligulate florets only
Floret Size	All florets about same-sized	Outer disk florets large, often bilateral	All florets about same-sized	All disk florets about same-sized	All florets about same-sized
Floret Sex	Each floret bisexual (rarely male-only with sterile ovary, e.g., male florets of *Baccharis*)	Each floret bisexual	Disk florets (bisexual or male) surrounded by tubular female filiform or naked florets	Disk florets generally bisexual, all fertile, sometimes male; ray florets female or sterile	Each floret bisexual
Examples	*Artemisia tridentata, Cirsium, Bebbia, Pluchea*; male florets of *Baccharis* have a small sterile ovary	*Chaenactis, Lessingia; Centaurea* has sterile outer florets	*Artemisia californica* and *A. douglasiana, Pseudognaphalium,* female heads of *Baccharis*	*Encelia* (rays sterile), *Eriophyllum, Helianthus, Pentachaeta aurea,* true asters	*Agoseris, Malacothrix, Microseris, Stephanomeria, Tragopogon*

Flowerhead Anatomy and Major Types

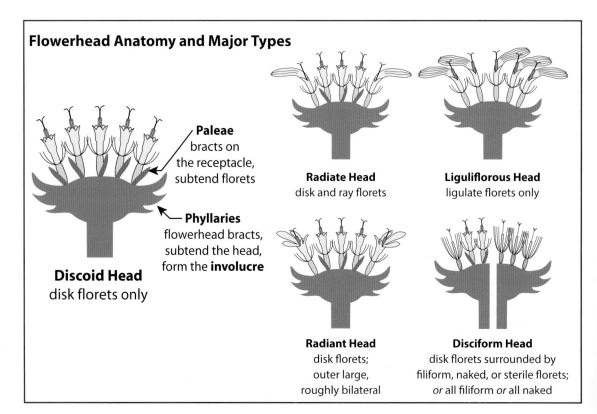

Discoid Head
disk florets only

Paleae bracts on the receptacle, subtend florets

Phyllaries flowerhead bracts, subtend the head, form the **involucre**

Radiate Head
disk and ray florets

Liguliflorous Head
ligulate florets only

Radiant Head
disk florets; outer large, roughly bilateral

Disciform Head
disk florets surrounded by filiform, naked, or sterile florets; or all filiform or all naked

DISCOID, RADIANT, AND DISCIFORM HEADS
Sacapellote
Acourtia microcephala DC.

An upright perennial with multiple main stems, 60-160 cm tall, covered with short sticky-tipped hairs. *Leaves 2.5-15 cm long, widely oval, edges fine-toothed, the base blunt or wrapped around the stem, alternate.* Its slender numerous flowerheads have 10-20 florets each, arranged in loose clusters on a branched inflorescence. *Florets two-lipped (bilabiate) with both lips strongly curved backward.* The inner-facing lip has two slender lobes, the outer-facing has a single broad three-lobed ray. *Corolla pink-purple, rarely white.* Filament cylinder orange, tipped with purple-pink. In flower May-Aug.

A common plant of both dry slopes and moist soils, sometimes in shade, among coastal sage scrub and chaparral. Widespread, found throughout our area from low to high elevations. Easily seen in places such as Santiago Oaks Regional Park, Limestone Canyon, Whiting Ranch Wilderness Park, Caspers Wilderness Park, San Juan Loop Trail, and near Blue Jay Campground.

The epithet was given in 1838 by A.P. de Candolle for its small flowerheads (Greek, mikros = small; kephale = head).

Similar. California everlasting (*Pseudognaphalium californicum*): similar only when senescent; strongly scented; leaves much smaller and narrower; heads disciform.

Sacapellote

When sacapellote senesces in late summer-fall, the plant turns golden brown. After its corollas fall away, its long white pappus bristles help loft the fruit on the wind.

Sacapellote

Crofton Weed, Sticky Eupatorium ✖
Ageratina adenophora (Spreng.) R.M. King & H. Rob.

An upright perennial 50-150 cm tall from an obviously woody base. *Stems and leaf undersides purple,* covered with sticky-tipped hairs. *Leaves opposite;* blade 4-10 cm long, *triangular* to oval. *Flowerheads discoid, crowded.* Phyllaries small, held flat. *Corolla white,* sometimes pink-tinged. In flower nearly all year, mostly Apr-July.

An extremely invasive weed of watercourses and seeps, native to Mexico, known to dominate shady riparian communities and crowd out other plants. Known here from north Bell Canyon on the Audubon Starr Ranch Sanctuary, Trabuco Canyon, waterside in San Juan Canyon, and in moist shady regions along Ortega Highway.

Named in 1826 by K.P.J. Sprengel for many glands that exude a substance that makes it sticky to the touch (Greek, adenos = gland; -phoros = bearing).

Similar. Canyon sunflower (*Venegasia carpesioides*): stem burgundy, brittle; *leaves triangular, alternate; flowerheads radiate;* phyllaries broad, often twisted; *disks yellow to orangish; rays bright yellow.*

Crofton Weed

Dwarf Coastweed
Amblyopappus pusillus Hook. & Arn.

A *small annual* 3-16 (up to 40) cm tall *with a strong sweet scent*, covered with sticky-tipped hairs, rarely hairless. Leaves 1-4 cm long, linear, undivided to pinnately lobed, generally fleshy, alternate, some lower leaves opposite. *Flowerheads discoid* or radiate with only a few tiny ray florets. Corolla yellow. In flower spring-summer.

Occurs in coastal dunes, beaches, and bluffs above the sea. Known from our sandy coastline, easiest to find at Crystal Cove State Park along the cliff base above the high tide mark. A charming little plant, often overlooked, commonly covered with particles of sand and salt.

The epithet was given in 1841 by W.J. Hooker and G.A.W. Arnott for its small size (Latin, pusillus = very small, weak).

Dwarf Coastweed

Ragweeds and Bur-sages: *Ambrosia*

Annuals or perennials. *Their leaves are generally oval or triangular in outline;* toothed, *pinnately lobed, or 1-4-pinnate.* The inflorescence is upright, often long. *Male- and female-only florets are in separate flowerheads on the same plant: males above, facing down; females below, facing up and/or out.* Each flowerhead of male-only disk florets is arranged in a shallow cup-like structure formed by fusion of the phyllaries; their corolla is pale yellowish. Each flowerhead of female-only florets is composed of 1-5 florets that have no corolla but do have very long style branches. They are all wind-pollinated; the pollen is a strong allergen for many people. *Fruit is a hard bur, often spiny.* We have 5 species in our area.

Beach Bur-sage

Males

Females

Right: inflorescence of beach bur-sage (Ambrosia chamissonis). *Male florets above, in down-facing green cups. Female florets below, with long style branches that face up and/or out.*

Sand-bur, Annual Bur-sage
Ambrosia acanthicarpa Hook.

An *upright gray-green annual*, 40-150 cm tall from a thin taproot. *The stems are covered with whitish hairs that lie flat and stiff bristles that can be painful to the touch.* Leaf blade 1-7 cm long, pinnately lobed, hairy or bristly. On the male flowerheads, the three longest phyllaries have a black stripe or patch. Styles of the female florets are pale pink. *Fruits 5-7 mm long, goldish when mature, surrounded by flat, straight, long spines* (paleae) that gradually narrow to a sharp point. In flower Aug-Nov.

Found in *sandy soils* in disturbed areas, roadsides, trailsides, and creekbeds. Known from *hotter, drier regions* of our area, *mostly away from the immediate coast.* Easily found along Santiago Canyon Road, Ortega Highway, and in most wilderness parks.

The epithet was named in 1833 by W.J. Hooker for its spiny fruit (Greek, akantha = a thorn or prickle; karpos = fruit).

Similar. Weak-leaved burweed (*A. confertiflora* DC.): low to upright; leaves 1-4-pinnate, petiole winged and lobed; fruits larger, *spines hooked at tip;* Elsinore Basin, scarce on coastal side of Santa Ana Mountains. **Western ragweed** (*A. psilostachya*): upright; green; leaves smaller; *fruits smaller and spineless;* grasslands, creeksides.

Sand-bur

Sand-bur

Weak-leaved Burweed

Beach Bur-sage, Bather's Delight
Ambrosia chamissonis (Less.) Greene

A *sprawling silvery perennial* from a central stem and root, the stems nearly 4 meters long. Stems and leaves are covered with rough or soft silvery hairs. Leaf blade 2-5 cm long, toothed to pinnately lobed. On the male flowerheads, the phyllaries usually have a black stripe or patch. Styles of the female florets are yellow to pale pink. *Fruits 5-10 mm long, brownish, with straight spines that are broad at the base and quickly narrow to a point, very painful to step on.* In flower July-Nov.

A common plant of coastal beaches and dunes, though widely extirpated by beach traffic and habitat alteration. Known from our entire coastline, easiest seen at the southern end of Huntington State Beach and the Bolsa Chica Ecological Reserve, where it is host to western field dodder (*Cuscuta campestris,* Convolvulaceae). Fairly common on coastal dunes at Balboa Peninsula and Trestles.

The epithet was given in 1831 by C.F. Lessing to honor L.K.A. von Chamisso.

Beach Bur-sage

Beach Bur-sage

Beach Bur-sage

Western Ragweed
Ambrosia psilostachya DC.

An *upright perennial* from creeping underground rhizomes, it grows 0.5-2 meters tall, *covered with short, stiff, bristly hairs*. Leaf blade 4-12 cm long, pinnately divided, opposite below, alternate above. *On the male flowerheads, the phyllaries are all green with no markings.* Styles of the female florets are pale yellow-green. *Fruits* to 5 mm long, greenish-brown, *spineless,* sometimes with a few teeth or bumps. In flower July-Nov.

A very common plant of grasslands, oak woodlands, riparian areas, creeksides, and disturbed soils throughout our area. Easily found in places such as Laurel and Limestone Canyons.

The epithet was given in 1836 by A.P. de Candolle for its spineless fruits (Greek, psilos = naked, smooth; stachys = an ear of grain or spike).

Similar. Sand-bur (*A. acanthicarpa*): upright annual; gray; leaves larger; *fruits large and spiny;* inland sandy soils. **Weak-leaved burweed** (*A. confertiflora* DC.); low to upright; leaves 1-4-pinnate, petiole winged and lobed; fruits larger, *spines hooked at tip;* Elsinore Basin, scarce on coastal side of Santa Ana Mountains.

San Diego Ambrosia ★
Ambrosia pumila (Nutt.) A. Gray

A small *soft-hairy gray perennial* 10-20(-50) cm tall, it grows in small patches and spreads via rhizomatous roots. Leaf blade 3-13 cm long, 2-4-pinnate. On the male flowerheads, the phyllaries are all gray-green with no markings. Styles of the female florets are pale yellow-green. *Fruits* under 2 mm long, brownish, *spineless or few-spined.* In flower June-Sept.

A charming plant of grasslands, vernal pools, and disturbed areas. Uncommon in our area, known from Nichols Road Wetlands Area in north Lake Elsinore, south toward Temecula/Murrieta, then south through San Diego County. Seriously threatened by development, non-native plants, road maintenance, trampling, and off-road vehicles. CRPR 1B.1. In cultivation, it makes a fine groundcover plant, especially hardy in sandy soils.

The epithet was given in 1840 by T. Nuttall for its small size (Latin, pumilus = diminutive).

Similar. Weak-leaved burweed (*A. confertiflora* DC.): much taller; leaves 1-4 pinnate, petiole winged and lobed; fruits larger, spines hooked at tip; Elsinore Basin.

Rayless Arnica
Arnica discoidea Benth.

An upright ephemeral perennial covered with soft long hairs and some shorter sticky-tipped hairs. It grows from extensive rhizomes by which it may form dense colonies. Generally has a single stem, 30-60 cm tall, that branches above the foliage. *Leaves mostly basal,* with a long petiole, the blade somewhat oval and tooth-edged, 2-12 cm long; upper leaves opposite with little or no petiole. *Heads discoid, disks at periphery sometimes expanded and ray-like. The inflorescence usually bears a single central flowerhead with a pair of opposing flowerheads just below it,* easily seen in the photograph. The corollas are yellow to orangish. In flower May-July.

An uncommon treat, it occurs in mixed coniferous forests, chaparral, and woodlands, more common in larger mountain ranges. Here known only from north-facing slopes below Maple Spring Saddle, down slope to Maple Spring in upper Silverado Canyon.

The epithet was given in 1849 by G. Bentham for its (mostly) discoid flowerheads; most other species of Arnica have radiate heads (Greek, diskos = a disk; -oidea = form of, type of).

Sagebrushes: *Artemisia*
Annuals or perennials with characteristic scents. Leaves often lobed. *Flowerheads discoid or disciform, arranged in terminal spikes.* Within each flowerhead, the outermost ring of florets are female-only. *The central disk florets are bisexual (rarely male-only) and have pale yellow corollas.* All flowerheads generally face downward, at least in fruit. Pollinated by bees and flies.

Our species are perennials that enter aboveground dormancy during periods of drought. In late fall-early winter, their roots start actively growing, followed by new stems and leaves.

Legend has it that before cowboys would head into town, they would rub California sagebrush on their clothes to mask the odor, giving the plant's scent the nickname of "cowboy cologne."

California Sagebrush
Artemisia californica Less.

An upright shrub, 1-2.5 meters tall, with a scent somewhat like menthol. *Its aromatic blue-green to silvery-gray threadlike leaves are 1-10 cm long, pinnately divided and lobed, rarely undivided, the edges rolled under. Flowerheads disciform.* The outermost ring of each flowerhead is made up of 6-10 females; the central disk florets have both male and female parts. Corolla pale yellow. In flower Aug-Feb.

An abundant plant throughout our area. Those living in areas that regularly receive a lot of fog tend to have more leaf divisions than plants in fogless areas. The additional divisions increase surface area to capture fog. Weighed down, the weak leaflets droop and the droplets coalesce, fall to the ground, and water the roots.

The epithet was given in 1831 by C.F. Lessing for the place of its discovery.

California Sagebrush

California sagebrush is usually blue-green but also comes in a silvery-gray form, most often seen near the coast.

SAGEBRUSH: AN IMPERILED HABITAT

California sagebrush and the coastal sage scrub plant community in which it occurs provide vital habitat for the coastal California gnatcatcher. Rampant destruction of coastal sage scrub habitat in southern California has reduced their numbers and fragmented their populations, essentially into islands separated by seas of commercial and industrial buildings, residences, and roadways. The bird is currently protected by law and some habitat has been set aside to encourage its survival. Their call sounds like that of a crying kitten. If you encounter them when wildflower-watching, be mindful not to disturb them.

Coastal California Gnatcatcher ★
Polioptila californica californica (Brewster) (Polioptilidae)

Gnatcatchers are charming little birds with slender bills and comically long tails that are always in motion. The coastal California gnatcatcher is about 11 cm long from bill-tip to tail-tip. *Its plumage is medium to dark gray with a gray belly. Males have a black cap during breeding season.* Its tail feathers are black, narrowly edged with white that is visible from below. A resident of coastal sage scrub, it flits from plant to plant gleaning insects and arachnids, often vocalizing during the process. Its populations are in serious decline.

The epithet was established in 1881 by W. Brewster in reference to the place of its discovery.

Coastal California Gnatcatcher

Robb Hamilton

Blue-gray Gnatcatcher
Polioptila caerulea (L.) (Polioptilidae)

The blue-gray gnatcatcher is also about 11 cm long and has similar behavior. *Its plumage is blue-gray with a whitish belly.* Males are a bit bluer than females. Its tail is black above; the outermost tail feathers are completely white, visible from above, easiest seen from below.

The epithet was established in 1776 by C. Linnaeus in reference to its bluish color (Latin, caerulus = dark-colored, dark blue).

Blue-Gray Gnatcatcher female

California Mugwort
Artemisia douglasiana Besser

An *upright perennial* from a creeping rhizome, it stands 0.5-2.5 meters tall. Its *leaves* are 1-15 cm long, linear to broadly elliptical, *smooth-edged or with 3-5 large lobes*. Leaves are *two-toned: dark green and often hairless above, white-hairy below*. *Flowerheads disciform*. The outermost ring of each flowerhead is made up of 6-9 all-female florets; the central disk region has 9-25 all-male florets. Corolla pale yellow. In flower June-Oct.

A common plant along creeks and streams. When the leaf is broken, it releases a sharp odor like turpentine, kerosene, or tar, sometimes with a hint of citrus. Rumored to be a remedy for poison oak dermatitis when crushed and added to bathwater, but its effectiveness has not been substantiated.

The epithet was given in 1833 by W.S.J.G. von Besser in honor of D. Douglas.

Dragon Sagewort, Tarragon
Artemisia dracunculus L.

A tall, slender, perennial from rhizomes and long wandlike stems, 0.5-1.5 meters tall. It is odorless or smells like licorice or dill. Leaves 1-7 cm long, slender and linear, *sometimes with linear lobes, mostly hairless*. Young leaves bright green, older leaves medium green. *Flowerheads disciform*. The outermost ring of each flowerhead is made up of 14-25 all-female florets; the central disk region has 8-20 all-male florets. Corolla pale yellow. In flower mostly Aug-Oct.

Found in meadows, creeks, and dry disturbed soils at many elevations within most plant communities. Known widely including Puente-Chino Hills, Santa Ana River channel, upper Newport Bay, San Joaquin Freshwater Marsh, San Joaquin Hills, Hot Springs Canyon, and San Juan Canyon. Especially common along Ortega Highway from Upper San Juan Campground to Long Canyon and from The Lookout down to Lake Elsinore. The form of this plant from Europe is the commercially grown tarragon used in cooking, often in the preparation of fish.

The epithet was named in 1753 by C. Linnaeus, who used the name coined by Pliny the Elder for the resemblance of its rhizomes to a small snake (Latin, draconis = dragon, serpent; -unculus = little).

Great Basin Sagebrush, Big Sagebrush
Artemisia tridentata Nutt. ssp. *tridentata*

An *upright, rounded, evergreen shrub* from a very thick *woody base*, up to 2 meters tall. Scent is like mild lavender, citrus, or pine. *Leaves* 1.2-4 cm long, shaped like a long triangle, narrow at the base, *with 3 broad teeth* at the tip (rarely to 5 teeth). Each narrow flowerhead is made up of 4-6 bisexual florets. Corolla pale yellow. In flower Aug-Oct.

A plant of the Great Basin and Mojave Desert, found sparingly here in areas where cold air settles. Known locally from the Santa Ana River in Yorba Linda and the lower mouth of Coal Canyon but most abundant in Cristianitos Canyon on the Rancho Mission Viejo, where it occurs among coastal sage scrub, oak woodland, and California junipers, perhaps a remnant of a juniper woodland community.

The epithet was named in 1841 by T. Nuttall for its three-toothed leaves (Latin, tri = three; dentatus = toothed).

Baccharis: *Baccharis*

Upright woody perennial or woody-based shrubs covered with sticky resin. Our species are hairless. Many have grooves or ridges along their stems. *Leaves* alternate, *undivided,* often toothed. *In addition to a thick central leaf vein, the leaves often have two narrow lateral veins that begin at the base and run up the leaf;* this condition is denoted as "main veins 3." The veins are best seen when a light is behind the leaf.

Plants dioecious; flowerheads unisexual, white to yellowish or pinkish. Phyllaries small, held against the florets; green, pink, or red.

Male flowerheads discoid, short and broad. Male floret with a short sterile ovary and a long style that protrudes above the corolla. The style helps disperse pollen and whithers soon after pollen is released. Corolla funnel-shaped, white to yellowish, sometimes pink-tinged, the lobes curve backward. Pappus bristles few, twisted, shorter than the corolla.

Female flowerheads disciform; long and slender. Female floret with a long fertile ovary. Corolla cylindrical, white to greenish. Pappus bristles threadlike, numerous, and longer than the corolla but shorter than the pistil. In fruit, these bristles elongate and form a plume that helps catch the wind to disperse the seed.

The genus was established in 1753 by C. Linnaeus who reused the name of a now-unknown plant (Greek, bakkaris = a plant with an aromatic root from which an oil was extracted, possibly *Gnaphalium sanguineum* L. of Egypt and Palestine). Some botanists cite the Roman God of wine, Bakkhos (Bacchus) as the origin of the genus.

Above: mule fat (Baccharis salicifolia.) Male flowerheads (left), female flowerheads (right).

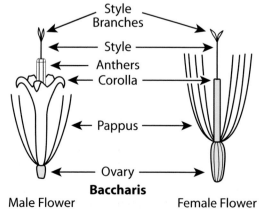

Baccharis
Male Flower — Female Flower

Marsh Baccharis
Baccharis glutinosa Pers. [*B. douglasii* DC.]

A perennial from a creeping rhizome, 1(-2) meters tall, most often single-stemmed with no substantial side branches. *Leaves* 3-10 cm long, lance-shaped, smooth-edged or large-toothed. Main veins 3. *Stems and leaves covered with resin glands that make the plant sticky to the touch.* Flowerheads several, in round- or flat-topped clusters. In flower July-Oct (-Dec).

A plant of permanent moisture, it occurs along watercourses and along both freshwater and saltwater marshes. The stems and first leaves appear in spring, the flowers in summer. Known here from Bolsa Chica Ecological Reserve (mostly along the trail that parallels Pacific Coast Highway), near the mouth of the Santa Ana River, in the Lomas de Santiago near the junction of Santiago and Limestone Canyons, and along the Santa Ana River in Corona. Reportedly common in Holy Jim Canyon in the 1920s, but not found there in recent years.

Named in 1807 by C.H. Persoon for its stickiness (Latin, gluten = glue; -osus = full of).

Marsh Baccharis

Male flowers — Female flowers — Marsh Baccharis

Marsh Baccharis — Coyote Brush

Coyote Brush — Male flowers — Female flowers

Coyote Brush, Chaparral Broom
Baccharis pilularis DC. **ssp. *consanguinea*** (DC.) C.B. Wolf

A rounded woody shrub 1-4 meters tall, often as wide, with several cone-shaped branches. Its *numerous stiff leaves* are 0.8-5.5 cm long, oval to reverse egg-shaped in outline, *often large-toothed, light green, the base often wedge-shaped,* petiole absent or very short. Main veins 3. It keeps its leaves, even in drought. Flowerheads appear at the branchtips both at the top of the plant and on side branches. Male florets are yellowish, female florets are white. In flower Aug-Dec.

A common plant of *coastal sage scrub,* it often grows in large groups. It occurs in all of our wilderness parks. The low-growing *B.p.* ssp. *pilularis* from the Monterey coastline is commonly used as groundcover.

The epithet was given in 1836 by A.P. de Candolle, possibly for its plant galls (Latin, pilula = a globule).

Similar. Willow baccharis (*B. salicina*): taller; grooved twigs; leaves variable but longer, larger-toothed, main veins 3; near water; less common. **Broom baccharis** (*B. sarothroides*): very round-topped; bright green; small thick leaves, main vein 1, edges rolled under, drop off in summer; in/near ephemeral watercourses and disturbed sites; uncommon, apparently recently expanding into Orange County.

Coyote Brush

Malibu Baccharis ★
Baccharis malibuensis R.M. Beauch. & Henrickson

A woody-based perennial with sprawling, arching, or upright stems to 2 meters long. Green stems grooved, with short hairs among the flowerheads. Petiole short; *leaf blade* (15)20-45(68) mm long, 1-4(-8) mm wide, *hairless or a little hairy; linear; the edges smooth or short-toothed.* Upper leaf surface dotted with sticky glands. *Main veins 1-3. Leaves remain on the plant at flowering time.* Flowerheads long and cylindrical. *Phyllaries basically linear.* In flower Aug-Sep.

Primarily a plant of coastal chaparral in the Santa Monica Mountains, a small population was found in Fremont Canyon on the Irvine Ranch in 2000 by Rick Riefner. The plants occur in an alluvial wash near an oak woodland, in sandy soil at the bases of rock outcrops, and along a few power line access roads. A very rare species with few populations left in the wild; threatened by urbanization. CRPR 1B.1.

The epithet was given in 1996 by R.M. Beauchamp and J.S. Henrickson for Stokes Canyon in Malibu, the place of its initial discovery.

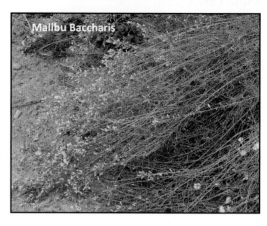

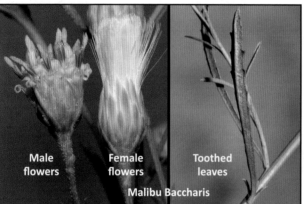

Encinitas Baccharis ★
Baccharis vanessae R.M. Beauch.

A woody-based hairless perennial with upright stems to 2 meters long. Petiole absent; *leaf blade bright green to yellow-green*, 1-45 mm long, 1-3 mm wide, *thread-like to linear*. Both leaf surfaces dotted with sticky glands. *Leaf edges smooth. Main vein 1. Leaves often fall off at flowering time.* Flowerheads funnel-shaped to rounded at base. *Phyllaries* lance-shaped, their *edges with tiny hairs;* rounded on back; tip pointed, *often curved backward in age.* In flower Aug-Nov.

Restricted to coastal chaparral in San Diego County. A small population was found in the southwestern edge of the San Mateo Wilderness Area in 1992 by Steve Boyd, Tim Ross, and Orlando Mistretta. The site is in lower Devil Canyon, about 1 mile upstream from San Mateo Canyon. CRPR 1B.1. Threatened by urbanization. A very rare species with few populations left in the wild. Once fairly common in Encinitas, the source of its common name and from where it was first collected. Now reduced to a few locations.

The epithet was given in 1980 by R.M. Beauchamp for his daughter, V.B. Beauchamp.

Mule Fat
Baccharis salicifolia (Ruiz & Pavón) Pers. [long ago, mistakenly called *B. glutinosa*, *B. viminea*]

An *upright woody tree-like shrub* with long wandlike main branches, *nearly 3 meters tall,* and few side branches. Most twigs are brown. *Leaves 5-15 cm long, linear to lance-shaped, the edges often toothed, light green. Main veins (1-)3.* Flowerheads are arranged in dense round-topped clusters. *Phyllaries green to red.* In summer, most flowerheads are at the tops of the branches and most leaves are toothed. In winter, most flowerheads are at the side branches and most leaves are smooth-edged. In flower all year, mostly Mar-Oct.

A common plant of waterways and seeps among riparian, marshes, coastal sage scrub, and coastal communities, often in large groups. Easily found in coast, hill, mountain, and inland sites. It occurs in all of our wilderness parks.

After creeks are flooded and the vegetation damaged or scoured, mule fat seeds are among the first to germinate and grow, taking advantage of the new territory. Once established, mule fat is a tough plant, as long as its roots have a reliable water source. Mule fat thickets are home to many species of insects, amphibians, reptiles, mammals, and birds. Its flowers are often visited by a host of insects such as thrips, beetles, bees, flies, moths, and butterflies.

The epithet was given in 1807 by H. Ruíz Lopez and J.A. Pavón for the similarity of its leaves to those of willows (Latin, salicis = willow; folium = leaf).

Similar. Willow baccharis (*B. salicina*): shorter; grooved twigs; leaves variable but shorter, larger-toothed, the base wedge-shaped, main veins 3; near water; less common. **Coyote brush** (*B. pilularis*): shorter stiff leaves, the base wedge-shaped, main veins 3; generally on drier sites, less water-dependent; abundant. **Willows** (*Salix* ssp., Salicaceae, not pictured): winter deciduous; leaves more pliable, main vein 1; flowers tiny, in upright catkins, never in flowerheads; often grow together.

Mule Fat

Male flowers

Mule Fat

Female flowers

Willow Baccharis
Baccharis salicina Torr. & A. Gray [*B. emoryi* A. Gray]
   An upright shrub, 1-3 meters tall, with grooved twigs. *Leaves 3.5-7 cm long, variable even on the same plant and twig*, surprisingly soft-textured. *Lower leaves large, toothed, the base wedge-shaped*; upper leaves smaller, narrower, not toothed. Petiole absent or short with winged edges. Main vein 3. Leaves sometimes fall off by flowering time. Flowerheads in clusters of 1-3 per stalk. In flower Aug-Dec.

   Sometimes mistaken for a hybrid between the more abundant mule fat and coyote brush, since its variable leaves are shaped like coyote brush but are often longer and not as stiff. Found in sandy soils along streams generally at low elevations. Known from Santa Ana Canyon, Anaheim Bay, Bolsa Chica Ecological Reserve, Upper Newport Bay, San Joaquin Marsh, south side of Irvine Regional Park, Bell Canyon on the Audubon Starr Ranch Sanctuary, in a seep along North Main Divide Road "Loop," south into the San Mateo Canyon Wilderness Area.

   Named in 1842 by J. Torrey and A. Gray for its similarity to willows (Latin, salicis = willow; -ina = likeness).

   Similar. Mule fat (*B. salicifolia*): taller; stems generally not grooved; leaves much longer, small-toothed, main veins (1-)3, base tapering to petiole; generally in/near water; abundant.
   Coyote brush (*B. pilularis*): shorter stiffer leaves, main veins 3; abundant.

Willow Baccharis

Willow Baccharis — Male flowers / Female flowers

Broom Baccharis — Male flowers / Female flowers

Broom Baccharis, Hierba del Pasmo
Baccharis sarothroides A. Gray
   A large rounded shrub 2-4 meters tall, *with dense bright green twigs and few leaves*. Leaves 1-3 cm long, *narrow; stiff, thick; linear*; edges smooth, rolled under; petiole absent. Main vein 1. The leaves drop off as summer approaches, most are gone by flowering time. Flowerheads appear on their own short peduncle, distributed along the stem, most numerous near the branchtips. There are generally no leaves or bracts among the flowerheads. In flower June-Oct.

   A plant of coastal sage scrub generally at low elevations from San Diego to Orange Counties, inland to western Riverside County, and east into the deserts. Generally found in or near ephemeral watercourses and seeps, in disturbed soils on hillsides, sometimes on irrigated road cuts. Known from Santa Ana Canyon, Irvine Regional Park, lower Baker Canyon, Limestone Canyon, Rose Canyon, and Lower Borrego Canyon wash near Trabuco Road.

   The epithet was given in 1882 by A. Gray for its leafless ("broom-like") appearance (Greek, saron = broom; -oides = like).

   Similar. Coyote brush (*B. pilularis*): leaves shorter, stiffer, broader, toothed-edged, main veins 3; generally on hillsides; abundant.

Broom Baccharis

Scabrid Sweetbush

Bebbia juncea (Benth.) Greene **var. *aspera*** Greene

A rounded, multi-branched, bush about 3 meters tall and wide, woody only at the base, hairless or covered with short bristles. Young stems are bright green and age to brown. *Leaves 1-3 cm long, linear, sometimes with small lobes or teeth, opposite, uppermost sometimes alternate.* Flowerheads solitary or in open branched clusters at the branch tips. Most flowerheads discoid; some heads disciform with 1-8 female florets in the outer ring. *Phyllaries broad to narrow; outer green and short white-hairy, inner straw-like, often red- or brown-tipped.* Corollas bright yellow, lobes short. In flower Apr-Sep.

A common plant of *dry rocky slopes, rock outcrops, and sandy soils along creekbeds.* Found in the Chino Hills, Lomas de Santiago, Santa Ana Mountains, Temescal Valley, and San Mateo Canyon Wilderness Area. Easy to find along Santiago Creek in Irvine Regional Park and San Juan Creek in Caspers Wilderness Park. *It drops its leaves in drought, leaving only its long slender green stems.* Pollinated by many types of bees, flies, butterflies, and moths.

The epithet was given in 1885 by G. Bentham for its rush-like appearance (Latin, junceus = rush-like). Rushes are plants (in the family Juncaceae) that have naked stick-like stems.

Similar. Scale-broom (*Lepidospartum squamatum*): stems grooved; leaves much smaller, scale-like, alternate; flowerheads in clusters; phyllaries green, hairless.

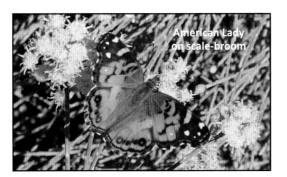

Left: An American lady butterfly (Vanessa virginiensis (Drury), Nymphalidae) nectaring from scale-broom. Note the tiny scale-like leaves along the stem. Scale-broom is a popular late-season nectar source for many types of insects.

Scale-broom

Lepidospartum squamatum (A. Gray) A. Gray

A *rounded shrub* 1-2 meters tall, often wider, it arises from creeping underground stems. *The young twigs grown in spring are covered with long white woolly hairs that fall off with age.* Stems have grooves that run down them. *Leaves 2-3 mm long, linear to spoon-shaped, and fall off with age.* Older twigs produce tiny scale-like leaves only 2-3 mm long, these may fall off as well. Flowerheads discoid, in open-branched sparse clusters near the tips of long naked side branches. *Phyllaries narrow, green, hairless.* Corollas yellow to orangish, the lobes long and strongly curved backward. In flower Aug-Nov.

A common plant of *sandy-gravelly soils* in washes and older stream terraces in sunny floors of all major canyons. Quite easy to find in Caspers Wilderness Park, O'Neill Regional Park and in Arroyo Trabuco upstream of the park. Well-adapted to its habitat, where it is known to push through as much as 10 meters of sediment after major storms. Pollinated by many types of bees, flies, butterflies, and moths.

The epithet was named in 1870 by A. Gray for the scale-like leaves on its older twigs (Latin, squamatus = scaly).

Similar. Scabrid sweetbush (*Bebbia juncea* var. *aspera*): stems not grooved; leaves much longer, linear, mostly opposite; flowerheads often solitary; outer phyllaries green, short-hairy, inner often red- or brown-tipped.

Brickellbush: *Brickellia*

Shrubs, upright or rounded in outline. *Leaves alternate or opposite, often heart-shaped or oval, the upper leaves much smaller than the lower.* The leaf and stem surfaces are covered with dot-like glands and short or long hairs. *Flowerheads discoid, phyllaries straight or curved.* Corolla cylindric, whitish, yellowish, or reddish. In fruit the phyllaries open outward 90 degrees to the stem, revealing the seeds, which are surrounded by many bristles.

The genus was established in 1824 by S. Elliott in honor of J. Brickell.

California Brickellbush

Brickellia californica (Torr. & A. Gray) A. Gray

An *upright many-branched shrub*, 0.5-2 meters tall, covered in tiny or long woolly hairs. *Leaves 1-6 cm long, gray-green, sometimes red-tinged, triangular to oval, the edge toothed, the base heart-shaped, alternate.* The leaf surface is covered with dot-like glands and short hairs. Flowerheads bisexual. The *slender florets* are whitish or tinged reddish. The phyllaries are hairless or nearly so; their tips are held upright. In flower primarily Aug-Oct, occasionally late spring and summer.

A plant of sandy washes, coastal sage scrub, and chaparral. Found in most natural areas in our area, more common away from the coast. Its leaves often hang down and present their blade surface directly to the sun. Its flowers release a pleasant scent in the evening, surely to attract nocturnal pollinators such as moths and beetles. Used in landscape as a partner to California buckwheat and California copperleaf, with which it grows in the wild.

The epithet was given in 1841 by J. Torrey and A. Gray for the state of its discovery.

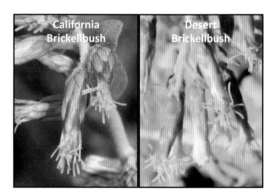

Desert Brickellbush

Brickellia desertorum Coville

A *light gray shrub*, 0.8-1.5 meters tall, often even wider, generally rounded in outline. It has *many stems that criss-cross each other.* Leaves 0.5-1.2 cm long, with short or no petiole, *edges toothed, covered with short woolly hairs, alternate or opposite. The uppermost leaves are quite tiny.* Flowerheads 8-10 mm long, in small clusters. Phyllaries about 21 in number, covered with minute hairs, especially near their tips; lower phyllaries shorter and especially hairy. Phyllary tips upright, not curved out- or backward. In flower primarily Sep-Dec.

A plant of rocky soils in our deserts and the Perris Basin, found sparingly here. Known in our area from dry north-facing slopes near Corona: along Skyline Drive in lower Tin Mine Canyon and Clay Canyon Drive in lower Brown Canyon.

The epithet was given in 1892 by F.V. Coville in reference to the place of its original discovery, in the Colorado Desert of California.

Similar. California brickellbush (*B. californica*): leaves much larger, alternate; flowerheads 12-14 mm long; phyllaries more numerous (21-35); common and widespread, co-occurs in Tin Mine Canyon. **Nevin's brickellbush** ✿ (*B. nevinii*): leaves alternate, slightly larger, densely covered with white hairs; flowerheads 1.5 cm, few in number; phyllaries about 30, densely covered with white hairs, tips curved out- or backward; uncommon, Silverado Canyon.

Nevin's Brickellbush ✿
Brickellia nevinii A. Gray

Nevin's Brickellbush

A low bushy shrub, 30-50 cm tall, *the stems densely covered with white woolly hairs, giving the entire plant a pale gray appearance.* The *small leaves* have few or no petioles, *the blade* 0.6-1.8 cm long, *oval to heart-shaped, alternate.* The leaf surface is densely covered with white woolly hairs and has no dot-like glands. Flowers all-disk, of both male and female, all white. The phyllaries are covered with white woolly hairs; their fuzzy tips curve outward or backward. In flower Sep-Nov.

A plant of dry slopes in coastal sagebrush scrub and chaparral. Here it occurs uncommonly in and above major canyons of the Santa Ana Mountains, such as Gypsum, Black Star, Tin Mine, Silverado, White, and Williams. In cultivation but reportedly difficult to grow.

The epithet was given in 1885 by A. Gray for J.C. Nevin, who discovered it in 1884 near Newhall.

Thistles: *Acroptilon, Carduus, Carthamus, Centaurea, Cirsium, Cynara, Silybum, Volutaria*

An easily recognized group, though its characteristics vary widely. *Sharp spines* make their presence known when you brush against them; one however, is spineless. Most are non-native, very invasive weeds that crowd out native vegetation and thus alter habitat. We have only 3 thistles native to our area, all in the genus *Cirsium*. Our natives are attractive, non-invasive, and important to wildlife as a food source. We have witnessed people cutting, digging, and poisoning all sorts of thistles, including our native species, because they assume all thistles to be unpleasant non-natives. We urge all land managers to have their thistles properly identified before taking control measures.

	Growth	Leaves	Inflorescence	Phyllaries	Corolla
Knapweed *Acroptilon*	Ephemeral perennial from rhizomes	Spineless; 1-2-pinnately lobed	Discoid	Broad, oval, spineless	White, pink, or purple
Italian *Carduus*	Annual (some species biennial or perennial)	Very spiny; pinnately lobed; stems winged	Discoid	Long, spine-tipped	White, pink, or purple
Distaff *Carthamus*	Annual	Very spiny; pinnately lobed	Discoid	Largest leaf-like, 1-pinnately lobed, the lobes spine-tipped	Yellow
Star- *Centaurea*	Annual or perennial	Spineless; 1-2 pinnately lobed	Disciform or radiant (some discoid)	Tips fringed to spiny; some pinnate; some with a spine from center	Yellow
Common *Cirsium*	Annual or perennial	Spiny; pinnately lobed; stems winged	Discoid	Long, spine-tipped	White, pink, or purple
Artichoke *Cynara*	Robust perennial from taproot	Very spiny; all pinnately lobed	Discoid	Very broad, spine-tipped	White to blue-purple
Milk *Silybum*	Annual or biennial from deep taproot	Spiny; pinnately lobed	Discoid	Broad, edges and tip spiny	Pink to purple
Morocco *Volutaria*	Annual	Spineless; pinnately lobed	Radiant	Narrow, spine-tipped	Pink to purple

Russian Knapweed ✖

Acroptilon repens (L.) DC. [*Centaurea repens* L.]

An *upright perennial* 30-100 cm tall that spreads by underground stems to form dense clonal colonies. *Unlike other thistles, the plant is covered with soft cobwebby hairs and has no spines.* Leaves alternate, petiole absent. Lower leaves 4-10 cm long and pinnately lobed, upper leaves smaller and toothed or smooth-edged. *Phyllaries spineless.* Corolla blue-purple or white. In flower June-Aug.

An invasive weed of cultivated fields, disturbed ground, and dry drainages; native to central Asia. Known here from Yorba Linda, Huntington Beach, Santa Ana, and Irvine (San Joaquin Freshwater Marsh, UCI Arboretum, and Turtle Rock).

The epithet was given in 1837 by C. Linnaeus for its creeping underground stems (Latin, repens = creeping, crawling).

Italian Thistle ✖

Carduus pycnocephalus L.

A *slender upright annual or biennial*, 0.2-2 meters tall, *hairless or a bit woolly, with narrow spine-tipped wings on its stems.* Basal leaves 10-15 cm long with 4-10 lobes, upper stem leaves often covered with white woolly hairs. *Flowerheads relatively small, narrow;* and occur *in tight clusters of 2-5 heads.* Phyllaries held upward, covered with white woolly hairs, their edges have short bristles that make them rough to the touch. *Corolla 10-14 mm long, pink to purple.* Pappus bristles have tiny barbs on them. Compare this with the similar genus *Cirsium*, which has plumose pappus bristles (each bristle has slender side branches on it). In flower May-July.

A weed of roadsides, disturbed soils, and many wild areas. Abundant and spreading throughout our area.

The epithet was given in 1763 by C. Linnaeus for its dense flowerheads (Greek, pyknos = dense, compact; kephale = head).

Similar. Bull thistle ✖ (*Cirsium vulgare*): *thick*-stemmed; spiny wings on stems; leaf veins sunken; *flowerheads large, rounded, solitary;* phyllaries bases held outward.

Smooth Distaff Thistle ✖

Carthamus creticus L. [*C. baeticus* (Boiss. & Reut.) Nyman]
 Upright annual, 0.4-1.0 m tall, stem whitish. *Stem leaves and outer phyllaries pinnately lobed, each lobe sharply spine-tipped.* Flowerheads urn-shaped, narrowed at tip. *Corolla yellow.* In flower Jun-Aug.

An uncommon weed of dry, disturbed, and overgrazed soils in the Chino Hills, such as in Brea and Tonner Canyons.

The epithet was named in 1763 by C. Linnaeus for the place of its origin (Latin, creticus = of the island of Crete).

Similar. **Safflower** (*C. tinctorius* L., not pictured): stem straw-colored; *leaves entire, toothed, tipped with small spines;* corolla yellow to red. One record for Santa Ana (1936); occasionally found in hot dry sites east of our area.

Star-thistles: *Centaurea*

Annuals or perennials, most upright, the *stems with raised ribs* along their length. *Involucre spheric, narrowed at its tip. Phyllaries with sharp spines.* Flowerheads disciform or radiant. *Disk florets bisexual, outermost florets often larger but sterile. Corolla yellow* in most of our species. About 10 species found here; we present only the 3 most common.

Blessed Thistle ✖

Centaurea benedicta (L.) L. [*Cnicus benedictus* L.]
 A *low-growing annual,* usually under 30 cm tall and twice as broad. *Stems short, many-branched, not winged, covered in long cobwebby hairs.* Leaves 10-20 cm long, elliptical in outline, lobed, toothed, and spine-tipped. *Flowerheads nestled among leaves. Phyllaries* green to yellowish, *uppermost with purple to golden spines, each with pinnately arranged lateral spines.* Corolla yellow. In flower Apr-June.

Uncommon, primarily in disturbed soils among coastal sage scrub, grasslands, and chaparral. *Sometimes a fire-follower;* occasionally a roadside weed. More often found in hot, dry, inland settings. Known from Santa Ana Canyon, Anaheim, Temecula Valley, Lake Elsinore, and Murrieta. Native to the Mediterranean region.

It was named in 1753 by C. Linnaeus, perhaps for its use in alcoholic beverages and medicines (Latin, benedictus = well-spoken of, blessed).

Malta or Maltese Star-thistle, Tocalote ✖
Centaurea melitensis L.

A *slender annual* 10 cm - 1 meter tall, *with scattered gray hairs and short bristles* that are rough to the touch. Lower leaves 2-15 cm long, undivided or lobed, the stem leaves are linear and have wings that continue a long way down the stem. Flowerheads one to a few, rounded, quite narrow around the yellow corollas. *Phyllaries* green, *upper ones purple toward the tip, topped with purple spines that fade to yellowish-brown in age.* The *large central spine* is 5-10 mm long and generally curves backward. In flower May-June.

A very invasive abundant weed thoughout our area. In Orange County, it is much more common and widespread than yellow star-thistle.

Named in 1753 by C. Linnaeus for its presumed country of origin (Greek, Melitaios = Malta; -ensis = from).

Similar. Yellow star-thistle ✖ (*C. solstitialis*): soft-haired; larger flowers; each phyllary has a huge, straight, yellow, central spine.

Right: The basal leaves of Malta and yellow star-thistle are hairy, spineless, and often pinnate. Their stem leaves are usually narrow and smaller.

Yellow Star-thistle ☠ ✖
Centaurea solstitialis L.

An *annual* 10 cm - 1 meter tall, *covered with fuzzy gray hairs.* Phyllaries green, *tipped with small, yellowish, lateral spines. The much larger central spine* is 10-25 mm long, *yellow, stout, and very straight*. In flower May-Oct.

One of the most problematic weeds in the western US, now increasingly common in our area. Mostly found in warm inland areas such as Temescal Valley and Elsinore Basin. In Orange County, it was virtually absent until the late 1990s, when small populations were discovered along Ortega Highway and in Silverado Canyon. Recently reported from Santiago Canyon Road and Elsinore Peak.

Yellow star-thistle is highly competitive and quickly develops thick stands that exclude native plants. It often first grows along roadsides but soon expands and invades grasslands, meadows, and other wildflower habitats. In addition to its devastating effects on native plants, horses that graze on it accumulate a toxin that causes chewing disease which kills brain cells that control fine motor movements such as lip and mouth movement. Unable to eat or drink, the disease is fatal. Infestations of this plant must be eradicated quickly and aggressively.

Named in 1753 by C. Linnaeus for its peak blooming period (Latin, solstitialis = belonging to midsummer).

Similar. Malta star-thistle ✖ (*C. melitensis*): bristly; smaller flowers; phyllaries green, purple-tipped, central purple spine curves backward.

Common Thistles: *Cirsium*

Very spiny, often large upright plants, from a stout taproot and/or runners. *Leaves largely basal,* stem leaves clearly tapered to tip, toothed or pinnately lobed, alternate. *Leaf and petiole edges usually wavy, spiny;* in some species, the petiole edges continue down the stem. Flowerheads discoid; clustered or solitary; *phyllaries spine-tipped, in 5-20 spirals.* Fruits oval, hairless. Many natives and non-natives, hybrids commonly encountered. The genus was named in 1754 by P.M. Miller (Greek, cirsion = a kind of thistle).

Cobwebby Thistle
Cirsium occidentale (Nutt.) Jeps. **var.** *occidentale*

Cobwebby Thistle

A *large upright biennial* 30-150 cm tall or more, from an obvious main stem and many upper branches, *covered with long hairs, especially on the lower leaf surfaces.* Lower leaves 10-40 cm long, reverse lance-shaped in outline, with lobes halfway in toward the midvein, the edges wavy and spine-edged. Upper leaves smaller, spinier, and clasp the stem or have spiny wings that continue a short way down the stem. *Flowerheads large and showy, 3-5 cm tall and about as wide, positioned far above the leaves. Corolla purple or red-purple.* Most *phyllaries* are held straight outward, their bases usually *densely covered with cobweb-like hairs.* Phyllaries halfway up the flowerhead have spines that are 1-2 cm long. Pappus bristles plumose (each bristle has slender side branches). In flower Mar-July.

Most often found at low elevations below 400 meters (about 1,310 feet) above mean sea level, *occasionally found along the Orange County coast and inland drainages; uncommon in Riverside County.* Known from upper Newport Bay, Laguna lakes, Temple Hill, Dana Point Headlands, Santa Ana Canyon, lower Santiago Canyon, Lomas de Santiago, Audubon Starr Ranch Sanctuary, Caspers Wilderness Park, and San Juan Canyon. We have two forms of *Cirsium occidentale* here; the pollen, nectar, and seeds of both are important food sources for many insects and birds.

The epithet was named in 1841 by T. Nuttall for its western distribution (Latin, occidentalis = western).

Similar. California thistle (*C.o.* var. *californicum*): phyllaries up- or outward; no or few cobwebby hairs; *corollas white to pink or pale purple; mid- to high elevations.*

The basal leaves of cobwebby and California thistle are pinnately lobed, the lobes wavy and spine-edged. They are covered in long gray or white hairs that may fall off with age.

Cobwebby Thistle / California Thistle

California Thistle
Cirsium occidentale (Nutt.) Jeps. **var.** *californicum* (A. Gray) D.J. Keil & C. Turner [*C. californicum* A. Gray]

Generally 50-200 cm tall. Most *phyllaries* are held upward or loosely outward, sometimes twisted, their bases *hairless or sometimes covered with cobweb-like hairs.* Phyllaries halfway up the flowerhead have spines that are up to about 1 cm long. *Corolla white to pink or pale purple.* In flower Apr-July.

Found at many elevations, less common at the coast. Known from major canyons such as Limestone, Silverado, Trabuco, Holy Jim, Santa Ana, and Claymine, also on the Audubon Starr Ranch Sanctuary. Common from foothills up to Modjeska and Santiago Peaks.

The epithet was given in 1857 by A. Gray for the state of its discovery.

Similar. Cobwebby thistle (*C.o.* var. *occidentale*): most phyllaries outward, dense cobwebby hairs; *corollas usually red-purple; coastal to low elevations.*

California Thistle

THE COMMON THISTLE GUILD

Metallic Mason Bee
***Osmia* sp.** (Megachilidae)

Small, *stout-bodied bees*, around 1.2 cm long. About 135 species of *Osmia* are known. *Many are metallic blue, green, or black. Like other leafcutter bees, they have long hairs on the underside of the abdomen.* Pollen collected from flowers is tightly packed between those hairs. When loaded with pollen, the abdomen is pushed upward.

Most species nest in abandoned wood cavities left by other bees. To prepare a nesting site, they cut a circular piece from a leaf, fly it back to the nest, and pack it in the back. This is repeated to make a leaf wall. Then they collect pollen, place it near the leaf wall, mix it with saliva, roll it into a ball, and lay an egg on it. More leaves are cut, packed, and a second leaf wall built. This is repeated until the nest is completed. The young hatch from the eggs, feed on the pollen, grow, pupate, emerge from the pupa as adults, crawl out of the nest, and fly off to begin their adult lives. They are abundant in spring in the Santa Ana Mountains, especially on **California thistle** and **poodle dog bush** (*Eriodictyon parryi*, Boraginaceae).

The genus was established in 1806 by G.W.F. Panzer, possibly for a scent it releases (Greek, osme = smell, scent).

Metallic Mason Bee

Pale Swallowtail
Papilio eurymedon Lucas (Papilionidae)

A *large butterfly* with wingspan of 7.6-9.5 cm, tails 1.1-1.4 cm long. *Wings are pale yellow with black stripes and borders.* In flight Mar-Aug.

A common sight in our mountains, canyons, and foothills, sometimes in adjoining neighborhoods. Like many adult butterflies, they are very fond of thistles, which they pollinate. They are one of the 3 pollinators of **ocellated Humboldt lily** (*Lilium humboldtii* ssp. *ocellatum*, Liliaceae). Larvae feed on **California coffeeberry** (*Frangula californica* ssp. *californica*), **spiny redberry** (*Rhamnus crocea*), **California lilacs** (*Ceanothus* spp., all Rhamnaceae), **holly-leaved cherry** (*Prunus ilicifolia* ssp. *ilicifolia*, Rosaceae), and sometimes domestic **peach** (*Prunus persica*, Rosaceae).

The epithet was given in 1852 by P.H. Lucas in honor of Eurymedon, son of Dionysus and Ariadne, one of the Argonauts.

Pale Swallowtail

Lesser Goldfinch
Spinus psaltria (Say) (Fringillidae) [*Carduelis psaltria* (Say)]

A *slender bird* about 11 cm long *with a dark conical bill*. Both sexes have a *white wing patch at the base of the primaries*, visible from above during flight or from the side when the bird sits. Males have a black cap that begins near the bill and continues up and over the head. The back is yellow-green, greenish, or black; the belly yellowish. Females have no cap, their back is greenish-yellow, their belly dull to pale yellowish. Young birds resemble adult females. The bird's song is a rapid series of twitterings; the call has strong notes. They often imitate other birds.

A common bird that often lives in flocks. They are very fond of thistle seeds and thrash the flowerheads to get them. Bird baths and home feeders filled with thistle seed will often attract them.

The epithet was given in 1823 by T. Say for its song (Greek, psaltria = a harper, a musician).

Lesser Goldfinch on California Thistle

Southern Meadow Thistle

Cirsium scariosum Nutt. **var. *citrinum*** (Petr.) D.J. Keil

  A *short biennial or perennial*, it flowers once, then dies off. *Most often stemless and low-growing*, rarely to 0.5 meter tall. *Leaves clustered; mostly or all basal; mostly hairless, cobwebby, or woolly;* often toothed to lobed, edged with long yellow spines. *Flowerheads large, usually nestled among leaves,* phyllaries flat, each tipped with a short spine. *Corolla white*, sometimes lightly blushed with purple. In flower May-Sep.

  A plant of moist soils among meadows, grasslands, and forest openings. Known here from the Santa Rosa Plateau, especially in grasslands near oak woodlands.

  The epithet was named in 1841 by T. Nuttall for the membranous edges of the outer phyllaries (New Latin, scariosus = thin, dry, membranous). The variety was named in 1917 by F. Petrak for its yellow leaf spines (Modern Latin, citrinum = lemon-colored).

Southern Meadow Thistle

Southern Meadow Thistle | Bull Thistle

Bull Thistle

Bull Thistle ✖

Cirsium vulgare (Savi) Ten.

  A *stout upright biennial* 20 cm – 2 meters tall, usually with *1 main stem with spine-tipped wings,* covered with scattered white woolly hairs and sometimes sticky-tipped hairs. *The spine-edged leaves have veins that appear to sink into the upper surface and pop out of the lower surface.* Lower leaves 10-40 cm long, 1-2 lobed; upper leaves a bit smaller, with spiny wings that continue down the stem. Leaf spines are very long and yellowish. *Flowerheads large,* with long straight phyllaries held outward. *Corolla purple to red-purple.* Like all *Cirsium* species, its pappus bristles are plumose (each bristle has slender side branches on it). In flower June-Sep.

  A native of Europe, common and widespread in many habitats but most often in disturbed soils and in moist drainages such as creeks and ponds.

  The epithet was given in 1836 by G. Savi for its abundance (Latin, vulgaris = common).

  Similar. Italian thistle ✖ (*Carduus pycnocephalus*): slender-stemmed, spiny leaf and petiole edges appear to continue down stems; leaf veins not sunken; *flowerheads slender, in dense clusters;* phyllaries held upward.

Bull Thistle

Cardoon, Globe Artichoke, Artichoke Thistle ✘
Cynara cardunculus L.

A *huge stout perennial* 50 cm – 2.5 meters tall. *Leaves 30 cm – 1 meter long*, 1-2 pinnately lobed or divided, very spiny, *covered with white woolly hairs that make the plant look grey. Flowerheads quite large*, 4-7 cm wide. Phyllaries broad, tipped with a massive spine. *Corolla blue-purple* (to white). In flower May-July.

Found throughout our area on dry slopes and moist depressions, most abundantly in overgrazed and disturbed clay soils. A native of Europe and perhaps the most invasive plant in southern California, where it occurs in dense colonies and crowds out native species. Cultivated artichoke is a nearly spineless form of it.

The epithet was named in 1753 by C. Linnaeus for its size, most certainly as a joke (Latin, carduus = thistle; -unculus = little).

Similar. Milk thistle (*Silybum marianum*): annual or biennial; leaves toothed, dark green with white blotches, mostly hairless; corolla purple.

Cardoon

Cardoon / Milk Thistle

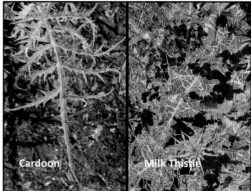
Cardoon / Milk Thistle

Milk Thistle ✘

Silybum marianum (L.) Gaertn.

A *stout perennial or biennial* (sometimes an annual), 20 cm – 3 meters tall, hairless or with some woolly hairs. *Leaves* are mostly basal, 15-60 cm long, toothed, edged with yellow spines, the blade *dark green with white in blotches and along the veins. Flowerheads large*, 2-6 cm across. Phyllaries very broad, tipped with a long narrow spine that stands outward. *Corolla purple*. In flower May-July.

Yet another very invasive thistle, all too commonly found beneath oaks, particularly at their dripline. Known throughout our area mostly in partial shade and most soils.

The common name refers to a European Christian legend that the white blotches on the leaves were from milk spilled by the Virgin Mary as she nursed her infant Jesus. The epithet was given in 1753 by C. Linnaeus in reference to that legend (New Latin, marianus = of Mary).

Similar. Cardoon ✘ (*Cynara cardunculus*): perennial; leaves pinnate, white-woolly, no white blotches; corolla blue-purple.

Milk Thistle

Pincushions: *Chaenactis*

Slender upright *hairy annuals* topped with attractive *pom-pom flowerheads*. Leaves alternate or basal, linear or 1-2-pinnate. *Heads discoid or radiant.* They are important sources of nectar and pollen for many types of insects.

The common name refers to their similarity to a pincushion used in sewing, especially the species with long stamens that appear like sewing pins stuck in the pincushion. We have 2 very common species, both more abundant after fires.

The genus was established in 1836 by A.P. de Candolle in reference to the outermost florets, the lobes of which are held wide open (Greek, chaino = to gape; aktis = a ray).

White Pincushion

Chaenactis artemisiifolia (Harv. & A. Gray) A. Gray

A *very tall annual*, often 1-2 meters in height. *Leaves* 15-20 cm long, largest blades 2-, 3-, or 4-pinnately lobed. Flowerhead discoid. *Corolla white*, sometimes pink-tinged. In flower Apr-July.

Found throughout our area among coastal sage scrub and chaparral, almost strictly during the first few years after fire. Known sparingly from Laguna Beach in the San Joaquin Hills, more common from the foothills and mountains.

The epithet was given in 1849 by W.H. Harvey and A. Gray in reference to the similarity of its leaves to European species of sagebrushes in the genus *Artemisia* (Latin, folium = leaf).

Similar. Yellow pincushion (*C. glabriuscula*): short; leaves 1-2-pinnate; bright yellow florets; abundant.

White Pincushion

Yellow Pincushion

White Pincushion

Yellow Pincushion

Chaenactis glabriuscula DC. **var. *glabriuscula***

A *thin-stemmed annual* to 50 cm tall, usually under 30 cm tall, even shorter and few-branched in very dry soils. *Leaves* are under 11 cm long, the largest undivided or *1-2-pinnately lobed*, the leaflets slender. *Flowerhead radiant. Corolla bright yellow.* In flower Mar-May.

A common widespread annual found in dry soils throughout our area. Especially frequent in sandy flats along San Juan Creek as it passes through Caspers Wilderness Park.

The epithet was given in 1836 by A.P. de Candolle for its small hairs that fall off as it ages (Latin, glaber = smooth or glabro = to make smooth; -cula = small).

Similar. White pincushion (*C. artemisiifolia*): tall; leaves 2-4-pinnate; white florets; mostly after fire. **Orcutt's pincushion** ★ (*C.g.* var. *orcuttiana* (Greene) Hall), stem spreading; largest leaf blades gen 2-pinnately lobed, fleshy; coastal bluffs of Laguna Beach to Dana Point, mostly extirpated, CRPR 1B.1. **Common beggar-ticks** ✘ (*Bidens pilosa* L.), leaves opposite, leaf lobes very broad; all disk florets slender, identical, symmetrical.

Yellow Pincushion

Brass-Buttons: *Cotula*

Annuals or perennials with alternate leaves. *Flowerheads disciform, shaped like buttons, broad and shallow; corolla short, yellow.* We have two species, both non-natives often seen in wild areas. The genus was named in 1753 by C. Linnaeus for the hollow cup-like space at the base of the larger leaves (Greek, kotyle = small cup).

Australian Brass-buttons ✖

Cotula australis (Sieber) Hook. f.

A *frail annual* to about 20 cm tall with sparse hairs and slender stems. *Leaves 2-3 pinnate, the lobes very thin.* Flowerheads 3-6 mm across, *corollas pale yellow.* In flower Jan-May.

An abundant widespread weed, especially in moist shaded soils of creekbeds, ponds, trailsides, and gardens. Native of Australia.

The epithet was given in 1826 by F.W. Sieber for its place of origin (Latin, australis = southern).

Similar. Australian brass-buttons ✖ (*C. coronopifolia*): stout fleshy perennial; flowerheads large, corollas yellow; coastal and saltmarshes.

Australian Brass-buttons

Australian Brass-buttons — Small flowerheads

African Brass-buttons — Large flowerheads

African Brass-buttons ✖

Cotula coronopifolia L.

A *perennial* from creeping rootstalks, the fleshy hairless stems are nearly 50 cm long. *Leaves 2-7 cm long, linear, undivided or pinnately lobed; the leaf base forms a sheath around the stem.* Flowerheads 6-15 mm across, *corollas yellow.* In flower Mar-Dec.

A plant of wet freshwater and saltwater soils, mostly along the coast. Most often found at Bolsa Chica Ecological Reserve, upper Newport Bay, San Joaquin Freshwater Marsh, Crystal Cove State Park, Doheny Beach at the San Juan Creek outfall, and Trestles Beach at the San Mateo Creek outfall. Still sold in the horticultural trade as a pond plant. Native of South Africa.

The epithet was given in 1753 by C. Linnaeus for the similarity of its leaves to crowfoot (Latin, coronopus = crowfoot plant; folium = leaf).

Similar. Australian brass-buttons ✖ (*C. australis*): weak annual; not fleshy; flowerheads tiny, corollas pale yellow; neither coastal nor saltmarshes, primarily a weed.

African Brass-buttons

Discoid-headed Goldenbushes: *Ericameria, Hazardia, Isocoma*

Shrubs with many stems and yellow to golden corollas. Those with discoid flowerheads are presented here; others with radiate flowerheads are placed in the Radiate section.

Great Basin Rabbitbrush

Ericameria nauseosa (Pall.) G.L. Nesom & G.I. Baird **var. *oreophila*** (A. Nelson) G.L. Nesom & G.I. Baird
[*Chrysothamnus nauseosus* (Pallas) Britton ssp. *consimilis* (E. Greene) H.M. Hall & Clements]

A *woody shrub* 0.5-2.5 meters tall *with hairy yellowish-green (or whitish) stems and a strong odor.* Leaves 2-6 mm long, *threadlike, with no petiole, undivided, alternate on the stem.* The narrow inflorescence occurs at the top of the plant. Flowerheads about 1 cm long, phyllaries with a brownish-colored ridge in the center. In flower Aug-Oct.

A plant that barely enters our area, more typical of disturbed soils in mountains at the desert edges. Known from only four places in our area: a landslide below Bigcone Springs along the Silverado Truck Trail, the mouth of lower Coal Canyon, foothills near Corona, and near Glen Ivy.

The epithet was given in 1814 by P.S. Pallas for its strong scent (Latin, nauseosus = that produces nausea). Our form was named in 1900 by A. Nelson for its occurance in mountains (Greek, oreos = a mountain; philos = loving, fond of).

Similar. Pine goldenbush (*E. pinifolia*): strong pine odor; hairless green stems; radiate flowerheads; not found together.

Great Basin Rabbitbrush

Great Basin Rabbitbrush | Wedge-leaved Goldenbush

Wedge-leaved Goldenbush

Wedge-leaved Goldenbush

Ericameria cuneata (A. Gray) McClatchie **var. *cuneata***

A *hairless deep green shrub* 10 cm - 1 meter tall, dotted with many glands that release a thick substance, making the plant *sticky to the touch. Leaves small,* the largest 3-14 mm long, 2-9 mm wide, with no petiole; their *base is wedge-shaped.* Flowerheads tiny, 8-11 mm long, 5-7 mm across, arranged in tight groups. The heads discoid, rarely with ray florets as well. In flower Sep-Nov.

An uncommon shrub here, usually *grows out of cracks in rocks at higher elevations,* more common inland and northward. Known from upper Fremont Canyon and among **canyon live oaks** (*Quercus chrysolepis*, Fagaceae) on the west-facing slope of Modjeska Peak.

The epithet was given in 1873 by A. Gray for its leaf shape (Latin, cuneatus = wedge-shaped).

Similar. Grassland goldenbush (*E. palmeri* var. *pachylepis*) and **Pine goldenbush** (*E. pinifolia*): *slender linear leaves; radiate flowerheads;* see entries under Radiate Goldenbushes.

Parish's Goldenbush

Ericameria parishii (E. Greene) H.M. Hall **var. *parishii***

A *stout tree-like shrub*, 1.5-5 meters tall, woody at the base, totally hairless, dotted with many glands that release a thick substance and make the plant *sticky* to the touch. *Leaves* are 2-6 cm long, 3-10 mm wide, *elliptical in outline. Flowerheads are in dense clusters at the top of the plant.* When in peak bloom, the entire top of the plant is a nearly-solid mass of flowers, allowing visiting insects to walk among them. Phyllaries are lance-shaped and whitish with a thick brown ridge down the center. Corolla yellow to yellow-orangish. In flower July-Oct.

A shrub of chaparral slopes and disturbed areas such as along roadcuts and after fires. Known in our area only from the Santa Ana Mountains such as in Coal Canyon, upper Silverado, Holy Jim Canyon Trail, many places along Main Divide Road; scarce at San Juan Hot Springs. Most easily seen along the North Main Divide "Loop." A very popular nectar and pollen source for several species of beetles, butterflies, flies, bees, and wasps.

The epithet was given in 1882 by E.L. Greene in honor of S.B. Parish.

Similar. Grassland goldenbush (*E. palmeri* var. *pachylepis*) and **Pine goldenbush** (*E. pinifolia*): *slender linear leaves;* see entries under Radiate Goldenbushes.

Parish's Goldenbush

Parish's Goldenbush

Saw-toothed Goldenbush

Hazardia squarrosa (Hook. & Arn.) E. Greene **var. *grindelioides*** (DC.) Clark

A *many-branched, generally open shrub,* 30 cm to 1 meter tall, rarely to 2 meters. *Leaves* 1.5-5 cm long, 1-2 cm wide, oblong to reverse-oval in outline, usually *bristly* and somewhat hairy on the upper surface, sometimes a bit sticky; *the base clasps the stem, and the edges have large sharp saw-like teeth.* The *slender flowerheads* are in tight clusters *along the branches,* never in a solid mass. *Phyllaries hairless or sticky-haired, tips curve backward. Corolla yellow-gold, streaked with red.* In flower July-Oct.

Found among coastal sage scrub and grassland edges but more commonly in chaparral. Widespread but generally not abundant, mostly lower to mid elevations but not along the coast. Known from Carbon and Tonner Canyons in the Chino Hills, Limestone Canyon especially near Bolero Spring, Niguel Hill, Mission Viejo, Rancho Mission Viejo, Audubon Starr Ranch Sanctuary, along San Juan Trail, in El Cariso, south into the San Mateo Canyon Wilderness Area.

The genus was established in 1887 by E.L. Greene to honor B. Hazard. The epithet was established in 1833 by W.J. Hooker and G.A.W. Arnott for its rough-surfaced leaves (Latin, squarrosus = rough). Our local form was named in 1836 by A.P. de Candolle for the similarity of its leaves to gumplants in the genus *Grindelia* (Latin, -oides = likeness).

Similar. Coastal goldenbush (*Isocoma menziesii*): hairless, hairy, or bristly; leaves in bundles, base narrower, generally not clasping, edges smaller-toothed; *flowerheads in clusters, phyllaries upright, fuzzy-tipped;* coastal to mountains; abundant. **Gumplant** (*Grindelia camporum*): stems hairless, shiny-white; *flowerheads large and round, radiate or discoid;* phyllaries outward or backward, covered with resin; more coastal.

Saw-toothed Goldenbush

Coastal Goldenbush
Isocoma menziesii (Hook. & Arn.) G.L. Nesom
var. *vernonioides* (Nutt.) G.L. Nesom

A variable shrub, hairless or bristly-haired, upright or low-spreading, 0.4-1.2 meters tall, many-branched from the base. Leaves 1-4.5 cm long, *oval or oblong, usually toothed, arranged in dense bundles*. Flowerheads about 1 cm long, in dense rounded clusters at the branch tips. Phyllaries upright to outward, the tips usually fuzzy-haired. *Corolla yellow to yellow-orangish*. In flower (Apr-)Aug-Dec.

A common plant of *sandy soils, grasslands, and coastal sage scrub* from the sandy coast inland to the foothills of the Santa Ana Mountains. Much less common in Temescal Valley. Especially good places to see it are Crystal Cove State Park, Laguna Coast Wilderness Park, Aliso and Wood Canyons Wilderness Park, Limestone Canyon, and Chino Hills. In some areas, such as Temple Hill, the plants are covered with long white hairs and the plants look gray. These populations are in need of study.

The epithet was given in 1838 by W.J. Hooker and G.A.W. Arnott for A. Menzies.

Similar. Saw-toothed goldenbush (*Hazardia squarrosa*): very bristly; leaves solitary, stiff, broad base clasps the stem, edges large-toothed; slender flowerheads along branches; phyllary tips curve backward; not coastal; less common.

Coastal Goldenbush

Coastal Goldenbush hairless form

Coastal Goldenbush hairy form

Prostrate Goldenbush ✿
Isocoma menziesii (Hook. & Arn.) G.L. Nesom
var. *sedoides* (Greene) G.L. Nesom

A low-growing, often mat-like form with stout stems. Leaves succulent, hairless to woolly. Flowerheads large and crowded. A very uncommon low-growing form *on sea bluffs along the immediate coast* in Corona del Mar, Crystal Cove State Park, and Laguna Beach. Easiest seen from a nice park bench on a point in Heisler Park. Threatened by coastal development. In Heisler Park, it is also threatened by competition from **Hottentot-fig ✖** (*Carpobrotus edulis,* Aizoaceae) which should be carefully removed in order to preserve this goldenbush.

The subepithet was given in 1887 by E.L. Greene for its growth habit (Latin, sedeo = to sit; -oides = like).

Prostrate Goldenbush

Prostrate Goldenbush

THE GOLDENBUSH GUILD

Coastal goldenbush is an important plant to many types of insects and the animals that feed upon them. This is a very small representation of its guild.

Spittlebugs

The nymphs (juvenile stages) of spittlebugs (Cercopidae) are commonly seen on coastal goldenbush and other plants such as baccharis (*Baccharis* ssp.) and sage (*Salvia* spp., Lamiaceae). *These harmless insects produce a white froth* (spittle) *to discourage predatory and parasitic insects from attacking them.* They reach adulthood in about 2-4 weeks, then exit the spittle and fly away.

Red-striped Goldenbush Beetle

Microrhopala rubrolineata (Mannerheim) (Chrysomelidae)

A *small, hard-bodied beetle* under 5 mm long. It is *dark blue-black, striped with red, and covered with tiny pits*. Its larvae tunnel into coastal goldenbush to feed and pupate. Adults emerge in summer-fall. They are most often found hidden among the leaf bases.

Named in 1843 by C.G. von Mannerheim for its red stripes (Latin, rubra = red; lineatus = streaked, marked).

Goldenbush Longhorn Beetle

Crossidius testaceus maculicollis Casey (Cerambycidae)

An elongate beetle, 1.2-1.5 cm long, covered with short yellowish hairs. Its head is black, thorax orange brown with 4 faint black patches, elytra orange-brown. As with most longhorn beetles, *its antennae are about as long as the body.* Young stages tunnel, feed, and pupate within roots of coastal goldenbush. Adults emerge in fall and feed on flowers and pollen.

The subepithet was given in 1912 by T.L. Casey, Jr., for the faint black patches on its thorax (Latin, macula = spot; collum = neck).

Squinting Blister Beetle

Epicauta straba Horn (Meloidae)

A soft-bodied beetle, 5-8 mm long. Like all blister beetles, the head is broader than the thorax. This species is all black, many other species are colorful. Adults are commonly seen in spring and summer eating pollen from many types of plants. As the beetles move about, pollen sticks to the body. Unlike bees that collect pollen for later use, pollen-feeding beetles ignore the pollen on their body and move on to the next flower to feed. At day's end, they are often covered in pollen.

Blister beetles have unusual life cycles. In some species, eggs are laid on flowers. The larvae hatch, lie in wait for a ground-nesting bee, and hitchhike a ride to the nest to feed on bee larvae. In other species, eggs are laid on the ground, the larvae hatch, and crawl into a bee nest on their own. Some species in this genus feed on the eggs of grasshoppers buried in soil. When disturbed, adults exude from their leg joints a fluid that can blister sensitive skin.

The epithet was given in 1891 by G.H. Horn for its eyes that narrow in front and appear to be squinting (Latin, strabo = a squinter).

Green Fruit Beetle
Cotinis mutabilis Gory & Percheron (Scarabaeidae)

A hefty but harmless beetle, 2-3 cm long, metallic green with light brown edging on its elytra. Larvae feed on decomposing fruit and leaves in soil, commonly in compost piles. Adults emerge in summer-fall and make buzzing sounds as they fly in search of pollen and fruit to eat. They are a member of the flower chafers, a group of scarabs that leave their elytra closed on their back and fly only with their hind wings, making them clumsy fliers. In the same group is one of the world's bulkiest insects, the **African goliath beetle** (*Goliathus goliatus* (L.)), which has the same shape but different coloration and size. The green fruit beetle is often mistaken by laypeople for a **Japanese beetle** (*Popillia japonica* Newman), which is small, 8-12 mm long, has a green thorax with brown elytra. In 1916, it was accidentally introduced from Japan to the east coast where it is now a serious garden pest; it is not known from our area.

Named in 1833 by H.L. Gory & A.R. Percheron for its variable colors (Latin, mutatus = change).

Green Fruit Beetle

Longhorn Bee
***Melissodes* sp.** (Apidae)

Medium-sized bees, 8-16 mm long, often gray, brown, or golden. Over 100 species of bees belong to this genus. *Females are generally larger and huskier than males. Antennae of the male are about one-half the length of its body.* Like nearly all of our native bees, they nest individually in the ground. An outpocket of the underground burrow (brood chamber) is provisioned with a ball of pollen, nectar, and saliva. The female lays a single egg on the ball. The larva hatches from the egg, feeds on the food ball, and pupates in the chamber. The adult emerges in spring, summer, or fall. Adults are commonly seen as they collect pollen and nectar from many plants, such as coastal goldenbush.

Longhorn Bee, male

Longhorn Bee, female

Slender Leafcutter Bee
Megachile angelarum (Megachilidae)

These leafcutter bees range in size from 8-11 mm long. *The dark abdomen has rings of white hairs.* Many related species are black, some others are metallic blue or green. Most pollen-collecting bees carry pollen on their hind legs; not so for leafcutter bees. *Pollen is incidentally collected on body hairs while visiting flowers.* They occasionally stop collecting, comb pollen with their legs, and *tightly repack the pollen between long hairs on the underside of the abdomen.* When loaded with pollen, the abdomen is pushed upward. They are important pollinators of many native plants such as manzanitas and crops such as alfalfa, almonds, and apples.

Leafcutter bees nest in unusual places, such as in beetle tunnels beneath bark and in holes in wood, often in nests used by other bees in past seasons. Unlike most other leafcutting bees, which separate their egg chambers with circular leaf pieces (see The Thistle Guild), this one separates its egg chambers with resin collected from plants. Their mandibles lack the leaf cutting surfaces needed to cut leaves.

The epithet was given in 1902 by T.D.A. Cockerell for its "angelic" appearance (angel; Latin, -ar = like; -um = adjectival ending).

Slender Leafcutter Bee

Common Pineapple Weed ✖
Matricaria discoidea DC.
[*M. matricarioides* (Less.) Porter, *Chamomilla suaveolens* (Pursh) Rydb.]

An *upright annual* usually 10-30 cm tall *with a strong scent of pineapple*. To check, crush a small amount in your hand and smell it. Its many branches arise from the base of the plant. Leaves under 5 cm long, alternate, 2-3-pinnate, with neither hair nor petiole. *Flowerheads* 1 cm across, *cone-shaped; fall apart when mature*. Corolla yellow to greenish. In flower May-Aug.

A common weed of *disturbed soils* throughout our area, most notably along trailheads, roadsides, and parking lot edges. Native to northwestern North America and northeast Asia.

The epithet was given in 1838 by A.P. de Candolle for its disc florets (Greek, diskos - a disc; oidea = form of).

Similar. Valley pineapple weed (*M. occidentalis*): no strong odor of pineapple; branches from upper plant; mature flowerheads remain intact.

Common Pineapple Weed

Valley Pineapple Weed
Matricaria occidentalis E. Greene
[*Chamomilla occidentalis* (E. Greene) Rydb.]

An *upright annual* 15-45 cm tall *with a faint sweet odor or odorless*. Its branches arise from the upper part of the plant. Leaves under 7 cm long, alternate, 2-3-pinnate, with neither hair nor petiole. *Flowerheads* up to 1.4 cm across, *conical or spherical*, and *do not fall apart when mature*. Corolla yellow to greenish. In flower Apr-July.

An uncommon native, often mistaken for common pineapple weed. Known here only from upper Newport Bay, though possibly more widespread and overlooked. Sometimes used in tea in place of **European chamomile** (*Anthemis nobilis* L.).

The epithet was given in 1886 by E.L. Greene for its western distribution (Latin, occidentalis = western).

Similar. Common pineapple weed ✖ (*M. discoidea*): strong odor of pineapple; branches from lower plant; mature flowerheads shatter.

Valley Pineapple Weed

Stinknet ✖
Oncosiphon piluliferum (L.f.) Källersjö
[*Matricaria globifera* (Thunb.) Fenzl ex Harv.]

An upright, *slender-stemmed, odorous annual*, usually 15-40 cm tall. *Leaves alternate, fern-like*, 2-3-pinnate. *Flowerheads discoid, spherical*, 5-8 mm wide. *Corolla bright yellow*. In flower Mar-July.

An aggressive weed of disturbed soils, roadsides, and nearby wild lands. Known from fields along Irvine Boulevard at the north end of the former MCAS El Toro base, Lake Elsinore near Nichols Road wetlands, and Camp Pendleton. Native to South Africa.

It was named in 1782 by C. Linnaeus "The Younger" for its spherical flowerheads (Latin, pilula = a globule, small spherical objects; fero = to bear).

Stinknet

Marsh-fleabane
Pluchea odorata (L.) Cass. **var. *odorata***

An *annual or ephemeral perennial*, 0.5-1.2 meters tall covered with a thick substance, *sticky to the touch*. *Leaves* 4-12 cm long, *oval*, often with a pointed tip, *the edges toothed*, alternate on the stem. Flowerheads broad, arranged into cone-shaped or flat-topped clusters at the top of the plant. Corolla purple. In flower July-Nov.

A charming plant that produces new aboveground growth in late spring, followed by attractive flowers. *Found in wet soils near watercourses and marshes.* Known from Featherly Park along the Santa Ana River, Oak Canyon Nature Center, Bolsa Chica Ecological Reserve, upper Newport Bay, San Joaquin Freshwater Marsh, Aliso Creek in Laguna Beach and Mission Viejo, lower San Juan Creek, San Juan Hot Springs, and moist areas along the coast.

The epithet was given in 1759 by C. Linnaeus for its scent (Latin, odoratus = sweet-smelling).

Similar. **Desert arrowweed** (*P. sericea*): upright willow-like shrub; silky hairs; neither sticky nor scented; near watercourses but not in wet soils.

Marsh-fleabane

Marsh-fleabane has much larger leaves than desert arrowweed. The two don't generally live near each other. Both are in cultivation.

Desert Arrowweed
Pluchea sericea (Nutt.) Coville

An *upright willow-like shrub* 1-5 meters tall *from upright stiff stalks, covered with soft silky hairs, neither sticky nor scented*. It spreads by underground rootstalks and often forms a dense thicket. *Leaves* 1-4 cm long, *linear to lance-shaped, smooth-edged*, crowded on the stems. Flowerheads in small groups at the branchtips. Corolla pink to deep rose. In flower Mar-July.

Widespread in California but known only from a few places here. Found in the Santa Ana Canyon channel near Green River Golf Course and Horseshoe Bend, canyons in Laguna Beach, the southwest shore of Irvine Lake, and lower Agua Chinon wash in the Lomas de Santiago.

Named in 1848 by T. Nuttall for the silky hairs that cover it (Latin, seris = silk).

Similar. **Marsh-fleabane** (*P. odorata* var. *odorata*): annual or ephemeral perennial; sticky; sweet scent; in wet soils.

Desert Arrowweed

Poreleaf, Odora

Porophyllum gracile Benth.

A low and rounded bushy shrub 30-70 cm tall, with many upright slender stems, an *unusual blue-green color, hairless, covered with whitish powder*. It has a *strong pungent scent* that is difficult to wash off of your hands. *Leaves* 1-5 cm long, *linear, very narrow,* alternate or opposite, with dot-like oil glands from which the scent is released. *Flowerheads solitary. Phyllaries long, straight, often purple, dotted with oil glands. Corolla whitish or purplish, the outside streaked with purple.* In flower Oct-June.

An unforgettable plant, its scent will remain on your skin and in your memory for days! Widespread but generally not abundant here. Typically a desert plant, found in our mountains and reaches the coast in San Diego and Orange Counties. Known from Temple Hill, Chino Hills, Lomas de Santiago, Black Star Canyon, Modjeska Canyon, Caspers Wilderness Park, Audubon Starr Ranch Sanctuary, Rancho Carrillo, Lucas Canyon, Hot Springs Canyon, and lower portions of San Juan Trail.

Named in 1844 by G. Bentham for its slender stems (Latin, gracilis = slender).

Cudweeds, Everlastings: *Gnaphalium, Pseudognaphalium*

Annuals, biennials, or ephemeral perennials, *covered with white woolly hairs*. Flowerheads small and narrow with tips that don't open up very wide, each with rings of noticeable phyllaries that impart color and luster. In fruit, the phyllaries dry and spread apart, releasing the individual fruits. Though some small flies, bees, and wasps visit the flowers, many species are suspected to be largely or entirely self-pollinated.

Larvae of the American lady butterfly (*Vanessa virginiensis*, Nymphalidae) feed on various species of cudweeds. They fold over the leaves at the branchtips and sew them into an upside-down cup-shaped shelter (hibernaculum). They pupate in or near the hibernaculum or crawl off the plant to pupate elsewhere.

Lowland Cudweed

Gnaphalium palustre Nutt.

A low-growing annual densely covered with white woolly hairs, generally 3-7(-30) *cm tall when in flower, it has many branches from the base. It has no scent.* Leaves 0.4-3 cm long, oblong to spoon-shaped. Flowerheads at the branchtips or among leaf bases. *Phyllaries brown, tipped with white.* In flower May-Oct.

A common plant of moist soils along watercourses, meadows, shallow soil depressions, and vernal pools, usually at low elevations. Easiest found in upper Newport Bay and the San Joaquin Hills such as at Laguna Coast Wilderness Park.

It was named in 1841 by T. Nuttall for its habitat (Latin, palustre = marshy).

Similar. Weedy cudweed ✖ (*Pseudognaphalium luteoalbum*): phyllaries fairly transparent, their tips whitish to yellowish; garden weed, not usually far into the wild, though sometimes after fire.

Weedy Cudweed ✖

Pseudognaphalium luteoalbum (L.) Hilliard & B.L. Burtt
 [*Gnaphalium luteo-album* L.]

An annual 15–60 cm tall, covered with white or yellow woolly hairs. *It has no scent. Widely branched from the base. Inflorescence long. Phyllaries fairly transparent, their tips whitish to yellowish.* In flower all year.

A common weed of both urban and wild areas, often at edges of ponds, creeks, and parking lots, rarely in vernal pools. Native to Eurasia.

The epithet was given in 1753 by C. Linnaeus for its yellow or white phyllary tips (Latin, luteus = yellow; albus = white).

Similar. Lowland cudweed (*Gnaphalium palustre*): phyllaries brown, tipped with white; not a garden weed, usually coastal.

Weedy Cudweed

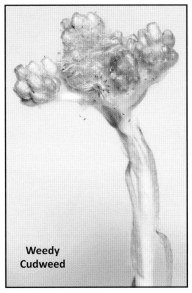
Weedy Cudweed

Phyllaries dry and spread apart

Weedy Cudweed

Cotton-batting Plant

Pseudognaphalium stramineum (Kunth) Anderb.
 [*Gnaphalium chilense* Spreng.]

An *annual or biennial* 8–70 cm tall with *no scent,* covered with white woolly hairs. *Leaves* 10–70 mm, mostly linear, with wavy edges that *continue as wings down the stem. Flowerheads very broad and rounded, in dense rounded fist-like clusters crowded into the branchtips.* Phyllaries are usually straw-colored with a pearly-luster. In flower Mar–Oct.

Found in moist sand and disturbed ground, mostly along the coast but also inland to Rancho Mission Viejo and Temescal Valley, south into the San Mateo Canyon Wilderness Area.

The epithet was given in 1820 by K.S. Kunth for the color of its phyllaries (Latin, stramineus = made of straw).

Cotton-batting Plant

Bioletti's or Bicolored Cudweed
Pseudognaphalium biolettii Anderberg
[*Gnaphalium bicolor* Bioletti]

An *upright perennial* 20 cm – 1.2 meters tall, *covered with gray-white woolly hairs, sweet-scented or with a strong odor* that smells different to each person. *Leaves* 2–8 cm long, linear or lance-shaped, wavy-edged, *two-colored: green and sticky above, gray-white woolly below.* Flowerheads broad, clustered at the branchtips. Phyllaries whitish with a pearly luster. In flower Jan-May.

Widespread but usually not abundant. Known from sandy soils along the coast, in creekbeds, and foothills, such as upper Newport Bay, Niguel Hill, Dana Point Headlands, Mission Viejo, and major canyons of the Santa Ana Mountains. Sometimes comes up after fire.

The epithet was given in 1991 by A.A. Anderberg to honor F.T. Bioletti.

Similar. California everlasting (*P. californicum*): differently scented; *leaves* narrower, *bright green* top and bottom; abundant.

California Everlasting
Pseudognaphalium californicum (DC.) Anderberg
[*Gnaphalium californicum* DC.]

An upright, *medium- to bright green annual or biennial,* 20–85 cm tall, with branches generally above the middle. Sometimes it has scattered gray-woolly hairs; it is always *entirely sticky to the touch.* It has a *characteristic scent* that smells different to each person, generally remniscent of maple syrup, pineapple, citrus, or curry. Leaves are 2–15 cm long, linear to lance-shaped, their edges continue a short bit down the stem. The inflorescence is broad and open with many branches. Flowerheads are rounded and tightly packed in clusters at the top. Phyllaries are stark white, rarely pink-tinged, aging to straw-colored. In flower Jan-July.

A plant of dry areas, often among openings in coastal sage scrub, oak woodlands, chaparral, on slopes and along creekbeds. Abundant throughout our entire area, easily seen in our wilderness parks and other wild areas. When it dries out, the surrounding air is filled with its wonderful scent.

The epithet was given in 1838 by A.P. de Candolle for the place of its discovery.

Similar. Bioletti's cudweed (*P. biolettii*): differently scented; *leaves* broader, medium to dark green above, *gray-white-woolly below,* much less common.

Fragrant Everlasting

Pseudognaphalium beneolens (Davidson) Anderberg
[*Gnaphalium canescens* DC. ssp. *beneolens* (Davidson) Stebb. & D.J. Keil, *G. beneolens* Davidson]

A biennial or short-lived perennial 40 cm – 1.1 meters tall, with many slender branches. The entire plant is covered with woolly gray hairs and *often appears greenish-yellow in color. It has a pleasantly sweet scent, though sometimes weak or absent.* Leaves 2–5.5 cm long, narrow, linear below, lance-shaped above, all mostly held upright against the stem, *the edges of the upper leaves continue down the stem, about halfway to the next lower leaf.* The inflorescence is many-stemmed and open. Flowerheads narrow, the phyllaries white to pale straw-colored. In flower June-Oct.

Found in openings among coastal sage scrub and chaparral, sometimes as understory in coniferous forests, generally at elevations higher than white everlasting (see below). Mostly known from the Santa Ana Mountains: San Juan Canyon up to Los Pinos Saddle, Holy Jim Canyon Trail, and Santiago Peak. Quite abundant and easily found near the Ortega Candy Store and Los Pinos Potrero.

The epithet was given in 1918 by A. Davidson for its scent (Latin, beneolens = smelling agreeably).

Similar. **White everlasting** (*P. microcephalum*): thicker stems, white-gray or blue-gray, no scent; larger leaves, leaf edges don't go down the stem; flower heads smaller; lower elevations and coastal. **Sonora everlasting** (*P. leucocephalum*): stems white-woolly, sticky; pine or lemon scent; leaves white-woolly below, *leaf edges continue a shorter way down the stem;* flowerheads spherical; sandy creeks, uncommon.

White Everlasting

Pseudognaphalium microcephalum (Nutt.) Anderberg
[*Gnaphalium canescens* ssp. *microcephalum* (Nutt.) Stebb. & D.J. Keil, *G. microcephalum* Nutt.]

Very similar to fragrant everlasting, sometimes considered a form of the same species. It is about the same size, 35 cm – 1 meter tall, but its *stems are usually thicker, it has no scent at all,* and it appears *completely white-gray or blue-gray in color.* Leaves 1.5-6 cm long but broader, 4-10 mm wide, reverse-lance-shaped, all mostly held outward; *their edges never continue down the stem.* In flower June-Aug.

Generally found in openings on slopes among chaparral and in washes at elevations *lower* than fragrant everlasting. Known from the Santa Ana River channel, Corona del Mar, San Joaquin Freshwater Marsh, Dana Point, Trabuco Creek, San Juan Creek, Rancho Mission Viejo, Limestone Canyon in the Lomas de Santiago, and major canyons of the Santa Ana Mountains such as along Maple Springs Road in Silverado Canyon, south into the San Mateo Canyon Wilderness Area.

The epithet was given in 1841 by T. Nuttall for its small flowerheads (Greek, mikros = small; kephale = head).

Similar. **Fragrant everlasting** (*P. beneolens*): thinner stems, greenish-yellow in color, sweet scent; smaller leaves, *leaf edges continue far down the stem;* flower heads larger; higher elevations, not coastal.

Sonora Everlasting, White Rabbit-tobacco ★
Pseudognaphalium leucocephalum (A. Gray) Anderberg
[*Gnaphalium leucocephalum* A. Gray]

An upright perennial 50–80 cm tall, branched mostly at the base, very sticky to the touch, and almost completely covered with white woolly hairs. It has a very pleasant scent like that of lemon and/or pine. The *numerous slender leaves* are 2.5–6.5 cm long, linear to lance-shaped, and *continue a short bit down the stem.* The leaves are deep to medium green and sticky above, densely white-woolly below. When viewed from above, the green leaves appear to be edged in white. Flowerheads are broadly spherical and densely clustered at the branchtips. *Phyllaries are stark white with a pearly luster.* In flower Aug-Sep.

A plant of *sandy soils along washes*, uncommon in our area. Known from the Santa Ana River channel, lower Trabuco Creek near Oso Parkway, Bell Canyon, San Mateo Canyon, Lucas Canyon, and San Juan Creek from Caspers Wilderness Park to San Juan Capistrano. CRPR 2.2.

The epithet was given in 1853 by A. Gray for its white flowerheads (Greek, leukos = white, bright; kephale = head).

Similar. Fragrant everlasting (*P. beneolens*): thinner stems; greenish-yellow in color; not sticky; sweet scent; leaf edges continue farther down the stem; flowerheads narrow; higher elevations, not coastal.

Sonora Everlasting

Sonora Everlasting

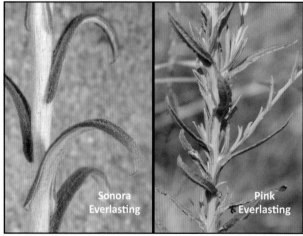

Sonora Everlasting

Pink Everlasting

Pink Everlasting
Pseudognaphalium ramosissimum (Nutt.) Anderb.
[*Gnaphalium ramosissimum* Nutt.]

An upright, *medium green*, biennial 50–120 cm, *with a very pleasant sweet-scent*, covered with sticky-tipped hairs. *Leaves 20–85 mm long, narrowly lance-shaped, green*, hairy but losing hairs with age. *Inflorescence large, broad, and open*, with many upright and spreading branches. *Flowerheads small and numerous, each appears pink-colored due to the pink phyllaries* (sometimes whitish or greenish). In flower July-Sep.

An uncommon plant mostly along the coast. Known from Corona del Mar, Moro Canyon in Crystal Cove State Park, Laguna Coast Wilderness Park, upper Newport Bay, and Dana Point Headlands. Uncommon at Dripping Springs in Limestone Canyon. This is a handsome plant that should be in cultivation.

The epithet was given in 1848 by T. Nuttall for its many-branched inflorescence (Latin, ramus = branch; -issimus = very much).

Pink Everlasting

Pink Everlasting

Valley Lessingia

Lessingia glandulifera A. Gray **var.** ***glandulifera***

An *upright open annual* 5-50 cm tall, the stems and leaf undersides are hairless or with scattered long hairs. Its basal leaves reach nearly 11 cm long, reverse lance-shaped, toothed or pinnately lobed, and dry up before flowering. Stem leaves much smaller, oblong to oval, sometimes tooth-edged or pinnately lobed. *Flowerheads radiant,* scattered among the branchtips. *Corolla* funnel-shaped, *yellow,* florets along the outermost edge are much larger than those in the middle. *Phyllaries densely covered with glands shaped like the head of a nail* that was not hammered all the way in, like one on which you would hang a framed picture. In flower May–Oct.

A plant of the western edge of the Mojave Desert, larger mountains, and westward into our area, usually away from the coast. Found in open, sunny spots with sandy soils among coastal sage scrub and chaparral. Known from Santa Ana Canyon and Hicks Canyon in the Lomas de Santiago (may have been extirpated from both sites). Possibly more widespread but not reported, probably because it flowers during very hot weather when most people are not out wildflower-watching.

The epithet was given in 1882 by A. Gray for the sticky glands on the upper leaves and phyllaries (Latin, glandula = gland; -fero = to bear).

Valley Lessingia

Valley Lessingia

Cotton-thorn

Tetradymia comosa A. Gray

A *small shrub* 30-120 cm tall, often much wider. *Stems and leaves are covered with white cotton-like hairs. Leaves* are 2-6 cm long, narrowly linear, and *tipped with a fine spine.* In some plants, the *leaves stiffen with age and act as long spines.* Flowerheads about 1 cm wide, of 5-9 disk florets. Each floret light yellow, sometimes brownish, the lobes long and curved backward. In flower June-Sep.

Occurs in scattered colonies in openings among coastal sage scrub and chaparral. Known from such places as the Chino Hills, Oak Canyon Nature Center, Coal Canyon, Temescal Valley, Lake Elsinore basin, and El Cariso. The easiest place to view it is at the Bear Canyon Trailhead near the Ortega Candy Store.

A common host of the **Pacific ambush bug** (*Phymata pacifica* Evans, Reduviidae) which waits on the flowerheads for its prey to arrive.

The plant was given its epithet in 1877 by A. Gray for its long cottony hair (Latin, comosa = with long hair). The insect received its name in 1931 from J.H. Evans in reference to its distribution along the Pacific coast of North America.

Cotton-thorn

Cotton-thorn

Pacific ambush bug

Spiny Cocklebur, Spiny Clotbur
Xanthium spinosum L.

A stiff-stemmed annual covered with groups of three spines that arise from the leaf nodes and flower bases. Leaf blades 3-10 cm long, linear to elliptical, entire to pinnately lobed, gray-green above, covered with white hairs below. Clusters of tiny male-only flowers appear at the branchtips, female flowers are below them in two-flowered heads. The fruit is a 1-cm-long *bur covered with hook-tipped spines* (elongated paleae); it contains 2 black seeds. In flower July-Oct.

A worldwide plant, perhaps native only to South America. Found primarily in moist disturbed soils in our area. Look for it in riparian areas, sandy creeks, and at pond edges. Known from scattered sites such as Moro Canyon in Crystal Cove State Park, Santiago Canyon, San Juan Creek, and the campus of Irvine Valley College.

The genus was established in 1753 by C. Linnaeus in reference to the yellow dye made from its fruit (Greek, xanthos = various shades of yellow). He gave it its epithet at the same time, for its spiny stems (Latin, spina = thorn, spine; -osum = full of).

Similar. Cocklebur (*X. strumarium*): no spines; stems green- or red-spotted; leaves triangular, green, hairless; fruits stout; common.

Cocklebur
Xanthium strumarium L.

Similar to spiny cocklebur but *without spines* on its *green- or red-spotted stems*. The *leaf blades* are up to 15 cm long, *triangular*, green, and hairless. Its *stout bur* reaches up to 2 cm long, *covered in hook-tipped spines* (elongated paleae); it contains 2 black seeds. In flower July-Oct.

A worldwide plant of wet, moist, and disturbed soils, much more common here than spiny cocklebur. Found in many of our wilderness parks and creeks. In the 1940s, a study of cocklebur fruit that had attached itself to both his dog's fur and his wool pants led Swiss inventor George de Mestral to invent hook-and-loop fasteners, often known today as Velcro®.

The epithet was given in 1753 by C. Linnaeus for its bulbous burs (Latin, struma = a lymph gland tumor; -arium = belonging to).

Similar. Spiny cocklebur (*X. spinosum*): spines on leaf nodes and flower bases; no spots; leaves linear, white-haired beneath; fruits narrower; uncommon.

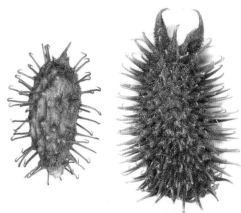

Spiny Cocklebur   **Cocklebur**

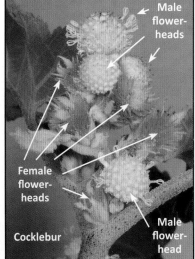

Cocklebur. Male florets in spherical flowerheads, female florets in barrel-shaped flowerheads below them.

RADIATE HEADS
White Yarrow
Achillea millefolium L.

An *upright perennial* 50 cm-1.2 meters tall with a strong pleasant scent, it grows from a stout taproot and spreads by creeping rhizomes. The plant is all green, sometimes grayish. *Leaves* 10-30 cm long, *mostly basal, 3-pinnate into small linear pointed leaflets that are tilted in different planes, which gives the leaves a ruffled, fern-like appearance.* Flowerheads 1-1.5 cm across, in dense clusters at the top of the plant; all *florets white, sometimes pink.* Disks 15-40 per head, 2-3 mm tall; rays 3-8, oval or round. In flower Mar-July.

A common plant of grasslands, coastal sage scrub, chaparral, and oak woodlands. Known from San Joaquin Freshwater Marsh, Laguna Coast Wilderness Park in the San Joaquin Hills, Rancho Mission Viejo, Audubon Starr Ranch Sanctuary, Limestone Canyon in the Lomas de Santiago, Puente-Chino Hills, and Temescal Valley.

The epithet was given in 1753 by C. Linnaeus for its numerous leaf divisions (Latin, mille = thousand; folium = leaf).

Similar. Golden yarrow (*Eriophyllum confertiflorum* var. *confertiflorum*): leaflets hairy, lobed or pinnate; florets golden.

White Yarrow

White Yarrow / Golden Yarrow

White Yarrow / Golden Yarrow

Golden Yarrow
Eriophyllum confertiflorum (DC.) A. Gray **var. *confertiflorum***

An *upright woody perennial* 20-70 cm tall, *densely covered with white woolly hairs.* Leaves oval in outline, *deeply lobed or pinnate, the edges rolled under, the upperside green and hairless or nearly so, underside very woolly.* Flowerheads 1-1.5 cm across, in dense clusters, *all florets golden.* Disks 10-75 per head, 2-4 mm tall; rays 4-6, rarely absent, oval or round. In flower Feb-Aug.

An abundant plant of coastal dunes, coastal sage scrub, and chaparral, common along trails and roadcuts in most of our wild areas. Easily overlooked before and after bloom but impossible to miss while in full flower. In cultivation as a garden accent in full sun with fast-draining dry soils.

The epithet was given in 1836 by A.P. de Candolle for its flowerheads, crowded at the top of the plant (Latin, confertus = crowded; floris = flower).

Similar. White yarrow (*Achillea millefolium*): leaflets hairless, 3-pinnate and ruffled-looking; all florets white.

Golden Yarrow

Blow-wives
Achyrachaena mollis Schauer

A *hairy, soft-leaved, upright annual* 2.5-40 cm tall with few branches. *Leaves undivided with no petiole;* their bases are partially wrapped around the stem; the blade 2-12 cm long, linear, sometimes tooth-edged; lower leaves opposite, their bases totally fused together, upper leaves alternate. The inflorescence stalk has leaf-like bracts along it. Flowerheads are radiate but never broadly open, so they appear discoid at first glance. *All corollas are yellow to red.* Phyllaries 3-8, upright, green with yellowish or clear edges, soft, sticky-tipped hairs. Disks 6-35 per head, tiny; *rays* 3-8, yellow to red, and *stand up (not outward)*. Each seed is directly topped with two rows of broad blunt-tipped white scales that spread out with age. In flower Apr-May.

An uncommon charming plant of heavy clay soils among native grasslands at fairly low elevations. Known from San Joaquin Hills (Quail Hill area, perhaps also in Shady Canyon), Lomas de Santiago, Rancho Mission Viejo, Oak Flats, and Nichols Road wetlands in Lake Elsinore.

The epithet was given in 1837 by J.C. Schauer for its soft leaves (Latin, molle = soft).

Similar. Mountain dandelions (*Agoseris* spp.): milky sap; liguliflorous flowerheads on a leafless stalk; phyllaries overlap, tip is long, narrow, curves outward, hairy; a slender stalk on the seed, topped with a plume of narrow white bristles.

Dog Mayweed, Stinkweed ✖
Anthemis cotula L.

A foul-smelling annual, 5-24 cm tall, sparsely covered with hairs. *Leaves* 2-6 cm long, carrot-like, *2-3-pinnate with linear lobes.* Flowerheads 1-2.5 cm across. *Disks bright yellow. Rays white,* fold back with age and contain no reproductive organs. In flower Apr-Aug.

An abundant weed of dry disturbed areas, roadways, and watercourses. Uncommon in Orange County but recently invaded Aliso and Wood Canyons Wilderness Park. Quite abundant in western Riverside County, especially around Lake Elsinore. Native of Europe.

The epithet was named in 1753 by C. Linnaeus probably in reference to the disk florets that sit on a conical receptacle and look like an upside-down teacup (Greek, kotyle = cup-shaped, small cup).

True Asters: *Eurybia, Symphyotrichum, Corethrogyne, Erigeron, Laennecia*

In our species, the *flowerheads are radiate* (three are disciform: *Erigeron bonariensis, E. sumatrensis, Laennecia coulteri*). Flowers solitary or in a sparse inflorescence. *Phyllaries overlapping, in many rows, the tips straight or curled outward. Rays long and narrow* (very short in *Erigeron canadensis*, absent in *Laennecia* and non-native *Erigeron*), *purplish in color* (some white). *Receptacle often flat* (some convex); *epaleate*. Pappus usually of several long bristles.

Roughleaf Aster ✿
Eurybia radulina (A. Gray) G.L. Nesom
[*Aster radulinus* A. Gray]

An *upright perennial* 20-70 cm tall, with hairy stems and a woody rhizome. *Leaves large*, 4-10 cm long, *elliptic to oval with large teeth, the underside covered with short hairs that are rough to the touch*. Flowerheads at the branchtips, few in number to several and crowded. Phyllaries upright, green, paler at the base, with purple edges. Disks yellow; rays white to pale violet. In flower June-Oct.

A plant of dry woodlands. Here it lives in the understory of chaparral at upper elevations. Very uncommon, known only from the northern Santa Ana Mountains near Oak Flats, where Skyline Drive meets Main Divide Road.

The epithet was named in 1872 by A. Gray for the bristly hairs on its leaves (Latin, radula = a scraper).

Roughleaf Aster

Roughleaf Aster | San Bernardino Aster

San Bernardino Aster ★
Symphyotrichum defoliatum (Parish) G.L. Nesom
[*Aster defoliatus* Parish; *A. bernardinus* H.M. Hall]

A *perennial* 10 cm–1 meter tall *covered with short white hairs*, it spreads by long rhizomes that enable it to form colonies. *Basal leaves* 4-12 cm long, 3-6 mm wide, *linear, often withered at flowering time*. Upper leaves smaller, sometimes in bundles at the nodes. Flowerheads solitary or in sparse groups at the branchtips. Phyllaries upright, green with pale edges near their base. Disks yellow; rays whitish to pale violet. In flower July-Nov.

Long ago it was known from the borders of swamplands of Orange County such as in Buena Park, Seal Beach, Newport Beach, and Tustin; most of those lands have been drained, filled, and built upon. *Today known only from grasslands in Los Pinos Potrero near Falcon Group Camp and Santa Rosa Plateau. Apparently extirpated from the Temescal Valley* CRPR 1B.2. In the Los Pinos Potrero, the leaves and buds are frequently eaten by **grasshoppers** (*Melanoplus* sp., Acrididae) during summerfall and do not produce additional flowers.

The epithet was named in 1904 by S.B. Parish for its lower leaves that die off before the flowers appear (Latin, de- = undoing; foliolum = leaf).

San Bernardino Aster
Usually leafless when in flower

Western Lanceleaf Aster
Symphyotrichum lanceolatum (Willd.) G. L. Nesom **ssp. hesperium** (A. Gray) G. L. Nesom
[*Aster lanceolatus* Willd. ssp. *hesperius* A. Gray]

  An *upright perennial* from a long rhizome from which it can form colonies. Stems 16-160 cm tall, hairless or with lines of hairs down the stem. *Leaves* 8-16 cm long, *lance-shaped; lower often toothed, upper usually smooth-edged;* basal leaves withered by flowering time. *Flowerheads solitary or grouped atop lateral branches, usually subtended by leafy bracts. Phyllaries curved outward; green, pale-edged from base to about halfway up.* Disks yellow-orange, rays lavender to blue or white. In flower July-Aug.

  A plant of wet soils in meadows and grasslands. Historical records from Newport Beach (probably upper Newport Bay) and Wintersberg Channel in Huntington Beach; not seen there in many years, probably extirpated. One record from Lake Elsinore, present status unknown. Currently known from Santa Rosa Plateau.

  The epithet was named in 1803 by C.L. Willdenow for its lance-shaped leaves (Latin, lanceolatus = lance-like, armed with a pointed weapon). The variety was named in 1884 by A. Gray for its western distribution (Greek, hesperos = western).

Slender Aster
Symphyotrichum subulatum (Michx.) G.L. Nesom
[*Aster subulatus* Michx., *A. exilis* Elliott]

  An *upright, slender, hairless annual,* 20-80 cm tall. *Leaves* 3-6 cm long, *linear to reverse lance-shaped, with a long slender point. Flowerheads tiny,* in an open inflorescence, at the top of the plant or along the stems. Phyllaries upright, slender, green to reddish, the edges paler. Disks yellow; rays pink to violet. In flower July-Oct.

  A delicate-looking plant of wetlands and other spots with permanent moisture such as along rivers, creeks, and bays. It can tolerate freshwater, saltwater, and alkaline soils. Known from the Santa Ana River channel, Anaheim Bay, Bolsa Chica Ecological Reserve, Huntington Beach Central Park, San Joaquin Freshwater Marsh, Aliso Creek near Laguna Beach, lower San Juan Creek in San Juan Capistrano, inland wetlands in San Clemente, and Lake Elsinore.

  The epithet was given in 1803 by A. Michaux for its awl-shaped phyllaries (Latin, subula = an awl).

Sand Aster, California Aster
Corethrogyne filaginifolia (Hook. & Arn.) Nutt.
  [*Lessingia filaginifolia* (Hook. & Arn.) M.A. Lane]

Sand Aster
Lower leaves often white, upper leaves often green

A *twiggy perennial* 20-80 cm tall with several branches *densely covered with soft white hairs* that give the plant a gray appearance. *Leaves* 1-7 cm long, spoon-shaped, oval, or linear, often with large teeth, uppermost leaves much smaller. Flowerheads arranged in a branched inflorescence at the top of the plant. *Phyllaries thick,* covered with *sticky-tipped hairs;* the tips curve backward in age. *Disk florets tall,* yellow, and have both male and female organs. Rays purple, sterile. In flower July-Oct.

A very common plant of grasslands, coastal sage scrub, and oak woodlands throughout our area but more common along the coast and its foothills. Pollinated by many kinds of beetles, butterflies, flies, bees, and wasps. Some native bees chew the white wool off the stems to provision their nests.

The epithet was given in 1841 by W.J. Hooker and G.A.W. Arnott for its narrow leaves, though they are not usually all that narrow (Latin, filum = thread; folium = leaf).

Similar. Leafy daisy (*Erigeron foliosus*): green mostly hairless stems; narrow hairless leaves; *short disks,* bluish rays; in flower late spring-summer.

Sand Aster

Sand Aster

Leafy Daisy

Leafy Daisy, Fleabane Daisy
Erigeron foliosus Nutt.

Leafy Daisy

An *upright perennial* 20 cm–1 meter tall from long slender green (rarely purple) branches. *Leaves* 2-5 cm long, *very narrow, often thread-like, hairless* or with scattered hairs. Flowerheads 1-1.6 cm across, in open groups at the top of the plant. *Phyllaries flat,* green with a yellowish raised ridge in the center, held upright. *Disk florets short,* yellow. Rays bluish, rarely whitish; sterile. In flower May-Aug.

An abundant plant of many habitats, known here along the coastline from Dana Point Headlands inland to our foothills, mountains, and inland valleys. Pollinated by beetles, butterflies, flies, bees, and wasps.

The epithet was given in 1840 by T. Nuttall for its many leaves (Latin, foliosus = full of leaves).

Similar. Sand aster (*Corethrogyne filaginifolia*): white-hairy stems; broader leaves; *tall disks;* purple rays; in flower summer-fall.

Bur-marigolds: *Bidens*

Annuals or perennials. *Leaves opposite,* simple to 1-pinnate. *Outer phyllaries often long, leaf-like.*

The genus was named in 1753 by C. Linnaeus for the 2 *barbed awns on the fruit* (Latin, di- = two; dens = tooth).

Smooth Bur-marigold

Bidens laevis (L.) Britton, Stearns, & Pogg.

A *hairless, rather fleshy annual or perennial herb* with sprawling or upright stems 20 cm-2.5 meters long; rooting at the nodes. Its *dark green leaves have no petiole,* the blade 5-15 cm long, lance-shaped, edges toothed, mostly opposite. *Flowerheads large,* about 6-8 cm across. Outer phyllaries 1-2 cm long, leaf-like. Disks several, yellow. *Rays 1.5-3 cm long, usually longer than outer phyllaries, bright yellow to orangish,* the tips sometimes pale. In flower Aug-Nov.

A stunning plant of freshwater wetlands and creeks at low elevations. Found abundantly along the Santa Ana River from Featherly to Yorba Regional Parks, sparingly along San Juan Creek. A fine plant for garden pond margins.

The epithet was given in 1753 by C. Linnaeus for its smooth hairless stems (Latin, laevis = smooth).

Similar. Nodding bur-marigold (*B. cernua* L., not pictured): leaves lance-shaped; flowerheads radiate (discoid); outer phyllaries 1-3 cm long, leaf-like; *rays longer, 8-15 cm long, longer than outer phyllaries,* bright yellow; one record for Santa Ana River channel in Anaheim (1973).

Smooth Bur-marigold

Smooth Bur-marigold

Common Beggar-ticks

Form with short white rays
Common Beggar-ticks

Sticktight has very long outer phyllaries

Common Beggar-ticks ✖

Bidens pilosa L.

An upright hairless or hairy annual, generally up to 1 meter tall, the stems simple or branched, square in cross-section. *Leaves 3-12 cm long, 1-pinnate* (simple), *edges smooth or toothed. Flowerheads discoid (radiate), often on long stalks.* Outer phyllaries very short, 4-5 mm. Disks several, small, yellow (white). Rays usually absent; when present they are short and white. Fruit narrow, 4-angled in cross-section. In flower year-round.

An abundant weed of moist disturbed ground in creeks, orchards, and gardens, more common in northern Orange County than southern. Recently discovered in Whiting Ranch Wilderness Park and Mission Viejo.

The epithet was given in 1753 by C. Linnaeus for its hairiness (Latin, pilosus = hairy).

Similar. Sticktight (*B. frondosa* L.): leaves 1-pinnate; flowerheads radiate or discoid, *outermost 5-8 phyllaries very long, 1-5 cm, leaf-like;* disks orange; rays usually absent (short, yellow). Uncommon in muddy soils and sand bars along watercourses. Santa Ana River channel in Yorba Linda and Whittier Narrows along the San Gabriel River. **Yellow pincushion** (*Chaenactis glabriuscula*): *leaves alternate,* lobes very slender; *flowerheads radiant,* outer disk florets large, bilateral.

Common Beggar-ticks

Lower leaves 1-pinnate

Tarplants: *Centromadia, Deinandra, Holocarpha, Madia, Lagophylla, Osmadenia*

These are *upright annuals* usually branched above the middle of the plant, often covered with sticky-tipped hairs. Most have a *strong aromatic scent* that some people liken to tar. *Their florets are bright-, pale-, or golden-yellow.* Most begin to flower in late spring, some in late summer to fall. Those in the genera *Centromadia, Deinandra,* and *Holocarpha* can be difficult to distinguish, so we provide this table to help.

	Southern *Cen. parryi* ssp. *australis*	Smooth *Cen. pungens* ssp. *laevis*	Fascicled *Deinandra* *fasciculata*	Kellogg's *Deinandra* *kelloggii*	Paniculate *Deinandra* *paniculata*	Graceful *Holocarpha* *virgata*
Leaves	Spine-tipped	Spine-tipped	No spines, leaves in bundles	No spines	No spines	No spines; upper leaves with pit glands
Flowerheads	Nestled among leaves	Nestled among leaves	Usually in pairs	Solitary	Solitary	Solitary; phyllaries with pit glands
Disk Florets	Numerous; anthers reddish to dark purple	Numerous; anthers yellow	6; anthers reddish to dark purple	6; anthers yellow to dark	8-13; anthers dark purple	Few to many; anthers red to dark purple
Ray Florets	9-30+	Numerous	5, ray 6-14 mm	5, ray 4-8 mm	(7)8(10), ray 3-7.5 mm	(3-)5(-7)
Notes	Rare; coastal	Rare; inland	Common; coastal and inland	Common; inland	Rare; coastal and inland	Rare; Santa Ana Mtn foothills

Southern Tarplant, Southern Spikeweed ★
Centromadia parryi (E. Greene) E. Greene **ssp. *australis*** (D.D. Keck) B.G. Baldwin
[*Hemizonia parryi* E. Greene ssp. *australis* D.D. Keck]

Southern Tarplant

Upright annual, 10-70 cm tall with *many stiff branches*, the stems greenish, purplish, or reddish with long pale hairs. Most upper stem and leaf surfaces are dotted with large yellow glands from which a mildly-scented, sticky fluid is secreted. *Lower leaves* 5-20 cm long, *1-2-pinnate*, often withered at flowering time. *Upper leaves* linear, much smaller, *tipped with a very sharp spine.* Flowerheads along the branches, nestled near leaves that seem to guard them. *Phyllaries green, long, and spine-tipped.* Disks numerous, yellow, most male-only; *anthers reddish to dark purple; pappus (calyx) of 3-5 long narrow scales. Rays 9-30 or more,* yellow, to orangish in age. In flower June–Oct.

A plant of wetlands, vernal pools, alkaline, and saline soils along the coast and foothills. Once much more widespread and common but largely extirpated by habitat destruction. Today known from coastal sites such as Seal Beach, Anaheim Bay, Bolsa Chica Ecological Reserve, upper Newport Bay, San Joaquin Freshwater Marsh, and San Diego Creek. Inland localities include Orange Park Acres, Peters Creek Channel, and both Cañada Chiquita and Cañada Gobernadora on Rancho Mission Viejo. CRPR 1B.1.

The genus was established in 1897 by E.L. Greene for its similarity to tarplants in the genus *Madia* and for its spines (Greek, kentron = spine; *Madia*). The epithet was given in 1882, also by Dr. Greene, to honor C.C. Parry. Our local form was named in 1935 by D.D. Keck for its southern distribution (Latin, australis = southern).

Southern Tarplant
Leaves and phyllaries are tipped with very sharp spines

Smooth Tarplant ★
Centromadia pungens (Hook. & Arn.) E. Greene
ssp. *laevis* (D.D. Keck) B.G. Baldwin
[*Hemizonia pungens* subsp. *laevis* D.D. Keck]

An upright or low-branched annual, 10-120 cm tall, its *stems are stiff, white-hairy, often bristle-covered*. Lower leaves 5-15 cm long, *2-pinnate*. Upper leaves smaller, linear, sometimes not divided, *tipped with a sharp spine*. Leaf faces (flat portions) are smooth and hairless. *Phyllaries keeled and spine-tipped*, often hidden by uppermost leaves. *Disk and ray florets numerous, bright yellow, as are the anthers. Disk florets have **no** pappus (calyx)*. In flower Apr-Sep.

A plant of soil depressions, streambeds, grasslands, and disturbed sites in warm-to-hot sunny inland areas. *Known here along Temescal Valley from Indian Canyon south through Lake Elsinore basin (easily found near Diamond Stadium), Murrieta, and Temecula*. More common eastward to Hemet. Quickly declining, seriously threatened by agriculture, road maintenance, urbanization, and flood control projects. CRPR 1B.1.

The epithet was established in 1894 by W.J. Hooker and G.A.W. Arnott in reference to the sharp spines on the leaftips (Latin, pungens = stinging). The subepithet was given in 1935 by D.D. Keck for its smooth leaf faces (Latin, laevis = smooth).

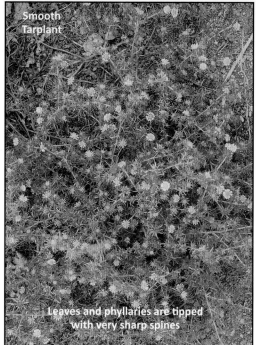
Leaves and phyllaries are tipped with very sharp spines

Fascicled Tarplant
Deinandra fasciculata (DC.) E. Greene
[*Hemizonia fasciculata* (DC.) Torr. & A. Gray]

An *upright annual* 40 cm-1 meter tall with sparse branches above. *Stems hairless, sometimes with short bristles*. Lower leaves 3.5-15 cm long, toothed or deeply lobed, covered with short bristles. Upper leaves much shorter, linear, and held closely to the stem. *Flowerheads are in bundles of 2 or more at tips of top- and side-branches, like pom-poms. Disks* yellow, generally male-only, with reddish to dark purple anthers that appear black; *only 6 disks per head*. Rays are yellow, broad, with 3 lobes; *only 5 rays in each head*; the strap is 6-14 mm long. In flower May-Sep.

A common plant of clay or sandy-clay soils, usually among native grasslands and open lots from the coastline, foothills, and canyons, to lower elevations of the Santa Ana Mountains and much of the Temescal Valley. Easily found in Crystal Cove State Park, Niguel Hill, Richard and Donna O'Neill Conservancy, Audubon Starr Ranch Sanctuary, Telegraph Canyon in the Chino Hills, near Lake Elsinore, and along Clinton Keith Road.

The epithet was given in 1836 by A.P. de Candolle for the arrangement of its flowerheads (Latin, fasciculus = a bundle).

Kellogg's Tarplant
Deinandra kelloggii (E. Greene) E. Greene
 [*Hemizonia kelloggii* E. Greene]

A *tall annual* 10 cm-1(-1.5) meter tall, *stems covered with soft hairs and/or bristles below, both bristles and sticky-tipped hairs above*. Lower leaves 3-9 cm long, oblong to reverse-lance-shaped, toothed to lobed. Upper leaves much smaller, neither toothed nor lobed, but still broad-based. Inflorescence open and branched. *Each flowerhead is atop a long, thin, naked peduncle that allows the flowerhead to flutter in even the slightest breeze. Disks yellow,* generally male-only. Anthers variable: yellow to brownish in some areas, reddish to dark purple or maroon in others; only 6 disks per head. *Rays are deep yellow,* never golden; very broad, with 3 lobes; *only 5 rays in each head;* the strap is 4-8 mm long. In flower Apr-July (-Nov).

A plant of grasslands and open patches, generally in decomposed granitic soils on and near rocky hills. Uncommon in Orange County; recently found in the Santa Ana Mountains along Long Canyon Road downhill to Acjachemen Meadow. In Riverside County, it is abundant in the Temescal Valley, south to Lake Elsinore, and eastward to Hemet; also in upper San Mateo Canyon near Tenaja Road.

The epithet was given in 1883 by E.L. Greene in honor of A. Kellogg, who first collected the plant in Alameda in 1870.

Kellogg's Tarplant

Kellogg's Tarplant

Paniculate Tarplant

Paniculate Tarplant   Fascicled Tarplant

Paniculate or San Diego Tarplant ★
Deinandra paniculata (A. Gray) Davidson & Moxley
 [*Hemizonia paniculata* A. Gray]

An *annual* 10-80 cm tall, rarely taller; *generally hairy*. Lower stems bristly, upper stems sticky to the touch. Lower leaves 1-10 cm long, linear to reverse-lance-shaped, toothed to pinnate. Upper leaves much shorter, linear. *The inflorescence has many branches that make the plant appear very wide.* Flowerheads are solitary, not in clusters. Disks yellow, male and female or male-only, with reddish to dark purple anthers that appear black; 8-13 disks per head. *Rays* yellow, broad, with 3 lobes; with *8-13* rays in each head, *the strap portion of each ray is 3.5-6 mm long.* In flower May-Nov.

Found in clay or sandy-clay soils of grasslands and open areas among coastal sage scrub, chaparral, and disturbed ground. Relatively common in southeastern Orange County and western Riverside County. Look for it in Aliso and Wood Canyons Wilderness Park, San Juan Canyon on the Rancho Mission Viejo, Richard and Donna O'Neill Conservancy, Arroyo Trabuco, Crow Canyon on Audubon Starr Ranch Sanctuary, and lower San Mateo Canyon. CRPR 4.2.

The epithet was given in 1883 by A. Gray for the arrangement of its flowerheads (Latin, paniculata = having tufts of flowers).

Paniculate Tarplant

Graceful Tarplant ★

Holocarpha virgata (A. Gray) Keck **var. *elongata*** Keck

An *upright annual* 20 cm–1.2 meters tall with *many long stems* that gracefully curve. It is *densely covered with glands* that secrete a thick substance and give off a *strong lemon- or pine-like scent*. Lower leaves 6-15 cm long, linear, often tooth-edged. *Upper leaves blunt ended, tipped with an obvious pit that houses a gland ("pit-gland")*. Flowerheads are scattered along branches. *Phyllaries green, their outer surface with 5-20 stalks, each tipped with a pit-gland.* Disks 9-25, yellow, anthers reddish to dark purple. Rays only 3-7 (usually 5), each rounded, yellow to orangish. In flower July-Sep(-Nov).

Occurs in *grasslands* and grassy openings among coastal sage scrub and oak woodlands. Not widely found here, more common in San Diego County. Known only from the former Lower Trabuco Canyon Campground about 1 mile upstream from Live Oak Canyon Road, Lookout Trail on the Audubon Starr Ranch Sanctuary, on mima mound fields in San Clemente State Beach, and San Mateo Canyon near both Bluewater and Tenaja Canyons in the San Mateo Canyon Wilderness. More common near Santa Rosa Plateau. It sometimes occurs with paniculate tarplant. CRPR 4.2.

The epithet was given in 1859 by A. Gray for its long twigs (Latin, virgatus = twiggy, made of twigs). The variety was named in 1958 by D.D. Keck for its stems, which curve away from the middle of the plant (Latin, elongatus = removed).

Graceful Tarplant

Graceful Tarplant

Elegant Madia

Elegant Madia

Elegant Madia

Madia elegans D. Don **ssp. *elegans***

A *slender-stemmed annual* 20-90 cm tall with a *strong scent* and sticky-tipped hairs. A few small leaves often form a basal rosette. *Stem leaves 2-20 cm long, lower ones crowded on the stem, soft and hairy, upper leaves very sticky.* Disks few or many, yellow to maroon. Rays 8-16, the strap long and fan-shaped, yellow, sometimes maroon near the base, held straight out. In flower May-June(-Aug).

An uncommon plant of dry slopes and meadow-edges. Known from Pruesker Peak on the Audubon Starr Ranch Sanctuary, Old Camp (where the Santiago and Joplin Trails meet), and scattered sites in the San Mateo Canyon Wilderness Area. It may be more common and widespread, but it is in flower when temperatures are high and most people are not out hiking.

The epithet was given in 1831 by D. Don for its elegant appearance (Latin, elegans = neat, elegant).

Elegant Madia

Thread-stem Madia
Madia exigua (Smith) A. Gray

A tiny aromatic annual 1-30(-50) cm tall with many slender stems, covered with soft or bristly hairs, some sticky-tipped. *Leaves linear,* 1-4 cm long, with bristles. *Flowerheads rounded but with pointed top and bottom like a toy top.* Phyllaries green, bulging outward, covered with long hairs, each tipped with a sticky gland. Disks only 1-2 per head, corolla and anthers yellow. *Rays 5 or 8, yellow, very short, and barely peek out of the flowerhead.* In flower May-July.

A low-growing plant of dry places, meadows, and post-fire chaparral sites. Small and easily overlooked, we find it among meadow plants and hidden beneath shrubs. Individual plants are in flower briefly, more commonly seen in May than July. Known from South Main Divide Road, Los Pinos Potrero, and San Mateo Canyon Wilderness, perhaps more widespread.

The epithet was given in 1815 by J.E. Smith for its short stature (Latin, exiguus = short, small).

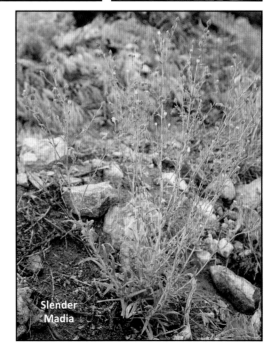

Slender Madia, Slender Tarplant
Madia gracilis (Smith) Keck

An upright annual 10 cm–1 meter tall with a pleasant scent, covered with soft or bristly hairs, upper parts sticky-tipped. Leaves 3-10 cm long, 5 mm wide, with soft hairs, and long sticky-tipped hairs. *Flowerheads oval.* Phyllaries green, covered with long hairs, each tipped with a sticky gland. Disks 2-12 per head, yellow, the anthers black. Rays 3-9, lemon yellow, stand up or outward. In flower Apr-Aug.

Usually common but sometimes absent from an area for years then reappears. Found in grasslands, meadows, and disturbed soils such as along trails. Known from Claymine Canyon near Sierra Peak, Harding Truck Trail, Trabuco Trail, Audubon Starr Ranch Sanctuary, Viejo and San Juan Canyons, Sitton Peak, Los Pinos Potrero, along North Main Divide Road "Loop," south into the San Mateo Canyon Wilderness Area.

The epithet was given in 1815 by J.E. Smith for its slender stems (Latin, gracilis = slender).

Common Hareleaf
Lagophylla ramosissima Nutt. **ssp. *ramosissima***
An upright annual 10 cm – 1.5 meters tall *from a skinny central stalk and slender side-branches.* Stem and leaf hairs are very soft. The stems are pale green and hairy when young, but in age they turn straw-colored or reddish and the hairs fall off. *Lower leaves* 3-12 cm long, linear, *dried up by flowering time.* Upper leaves smaller, very hairy, and few in number. Uppermost leaves mostly in small tufts. Flowerheads oval, the phyllaries green, very hairy, with dot-like yellow glands at the tips of tiny stalks. Disks 6 per head, male-only, anthers black, corolla yellow. *Rays 5, each 3-lobed, the central lobe is smaller, each lobe pale yellow with a reddish stripe on the outside.* In flower May-Oct.

Its flowers open in the evening and close by mid-morning, so you must get up early to see them. Found in grasslands and meadows, easily overlooked. Known from north Bell Canyon and northwest of Pruesker Peak, both on the Audubon Starr Ranch Sanctuary. Also in Oak Flats and in the meadow along South Main Divide Road just past the community of Rancho Capistrano, and in Whittier Hills and Santa Rosa Plateau. Within the Los Pinos Potrero it is sometimes common and easy to access, especially opposite the entrance to Falcon Group Camp.

The epithet was given in 1841 by T. Nuttall for its many branches (Latin, ramus = branch; -issima = very much, most). The common name refers to its soft-hairy leaves.

Common Hareleaf

Common Hareleaf

False Rosinweed

False Rosinweed

False Rosinweed, Three-spot
Osmadenia tenella Nutt.
[*Calycadenia tenella* (Nutt.) Torr. & A. Gray]

An *upright slender annual* 15-50 cm tall, often many-branched, hairy and sticky, with a *strong but pleasant resin scent. Leaves* 1.5-5 cm long, *slender,* usually toothed, hairy. Flowerheads at the branchtips, about 1 cm across. *All florets are white, often with a red splotch, the white fades to pinkish with age.* Disks 3-10 per head, deeply 5-lobed, anthers yellow, pistils white and very long. *Rays 3-5, each deeply 3-lobed and held apart from each other.* The 3 ray lobes overlap those of adjacent florets and make it difficult to recognize that the heads are radiate. In flower May-July, rarely to Oct.

An abundant plant of late spring and summer thoughout our area. It grows in barren sandy or rocky soils, in openings among grasslands and coastal sage scrub. Easily found in all of our wilderness parks and preserves, especially in Limestone Canyon, Richard and Donna O'Neill Conservancy, and Caspers Wilderness Park.

The epithet was given in 1841 by T. Nuttall for its delicate appearance (Latin, tenellus = quite delicate).

False Rosinweed

Common Yellow Carpet
Blennosperma nanum (Hook.) S.F. Blake

  A *slender upright annual*, 3-12 cm tall. *Leaves alternate, linear to pinnately divided, hairless to sparsely woolly. Phyllaries of 2 lengths, fused at their bases, curved upward or backward, tips purplish. Disks 20-60 or more, male-only* (style present but ovary is sterile), *yellow; pollen white. Rays 5-13 or more; yellow; sometimes purplish below;* stigma yellow. In flower Jan-May.

  Lives around vernal pools on the Santa Rosa Plateau, often abundant. Pollinated by native bees and flies, some of which specialize on it.

  The epithet was given in 1833 by W.J. Hooker for its small size (Latin, nanus = dwarf).

Common Yellow Carpet

Garland Daisy ✖
Glebionis coronaria (L.) Spach.
  [*Chrysanthemum coronarium* L.]

  A *stiff-stemmed upright annual* to 1 meter tall. *Leaves 2-6 cm long, 2-pinnate, alternate on the stem.* Flowerheads 2-4 cm across, the numerous disk and ray florets have both sexes and produce seed. Disks yellow; rays yellow, rarely white-tipped to mostly white. In flower Mar-Aug.

  An overly abundant weed of roadsides and disturbed soils, it escaped from cultivation, made its way into wild areas, and is spreading. Find it commonly behind the former Marine Corps Air Station El Toro, in upper Newport Bay, Crystal Cove State Park, lower Aliso Canyon, Niguel Hill, Arroyo Trabuco, and along lower Ortega Highway. Also around Nichols Road near Lake Elsinore. Native of southern Europe.

  The epithet was given in 1753 by C. Linnaeus for its flowers, used in floral displays (Latin, coronarius = crown, wreath, or garland).

Garland Daisy

False Daisy
Eclipta prostrata (L.) L. [*E. alba* (L.) Hassk.]

  A *low-growing to partially upright annual,* 10 cm-1 meter long, sometimes rooting at the nodes. Leaves linear, lance-shaped, or elliptical, smooth-edged or small-toothed, opposite. *Flowerheads at the leafy branchtips,* about 1 cm across, solitary or in small groups. Phyllaries green, broad, the inner narrower. *Disks white, anthers brown. Rays numerous, very narrow, white.* In flower all year.

  A plant of *watercourses*, less commonly along standing water. Known from Santa Ana River channel, Huntington Beach, San Joaquin Freshwater Marsh, San Diego Creek, Laguna Canyon, Arroyo Trabuco, San Juan Creek, Horsehoe Bend on the Santa River, and Elsinore Basin.

  The epithet was given in 1753 by C. Linnaeus for its low-growing habit (Latin, prostratus = prostrate).

False Daisy

Sunflowers: *Encelia, Helianthus, Venegasia, Verbesina, Bahiopsis, Wyethia*

These are the big beautiful sunflowers most people typically associate with the name. They all have a fairly *large disk area* with bright yellow to orangish rays. Some are large annuals that stand upright, while others are rounded shrubby perennials.

California Bush Sunflower, California Brittlebush
Encelia californica Nutt.

A *rounded shrub* 50 cm–1.5 meters tall, as wide or wider. The many stems arise mostly from below. Its *green leaves* are covered with very short hairs, long-petioled, blades 3-6 cm long, often heart-shaped at the base, elliptic to oval, smooth-edged or toothed, the tip narrow, alternate. *Flowerheads on long peduncles,* solitary or in groups. Phyllaries numerous, all similar, lance-shaped, covered with white hair. *Disks brown-purple,* giving the flower a dark "face." *Rays 15-25, 3-lobed or blunt-tipped, bright yellow.* In flower Feb-June.

Common throughout our area from the moist coastline inland to the foothills and mountains. Absent from higher elevations, less common in the Temescal Valley. In cultivation, often planted for slope stabilization, it will continue to flower through summer and fall if irrigated. Under water stress, it drops its leaves.

The epithet was given in 1841 by T. Nuttall for the place of its discovery.

Similar. Desert bush sunflower (*E. farinosa*): fuzzy gray leaves; disks orange, yellow, or brown-purple; more common in dry inland areas.

California Bush Sunflower

Below left: California bush sunflower, disk florets brown-purple.
Below right: Desert bush sunflower, disk florets usually yellow to orange, more often brown-purple in deserts.

Desert Bush Sunflower, Incienso, Desert Brittlebush
Encelia farinosa A. Gray ex Torr.

A *fragrant shrub* 30 cm–1.5 meters tall, from one to many obvious trunks. Most of the plant looks as if it were covered with whitish powder; actually it is covered in white, silver, or gray hairs. The *gray leaves* have a long petiole, blade 2-7 cm long, oval to lance-shaped, smooth- and often wavy-edged, the tip rounded or narrow, alternate. *Flowerheads on very long, yellowish, hairless peduncles that stand far above the leaves.* Phyllaries numerous, all similar, lance-shaped. *Disks variable in color: generally yellow or orange in our area, typically also brown-purple in the deserts. Rays 11-21, bright yellow.* In flower Mar-May.

California Bush Sunflower

Desert Bush Sunflower

A plant of the open desert, found in rocky to sandy soils in our area. Probably not naturally occurring in Orange County but used in hydroseed mixes on roadcuts and altered land, reproducing in some of those areas. Widely used in Mission Viejo during the 1970s-1980s, especially along Oso Parkway just inland from Marguerite Parkway, though much of it had been eliminated by housing developments by 1990. Also known from Santiago Canyon, Hicks Canyon, and roadcuts along Ortega Highway. More common inland of the Santa Ana Mountains, such as along Ortega Highway near The Lookout, downhill to Lake Elsinore, near Corona, and in many spots in Temescal Valley. In Mexico, the stems are scraped, the sap allowed to dry, and then burned for incense (incienso). Its stems break easily, thus the common name *brittlebush*.

The epithet was given in 1848 by A. Gray for the whitish "meal" (powder) that covers it (Latin, farinosus = mealy).

Similar. California bush sunflower (*E. californica*): nearly-hairless green leaves; disks always brown-purple; more common in coastal areas.

Desert Bush Sunflower

Western Sunflower
Helianthus annuus L.

A *large upright annual* to 3 meters tall, nearly as wide, with *many branches mostly above its center,* the leaves covered with short hairs rough to the touch. *Leaves with a long petiole,* the blade 10-40 cm long, broad, often tooth-edged, *lower leaves generally triangular or oval,* upper leaves lance-shaped, alternate. *Flowerheads numerous and huge,* often to 10 cm across. *Phyllaries white-hairy, oval with a very long narrow tip.* The disk area is broad, 2-3.5 cm across, florets yellow to reddish-brown. Rays large, the strap about 2.5 cm long, narrowly pointed, bright yellow. In flower Feb-Oct.

A common plant of dry clay or sandy soils, south-facing hillsides, sometimes along creeks, and disturbed soils, especially landslides. Known throughout our entire area. Particularly abundant in the hills directly above San Clemente, near Ladera Ranch, Rancho Mission Viejo, and Whiting Ranch Wilderness Park in the Lomas de Santiago. This is the wild ancestor of the common sunflower grown for snack seeds, oil, and horticulture.

The epithet was given in 1753 by C. Linnaeus for its annual growth habit (Latin, annuus = yearly, annual).

Similar. Slender sunflower (*H. gracilentus*): slender perennial; branched at base; lower leaves opposite; flowerheads ≤5 cm across; phyllaries without a long tip, short-hairy.

Western Sunflower

Western Sunflower

Slender Sunflower

Western Sunflower

Slender Sunflower

Slender Sunflower
Helianthus gracilentus A. Gray

An upright slender perennial 60 cm–2 meters tall, *branched mostly from the base,* the stems and leaves covered with short hairs, rough to the touch. *Leaves with short or no petioles;* the blade 5-11 cm long, *lance-shaped to ovalish,* sometimes toothed; *opposite, uppermost sometimes alternate.* Flowerheads few, up to about 5 cm across. *Phyllaries with very short hairs, lance-shaped, pointed.* The disk area 1-1.5 cm wide, florets yellow to red. Rays with a strap 15-25 mm long, bright yellow to orangish. In flower May-Oct.

A common plant of dry hillsides and creekbed margins, often growing from cracks in granitic boulders or in decomposed granitic or sandstone soils. Known widely from the Santa Ana Mountains and Lomas de Santiago, apparently absent from the coast. Common along nearly the entire route of Main Divide Truck Trail, easily found along the paved section both north and south of El Cariso.

The epithet was given in 1876 by A. Gray for its slender stems (Latin, gracilis = slender).

Similar. Western sunflower (*H. annuus*): large annual, branched above the middle; leaves alternate; flowerheads to 10 cm across; phyllaries with a very long tapered tip, long-hairy.

Slender Sunflower

Canyon Sunflower
Venegasia carpesioides DC.

An *upright shrub*, 50 cm-1.5 meters tall, nearly hairless, with *numerous weak burgundy branches* that often break off from the weight of its own flowers. *Leaves alternate; blade 3-15 cm long, triangular to oval,* smooth-edged or toothed. *Flowerheads roughly 4 cm in diameter. Phyllaries large and leaf-like; some very broad, others narrow; often twisted.* Disks and their anthers are yellow to orangish. Rays bright yellow. In flower mostly Jan-June (-Sep).

An uncommon plant of coastal sage scrub, oak woodlands, and riparian communities, *usually in partial to complete shade on moist north-facing slopes.* Sometimes numerous when found, especially after fire. Known sparingly from lower Aliso Canyon in Laguna Beach and along Pacific Island Drive up Niguel Hill. More abundant in upper Lucas Canyon within both Caspers Wilderness Park and San Mateo Canyon Wilderness Area. A good performer in the garden with partial shade and adequate water.

The epithet was given in 1838 by A.P. de Candolle for its similarity to an Asian plant (Greek, carpesi = an aromatic medicinal plant from Asia; Latin, -oides = like).

Similar. Big-leaved crownbeard ★ (*Verbesina dissita*): bristly; leaves narrower, opposite below; *phyllaries oval, flat; coastal.* **Crofton weed** ✖ (*Ageratina adenophora*): *leaves opposite; flowerheads discoid,* phyllaries small, flat; *corolla white*.

Canyon Sunflower

Phyllaries leafy, often twisted

Big-leaved Crownbeard ★
Verbesina dissita A. Gray

An *upright narrow perennial*, 50 cm-1 meter tall, *covered with short bristly hairs.* Spreads by rhizomes and forms small clonal colonies. Its *few branches* are pale yellow-green. The *bright green leaves have no petiole*, the *ovate blade* is 4-8 cm long, smooth-edged or with small teeth. Lower leaves *opposite and held straight out; upper leaves alternate.* The *large flowerheads* are 5 cm in diameter; its flowers do not remain open for very long. *Phyllaries roughly identical, all oval and flat.* Disks yellow with dark brown anthers. Rays yellow-orange and sterile. In flower Apr-Aug (-Sep).

Occurs among dense stands of other plants, and is sometimes difficult to locate. Typically lives with bigpod lilac, bushrue, spiny redberry, laurel sumac, and California bush sunflower. Known only from Laguna Beach and in Baja California. Found on Goff Ridge (south of Nyes Place), Temple Hill (as on the slopes of Hobo Canyon), and Niguel Hill (within view of Pacific Island Drive).

A very rare plant, listed as CRPR 1B.1 and as Threatened by both State and Federal agencies. Its small population in Orange County is the only one in the United States. This beautiful plant is in very serious danger from rampant development and brush-clearing operations in the hills of Laguna Beach and Laguna Niguel. It makes an attractive garden plant. Away from the immediate coast, it does best in partial shade.

The epithet was given in 1885 by A. Gray for its leaf pairs, which are far apart, along the stem (Latin, dissitus = lying apart).

Similar. Canyon sunflower (*Venegasia carpesioides*): *smooth throughout; very large triangular leaves, alternate; phyllaries very leaf-like,* at least some very broad; coastal to foothills, infrequently co-occurring; uncommon. **California bush sunflower** (*Encelia californica*): *leaves alternate, petiole long;* disks brown-purple; abundant, co-occurs in Laguna Beach.

Big-leaved Crownbeard

Phyllaries oval, flat

Big-leaved Crownbeard

Upper leaves alternate, lower leaves opposite

Golden Crownbeard ✖

Verbesina encelioides (Cav.) A. Gray

A *bushy annual*, most often 0.5-1 meter tall, *with an unpleasant scent* similar to that of a goosefoot (*Chenopodium* spp., Chenopodiaceae). Leaves 4-10 cm long; *lower leaves usually opposite, upper alternate*. Leaf blade upper surface green and soft-hairy, lower surface grayish and slightly bristly. The blade triangular to lance-shaped, with 3 main veins from the base, the edges toothed. *The petiole of larger leaves is often narrowly winged.* The plants we have observed in Orange County usually have broad tooth-edged ear-like wings at the base of the petiole, though in California this species is reported to usually lack them. *Phyllaries linear, very long, sparsely covered with hairs.* Disk florets numerous; corolla orange-yellow, anthers purple (yellow to brown). Ray florets fertile, style present; corolla yellow, narrow to broad, with 3 obvious lobes. At maturity, fruits from disk florets are flattened and winged. In flower May-Dec.

A plant of dry disturbed soils and roadsides; recently established at San Joaquin Wildlife Sanctuary. In our area, found most often in Temescal Valley such as in Corona. Also known from northern Orange County south into Lake Forest. More common in the interior valleys, outside of our area. Native to Arizona, eastward through central and eastern U.S., south into Mexico. Naturalized in California, spreading toward the coast as the drying trend continues.

The epithet was given in 1793 by A.J. Cavanilles for its similarity to the genus *Encelia* (New Latin, -oides = likeness).

Right: The leaves of golden crownbeard are large and often have large "ears" at their base. Far right: The leaves of San Diego sunflower are small, wavy, and bristly to the touch.

San Diego Sunflower ★

Bahiopsis laciniata (A. Gray) E.E. Schilling & Panero
[*Viguiera laciniata* A. Gray]

A *dense, rounded shrub* 1.2 meters tall and up to 2 meters wide, *covered with short bristly hairs and shiny resin*. The *leaf blade is 1-5 cm long*, oblong to lance-shaped, the base flat or spear-shaped, the *edges wavy, sometimes toothed*. Flowerheads appear at the stem tips, alone or in clusters. *Phyllaries linear, covered with short bristly hairs.* Disk flowers are orangish, the 5-13 ray flowers yellow. In flower Feb-June.

A native of San Diego County, where uncommon and threatened by development. Ranked CRPR 4.2. It is used locally in hydroseed mixes on road cuts, where it persists and occasionally spreads, as it is doing in Orange and Riverside Counties. Find it along Ortega Highway through Caspers Wilderness Park, South Main Divide Road, Black Star Canyon Road, Santiago Canyon, and in some developed areas. A wonderful performer in cultivation, tolerant of many soil types and harsh sun exposure.

The epithet was given in 1859 by A. Gray for its leaf edges, which appear ragged and torn (Latin, lacinia = a thing torn).

Southern Mule's Ears
Wyethia ovata Torr. & A. Gray
[*Agnorhiza ovata* (Torr. & A. Gray) W.A. Weber]

  A low-growing perennial from a stout taproot and underground stems, it stands 10-50 cm tall. Usually covered with soft hairs that may fall off with age. The entire plant is entirely coated with a sticky substance and has a *sweet scent,* the flowers smell like honeysuckle. The *large leaves* with *long petioles* are alternate on the stem and often stand straight up; the *blade* is 8-20 cm long, *elliptic to ovate.* The *bright yellow flowerheads* are 5-6 cm across, *hidden among the leaves,* each with several yellow disk florets and 5-8 yellow ray florets that have very long styles. In flower May-July, rarely into Aug.

  A charming plant, thrilling to discover in the wild. Its leaves begin to appear in late March and continue through spring and summer. In summer it sets seed and its aboveground parts die off as it enters dormancy. It grows in colonies of several individuals to an area. Uncommon here, known only from openings among grasslands at Los Pinos Potrero, along San Juan Trail through upper Lion Canyon, and within oak woodlands and chaparral in upper Devil Canyon of the San Mateo Canyon Wilderness Area. During summers at Los Pinos Potrero, it is routinely devoured by **grasshoppers** (*Melanoplus* sp., Acrididae).

  The genus was established in 1834 by T. Nuttall in honor of his friend N.J. Wyeth. The epithet was assigned in 1848 by J. Torrey and A. Gray in reference to its oval leaf blades (Latin, ovatus = egg-shaped).

Radiate-headed Goldenbushes: *Ericameria*

  Shrubs with many stems, small leaves, and yellow to golden corollas. Those with *radiate* flowerheads are presented here. Our two species with *discoid* flowerheads are placed in the *Discoid* section.

Grassland Goldenbush
Ericameria palmeri (H.M. Hall) H.M. Hall **var. pachylepis** (H.M. Hall) G.L. Nesom

  Frequently a *low-growing green shrub* with several arching branches sparsely covered with short hairs, 50 cm-1.5 meters tall. *Covered with dot-like glands that give off a sticky resin with faint resin scent.* Numerous *slender linear leaves* are 5-16 mm long, *in bundles on the stem.* Its bright yellow flowerheads appear along the upper branches. Disks 5-20 per head, 5-8 mm long; rays 1-6, 4-6 mm long. In flower Aug-Dec.

  A common shrub of *grasslands and coastal sage scrub,* sometimes in disturbed soils. Widely distributed, often overlooked when not in flower. Known from UCI Ecological Preserve, Chino Hills, throughout the Lomas de Santiago such as Limestone Canyon (abundant), Los Pinos Potrero, Temescal Valley, and Elsinore Basin.

  The epithet was given in 1876 by H.M. Hall in honor of E. Palmer. Our local variety was named in 1928 by H.M. Hall for its thick phyllaries (Greek, pachys = thick; lepis = scale).

  Similar. Pine goldenbush (*E. pinifolia*): taller, woodier stems, upright branches; pine scent; leaves narrower, in tight bundles; growing in/near sandstone. **Great Basin rabbitbrush** (*E. nauseosa* var. *oreophila*) and **wedge-leaved goldenbush** (*E. cuneata* var. *cuneata*): broader leaves; discoid flowerheads; see entries under Discoid Goldenbushes. **California sagebrush** (*Artemisia californica*): stems not sticky; blueish to grayish; leaves often lobed, soft, sage-scented; flowerheads discoid; may grow nearby.

Pine Goldenbush
Ericameria pinifolia (A. Gray) H.M. Hall

An *upright hairless green shrub,* 60 cm–2.5 meters tall, from *woody* lower stems. Its *needle-like leaves* are 10-40 mm long, under 1 mm wide, with *clusters of smaller leaves* at their base. Leaves are covered with dot-like glands that give off a sticky resin with a *pine-like scent.* Flowerheads bright yellow, disk florets 12-many per head, 6-7.5 mm long; rays 3.5-5 mm long. *It has two flowering periods per year.* In spring, the flowerheads are solitary at the branchtips, each with 15-30 rays. In fall, the flowerheads are clustered and smaller, each with only 5-10 rays. In flower Apr-July, Sep-Jan.

Pine Goldenbush

A plant of *coastal sage scrub and chaparral,* generally grows from cracks in sandstone or granitic rocks or in adjacent soils. Known from Chino Hills bordering Santa Ana Canyon, throughout the San Joaquin Hills, Talega Canyon in San Clemente, Bear Canyon Trail, northern Tenaja Trail, and inland foothills of the Santa Ana Mountains bordering Temescal Valley to Elsinore Basin. Apparently absent from the Lomas de Santiago and most of the coastal side of the Santa Ana Mountains. Easiest to see near the parking area at Laguna Coast Wilderness Park and near Ortega Candy Store.

The epithet was given in 1873 by A. Gray for its pine-scented leaves (Latin, pinus = pine; folium = leaf).

Similar. Grassland goldenbush (*E. palmeri* var. *pachylepis*): shorter, less woody, arching branches; faint resin scent; leaves shorter, not in such tight bundles; in deeper soils, not rock. **Great basin rabbitbrush** (*E. nauseosa* var. *oreophila*): strong odor but not pine-like; stems with yellowish-green or whitish hairs; discoid flowerheads; not found with pine goldenbush.

Gumplant
Grindelia camporum E. Greene
[*G. hirsutula* Hook. & Arn.; *G. robusta* Nutt.]

An *upright perennial* 60 cm–1.5 meters tall with a variety of growth forms, hairless or sparsely covered with short hairs. Its stems are white and look varnished. *Most leaves have no petiole,* the base often clasps the stem; blade 1-10 cm long, oblong to lance-shaped, usually tooth-edged, light green to yellowish. *Flowerheads* large, heavy, *spherical,* the buds covered in a white gummy resin. *Outermost phyllaries held outward, curved backward, or strongly coiled backward into a loop.* Disk flowers numerous, yellow. *Rays yellow,* generally 25-39 per head, held straight out, *sometimes absent.* In flower May-Oct but may stop blooming early in very dry years.

Gumplant

Rayless form, cut-away

Gumplant

A plant of clay soils within coastal sage scrub, grasslands, and trail edges, primarily near the coast but not on open beaches. Found commonly in the San Joaquin Hills such as in Corona del Mar, Crystal Cove State Park, south through Dana Point, its headlands, and San Clemente. Also from Chino Hills, Limestone Canyon, Fremont Canyon, Rancho Mission Viejo, and the Richard and Donna O'Neill Conservancy, and near Pruesker Peak on the Audubon Starr Ranch Sanctuary.

The genus was established in 1807 by C.L. Willdenow for D.H. Grindel. The epithet was given in 1833 by W.J. Hooker and G.A.W. Arnott for hairs observed on the first plants studied (Latin, hirsutus = hairy, bristly).

Similar. Saw-toothed goldenbush (*Hazardia squarrosa*): stems hairy to hairless, not shiny-white; leaves shorter; *flowerheads very slender,* discoid; phyllaries outward or backward, *no resin;* inland foothills to mountains; not found together.

California Matchweed
Gutierrezia californica (DC.) Torr. & A. Gray

A *small shrubby perennial* 20-60 cm tall with *many slender branches*. *Leaves* 2-5 cm long, about 1 mm wide, *needle-like,* alternate, dotted with glands and *sticky to the touch*. Flowerheads small, bright yellow, the style very long and held way above the flower. Disks 4-13 per head; rays 4-13, held outward, commonly curved backward. In flower May-Oct.

A common plant of *sandy soils* in full sun among coastal sage scrub, grasslands, and along trailsides. Known from the San Joaquin Hills, Chino Hills, Lomas de Santiago, Santa Ana Mountains, and Temescal Valley.

The common name refers to the stems being as thin as a matchstick. The epithet was given in 1836 by A.P. de Candolle for the place of its discovery.

Rosilla, Sneezeweed
Helenium puberulum DC.

An *upright annual or perennial,* 50 cm–1.6 meters tall, covered with short downy hairs or hairless. *Leaves* alternate, dotted with glands, *their edges continue down the stem.* Basal leaves reverse-lance-shaped, stem leaves similar or linear, 3-15 cm long. Many older leaves die off before flowering. *Flowerheads spherical,* up to 2 cm across, 4-20 per plant. Disks several per head, yellow, the lobes yellow, brown, or purplish. *Rays 5-10 per head,* the strap 3-8 mm long, *folded backward,* sometimes hidden or rays totally absent. In flower June-Oct.

A plant of streams, creeks, and other wetlands. Uncommon here; old records are from Santa Ana River channel, Bolsa Chica Ecological Reserve, lower Aliso Canyon, upper Santiago Canyon, and lower Hot Springs Canyon. Currently known only from lower Moro and Los Trancos Canyons in Crystal Cove State Park and moist areas in Falls, lower San Mateo, and Lucas Canyons.

The epithet was given in 1836 by A.P. de Candolle for its short downy hairs (Latin, puber = downy).

Telegraph-weeds: *Heterotheca*

Stout perennials and annuals with a strong taproot, *stems bristly and hairy, often sticky*. They give off an *odor of camphor* when cut or disturbed. The inflorescence is somewhat flat-topped. Flowerheads numerous, corollas yellow to golden. Pollinated by several species of bees, flies, butterflies, and moths.

Telegraph-weed
Heterotheca grandiflora Nutt.

An *upright, unbranched, annual or short-lived perennial,* 50 cm-2 meters tall. *Leaves of 3 types:* lowest leaves have a petiole, their ear-like basal lobes clasping the stem; middle leaves have a petiole, are linear to elliptical and densely hairy; upper leaves are linear, have no petiole, and are less hairy but stickier. *Flowerheads are in a sticky open-branched cluster at the top of the plant.* Corollas yellow. Disks 30-75 per head, each 4-6 mm long; rays 20-40, 5-8 mm long. Fruits from disk and ray florets are dissimilar. In flower year-round.

A *common* plant throughout our area, in sandy and disturbed soils, grasslands, coastal sage scrub, and the edge of riparian communities. Often quite tall, said to be named for its similarity to a telegraph pole or for its habit of growing in soils that were disturbed when telegraph poles were installed.

Named in 1840 by T. Nuttall for its abundant flowerheads (Latin, grandis = large, great, full, abundant; floris = flower).

Similar. Sessileflower golden aster (*H. sessiliflora*): plant shorter; scent different; leaves green to gray, *hairs very long;* corollas more yellow-gold; much less common.

Telegraph-weed

Telegraph-weed

Telegraph-weed

Bristly Golden Aster

Bristly Golden Aster

Bristly Golden Aster
Heterotheca sessiliflora (Nutt.) Shinners **ssp. *echioides*** (Benth.) Shinn.

An *upright perennial* 10 cm-1.3 meters tall, its lower region covered with bristly hairs, *upper with long soft woolly hairs* about 2 mm long. Leaves 1-5 cm long, linear; lower tapered, upper 1-2 cm long and with no petiole. Flowerheads arranged in an openly branched cluster at the top of the plant. *Corollas yellow to goldish.* Disks 30-50 per head, each 3-10 mm long; rays 3-30, 3-10 mm long. Fruits from disk and ray florets are identical. In flower July-Nov.

Occurs in *sandy soils* among coastal sage scrub, oak woodlands, and grasslands. Known from the Chino Hills, Santa Ana Canyon, Oak Canyon Nature Center, Lomas de Santiago, Trabuco Canyon, Rancho Mission Viejo, Lower Hot Springs Canyon, San Juan Canyon, the southern end of Verdugo Potrero and Oak Flats, Temescal Valley, and San Mateo Canyon. In Limestone Canyon, it grows with telegraph weed, making their comparison easy.

The epithet was given in 1840 by T. Nuttall for the short flower stalks of his original specimens (Latin, sessilis = low, dwarf; floris = flower).

Similar. Telegraph-weed (*H. grandiflora*): taller; scent different; leaves green; corollas yellow; abundant.

Bristly Golden Aster

Red-rayed Alpinegold
Hulsea heterochroma A. Gray

An *upright annual or short-lived perennial with a sweet scent,* 50 cm-1.5 meters tall with few main stems, *densely covered with sticky-tipped hairs.* Leaves oblong, the tip pointed, the margins toothed, without a petiole, alternate on the stem. Basal leaves often large, 10-20 cm long, and crowded, smaller and spaced farther apart up the stem, sometimes withered when flowers appear. *Disks florets yellow, tipped with red-purple. Ray florets* slender and numerous, *with a yellow tube, purplish-red strap,* and 3-toothed tip, sometimes yellowish, their outer surface hairy. In fruit, it forms a round head of bristle-tipped fruits (cypselae) that fall to the ground when they mature. In flower late May-Aug.

An uncommon plant *generally found only after mountain fires or rock slides,* sometimes on road cuts and in open areas among chaparral; also in similar habitats among coniferous forests in taller mountain ranges. Here known only from the fire road that ascends the west side of Modjeska Peak and in upper Silverado Canyon on the slopes just below Maple Springs Saddle. Pollinated by native bees, soft-winged flower beetles (Melyridae), and metallic wood-boring beetles (Buprestidae).

The genus was established in 1857 by J. Torrey and A. Gray in honor of G.W. Hulse. The epithet was given in 1868 by A. Gray for the different colors of the disk and ray flowers (Greek, heteros = other, different; chromos = color). In most species of *Hulsea,* the disks and rays are of the same color.

Red-rayed Alpinegold

Red-rayed Alpinegold

Red-rayed Alpinegold

Red-rayed Alpinegold

Fleshy Jaumea

Fleshy Jaumea
Jaumea carnosa (Less.) A. Gray

A *low-growing, fleshy, hairless perennial* that spreads from creeping underground rhizomes. The *narrow leaves* are 1.5-5 cm long, *linear to oblong, opposite.* Flowerheads 1.2-2 cm across, solitary at the branchtips, *corollas bright yellow.* Phyllaries oval, often tipped with purple or red. Disks 6-7 mm long, rays 6-10 per head, the strap 3-5 mm long. In flower May-Oct.

Common in *salt marshes* along the coast. Here known from Anaheim Bay in Seal Beach, Bolsa Chica Ecological Reserve, upper Newport Bay, and San Mateo Creek estuary. A common host of **goldenthread dodder** (*Cuscuta pacifica* var. *pacifica,* Convolvulaceae).

The epithet was given in 1831 by C.F. Lessing for its fleshy stems and leaves (Latin, carnosus = fleshy, pulpy).

Fleshy Jaumea
Orange strands are Goldenthread Dodder

Annual Daisies: *Lasthenia, Layia, Pentachaeta*

These are slender-stemmed annuals with yellow and white flowers, their identities sometimes confused with each other.

Goldfields (*Lasthenia* spp.) are *delicate-stemmed* upright annuals often found in great numbers *in seasonally moist soil depressions. Leaves are opposite,* usually undivided, rarely pinnate. Pollinated by beetles, moths, flies, and bees. *They appear nearly identical face-on, but their phyllaries are quite different.*

Southern Goldfields

Southern Goldfields, Royal Goldfields
Lasthenia coronaria (Nutt.) Ornduff

A *sweet-smelling annual* nearly to 40 cm tall with many or no branches, generally *hairy,* often with sticky-tipped hairs. Leaves 1.5-6 cm long, linear, *undivided or 1-pinnate, often hairy and sticky. This is our only goldfields that may have pinnate leaves.* Phyllaries 6-14, with sticky-tipped hairs, *not fused.* In flower Mar-May.

A generally uncommon plant of coarse sand. Widespread in the Lomas de Santiago, such as in Bee Canyon, Limestone Canyon, and Whiting Ranch Wilderness Park. Long ago known from Chino Hills above the Santa Ana River channel, upper Newport Bay, and near the Newport Beach Country Club. Inland sites include Alberhill and lower Horsethief Canyon.

The epithet was given in 1841 by T. Nuttall for its "small crown of diminutive leaflets," today called phyllaries (Latin, coronarius = pertaining to a crown).

Similar. Coulter's goldfields (*L. glabrata* ssp. *coulteri*): hairless, not sticky; leaves linear; basal two-thirds of the phyllaries are fused together; rare. **Coastal goldfields** (*L. gracilis*): hairy, not sticky; leaves linear; phyllaries not fused; common.

Southern Goldfields | Coulter's Goldfields

Coulter's Goldfields ★
Lasthenia glabrata Lindl. **ssp. *coulteri*** (A. Gray) Ornduff

Leaves 4-14 cm long, *linear, undivided, hairless.* Phyllaries 10-14, *all fused together for the first two-thirds of their length, hairless.* In flower Mar-May.

A plant of salt marshes, alkali flats, and vernal pools. Formerly widespread in Orange County, now mostly extirpated. Old sites include West Coyote Hills, the former Bryant Ranch (site of present-day Los Alamitos, Cypress, and Garden Grove), Bolsa Chica Ecological Reserve, and Costa Mesa. Most recent records (Pelican Point in Crystal Cove State Park, Weir Canyon, Lake Forest, and Mission Viejo) appear to have originated from hydroseeding. Almost completely wiped out of its already small range along the coast, it now occurs almost exclusively in western Riverside County. Known from the Lake Elsinore area, primarily near Nichols Road wetlands adjacent to Alberhill Creek. CRPR 1B.1.

The epithet was given in 1835 by J. Lindley because it is hairless (Latin, glabratus = to make smooth). Our form was named in 1884 by A. Gray to honor T. Coulter.

Similar. Southern goldfields (*L. coronaria*): hairy and sticky; leaves pinnate; phyllaries not fused; uncommon. **Coastal goldfields** (*L. gracilis*): hairy, not sticky; leaves linear; phyllaries not fused; common.

Coulter's Goldfields

Coastal Goldfields

Lasthenia gracilis (DC.) Greene
[*L. californica* Lindl., *L. chrysostoma* (Fischer & C. Meyer) Greene]

An *upright annual* nearly to 40 cm tall with few or no branches, *generally hairy. Leaves* 0.8-7 cm long, *linear, hairy. Phyllaries 4-13, not fused, hairy.* In flower Mar-May.

Abundant in moist soils among grasslands and coastal sage scrub throughout our area except higher elevations. In many areas, however, it is being eliminated by development and unnecessary plowing. Scenic populations once adorned our sea bluffs from Corona del Mar to Dana Point Headlands and San Clemente, all gone now. Known from upper Newport Bay, San Joaquin Hills, UCI Ecological Preserve (formerly common throughout the UCI campus), Temple Hill, Carbon Canyon in the Chino Hills, upper Black Star Canyon, Limestone Canyon, Arroyo Trabuco, Trabuco Canyon, and San Juan Canyon. Still common in many parts of Caspers Wilderness Park and Audubon Starr Ranch Sanctuary, especially after fire.

The epithet was given in 1836 by A.P. de Candolle for its slender stems (Latin, gracilis = slender).

Similar. Southern goldfields (*L. coronaria*): hairy and sticky; leaves pinnate; phyllaries not fused; uncommon.
Coulter's goldfields (*L. glabrata* ssp. *coulteri*): hairless, not sticky; leaves linear; basal two-thirds of the phyllaries are fused together; rare.

Coastal Goldfields

Coastal Goldfields

Coastal Goldfields

Common Tidy-tips

Common Tidy-tips

Common Tidy-tips

Layia platyglossa (Fischer & C. Meyer) A. Gray

A *many-branched annual* 3-70 cm tall *with soft hairs and sticky stems. Leaves* nearly 10 cm long, *linear, hairy, sticky, lower leaves lobed, alternate. Flowerheads large,* 2-3 cm across. *Phyllaries all-green, white-hairy at the base, not spine-tipped.* Disks 6-120 or more per head, 3.5-6 mm long, yellow, the anthers purple. *Rays* 5-18, the strap 3-20 mm long, *very wide, yellow with white tips.* In flower May-June.

A stunning beauty of grasslands and openings among coastal sage scrub and chaparral. Once abundant along the coast and its sea bluffs, largely eliminated by urbanization. Known from upper Newport Bay, Dana Point Headlands, Telegraph Canyon in the Chino Hills, lower Santa Ana Canyon, throughout the Lomas de Santiago, north of Irvine Regional Park, near O'Neill Regional Park, on the slopes of San Juan Canyon, and in grasslands below Elsinore Peak. Inland, it is known from the Bedford Canyon forest gate and within Temescal Valley, such as in Horsethief Wash, Glen Ivy Hot Springs, and lowlands near Lake Elsinore.

The epithet was given in 1836 by F.E.L. von Fischer and C.A.A. von Meyer for its broad ray florets (Greek, platys = flat, broad, wide; glossa = tongue).

Similar. Allen's daisy ★ (*Pentachaeta aurea* ssp. *allenii*): *leaves very narrow; rays very narrow,* their tips with (0-)3 tiny teeth; once common, now rare.

Common Tidy-tips

Allen's Daisy ★
Pentachaeta aurea Nutt. **ssp. *allenii*** D.J. Keil

An *upright few-branched slender annual* 5-36 cm tall, with short hairs. *Leaves narrow,* 1-5 cm long, 2-3 mm wide, *linear, often hairy.* Flowerheads 1-2.5 cm across. *Phyllaries greenish to reddish with a darker central line, their edges transparent, tipped with a slender spine.* Disks 30-90 per head, yellow. Rays 14-52, the tube yellow, the strap 3-12 mm long, narrow, tipped with 3 very tiny teeth. *The strap is bright yellow at the base; near our coast they are white at the tips, in our foothills nearly the entire strap is white.* Each brown fruit is topped with 5 (to 8) easily seen broad-based pappus bristles. In flower Mar-June.

This is a local form of **Golden daisy** ★ (*Pentachaeta aurea* Nutt. ssp. *aurea*), which has totally golden yellow rays. It is found in the counties of Los Angeles, San Bernardino, Riverside, San Diego, and Baja California, Mexico. Allen's daisy is found only in Orange County.

A plant of heavy clay soils on coastal bluffs, open mesas, and dry hillsides at fairly low elevations, most common after fire, now mostly eliminated by urbanization. It was first discovered by Leroy Abrams in 1901 near the train station in historic El Toro (southwest of El Toro Road at Muirlands Boulevard in present-day Lake Forest). Others were known from the southern Chino Hills overlooking Santa Ana River channel (last reported in 1935, probably extirpated by development). A population from Sierra (= Coal) Canyon in 1920 has not been relocated. Those on the Dana Point Headlands appear to have been eradicated by development in 2005-2006. It is still known from the San Joaquin Hills, such as in Shady Canyon, Camarillo Canyon, Sycamore Hills, and Laguna Coast Wilderness Park. Also found in Fremont Canyon, lower Borrego Canyon behind former Marine Corps Air Station El Toro, Glenn Ranch in Canada de Los Alisos, Santiago Canyon, Limestone Canyon just west of "The Sinks", and atop Pruesker Peak on Audubon Starr Ranch Sanctuary. CRPR 1B.1.

The epithet was given in 1840 by T. Nuttall for its golden yellow rays (Latin, aureus = golden). Our form was named in 2007 by D.J. Keil in honor of R.L. Allen, one of his former students and lead author of this book.

Similar. Golden daisy ★ (*Pentachaeta aurea* ssp. *aurea*): *rays all golden yellow;* fruits yellow; known from Lake Elsinore (last reported in 1920, probably extirpated), western summit of Miller Mountain (San Diego County), and Santa Rosa Plateau; CRPR 4.2. **Common tidy-tips** (*Layia platyglossa*): leaves broader, often pinnately lobed; flowerheads large; *rays broad* and clearly 3-lobed; widespread, common.

Lower left: Allen's daisy (smaller), common tidy-tips (larger). Lower center and right: Dana Point Headlands population with gentle transition from yellow to white. Note the phyllaries: red or green with transparent edges and spiny tips.

Allen's Daisy

Allen's Daisy

Golden Daisy, Santa Rosa Plateau

Allen's Daisy / Common Tidy-tips

Allen's Daisy, Dana Point

Allen's Daisy, Dana Point

Spanish Sunflower ✖
Pulicaria paludosa Link
[*P. hispanica* (Boiss.) Boiss.]

An *upright hairy perennial or annual* up to 1 meter tall. It grows from a gnarled underground tuber that is difficult to remove. *Leaves* 1-3 cm long, *linear, the base clasps the stem, alternate.* Its *small flowerheads* are at the top of the plant. Disks and rays are numerous, yellow, the *rays relatively short*. In flower all year.

A very *invasive* plant of waterways and standing water in wild and urban areas. Native to southern Europe, rapidly spreading throughout our area.

The epithet was given in 1806 by J.H.F. Link for its semi-aquatic habitat (Latin, paludosus = swampy).

Spanish Sunflower

Numerous very short ray florets

California Tickseed

California Tickseed

California Tickseed
Leptosyne californica Nutt.
[*Coreopsis californica* (Nutt.) H.K. Sharsm. var. *californica*]

An *upright, weak-stemmed, hairless annual* 5-30 cm tall. *Leaves* 2-10 cm long, 0.5 cm wide, *mostly basal, linear, red-tipped,* undivided or with small pinnate lobes. *Flowerheads* 1-3.5 cm across, *solitary, on tall leafless stalks. Phyllaries of two types: outermost narrow and green, innermost broad and green to yellow.* Disks 10-30 per head, yellow to orangish. Rays 5-12, broad, oval, bright yellow. In flower Mar-May.

Found in rocky or sandy soils, in openings among chaparral, grasslands and washes, sometimes after fire. *Scattered and generally uncommon here, mostly a desert plant.* Known sparingly from the coastal side of the Santa Ana Mountains: Cristianitos and Gabino Canyons on the Rancho Mission Viejo, the ridge overlooking San Juan Creek in Caspers Wilderness Park, and Fisherman's Camp/Bluewater Canyon in the San Mateo Canyon Wilderness Area. More common in the Temescal Valley, such as near Corona, lower Indian Canyon wash, Glen Ivy, Alberhill, and Lake Elsinore.

The epithet was given in 1841 by T. Nuttall for the place of its discovery.

Similar. Giant tickseed, Dr. Seuss plant (*L. gigantea* Kellogg [*Coreopsis g.* (Kellogg) H.M. Hall]): *Upright perennial with a thick trunk, 30 cm-2 meters tall;* leaves 3-4-pinnate, *leaflets narrow, carrot-like;* flowers large, numerous. In flower Jan-May. San Onofre bluffs, introduced by revegetation projects, now established and reproducing. **California butterweed** (*Senecio californicus*): annual; inflorescence branching; *phyllaries all identical, numerous, green, tipped with black; rays slender.*

Giant Tickseed

California Tickseed

Giant Tickseed

California Butterweed
Senecio californicus DC.

An *annual* 10-40 cm tall, *hairless or with scattered long hairs.* Stems slender, branched usually near the base. *Leaves* 2-7 cm long, *linear to oval, lowermost usually pinnate, upper leaves toothed or smooth-edged, with a clasping or ear-like base. Flowerheads large, on branched stalks,* rarely solitary. *Phyllaries upright, all similar, green, tipped with black.* Disks several per head, yellow to orangish. Rays 13, about 1 cm long, often curved backward, slender, bright yellow. In flower Mar-May.

  Uncommon here, found in openings among coastal sage scrub and chaparral along the coast, inland to mountain foothills. Its occurrence is not predictable; it generally does not show up in the same place every year. Known from Moro Canyon in Crystal Cove State Park, Dana Point Headlands, Chino Hills above Santa Ana River channel, Weir Canyon, upper Blind and Baker Canyons, Lomas de Santiago, lower Santiago Canyon, Rancho Mission Viejo, Caspers Wilderness Park, and Audubon Starr Ranch Sanctuary.

  The epithet was given in 1838 by A.P. de Candolle for the place of its discovery.

  Similar. California tickseed (*Leptosyne californica*): flowerheads on individual leafless stalks; *phyllaries of two types: outermost narrow and green, innermost broad and green to yellow; none black-tipped;* rays broad, oval.

California Butterweed

California Butterweed

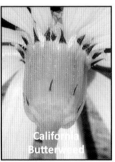
California Butterweed

Sand-wash Butterweed
Senecio flaccidus Less. **var. *douglasii*** (DC.) B. Turner & T. Barkley

An *open-branched shrub* 1-1.6 meters tall, covered with soft woolly hairs, blue-gray to blue-green in color. *Leaves* 3-10 cm long, *usually drooping, narrowly linear, often pinnate,* the lobes long and slender. Flowerheads often in dense clusters, each about 4 cm across. *Phyllaries* upright, 16-21 long ones, some smaller ones; all are green or grayish, *none black-tipped.* Disks yellow to orangish, rays 8-13, the strap 1-2 cm long, bright yellow. In flower June-Oct.

  A common plant of hot sunny locations, typically among coastal sage scrub, chaparral, oak woodland, and coniferous forests, *commonly growing from cracks in boulders, in sandy washes, and sometimes on road cuts.* Known from the West Coyote Hills, Chino Hills above and within the Santa Ana River channel, lower Santiago Creek, Audubon Starr Ranch Sanctuary, San Juan Canyon, along Ortega Highway above Lake Elsinore, Indian Truck Trail, many places along Main Divide Truck Trail, south into the San Mateo Canyon Wilderness Area. It is easiest to see in Silverado Canyon right along the Maple Springs Motorway, within 3 miles of the forest gate.

  The epithet was given in 1830 by C.F. Lessing for its drooping leaves (Latin, flaccidus = relaxed). Our form was named in 1838 by A.P. de Candolle to honor D. Douglas.

Sand-wash Butterweed with Virginia Lady Butterfly

Sand-wash Butterweed

Goldenrods: *Euthamia, Solidago*

Perennials that spread from underground creeping rhizomes or stems. *Leaves alternate,* the edges smooth or toothed. Their *many flowerheads* appear in a cone-shaped inflorescence at the upper branches. Phyllaries are of different lengths, each narrow-pointed. Corollas bright golden yellow. Generally bloom in late summer to fall.

Western Goldenrod

Euthamia occidentalis Nutt. [*Solidago occidentalis* (Nutt.) Torr. & A. Gray]

A *hairless perennial that stands straight up* to nearly 2 meters tall. *Leaves 4-10 cm long, narrowly linear* or lance-shaped, hairless, often sticky. *Flowerheads arranged in sparse many-branched clusters.* In flower July-Oct.

A delightful plant to encounter, it emerges from the soil in late spring and grows through summer until *it towers near or over your head.* It occurs in creekbeds, meadows, marshes, and willow thickets. Known from Bolsa Chica Ecological Reserve, San Joaquin Freshwater Marsh, San Diego Creek, San Juan Canyon, much of the Santa Ana River channel, and San Mateo Creek.

The epithet was given in 1840 by T. Nuttall for its western distribution (Latin, occidentalis = western).

Similar. California goldenrod (*Solidago velutina* ssp. *californica*): generally much shorter; green-grayish, covered with short hairs, often leans over; leaves spoon-shaped; rays broader; abundant in drier soils.

Western Goldenrod

California Goldenrod

California Goldenrod

Solidago velutina D.C. **ssp. *californica*** (Nutt.) Semple [*S. californica* Nutt.]

An *upright perennial* from short rhizomes, it stands 20 cm-1.5 meters but often leans over and looks shorter. *Covered with short hairs that often make the stems and leaves look gray. Lower leaves 5-12 cm long, spoon-shaped or oval, bases tapered.* Upper leaves similar. *Flowerheads arranged in a one-sided cluster at the branch tips.* In flower July-Oct.

Common and widespread in our area; usually appears partly under other plants, often near the border between grasslands and other communities such as oak woodlands, coastal sage scrub, and chaparral, even right atop Santiago Peak. Easily found in Laguna Coast Wilderness Park, Limestone Canyon, Silverado Canyon, Caspers Wilderness Park, and Blue Jay Campground. In the Lomas de Santiago, it is parasitized by goldenrod broom-rape (*Orobanche* sp., Orobanchaceae).

The epithet was given in 1840 by T. Nuttall for the state of its discovery.

Similar. Western goldenrod (*Euthamia occidentalis*): generally much taller; green, hairless, straight up; leaves linear; inflorescence crowded at branchtips; rays narrow; less common, in wetter soils.

California Goldenrod

LIGULIFLOROUS HEADS

All of our species ooze a milky white sap, easily seen when cut or broken. Characteristics of the phyllaries and seeds are quite helpful in their identification.

Mountain Dandelions: *Agoseris*

Annuals or perennials from a stout taproot, with long basal leaves and milky sap. The large flowerheads are atop a long leafless stalk. Phyllaries overlap, with a long narrow tip that often curves outward, hairy. All corollas are yellow. The flowers are open only in the early morning. You are more likely to see their *large fluffy fruiting heads* than open flowers. The *seed has a slender stalk, topped with a plume (pappus bristles) of soft narrow white bristles* like **common dandelion** ✖ (*Taraxacum officinale* F.H. Wigg., not shown). Our 3 species are certainly more common and widespread than the sparse field and herbarium records show; they are often overlooked.

The genus was established in 1817 by S.C. Rafinesque-Schmaltz, who used the Greek name for goat chicory, presumably a similar-looking European species (Greek, agos = chief; series = endive or lettuce).

In *Agoseris*, each fruit (cypsela) bears a long slender stalk, topped with several narrow white pappus bristles. Compare these fruits with those of *Microseris* (including *Uropappus* and *Stebbinsoseris*); their fruits *lack* the slender stalk. Instead, their pappus bristles are broad and attached directly to the top of the cypsela.

Spear-leaved Mountain Dandelion

When in fruit, the pappus bristles of mountain dandelions form a large sphere, soft to the touch and easily taken by the wind.

Large-flowered Mountain Dandelion

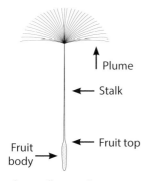

Large-flowered
Agoseris grandiflora
Fruit top is *gradually narrowed* to the stalk.

Woodland
Agoseris heterophylla
Fruit top is *fairly broad-shouldered;* the fruit body can be very broad.

Spear-leaved
Agoseris retrorsa
Fruit top is *abruptly broad-shouldered.*

Large-flowered Mountain Dandelion
Agoseris grandiflora (Nutt.) E. Greene

A *large perennial* 25-85 cm tall, completely covered with soft white non-sticky hairs. *Leaves highly variable in size and shape,* linear to reverse-lance shaped, undivided or deeply lobed, *most lobes pointing toward the* **tip** *of the leaf. Petiole generally purple.* Flowerheads 15-40 mm tall. When mature, the *fruit top is gradually tapered* up to the stalk. Fruit body slender. Plumes 7-15 mm long. In flower May-July.

An occasional plant of slopes, meadows, openings in chaparral, and oak woodlands. Known from the slopes of Santiago Peak, Silverado Canyon, Los Pinos Potrero, San Juan Trail, Sitton Peak Truck Trail, and southward in the San Mateo Canyon Wilderness Area.

The epithet was given in 1841 by T. Nuttall for its large flowers (Latin, grandis = large; floris = flower).

Large-flowered Mountain Dandelion

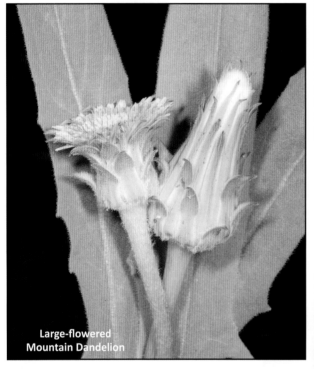
Large-flowered Mountain Dandelion

Woodland Mountain Dandelion
Agoseris heterophylla (Nutt.) E. Greene **var. *heterophylla***

An *annual* 5-40 cm tall, *covered with white hairs. Leaves* are *reverse-lance-shaped, undivided or few-toothed;* sometimes with 2-3 pairs of narrow lobes, each lobe 1.5-16 mm wide. *Petiole not purple. Flowerheads are only 8-25 mm tall.* The base of the flowerhead is densely covered with yellow to reddish sticky-tipped hairs. Flowers open only in the morning. When mature, *the fruit top is fairly broad-shouldered* where it meets the stalk (sometimes tapered). *The fruit body can be wide or narrow.* Outermost fruits are generally quite different from the innermost fruits. Plumes 4-9 mm long. In flower Apr-July.

An uncommon plant of grasslands, burns, and openings in coastal sage scrub and oak woodlands. Known from Mission Viejo, Caspers Wilderness Park, San Juan Canyon, and Los Pinos Potrero, certainly elsewhere.

The epithet was given in 1841 by T. Nuttall for its variable leaf shapes (Greek, hetero = other, different; phyllon = leaf).

Woodland Mountain Dandelion

Spear-leaved Mountain Dandelion

Agoseris retrorsa (Benth.) E. Greene

A *perennial* 10-50 cm tall, completely covered with soft white non-sticky hairs. *Leaves highly variable in size and shape,* linear to reverse-lance shaped, undivided or deeply lobed, *most lobes point toward the **base** of the leaf. Petiole often purple.* Flowerheads 20-60 mm tall. When mature, the *fruit top is broad-shouldered* where it meets the stalk. Fruit body slender. Plumes (11-) 15-20 mm long. In flower May-Aug.

Found in openings among chaparral, oak woodlands, coniferous forests, and talus slopes, generally at elevations higher than our other mountain dandelions. Known from many places along Main Divide Truck Trail, Modjeska and Santiago Peaks.

The epithet was given in 1849 by G. Bentham for its backward-pointing leaf lobes (Latin, retrorsus = bent or turned backward).

Spear-leaved Mountain Dandelion

Spear-leaved Mountain Dandelion

Crete Weed

Crete Weed ✖

Hedypnois cretica (L.) Dum.-Courset.

A *low-growing annual* up to 40 cm tall and wide, with many spreading branches and milky sap. *The entire plant is covered with short rough bristles.* Leaves 5-18 cm long, roughly oblong. Flowerheads solitary or in clusters. Phyllaries linear and strongly curved in toward the center of the flowerhead. *Corollas yellow, often tipped with brown.* In fruit, the pappus on each cypsela consists of short scales. Cypselae in the middle of the head have a single very long bristle atop each scale. In flower Mar-May.

An *abundant weed* of open areas, fire roads, and trails throughout our area, especially in heavy clay soils. Even found on the beaches of Balboa Peninsula. *Since the pappus are not plumed, they cannot fly to disperse seeds; fruits attach themselves to passing animals and hikers.* Help reduce the spread of this weed by removing fruits from your shoes and clothing and disposing fruits in the trash.

The epithet was given in 1753 by C. Linnaeus for the plant's original home (Latin, creticus = of the island of Crete).

Crete Weed

Smooth Cat's Ear ✖

Hypochaeris glabra L. [*Hypochoeris glabra* L.]

A *hairless* (rarely hairy) *annual* with milky sap and a deep taproot. *Leaves 2-10 cm long, all basal, edges smooth or shallow-lobed. Small flowerheads atop a long leafless stalk.* Phyllaries small, few. Ligules short, about equal to the longest phyllaries. *Fruits from the outermost florets have bristles directly attached; those from the inner florets have bristles attached to a long stalk ("beak").* In flower Mar-June.

An abundant weed of all wild lands, particularly grasslands, disturbed soils, trailsides, home gardens. Native to Europe.

The epithet was given in 1753 by C. Linnaeus for its hairless leaves (Latin, glabro = smooth, hairless).

Similar. Hairy cat's ear ✖ (*H. radicata* L., lower two flowerheads): low perennial or geophyte; *covered with rough hairs;* flowerheads and ligules larger; in flower Mar-June; uncommon, mostly in sandy soils near the coast, sometimes an urban weed.

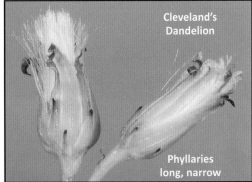

Cleveland's Dandelion

Malacothrix clevelandii A. Gray

An *upright, hairless, delicate annual,* 4–36 cm tall. *Basal leaves 3-10 cm long, crowded, long and narrow, with teeth or short lobes;* often withered at flowering time. Stem leaves much smaller, sparse. *Flowerheads small,* 4-8 mm long, 2.5-3.5 mm in diameter; open only briefly, sometimes not fully opening, possibly self-pollinated. Phyllaries often tipped with brown or purple. Outermost phyllaries are half the length of the inner ones. *Ligules and anthers yellow.* In flower Apr-June.

An uncommon plant of *chamise chaparral,* generally *in decomposed granitic soils.* More common after fires or landslides. Known locally from slopes above Lake Elsinore, Bear Canyon Trail, Round Potrero, northern Teneja Trail, and Bluewater Trail.

The epithet was given in 1876 by A. Gray in honor of D. Cleveland, who collected it "near San Diego" in 1875.

Cliff Aster ★

Malacothrix saxatilis (Nutt.) Torr. & A. Gray **var. *saxatilis*** (Nutt.) Torr. & A. Gray.

A *perennial* 30-90 cm tall with milky sap, hairless to minutely haired. *Basal leaves lobed or not, withered before flowering.* Stem leaves 3-10 cm long, lance-, spatula-shaped, or oval, usually smooth-edged, sometimes toothed, *succulent and leathery*. The inflorescence has several wiry branches. Flowerheads 3-4 cm across, *corollas white tinged with pink, outer surface streaked with pink*. In flower Mar-Sep.

A rare plant of sea bluffs. Known only from Dana Point, near Salt Creek Park and Ritz Carlton Hotel. CRPR 4.2. Sometimes visible from Dana Point Marine Reserve.

The epithet was given in 1841 by T. Nuttall for its association with rocks (Latin, saxatile = growing among rocks).

Similar. Slender-leaved cliff aster (*M. saxatilis* var. *tenuifolia*): leaves with long slender lobes; corollas white to pinkish; coastline to mountains, *not sea bluffs;* abundant.

Cliff Aster

All three photos: Fred M. Roberts, Jr.

Slender-leaved

Slender-leaved

Cliff Aster

Slender-leaved Cliff Aster

Malacothrix saxatilis (Nutt.) Torr. & A. Gray **var. *tenuifolia*** (Nutt.) A. Gray

A *stout perennial,* 30 cm *to about 2 meters tall* with milky sap, mostly hairless. In its early stages, its basal leaves are often round-lobed and edged with red, these wither by flowering time. *Stem leaves more or less linear, with long narrow lobes and a pointed tip.* The inflorescence has several wiry branches. Flowerheads 3-4 cm across, corollas white to pinkish, the outer surface is sometimes streaked with pink or red. In flower all year, mostly summer.

An abundant plant of hillsides, disturbed soils, and road cuts. Known from Carbon Canyon in the Chino Hills, Santa Ana River channel, Bolsa Chica Ecological Reserve, Upper Newport Bay, throughout the San Joaquin Hills such as Crystal Cove State Park and Niguel Hill, Aliso Viejo, Maple Springs Road, and along Main Divide Truck Trail above Falcon Group Camp.

The subepithet was given in 1841 by T. Nuttall for its slender leaf lobes (Latin, tenuis = thin; folium = leaf).

Similar. Cliff aster ★ (*M. saxatilis* var. *saxatilis*): leaves linear, not lobed; corollas pinkish; *sea bluffs only,* known from Dana Point Headlands, near Salt Creek Park, and Ritz Carlton Hotel.

Slender-leaved Cliff Aster

Silverpuffs and Brownpuffs: *Microseris* [including *Stebbinsoseris* and *Uropappus*]

Small annuals or perennials with milky sap. They are covered with small white hairs that dry out and look like tiny white scales. *Leaves are mostly basal, often pinnate.* Most flowerheads droop ("nod") when in bud. *Phyllaries green, often with a central pink or red stripe and/or tip; of two lengths: the outer much shorter, the inner longer and often edged with black hairs.* Corollas yellow to white, quite similar. *Each cypsela is directly topped with several slender pappus bristles that are broad at their base.* In the wild, you more often see the seedheads than the short-lived flowers. Most are fairly common plants of grasslands and vernal pools.

Some botanists place two of our species into the genera *Stebbinsoseris* and *Uropappus* on the basis of minor characteristics, a move we feel is unnecessary.

Similar. Mountain dandelions (*Agoseris* spp.): flowerheads upright in bud; a slender stalk on each cypsela, topped with a plume of narrow white bristles.

	Douglas's *Microseris d.* ssp. *douglasii*	San Diego *Microseris d.* ssp. *platycarpha*	Elegant *Microseris elegans*	Grassland *Microseris heterocarpa*	Lindley's *Microseris lindleyi*
Corolla	Yellow or white	Yellow or white	Yellow or orange	Yellow or white	Yellow
Fruit body	4-10 mm long; even width or slightly tapered to tip; brown	3-4.5 mm long; even width or broader at tip; brown	1.5-3 mm; even width or broader at tip; brown to tan	4.5-12 mm long; even width; dark purplish-brown	7-17 mm long, tapered to tip, black to dark brown
Pappus scales	Short, only 25% the length of the central bristle; silvery to blackish	Long, over 50% the length of the central bristle (can be longer or shorter); silvery to blackish	Short, only 25% the length of the central bristle; white or brownish	Long; light brown	Long; silvery
Scale Body Tip	Tapered	Tapered	Tapered	Notched	Notched
Central Bristle	3-8 mm long; stout and barbed	1-4 mm long; stout and barbed	3-5 mm long; narrow, barely barbed	3-8 mm long; narrow, finely barbed near tip	4-6 mm long; smooth, not barbed
Notes	Buds *upright* before opening	Buds *nod* before opening	Buds *nod* before opening	Buds *nod* before opening	Buds *upright* before opening

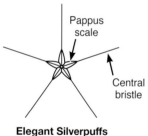

Elegant Silverpuffs
Microseris elegans

Douglas's Silverpuffs
Microseris douglasii
ssp. *douglasii*

San Diego Silverpuffs
Microseris douglasii
ssp. *platycarpha*

Grassland Brownpuffs
Microseris heterocarpa
[*Stebbinsoseris heterocarpa*]

Lindley's Silverpuffs
Microseris lindleyi
[*Uropappus lindleyi*]

San Diego Silverpuffs ★
Microseris douglasii (DC.) Sch.-Bip. **ssp. *platycarpha*** (A. Gray) K.L. Chambers

An annual, 5-60 cm tall. Leaves 3-25 cm long. Corollas yellow or white. *Fruit body short,* only 3-4.5 mm long. Pappus scales broad-based, upward nearly half of the central bristle's length. The broad portion is generally half of the central bristle's length but can be up to three-quarters its length, rarely less than half. Central bristles stout and barbed. In flower Mar-May.

Overall, a rare species though relatively common here. Found in clay soils among grasslands, vernal pools, and openings in coastal sage scrub. Known from the Fairview Park vernal pool complex in Costa Mesa, throughout the San Joaquin Hills such as on the UC Irvine campus, Quail Hill, the Dana Point Headlands, and Aliso and Wood Canyons Wilderness Park. Also in the Chino Hills (between Telegraph Canyon and Blue Mud Canyon), Lomas de Santiago (Limestone Canyon), Blind Canyon, Fremont Canyon, North Mission Viejo, Richard and Donna O'Neill Conservancy, and Alberhill in Temescal Valley. Ranked CRPR 4.2, threatened by urban development and non-native plants.

The epithet for this form was given in 1857 by A. Gray for its broad pappus scales (Greek, platys = flat, broad, wide ; karphos = twig or chaff).

Similar. Douglas's silverpuffs (*M. douglasii* ssp. *douglasii*, not shown) and **Elegant silverpuffs** (*M. elegans*): pappus scales broad-based only to about one-quarter of the central bristle's length; Douglas's silverpuffs and elegant silverpuffs are uncommon here.

San Diego Silverpuffs

Buds nod before opening

San Diego Silverpuffs

San Diego Silverpuffs

Elegant Silverpuffs
Microseris elegans A. Gray

An upright annual, 5-35 cm tall. Leaves 2-20 cm long, usually lobed. Corollas yellow or orange. *Fruit body very short,* only 1.5-3 mm long, widest at tip, brown to blackish. Pappus scales broad-based, upward only to about one-quarter of their length. Central bristles narrow, barely barbed. In flower Apr-June.

A plant of heavy clay soils in native grasslands and near vernal pools. Uncommon here, recorded from the summit and upper south flank of Miller Mountain.

The epithet was given in 1884 by A. Gray perhaps for the elegant appearance of its seedheads or its slender pappus bristles (Latin, elegans = neat, elegant).

Similar. Douglas's silverpuffs (*M. douglasii* ssp. *douglasii*): pappus scales similar to elegant but *fruit body larger,* 4-10 mm long; central pappus bristles stout and barbed; not confirmed from our area.

Elegant Silverpuffs

Buds nod before opening

Grassland Brownpuffs
Microseris heterocarpa (Nutt.) K.L. Chambers
[*Stebbinsoseris heterocarpa* (Nutt.) K.L. Chambers]

An annual, 8-60 cm tall. Leaves 5-35 cm long, pinnately toothed or lobed. *Nodding in bud, upright in flower.* Corollas white or yellow. *Each dark purplish-brown fruit is topped with 5 broad-based light brown pappus bristles*, each notched inward at the tip and topped with a long (sometimes short) slender central bristle. In flower Mar-May.

At times abundant in sandy-clay soil among grasslands. Known from San Joaquin Freshwater Marsh, UCI Ecological Preserve, hills of Dana Point, blufftops in San Clemente, Richard and Donna O'Neill Conservancy, Riley Wilderness Park, Audubon Starr Ranch Sanctuary, Arroyo Trabuco, Fremont Canyon, Chino Hills, Temescal Valley, Elsinore Basin, Santa Rosa Plateau, and the San Mateo Canyon Wilderness Area. The narrow lobes produced by the notched pappus tip are often held against the central bristle, especially when wet, making them difficult to see. Gently brush the pappus tip to make them pull away from the bristle.

The epithet was given in 1841 by T. Nuttall for its unusual pappus bristles, different from related plants known at the time (Greek, heteros = other, different; karpos = fruit).

Similar. Lindley's silverpuffs (*M. lindleyi*): fruit black or dark brown, pappus bristles silvery, also notched.

Grassland Brownpuffs

Buds nod before opening

Grassland brownpuffs photos: Fred M. Roberts, Jr.

Grassland Brownpuffs | Lindley's Silverpuffs

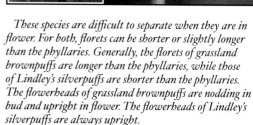

These species are difficult to separate when they are in flower. For both, florets can be shorter or slightly longer than the phyllaries. Generally, the florets of grassland brownpuffs are longer than the phyllaries, while those of Lindley's silverpuffs are shorter than the phyllaries. The flowerheads of grassland brownpuffs are nodding in bud and upright in flower. The flowerheads of Lindley's silverpuffs are always upright.

Lindley's Silverpuffs
Microseris lindleyi (DC.) Nutt.
[*Uropappus lindleyi* (DC.) A. Gray]

An annual 5-70 cm tall. Leaves 5-30 cm long, smooth-edged or pinnate with narrow lobes. *Upright in bud and in flower.* Flowerheads narrow, corollas yellow or white. *Each blackish (or dark brown) fruit is topped with 5 broad-based silvery pappus bristles*, each notched inward at the tip and topped with a long slender central bristle. In flower Apr-June.

Found in clay soils among grasslands, coastal sage scrub, and oak woodlands, sometimes after fire. Known from the San Joaquin Hills as on the UCI campus, and in Trabuco Canyon, Caspers Wilderness Park, Audubon Starr Ranch Sanctuary, Verdugo Canyon, San Juan Canyon, and the San Mateo Canyon Wilderness Area. As with grassland brownpuffs, you may need to gently brush the pappus tip in order to make the narrow pappus lobes pull away from the central bristle.

The epithet was given in 1838 by A.P. de Candolle in honor of J. Lindley.

Similar. Grassland brownpuffs (*Microseris heterocarpa*): fruit dark purplish-brown, pappus bristles light brown, also notched.

Lindley's Silverpuffs

Buds upright before opening

California Chicory
Rafinesquia californica Nutt.

An upright *hairless annual*, 20 cm to about 1.5 meters tall, with a rubbery texture and milky sap. *Lowest leaves often in a basal rosette*, not always withered at flowering time. Leaves 3-15 cm long, oblong in outline, toothed or with several wide lobes. *Upper leaves clasp the stem*. Inflorescence 1- to few-branched. Phyllaries green, often with a central pink or red stripe; of two types: innermost very long; outermost very short, often curved out- or backward. Corolla white to cream, often pink-tinged, sometimes pink-striped on the lower surface. In flower Apr-July.

A common plant of burns and disturbed soils along trails and fire roads, mostly in coastal sage scrub and chaparral. Known throughout the San Joaquin Hills, Santa Ana Mountains, Temescal Valley, and Elsinore Basin.

The epithet was given in 1841 by T. Nuttall for the place of its discovery.

Similar. **Wreath-plants** (*Stephanomeria* spp.): slightly or very hairy; leaf base does not clasp the stem; inflorescence with long slender wiry branches.

California Chicory

Wreath-plants, Wire Lettuce: *Stephanomeria*

Upright annuals or perennials with *long, wiry, wand-like stems*. Leaves are alternate and undivided; smooth-edged, toothed, or pinnately lobed. *Leaf bases generally do not clasp the stem*. Flowerheads numerous. Fruit 5-sided, topped with plumed bristles. Common in summer and fall.

We have 4 species here. **Chicory-leaved wreath-plant** (*Stephanomeria cichoriacea*) is a perennial that lives in rocky soils, mostly at higher elevations of the Santa Ana Mountains. It has a large pink flower and is easy to identify. The other 3 species resemble each other and differ mostly in their fruits. Each branch of the inflorescence usually has open flowers toward the tip and older flowers near the base. Examine older flowers to find those in fruit. You'll need magnification to study the fruit's pappus bristles and plumes and the fruit body's grooves and surfaces.

Deane's Wreath-plant *S. exigua* ssp. *deanei*	San Diego Wreath-plant *S. diegensis*	Tall Wreath-plant *S. virgata* ssp. *virgata*
Fruit body 5-sided; each side roughly flat and has a narrow groove; each edge has a narrow ridge.	Fruit body 5-sided; each side has a long narrow groove, often interrupted by bumps; each edge is rounded.	Fruit body 5-sided; each side roughly flat (no groove); each edge has a narrow ridge.
Pappus bristles have fallen off, but their bases remain as short scales attached to the fruit body.	Pappus bristles are still attached; after they fall off, no scales will be left on the fruit body.	Pappus bristles are still attached; after they fall off, no scales will be left on the fruit body.

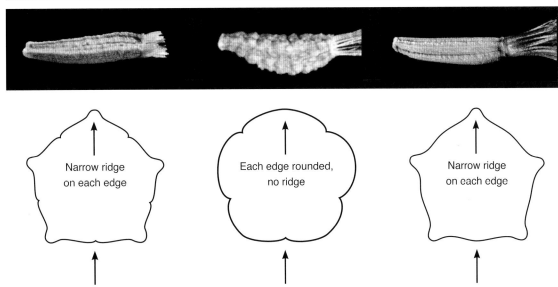

Chicory-leaved Wreath-plant
Stephanomeria cichoriacea A. Gray

A blue-green perennial from an obvious basal rosette. Leaves 5-18 cm long, lance-shaped, usually toothed, covered with white hair. The inflorescence stands 50 cm-1.5 meters tall. *Phyllaries green to pink; of two types: innermost very long; outermost very short, often curved out- or backward as they age.* Each flowerhead with 10-21 florets, *corollas pink.* In flower Aug-Oct.

A common plant of rock outcrops, mostly in higher elevations of the Santa Ana Mountains. Known from Silverado Canyon, San Juan Canyon, many places along North Main Divide Truck Trail, and atop Santiago Peak.

The epithet was given in 1865 by A. Gray for the similarity of its leaves to chicory (*Cichorium intybus* L.), a European plant used as a coffee substitute or flavoring.

Chicory-leaved Wreath-plant

Chicory-leaved wreath-plant has bold pink corollas. Its outermost phyllaries are short and curved backward; they are often dried up by peak flowering time.

Deane's Wreath-plant
Stephanomeria exigua Nutt. **ssp. *deanei*** (J.F. Macbr.) Gottlieb

An upright *many-branched short annual,* generally only 20-60 cm tall. Leaves are lance-shaped, often toothed, uppermost leaves small. Both basal and stem leaves usually wither before flowering time. The inflorescence is covered with a sticky substance. *Phyllaries generally green; of two types: innermost very long; outermost very short, held tight against the innermost, never curved backward.* Flowerheads with only (5-)7-9 florets each. *Corolla white to pink, sometimes blue-tinged; the underside often pink-striped.* In flower May-Sep.

A plant of open sites with dry sandy soils. Generally uncommon in Orange County, known from Crow and upper Bell Canyons on the Audubon Starr Ranch Sanctuary. Once abundant on the Dana Point Headlands. In Riverside County, known from Temescal Valley and foothills above Corona. In the Santa Ana Mountains, it appears north of the larger peaks along North Main Divide Road, then along South Main Divide Road and south into the San Mateo Canyon Wilderness Area.

The epithet was given in 1841 by T. Nuttall for its short stature (Latin, exiguus = short, small). The subepithet was given in 1918 by J.F. Macbride for G.C. Deane, who collected the type specimen in Sweetwater near San Diego, possibly on his property.

Deane's Wreath-plant

Deane's wreath-plant has white to pink corollas, sometimes blue-tinged. Its outermost phyllaries are short and remain flat against the long innermost phyllaries.

San Diego Wreath-plant
Stephanomeria diegensis Gottlieb

A *large green-stemmed annual* 1-2 meters tall, from a stout tap root. The *wiry stems* have many branches and are covered in tiny hairs. The basal leaves wither before flowering time, as may the stem leaves. Leaves small and linear. Phyllaries green, sometimes with pink or red; of two types: innermost very long; outermost very short, often curved out- or backward as they age. *Flowerheads with 11-13 florets each.* Corolla pale pink to white above, often streaked in pink or purple below. In flower July-Oct.

Mostly a coastal plant, often in sandy soils and dunes, occasionally in disturbed soils of the Santa Ana Mountains. Nearly absent from the Temescal Valley. Known from the Santa Ana River channel, Niguel Hill, Dana Point, Riley Wilderness Park, Audubon Starr Ranch Sanctuary, San Juan Creek, and San Mateo Canyon.

The epithet was given in 1972 by L.D. Gottlieb for the place of its discovery, in San Diego (Latin, -ensis = belonging to).

Similar. Tall wreath-plant (*S. virgata* var. *virgata*): taller, 0.5-2 meters tall; 5-9 florets per flowerhead; corolla white to pink above, purple-pink below; mostly inland and mountains.

Tall Wreath-plant
Stephanomeria virgata Benth. **ssp. *virgata***

A *large annual* 0.5-2(3) meters tall from a stout tap root. The long *wiry stems* have many branches that are hairless, sparsely covered in tiny hairs, or quite hairy. Basal leaves are usually withered by flowering time. *Phyllaries* green, sometimes with pink or red; of two types: innermost very long; *outermost very short, strongly curved backward as they age. Flowerheads with 5-9 florets each.* Corolla white to pink above, purple-pink below. In flower July-Oct.

A common plant of open dry areas, *mostly away from the coast.* Usually along road- and trailsides and in openings among chaparral. Known from the San Joaquin Hills, Lomas de Santiago, and Santa Ana Mountains.

The epithet was given in 1844 by G. Bentham for its long twigs (Latin, virgatus = twiggy).

Similar. San Diego wreath-plant (*S. diegensis*): shorter, 1-2 meters tall; 11-13 florets per flowerhead; corolla to pale pink to white; mostly coastal.

Yellow Salsify ✖
Tragopogon dubius Scopoli

A *stiffly upright hairless annual or biennial*, 30-100 cm tall, with *grass-like leaves* and milky sap. The flowerheads have several *yellow flowers* that generally open up between 8:30 a.m. and close by noon. Fruit head spherical, about 2.5-3.5 cm across; *pappus white*. In flower Apr-July.

Both of our salsifies are noted for their large flowers and seedheads that attract attention. Native to Europe, they were originally garden plants but are now wildland weeds. Known here from disturbed soils and meadows along South Main Divide Road and trailsides in adjacent Morrell Potrero. Common along Clinton Keith Road and on Santa Rosa Plateau.

Named in 1772 by J.A. Scopoli perhaps for its uncertain place of origin (Latin, dubius = uncertain).

Similar. Purple salsify ✖ (*T. porrifolius*): a bit taller; purple corollas; mostly in urban and suburban areas.

The aptly-named yellow salsify has yellow ligules. In fruit, the style lengthens and is tipped with white pappus. Purple salsify has purple ligules and brown pappus. Originally garden plants, both escaped from cultivation and are now invasive weeds.

Purple Salsify ✖
Tragopogon porrifolius L.

Upright hairless biennial, 40-100 cm tall, the *leaves grass-like*, with milky sap. Flowerheads are full of *purple flowers* that open up around noon. Fruit head spherical, about 2.5-4 cm across; *pappus brown*. In flower Apr-July.

An aggressive weed. Known mostly from urban and suburban areas such as Mission Viejo but also found in Lucas Canyon within the San Mateo Canyon Wilderness Area.

It was named in 1753 by C. Linnaeus for its leaves, which resemble those of leeks (Latin, porrus = leek [*Allium porrum* L.]; folius = leaf).

Similar. Yellow salsify ✖ (*T. dubius*): a bit shorter; yellow corollas; South Main Divide Road, Morrell Potrero, Santa Rosa Plateau.

BERBERIDACEAE • BARBERRY FAMILY

Herbaceous perennials, shrubs, rarely trees. Some grow from *creeping rhizomes*, thus forming dense thickets. *Leaves* alternate or basal, entire or pinnate, *with spines on the edges* of some species. Sepals 6-18 or 0. *Petals 6, arranged in two whorls of 3, each bearing a nectary.* Stamens 6 (rarely 4-18). Ovary superior. Fruit is a berry.

Our two species have *pinnate leaves with very spiny margins.* Their *flowers have 9 bright yellow sepals in 3 equal whorls and 6 bright yellow petals in 2 whorls of 3.* The fruit is a blue-purple berry. These plants are toxic, especially their roots.

The genus was named in 1753 by C. Linnaeus, who used an ancient name (Arabic, berberys = barberry fruit).

California Barberry
Berberis aquifolium Pursh **var. *dictyota*** (Jeps.) Jeps.
 [*B. dictyota* Jeps.]

An upright shrub, 0.5-1(-2) meters tall. *Leaves are gray-green (rarely bright green), paler at the edges, and thick, with large spines on its wavy leaf margins; the spines are 2-5 mm long.* In flower late Mar-May.

A chaparral plant, known from upper Cold Spring Canyon within the San Mateo Canyon Wilderness, upper Hot Springs Canyon just below Falcon Group Campground, Coldwater Trail, Pine Canyon near Silverado Canyon, lower Brown and Hunt Canyons in Corona, Bald Peak, and right along the Bedford Truck Trail 0.8 km below its intersection with North Main Divide Road. It's easiest to access at the last spot, but it is quite entangled with poison oak so be careful what you touch! There it grows on a rocky talus slope bisected by the roadway.

Named in 1891 by W.L. Jepson in reference to the lighter-colored net-like veins on its leaves (Greek, diktyon = net).

Similar. Shinyleaf barberry (*B. pinnata* ssp. *pinnata*): leaves greener, mostly flatter, more flexible; spines smaller but greater in number.

California Barberry

California Barberry

Leaves usually dull gray (background photo), rarely dark green (inset photo), but always stiff and long-spined

Shinyleaf Barberry
Berberis pinnata Lag. **ssp. *pinnata***

An upright shrub, 0.4-2 meters tall. *Leaves are glossy dark green and thin with small spines on the leaf edges; spines are 1-2 mm long.* In flower Mar-May.

Known here only from dense chaparral on steep north-facing slopes of Sitton Peak and hills to its west. Most of the population is inaccessible, but some plants grow along a short section of the Sitton Peak Truck Trail at 2.62 air miles west-northwest of Sitton Peak summit and 1.36 air miles east-southeast of San Juan Fire Station.

Named in 1803 by M. Lagasca y Segura for its pinnate leaves (Latin, pinnatus = plumed, with feathers, winged).

Similar. California barberry (*B. aquifolium* var. *dictyota*): leaves grayer, very wavy, much stiffer; spines larger but fewer in number.

Barberries are pollinated by carpenter bees such as this one (Xylocopa sp., Apidae).

Shinyleaf Barberry

Leaves flatter, more pliable, short-spined

BORAGINACEAE • BORAGE AND WATERLEAF FAMILY

Most local members of this family *are annuals, covered with stiff bristly hairs* that stand straight out (1 is a hairless perennial). Most grow upright but a few lie on the ground. Leaves are variable: alternate, opposite, and/or basal; simple or divided. *Flowers* are usually *arranged in a coil that resembles the fiddlehead of a violin or curved tail of a scorpion* (technically called a scorpioid cyme). Flowering proceeds from lower to upper flowers; the inflorescence uncoils as this progresses. By the time fruits mature, it is often entirely uncoiled. The *5 (6-10) sepals* are fused at their bases, but their tips are free. The *5 (5-8) petals* are fused at the base into a *shallow cup or deeper bell-shape*, with their tips free and facing outward. Some species have a small appendage at the top of the throat that serves to close the throat. The *ovary is superior.* In some species, the *ovary is divided into 4 lobes,* each lobe ("nutlet") housing a single seed; these have a single style that originates from the center of the 4 ovary lobes. Other species have a single undivided ovary full of many seeds; many of these have 2 separate styles or a single one that is forked.

Many of our borages are "belly plants," tiny enough that you must squat or lie down on your belly and use a hand lens to see and enjoy their flowers. Their beauty is well worth the effort.

Recently, two closely related plant families, the **waterleafs** (Hydrophyllaceae) and **lennoas** (Lennoaceae), were merged into the borage family. Though closely related, each can be recognized as a distinct group, so we present them in separate sections of this chapter.

	Classic Borages (Boraginaceae in the strict sense)	**Waterleafs** (formerly Hydrophyllaceae)	**Lennoas** (formerly Lennoaceae)
Leaves	Mostly alternate, linear, undivided	Alternate or opposite, often divided	Alternate, undivided
Inflorescence	Scorpioid cyme	Coiled cyme or solitary	Panicle, spike, or head
Corolla Throat	Narrowed at the throat	Open at the throat	Narrowed at the throat
Throat Appendage	Often present	Never present	Never present
Ovary	Superior, deeply 4-lobed	Superior, not lobed	Superior, not lobed
Style Origin	From center-base of the ovary	From top of the ovary	From top of the ovary
Style	Style 1, not forked	Styles 2, or 1 that is forked	Style 1, not forked
Fruit	4 nutlets with 1 seed each	A single capsule, often filled with many seeds	A single capsule, often filled with many seeds
Nutrition	Not parasitic	Not parasitic	Parasitic
Local Genera	Amsinckia, Cryptantha, Harpagonella, Heliotropium, Pectocarya, Plagiobothrys	Emmenanthe, Eriodictyon, Eucrypta, Nama, Nemophila, Phacelia, Pholistoma	Pholisma

Amsinckia, *with scorpioid cyme*

Phacelia minor *and* Phacelia parryi, *with open throats*

Pholisma arenarium, *with an upright tight cluster of flowers*

CLASSIC BORAGES
Common Fiddleneck
Amsinckia intermedia Fisch. & C.A. Mey.
  [*A. menziesii* (Lehm.) A. Nelson & J.F. Macbr. var. *intermedia* (F. & M.) Ganders]

A *bristly annual*, 20-120 cm tall, upright or spreading with the stem tips upright. Its *stems and leaves* are generally *bright to medium green* (stems sometimes yellowish or brownish), *covered with stiff bristles that stand outward, with no smaller hairs between them*. Leaves are 2-15 cm long, linear or oblong. The *trumpet-shaped corolla* is 7-11 mm long, 4-10 mm wide, *orange with 5 red-orange blotches* high up in the throat. The *corolla clearly extends far above the calyx*. In flower Mar-June.

Abundant, especially in meadows, grasslands, fallow pastures, open fields, and disturbed soils throughout our area, though generally not at high elevations. Look for it in all of our wilderness parks, in lower San Juan Canyon along Ortega Highway, in the San Mateo Canyon Wilderness, and throughout the Temescal Valley.

In 1831, J.G.C. Lehmann established this genus in honor of W. Amsinck. The epithet was given in 1836 by F.E.L. von Fischer and C.A. von Meyer, probably in reference to features of the plant that lie between other known species (Latin, inter = between; medius = middle).

Common Fiddleneck

Gray Fiddleneck

Nutlets, ventral views

Scale bars are 1mm long

Nutlet photos: Michael G. Simpson, San Diego State University, http://www.sci.sdsu.edu/plants/amsinckia

Gray or Rigid Fiddleneck
Amsinckia retrorsa Suksd.

An upright annual to about 60 cm tall. It has a single main stem with few side branches. Like common fiddleneck, the plant is *covered with long bristles* that stand outward. *Between those bristles are soft gray hairs that give the plant a gray appearance. On the stems those gray hairs bend toward the ground.* On the leaves those gray hairs are quite numerous and bend toward the leaf tip. Leaves are similar to common fiddleneck but are often narrower. The *corolla is very small, scarcely longer than the calyx lobes, pale yellow with no orange markings*. In flower Apr-June.

Often grows with common fiddleneck *away from the coast* as at Elsinore Peak. Common from Temescal Valley through Elsinore Basin and Temecula-Murrieta. Sparse on the coastal slope, such as in Limestone Canyon.

The epithet was given in 1900 by W.N. Suksdorf for its bent gray hairs (Latin, retrorsus = bent, turned backward).

Common Fiddleneck

Gray Fiddleneck

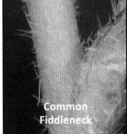
Common Fiddleneck

Gray Fiddleneck

THE FIDDLENECK GUILD
Little Bear Scarab Beetle
Paracotalpa ursina (Horn) (Scarabaeidae)

  Stout scarab beetles, 10-23 mm long, *head and thorax metallic blue, elytra a rich reddish brown*. Long yellowish hairs clothe the body, giving it a fuzzy appearance. In flight during spring when fiddlenecks are in bloom and fruit.

  Usually found among fields of fiddleneck. They fly low to the ground then land and feed on the plants, often several to a plant. Once more common and widespread in our area, now greatly diminished with the elimination of open wildflower fields. They used to be an especially common sight in Dana Point, now uncommonly found. Recently found in Laguna Coast Wilderness Park near the Nix Nature Center. More common inland but also diminishing there because of development.

  The epithet was given in 1867 by G. Horn for its fuzzy, bear-like appearance (Latin, ursinus = resembling a bear).

Little Bear Scarab Beetle

Cat's-eyes (*Cryptantha*) and Popcorn Flowers (*Plagiobothrys*)

Small white-flowered annuals or perennials found commonly throughout California, especially during the first 1-2 years after a fire. They are *covered with obvious whitish or translucent hairs.* Some species lie flat on the ground, others stand upright. Leaves undivided and linear, those on upper stems alternate. Petals fused at the base into a short tube, separated into 5 rounded lobes above. *Flower throat with 5 appendages,* easily seen. For most species, the corolla falls off easily and does so as a single unit, not as individual petals. Fruits are 1-4 dry nutlets; each nutlet contains a single seed. The nutlet surface is often rounded or teardrop-shaped, smooth or covered with bumps, prickles, or spines. To identify them positively, you must examine the mature nutlets under a good hand lens or microscope. Species in these two genera are similar and difficult to distinguish, especially in the field. Some people call members of both genera by the common name of popcorn flower.

	Cat's-eyes *Cryptantha*	Popcorn Flowers *Plagiobothrys*
Hairs	*Stiffer,* thicker hairs held away from the plant; often painful to the touch	*Softer,* narrower hairs held against the plant; generally *not* painful to the touch
Leaves	Lower leaves *not* crowded; opposite	Lower leaves crowded; opposite or in a basal rosette
Sap	Sap *clear,* never purple	Sap of root, stem, and petioles *may be* purple
Flower size	*Smaller* flowers	*Larger* flowers
Seeds	*Smoother* seeds	*Rougher* seeds
Nutlet scar	Nutlet scar recessed *inward* ("grooved")	Nutlet scar elevated *outward* ("keeled")
Flowering	Generally in flower a little *later* than *Plagiobothrys*	Generally in flower a little *earlier* than *Cryptantha*

Cat's-eyes (*Cryptantha* spp.) are *covered with stiff bristly hairs,* often held outward from the plant. The lowest leaves are not crowded and are opposite each other on the stem. Flowers are tiny (less than 7 mm across). Fruits are smooth or covered with bumps, prickles, or spines. We have 6 local species, only 4 of which are commonly found.

	Cleveland's *Cryptantha clevelandii*	Common *Cryptantha intermedia*	Coast *Cryptantha leiocarpa*	Guadalupe *Cryptantha maritima*	Tejon *Cryptantha microstachys*	Jones's *C. muricata* var. *jonesii*
Stature	Stout plant; upright central stem, sometimes branched	Slender plant covered with many long bristles; upright central stem	Moderate plant; upright or spreading, branched from the base	Slender plant covered with long, white, spreading hairs; upright central stem	Slender plant; stem unbranched or with few branches; very hairy	Stout plant; upright central stem; branches along top to bottom or just near the top
Nutlets	1 (2-4); seed a teardrop, smooth, shiny, black when mature	1-4; seed conical, shiny, rough bumps	3-4; seed teardrop-shaped, smooth, pale tan with brown mottling	1-2; seed a long narrow teardrop, smooth and shiny or with tiny bumps	1; seed a long narrow teardrop, smooth, shiny, pale brown with medium brown mottling	4; seed oval to triangular, with a dorsal ridge, covered with short bumps
Habitat	Grasslands, coastal sage scrub, chaparral; coast to mountains	Coastal sage scrub, chaparral	Coastal sandy soils and dunes	Dry areas, mostly deserts; once found in upper Newport Bay	Sandy to clay soils; grasslands, coastal sage scrub, chaparral; foothills to mtns	Dry, open sandy soils; coastal sage scrub, chaparral

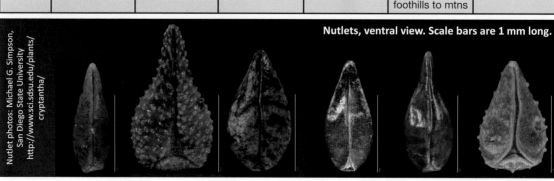

Nutlets, ventral view. Scale bars are 1 mm long.

Nutlet photos: Michael G. Simpson, San Diego State University http://www.sci.sdsu.edu/plants/cryptantha/

Cleveland's Cat's-eyes
Cryptantha clevelandii Greene

Upright, single-stemmed or branched, 10-60 cm tall, the *stem covered with hairs that mostly hold to the plant*. Leaves 1-5 cm long, linear. Flowers on long, leafless stalks. Lowest flowers far apart, upper flowers crowded. Sepals covered with bristles; lower ones curve backward. Each *white corolla* is 1-6 mm across and has a *yellow (rarely white) throat*. Nutlets *1* (2-4), about 1 mm long, shaped like a long narrow teardrop, smooth, and shiny black when mature. In flower Apr-June.

In sandy to rocky soils, open areas and chaparral, from coast to lower elevations in the mountains. Fairly common in Caspers Wilderness Park, especially after fire.

Named in 1887 by E.L. Greene in honor of D. Cleveland.

Similar: **Tejon cat's-eyes** (*C. microstachys*): generally smaller; most hairs stand straight out, giving the plant a very bristly feel; flowers minute, calyx mostly under 1 mm long, with very long straight bristles; more often in foothills and mountains.

Cleveland's Cat's-eyes

Common Cat's-eyes
Cryptantha intermedia (A. Gray) Greene

An upright slender annual, 10-60 cm tall, *covered with sharp hairs that spread away from the foliage*. It often grows in dense colonies. Leaves 1.5-5 cm long, linear, and sparse on the stem. Each *white corolla* is 3-6 mm across and has an obvious *bright yellow throat*. Nutlets 1-4, 1.5-2 mm long, conical, shiny, and covered with rough bumps. In flower Mar-July, earlier with good rains.

Abundant and widespread in chaparral, from coastline to mountains. Common along trails and roadsides, often in loose or disturbed soils. Easily found on the Dana Point Headlands and near the Ortega Candy Store.

Named in 1882 by A. Gray for the line of stiff hairs down the center of each sepal (Latin, inter = between; medius = middle).

Similar. **Jones's cat's-eyes** (*C. muricata* var. *jonesii*): upright, stout, central stem thicker; seeds fatter; generally less abundant.

Common Cat's-eyes

Coast Cat's-eyes ✿
Cryptantha leiocarpa (Fisch. & C.A. Mey.) E. Greene

An *upright or spreading* annual, the *stems branch from the base* and reach 5–30 cm long. Its stems are sparsely covered with spreading hairs; some hairs may be rough to the touch. Leaves 1–3.5 cm, linear, often held upward, *covered with soft silvery hairs,* some hairs with a swollen bulb at their base. Sepals densely covered with long bristles. The *white corolla* is 1–2.5 mm wide. Nutlets 3-4; 1.5-2 mm long, teardrop-shaped, smooth, pale tan with brown mottling. In flower Mar-Apr.

A scarcely found coastal plant of sandy soils, right on the beach. It occurs along the coast from southern Oregon to Orange County. Recently found on backdunes along the seaward side of Balboa Peninsula in Newport Beach, where it grows with **beach bur-sage** (*Ambrosia chamissonis,* Asteraceae), **beach morning-glory** (*Calystegia soldanella,* Convolvulaceae), **beach sand-verbena** (*Abronia umbellata,* Nyctaginaceae), and **beach evening-primrose** (*Camissoniopsis cheiranthifolia* ssp. *suffruticosa,* Onagraceae). Extensive searches, especially in Crystal Cove State Park, San Clemente State Beach, and San Onofre State Beach, may turn up additional populations.

The epithet was given in 1835 by F.E.L. Fischer and C.A.A. von Meyer for its smooth nutlets (Greek, leios = smooth; karpos = fruit).

Coast Cat's-eyes

Guadalupe or White-haired Cat's-eyes
Cryptantha maritima (E. Greene) E. Greene

Upright slender annual 10-40 cm tall, *covered with very long, white, spreading hairs.* Densely branched in the upper half with *several short flower stalks.* Leaves are 1-4 cm long, linear, with flattened hairs. Some or all hairs on the leaves have a tiny swollen bulb at their base, visible with a hand lens. Flowers are very tiny, 0.5-1 mm wide. *Nutlets only 1-2,* 1.5-2 mm long, a long narrow teardrop, smooth and shiny or with tiny bumps. In flower Mar-May.

Known here only from the hills surrounding upper Newport Bay, perhaps more widespread. Primarily a species of our deserts and the channel islands.

Named in 1885 by E.L. Greene from plants collected near the sea, on Guadalupe Island, Baja California (Latin, maritimus = of the sea).

Tejon Cat's-eyes
Cryptantha microstachys (Greene ex A. Gray) Greene

An *upright annual,* 10-50 cm tall. *Stem is unbranched or with few branches,* covered with flattened or standing hairs. Leaves 0.5-4 cm long, linear or oblong, held upward and bristly. Sepals are covered with short hairs and 1-2 very long bristles each as long as the sepal itself. Its *tiny flowers* are 0.5-1 mm wide. *Nutlets 1,* about 1.5 mm long, shaped like a long, narrow, smooth, shiny teardrop, pale brown with medium brown mottling. In flower Apr-June.

Common from coastline to lower elevations of the mountains, usually after fire. Most often found along South Main Divide Road, about Elsinore Peak, and in the San Mateo Canyon Wilderness Area.

Named in 1885 by E.L. Greene for its very slender flower spike (Greek, mikros = small; stachys = an ear of grain, spike). It was first collected at Fort Tejon in Kern County, the basis of the common name.

Similar. Cleveland's cat's-eyes (*C. clevelandii*): generally taller; most hairs hold to the plant; corolla larger; calyx longer; sepal bristles more numerous, shorter, the lower ones curving backward; generally more coastal and foothills.

Jones's Cat's-eyes
Cryptantha muricata (Hook. & Arn.) Nels. & Macbr. **var. jonesii** (A. Gray) I.M. Johnst.

A *stout thick-stemmed annual,* 10-100 cm tall. *It branches from top to bottom or just near the top.* Leaves are 0.5-4 cm long, linear, and covered with sharp bristles, some with a tiny swollen bulb at their base, visible with a hand lens. *Inflorescence spherical and tight, like a clenched fist.* Flowers are 2-6 mm wide. *Nutlets 4,* 1-2 mm long, oval to triangular, with a dorsal ridge, covered with short bumps. In flower Apr-June.

Found in sandy, gravelly areas in *dry places* among our foothills, canyons, and mountains. Fairly regular around Elsinore Peak and among talus slopes along Holy Jim Canyon Trail.

The epithet was given in 1838 by W.J. Hooker and G.A.W. Arnott for the sharp bristles that cover the plant (Latin, muricatus = pointed, full of sharp points).

Similar. Common cat's-eyes (*C. intermedia*): same upright habit, much less stout, central stem much thinner; seeds slender; generally much more abundant and widespread.

POPCORN FLOWERS: PLAGIOBOTHRYS

Popcorn flowers (*Plagiobothrys spp.*) are covered with slender bristly hairs that are often held against the plant. The lowest leaves are numerous and crowded, often arranged into a basal rosette or simply opposite each other. The corolla is slightly tubular, from 2-6 mm deep and 1-9 mm across, white with a yellow throat. Fruits are covered with bumps, prickles, or spines, rarely smooth. Some contain a purple sap that may color the plant purplish; when a piece is broken off, purple sap will flow from the plant and stain your fingers. We have 5 local species (a sixth is rarely seen).

	Adobe *Plagiobothrys acanthocarpus*	**Valley** *Plagiobothrys canescens*	**California** *P. collinus* var. *californicus*	**Rusty** *Plagiobothrys nothofulvus*	**Pacific** *Plagiobothrys tenellus*
Stature	Upright or spreading, stems 10-40 cm long; light green, sparsely haired; no basal rosette	Prostrate, many-branched, stems 10-60 cm long; purple sap; roots, stems, and leaves often purplish; rough haired; basal rosette	Low, spreading, sometimes tufted or upright, stems 10-40 cm long; lower leaves opposite, upper leaves alternate	Upright, tall, stems 20-70 cm, covered with rough sharp hairs; strong basal rosette with long leaves; purple sap; flowers crowded; calyx hairs golden, calyx circumscissle	Upright, short, slender, stems 5-30 cm tall, densely covered with spreading hairs; strong basal rosette with short leaves; clear sap; flowers crowded at tips
Nutlets	4; teardrop-shaped; bumpy, long-spines; attachment scar deeply concave	4; strongly arched; midrib and lateral ribs, no cross-ribs; attachment scar flat	4; oval; sharp ribs; on a very long stalk	1-3(4); slightly arched in profile, teardrop shaped, pointy tip; ribs narrow; on a short stalk	3; thick, arched in profile, "Maltese cross"; small tubercles on ribs; short attachment scar near middle
Habitat	Vernal pools and moist clay soils of grasslands	Grasslands and coastal sage scrub, foothills to mountains	Dry to clay soils; grasslands and open coastal sage scrub	Grasslands, meadows, openings in coastal sage scrub	Moist clay soils and flat open areas among grasslands and chaparral

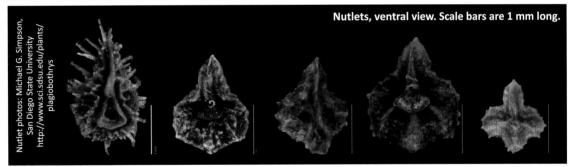

Nutlets, ventral view. Scale bars are 1 mm long.

Nutlet photos: Michael G. Simpson, San Diego State University http://www.sci.sdsu.edu/plants/plagiobothrys

Adobe Popcorn Flower ✿
Plagiobothrys acanthocarpus (Piper) I.M. Johnst.
Upright or spreading, 10-40 cm tall, its foliage is *light green, sparsely covered with hairs. Leaves are not in a basal rosette.* Lower leaves are linear to spoon-shaped, 2-6 cm long. *Corolla tiny and short-lived;* 3-6 mm deep, 1-2.5 mm wide; white with yellow throat. *Nutlets 4,* teardrop-shaped, covered with low rounded bumps and long spines, the spines often bear backward-pointing barbs at their tips. In flower Mar-May.

Found in vernal pools and in clay soils of grasslands. Formerly known from UCI Ecological Preserve (not found recently) and in the hills of northern Dana Point, directly west of Marco Forster Middle School (an area now occupied by homes). Recent records include only Quail Hill in Irvine and coastal bluffs above Surfers Beach in San Onofre. Its bumpy, spine-covered fruits hold onto the fur of small mammals that unknowingly disperse them.

It was named in 1920 by C.V. Piper in reference to its prickly nutlets (Greek, acantha = a thorn or prickle; karpos = fruit).

Adobe Popcorn Flower

Valley Popcorn Flower
Plagiobothrys canescens Benth. **var. *canescens***

Usually *a prostrate plant with many branches,* 10-60 cm long, *covered with long rough white hairs.* Lowest leaves are 1.5-5 cm long, in a basal rosette. *Sap purple;* roots, stems, and leaves often purplish. There are *leafy bracts all along the inflorescence.* The corolla is 4-6 mm deep, 2-3 mm wide; it generally does not fall off in fruit but instead dries up and stays on the plant. *Nutlets 4,* strongly arched in profile. Each is *ribbed:* the outer surface of each has a central vertical midrib with a series of horizontal lateral ribs, no vertical cross-ribs; the spaces between the ribs are wide and flat. In flower Mar-May.

Generally common in grasslands and coastal sage scrub, from the foothills up into the mountains. Known from Riley and Caspers Wilderness Parks, Rancho Mission Viejo, along upper San Juan Trail in Chiquito Basin, Puente Hills, Temescal Valley, Elsinore Basin, and Santa Rosa Plateau.

Named in 1849 by G. Bentham in reference to its covering of long white hairs (Latin, canescens = becoming gray, white, or hoary).

Similar. California popcorn flower (*P. collinus* var. *californicus*): less hairy; nutlet atop a narrow stalk.

California Popcorn Flower
Plagiobothrys collinus (Phil.) I.M. Johnst. **var. *californicus*** (A. Gray) Higgins

Low-growing and spreading, sometimes upright, stems 10-40 cm long. *Stem hairs slender, held against the stem.* The plant often grows as a small tuft, flat on the ground, and produces flowers before the long stems begin to grow and elongate. Lower leaves are 1-4 cm long, opposite; upper leaves alternate. There are *leafy bracts all along the long wand-like inflorescence.* Corolla about 3 mm deep, 4-7 mm wide. *Nutlets 4,* 1-2 mm long, oval, on a very long stalk. Each is *ribbed:* a sharply defined central vertical midrib, horizontal lateral ribs, and vertical cross-ribs. In flower Mar-May but with good rains, it begins to flower in Jan.

Very common and widespread throughout our area, common on grasslands and open coastal sage scrub, usually on clay soils. Look for it on the Dana Point Headlands and in parks such as Crystal Cove State Park, Laguna Coast Wilderness Park, and Caspers Wilderness Park. Also known from Limestone Canyon, Puente Hills, Temescal Valley, and Elsinore Basin.

The epithet was given in 1857-58 by R.A. Philippi for its general distribution on hillsides (Latin, collinus = hill-loving). Our form was named in 1877 by A. Gray for the place of its discovery.

Similar. Two other varieties of *P. collinus* exist here, one possibly extirpated when UC Irvine was expanded, the other found only in the San Joaquin Hills. Both of them intergrade with California popcorn flower. **Valley popcorn flower** (*P. canescens*): white-hairy; nutlet attachment scar flat.

Rusty Popcorn Flower
Plagiobothrys nothofulvus (A. Gray) A. Gray

An erect annual, rarely biennial or perennial. A *tall plant* that often towers over surrounding wildflowers, it stands upright 20-70 cm and is covered with rough sharp hairs. Its *sap is red-purple,* which gives the roots, stems, and leaves that color. Lowest leaves are 3-10 cm long in a *large basal rosette.* Stem leaves are alternate and few in number. Flowers are crowded at the ends of the flowering stalks. *Hairs on the calyx are yellow-gold in color.* Petals are 2-3 mm deep, 3-9 mm wide. *Calyx lobes* often *curve inward over the fruit* and pop off as a single cap-like unit when the fruit is mature (circumscissle). *Nutlets 1-3(4),* slightly arched in profile, on a short stalk, teardrop shaped, with a pinched-in region just below the tip. Nutlet scar convex to shallowly concave. *Ribbed;* midrib, lateral ribs, and cross-ribs narrow. In flower Mar-May.

Generally found in *moist soils* of grasslands, meadows, and openings in coastal sage scrub. Once abundant and widespread, formerly known from many locations near the coast and foothills, especially Dana Point; now mostly extirpated from Orange County by massive development. Currently abundant only in the Richard and Donna O'Neill Conservancy, portions of Rancho Mission Viejo, and pockets in the Lomas de Santiago such as Limestone Canyon, particularly after fire and good winter rains. Also known from Laguna Coast Wilderness Park, Blue Jay Campground, South Main Divide Road, Elsinore Peak, San Mateo Canyon Wilderness Area, Temescal Valley, and Lake Elsinore region.

Named in 1882 by A. Gray in reference to its gold calyx hairs, which are quite unlike its relatives' (Greek, nothos = unlike the others; fulvus = gold-colored).

Similar. Field popcorn flower (*P. fulvus* (Hook. and Arn.) I.M. Johnst. var. *campestris* (E. Greene) I.M. Johnst., not pictured): *sap clear or green;* calyx lobes upright or spread out; *calyx hairs brown; fruit not circumscissle; nutlet scar deeply concave.* Found in *remote areas,* such as grasslands south of Oak Flats and on the southern slopes of Miller Mountain; one old vague record for "a meadow" along Ortega Highway. The two species co-occur on Santa Rosa Plateau.

Pacific or Slender Popcorn Flower
Plagiobothrys tenellus (Nutt.) A. Gray

An *upright slender-stemmed* annual 5-30 cm tall, *densely covered with spreading white hairs.* Short basal leaves in an obvious rosette, each leaf only 1-5 cm long; stem leaves smaller, fewer, alternate on stem. Inflorescence tip tightly coiled. Calyx with whitish and/or reddish hairs. Corolla 1-3 mm wide. *Nutlets generally 3;* each 1-2 mm long, thick, arched in profile, shaped like a Maltese cross, short attachment scar near middle. *Ribbed;* central vertical midrib and horizontal lateral ribs with small tubercles, no vertical cross-ribs. In flower Mar-May.

Generally uncommon in our area, though sometimes occurs in great numbers in limited areas, mostly in clay soils and flat open areas among grasslands and chaparral. Known from Chiquito Basin along San Juan Trail and along South Main Divide Road from the USFS gate near Rancho California southward to Elsinore Peak and scattered sites in the San Mateo Canyon Wilderness Area.

The epithet was given in 1885 by T. Nuttall in reference to its slender, delicate stems (Latin, tenellus = quite delicate).

Palmer's Grappling Hook ★
Harpagonella palmeri A. Gray

A *delicate little annual* that looks like a *Cryptantha* and is often overlooked, even by experienced botanists. It grows outward or upward, 3-30 cm tall, with slender branches that arise from the base. Leaves are 0.5-3.5 cm long, narrow, with sparse hairs that lie flat against the foliage. Corolla is white, minute, nearly 2 mm long, barely longer than the sepals, and short-lived. *In fruit, the upper 2 sepals grow to form a gnarled, bur-like "fist" with 5-10 stout spines that bear hooked bristles.* This "grappling hook" surrounds the single nutlet. The other 3 sepals are normal in appearance. In flower Mar-Apr.

Found primarily in clay soils among dry grasslands and openings in coastal sagebrush scrub and chaparral at places such as Dana Point Headlands, Mission Viejo, Christianitos and Bell Canyons, Elsinore Peak, and Miller Mountain. Good sites to see it are at the Richard and Donna O'Neill Conservancy, Caspers Wilderness Park, and Elsinore Peak. CRPR 4.2.

It was first found on Guadalupe Island, Baja California, Mexico, by E. Palmer. In 1876, A. Gray established a new genus for it, named in reference to the flower's small hook-like calyx spines (Latin, diminutive form of harpago = grappling hook). The specific name is in honor of Palmer. It is the only member of the genus.

Palmer's Grappling Hook
Nutlet
Fruits

Nutlet: Michael G. Simpson, http://www.sci.sdsu.edu/plants/pectocarya/

Combseed: *Pectocarya*

These are *dainty annuals,* easily overlooked but abundant on open ground during spring. They are sparsely covered with hairs that lie flat against the stems and leaves. The plants are bright green and age yellowish then light brown before senescing and shattering into stick-like pieces at each node. All leaves are linear, the upper ones arranged alternately on the stem.

Their *flowers are tiny,* arranged along the stem, each on a curved pedicel. Petals are 0.8-3 mm long, bell-shaped or more flattened, and bright white. *Their most diagnostic and charming feature is their fruit. Each is composed of 4 nutlets, their arms either bilateral (arranged like toes of a woodpecker: 2 forward, 2 backward) with curved prickles at their edges or radiating outward like a Maltese cross with straight spines.* These fruits catch on a mammal's fur as it passes by, pulling them off the plant or from the ground, thus giving the seeds a free ride to a new location. California has 8 species, 3 of them are in our area. You will need to observe the fruit under a hand lens to identify them.

The genus was established in 1840 by A.P. de Candolle for the comb-like teeth on the fruits (Latin, pecto = to comb; Greek, karyon = a nut).

Pectocarya linearis ssp. ferocula
Large, triangular teeth along nutlet margin

Pectocarya penicillata
Narrow, hook-shaped bristles only near the tip of each nutlet arm

Pectocarya setosa
Nutlet arms flat and rounded, slender bristles all around

Narrow-toothed or Slender Combseed

Pectocarya linearis (Ruiz & Pavón) DC **ssp. *ferocula*** (I.M. Johnst.) Thorne

The *stems* of this combseed are 6-26 cm long, *lie on the ground, and curve upward at the ends.* Leaves are narrowly linear, 0.5-2.5 cm long. Each nutlet ("toe") in its "woodpecker-foot" fruit is 2-3.8 mm long, straight or slightly curved backward near the tip. *Wide triangular-based sawteeth arise from base to tip of fruit, each ending in a curved prickle.* In flower Mar-May.

Our most common and widespread combseed, it lives in sandy open areas within grasslands, coastal sagebrush scrub, chaparral, and recovering burn sites. Find it along our coast in UCI Ecological Preserve, Laguna Beach, and Dana Point, foothills of Caspers Wilderness Park, lower Trabuco Canyon, and along many trails in the Santa Ana Mountains.

The species, which also lives in Chile, was named in 1799 by H. Ruíz Lopez and J.A. Pavón for its linear leaves (Latin, linearis = pertaining to a line or lines). Our subspecies was named in 1932 by I.M. Johnston in reference to the ferocious-looking teeth on its fruit (Latin, feroculus = fierce, wild, bold).

Winged or Northern Combseed

Pectocarya penicillata (Hook. & Arn.) A. DC.

Much like slender combseed, this species has *stems* 2-25 cm long, and grows *flat or upright.* Its leaves are 1-3 cm long, narrowly linear, the edges often rolled under. Margins of the fruit are folded over unevenly, broadest near base and tip. *Narrow hook-shaped bristles arise only from the tip of each nutlet.* In flower Mar-May.

Recorded in our area from the Chino Hills near the Santa Ana River channel, Santiago Canyon wash within Irvine Regional Park, Audubon Starr Ranch Sanctuary, South Main Divide Road, Elsinore Peak, and sparingly in Devil Canyon within the San Mateo Canyon Wilderness Area.

Named in 1838 by W.J. Hooker and G.A.W. Arnott in reference to the fruits that end in fine bristles (Latin, penicillatus = having the form of a brush, ending in a tuft of fine hairs).

Stiff-stemmed, Round-nut, or Bristly Combseed

Pectocarya setosa A. Gray

An *upright* plant, 2-23 cm tall, *covered with long bristly hairs.* Small individuals unbranched, larger ones with numerous zig-zagging branches. Leaves 0.5-2.0 cm long, lower ones opposite, upper ones alternate. Sepals are curved, longer than the nutlets, and bear 3-6 straight spines at their tips. *Nutlets 1.5-4 mm long, rounded, arranged like a Maltese cross in fruit, and covered with fine hook-tipped bristles.* In flower Apr-May.

Usually a desert-dwelling plant, found uncommonly in our area on clay soil between chaparral and grassland on dry northeast-facing slopes of Elsinore Peak. Look for it in similar places elsewhere in the Santa Ana Mountains and the valley below.

Named in 1876 by A. Gray for the bristles on its foliage and fruit (Latin, setosus = full of hairs, hairy).

Photo of Bob: Fred M. Roberts, Jr.

Salt or Alkali Heliotrope
Heliotropium curassavicum L. **ssp.** *oculatum* (Heller) Thorne

A *hairless, fleshy perennial* that grows from simple or rhizome-like roots from which additional stems may sprout, thus vegetatively spreading the plant. *Often thinly coated with a fine white powder.* Branches 10-60 cm long, sprawling to partly upright. Leaves 1-6 cm long, lance-shaped, a bit wider toward the tips, with little to no petiole. *Corolla bell-shaped, white to bluish with a yellow or purplish throat,* 3-5 mm long, 3-4 mm wide, and arranged in long curved cymes. In flower Mar-Oct.

Salt Heliotrope

Grows in *moist to wet alkaline ground* along marshes, creeks, and reservoirs. Quite widespread along the coast as at Bolsa Chica Ecological Reserve, Upper Newport Bay, and Aliso Canyon; inland along watercourses and on alkali flats to the Chino Hills, foothills of the Santa Ana Mountains, in Temescal Valley, and in Elsinore Basin.

Heliotropes contain pyrrolizidine alkaloids (PAs), double-ringed chemicals that taste bad to us and cause liver damage in some vertebrate animals if eaten. Insects who feed on plants that contain PAs store the chemicals, making them distasteful to predators that eat them. Since heliotropes are unique in structure and chemistry, some botanists place them in their own family, the Heliotropaceae.

The genus was named in 1753 by C. Linnaeus in reference to the summer solstice, supposedly when a related species blooms (Greek, helios = the sun; tropos = a turn). He gave the epithet for the island of Curaçao, where an early collection was made (Latinized from Curaçao; -icum = belonging to). Our form was named in 1904 by A.A. Heller because its flower has a purplish center like an eye (Latin, oculus = eye; -atus = likeness).

Buckeye Butterfly

At Whiting Ranch Wilderness Park, a buckeye butterfly (Junonia coenia Hüebner, Nymphalidae) sips nectar from alkali heliotrope. The plant is irresistible to many butterflies.

WATERLEAFS
Whispering Bells
Emmenanthe penduliflora Benth. **var.** *penduliflora*

A *soft annual* covered with sticky-tipped hairs, it stands upright 5-85 cm tall. *Leaves* 1-12 cm long, *toothed or pinnate,* mostly basal but some also up the stem. Its *bell-shaped flowers* are 6-15 mm long, yellow to whitish. In flower Apr-July.

A *fire-follower* that mostly comes up one year after a fire and often persists for a few more years. Known from the southern end of the Chino Hills and throughout the Santa Ana Mountains such as at Caspers Wilderness Park, Sitton Peak, San Juan Canyon, San Juan Trail, and the higher peaks. Probably also occurs in the San Joaquin Hills but has not yet been reported from there.

Its young flower buds stand upright but begin to hang down as they open and mature, looking very much like hanging bells. Listen closely to the plant as the wind blows older and dried flowers against each other, making a very soft sound like... whispering bells. They are pollinated by bees.

The genus was established in 1835 by G. Bentham in reference to the flowers that remain on the plant long after flowering (Greek, emmemo = to be faithful; anthos = flower). At the same time, he gave the epithet for the flowers that hang down as they mature (Latin, pendulus = hanging down; fero = to bear). This is the only member of its genus.

Whispering Bells

Yerba Santas: *Eriodictyon*

These are *pleasantly scented,* upright *evergreen shrubs that often come up after fire.* Ten of the 11 species in the genus are found in California, 3 of them in our area.

The genus was established in 1844 by G. Bentham in reference to the netlike veins and dense hairs of the leaf undersides (Greek, erion = wool; diktyon = net).

Thick-leaved Yerba Santa
Eriodictyon crassifolium Benth. **var.** *crassifolium*

A *tough upright shrub* 1-3 meters tall, *its twigs and leaves are densely covered with white hair, giving them a grey appearance.* The leaf is 3-17 cm long, 1-6 cm wide, linear to oblong, usually toothed. The *lavender bell-shaped corolla* is 5-16 mm long and densely hairy. In flower Apr-June.

A *very common plant of chaparral and along washes in canyons below, often in sandy or decomposing granitic soils.* Known from Niguel Hill, Lomas de Santiago, Santiago Canyon, and Santa Ana Mountains. Easily seen at Tucker Wildlife Sanctuary and Caspers Wilderness Park. Especially abundant after fire. In some areas, such as Coal Canyon, it is parasitized by **clustered broomrape** (*Orobanche fasciculata,* Orobanchaceae). Its lovely flowers are visited by a wide assortment of bees and butterflies.

The epithet was given in 1844 by G. Bentham for its thick leaves (Latin, crassus = thick; folium = leaf).

Thick-leaved Yerba Santa

Hairy Yerba Santa
Eriodictyon trichocalyx A. Heller **ssp.** *trichocalyx*

An *upright shrub* to 0.65-1 meter tall (rarely to just under 2 meters), *its twigs are generally hairless and sticky.* The *leaf* is 3-14 cm long, 0.5-4 cm wide, *linear,* usually toothed, the edge sometimes rolled under. *The leaf upperside is mostly hairless and sometimes sticky, so it appears green and shiny (not grey and fuzzy).* The underside has net-like veins and dense white hairs. The *corolla* is 4-13 mm long, densely hairy, *white to lavender in color.* In flower May-Aug.

An uncommon plant of sandy soils, usually away from the active water channel. Known from the Santa Ana River bed near Yorba Linda, upstream through Riverside. More common in the San Bernardino Mountains and Cajon Pass.

It was named in 1904 by A.A. Heller for its hairy calyx (Greek, trichos = hair; kalyx = bud cup of a flower).

Hairy Yerba Santa

Hairstreak Butterfly

THE YERBA SANTA GUILD

Southern Yerba Santa Leaf Beetle
Trirhabda eriodictyonis Fall (Chrysomelidae)

Larvae metallic blue-green, nearly 1 cm long. *Adults* soft-bodied, 7-10 mm long; *all-yellowish,* each elytron with a dark stripe, a third stripe where the two elytra meet. Head and prothorax with small dark spots. Larvae emerge from the egg in spring, feed on flower buds and young leaves of yerba santa. They pupate in soil, emerge as adults in summer, and feed and on mate leaves. When approached, they often hide or drop to the ground to avoid capture.

The epithet was given in 1907 by H.C. Fall for its host plants, all yerba santas in the genus *Eriodictyon.*

Similar. At least 4 other species of *Trirhabda* occur in our area, each specific to its host plant. The larvae are all metallic blue-green, but adults are a bit different. One other species feeds on yerba santas, 2 on **California sagebrush** (*Artemisia californica,* Asteraceae), and 1 on **desert bush sunflower** (*Encelia farinosa,* Asteraceae).

Larva of Southern Yerba Santa Beetle

Poodle-dog Bush

Eriodictyon parryi (A. Gray) Greene
[*Turricula parryi* (A. Gray) J.F. Macbr.]

A *short-lived perennial* 1-3 meters tall, *covered throughout with stiff sticky-tipped hairs*. The *large leaves* are 4-30 cm long, linear, smooth-edged or toothed, the edges sometimes rolled under. The dense cluster of flowers arises on long stalks, far above the other growth. The calyx is covered with long sticky hairs. The hairy *corolla* is 10-20 mm long, trumpet-shaped with shallow rounded lobes, and *blue, purple, or lavender in color*. In flower June-Aug.

A generally uncommon plant of chaparral and coniferous forests in the Santa Ana Mountains, sometimes found in large numbers, especially in the years following fire, but later dying out. Known from upper Silverado and Trabuco Canyons but more common along Main Divide Truck Trail from Falcon Group Camp to Los Pinos Saddle and right on top of Santiago Peak. Rarely found in the Santa Ana River channel near Yorba Linda.

The plant gives off a thick sweet scent that is foul to some people and pleasant to others; you have to experience it and decide for yourself. It is pollinated by several types of bees, tumbling flower beetles, butterflies, and hummingbirds. *Some people suffer severe dermatitis from contact with it, so play it safe and do not to touch it.*

The epithet was given in 1876 by A. Gray in honor of C.C. Parry, who first collected it in the San Bernardino Mountains the previous year. The common name poodle-dog bush refers to the leaves, which are often crowded in tufts like those of a poodle-dog groomed for showing.

> *Right: Pale swallowtail (Papilio eurymedon, Papilionidae) nectaring from the plant, atop Santiago Peak.*
> *Far right: native mason bee (Osmia sp., Megachilidae), quite common on the plant, also atop Santiago Peak.*

Common Eucrypta

Eucrypta chrysanthemifolia (Benth.) E. Greene
var. *chrysanthemifolia*

This *delicate, brittle-stemmed annual* has few branches and stands upright about 20 cm tall or has multiple spreading branches nearly 1 meter long and sprawls over the ground. It is *covered with a thick sticky fluid and has a pleasant odor*. The lowest leaves are 2-10 cm long, oblong to ovate in outline, 1-2-pinnate, their divisions often lobed, mostly opposite on the stem. Upper leaves are smaller, generally alternate. The base of the petiole clasps the stem. The *corolla* is 2-6 mm long, 6-8 mm wide, bell-shaped with 5 free lobes, *white to yellowish in color, often with purple spots and lines*. The round fruit contains two types of seeds: the outermost are larger and wrinkled, the innermost smaller and smooth, somewhat hidden by the larger seeds. In flower Mar-June.

A delightfully abundant plant of coastal sagebrush scrub, chaparral, oak woodlands, and moist drainages, where it often grows in full or partial shade. Especially common in the years after fire and plentiful winter-spring rains. Known from coastline, canyon, valley, foothills, and mountains throughout our area.

The genus was established in 1848 by T. Nuttall in reference to its secret, hidden seeds (Greek, eu = true; kryptos = secret, hidden). The epithet was given in 1834 by G. Bentham in reference to the similarity of its leaves to those of plants in the genus *Chrysanthemum* (Greek, folium = leaf).

Mud Nama ★
Nama stenocarpum A. Gray

A *low-growing, often sprawling annual,* with stems 8-40 cm long. Covered with hairs, mostly short, soft, and sticky-tipped, some stiff and swollen at their base, visible with a hand lens. Leaves are 5-30 cm long, *oblong, the edges wavy but not divided.* The *corolla* is 4-6 mm long, *funnel-shaped, white or cream to pale violet.* In flower Mar-May (-July).

An uncommon plant of muddy patches, drying ponds, and vernal pools. Known locally from Fairview Park vernal pools, Chiquita Ridge, and portions of Rancho Mission Viejo. With widespread loss of open, muddy habitats, this plant is in serious trouble in our area. CRPR 2.2.

The epithet was given in 1875 by A. Gray for the shape of its fruit (Greek, stenos = narrow, straight; karpos = fruit).

Mud Nama

Blue-eyes: *Nemophila*

Brittle, weak-stemmed annuals, generally low-growing, stems often angular in cross-section or with wings down their lengths. Leaves pinnately toothed to lobed, often bristly. Calyx with 5 free hairy lobes; between lobes is a backward-pointing calyx appendage. The corolla is hairy, funnel-shaped, fused at the base, with 5 large free lobes. The ovary is hairy but not bristly. Generally live in partial shade of oak woodlands and adjacent chaparral. The genus was established in 1822 by T. Nuttall in reference to the plants' occurance in woodlands (Greek, nemos = pasture or woodland glade; philos= loving).

Menzies's Baby Blue-eyes

Menzies's Baby Blue-eyes
Nemophila menziesii Hook. & Arn. **var.** *menziesii*

A *slender annual,* it sprawls on the ground or stands upright. Its *brittle, slender stems* are 10-30 cm long, hairy but not prickly. The *lower leaves* are 1-5 cm long, *opposite,* oval in outline, *pinnate,* generally with 6-13 rounded lobes. Upper leaves have a short petiole, with fewer lobes or undivided, sometimes just toothed. *The calyx has 5 free hairy lobes, between lobes is a backward-pointing calyx appendage.* The corolla is hairy, funnel-shaped, fused at the base, with 5 large free lobes. Each *corolla* measures 5-20 mm long, 10-40 mm wide; its color is *bright baby blue (rarely white) with a white center that is sometimes black-dotted, the veins often blue.* Its 5 stamens are on fairly short filaments. The style is 2-7 mm long. In flower Feb-June.

A delicate wildflower of shady north-facing slopes and moist soils among grasslands, coastal sagebrush scrub, chaparral, and oak woodlands throughout our area. Look for it at upper Newport Bay, San Joaquin Hills, Laguna Canyon, Chino Hills, Rancho Mission Viejo, Trabuco Canyon, Santiago Canyon, Silverado Canyon, Maple Springs Saddle, Temescal Valley, most of our wilderness parks, and major canyons of the Santa Ana Mountains.

The epithet was given in 1833 by W.J. Hooker and G.A.W. Arnott in honor of A. Menzies, who first collected it in California.

Menzies's Baby Blue-eyes

Pale Baby Blue-eyes
Nemophila menziesii Hook. & Arn. **var. *integrifolia*** Parish

A *smaller form* of baby blue-eyes that occurs mostly at higher elevations. *The lower leaves generally have only 5-7 lobes, each lobe smooth-edged or toothed.* Upper leaves have no petiole and are undivided or shallow-toothed. The *corolla is only 5-10 mm long, 6-15 mm wide, mostly paler blue.* In flower Feb-June.

Known from Pine, Silverado, Trabuco, and Hot Springs Canyons, Harding Truck Trail, Maple Springs Saddle (with the typical variety), and atop Santiago Peak.

This variety was named in 1898 by S.B. Parish in reference to its entire (not divided) upper leaves (Latin, integ = entire; folium = leaf).

Similar. Meadow baby blue-eyes (*N. pedunculata*): corolla smaller (2-8 mm across), whitish to pale blue with dark to purple spots; difficult to distinguish, possibly not distinct.

Pale Baby Blue-eyes

A NOTE ABOUT BABY BLUE-EYES

Baby blue-eyes is a single species that exists as two different varieties. The status of those varieties is a topic of discussion. Some botanists cite their similarities, eliminate the "variety" status, and treat them all as Nemophila menziesii. Other botanists, however, cite their differences and treat them as two varieties (as we elected to do here) or even two separate species, as first proposed by LeRoy Abrams in 1904.

Evidence that they are speciating (diverging into separate species) includes ultraviolet reflectance patterns that influence their appearance to pollinating insects. In southern California, Menzies's baby blue-eyes **reflects** ultraviolet light (so it appears **lighter** to an insect) while pale baby blue-eyes **absorbs** it (and appears **darker** to an insect). This may explain why they are pollinated by different species of Andrenid bees, certainly a factor in their divergence. See Cruden (1972).

Meadow Baby Blue-eyes
Nemophila pedunculata Benth.

A *low-growing, spreading annual* with sparsely hairy stems 10-30 cm long. The leaves are all opposite, the blade 5-35 mm long, oblong to oval in outline, with 5-9 lobes that are sometimes toothed. *The calyx is 1-4 mm long, the calyx appendages are almost as long.* The corolla is 2-5 mm long, 2-8 mm wide, white or blue with darker veins and dark dots in the throat. The petal tips sometimes with a purple spot. The style is less than 2 mm long. In flower Mar-May.

Not widely reported but certainly more common and probably often mistaken for stunted baby blue-eyes. Grows in grassy openings among chaparral and woodlands. Known from the area near Rancho Carrillo Road in the foothills east of Caspers Wilderness Park.

The epithet was given in 1835 by G. Bentham for its long peduncles (individual flower stalks, sometimes called pedicels) or the fact that it is small and under your feet when hiking (Latin, pedunculus = small foot).

Meadow Baby Blue-eyes

Fiesta Flowers: *Pholistoma*

Succulent brittle-stemmed annuals that sprawl over the ground and clamber up nearby plants and objects, aided by bristles and backward-pointing prickles. Like *Nemophila*, the *calyx has 5 free lobes, with 1 backward-pointing calyx appendage between them*. In the corolla throat are *5 scale-like corolla appendages that appear to close it.* The ovary is hairy *and* bristly. Common plants of early spring, often seen along shady hillsides, trails, and roadsides. They are pollinated by Andrenid bees.

Blue Fiesta Flower

Pholistoma auritum (Lindl.) Lilja **var. *auritum***

A *straggling annual* with *hairy, prickly, succulent stems* from 10 cm to about 1.5 meters long. The *individual lobes of the pinnate leaves* usually *point backward* toward the base. *Where it meets the stem, the leaf petiole has broad ear-like flaps that wrap around the stem.* The calyx lobes are 3-9 mm long, the calyx appendages 1-4 mm long. The *corolla* has a broad funnel shape, 3-15 mm deep, 5-20 mm wide, its *color lavender to purple, rarely bluish*. The corolla appendages are 2-3 mm long and purple. In flower Mar-May.

An abundant plant that grows among coastal sagebrush scrub and oak woodlands, especially in moist soils on shady, north-facing slopes. With adequate water and shade, it can grow in very dense mats and put on quite a show. Found from the coast as at upper Newport Bay and the San Joaquin Hills, inland to Chino Hills, Rancho Mission Viejo, Caspers Wilderness Park, and the Santa Ana Mountains.

Named in 1833 by J. Lindley for its ear-like leaf bases (Latin, auritus = eared).

Right: Blue fiesta flower. Note the backward-pointing calyx appendages between the calyx lobes. The corolla is a soft lavender to purple, sometimes bluish. The flower stalks and stems also have backward-curved prickles with which it clambers over other plants and fences.

Far right: San Diego fiesta flower. It too has backward-pointing calyx appendages and climbing prickles. Its corolla is white, rarely blue-tinged.

San Diego Fiesta Flower

Pholistoma racemosum (A. Gray) Constance

A *small low-growing plant* with stems 20-60 cm long, covered with hairs and prickles. Its *lowest leaves* are 3-10 cm long, 2-8 cm wide, *oval to triangular in outline, pinnately divided*. The leaflets point outward, generally not backward. *The petioles have wide flaps that only slightly wrap around the stem.* Calyx lobes are 2-4 mm long, linear, its appendages <2 mm long. *Corolla* is 4-10 mm long, 5-15 mm wide, *white* (rarely bluish) in color. Corolla appendages are small or absent. In flower Mar-May.

An uncommon plant, it grows in shady places at low elevations among coastal sagebrush scrub, chaparral, and oak woodlands. Locally known only from the Gavilan Hills (barely outside our area), and Tenaja and Cold Spring Canyons in the San Mateo Wilderness. More common in San Diego County.

The epithet was given in 1875 by A. Gray in reference to its rounded fruits (Latin, racemulus = the stalk of a cluster or bunch of berries or grapes).

Cluster Flower, Wild Heliotrope, Scorpionweed, Phacelia: *Phacelia*

A large genus of beautiful annuals and perennials, very popular among wildflower-watchers. *Most are hairy, those hairs often sticky-tipped.* Their leaves are usually alternate, undivided, toothed, or lobed, but many are 1-2-pinnate. *The inflorescence is a dense, one-sided coil like the head of a violin.* The calyx lobes are generally thin and free most of their lengths. The corolla is fused at its base, with 5 free rounded lobes. Generally, the corolla is shaped like a bowl, funnel, or bell and is white, blue, or purple in color. The 5 stamens are attached to the corolla, each sometimes flanked by a pair of short flaps called "scales."

Nearly all species are much more common in the years following fire. We have 13 types in our area; most are annuals, 2 are perennials.

Some have *stiff hairs* that jab the skin when touched and inject chemicals that produce severe dermatitis in sensitive-skinned individuals. An itchy rash and raised pustules appear at the contact site, and the toxins can translocate and affect the eyelids and other sensitive skin. They are pollinated by bees and flies.

The genus was established in 1789 by A.L. de Jussieu in reference to the dense cluster of flowers on each coiled inflorescence (Greek, phakelos or phakellos = a cluster or bundle).

Santiago Peak Phacelia

	Leaves	Corolla	Notes and Range
Annuals			
Short-lobed *P. brachyloba*	Deeply lobed to compound	White to lavender, yellow throat	Mostly found only after fires. Santa Ana Mountains.
Caterpillar *P. cicutaria*	Deeply lobed to compound	Whitish to pale lavender	Abundant. Coastline, foothills, and mountains.
Hubby's *P. hubbyi*	Deeply lobed to compound	Whitish to pale lavender	Long wavy hairs on inflorescence stems; inflorescence dense and spherical. Uncommon. In Limestone, Santiago, and Hicks Canyons, mostly on steep slopes.
Great Valley *P. ciliata*	Lobed to pinnate	Blue	Very broad sepals. Uncommon. Loma Ridge, Temescal Canyon, and Gavilan Hills.
Washoe *P. curvipes*	Entire	White with purple tips	Small, many-branched. Santa Ana Mountains: Modjeska Peak, Indian Truck Trail.
Davidson's *P. davidsonii*	Lower deeply lobed to compound, upper entire to lobed	White with purple tips	Small, many-branched. Santa Ana Mountains: Santiago and Modjeska Peaks.
Common *P. distans*	Pinnate	Blue	Coastline, foothills, mountains. Abundant in years of plentiful winter-spring rain.
Tansy *P. tanacetifolia*	Pinnate	Blue	Uncommon, coastal to foothills, and Corona foothills.
Large-flowered *P. grandiflora*	Entire	Lavender, large!	Only after fire. Locally rare: Caspers and Whiting Ranch Wilderness Parks and upper San Juan Canyon.
Desert C. Bells *P. campanularia*	Entire	Bright blue	A desert native, common in seed mixes, often hydroseeded on road cuts.
Wild C. Bells *P. minor*	Entire	Purple to bluish	Corolla throat slightly pinched-in at the top. Abundant, mostly in San Juan Canyon and above Temescal Valley.
Parry's *P. parryi*	Entire	Blue to purple, 5 white dots; lighter throat	Coastline, foothills, mountains. Abundant post-fire and in years of plentiful winter-spring rain.
Santiago Peak *P. keckii*	Entire, toothed to slightly lobed	Lavender to red-purple, throat yellow	Strict fire-follower, quite rare. Santa Ana Mountains: Santiago, Modjeska, and Pleasants Peaks, plus upper Holy Jim and Coldwater Trails.
Perennials			
Imbricate *P. imbricata*	Primarily basal, pinnate	White to pale lavender	Often grows from rock cracks. Santa Ana Mountains: foothills and major canyons.
Branching *P. ramosissima*	Pinnate	Pale to medium blue	San Joaquin Hills, Chino Hills, Santa Ana Mountains.

CLUSTER FLOWERS: **PHACELIA**   EUDICOTS: **BORAGINACEAE • 161**

Short-lobed
Phacelia brachyloba
Shallow bowl; white to
pink, throat yellow;
leaves linear
to 1-pinnate

Caterpillar
Phacelia cicutaria
Shallow bowl; pale lavender,
throat often spotted;
leaves deeply lobed
to 1-pinnate

Hubby's
Phacelia hubbyi
Shallow bowl; pale lavender,
throat often spotted;
inflorescence densely hairy;
leaves deeply lobed or 1-pinnate

Great Valley
Phacelia ciliata
Shallow bowl; pale blue
to lavender, throat pale;
leaves deeply
lobed to divided

Washoe
Phacelia curvipes
Shallow bowl; white with
blue to violet lobes; leaves
roughly elliptical
(rarely lobed or 1-pinnate)

Davidson's
Phacelia davidsonii
Shallow bowl; white with
violet lobes; lower leaves
deeply divided
or 1-pinnate

Common
Phacelia distans
Shallow bowl; blue to
blue-lavender;
long filaments;
leaves 1-2-pinnate

Tansy
Phacelia tanacetifolia
Shallow bowl; blue to
blue-lavender; *very
long* filaments; leaves
1-2-pinnate

Tansy: Genevieve Walden

Large-flowered
Phacelia grandiflora
Large shallow bell,
25-40 mm across;
blue to violet, dark spots;
leaves oval, toothed

Desert Canterbury Bells
Phacelia campanularia
Deep vase-like bell, *throat
broad,* not narrowed;
bright blue; leaves
oval-round, toothed

Wild Canterbury Bells
Phacelia minor
Deep bell, *narrowed at
upper throat;* purplish
to bluish; leaves
oval-round, toothed

Parry's
Phacelia parryi
Shallow bowl; purple to
blue, *5 white patches
near throat;* leaves
oval-round, toothed

Santiago Peak
Phacelia keckii
Deep narrow bell; lavender
to purplish, throat yellow;
leaves elliptical to oval,
toothed or lobed

Northern Imbricate
P. imbricata ssp. *imbricata*
Narrow cylinder; white;
leaves pinnate, *7-15 lobes;*
upright; medium
to high elevations

Southern Imbricate
P. imbricata ssp. *patula*
Narrow cylinder; white;
leaves pinnate, *3-7 lobes;*
mounded; higher
elevations

Branching
Phacelia ramosissima
Medium-depth funnel or
bell; dirty-white to bluish
to lavender; leaves
1-2-pinnate; mounded

Annual Phacelias
Short-lobed Phacelia
Phacelia brachyloba (Benth.) A. Gray

An annual, 8-60 cm tall, the *stems covered with short hairs, often sticky-tipped. Leaves* 1.5-4.5 cm long, linear, narrow, with rounded lobes, sometimes pinnate, *often in a basal rosette. Corolla* 7-10 mm long, bowl-shaped with 5 rounded *white* lobes (*rarely pinkish*) and a *yellow throat*. In flower Mar-June.

Found almost exclusively during the first few years following fire, in the foothills and higher elevations of the Santa Ana Mountains. Pollinated by many species of small flies and bees.

The epithet was given in 1835 by G. Bentham in reference to its short leaf lobes (Greek, brachys = short; lobos = lobe).

Caterpillar Phacelia
Phacelia cicutaria E. Greene **ssp. *hispida*** (A. Gray) Thorne [*P.c.* E. Greene var. *hispida* (A. Gray) J.T. Howell]

An annual, 18-60 cm tall, the *stems upright or clambering* up adjacent plants, tree trunks, or rocks. The sticky stems, leaves, and calyx lobes are *covered with long stiff white hairs.* Each hair has a swollen bulb at its base. *Leaves* 2-15 cm long, oval to oblong in outline, *deeply lobed, usually pinnate with toothed lobes.* Calyx lobes are grayish, very long and narrow. *Corolla* 8-12 mm long, bell-shaped, *pale lavender,* often spotted above the throat. *The flowers are in long coiled groups.* As each coil unfurls and its last flowers open, the group looks like a hairy caterpillar, especially after it has dried. In flower Mar-May.

Our most widespread and abundant phacelia. Found in grasslands, coastal sagebrush scrub, chaparral, and oak woodlands. Common in the San Joaquin Hills, Puente-Chino Hills, Temescal Valley, Lomas de Santiago, Santa Rosa Plateau, and major canyons and hillsides throughout the Santa Ana Mountains.

The name was given in 1902 by E.L. Greene for the similarity of its leaves to *Cicuta,* a genus of plants in the carrot family, Apiaceae (Latin, cicutarius = similar to *Cicuta*). Our form was named in 1878 by A. Gray for the stiff bristly hairs on its calyx (Latin, hispidus = spiny, shaggy, rough).

Hubby's Phacelia ★

Phacelia hubbyi (J.F. Macbr.) L.M. Garrison
[*P. cicutaria* E. Greene var. *hubbyi* (J.F. Macbr.) J.T. Howell]

An annual, quite similar to caterpillar phacelia (*Phacelia cicutaria,* above). Its long gray hairs give the plant a gray appearance. Branches of the inflorescence are covered with very long wavy hairs. *Flowers are concentrated in very large fuzzy spheres.* Calyx lobes are grayish and shaggy-haired. *Corolla* 8-12 mm long, bell-shaped, *pale lavender,* often spotted above the throat. In flower Apr-June.

It occurs on steep cliff-faces, only in the counties of Kern, Los Angeles, Santa Barbara, Ventura, and Orange. Known here from Aqua Chinon, Hicks, Limestone, and Santiago Canyons, plus Santiago Trail to Old Camp. CRPR 4.2.

The epithet was given in 1917 by J.F. Macbride in honor of F.W. Hubby, Sr., who collected the first specimen on 20 May 1896 in Ojai, California.

Hubby's Phacelia

Inflorescence white-fuzzy and clustered like a large fist

Hubby's Phacelia | Caterpillar Phacelia | Hubby's Phacelia | Great Valley Phacelia

Great Valley Phacelia ✿

Phacelia ciliata Benth.

An upright annual, 10-55 cm tall, *often hairy and sticky.* Leaves 3-15 cm long, oblong to oval, lobed to divided. *Calyx lobes very broad and hairy, major veins raised, the edges with small stiff hairs. Calyx lobes broaden with age. Corolla* 8-10 mm long, *pale blue to lavender, throat pale.* In flower Feb-June.

In clay and gravel soils among grasslands and on slopes. Mostly a plant of the Central Valley. Known here from Huntington Beach and Peters Canyon in 1929 and "hill slopes near Handy Creek (on road to Irvine County Park, 3.5 mi e of El Modeno)" in 1965. No recent Orange County records until it was rediscovered by Fred Roberts on Loma Ridge in late April 2012. Also known from near Cajalco Road in the Gavilan Hills, just east of our area. An old 1897 record for "Temescal" may refer to a population in the Temescal Valley.

Named in 1835 by G. Bentham for the small hairs on its calyx lobes (Latin, ciliatus = furnished with cilia).

Great Valley Phacelia

Washoe Phacelia
Phacelia curvipes S. Watson

An annual 4-15 cm tall, its *many spreading branches* are covered with short hairs, the upper stems also with *sticky-tipped hairs*. Leaves are *mostly basal*, 1-4 cm long, elliptical or wider near the tip, and *undivided* (sometimes unevenly lobed or pinnate). The individual flower stalks curve with age. Calyx lobes with hairs of two types: very short hairs mixed with longer stiff hairs. Corolla 4-8 mm long, broad to bell-shaped, *white tube and throat with blue to violet lobes*. Stamens 2-6 mm long, each has a pair of linear "scales" at its base. The style is 2-4 mm long. Seeds are generally 1 mm long, rarely to 1.5 mm. In flower Apr-June.

Grows in open areas among chaparral and coniferous forests. Known from the summit of Modjeska Peak and near the junction of Main Divide Road and Indian Truck Trail.

The name was given in 1871 by S. Watson for its curved flower stalks (Latin, curvus = curved or bent). The common name refers to Washoe, Nevada, where it was first found.

Similar. Davidson's phacelia (*P. davidsonii*): leaves lobed; corolla, stamens, and style larger.

Davidson's Phacelia
Phacelia davidsonii A. Gray

An annual, 5-20 cm tall, its *few-branched stems spread out or stand upright. It is covered with short hairs or sparse stiff hairs; none are sticky-tipped*. Leaves 0.8-7 cm long, elliptical or wider near the tip, the *lower leaves deeply divided or pinnate*, upper leaves undivided or unevenly lobed. Calyx lobes with short hairs only. Corolla 7-15 mm long, broad to bell-shaped, *white with violet lobes*. Stamens 3-6 mm long, each has a pair of linear "scales" at its base. The style is 4-8 mm long. Seeds are generally 2 mm long, sometimes 1-2 mm. In flower May-June.

Found in open areas among chaparral and coniferous forests. In our area, known only from the summits of Santiago and Modjeska Peaks, most often after fire or other soil disturbance. It may also occur on other peaks such as Trabuco, Pleasants, or Sierra. In Spring 2008, we found it growing with *Phacelia keckii* atop Modjeska Peak.

Named in 1875 by A. Gray to honor G. Davidson, who first collected it in Kern County, California.

Similar. Washoe phacelia (*P. curvipes*): most leaves not lobed; corolla, stamens, and style smaller.

Common Phacelia
Phacelia distans Benth.

An annual, 15-80 cm tall, *sprawling or upright,* sparsely covered with stiff hairs, the upper parts with sticky-tipped hairs. *Leaves 2-10 cm long, 1-2-pinnate, each segment rounded or toothed, very fern-like.* Calyx lobes oblong to lance-shaped. Corolla 6-9 mm long, funnel-shaped. *Stamen filaments, pistil, and corolla are all the same blue to blue-lavender color, rarely white. The stamens are only a little longer than the corolla. In fruit, the corolla and stamens fall off as a single unit.* In flower Mar-June.

Quite common in sandy, clay, and rocky soils. Widespread, known from upper Newport Bay, Crystal Cove State Park, San Joaquin Hills, Laguna Canyon, Dana Point Headlands, Rancho Mission Viejo, Chino Hills, and in major canyons of the Santa Ana Mountains.

Named in 1844 by G. Bentham for its widely spaced stem hairs (Latin, distans = separated, apart).

Similar. Tansy phacelia (*P. tanacetifolia*): fewer-branched; leaves also fern-like; *stamens much longer than corolla;* much less commonly found.

Common Phacelia

Common Phacelia — Stamens about as long as corolla, flower size identical
Tansy Phacelia — Stamens *much longer* than corolla, flower size identical

Tansy Phacelia
Phacelia tanacetifolia Benth.

An annual, 15-100 cm tall, *upright with few branches.* It is covered with short, stiff, sticky-tipped hairs. The *fern-like leaf* is 2-20 cm long, oblong to oval in outline, 1-2-pinnate with many small toothed or rounded lobes. Calyx lobes slender. The *light to medium blue or lavender corolla* is 6-9 mm long, bell-shaped. *Its very long stamens protrude far out of the flower. In fruit, the corolla and stamens dry up and remain on the flower.* In flower Mar-May.

Primarily an inland plant of sandy soils among coastal sage scrub. In Orange County, more common near the coast. Known from Newport Beach, upper Newport Bay, Laguna Hills, Silverado Canyon, and San Juan Canyon. Also found sparingly in the foothills near Corona and at Glen Ivy above Temescal Valley. In cultivation.

Named in 1835 by G. Bentham for the similarity of its leaves to that of tansy (*Tanacetum vulgare* L., Asteraceae), a common European plant (Latin, tenacetum = tansy; folium = leaf).

Similar. Common phacelia (*P. distans*): more branched; leaves also fern-like; *stamens only slightly longer than corolla;* quite abundant.

Tansy Phacelia

Large-flowered Phacelia
Phacelia grandiflora (Benth.) A. Gray

A *stout upright annual* 50-100 cm tall, sparsely covered with long stiff hairs and short sticky-tipped hairs. *Leaves* 3-20 cm long, *oval or round in outline*, edged with uneven teeth. The *large corolla* is 25-40 mm across and broadly open like a shallow bowl. *The throat is white and the lobes are blue to violet, the base of each lobe usually blue- or violet-spotted.* Flowers give off a sweet-smelling perfume that fills the air. In flower Apr-June.

Appears only during the first few years after fire, among chaparral and riparian vegetation. Very uncommon in our area. To date, known here only from Hicks Canyon, Whiting Ranch Wilderness Park, Caspers Wilderness Park, and portions of San Juan Canyon.

Named in 1875 by G. Bentham for its large flowers (Latin, grandis = large; floris = flower).

Large-flowered Phacelia

Large-flowered Phacelia

Desert Canterbury Bells

Desert Canterbury Bells, Desert Bluebells
Phacelia campanularia A. Gray

An *upright annual*, 18-55 cm tall. *Leaves ovate to round*, the edges toothed. *Corolla* 15-40 mm long. Its *deep tubular flowers* are *bright blue*, shaped like a straight-sided bell. A native of our deserts, it is often included in wildflower seed mixes used in gardens and along roadsides; it sometimes appears in or near our wildlands. In flower Feb-Apr.

It was named in 1878 by A. Gray for its bell-shaped flowers (Latin, campanula = a bell).

Similar. Wild Canterbury bells (*P. minor*): corolla purple to blue-purple; bell is much deeper and constricted just before the free lobes flare outward; much more abundant.
Parry's phacelia (*P. parryi*): corolla purple to blue, shallow, bowl-shaped, a pale patch at the base of each lobe.

Desert Canterbury Bells

Wild Canterbury Bells, California Bluebells
Phacelia minor (Harvey) Thell

An annual, 20-60 cm tall with *upright stout stems*, covered by both short sticky-tipped hairs and stiff longer hairs. *Leaves 2-11 cm long, oval or rounded in outline,* the edges with uneven teeth. *Corolla* is *a very deep bell* 10-40 mm long, *purple to bluish, slightly constricted at the throat.* In flower Mar-June.

Very common throughout our area, especially after fire, when it dominates slopes, turning them vivid purple. Known from canyons and foothills of the San Joaquin Hills and throughout the Santa Ana Mountains. Some of those from Indian Truck Trail are more blue-purple than those in other places (as at Caspers Wilderness Park).

Named in 1846 by W.H. Harvey for its corolla lobes, which are disproportionately small for its large flowers (Latin, minor = smaller).

Similar. Desert Canterbury bells (*P. campanularia*): corolla bright blue, vase-shaped, throat not constricted; **Parry's phacelia** (*P. parryi*): corolla purple to blue, shallow, bowl-shaped, a pale patch at the base of each lobe.

Parry's Phacelia
Phacelia parryi Torr.

An annual, 10-70 cm tall, *upright, sticky, and hairy. Leaves* 1-12 cm long, *oblong to oval in outline,* with uneven teeth along the edges. *Corolla* is *shallow, bowl-shaped,* 10-20 mm long, *purple to blue, the throat paler or whitish, usually with a white or yellowish patch at the base of each lobe.* In flower Mar-May.

A plant of canyons, slopes, and foothills of the San Joaquin Hills and Santa Ana Mountains. Generally more common along the coast than on inland hills and mountains. Look for it at Crystal Cove State Park, Laguna Coast Wilderness Park, Limestone Canyon, and sparingly at Caspers Wilderness Park.

Named in 1859 by J. Torrey in honor of C.C. Parry.

Similar. Desert Canterbury bells (*P. campanularia*): corolla bright blue, vase-shaped, usually no pale patch at the base of each lobe. **Wild Canterbury bells** (*P. minor*): corolla purple to blue-purple; bell is much deeper and constricted just before the free lobes flare outward; much more abundant.

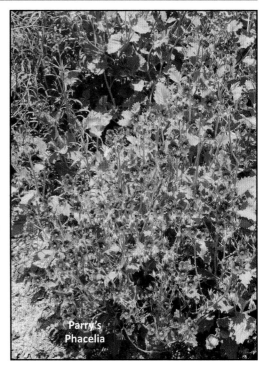

Santiago Peak Phacelia ★
Phacelia keckii Munz & I.M. Johnst.
[*P. suaveolens* E. Greene ssp. *keckii* (Munz & Johnston) Thorne]

Santiago Peak Phacelia

An annual, 5-40 cm tall, *multiple-branched, upright or spreading*, covered with short sticky-tipped hairs. *Leaves* 1-7.5 cm long, *elliptical to oval in outline, toothed or lobed*. The *lavender to rose-purple corolla is a long narrow bell, 10-14 mm long, the tube and throat yellow, streaked with purple*. It gives off a mild sweet scent. In flower May-June(-Sep).

Found on dry slopes, ridges, and peaks among chaparral and coniferous forests almost exclusively in the Santa Ana Mountains. A very rare plant, it appears during the first year or so after fire and sometimes after severe soil disturbance. Known from the Coldwater Canyon Trail at 4700 feet elevation (the type locality, first found there in 1923, then again in 1931, 1935, and 1968), Holy Jim Canyon Trail at 4800 feet (in 1923), Pleasants Peak at 3860 feet (in 1981), and Modjeska Peak at 5059 feet (in 1988). In 1990, a single plant was found in Arroyo Seco Creek in the Agua Tibia Mountains of western Riverside County (about 2000 feet). CRPR 1B.3.

Our repeated attempts to find the plant failed for many years, then finally paid off. In October 2007, the Santiago Fire (USFS Fire CA-ORC-068555) broke out in the Lomas de Santiago, charring the foothills before traveling upward into the Santa Ana Mountains. Fire agencies responded by clearing numerous new fire breaks and safety zones in the forest. The fire was terminated a few days later. Harding Canyon was extensively burned, as was Maple Springs Saddle and a small portion of upper Silverado Canyon. Neither Santiago nor Modjeska Peak was burned. The following spring, on 16 May 2008, while documenting the status of the burned and otherwise disturbed areas for the US Forest Service, Bob Allen and Reggie Durant discovered a total of 4 individuals of the plant growing in 2 firebreaks on the west- and south-facing regions of Modjeska Peak near Main Divide Road. A few hours later, they found a third location, in a large firebreak on the north-facing slope of Modjeska Peak, which hosted about 100 plants. The following spring, however, very few of the plants were found in that location.

Santiago Peak Phacelia

The epithet was given in 1924 by P.A. Munz and I.M. Johnston in honor of D.D. Keck, a student of Munz's and co-collector of the first specimen, discovered as they hiked Coldwater (Glen Ivy) Trail on 14 June 1923.

Santiago Peak Phacelia

Santiago Peak Phacelia

Perennial Phacelias

Northern Imbricate Phacelia

Phacelia imbricata E. Greene **ssp. *imbricata***
[*P.i.* ssp. *bernardina* (Greene) Heckard]

A perennial from a woody taproot. *Its stems are upright or spreading out, covered with stiff hairs, some sticky-tipped.* Most of its *grayish leaves* are in a *basal tuft*. Each blade 5-15 cm long, linear to oval in outline, pinnate, *generally with 7-15 smooth-edged lobes.* Some upper leaves are not divided. The inflorescence is tightly crowded. *Calyx lobes overlap each other, outermost lobes narrowly oval, usually sticky.* Corolla 4-7 mm long, *a narrow cylinder or bell, usually white,* sometimes lavender. Stamens very long and protrude far from the flower. Each stamen has a pair of oblong "scales" at its base. In flower Apr-July.

A plant of foothills and mountains, found mostly among coastal sagebrush scrub, chaparral, and oak woodlands, often in partial shade. Known from Chino Hills, Audubon Starr Ranch Sanctuary, our wilderness parks, up to higher elevations in the Santa Ana Mountains, often shaded by oak trees. Easily found along North Main Divide Truck Trail on the flanks of Modjeska Peak, where it is commonly pollinated by bumble bees and other native bees.

Named in 1893 by E.L. Greene for its overlapping calyx lobes (Latin, imbricatus = covered with tiles or scales).

Southern Imbricate Phacelia

Phacelia imbricata E. Greene **ssp. *patula*** (Brand) Heckard

Similar to northern imbricate phacelia (*P. imbricata* ssp. *imbricata*) but *grows mostly flat on the ground, spread out, with stem tips upward,* sometimes entirely upright. *Leaves with only 3-7 lobes. Outer calyx lobes lance-shaped, not sticky.* In flower May–Aug.

Found at *higher elevations* among chaparral and woodlands. Known here only from Modjeska Peak, possibly more widespread.

The epithet was given in 1912 by A. Brand for its low-growing, spreading growth habit (Latin, patulus = spread out).

The flowers of imbricate phacelia don't open up very wide, so they're just right for this slender-bodied Andrenid bee to slip inside for nectar.

Branching Phacelia
Phacelia ramosissima Douglas ex Lehm.

A large perennial, its many-branched stems spread out, 30-150 cm long. *It often forms large intertwined mats and mounds.* Long stiff hairs on the stem and leaves are rough to the touch and have swollen bulbs at their base. On the stems, below the inflorescence, the hairs are sticky-tipped. Stem hairs are numerous, very stiff, often painful to the touch. The *leaf blade is 40-200 mm long, oblong to oval in outline, pinnate with lobed leaflets or 2-pinnate.* The rough-edged leaflets are alternately arranged on the leaf. *Corolla* is 5-8 mm long, funnel- or bell-shaped, *dirty-white to bluish or lavender.* Its long stamens protrude far out of the flower. *Pollen is lavender, easily seen on bumble bees as they gather it.* In flower May-Aug.

Widespread and abundant, found from coastline, valleys, foothills, and mountain slopes. Known from San Joaquin Hills, Laguna Canyon, Carbon Canyon in the Chino Hills, Silverado Canyon, Bell Canyon, Hot Springs Canyon, San Juan Canyon, Bear Canyon Trail, all the way up to the summit of Santiago Peak, and in the Elsinore Basin. It makes a great garden plant, easily grown from seed raked into the soil.

The epithet was given in 1830 by D. Douglas for its many branches (Latin, ramus = branch; -issimus = very much).

Branching Phacelia

Branching Phacelia

Branching Phacelia

Bumble Bee on Branching Phacelia

LENNOAS
Scaly-stemmed Sand Plant
Pholisma arenarium Nutt. ex Hook.

Perennial herb, parasitic on the roots of nearby perennial plants. The *stout stem* stands about 5-10 cm aboveground, nearly one meter belowground. Leaves small, 5-25 mm long, linear to triangular, sticky, generally underground only, suspected to absorb water. *Inflorescence conical in outline and densely flowered. Corolla* 4-10 lobed, *lavender to deep blue-purple with white frilly edges.* Stamens equal in number to corolla lobes. In flower Apr-Jul(-Aug), Oct.

An unusual plant of *sandy soils,* found most often along the western edge of the Sonora and Mojave Deserts. Also in coastal dunes of San Diego, Los Angeles, Santa Barbara, and San Luis Obispo Counties. On 20 May 1991, Steve Boyd found it in decomposing granite soil among chaparral in lower Rice Canyon, northwest of Lake Elsinore. It may once have occurred on our coastal dunes, but foot traffic and sand manipulation may have eradicated it. We remain hopeful for its discovery on the coast.

Named in 1844 by T. Nuttall for its growth in sand (Latin, arenarius = pertaining to sand).

Scaly-stemmed Sand Plant

BRASSICACEAE • MUSTARD FAMILY

Annuals, biennials, perennials, or shrubs. Most of our species are herbaceous annuals, but some are biennials or herbaceous perennials. Mustards enjoy cooler months and flower primarily during winter and spring. Look for our native mustards in moist soils, sometimes in partial shade, often growing among other plants that protect and cool the soil above their roots.

For many species, the leaves are larger at the plant's base and smaller upward on the stems. They are *alternately arranged* and entire or dissected. *Corollas are usually yellow or white* though a few are pink, purple, or orange. *Sepals and petals are each 4 in number and free* (not fused together) though petals are lacking in a few tiny-flowered species. The petals are very narrow at the base and held upward, parallel to the style, then often turn 90 degrees outward and widen; some curve backward and remain narrow. *The petals are arranged in 2 perpendicular groups that form a figure like a Maltese cross or slightly offset like the letter "X"* (2 petals high, 2 low). *Stamens vary but are generally 6 (4 long, 2 short), rarely only 2 or 4*. The *ovary is superior* and sports a single short style that ends in a simple or two-lobed stigma. The *fruit is a 2-chambered capsule called a silique* that opens when the two sides fall away, exposing the seeds, which are attached to a paper-thin partition (false septum) within. Siliques are typically long and narrow. Those that are round or only slightly longer than wide, often flattened, are called *silicles*. The style often persists at the far end of the fruit and may grow with it, forming a beak. The length of the beak is important in identification, as are leaves, pedicel, flower structure, color, and fruit shape.

Members of the family have a characteristic odor and taste. They produce a group of organic compounds called *glucosinolates* (mustard oils) that contain both sulfur and nitrogen, manufactured by shuffling and recombining atoms from glucose and an amino acid.

This is a large worldwide group with over 3,000 species, about 250 in California and nearly 50 in our area. Many non-native mustard species are troublesome weeds in both natural and urban areas.

Mustards are all quite similar and can be surprisingly difficult to identify. When doing so, note the divisions and shape of the lower and upper leaves; shape and color of the flower; length and shape of the pedicel, fruit, and beak; how the mature fruits are held (downward, outward, upward); and arrangement of the seeds within the fruit.

	Linear-pod Mustards	Oblong-pod Mustards	Flat-pod Mustards
Fruit	A linear silique, at least 4 times longer than wide.	An oblong silique, less than 4 times longer than wide, fleshy	A silicle: rounded, flattened, generally wider than long
Subgroups and Genera	Cresses: *Barbarea, Cardamine, Planodes, Sibaropsis, Turritis* Jewelflowers: *Caulanthus* Tansy Mustards: *Descurainia* Desert Whitlow: *Draba* Wallflowers: *Erysimum* Water Cresses: *Nasturtium, Rorippa* Keeled Fruits: *Tropidocarpum* Non-Native Weeds: *Brassica, Diplotaxis*, Hirschfeldia, Matthiola, Raphanus, Sisymbrium*	Sea Rockets: *Cakile*	Lacepods and Fringepods: *Athysanus, Thysanocarpus* Shepherd's-purse: *Capsella* Peppergrasses: *Hornungia*, Lepidium*

* Not presented in this book

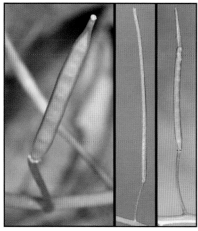

Linear Silique:
Cardamine, Sisymbrium, Brassica

Oblong Silique:
Cakile

Silicle:
Thysanocarpus, Capsella

LINEAR-POD MUSTARDS

Cresses: *Barbarea, Cardamine, Planodes, Sibaropsis, Turritis*

American Winter Cress
Barbarea orthoceras Ledeb.

A stout, stiffly branched mustard with ribbed, angular stems. A biennial or perennial, it grows 10-60 cm tall. *Leaves* are *pinnately lobed, dark green, often tinged with purple*. Basal leaves < 12 cm long, 2-3 pairs of lateral lobes; terminal lobe large and rounded. Stem leaves smaller, often with fewer lobes, and may clasp the stem. *Flowers small with bold yellow petals, sometimes paler.* Fruits 1.5-5 cm long, 4-sided, square in cross-section, firmly held upward or outward. There is one row of seeds on each side of the septum. Seeds are brown and oval, not spherical. In flower Mar-Sep.

Find it in areas of *moist soil* in meadows and along seasonal drainages, often in partial shade. Known from Hot Springs Canyon, San Juan Canyon, meadows near Blue Jay and Falcon Campgrounds, in the San Mateo Canyon Wilderness Area, and along Trabuco Trail.

The epithet was established in 1824 by C.F. von Ledebour for its straight fruits (Greek, orthos = straight; keras = horn).

Similar. Marsh yellow-cress (*Rorippa palustris* (L.) Besser, not shown): fruit much shorter, only 7-15 mm long, slightly curved and held upward. **Western yellow-cress** (*R. curvisiliqua*): fruits also shorter, only 6-15 mm long, usually strongly curved upward. The fruits of both have 2 rows of seeds per chamber.

American Winter Cress

California Toothwort or Milkmaids
Cardamine californica (Nutt.) Greene
[*Dentaria californica* Nutt.]

An *early-winter* delight, this plant appears during our rainy season and leads the way for the season's wildflowers. It stands 20-70 cm tall, often leans over or sprawls outward, and grows from a short, tuberous, perennial rhizome. Its main stalk and basal leaves grow directly from the rhizome. The *basal leaves* appear after Dec or Jan rains and are *divided into 3 linear or rounded leaflets* that may have toothed margins. Farther north in the state, the basal leaves are entire and rounded or heart-shaped. *Stem leaves 1-pinnate,* the lower divided into 3 rounded leaflets, the upper into 3-5 linear sharp-pointed, toothed leaflets. *Petals* 9-14 mm long, *widely rounded, white to pale rose.* Fruits upright, 20-50 mm long. In flower Feb-May.

It *grows in partial or full shade, often on north-facing slopes,* generally among grasslands and *oak woodlands*. Look for it in the hills surrounding upper Newport Bay and along trails in Laguna Beach, Laguna Coast Wilderness Park, Aliso and Wood Canyons Wilderness Park, Black Star Canyon, Silverado Canyon, Holy Jim Canyon, Trabuco Canyon, Caspers Wilderness Park, Audubon Starr Ranch Sanctuary, Hot Springs Canyon, San Juan Canyon, Santa Rosa Plateau, and the San Mateo Canyon Wilderness Area.

The epithet was given in 1838 by T. Nuttall in reference to the place of its discovery.

Similar. Little bitter-cress (*C. oligosperma* Nutt., see inset): *small* upright annual or biennial, 8-30 cm tall; *strong basal rosette of leaves,* each 1-pinnate; *2-5 pairs of lateral leaflets,* terminal leaflet large; *stem leaves similar, very narrow leaflets;* petals about 2-4 mm long. In flower Mar-July. Moist areas along creeks and trails; also a nursery and garden weed.

California Toothwort

California Toothwort

Little Bitter-cress

Often much taller

Side branches and basal leaves sometimes trail on the ground

Leaves variable

Virginia Rock Cress ✿
Planodes virginicum (L.) Greene
[*Sibara virginica* (L.) Rollins]

An *annual or biennial*, 10-30 cm tall, that *lies flat on the ground or stands upright*. Leaves 2-6 cm long, *pinnately lobed, in an obvious basal rosette*, their terminal segment is broad. *Flowers are small and short-lived*. Sepals are tiny, to 1.5 mm long, purplish. *Petals* slightly longer, *spoon-shaped, white to faintly pink*. Fruits 2-2.5 cm long, to 2 mm wide, flat, straight, held upward. *Seeds are round and flattened, each with a narrow wing around it*. In flower Apr-May.

A *wetland plant, found at the edges of vernal pools and seasonally moist soil*. Quite uncommon here, known only from vernal pools at Fairview Park in Costa Mesa and Rancho Mission Viejo. Also known from Skunk Hollow, east of Murrietta, not far from our area.

The plant was originally assigned to the genus *Cardamine* (1753), then bounced between the genera *Arabis* (1811), *Planodes* (1912), and *Sibara* (1941). In 2007, I.A. Al-Shehbaz moved it back into *Planodes*, a genus established by E.L. Greene in 1912, in reference to its characteristics, which tricked other botanists into placing it into the wrong genus (Greek, planos = deceiving). The epithet was established in 1753 by C. Linnaeus in reference to the state of Virginia, where it was originally discovered.

Hammitt's Clay-cress ★
Sibaropsis hammittii Boyd & Ross

This *delicate herbaceous annual* was discovered by local botanists Steve Boyd and Tim Ross near Elsinore Peak in 1992 and formally described in 1997. The plant was so different that they had to establish a new genus for it. *It stands 5-15 cm* (rarely to 20 cm) *tall* and is branched from the base. The branches are hairless, light green, and sometimes have a faint purplish color. *Leaves* are 10-45 mm long, 0.5-1.0 mm wide, *very narrow, rounded in cross-section, fleshy, and basically hairless*. The inflorescence has fewer than 10 flowers. Sepals are 3 mm long, 0.5-1.0 mm wide, purplish to greenish-purple, paler at the base, with a whitish margin. Petals are arranged in two groups, those in one group slightly longer than those in the other. *Each petal* is somewhat *spoon-shaped, often indented at the tip*, 8.5-10 mm long, 2-2.5 mm wide at its broadest, *light purplish- or pinkish-lavender with darker purplish veins, aging purplish*. Fruits are held upward and resemble leaves. They are 15-25 mm long, 0.6-0.8 mm wide, and terminate with a beak 1.5-4.5 mm long. Seeds are elongate and arranged in only one row on each side of the septum. In flower Mar-Apr.

In our area, known only from *clay soils among grasslands* near Elsinore Peak. It also occurs on Poser and Viejas Mountains in the southern end of the Cuyamaca Range in San Diego County. The plant always occurs in low numbers at only a few locations; it is never common. It is listed as a Sensitive Species by the US Forest Service. CRPR 1B.2. If you find it, admire, but please do not touch or otherwise disturb it.

The genus was named in 1997 for its similarity to members of the genus *Sibara* (Greek, -opsis = resembling in appearance). The epithet honors M. Hammitt, a friend of Boyd and Ross.

Tower Mustard
Turritis glabra L. [*Arabis glabra* (L.) Bernh.]

A tall stout mustard, 0.4-1.2 meters in height, usually with an *unbranched single main stem*, biennial or rarely perennial. Its *basal leaves* are 6-15 cm long, hairy and elongate, often with teeth that point backward. *Stem leaves* are about the same length, nearly hairless; their base looks like two rounded earlobes that *clasp the stem*. Sepals are 3-5 mm long and yellowish. *Petals* are elongate or spoon-shaped, *yellow-cream*, rarely purplish. The *slender fruits are held upright* and close to the stem, 4-10 cm long, 1 mm wide, with a very short and stout beak. Seeds are in two rows on each side of the septum. In flower Mar-July.

Known from shaded sites along the Maple Springs Truck Trail in Silverado Canyon, along Main Divide Truck Trail southeast of Santiago Peak, and along San Juan Trail south of Blue Jay Campground.

The epithet was established in 1753 by C. Linnaeus in reference to its hairless, smooth upper leaves (Latin, glabro = hairless, smooth). The common name refers to the height of this species.

*Right: Plants on the southeast flank of Santiago Peak, in shade of coast live oaks and canyon live oaks (*Quercus agrifolia *and* Q. chrysolepis, *Fagaceae).*
Insets: Stem leaves, note the ear-like base. A small metallic sweat bee in the genus Lasioglossum (Halictidae) *uses its mandibles to coax and collect pollen from an anther.*

Jewelflowers: *Caulanthus*

These plants are so named for their small delicate flowers with intricate often wavy-edged petals, like little jewels presented to the careful observer willing to have a closer look. They are annuals or perennials, some with "pouched" sepals that curve outward at the base and back inward near the top, like a full cloth sack set on the ground.

San Diego Jewelflower
Caulanthus heterophyllus (Nutt.) Payson
[*Streptanthus heterophyllus* Nutt.]

A very slender upright annual. The main stem is bristly near the ground, less so above. The basal leaves are 3-12 cm long, oblong, narrow, toothed or lobed. *Upper leaves are less lobed and clasp the stem.* Flowers are held upright. Sepals 8-9 mm long, barely pouched, purplish or greenish. *Petals 12-14 mm long with straight (not wavy) membranous edges, pale with purple veining, linear, curved backward.* Fruit is slender, 5-8 cm long, and hangs downward. In flower Mar-May.

Found on open slopes in coastal sage scrub, chaparral, and mixed evergreen woodland, especially common following fire. Look for it along the Harding Canyon Trail, in Limestone Canyon, Caspers Wilderness Park, along Main Divide Truck Trail, Temescal Valley, in the San Mateo Canyon Wilderness Area, and in the Chino Hills along Santa Ana Canyon.

The epithet was given in 1838 by T. Nuttall for its different basal and stem leaves (Greek, heteros = other; phyllon = leaf).

California Mustard

Caulanthus lasiophyllus (Hook. & Arn.) Payson
 [*Guillenia lasiophylla* (Hook. & Arn.) E. Greene, *Thelypodium lasiophyllum* (Hooker & Arn.) E. Greene]

Often a *tall annual*, 20-180 cm in height. *Basal leaves* 3-21 cm long, *lance-shaped to oblong, with wavy edges or pinnate.* The lowest stem leaves are usually larger than the basal leaves, upper stem leaves much smaller. *Flowers are held upright.* Sepals are greenish to pinkish, 1.5-4 mm long, not pouched. *Petals yellowish-white, rarely pinkish,* 3-6 mm long, not wavy-edged. *Fruit hangs down when mature,* 1-7 cm long, straight or curving outward near the tip. In flower Mar-June.

Occurs infrequently within coastal sage scrub, on slopes, along washes and canyons, sometimes after a burn. Known from the San Joaquin Hills, Lomas de Santiago, Verdugo Canyon, and scattered sites within the San Mateo Canyon Wilderness Area.

The epithet was given in 1841 by W.J. Hooker and G.A.W. Arnott for its hairy leaves (greek, lasios = hairy; -phyllon = leaf).

California Mustard

Tansy Mustard

Descurainia pinnata (Walter) Britton

A dainty annual with fine, feathery, bright green leaves. Grows 10-70 cm tall, upright with an obvious main stem, often with shorter side branches. *Basal leaves* 3-9 cm long, *2-pinnate,* the lobes oval or linear. Upper leaves 1-10 cm long and 1-2-pinnate. *Petals* 2-3.5 mm long, *bright yellow.* Fruit 4-20 mm long, oblong to somewhat club-shaped, held upward or outward. Two rows of seeds on each side of septum. In flower Mar-June.

It appears on chaparral slopes in alkaline and/or rocky soils, along trails, and in other slightly disturbed sites. Known from Newport Beach, Laguna Beach, Costa Mesa, Irvine, Dana Point, San Clemente, and Santa Ana Canyon near Chino Hills. Formerly more common but still easily found along trails in places such as Laguna Coast Wilderness Park.

The epithet was given in 1888 by T. Walter in reference to its pinnate leaves (Latin, pinnatus = plumed, with feathers).

Tansy Mustard

Wedge-leaf Draba, Desert Whitlow

Draba cuneifolia Nutt. ex Torr. & A. Gray

An annual, 5-25 cm tall. *Basal leaves* 0.5-2 cm long, *oval to linear,* toothed at least at their lower half. *Stem leaves* are few but of similar shape, each *with a wedge-shaped base.* Each inflorescence is on the end of a long, naked stalk. *Petals* are 3.5-5 mm long, *white, notched down the center near the tips.* Fruits are 3-12 mm long, football-shaped and flattened parallel to the septum, with a minute beak, held upward or outward; each contains 2 or more seeds. In flower Feb-May.

Grows in rocky or sandy-clay soils, often in barren or grassy patches within coastal sage scrub. Found in the Chino Hills above Santa Ana Canyon, on Goff Ridge in Laguna Beach, Lomas de Santiago, and Caspers Wilderness Park. More common in coastal San Diego County and our deserts.

The epithet was given in 1838 by T. Nuttall for its wedge-shaped leaf base (Latin, cuneus = wedge; folium = leaf).

Wedge-leaf Draba

Wallflowers: *Erysimum*

Western Wallflower

Erysimum capitatum (Douglas) E. Greene **var. *capitatum***

This lovely plant is a *biennial or short-lived perennial*. It stands upright, from 50-100 cm tall, sometimes taller. *Leaves are linear*, entire or toothed. Lower leaves 2-10(-25) cm long, covered with flattened many-branched gray hairs. *Petals are bold school-bus orange to bright yellow*. In flower May-July.

Grows in dry rocky areas among chaparral, sometimes springing up from the shelter of a shrub. Found in upper elevations of the Santa Ana Mountains, mostly along the Main Divide Truck Trail on both north and south flanks of Santiago Peak, and in Hagador Canyon.

The related European species were called *wallflowers* because the plants often grow on old rock walls.

The epithet was given in 1829 by D. Douglas for its flowers which are crowded into a head-like cluster (Latin, capitatus = having a head).

Western Wallflower

Water Cresses: *Nasturtium, Rorippa*

White Water Cress

Nasturtium officinale W.T. Aiton
[*Rorippa nasturtium-aquaticum* (L.) Hayek]

A *hairless perennial* that grows *in water*, often submerged, stems 10-60 cm long, root at the nodes. *Leaves 1-pinnate*, 1-10 cm long; *leaflets oval to linear, the tip round or pointed*, the terminal segment larger than the others. Flowers 3-4 mm long, *petals white*. Fruit 8-15 mm long, held outward or upward. In flower Mar-Nov.

Common along *slow-moving wet streamsides* throughout our area. This is the cultivated water cress, now found worldwide. There is some debate among botanists whether this plant is actually native to North America.

The epithet was given in 1812 by W.T. Aiton for its past use in medicine (Latin, officinalis = used in medicine).

Similar. Pacific yellow-cress (*R. palustris*, not pictured) and **Western yellow-cress** (*R. curvisiliqua*): petals yellow; both much less common here.

White Water-cress

Western Yellow-cress

Rorippa curvisiliqua (Hook.) Britton

An annual or biennial with many sprawling or upright stems, 10-40 cm long. *Leaves 2-7 cm long, oblong in outline, undivided to 1-pinnate*, the lobes and tip often sharply tooth-edged, the terminal segment larger than the others. Its *tiny flowers* are 1-2 mm long, petals yellow. *Fruit 6-15 mm long, <1.5 mm wide, curves upward* (rarely straight), held upward or outward, 2 rows of seeds per chamber. In flower Apr-Sep.

An uncommon plant, recorded long ago from the shore of Laguna Lakes in the San Joaquin Hills near Laguna Beach, more recently in portions of Temescal Valley.

The epithet was given in 1830 by W.J. Hooker for its curved fruit (Latin, curvus = curved; siliqua = pod).

Similar. Pacific yellow-cress (*R. palustris*, not pictured): petals also yellow but longer, terminal leaflet very much longer than the others; fruits straight; uncommon. **White water cress** (*Nasturtium officinale*): petals white, common. **American winter cress** (*Barbarea orthoceras*): fruits much larger, 1.5-5 cm long, square in cross-section, firmly held upward or outward, 2 rows of seeds per chamber.

Western Yellow-cress

Keeled-Fruit Mustards: *Tropidocarpum*
Slender Dobie-pod
Tropidocarpum gracile Hook.

A sprawling to upright annual, stems (3-)10-45 cm long. *Leaves* 1-10 cm long, 1-pinnate, *terminal segment narrow.* Sepals greenish-yellow to purplish. *Petals* 3-6 mm long; *yellow, sometimes purple-tinged, especially as they age. Fruit flattened,* 2.5-6 cm long, *held upright.* In flower (Jan-)Mar-May.

Lives in small soil depressions and partial shade among grasslands, scrub, and oak woodlands; also post-fire. Uncommon. Known from San Joaquin Hills (Temple Hill, Aliso and Wood Canyons Wilderness Park), San Mateo Canyon Wilderness Area, and shaded spots on sandy benches along lower Indian Canyon wash.

The epithet was given in 1836 by W.J. Hooker, probably for its slender fruits (Latin, gracilis = slender).

Slender Dobie-pod

Non-Native Weeds: *Brassica, Hirschfeldia, Matthiola, Sisymbrium, Raphanus*

BEAUTY OR THE BEAST?

A majority of plants native to California have an interesting relationship with soil fungi called **mycorrhizal fungi**. The fungi serve as conduits for water, nutrients, and other chemicals. In return, they get to live among or within cells of the roots. Many plants grow poorly or not at all if these fungi are lacking. Aware of this, native-plant nurseries inoculate their planting soils with these fungi to promote healthy growth of their stock.

Most members of the mustard family have no relationship with mycorrhizal fungi and therefore they can grow in areas with damaged soils where the fungi no longer exist. This gives non-native mustards an advantage over native plants, many of which *require* mycorrhizal fungi for vigorous growth. This explains why mustards are often first to grow in areas after a catastrophe: they can live in poor fungi-less soils, soak up soil moisture and nutrients, and crowd or shade out other plants. If the mycorrhizal fungi are not reestablished, mustards may continue to flourish and prevent recolonization by native species.

During the second and third winter-spring after a wildfire, our hills are often blanketed with bright yellow blossoms of non-native annual mustards, especially the ubiquitous black mustard (*Brassica nigra*). Uninformed citizens see only the beautiful yellow hills, unaware of the damage done by these mustards and the possibility that native vegetation may never reclaim the area. Well-meaning fire agencies, in their single-minded efforts to

Black Mustard infestation in Peters Canyon

"control" wildfire before, during, and after the event, routinely clear native vegetation and severely disturb the soil. This promotes the establishment of non-native mustards. After the mustards die off, their dry skeletons become mustard matchsticks, fodder for future fires.

A wise choice would be to limit native vegetation removal and soil disturbance, control mustards and other non-native plants, and vigorously restore native vegetation. We currently have the technology to carry this out.

Field Mustard ✖
Brassica rapa L. [*B. campestris* L.]

An *upright annual,* 0.2-1 meter tall. *Basal leaves pinnately lobed,* with 2-4 lateral lobes and a large rounded terminal lobe that is wavy-edged or toothed. Stem leaves linear, *the base of each ear- or arrowhead-shaped and surrounds the stem.* Petals 6-11 mm long, bright yellow. Fruits are 3-7 cm long, stout, held outward or upward; the pedicel is 7-25 cm, beak 8-15 mm, narrowed toward the tip. Seeds are dark brown or black. In flower Jan-May.

Found mostly in moist soils among disturbed areas and citrus orchards. The turnip is a cultivar of this plant. Native to Europe.

The epithet was given in 1753 by C. Linnaeus, who used the Latin name for turnip (Latin, rapa = turnip).

Field Mustard

Black Mustard ✖

Brassica nigra (L.) Koch

A very tall annual, 0.4-2.5 meters in height, *with an obvious main stem* and scattered branches above. Stems are usually smooth but are sometimes covered with stiff hairs. *Basal leaves are pinnately lobed* and variously toothed. Stem leaves similar but smaller with a tapered base. *Petals* 7-11 mm long and *bright yellow. Fruits* 1-2 cm long, upright, *held tightly to the stem, with a slender seedless beak* 1-3 mm long; pedicel thin. Seeds black. In flower primarily Mar-July, longer with summer rains.

Found throughout the region in disturbed areas and wildlands, from coastlines and canyons to foothills on all sides of the Santa Ana Mountains and the Puente-Chino Hills. After it dies off in spring, its old growth is left standing in place for the rest of the year and is finally knocked over by the next winter's rains.

This is the largest mustard brought by man from Eurasia to North America. Legend has it that during the late 1700s, Spanish missionaries scattered the seeds along routes between the California missions so that a weary traveler could follow the trail of mustard plants to the shelter of a mission. Unfortunately, these plants spread everywhere and are now among the most invasive of weeds that crowd out our native vegetation. With adequate soil moisture they can grow quite tall, often far over your head.

The epithet was given in 1753 by C. Linnaeus in reference to its black seeds (Latin, niger = black).

Similar. Shortpod mustard (*Hirschfeldia incana*): perennial; shorter in stature; smaller, paler flowers; holds its sepals outward. It too has short fruits held to the stem, but its fruit has a swelling at the base of the beak, 1-2 seeds within, sometimes only one seed per fruit body.

Black Mustard

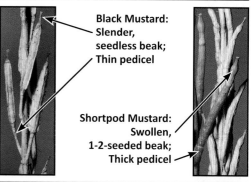

Black Mustard: Slender, seedless beak; Thin pedicel

Shortpod Mustard: Swollen, 1-2-seeded beak; Thick pedicel

Shortpod Mustard, Summer Mustard ✖

Hirschfeldia incana (L.) Lagr.-Fossat
[*Brassica geniculata* (Desf.) Benth.]

A biennial or perennial, covered with stiff, bristly, grayish-white hairs. It stands 0.2-1 meter tall. Its *basal leaves are pinnate* with a large terminal lobe and wavy or toothed edges. Uppermost stem leaves are very small, linear, and not lobed. Flowers are in short tight clusters at the end of the stems. Sepals 3 mm long, yellow-green, and curve outward. *Petals* 5-6 mm long, and *light to bright yellow. Fruits are very short,* 1-1.5 cm long, with a 3-4 mm long *thick pedicel,* and are *held upright, very tightly to the stem.* The *rocket-like beak* is 3-6 mm long, *abruptly swollen at the base; contains 1-2 seeds;* the tip roughly flattened. There is 1 row of seeds on each side of the septum but sometimes only 1 seed in the entire pod. Seeds red-brown. In flower May-Oct.

A common weedy species, found in disturbed areas and roadsides throughout the region. It usually appears later in the year than other mustards and continues flowering through summer, long after other species have faded. Native to the Mediterranean region.

The epithet was given in 1755 by C. Linnaeus for its grayish-white hairs (Latin, incanus = covered with grayish-white hairs).

Similar. Black mustard (*Brassica nigra*): annual; much taller; larger bold yellow flowers; sepals held upright. It too has short fruits held to the stem, but its fruit has a tapered seedless beak and contains several seeds per fruit body.

Shortpod Mustard

Sahara Mustard ✖

Brassica tournefortii Gouan

A *stiff, upright, many-branched annual*, 10-70 cm tall. It *often forms a rounded mass of intertwined stems* and fruits. *Leaves in a basal rosette, pinnately lobed with numerous toothed lobes.* Stem leaves few, narrowed at the base, the highest leaves tiny. Flowers crowded at tips when first opening. *Sepals often purple.* Petals 3-10 mm long, *pale yellow.* Fruits 3-7 cm long, pedicel 8-15 mm, with a stout beak 1-1.5 cm long, held outward. Seeds brown-purple. In flower Jan-June.

A native of Europe, this *extremely invasive plant* is currently spreading in the Mojave and Sonora Deserts. Fortunately, at present it is not common in our area. Known from upper Newport Bay, Garden Grove, near Saddleback College in Mission Viejo, Laguna Coast Wilderness Park, Dana Point Headlands, Lomas de Santiago, and the Temescal Valley.

The epithet was given in 1773 by A. Gouan in honor of J.P. de Tournefort.

Stock ✖

Matthiola incana (L.) R.Br.

An upright perennial, 10-80 cm tall, usually from a single stalk. Leaves 5-15 cm long, thick, linear, the edges smooth, rarely lobed or toothed. Flowers fragrant. Petals 6-16 mm long, purple to white, sometimes of one color and streaked with the other. Fruit a long silique, 8-16 cm long, 4-5mm thick, narrowed between seeds, tipped with a tiny beak. In flower Mar-May.

Native to Europe, this plant is popular in the cut-flower and gardening trades. It has naturalized and is now found along the coast of California and Mexico. Look for it in sandy soils and coastal bluffs in Seal Beach, Huntington Beach, Bolsa Chica Ecological Preserve, Newport Beach, Balboa Peninsula, upper Newport Bay, and on the cliffs near the mouth of Aliso Canyon in Laguna Beach.

The epithet was given in 1753 by C. Linnaeus for the hairs that cover it (Latin, incanus = hoary [covered with gray or whitish hairs]).

Hedge and Tumble Mustards, Rockets: *Sisymbrium*

Tumble Mustard ✖

Sisymbrium altissimum L.

An *upright annual*, 0.3-1.5 meters tall, most side-branches from above the middle. *Basal leaves pinnately lobed or pinnate. Stem leaves pinnate with thread-like lobes,* the terminal lobe about equal in length to the lateral lobes. *The 2 outermost sepals have an upright "horn" near their tip.* Petals 6-8 mm long, yellowish-white. *Fruit 5-10 cm long, the pedicel as broad as or broader than the fruit.* In flower May-July.

A plant of disturbed soils and watercourses. Uncommon here, known from Fullerton, Santa Ana River channel, upper Newport Bay, atop Santiago Peak, and sparingly in the San Mateo Canyon Wilderness Area. More common east of our area, especially in the Perris Basin. Native to Europe.

The epithet was given in 1753 by C. Linnaeus for its height (Latin, altus = high; -issimmus = very much).

Similar. London rocket (*S. irio*): plant branched at base; upper leaves pinnate or oblong with 2 spreading lobes at base; fruit stalk much narrower than the fruit. **Hedge mustard** (*S. officinale*): upper stems wiry, outspread; fruits very short, held tightly against the stem. **Oriental tumble mustard** (*S. orientale*): upright; fruits long, held away from stem, fruit stalk about the same width as the fruit.

London Rocket ✖
Sisymbrium irio L.

An *upright annual,* 15-50 cm tall, branched from near the base. Early winter and well-hydrated plants are more pliable and leafier than others. *Lower leaves* to 15 cm long, *pinnately lobed, the terminal lobe is longer than the lateral lobes and is usually arrowhead- or spear-shaped at the base.* Upper leaves similar but smaller, or oblong with 2 lobes at base, sometimes entire. *Petals tiny,* 2.5-4 mm long, *yellow. Young fruits soon overtop the flowers.* Mature fruit 3-4 cm long, the stalk (pedicel) much narrower than the fruit. In flower Jan-Apr.

The most common of our *Sisymbrium* mustards, found in vacant lots and non-native grasslands, along roadsides and trails, after fire, sometimes a garden weed. Native to Europe.

The epithet was given in 1753 by C. Linnaeus, who used an ancient name for a mustard (Latin, irio = a kind of cress). The common name refers to the fruits (Italian, rocchetto = cylindrical shape), though these days it is often attributed to the "rocket-shaped" leaves.

Similar. Tumble mustard (*S. altissimum*): upright, narrow plant; upper leaf lobes thread-like; sepals horned; fruit stalk as broad as or broader than the fruit. **Hedge mustard** (*S. officinale*): upper stems wiry, outspread; fruits very short, held tightly against the stem. **Oriental tumble mustard** (*S. orientale*): upright; fruits long, held away from stem, fruit stalk about the same width as the fruit.

London Rocket
Hairless leaves

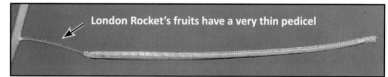
London Rocket's fruits have a very thin pedicel

Hedge Mustard ✖
Sisymbrium officinale (L.) Scop.

An *upright stiff-stemmed annual,* 30-80 cm tall, *covered with sharp hairs,* most side branches are from *above* the middle; those branches often twist. *The plant often has a blue-green tinge.* The *basal leaves* are up to 20 cm long, *pinnately lobed, the terminal lobe is about equal to or longer than the lateral lobes.* Stem leaves much smaller, with narrow lobes, often with no petiole. *Petals* 3-4 mm long, *pale yellow. Fruit short,* 8-15 mm long, *held tightly against the stem.* In flower Apr-July.

Commonly found among grasslands, along creekbeds, and *under oaks* in the San Joaquin Hills, Chino Hills, Limestone Canyon, foothills of the Santa Ana Mountains, and the San Mateo Canyon Wilderness Area. Native to Europe.

The epithet was given in 1772 by C. Linnaeus for its past use in medicine (Latin, officinalis = used in medicine). The common name refers to the branches which are so dense that the plant seems to form a hedge.

Similar. Tumble mustard (*S. altissimum*): upright, narrow plant; upper leaf lobes thread-like; sepals horned; fruit stalk as broad or broader than the fruit. **London rocket** (*S. irio*): plant branched at base; upper leaves pinnate or oblong with 2 spreading lobes at base; fruit stalk much narrower than the fruit. **Oriental tumble mustard** (*S. orientale*): upright; fruits long, held away from stem, fruit stalk about the same width as the fruit.

Hedge Mustard
Fruits short, held against the stem
Broad terminal lobe

Oriental Tumble or Hedge Mustard ✖

Sisymbrium orientale L.

An *upright annual* to 30 cm tall, *covered with soft hairs*, variously branched. *Basal leaves pinnately lobed or pinnate;* stem leaves smaller, with 2 spreading lobes at their base. Petals 8-10 mm long, *pale yellow.* Fruit 3-10 cm long, *pedicel 3-6 mm long, about the same width as the fruit.* In flower late Mar-May.

An occasional weed of coastal hillsides in Newport Beach and Corona del Mar and on the UCI campus. Easily found in Hot Springs Canyon and along Ortega Highway through San Juan Canyon. A common roadside weed in inland portions of our area. Native to Eurasia.

The epithet was given in 1756 by C. Linnaeus for its presumed place of origin, in eastern Europe or Asia (Latin, orientalis = of the East).

Similar. Tumble mustard (*S. altissimum*): upright, narrow plant; upper leaf lobes thread-like; sepals horned; fruit stalk as broad or broader than the fruit. **London rocket** (*S. irio*): plant branched at base; upper leaves pinnate or oblong with 2 spreading lobes at base; fruit stalk much narrower than the fruit. **Hedge mustard** (*S. officinale*): upper stems wiry, outspread; fruits very short, held tightly against the stem.

Oriental Hedge Mustard

Stem leaves with 2 basal lobes or unlobed

Fruit stalk very thick

Radishes: *Raphanus*

Upright annuals, sometimes perennials, *from a strong taproot.* Stems are branched mostly from above the middle, the plants *covered with stiff, bristly hairs.* Lower leaves pinnate, the terminal lobe often wide and round. *The 2 innermost sepals have a sac-like bulge at their base.* The *hard fruits* are needed for positive identification.

Jointed Charlock ✖

Raphanus raphanistrum L.

Upright, 30-80 cm tall. *Sepals often red or purplish.* Petals 15-20 mm long, *yellowish,* often with gray or purplish veins, age to pale or whitish. *Fruit strongly narrowed between seeds.* In flower Apr-June.

A common weed of vacant lots, especially along Santiago Canyon Road. Native to the Mediterranean region.

The epithet was given in 1753 by C. Linnaeus for its similarity to both cabbage and wild turnip (Greek, rhaphanos = cabbage; Latin rapistrum = wild turnip).

Similar. Wild radish (*R. sativus*): *petals generally purple or pink; mature fruit not strongly narrowed between seeds;* much more abundant. The two rarely grow together but are known to hybridize when they co-occur. Hybrids have salmon-colored petals.

Jointed Charlock

Wild Radish ✖

Raphanus sativus L.

Upright, 40-120 cm tall. *Sepals mostly yellow-green to green.* Petals 15-25 mm long, *purplish, pink, yellowish, or white with purplish or rose veins. Mature fruit not strongly narrowed between seeds.* In flower Feb-July.

Found most often in vacant lots, open fields, and grazed sites throughout our area, sometimes in wild lands. A very common weed worldwide, native to the Mediterranean region. This is the wild ancestor of the common radish.

The epithet was given in 1753 by C. Linnaeus for its use in cultivation (Latin, sativus = that which is sown).

Similar. Jointed charlock (*R. raphanistrum*): *petals generally yellow; fruit strongly narrowed between seeds;* much less abundant.

Wild Radish

Fruits: Jointed Charlock (young, constricted between seeds); Wild Radish (mature, not constricted)

OBLONG-POD MUSTARDS

Horned Sea Rocket ✘
Cakile maritima Scop.

A *fleshy annual or perennial* that is sprawling, rounded, or upright to about 60 cm long or tall. *Leaves* 4-8 cm long, *deeply pinnate* with rounded lobes. *Petals* 9-10 mm long, *pink, purplish, or rarely white. Fruit* 1.5-3 cm long, *lower half with 2 horns.* In flower Apr-Nov.

Found only on sandy beaches and dunes along our immediate coast in places such as Huntington Beach, Newport Beach, Capistrano Beach, and Trestles. The name "sea rocket" refers to the rocket-shaped fruit that breaks into 2 halves with only 1 seed in each. The outer half (the beak) is blown away by wind or carried out to sea, distributing the seed to new locations. The basal half remains on the plant and drops its seed nearby to maintain the species in its habitat. Native to Europe.

The epithet was given in 1772 by J.A. Scopoli in reference to its habitat near the sea (Latin, maritimus = of the sea).

Similar. Oval sea rocket ✘ (*C. edentula* (Bigelow) Hooker, not pictured): more sprawling; leaf shallowly lobed; lower half of fruit without horns; native to Eastern U.S.; not found here in many years, possibly eliminated.

Horned Sea Rocket

FLAT-POD MUSTARDS

Lacepods, Fringepods: *Athysanus, Thysanocarpus*

Dwarf Athysanus, Dwarf Fringepod
Athysanus pusillus (Hook.) Greene

A *small annual* 5-35 cm in height, and sometimes lies flat. *Leaves* 6-30 mm long, 2-10 mm wide, *mostly basal,* entire or shallow-lobed. The *tiny short-lived flowers* are held upright and appear on only one side of the stalk. Sepals about 1 mm long and fall away soon. The *small white petals* are the same length or twice as long as the sepals; sometimes lacking. *Fruit* is *circular, flat, hairy,* and droops from the stalk; it contains only one seed. In flower Feb-June.

Found mostly on gentle north-facing shaded slopes in moist soils, often under shrubs. Recorded from Chiquito Basin, Temescal Valley below Horsethief Canyon, and San Mateo Canyon Wilderness Area, certainly elsewhere as well.

The epithet was given in 1837 by W.J. Hooker for its size (Latin, pusillus = very small, weak).

Dwarf Athysanus

Hairy Fringepod or Lacepod
Thysanocarpus curvipes Hook.

An *annual* 10-60 cm tall. *Leaves* are *elongate* and *mostly basal,* 1.5-7 cm long, *entire or toothed. The base of each stem leaf is arrowhead-shaped and surrounds and/or clasps the stem.* Sepals about 1 mm long, often purplish, with white edges. Petals similar, narrow, shaped like a tongue depressor, white to purplish. At opening, the anthers are purple, aging yellowish as the pollen is released. *Fruit* 5-8 mm across *with a smooth- or wavy-edged wing;* it contains only one seed. In flower Mar-May.

Known from hills surrounding upper Newport Bay, Lomas de Santiago, Harding Canyon, upper Modjeska Canyon, San Juan Canyon, Hot Springs Canyon, San Juan Loop Trail, Blue Jay and Falcon Campgrounds, along Main Divide Truck Trail above Maple Springs, Alberhill, Elsinore Peak, and the San Mateo Canyon Wilderness Area.

The epithet was given in 1833 by W.J. Hooker for the leaf base, which curves around the stem (Latin, curvo= to curve).

Hairy Fringepod

Southern Fringepod or Lacepod
Thysanocarpus laciniatus Nutt.

A *slender annual*, 10-60 cm tall, glaucous throughout. *Leaves* 1-4 cm long, *linear, and usually do not clasp the stem. Leaf margins entire or deeply lobed.* Petals white to purplish. Fruit 3-6 mm across, smooth- or wavy-edged; it contains only one seed. In flower Mar-May.

Found in grassy meadows and coastal sagebrush scrub in foothills and mountains, such as Mission Viejo, Hot Springs Canyon, San Juan Canyon, Upper McVicker Canyon, and in the San Mateo Canyon Wilderness Area.

The epithet was given in 1838 by T. Nuttall for its lacy-edged fruits (Latin, lacinia = a thing torn).

Southern Fringepod

Southern Fringepod

Shepherd's-purse and Peppergrasses: *Capsella, Lepidium*

Fruits round or triangular in shape, flattened perpendicular to the septum, held upward or outward, and contain 2 or more seeds.

Shepherd's-purse ✖
Capsella bursa-pastoris (L.) Medik.

A *small upright annual*, 10-50 cm in height. *Basal leaves* 3-6 cm long, often pinnately lobed, *arranged in a rosette.* Stem leaves smaller, less lobed, often with 2 rounded basal lobes that surround the stem. *Flowers small*, 2 mm long, with *white petals. Fruit* is a *triangular or heart-shaped* silicle, 4-8 mm long, *flattened with an indented tip,* held upward or outward. In flower most of the year.

Native to Europe and widespread in our area. Prefers moist soils and partially shaded spots. Common along trails and creekbeds, under oaks, and sometimes in home gardens.

The epithet was given in 1753 by C. Linnaeus for its fruit (Latin, bursa = purse; pastor = shepherd).

Shepherd's-purse

Peppergrasses: *Lepidium*

Most of our species are annuals, though one is a perennial. All have a tuft of basal leaves. *Flowers* of most are *small and short-lived. Fruits* are held upward or outward, rounded or oval in shape, *flattened, usually notched at the tip,* often surrounded by 2 points; each contains only 1 seed. Common elements within coastal sage scrub and chaparral, and along trails and roadsides. *To identify them, you must examine the fruits.*

	Alkali	Hairy-pod	Shining	Robinson's	Broad-leaved	Pinnate-leaved
	L. acutidens	L. lasiocarpum	L. nitidum	L.v. robinsonii	L. latifolium	L. pinnatifidum
Growth	Annual; low and spreading, to 20 cm long	Annual; upright, to 30 cm tall	Annual; upright or spreading, to 40 cm tall	Annual; upright, to 20 cm tall	Perennial; upright, to 1-1.5 meters	Annual, biennial, or perennial; to 50 cm tall
Basal leaves	Pinnate	Linear	Pinnate	Pinnate	Roughly oval, oblong	Pinnate
Stamens	4	2	4	2	6	4
Fruit	Hairy, notched, 2-horned	Roughly round, shallow-notched	Hairless, winged; tip notched	Round, tip shallow-notched	Round, sparse-haired, not notched	Round, mostly hairless, shallow-notched

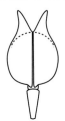 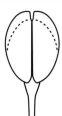    

Alkali Peppergrass
Lepidium acutidens (A. Gray) Howell
[*L. dictyotum* A. Gray var. *acutidens* A. Gray]

A *low-growing, spreading annual,* with many branches that arise from its base and grow to 2-20 cm long. Basal leaves are pinnate with long narrow lobes. *Petals absent. Stamens 4. Fruits are mostly over 4 mm long, hairy, notched, and have two large pointed horns that curve away from each other.* In flower Mar-May.

Uncommon here, known only from saline soils at Nichols Road and a historical collection from near Buena Park.

The epithet was given in 1867 by A. Gray for the net-like veins on its fruit (Latin, diktyon = a net). The variety name was given in 1876 by A. Gray for the pointed horns of the fruit (Latin, acutus = sharp, pointed; dens = tooth).

Alkali Peppergrass
Aaron Schusteff

Hairy-pod Peppergrass
Lepidium lasiocarpum Nutt.

An *annual* 10-30 cm tall *with hairy branches. Basal leaves linear and broad,* margins toothed or with oblong lobes. Uppermost stem leaves nearly entire. *Pedicels are distinctly flattened.* Sepals 1-1.5 mm long, wide, purplish. Petals white or absent, *shorter* than sepals. *Stamens 2. Fruits sparsely haired,* 3-4.5 mm wide, *round or slightly longer than wide, the tip with a shallow notch.* The beak is minute or absent. In flower Feb-May.

Occurs in sandy and sandy-clay soils among grasslands, coastal sage scrub, and chaparral. Known from upper Newport Bay, Corona Del Mar, Costa Mesa, hillsides near UC Irvine, Laguna Canyon and the beaches in Laguna Beach, Dana Point Headlands, Santiago Creek, Trabuco Canyon, Whittier Hills, Temescal Valley, and the San Mateo Canyon Wilderness Area.

The epithet was given in 1838 by T. Nuttall for its hairy fruits (Greek, lasios = hairy; karpos = fruit).

Hairy-pod Peppergrass
Fred M. Roberts, Jr.

Shining Peppergrass
Lepidium nitidum Nutt.

An upright or spreading annual 10-40 cm tall. Its *basal leaves are pinnate with long narrow lobes,* stem leaves pinnate with linear pointed lobes. *Pedicels are flattened in cross-section.* Petals white, 1.5 mm long, slightly *longer* than the sepals. *Stamens 4. Fruits hairless, smooth, and shiny, completely surrounded by a thin "wing"; the tip has a narrow notch.* In flower Feb-May.

Grows in saline soils of grasslands and vernal pools and openings in coastal sage scrub and chaparral. Commonly found in most of our coastal cities, UC Irvine Ecological Preserve and grasslands near campus, the Chino Hills above Santa Ana River channel, Caspers Wilderness Park, Audubon Starr Ranch Sanctuary, San Juan Canyon, Whittier Hills, Temescal Valley, Santa Rosa Plateau, and the San Mateo Canyon Wilderness. This is our most commonly encountered peppergrass.

The epithet was given in 1838 by T. Nuttall for its shiny fruits (Latin, nitidus = bright, glittering).

Shining Peppergrass
Plant: Fred M. Roberts, Jr.

Robinson's Peppergrass ★
Lepidium virginicum L. **var. robinsonii** (Thell.) C.L. Hitchc. [*L. virginicum* ssp. *menziesii* (DC.) Thell.]

An *upright annual,* 10-20 cm tall, *densely covered with pointed hairs.* Main stem branches above, not from the base. *Basal leaves* 5-15 cm long, oval, *pinnate. Stem leaves divided or lobed with narrow segments only 1-2 mm wide. Pedicel hairy, roughly rounded in cross-section.* Petals 1-2 mm long, white, equal to or longer than the sepals. *Stamens 2. Fruits round with a shallow notch.* In flower Jan-Apr.

Occurs sparingly in dry soils of coastal sage scrub and chaparral. Old records exist from Yorba Linda and Anaheim. More recently found in the Sycamore Hills of Laguna Beach, Loma Ridge in Lomas de Santiago, and San Juan Canyon. CRPR 1B.2.

The epithet was given in 1753 by C. Linnaeus for the place of its discovery. Our form was named in 1906 by A. Thellung in honor of B.L. Robinson.

Similar. Hairy peppergrass (*L.v.* var. *pubescens* (E. Greene) Thell., not pictured): upright, 20-70 cm tall, covered with rigid hairs; stem leaves not or shallowly dissected; fruit 3-4 mm long; in flower Mar-Aug; mostly wet places and moist soils in coastal sage scrub, chaparral, roadsides; uncommon, known from upper Newport Bay, UC Irvine Ecological Preserve, Costa Mesa, Caspers Wilderness Park, Chino Hills near Santa Ana River channel.

The current treatment in Baldwin et al. (2012) considers both Robinson's and hairy peppergrass as poorly defined forms of Menzies's peppergrass (*L. virginicum* ssp. *menziesii* (DC.) Thell., not shown). We follow the traditional arrangement and recognize the two as valid varieties.

Robinson's Peppergrass

Broad-leaved Peppergrass ✖
Lepidium latifolium L.

A *very invasive non-native perennial, it grows from rhizomes and forms dense colonies.* The tallest peppergrass, it grows 0.4-1 meter (-2 meters!). Basal leaves are up to 30 cm long and 6-8 cm wide, entire and toothed. Petals white and rounded, large and easily seen, unlike those of other peppergrasses. *Stamens 6. Fruit is round,* 2 mm across, *sparsely haired, and has no notch.* In flower May-June, sometimes later.

It occurs near beaches and in saline soils such as on the Bolsa Chica Ecological Reserve. Sporadically appears along roadsides, freeways, and railroad rights-of-way. Native to Eurasia.

The epithet was given in 1753 by C. Linnaeus for its broad leaves (Latin, latus- = broad, wide; folium = leaf).

Broad-leaved Peppergrass

Pinnate-leaved Peppergrass ✖
Lepidium pinnatifidum Ledeb.

Annual, biennial, or perennial, 20-50 cm tall. Branches above, usually not from the base. *Basal leaves pinnate;* stem leaves entire, pinnate, or dentate. *Pedicel round in cross-section.* Petals white or absent. *Stamens 4. Fruits* 1.8-2 mm long, *round, nearly hairless, with a shallow notch.* In flower May-June.

Uncommon here, known from salty soils in Huntington Beach and along San Diego Creek in Irvine. Native to Eurasia.

The epithet was given in 1841 by C.F. von Ledebour for its basal leaves, which are always pinnate (Latin, pinna or penna = wing or feather; fidus = trustworthy, sure).

Pinnate-leaved Peppergrass

CACTACEAE • CACTUS FAMILY

Traditional symbols of the Old West, cacti are one of the most distinctive plants in the New World. Their fleshy stems store water, but their covering of sharp spines says, "Stay back, you can't have it!" Their spiny appearance contrasts sharply with the unrivaled beauty of their large colorful cup-like flowers.

Herbaceous perennials, shrubs, trees, generally fleshy. Ours are *shrubs* with *succulent, spiny-covered, green stems*. Photosynthetic leaves are generally lacking but when present are tiny and short-lived. Other leaves are modified into spines. Cactus stems have regions of specialized tissue called **arcoles** that may bear leaves, spines, and/or hairs. Their **spines** are of two types: **long spines** and short spines called **glochids**. Long spines can be flat in cross-section like a spatula or round like a toothpick. Long spines usually splay out in a radial pattern. The longest among them often face mainly one direction, useful for identification. The tips are either smooth and removed easily or barbed, making removal more difficult and painful. Glochids are short slender spines that are extensively barbed and difficult to see and remove. The common name *cactus* refers to the spines (Greek, kaktos = a prickly plant).

Flowers appear at the upper ends of the stems and are held firmly upright. Instead of traditional sepals and petals, those parts begin as small green scales at the flower base and enlarge upward, gaining color and appearing like petals; these parts are called **tepals**. *Stamens* are too numerous to count; anthers and pollen are yellow or golden. The pistil has a *single style* with a *stigma that is divided into 3 or more club- or finger-like lobes*. The *ovary is inferior,* surrounded by a substantial *nectary* that attracts bees, wasps, flies, and beetles. *Fruit is a fleshy berry,* often red in color, sometimes covered with spines, and edible (after the spines are removed). In some species the fruit is dry, not fleshy, and certainly inedible.

> **PRETTY CAN BE PAINFUL**
>
> *Long spines easily enter your skin and are often painful to remove. Glochids, however, are far more dangerous because you can unknowingly get them into a finger and transfer them if you later rub your eye or touch other body parts. For spine and glochid removal, a handy field tool is a set of fine-tipped tweezers available at REI Coop and other outdoor stores.*
>
> *Cactus stems can break off at constricted areas (called joints), take root, and grow as a new plant. Sometimes a piece breaks off into your skin as you maneuver around it. Don't try to swat the piece away - it will just roll down your body and remain stuck. For this reason, you should carry a stiff comb as part of your field gear. Maneuver the comb's teeth among the spines between you and the cactus and pull the piece straight out and away from your body.*

Cactus in Bob's leg. Photo by Alison Linden

Two genera of cactus live in our area; they were once considered members of a single genus. The *chollas* (or cylindrical-stemmed opuntias), in the genus *Cylindropuntia*, have stems that are rounded in cross-section and bumpy-surfaced. The *prickly-pears* (or flat-stemmed opuntias), in the genus *Opuntia*, have paddle-shaped stems (called pads) that are flat in cross-section and generally smooth-surfaced.

Identification is best done when cacti are in flower. You'll need to note the color of the innermost tepals, stamen filaments, style, and stigma and the shape of the stems and pads. Also important are features of their spines such as number of long spines per areole, shape in cross-section, and color. Be sure to take tweezers and comb with you into the field, in case you get too close.

Cholla
Cylindropuntia
Stems round in cross-section

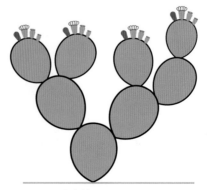

Prickly-pear
Opuntia
Stems flat in cross-section

Cactus Leaves
Opuntia littoralis
Fleshy, small, and short-lived

Cholla: *Cylindropuntia*

Cacti with bumpy-surfaced cylindrical stems, rounded in cross-section. We have 2 species, easily identified.

Cane or Valley Cholla
Cylindropuntia californica (Torr. & A. Gray) F.M. Knuth **var. *parkeri*** (J.M. Coult.) Pinkava
  [*Opuntia parryi* Engelm.]

A *spreading or upright cactus,* up to 3 meters tall, but *without a main trunk.* It *generally branches from far below the middle.* Stems are generally more than 30 cm long, green to purplish. The *innermost tepals are yellow to green-yellow, outer ones generally tipped with red or purple.* Filaments pale green. Fruits are not fleshy but leathery or dry, with a variable number of spines. In flower May-June.

Not a common plant in our area, mostly found in riparian habitats with California sycamores. Old records are from Scully Hill and the former site of Rancho Santa Ana Botanic Garden at the edge of Santa Ana Canyon (both in the southern Chino Hills). Today, known from the Arroyo Trabuco south of the developed portion of O'Neill Regional Park and on the first creekside mesa past the entry booth at Caspers Wilderness Park (also planted in the garden just before the booth).

The epithet was given in 1840 by J. Torrey and A. Gray in honor of the place of its discovery. Our form was named in 1896 by J.M. Coulter and honors J.C. Parker, who first collected it.

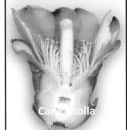

Coast Cholla
Cylindropuntia prolifera (Engelmann) F.M. Kunth
  [*Opuntia prolifera* Engelmann]

A *tree-like cactus* most often 1-2 meters tall. It *generally branches from above the middle.* Stems are generally less than 30 cm long, green-gray. The *innermost tepals are rose to magenta.* Filaments yellow-green, often tinted purple near the tips. The fruits are fleshy, green, and spineless. In flower Apr-June.

Found along the immediate coast and inland to the base of hot south-facing slopes among chaparral and coastal sage scrub in the San Joaquin Hills, Laguna Hills, Lomas de Santiago, and lower elevations of the Santa Ana Mountains. You'll find it on the bluffs below Hoag Memorial Hospital in Newport Beach, Crystal Cove State Park, Turtle Rock, and both Laguna Coast and Whiting Ranch Wilderness Parks. Midway along the Borrego Trail in Whiting Ranch it grows with **chaparral beargrass** (*Nolina cismontana,* Ruscaceae).

The epithet was given in 1852 by G. Engelmann for its abundance (Latin, proliferus = reproducing freely).

Prickly-pears: *Opuntia*

Cacti with smooth-surfaced paddle-shaped *pads that are flat in cross-section*. We have 6 species. Identification can be difficult and requires you to study their pads.

Beavertail Prickly-pear *Opuntia basilaris* **var.** *basilaris*	Mission Prickly-pear *Opuntia ficus-indica*	Chaparral Prickly-pear *Opuntia oricola*
Growth: Very low, < 0.5 meter tall	Growth: Very tall, tree-like, up to 5 meters tall	Growth: Generally low to medium, up to 2 meters tall
Pads: Oval; green to blue-green; thin Length: 8-21 cm Width: 5-13 cm	Pads: Oval, elliptic, oblong; green; thick Length: 25-43 (-60) cm Width: > 25 cm	Pads: Circular to oval; dark green Length: 16-25 cm Width: >15 cm
Long spines: None (rarely 1-8 per areole)	Long spines: 0 (1-6 per areole, flat in cross-section, short)	Long spines: 5-13 per areole; yellow, age red-brown; flat in cross-section; curved; mostly downcurved
Inner tepals: Pink-magenta (white)	Inner tepals: Yellow to orange	Inner tepals: Yellow or slightly tinged pink to orange
Style: White to pink	Style: White (pale pink)	Style: Red

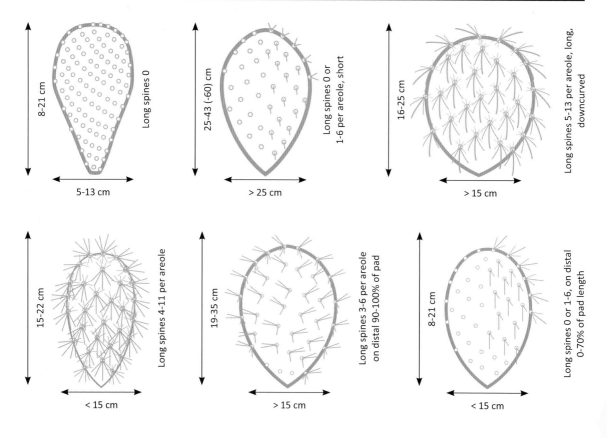

Coast Prickly-pear *Opuntia littoralis*	Western Prickly-pear *Opuntia x occidentalis*	Mesa Prickly-pear *Opuntia x vaseyi*
Growth: Generally < 1.5 meters tall	Growth: Low, <1 meter tall	Growth: Low, generally < 1 meter tall
Pads: Long oval; gray-green; often thick Length: 15-22 cm Width: < 15 cm	Pads: Oval; green Length: 19-35 cm Width: >15 cm	Pads: Oval; green; often thick Length: 9-22 cm Width: <15 cm
Long spines: 4-11 per areole; yellow; round in cross-section (square); straight; held outward	Long spines: 3-6 per areole; yellow; flat in cross-section; straight; upper 1-2 longest	Long spines: 0-4 (-6) per areole; yellow, age dark red; flat in cross-section; straight
Inner tepals: Yellow to dull red	Inner tepals: Yellow (-orange) to deep pink	Inner tepals: Yellow, orange, or dull red
Style: Pink to red	Style: Pink to white	Style: Pink to red (white)

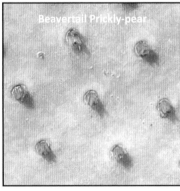

Beavertail Prickly-pear
Opuntia basilaris Engelm. & J.M. Bigelow **var. *basilaris***
   A *short cactus,* under 0.5 meter tall, it forms dense clumps. The *pads* are 8-21 cm long, 5-13 cm wide, *green to blue-green and often tinged purplish,* reverse-oval in shape, and not very thick. They are often covered with fine hairs. It has *no long spines but plenty of glochids* (it rarely has up to 8 long spines per areole). The *inner tepals are pink-magenta* (rarely white). Filaments deep magenta-red, anthers yellowish. Style white to pink, stigma white to cream. The *spineless fruit* is tan and dry when ripe. In flower Mar-June.

   Grows on dry slopes among coastal sage scrub and chaparral in Temescal Valley but is quite uncommon. This is primarily a desert plant barely found in our area.

   The epithet was given in 1856 by G. Engelmann and J.M. Bigelow for its low-growing stems (Latin, basilaris = at the base).

   Top: Beavertail prickly-pear with green pads. Note the lack of long spines but an abundance of glochids at each areole. Above, right: Blue-green pads. Bottom, right: Flower with pink tepals and style.

Mission Prickly-pear ✖
Opuntia ficus-indica (L.) Mill.

A *large tree-like cactus*, 4-5 meters tall, *often spineless*. Its pads are 25-43 cm long, >25 cm wide, and quite thick. When spines appear, they cover 0-100% of the pad. *Each spine is white-tan or brown-gray, flat in cross-section, 1-6 per areole. Inner tepals yellow to orange.* Filaments yellow (pale green to pale pink), anthers yellow. Style white to pale pink, stigma green to yellowish. The *large fruit is fleshy*, yellow to orange to purple in color, often roughly spherical in shape. In flower May-June (-July).

The exact origin of this plant is unknown but is probably Mexico. It is likely that the native peoples selected for fewer-spined individuals on which to cultivate cochineal scale insects (see The Cactus Guild). Mexican foods called *nopales* or *nopalitas* are made from the young pads and *tunas* from the fruits. It is now cultivated worldwide for food. Locally it is uncommon and grows in coastal to foothill areas, primarily as a garden escape or persisting from an old settlement. Look for it in the hills surrounding upper Newport Bay, settlements along Santiago Canyon and its tributaries such as in Modjeska Canyon, Tucker Wildlife Sanctuary, Tracubo Canyon, and near old mining homesteads in Lucas Canyon.

Legend has it that the horticulturalist Luther Burbank produced a spineless variety of this plant simply by asking it to stop making spines. In reality, he cross-bred other types of cacti with it. The results, however, are a mixed bag of spineless and spiny cacti.

The epithet was given in 1768 by C. Linnaeus for the similarity of its fruits to figs from India (Latin, ficus = fig; indicus = of India).

Above: Mission prickly-pear near Cook's Corner. It is often tree-like, tall, and broad-crowned. Upper right: When in flower, its ovary is long and sometimes narrow. The flowers grow from the edges of the pads and sometimes also from the face of the pad. Right: Long spines, when present, are short.

Chaparral Prickly-pear ✤
Opuntia oricola Philbrick

A spreading, medium-sized cactus that often forms dense clumps, 1-2 meters tall, occasionally tree-like. *Its smooth, dark green pads* are 16-25 cm long and *quite rounded* (sometimes reverse-oval). *Spines 5-13 per areole, translucent yellow-gold, flat in cross-section, usually curved, and often point downward.* The innermost tepals are yellow. Filaments yellow to orange-yellow, anthers yellow. *Style red*, stigma green. The fleshy fruits are red to purple-red, pale yellow inside. In flower Apr-May (-July).

Widespread but generally not common. *Found most often on the immediate coast, below 500 feet in elevation,* among coastal sage scrub and coastal bluff scrub. Reported to grow in chaparral, but not observed to do so in our area. Known from just north of Sunny Hills in the Coyote Hills in Fullerton, on south-facing slopes in upper Newport Bay just east of the Muth Interpretive Center, U.C. Irvine Open Space Preserve, San Joaquin Hills, Laguna Beach, Dana Point Headlands, Ken Sampson Overview in Dana Point, Rancho Mission Viejo, and San Clemente.

The epithet was given in 1964 by R.N. Philbrick for its coastal habitat (Latin, ora = coast; colere = to inhabit).

Above, left: Individuals are visible from the nature trail on the Dana Point Headlands above Dana Point Harbor. Above, right: innermost tepals are yellow or slightly tinged pink to orange. Middle, right: Long spines are flat in cross-section and glow golden in afternoon sun. Right: Pads are very spiny and clearly round.

Coast Prickly-pear
Opuntia littoralis (Engelm.) Cockerell

This cactus forms a *low-growing clumping mass,* up to 9 meters wide and 1 meter tall. Its gray-green to blue-green pads are 15-22 cm long, *< 15 cm wide, elongated or egg-shaped,* lightly coated in a whitish powder. *Long spines 4-11 per areole, yellow, coated whitish, yellow or brown at the base, and round in cross-section; uppermost spines are square in cross-section.* The innermost tepals are yellow to dull red. Filaments yellow to orange-yellow, anthers yellow. *Style pink to red,* stigma yellow-green to green. The *spineless fruits* are dark red-purple and fleshy. In flower Apr-May (-July).

Primarily a coastal plant, it grows among grasslands, coastal sage scrub, and chaparral. Look for it on the ocean bluffs near Crystal Cove State Park, on Scully Hill in the Puente-Chino Hills, throughout Rancho Mission Viejo, in Whiting Ranch and Caspers Wilderness Parks, near Pruesker Peak on the Audubon Starr Ranch Sanctuary, in lower elevations of the Santa Ana Mountains, and in Temescal Valley. Present but uncommon in the San Mateo Canyon Wilderness Area. It often thrives in overgrazed cattle pasturelands.

The epithet was given in 1905 by G. Engelmann for its occurrence near the ocean (Latin, littoralis = belonging to the seashore).

Similar. Mesa prickly-pear (*O.* x *vaseyi*): spines flat in cross-section; innermost tepals yellow, orange, or dull red; both are coastal.

Western Prickly-pear

Opuntia* x *occidentalis Engelm. & J.M. Bigelow

A *spreading mass* up to 1 meter tall. Each *oval green pad* is 19-35 cm long and over 15 cm wide. *Long spines 3-6 per areole* (the distal 90-100% of each pad has long spines), each spine flat, straight; the upper 1-2 are longer than the lower 2-4. Innermost tepals are yellow to deep pink. Filaments yellow or white, anthers yellow. *Style pink to white,* stigma green. The fruit is red-purple and fleshy. In flower Mar-May.

An uncommon hybrid in coastal sage scrub and chaparral communities at elevations below 500 meters (1640 ft). Not likely to be recognized by the casual wildflower-watcher. Known from Trabuco Canyon, the western edge of the San Mateo Wilderness along Lucas Canyon Trail near Aliso Canyon, San Juan Canyon (20 miles east of San Juan Capistrano), and the Temecula wash along Highway 79 (just outside our area). Possibly more widespread here. It is a stable hybrid between **coast prickly-pear** (*O. littoralis*) and the hybrid offspring of **Engelmann's prickly-pear** (*O. engelmannii* Engelm.) and **brown-spined prickly-pear** (*O. phaeacantha* Engelm.).

The epithet was given in 1856 by G. Engelmann for its western distribution (Latin, occidentalis = western).

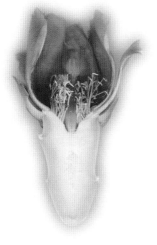

Right and center: Mesa prickly-pear usually has two-toned filaments, yellow at the base and transitioning to orange toward the anthers. Far right: Western prickly-pear has yellow or white filaments.

Mesa Prickly-pear

Opuntia* x *vaseyi (J.M. Coult.) Britton & Rose

Similar to *coast prickly-pear,* this cactus grows as a *spreading mass* up to 1.2 meters tall. Its *green pads* are 9-22 cm long, elongated or egg-shaped. *Long spines 0-4(-6) per areole,* covering 0-70% of the apical portion of the pad; each is yellow with a whitish-coating, brown or yellow at the base, and *flat in cross-section*. The *innermost tepals* are *completely yellow, orange, or dull red*. Filaments orange-yellow, anthers yellow. *Style pink to red* (rarely white), stigma green to yellow-green. The fruit is red-purple and fleshy. In flower Apr-June.

This is a stable hybrid that originated as a natural cross between **coast prickly-pear** and **Englemann's prickly-pear** (*O. phaeacantha* Engelmann). It grows among chaparral, coastal sage scrub, and in disturbed areas. Mostly a coastal plant, it is known from the San Joaquin Hills, Turtle Rock, Dana Point Headlands, Laguna Hills, Coyote Hills, and Loma Ridge in the Lomas de Santiago.

The epithet was given in 1908 by J.M. Coulter in honor of G. Vasey or his son, G.R. Vasey.

Similar. Coast prickly-pear (*O. littoralis*): spines round in cross-section; innermost tepals yellow to dull red; both are coastal.

THE CACTUS GUILD

Prickly-pear Bug
Narnia femorata Stål (Coreidae)

Brownish true bugs, 14-19 mm long, with unusual hind legs. Each *hind femur* (upper section of the hind leg) is *inflated* and bears *stout spines.* The *hind tibia* (lower section of the hind leg) has an *expanded leaf-like region,* typical of most members of this family, the leaf-footed bugs. The forewings often have a white line across them.

They feed on cactus fruit and pads in spring, summer, and early fall. Females lay eggs on cactus spines. There are two generations per year, so you may see young and adults feeding together on the same cactus. Look for them anywhere you find prickly-pear cactus, such as Laguna Coast Wilderness Park, the Dilley Reserve, and Peters Canyon. They are completely harmless; if you pick them up, they'll probably crawl on you and then fly away.

The epithet was given in 1862 by C. Stål for the insect's spiny femur (Latin, femur = the thigh).

Prickly-pear bug

Cochineal Scales
Dactylopius spp. (Dactylopiidae)

Small insects with an unusual life cycle. They have piercing-sucking mouthparts with which they remove water and nutrients from prickly-pear cacti. *Adult females are tiny* (up to 0.3 mm long), *sack-like, wingless,* and covered with a *thick coating of wax* that they secrete from hollow bristles on their back and sides. Males are smaller when young and secrete no cottony wax. As they grow, males develop a single set of wings externally (not internally, as beetles, butterflies and bees do). When mature, males fly off to another colony to mate. They have no mouthparts, cannot feed, and die soon. After mating, the female's body swells to a pear shape and outgrows her legs, which prevents her from walking. She soon gives birth to nymphs that remain nearby or move to another section of the cactus.

Females contain *carminic acid,* a compound that repels ants and other predatory insects; it also gives her body its deep red color. Early native Americans such as the Aztecs carefully collected the females, dried them, and crushed them into a powder that, when mixed with water, dyed clothing a rich carmine red color. The Spanish explorer Hernán Cortés discovered this use when he visited the Aztecs in 1519. By 1600, export of cochineal dye had become a major industry for Mexico. It was not until the mid-1800s, when the first aniline dye was chemically produced, that the cochineal dye industry collapsed. Cochineal is still used to dye some fabrics and to color some food and drinks, such as Sobé Lizard Fuel™, fruit juices, and frozen fruit bars.

The most common local species is *Dactylopius opuntiae* (Cockerell), though *D. confusus* (Cockerell) is also here. Both species produce wax threads that look like spiderwebs. *D. coccus* Costa occurs from Mexico to South America; it is the species most often used for cochineal dye. Its wax does not resemble spiderwebs.

The genus was named in 1835 by O.G. Costa, probably for the hollow, finger-like bristles on the female's body (Greek, daktylos = finger).

Cochineal Scales

The blood of the cochineal scale insect is carmine red, a valuable dye

Ladybird Beetles
(Coccinellidae)
  Ladybirds have high-domed rounded or oval bodies. Most species in our area are black or tan in color, though people often assume all ladybirds are red. They all use their biting mandibles to feed on insects that have piercing-sucking ("drinking-straw") mouthparts such as mites, aphids, mealybugs, and scales. Pictured here are three of our many local species, two of them found almost exclusively on cacti.

  The **Three-forked Ladybird** (*Hyperaspis trifurcata* Schaeffer) is tiny, only 2.3-3.0 mm long. Its front wings (*elytra*) are *black, marked with yellow,* though the yellow is sometimes greatly reduced. It is known from coastal southern California to southern Arizona, southern Texas, and Mexico. Its larvae feed on cochineal scales. The carminic acid from the scales is absorbed into the young ladybird's bloodstream, turning it pink. When disturbed, the ladybird larva bleeds (harmlessly) carminic acid through its body wall, which repels ants and other predators.
  The epithet was given in 1905 by C. Schaeffer for the three black bands on the elytra that resemble a 3-pronged fork (Latin, tri = three; furcatus = forked).

  The **Banded Ladybird** (*Exochomus fasciatus* Casey) is 2.6-3.7 mm long. Its *elytra* are *red with 1-2 bands of black.* It also feeds on cochineal scales. Its larvae may also incorporate carminic acid in their bodies. It is found only in coastal southern California.
  The epithet was given in 1899 by T.L. Casey for its banded elytra (Latin, fasciatus = bundled, banded).

  The **European Seven-spotted Ladybird** ✖ (*Coccinella septempunctata* L.) is 6.5-7.8 mm long, the *elytra orange with 3.5 black spots each.* Its black head has two tiny white spots in front of its eyes. It was introduced here in 1979 to control agricultural pests but has expanded its range beyond crop fields and is now found throughout much of the U.S.
  The epithet was given in 1758 by C. Linnaeus for the seven black spots on its back (Latin, septem = seven; punctatus = spotted as with punctures).

Cactus Flower Beetles
Carpophilus pallipennis (Say) (Nitidulidae)
  Small, flattened, dark brown beetles, 2.5-4 mm long, often found in abundance among the stamens and tepals of cactus flowers. Their front wings (elytra) are short and expose the last 2-3 segments of their abdomen. The elytra are mostly dull yellow-brown, except for a darker brown triangle near the center of the body. The other species of *Carpophilus* known from our cacti are all-brown and much less common. The larvae feed on decomposing cactus fruits, the adults on nectar and sap.

  The epithet was given in 1823 by T. Say for its pale elytra (Latin, palli, pallium = mantle, cover; penn = wing).

Mexican Cactus Fly
Copestylum mexicanum Macquart (Syrphidae)
  A large, smooth, shiny, purplish-black fly (15-20 mm long) with black at basal half of the leading edge of its wings. Larvae feed in decomposing tissues of dead cacti; they do not cause the death of cacti. Adults are commonly seen in spring and summer as they feed on many types of flowers.
  The epithet was given in 1842 by P.J.M. Macquart for the country of its initial discovery.

Cactus Bees
Diadasia spp. (Apidae)

These bees visit cactus flowers to collect pollen and nectar to eat and to provision their ground nests. They are *fast flyers* and *quite wary,* quick to depart the flowers if disturbed. To watch them, use close-focusing binoculars or camera lens and move slowly. We have several species here.

The genus was established in 1879 by W.H. Patton for the shaggy hairs that occur all around its abdomen (Greek, dia- = throughout; dasys = hairy, shaggy).

Cactus Bee

Cactus Wren
Campylorhynchus brunneicapillus (Lafresnaye) (Troglodytidae)

A *large wren,* brownish in color with black spots and a *bold white "eyebrow."* It mostly *keeps its tail outward or downward,* not upward like most wrens'. It lives in deserts of California, Arizona, and Mexico, but populations also exist in coastal southern California. Here, we find it in the San Joaquin Hills, Lomas de Santiago, and lower elevations of the Santa Ana Mountains. It nests among the stems of cacti, primarily coast cholla and chaparral prickly-pear, at a height >1.2 meters above the ground, presumably to avoid predators.

The epithet was given in 1835 by N.F.A.A. de Lafresnaye for its brown color (Latin, bruneus= dark brown; capillus = hair).

Cactus Wren

Greater Roadrunner
Geococcyx californianus Lesson (Cuculidae)

An *attractive bird* with a *very long tail,* it can reach a total length of about 58 cm (23 inches). Its feathers are *streaked with brown and white.* Its head sports a tall dark crest that it can raise and lower. Behind the eye is a streak of blue, followed by red. The dark bill is disproportionally long and heavy for its slender head and neck. Like other cuckoos and the woodpeckers, it has two toes that point forward, two that point backward; thus it leaves interesting tracks as it walks on the ground.

Primarily a bird of the southwestern U.S. but ranges east to Arkansas and south into Mexico. Though it can fly, it is more often seen running in short bursts, a technique it uses to capture insects, lizards, snakes, rodents, and small birds. It is indelibly linked in the American mind to cactus, coyotes, and cartoons (*beep! beep!*).

It uses behavioral tricks to cool its body: panting to release heat by evaporation, drooping its wings to increase surface area and air circulation near its body, perching on something above the hot ground, and standing in shade.

The epithet was established in 1829 by R.P. Lesson in reference to the place (then a Spanish territory) of its discovery.

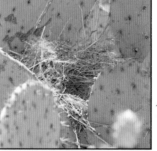
Cactus wren nest. If you find this in a cactus, please keep away so you don't disturb the family within.

Greater Roadrunner

> *I'm looking for an open trail to steal*
> *away on, steep, with rocks and poison oak,*
> *and around the next turn, a roadrunner.*
>
> – Thea Gavin

CAMPANULACEAE • BELLFLOWER FAMILY

Annuals, perennials, shrubs, rarely trees. *Ours are herbaceous annuals or perennials,* often small and delicate-looking. Leaves usually simple and alternate, stipules always absent. *Calyx and corolla are each fused into a tube below with 5 free lobes above. Stamens 5,* free or variously fused to each other; their anthers open inward toward the pistil. Ovary inferior or half-inferior. Style 1, sometimes branched 2, 3, or 5 times at the tip which curls backward as the flower ages (as in the sunflower family, a close relative). Fruit is a capsule.

Flowers are of three general types. Those of the **Campanula type** *have a bell-shaped corolla* fused only at the base with 5 free equal-sized lobes above. The stamens are not fused to each other. California genera of this type are *Campanula, Githopsis, Heterocodon, Legenere,* and *Triodanis.*

Those of the **Lobelia type** *have a corolla that is fused into a tube below, free above, and is strongly two-lipped above,* with 2 small lobes held upward and 3 larger lobes downward. The filaments are free at their base but fused above into a tube that surrounds the pistil. The anthers are fused or free. In most species, the *flower is inverted.* Before the flower opens, the pedicel twists 180 degrees, and so the flower we see is actually upside-down (most orchids do the same thing). Some flowers never open and undergo in-bud fertilization. They live in wet areas. California genera are *Downingia, Lobelia,* and *Porterella*. They are sometimes placed in their own family, the Lobeliaceae.

Those of the **Nemacladus type** *have a bell-shaped or 2-lipped corolla,* fused at the base, free above, with 2 small lobes held upward and 3 larger lobes downward. The filaments are free at their base, fused into a tube above the middle, the anthers free. In some species, the flower is inverted. They live in dry areas. The only California genus is *Nemacladus.* They are sometimes placed in their own family, the Nemacladaceae.

Several species are cultivated as ornamentals, valued for their beautiful flowers often in shades of purple, blue, or white, rarely red. There are about 2,000 species worldwide, 290 in the U.S., 49 in California, 10 in our area.

CAMPANULA-TYPE FLOWERS

Blue-flowered Southern Bluecup
Githopsis diffusa A. Gray **ssp. *diffusa***

An annual 3-30 cm tall, most often *short and single-stemmed* but taller and branching in rainy years. Leaves 4-10 mm long, linear, often with small teeth at their margins. Sepals held up or out. *Corolla* (1.5-)3-5 mm long, *light to deep blue. Each petal with 3 obvious veins and a pointed tip.* In flower Apr-June.

A delightful reward for sharp eyes, it is widespread but never abundant and is easily overlooked. More common in spring seasons after fire. Look for it among grasslands, coastal sage scrub, chaparral, and oak woodlands. Scarcely found among grasslands bordering the San Juan Trail in Hot Springs Canyon, and on shaded slopes among chaparral in the San Mateo Canyon Wilderness Area.

The epithet was named in 1882 by A. Gray for its openly branching growth form (Latin, diffusus = diffuse or loosely branched).

Blue-flowered Southern Bluecup

Upper: Blue-flowered southern bluecup, along the lower section of San Juan Trail, just above the trailhead in Hot Springs Canyon.

Lower: White-flowered southern bluecup, along the upper section of San Juan Trail, just below Blue Jay Campground.

White-flowered Southern Bluecup
Githopsis diffusa A. Gray **ssp. *candida*** (Ewan) N. Morin

Upright, 4-20 cm tall. Leaves 6-13 mm long. *Corolla 5-7.5 mm long, stark white.* In flower Apr-June.

Larger, more widespread, and more common than the blue-flowered form. Found among coastal sage scrub, chaparral, oak woodlands, grasslands, and meadows. Look for it on Sitton Peak, upper San Juan Trail near Blue Jay Campground, Los Pinos Potrero, upper Hot Springs Canyon near Falcon Group Camp, Modjeska Peak, and along the Cold Spring Trail and Oak Flats area in the San Mateo Canyon Wilderness Area.

The white-flowered form was named in 1939 by J.A. Ewan for the white color of its flowers (Latin, candidus = white, clear, shining).

White-flowered Southern Bluecup

Rareflower Heterocodon
Heterocodon rariflorum Nutt.

A delicate annual with spreading or upright stems 5-30 cm long. *Leaves* 2-10 mm long, *round to heart-shaped, tooth-edged, held against the stem. Flowers solitary at the nodes.* Sepals widely triangular, leaf-like, tooth-edged. *Corolla* 3-5 mm long, *clearly bell-shaped, the lobes very short; blue to lavender at the lobes, paler toward the corolla base.* Only the uppermost flowers open; the others remain closed and undergo in-bud pollination. Sometimes none of the flowers open. In flower May-July.

Uncommon, generally found in moist places such as sandy soils along streams and in shady understory. Reported from San Juan Canyon by Pequenaut (1951). Known from the vicinity of Elsinore Peak. Also in the San Mateo Canyon Wilderness Area: Wildhorse Canyon above Los Alamos Canyon, near the junction of Tenaja and Morgan Trails, Miller Canyon, and Devil Canyon. Probably more widespread.

The epithet was given in 1843 by T. Nuttall for its rarely seen flowers (Latin, rarus = infrequent, rare; floris = flower).

Small Venus's Looking-glass
Triodanis biflora (Ruiz & Pav.) E. Greene

An upright annual, 5-40 cm tall, upper parts with sparse stiff hairs that point backwards. *Leaves* are 5-15 mm long, *linear or oval, short and cup-like, held against the stem. Flowers 1-4 at a node.* Sepals narrowly triangular, not leaf-like. *Corolla* 5-9 mm long, *open, the lobes free for the last 4-7 mm, deep blue to blue-violet.* They have 3, 4, or 5 corolla lobes. Only the uppermost flowers open; the others remain closed and undergo in-bud pollination. Sometimes none of the flowers open. In flower Apr-June.

Found in dry meadows, burns, and disturbed areas. Known from the San Joaquin Hills (Moro and Laurel Canyons, also the Sycamore Hills), Trampas Canyon on Rancho Mission Viejo, Caspers Wilderness Park, Potrero Los Pinos, and scattered sites within the San Mateo Canyon Wilderness Area.

The epithet was given in 1799 by H. Ruíz Lopez and J.A. Pavón, probably in reference to a specimen that had only 2 flowers or had 2 flowers per node (Latin, bi- = 2; flora = flower).

All photos: Caspers Wilderness Park, Spring 2005. Lower left, note the growth form: stems radiate out from a central point, sprawl on the ground, then turn upward. The plant often grows entwined among other annual wildflowers and grasses, which makes it difficult to locate.

LOBELIA-TYPE FLOWERS

Calicoflower: *Downingia*

Small hairless annuals, spreading to upright, with stems 20-40 cm long. Leaves linear or wider. Flower Lobelia-type, twisted 180 degrees when open. Corolla usually blue, lower lip with a large white patch, 1-2 yellow or orange patches at its center; upper lip with 2 small lobes; lower lip larger, 3-lobed. Rarely, the flowers are nearly or entirely white. Ovary inferior. Plants of wet soils, often vernal pools.

Hoover's Calicoflower
Downingia bella Hoover

Corolla length 10-12 mm from base of tube to tip of longest lobe. *Corolla blue; lower lip yellow region surrounded by white, the white is broken up by blue near junction of lower and upper lips; the lower lip has 2-3 purple spots at that junction.* Each seed has longitudinal lines on it. In flower Mar-May.

Uncommon in vernal pools of the Santa Rosa Plateau, where it grows in deeper water than toothed calicoflower. Pollinated by native bees.

The epithet was given in 1937 by R.F. Hoover for its beauty (Latin, bellus = neat, charming, handsome).

Hoover's Calicoflower

Toothed Calicoflower
Downingia cuspidata (Greene) Rattan

Corolla length 7-15 mm from base of tube to tip of longest lobe. *Corolla pale to bright blue or lavender; lower lip yellow region is completely surrounded by white; no purple near the throat.* Seeds have spiral lines that make them look twisted. In flower Mar-June.

Found in vernal pools of the Santa Rosa Plateau, where it grows in shallower water than Hoover's calicoflower. Sometimes quite abundant, especially as the vernal pools begin to dry, a spectacular sight. Pollinated by native bees.

The epithet was named in 1895 by E.L. Greene for the small tooth at the tip of each lower corolla lobe (Latin, cuspidatus = made pointed).

Hoover's Calicoflower

Both images: Wayne Armstrong

Toothed Calicoflower

Toothed Calicoflower

Both images: Fred M. Roberts, Jr.

Rothrock's Lobelia

Lobelia dunnii Greene **var. *serrata*** (A. Gray) McVaugh

Upright or sprawling, 20-85 cm tall. A perennial though its stems die back annually. *Leaves* are 3-7 cm long, *linear,* most with no petiole, the *edges toothed* and sticky-tipped. *Flowers* are arranged *in a cluster* at the stem tips. *The pedicel of each flower twists around, turning the flower upside-down.* Corolla is 2-lipped, the upper with 2 small slender horn-like lobes that curve up and back, the lower with 3 much larger lobes held out- or downward. *Petals are light blue to pale lavender, rarely white.* The 5 stamens are fused at their anthers into a tube or a ring around the style. In flower July-Oct.

The stems are very brittle and easily broken, so it should not be touched. In addition, some species of *Lobelia* contain toxic alkaloids such as lobeline, sometimes fatal if ingested. Lobelias are best admired and neither touched nor tasted.

Found sparingly here at the edges of waterfalls and seeps, such as Dripping Springs in Limestone Canyon, upper Santiago Canyon, Falls Canyon, Holy Jim Falls, and Coldwater Canyon above Glen Ivy. It probably occurs at other falls and seeps in the Santa Ana Mountains.

The epithet was named in 1889 by E.L. Greene in honor of G.W. Dunn, who first collected the plant in Baja California, Mexico. In 1876, A. Gray named our form for its for its toothed leaves (Latin, serratus = saw-shaped, serrated). The common name is in honor of J.T. Rothrock.

*Clockwise from top right: A single flower; pollen is being released inside the stamen tube, the style and stigma have not yet punched their way through. Plants at Dripping Springs. Inset, small bee gathering pollen at Holy Jim Falls. Blue- and white-flowered forms growing together in upper Santiago Canyon. Inset, metallic sweat bee (*Lasioglossum* sp.), a likely pollinator at Dripping Springs.*

NEMACLADUS-TYPE FLOWERS

Threadplants: *Nemacladus*

Very slender tiny annuals with thin, wispy stems. Leaves are in a compact basal rosette, each leaf linear, pinnate in some species. Flowers face upward or hang downward. Corolla tubular with 5 free tips or 2-lipped, inverted (twisted upside-down) in most species. *Stamen filaments united into a tube that surrounds the pistil and curves back toward the flower at the tip. Stigma 2-lobed, covered with tiny bumps, and tightly surrounded by the anthers.* Several species occur in our deserts, 3 in our area.

The genus was named in 1843 by T. Nuttall in reference to its thin thread-like branches (Greek, nema, nematos = thread; klados = branch).

Long-flowered Threadplant

Nemacladus longiflorus A. Gray **var. *longiflorus***

Upright, 2-21 cm tall, *purplish, usually without side branches.* Leaves 3-12 mm long, oval, hairy. *Inflorescence zigzagged, pedicels long, curved upward at tip.* Flower inverted. *Corolla 5-8 mm long, 2-lipped, throat narrow, face white with yellow spots, candycane-striped behind.* Mature ovary superior. In flower Apr-June.

Uncommon, found along sandy benches in Temescal Valley and lower Indian Canyon wash. Also found "near Santa Ana" in 1907 but probably extirpated by channelization of the Santa Ana River and nearby urbanization.

The epithet was given in 1876 by A. Gray for its long flowers, perhaps specifically for its long pedicels (Latin, longus = long; floris = a flower).

Comb-leaved Threadplant

Nemacladus pinnatifidus Greene

Upright, 6-20 cm tall, branches green, often brownish or purplish near their base. *Leaves 5-20 mm long; hairless; gently narrowed to the long petiole; pinnately lobed or toothed,* rarely undivided. *Inflorescence strongly zigzagged; bracts are held tightly against the pedicel; lower bracts pinnately lobed. Each pedicel is roughly straight, but its tip is sharply hooked upward.* Corolla about 2 mm long, star-shaped (rotate, not 2-lipped); *upper 2 lobes folded backward, lower 3 lobes outward or backward. Corolla whitish, pale yellowish, or pinkish.* Filaments strongly arched back toward the ovary. In flower May-June.

A plant of open spots in chaparral and dry sandy washes, often found after fire. Uncommon here, known from Bear Canyon Trail, North Main Divide Truck Trail near west fork of upper Long Canyon, Coldwater (Glen Ivy) Trail, and Miller Canyon.

The epithet was given in 1886 by E.L. Greene for its leaves (Latin, pinnatifida = pinnately divided).

Similar. Nuttall's threadplant (*N. ramosissimus* Nutt.): upright, 5-32 cm tall, branched only from the base; leaves 3-18 mm long, toothed or pinnately lobed; inflorescence axes straight; pedicel-base bracts linear, never toothed, long, and often stand up like flags; pedicel curved or straight, its tip not sharply hooked upward; *corolla white to pinkish with a central pink stripe, lobes upward or outward (not backward);* filaments strongly arched back toward the ovary; mature ovary half-inferior; in flower Apr-May; similar habitats; Hicks Canyon, between Santiago and Rose Canyons, Holy Jim Trail, North Main Divide Truck Trail near upper Long Canyon, lower Bedford Canyon, Glen Ivy, Temescal Valley (Indian Wash and south flank of Alberhill Mountain), and San Mateo Canyon Wilderness Area (Bluewater Trail and San Mateo Canyon); named in 1843 by T. Nuttall for its branching habit (Latin, ramus = branch; -issimus = very much).

Above: A soft-winged flower beetle, suspected pollinator of long-flowered threadplant. As the beetle passes beneath the arched filament/style tube, pollen rubs on its back. Its movement causes pollen transfer to the stigma of this or another flower, effecting pollination.

CAPRIFOLIACEAE • HONEYSUCKLE FAMILY

Shrubs or vines, often clambering over nearby plants. *Leaves opposite,* somewhat oval, simple or divided. *Calyx tube fused to the ovary at the base,* with *5 free lobes* at the tip. The *corolla is tubular or two-lipped, with 5 free lobes.* Stamens 5, filaments fused to the corolla. Ovary inferior. Fruit is a berry, drupe, or capsule. Worldwide 220 species, 11 in California, 4 in our area. Elderberries were once considered members of this family but were found to constitute their own group and placed into a separate family, the Adoxaceae.

Southern Honeysuckle
Lonicera subspicata Hook. & Arn. **var. *denudata*** Rehder

A *shrub,* somewhat upright but more *commonly sprawling over other plants.* Stems generally 1-2.5 m long, woody at the base, less so near the growing tips, often hairy. *Leaves* 1-4 cm long, elliptic, simple, not toothed. Inflorescence a short spike, sometimes sticky, held upright, flowers in opposing groups that arise from the branch tips. *Corolla* 8-12 mm long, *strongly 2-lipped, pale creamy-yellow.* The upper 4 lobes roll up and back together, the lower lobe spirals down or rolls back under the flower. The stamens have whitish filaments and end in large bright yellow anthers. Its *yellow or red fruit* is spherical and about 8 mm across. In flower Apr-June.

Widespread and common throughout our area, primarily in coastal sage scrub, dry chaparral-clad slopes, and shady riparian areas. Look for it in undisturbed canyons, all of our wilderness parks, and along popular trails such as San Juan Loop and Bear Canyon.

The genus was named in 1753 by C. Linnaeus in honor of A. Lonitzer. The epithet was given by W.J. Hooker and G.A.W. Arnott in reference to the arrangement of its flowers in short spikes (Latin, sub- = under, below, almost, somewhat, near; spicatus = spiked). Our form was named by A. Rehder for its leaves, which are much less hairy than those of other varieties (Latin, denudatus = stripped, made bare).

Southern Honeysuckle

Southern Honeysuckle

Southern Honeysuckle fruits

THE HONEYSUCKLE GUILD
Honeysuckle Bud Gall Midge
Rhopalomyia lonicera Felt (Cecidomyiidae)

These galls appear on axillary buds of southern honeysuckle. When occupied, these *spherical, fleshy galls* are green, sometimes red-tinted, and grow to about 10-15 mm wide. An *adult fly* emerges from the tip of the gall in Feb-Mar. After mating, the female lays one egg in a young axillary bud. The larvae emerge and spend summer, fall, and winter in the growing gall, pupate within, and emerge the following year.

A common sight on honeysuckle plants throughout cismontane southern California. For such a common insect, very little is known about it.

The epithet was given in 1925 by E.P. Felt for the genus of its host plant.

Honeysuckle Bud Gall Midge in Southern Honeysuckle

Honeysuckle Bud Gall Midge adult

Snowberries: *Symphoricarpos*

Upright or creeping shrubs with slender, woody, brittle stems. Leaves are small, dark green, round to elliptical, sometimes lobed, mostly winter deciduous. *Clusters of tubular flowers* are near the branch ends and hang downward. The *short calyx* has 5 narrow lobes. *Corolla* pink or white, *bell-shaped*, with 5 rounded lobes. *Fruit is a white or pinkish spherical berry with 2 seeds.* As the fruits grow, they swell and crowd each other. Common understory shrubs, especially among shady oak woodlands. The berries are not eaten by wildlife. Those of common snowberry are toxic to humans.

The genus was established in 1755 by H.L.D. de Monceau in reference to the fruits, which grow (swell) together into a crowded cluster (Greek, sym = with, together; phoros = bearing; karpos = fruit).

Common Snowberry

Common Snowberry ☠
Symphoricarpos albus (L.) S.F. Blake **var. *laevigatus*** (Fern.) S.F. Blake

An *upright plant,* 60-180 cm tall; *our other snowberries generally creep or trail.* Its branches are stiff, the young shoots held upright. Leaf blades 1-3(-6) cm long, sometimes lobed, usually hairless. *Inflorescence crowded,* usually with 8-16 flowers. Calyx divided into lobes at about one-half its length. *Corolla* 4-6 mm long, bell-shaped, *pink. The corolla is noticeably swollen on the side that faces the ground; five nectar glands exist within that swelling; dissect a flower to see it.* The corolla lobes are about one-half the length of the entire corolla; each lobe is very hairy inside. Fruits are round, 8-12 mm across, and bright white; toxic to humans. In flower May-July (-Aug).

Known here sparingly, to date only from Lucas Canyon in the San Mateo Canyon Wilderness Area, where it grows beneath **coast live oaks** (*Quercus agrifolia,* Fagaceae). It may be more widespread; Caspers Wilderness Park is a likely place for it.

The epithet was given in 1753 by C. Linnaeus in reference to its white fruits (Latin, albus = white). The subepithet was given in 1905 by M.L. Fernald, perhaps for its hairless leaves or its smooth stems and fruits (Latin, laevigatus = slippery, smooth).

Common Snowberry

Spreading Snowberry
Symphoricarpos mollis Nutt. in Torr. & A. Gray

Of various growth forms, usually low-growing, sometimes creeping on the ground and rooting, rarely upright and arching, its branches 15-60 cm long. Leaf blades are 0.5-3 cm long, dark green, hairless to soft-hairy. *Flowers in groups of 2-8.* The calyx is divided about halfway, the free lobes held outward away from the flower. *Corolla* 4 mm long, bell-shaped, *pink, sometimes red outside. The corolla is symmetrical; each lobe has a single rectangular nectar gland at its inner base; dissect a flower to see it.* The corolla lobes are about one-half the length of the entire corolla; each lobe is hairy inside. Fruits are roughly spherical, white or pinkish, 8 mm wide. In flower Apr-June.

Common among oak woodlands and riparian habitats, often in partial or full shade of oaks. Found in all of our foothills, major canyons, and mountains. Easily found in Carbon Canyon, Laguna Coast Wilderness Park, and Caspers Wilderness Park.

Named in 1841 by T. Nuttall for its soft leaves (Latin, mollis = soft).

Spreading Snowberry

Parish's Snowberry

Symphoricarpos rotundifolius A. Gray **var. *parishii*** Rydb.

A low-growing shrub with branches 30-60 cm long, arching or trailing on the ground and rooting at the tips. Its old bark shreds away in strips. Leaf blades are about 1-2 cm long, gray-green but paler beneath, covered with a fine white powder (glaucous), round (oval to elliptic), the young leaves often lobed. *Only 1-2 flowers per group.* Calyx deeply divided into lobes that flare out away from the flower; their margins are transparent. *Corolla 6-9 mm long, pink or white, shaped like a long narrow bell, sparsely hairy inside. The corolla is symmetrical; each lobe has a single long nectar gland at its inner base; dissect a flower to see it.* Fruit is oval, 8-12 mm across, white to pinkish, and glaucous. In flower June-Aug.

Found high up on Santiago Peak, primarily on the north-facing side, along Main Divide Truck Trail. This species is much more common in taller, wetter mountain ranges.

Named in 1850-1852 by A. Gray in reference to its round leaves (Latin, rotundus = round; folium = leaf). Our local form was named in 1899 by P.A. Rydberg in honor of S.B. Parish.

Parish's Snowberry

Parish's Snowberry

Parish's Snowberry

THE SNOWBERRY GUILD
Snowberry Clearwing Sphinx Moth

Hemaris diffinis (Boisduval) (Sphingidae)

A large day-flying moth, its larvae feed on honeysuckle and snowberry, pupate in cocoons spun in leaf litter on the ground, and emerge in spring. Adults first appear in May-June and sip nectar from plants such as **California thistles** (*Cirsium occidentale* var. *californicum,* Asteraceae) high in the Santa Ana Mountains and from other plants at lower elevations. These adults mate and lay eggs on the host plants; their larvae feed, grow, mature quickly, and soon a second batch of adults appears in July-Aug. *Adults rarely land to feed; they usually hover like a hummingbird* and extend their long proboscis into the corolla to reach its nectar. Find them along Main Divide Road on Modjeska and Santiago Peaks, in San Juan Canyon, in foothills, and in places such as Caspers Wilderness Park.

The epithet was given in 1836 by J.B.A.D. de Boisduval for the two yellow lines on its abdomen (Greek, di- = two or double; Latin, finis = boundary, limit).

Snowberry Clearwing Sphinx Moth on California Thistle

CARYOPHYLLACEAE • PINK OR CARNATION FAMILY

Delicate-looking annuals, biennials, or perennials, often with slender *stems* that are *jointed at the nodes. Leaves* are *undivided, usually opposite,* often lacking stipules and petioles. Flowers in an open group at branch tips or solitary from the nodes. *Sepals 5,* free or fused into a tube. *Petals 5* or lacking, entire or lobed, held up and often curved outward. Each petal is often "pinked" (cut into, as with pinking shears), hence the common name. *Stamens 10,* usually held straight, sometimes long and conspicuous. Ovary superior. Fruit is usually a capsule. Cultivated species include carnation, sweet William, and baby's breath.

Sandmat

Cardionema ramosissimum (Weinman) Nelson & J.F. Macbr.

A *small low-growing perennial* with a very stout taproot. The numerous branches are 5-30 cm long and appear woolly because they are crowded with leaves, stipules, and bundles of secondary leaves. *Leaves* are 5-13 mm long, *quite narrow, and end in a tiny spine.* Each node has 2 thin papery stipules, 4-8 mm long, and often 2-lobed. Its *tiny flowers* appear along the stems. Sepals 5, 4-6 mm long, woolly below, ending in a long narrow spine above. Petals 5, reduced to minute green scales less than 1 mm long. Stamens 3-5, alternating with 5 white staminodes. Each anther shaped like an upside-down heart. In flower Apr-Aug.

The spine-tipped leaves, stipules, and sepals make this plant prickly to the touch. Easily and often overlooked. Grows in sandy soils on coastal dunes and bluffs, among coastal strand and coastal sage scrub communities. Once known from upper Newport Bay, Bonita Canyon in Irvine, Newport Beach, and coastal Laguna Beach but not reported recently. Still common on portions of Temple Hill, in Badlands Park in Laguna Niguel, and on the Dana Point Headlands. This plant is seriously declining in our area because of coastal development and habitat disturbance.

The genus was named in 1828 by A.P. de Candolle for the heart-shaped anthers on narrow filaments (Greek, kardia = heart; nema = thread). This species was named in 1820 by J.A. Weinmann for its abundant branches (Latin, ramus = branch; -issimus = very much).

Sandmat

Douglas's Stitchwort

Minuartia douglasii (Fenzl ex Torr. & A. Grey) Mattf. [*Arenaria douglasii* Fenzl ex Torr. & A. Grey]

A *delicate annual,* 4-30 cm tall, unbranched or branched from the base, *with green or purple threadlike stems. Leaves* are 0.5-40 cm long, *threadlike, curled back in age.* Sepals are 2.5-3.7 mm long, green, narrow to broad, and pointed. *Petals white, sometimes with a notch in their tip, about twice as long as the sepals and alternating with them.* Stamens 10, nearly as long as the petals. Styles 3. In flower Apr-June.

Its stiff but slender stems allow it to blow easily in the wind, making photography and observation a challenge. Pollinated by small bees and flies. Prefers hot, dry hillsides. Uncommon overall but sometimes grows in dense populations. Known from the Chino Hills above Santa Ana Canyon, Ashbury Canyon in O'Neill Park, San Juan Trail just above Chiquito Basin, along Main Divide Truck Trail, on slopes of Modjeska Peak, and in the San Mateo Canyon Wilderness Area.

The genus was established in 1753 by C. Linnaeus in honor of J. Minuart. This species was named in 1840 by E. Fenzl for D. Douglas.

Douglas's Stitchwort

Catchflies, Campions, Pinks: *Silene*

Lovely annuals or perennials, often covered with sticky-tipped hairs. Leaves are linear, narrow, with a single obvious vein down the center. Flower clusters vary from a loose bunch at the top of the plant to single flowers that arise from the nodes. Pedicels often long. *Calyx tubular* with 1-13 free tips. *Petals 5, not fused, often with lobed tips.* Some have an appendage at the flower throat. Styles 3-5. We have 5 native and 2 non-native species in our area.

The genus name is from the Greek myth of Silenus, leader of the satyrs (part man, part horse or goat), often intoxicated and covered with foam. The name was given in 1735 by C. Linnaeus, who was reminded of the legend by the sticky, foamy secretion on the surface of some of these plants.

Snapdragon Catchfly

Snapdragon Catchfly
Silene antirrhina L.

An *upright annual,* 12-80 cm tall. Stems are hairless or nearly so; lower stems have tiny backward-pointing barbs, the upper stems are sticky. Leaves are 1-3 (rarely up to 6) cm long, linear. *Flowers generally open only at night and early morning.* Calyx 4-9 mm long, tubular, *with* short lobes 1-2 mm long, and *10 veins.* Petals are 2-lobed, about the same length as the calyx. *The inner surface of the petals is white to pink, the outer pink to magenta.* In flower Apr-Aug.

Widespread but generally uncommon here, sometimes locally abundant after fire. Known from the southern Chino Hills, Dana Point Headlands, Santiago Canyon, lower Arroyo Trabuco, Caspers Wilderness Park, Modjeska Peak, and the San Mateo Canyon Wilderness Area.

Named in 1753 by C. Linnaeus for its similarity to snapdragons (*Antirrhinum* spp., Plantaginaceae).

Similar. **Many-nerved catchfly** ✘ (*S. coniflora*): buds and flowers much larger; calyx with 20-25 veins; uncommon, usually at high elevations.

Many-nerved Catchfly

Many-nerved Catchfly ✘
Silene coniflora Nees ex Otth
[*S. multinervia* S. Watson]

A *slender annual,* 20-65 cm tall. Stems are upright, covered with short sticky-tipped hairs. Lower leaves 3-8 (rarely up to 10) cm long, lance-shaped; upper leaves 0.8-3 cm long, narrow. *Its short-lived flowers are held upright and are rarely open during daytime.* Calyx 8-12 mm long, the free lobes 1-3 mm long, *with 20-25 veins.* Petals about as long as the calyx, *white to pink, usually notched at the tip.* In flower Apr-May.

Not common, it appears only in the first few years after a fire. Known from the Santa Ana Mountains in the region of Coal, Gypsum, Fremont, and Black Star Canyons, on slopes along North Main Divide Road at Upper Long Canyon, the Santa Rosa Plateau, and the San Mateo Canyon Wilderness Area. Native to Asia.

It was named in 1824 by C.G.D. Nees von Esenbeck for its conical calyx and fruit (Greek, konikos = cone-shaped; Latin, floris = flower).

Similar. **Snapdragon catchfly** (*S. antirrhina*): buds and flowers much smaller; calyx with only 10 veins; much more common and widespread.

Flowers: Fred M. Roberts, Jr.

Windmill Pink ✖
Silene gallica L.

An annual, 10-40 cm tall, sprawling or upright, *covered with rough hairs or tiny bristles.* Lower leaves 1-3.5 cm long, upper about 1-2 cm. Flowers held outward or upward. Calyx 6-10 mm long with tiny free lobes. *Petals white, pink, or lavender, much longer than the calyx, entire or notched at the tip, each slightly twisted like blades of a windmill.* Each petal has 2 appendages at the flower throat. It has 3 styles. In flower Feb-June.

A common and widespread weed of grasslands, disturbed soils, and post-burn sites. Found throughout our area in abundance. A native of Europe.

Named in 1753 by C. Linnaeus for Gaul (part of that area is now occupied by France), where his specimens were probably collected (Latin, gallicus = from Gaul).

Mexican Pink, Southern Pink
Silene laciniata Cav. **ssp.** *laciniata*
[*S. laciniata* Cav. ssp. *major* Hitchc. & Maguire]

Our *largest member of this genus* and always a treat to discover in the field. *An upright or straggling perennial* 30-70 cm tall, it grows from a deep fleshy taproot. Upper stems are sticky and a little hairy. Lower leaves are 1.5-10 cm long, upper leaves slightly shorter. Flowers are held outward or upward. Calyx 12-26 mm long, its free lobes often irregularly shaped, 3-5 mm long. *Petals are large, bent 90 degrees outward from the calyx, and bright red with 4-6 long lobes. Each petal has 2 appendages at the flower throat.* Stamens are about as long as the petals. Styles 3, held above the petals among the long stamens. In flower May-July, starting earlier in rainy years.

Very common among coastal sage scrub, chaparral, and on cliff-faces in many locations, often in partial or full shade. Consistent localities include Laguna Coast Wilderness Park, Whiting Ranch Wilderness Park, Limestone Canyon, Caspers Wilderness Park, upper Santiago Canyon, along both Main Divide and Indian Truck Trails, the Whittier Hills, Nolina Canyon in the Chino Hills, and in the San Mateo Canyon Wilderness Area.

The epithet was given in 1801 by A.J. Cavanilles for its cut petal edges (Latin, lacinia = a thing torn, the edge of a garment).

Mexican pink varies in its color and pinking.

Lemmon's Campion
Silene lemmonii S. Watson

A sprawling or upright perennial 15-45 cm tall, *from a woody stem-like root*. Several basal leaves, fewer stem leaves. Its delicate *flowers hang downward* like bells when they open. Calyx 6-10 mm long, the free lobes 1-2 mm long, with 10 veins. *Petals pink to yellowish-white, with 4 long slender lobes, the lobes held outward or backward, sometimes twisted*. Each petal has 2 appendages at the flower throat. Styles 3, about equal to the petals. In flower June-Aug.

Never abundant, found among oak woodland and mixed coniferous forests high up in the Santa Ana Mountains. *Look for it in shade* below Bear Spring along the Holy Jim Canyon Trail, in upper McVicker Canyon near Main Divide Truck Trail, and on rock faces along Indian Truck Trail.

It was named in 1875 by S. Watson for J.G. Lemmon.

Lemmon's Campion

Below, left two: The flowers of Lemmon's campion hang down; each petal has 4 long slender lobes that curl and twist with age. Below, right two: The flowers of San Francisco campion face upward or outward; each petal has 2 broad lobes that curl back with age.

Lemmon's Campion

Lemmon's Campion

San Francisco Campion

San Francisco Campion

San Francisco Campion
Silene verecunda S. Watson

An upright perennial 10-55 cm tall *from a branched woody root*. Several basal leaves, fewer stem leaves. *Flowers are held upward or outward when they first open*. Calyx 10-15 mm long, the lobes 2-5 mm long, with 10 veins. *Petals white to rose, held outward, 2-lobed with broad edges that may have tiny lower side lobes*. Each petal has 2 appendages at the flower throat. Styles 3 or 4, about equal to the petals. In flower June-Aug.

Grows in more open, sunnier areas and at higher elevations than Lemmon's campion, often within oak woodland and coniferous forest. Known from rocky soil on the summit of Santiago Peak and north-facing outcrops and talus slopes in the Santa Ana Mountains.

The species was first collected near Mission Dolores in San Francisco and named in 1869 by S. Watson for its shy appearance or its soft, muted flower color (Latin, verecundus = shy, unassuming).

San Francisco Campion

Sand-spurreys: *Spergularia*

Brittle-stemmed, upright to sprawling annuals or perennials with an obvious taproot. Leaves long, very narrow with papery stipules at their base. *Flowers white to pink* at stem tips. Sepals and petals 5, not divided. Their petals fall off easily when disturbed. Stamens 2-10, styles 3.

Sticky Sand-spurrey
Spergularia macrotheca (Hornem.) Heynh. **var. *macrotheca***

A stout perennial, upright or sprawling, 5-35 cm tall. *Leaves* 1-3.5 cm long, *fleshy, and mostly in bundles (fascicles) along the stem. Stipules* 4.5-11 mm long, *very obvious, triangular with a long sharp point.* The inflorescence is covered with sticky-tipped hairs. Sepals 5-10 mm long, fused below, free above. *Petals* 4-7 mm long, *oval, and pink or rose. Stamens 9-10.* In flower most of the year.

Found in alkaline soils of salt flats and marshes in Seal Beach, Newport Bay, and Laguna Canyon, on coastal bluffs in San Clemente, and in foothills.

Named in 1826 by J.W. Hornemann, perhaps in reference to the large stipules that cover the leaf nodes (Greek, makros = long, large; theca = cover, case, container).

Sticky Sand-spurrey

Salt Marsh Sand-spurrey
Spergularia marina (L.) Besser
[*S. salina* J. Presl & C. Presl]

A delicate, succulent annual, 5-30 cm tall. *Leaves* 2-4 cm long, *linear, narrow, and fleshy. Stipules small,* 1-3 mm long, *broadly triangular, and narrowed to a point.* Sepals 2.5-5 mm long, oval, free at the tips. *Petals generally slightly shorter than sepals; blunt-tipped; white, pink, or rose. Stamens 3* (2-5). In flower Mar-Sep.

Common in sandy soil along the seashore and in alkaline soils of salt marshes in Seal Beach, Bolsa Chica, upper Newport Bay, San Joaquin Marsh, Laguna Canyon, Dana Point Headlands, and Elsinore Basin. Uncommon in Devil Canyon in the San Mateo Canyon Wilderness Area.

The epithet was given in 1819 by J.S. and C.B. Presl for its salty habitat (Latin, salinus = salty).

Salt Marsh Sand-spurrey

Michael Charters

Common Chickweed ✖
Stellaria media (L.) Vill.

A straggling annual with weak stems, 7-50 cm long. *It has a single line of hairs down the stem,* easily seen with naked eye or hand lens. *Leaves* 8-45 mm long, *broad, oval, evenly spaced along the stem.* Pedicels grow outward or upward, then curve backward when in fruit. Sepals 5, free not fused, 3-5 mm long. *Petals* are slightly shorter, *2-lobed, white, sometimes absent.* In flower Feb-Sep.

An abundant weed of moist, shady places, commonly found in the understory of riparian and oak woodlands; also a garden weed. Native of Europe.

Named in 1753 by C. Linnaeus (Latin, medius = middle) for the line of hairs on the middle of the stem.

Line of hairs along stem
Common Chickweed

CHENOPODIACEAE • GOOSEFOOT FAMILY

Annuals or perennials; some are large shrubs. Leaves simple, usually alternate. Flowers solitary or mostly in clusters; small to minute. In some species, flowers of both sexes appear on the same plant, while in others each plant has flowers of only one sex. Calyx 1-5 lobed, often absent in female-only flowers. Petals absent. Stamens 1-5, opposite the calyx lobes. Single pistil with 1-4 branches, ovary superior, fruit a 1-seeded achene or utricle. Familiar cultivated species include beet, spinach, and cockscomb. Also in this family is the movie industry's traditional symbol of the West, the tumbleweed or Russian thistle, which was introduced here in the 1800s. We have numerous species here, most of which are introduced weedy annuals. Of our 45 recorded species, 24 are native; we present here only our showiest 4.

Pickleweeds: *Arthrocnemum, Salicornia*

Annuals or shrubby perennials, all are hairless. Their fleshy green stems are variously jointed and constricted, covered with a fine powder. Leaves are reduced to tiny triangular or linear scales, opposite on the stem. The flowers are tiny and embedded in the stem, mostly noticeable as a pair of stamens or forked styles that emerge from the stem. Three species live in our area, distinguished by their stems and seeds.

Goldenthread dodder (*Cuscuta pacifica* var. *pacifica*, Convolvulaceae) is commonly found parasitizing Parish's and Pacific pickleweeds.

Parish's Pickleweed
Arthrocnemum subterminale (Parish) Standl.
[*Salicornia subterminalis* Parish]

A dense shrub with vibrant green stems, 7-30 cm tall. The flowers generally do not appear on or near the ends of the stems; the uppermost ones only occur on 5-14 segments below it. Seeds are hairless. In flower Apr-Sep.

It grows farther inland, in drier soils, and at higher elevations than our other pickleweeds. Very common in upper Newport Bay from the upper reaches of the marsh, right along Back Bay Drive (especially near San Joaquin Hills Road), and onto the lower reaches of the cliffs. Also found in Anaheim Bay at Seal Beach, Bolsa Chica Ecological Reserve, Little Corona Beach in Corona del Mar, and a few places in Laguna Beach such as Abalone Point.

Named in 1898 by S.B. Parish for the position of its flowers, below the terminus (end) of the stem (Latin, sub- = under, below; terminalis = terminal).

Parish's Pickleweed

Pacific Pickleweed
Salicornia pacifica Standl.
[*Salicornia virginica* L.; *Sarcocornia pacifica* (Standl.) A.J. Scott]

A woody pickleweed that grows as a multi-branched upright or rounded shrub 20-70 cm tall. *Stems are blue-green.* Its growth is not as dense as that of Parish's pickleweed. Its flowers are mostly on the central stem, near the top. Seeds are hairy. In flower Aug-Nov.

Grows in the water, often with Bigelow's pickleweed, but *mostly in soils that are a bit higher and drier, and all the way up to the margins of the estuary,* where it often grows with Parish's pickleweed. Found in Anaheim Bay at Seal Beach, Bolsa Chica Ecological Reserve, upper Newport Bay (along Back Bay Drive) and upstream at the San Joaquin Marsh in Irvine; also on ocean bluffs in Laguna Beach, Capistrano Beach, and San Clemente.

Named in 1916 by P.C. Standley for the region of its discovery.

Parish's Pickleweed

Pacific Pickleweed

Pacific Pickleweed

Bigelow's Pickleweed
Salicornia bigelovii Torr.

A slender annual 9-45 cm tall, it grows straight up from the ground and branches only above its middle. Stems are lighter green than our other pickleweeds. Seeds are hairy. In flower July-Nov.

Grows in open mud flats within low reaches of the estuary in wet soils that are inundated by high tides, exposed at low tides. Look for it in Seal Beach, Bolsa Chica Ecological Reserve, and upper Newport Bay at Big Canyon and North Star Beach.

Named in 1859 by J. Torrey in honor of J. Bigelow.

Bigelow's Pickleweed · Bigelow's Pickleweed

Sea-blites: *Suaeda*

Annuals or small perennial shrubs, hairless or hairy. Leaves are alternate, linear with a slender tip, arch upward, and are covered with a fine white powder. Their flowers have 5 sepals and 2-4 stamens.

Woolly Sea-blite ★
Suaeda taxifolia (Standl.) Standl.
[*S. californica* S. Watson var. *pubescens* Jeps.]

This *perennial* grows as a *rounded shrub* up to 1.5 m tall and wide. It is *densely covered with whitish hairs giving it a gray appearance,* rarely hairless. New stems are light green. *Leaves* are under 3 cm long, *linear, in cross-section rounded or flat.* Flowers are spread out on the plant, in clusters of 1-3 flowers each, 1-3 mm across. All calyx lobes are of equal size. *This is the only sea-blite with long stamens that stand out of the flower.* Stigmas 3-4. In flower all year.

Occurs on the upper borders of salt marshes just above the high tide line and on nearby cliff faces in moist silty-sandy or sandy-clay soils. Known from Anaheim Bay in Seal Beach, Bolsa Chica Ecological Reserve, and upper Newport Bay inland to the San Joaquin Marsh in Irvine. Also on coastal bluffs along the Newport Beach coastline, south to Crystal Cove State Park, Laguna Beach, Capistrano Beach, and San Clemente. CRPR 4.2.

Named in 1916 by P.C. Standley for the similarity of its leaves to that of a yew tree (*Taxus* spp., Taxaceae) (Latin, taxus = the yew tree; folium = leaf).

Similar. Horned sea-blite (*S. calceoliformis* (Hook., not pictured) Moq.): hairless annual; striped stems; leaves triangular to linear, crowded; uppermost calyx lobe larger than all 4 others, all 5 margined with a thin membranous "wing." Stigmas 2. In flower July-October. Uncommon. **Estuary sea-blite ★** (*S. esteroa* Ferren & S.A. Whitmore, not pictured): hairless perennial; leaves linear, green to reddish; uppermost calyx lobe larger than all 4 others, but none are winged. Stigmas 2. In flower July-October. CRPR 1B.2.

Woolly Sea-blite

Woolly Sea-blite

Tumbleweed, Russian Thistle ✖

Salsola tragus L.

An upright, *many-branched, rounded annual* to 1.5 meters tall and broad. Quite variable, its stems and fruits vary from green to red. *Leaves narrow, linear, spine-tipped,* painfully sharp and stiff when dry. *Flowers solitary at nodes,* surrounded by 1-2 spine-tipped leaf-like bracts. Stamens 5, anthers yellow. Ovary superior with 2 style branches. *In fruit, the 5 calyx lobes spread out and develop translucent "wings."* In flower July-Oct.

This is the familiar tumbleweed of the western plains as portrayed in movies. In truth, however, it is native to Eurasia, probably introduced here with the cattle industry in the 1800s or earlier. After it dries, the main stem easily breaks off and the plant rolls away, spreading seeds as it tumbles. Known throughout our area, especially in empty lots and disturbed soils.

The epithet was given in 1756 by C. Linnaeus possibly to honor J. Bock (Hieronymous Tragus, Greek name of Jerome Bock).

Tumbleweed

CISTACEAE • ROCK-ROSE FAMILY

Annuals, perennials, and shrubs. *Ours are small shrubs* with slender stems and simple undivided leaves. *Flowers with 5(3) sepals, 5 petals that fall off easily, and many stamens.* Ovary superior, the style very short, generally much shorter than the ovary. *Fruit is an oval capsule.* A small family of 165 species, 2 native to California, only 1 in our area.

California Rush-rose

Helianthemum scoparium Nutt.

A *small shrub* 12-45 cm tall with *many slender wiry green to reddish stems that stand upright or spread out.* Leaves 5-40 mm long, *linear,* usually alternate, *often fall off.* Sepals 5, the outer 2 very narrow. *Petals 5, bright yellow, held straight out.* Stamens 10 or more, the anthers short and round. *Fruit an oval capsule.* In flower Feb-Sep.

Very common plant of *bright sunny places,* mostly in sandy soils among coastal sage scrub and chaparral, and also in rock cracks on hilltops. Especially common after fire. Known throughout our area, such as in the San Joaquin Hills, along the Harding Canyon Truck Trail, in San Juan Canyon, and within all of our wilderness parks.

The epithet was given in 1838 by T. Nuttall for its broom-like stems (Latin, scopula = thin branches, twigs, a broom).

Similar. Poppies (Papaveraceae): sepals 2, fall off easily; petals 4 or 6; anthers very long; fruit a capsule that elongates as it matures.

California Rush-rose

CLEOMACEAE • SPIDERFLOWER FAMILY

Perennial herbs and woody shrubs, often with a foul odor. Most have *alternate leaves,* usually *divided palmately into 3 leaflets.* Sepals and petals 4 (0, 4-16), *petals yellow, rarely purple or white.* Stamens generally 6 (4-many), much longer than the petals; their anthers coil as they release pollen. *The ovary is superior, often with a stalk at its base.* Sometimes stamens or pistils do not form, leaving the flower with parts of only one sex. Fruit is a capsule.

Many species are used in gardens for their colorful flowers and graceful long stamens. The family has about 900 species worldwide, 12 in California, mostly in our deserts, only 1 in our area.

Bladderpod
Peritoma arborea (Nutt.) H.H. Itis
[*Cleome isomeris* Greene, *Isomeris arborea* Nutt.]

A *rounded shrub* 0.5-2 m tall, with a multitude of interwoven branches. Sometimes covered with a fine whitish power, making it appear grey-green. Its *3 oblong leaflets* are 15-45 mm long. The lower half of its green calyx is fused, the upper half divided into 4 free lobes. The *bright yellow corolla* is 8-14 mm long and has 4 free petals. *Its 4 long stamens and single pistil are longer than the petals. The ovary is on a long stalk.* The *fruit inflates with age* and is the basis for its common name. In flower most of the year.

Bladderpod

A conspicuous plant of our coastline, it occurs among coastal sage scrub on coastal bluffs, canyons, and gullies. Find it on the bluffs of Bolsa Chica, Newport Beach, upper Newport Bay, Corona del Mar, UCI Ecological Preserve, Laguna Beach, Dana Point (harbor, headlands, and canyons), San Clemente, the mouth of San Mateo Creek, and Santa Rosa Plateau. Also known inland along Santiago Canyon near Irvine Regional Park, and in Silverado and Modjeska Canyons, where it was possibly introduced. Interestingly, it also grows in our deserts. It is used in hydroseed mixes in places where it is not naturally found, such as along South Main Divide Road not far from El Cariso. Valued in horticulture for its bright yellow flowers.

Bladderpod

The epithet was given in 1838 by T. Nuttall for the tree-like growth form of older plants (Latin, *arboris* = tree).

THE BLADDERPOD GUILD
Harlequin Bug
Murgantia histrionica (Hahn) (Pentatomidae)

This *colorful, shiny insect* displays a *variable amount of black, orange, and white.* It has *broad, squared, "shoulders" and a shield-shaped outline* and grows to about 12 mm long. Both young and adults are found together year-round on spiderflower and mustard plants. They use their piercing-sucking mouthparts to feed on sap. Though they may let off a mild odor when disturbed, they are not harmful to humans.

The epithet was given in 1834 by C.W. Hahn for its resemblance to the red and black mask worn by the Harlequin character of the theater (Etruscan, *histrionicus* = relating to an actor).

Harlequin Bug

CONVOLVULACEAE • MORNING-GLORY FAMILY

These are annuals or perennials, *many species growing as twining vines*. Leaves are alternate, *commonly triangular or arrowhead-shaped*. Flowers generally arise from the nodes. Sepals 5, overlapping, sometimes fused at their base. *The 5 petals are fused into a tube or funnel; in some species the petals are free at the tips*. Stamens 5, attached to the petals. The ovary is superior, styles 1-2. Fruit is a capsule filled with a few large seeds. To identify them, you must examine their growth form, leaves, and female flower parts. You will need a hand lens.

True Morning-glories: *Calystegia*

Members of this genus are similar to the bindweeds (*Convolvulus* spp.) In *Calystegia*, the calyx is over 7 mm long, the style is 2-lobed near the top; the stigma on each lobe is swollen, cylindrical to oblong, often flattened, obviously broader and different in texture from the style. The name refers to a pair of leaf-like bracts that often cover the sepals (Greek, kalux = cup; stege = a covering or roof). We have 4 species in our area.

Large-bracted Morning-glory

Calystegia macrostegia (E. Greene) Brummitt

A *green perennial*, its *long stems* generally exceed 1 m long, climb up, and are *woody at the base. The calyx is surrounded by a pair of large green to purplish bracts that look like giant sepals* (it looks as if they had 7 sepals instead of the normal 5; count them to be sure). *Corolla funnel-like, white*, sometimes with a pink tinge. In flower Mar-Aug.

Widespread, it grows in dry soils among coastal sage scrub and chaparral, from coastline to mountaintop and valleys beyond. Especially common following fires.

Named in 1885 by E.L. Greene for the large bracts that cover its sepals (Greek, makros = long, large; stege = a covering or roof).

Sonora Morning-glory

Calystegia occidentalis (A. Gray) Brummitt **ssp. *fulcrata*** (A. Gray) Brummitt [*Calystegia fulcrata* (A. Gray) Brummitt]

A *grayish perennial*, its *short stems* grow to about 1 meter long, usually trailing on the ground. *Leaves* arrowhead-shaped and *wider, stiffer, and fleshier* than those of other morning-glories. *Below the calyx and clearly separate from it is a pair of large, lobed, arrowhead-shaped bracts that look like a pair of opposite leaves*, each 5-30 mm long, 2-7 mm wide. *Corolla creamy-yellow* (not white), sometimes pink-tinged. In flower May-Aug.

Grows at *higher elevations* among chaparral and coniferous forests. Best seen on talus slopes along the high reaches of Holy Jim Canyon Trail. Also known from upper Trabuco Trail and the west slope of Modjeska Peak along Main Divide Truck Trail but certainly more widespread.

Named in 1876 by A. Gray, probably for the pair of long bracts that hold up the flower (Latin, fulcrum = a prop).

Native bees nap, sleep, and chew on morning glories!

Large-bracted Morning-glory

Sonora Morning-glory

Arrows point to bracts

Arrows point to bract tips

Beach Morning-glory ✿
Calystegia soldanella (L.) R. Br.

Low-growing and hairless, its roots are very deep and its *stems are fleshy,* 10-50 cm long. *Leaves are fleshy, broad, and kidney-shaped or triangular in outline.* The calyx is enclosed by a pair of large bracts. Corolla pink, each petal with a white stripe down its center. In flower Apr-May.

Lives near the ocean in sandy soils, from Sunset Beach south to Bolsa Chica Ecological Reserve, Huntington Beach State Park, Balboa, Newport Beach, Laguna Beach, and Trestles. Once very common, now difficult to find because of beach grading and trampling by coastal visitors.

Named in 1753 by C. Linnaeus for the resemblance of its leaves to European primroses in the genus *Soldanella.* Both have rounded leaves that lie on the ground like a coin fallen in sand (Latinized from Italian, soldo = a coin).

Beach Morning-glory

Beach: 2 large fleshy bracts surround the calyx

Bindweeds, Morning-glories: *Convolvulus*

Similar to the true morning-glories (*Calystegia* spp.). The calyx is shorter (under 5 mm long). The style is also 2-lobed near the top, but the stigma on each lobe is linear to thread-like, not flattened, and not very different from the style. The name refers to their habit of rolling together or entwining (Latin, convolvo = to roll around). We have 1 native and 1 introduced species here.

Field Bindweed ✖
Convolvulus arvensis L.

A low-growing deep-rooted *perennial* with stems up to 1 m long. Leaf 2-3 cm long, arrowhead-shaped. *A pair of tiny bracts grows well below the flower.* Calyx 5 mm long, corolla 2-2.5 cm long. *Corolla white, often pink-striped,* sometimes purplish on the outside. In flower May-Oct.

A very invasive weed, native to Eurasia. Mostly known near human settlements, along unkept roadways, on ranchlands, in vacant lots, and sometimes also in nature reserves and parks.

Named in 1753 by C. Linnaeus for its usual place of occurrence in cultivated fields (Latin, arvensis = of or belonging to a field).

Field Bindweed; 2 tiny bracts far below the flower

Small-flowered Morning-glory ★
Convolvulus simulans L.M. Perry

Quite unlike other members of its family, this *delicate annual* has spreading or upright branches 10-40 cm long. Leaves are oblong or linear, 1.5-4 cm long, narrow at the base. *Corolla* is about 6 mm long, *pinkish or bluish, clearly divided into 5 lobes.* The developing fruits are held downward (said to "nod"). In flower Mar-Apr.

Found in clay and serpentine-derived soils. Uncommon here, long ago known only from 3 sites in the Chino Hills (Bee Canyon and Brea) and on the UC Irvine campus, all likely destroyed by development. Today, known from San Joaquin Marsh at UC Irvine, Hicks Canyon, the Richard and Donna O'Neill Conservancy, the Joplin Youth Center lands above Trabuco Canyon, several sites on the Rancho Mission Viejo, one site in Caspers Wilderness Park, in the Temescal Valley, and on Alberhill. CRPR 4.2.

Named in 1931 by L.M. Perry, perhaps because it is different from most of its family and appears to imitate other morning-glories and some species of phlox (Latin, simulans = imitative).

Similar. Slender phlox (*Microsteris gracilis,* Polemoniaceae): sticky-tipped hairs; leaves lance-shaped, opposite (alternate above); corolla lobes pink to white, tips blunt or notched inward; 3 style branches; fruits held upright.

Small-flowered Morning-glory

Alkali Plant
Cressa truxillensis Kunth

A small perennial 7-25 cm tall, with many branches and covered in silky grey hairs. *Unlike most members of its family, it grows upright and is not a vine.* Leaves are under 1 cm long, undivided, somewhat elliptic. Calyx 4-5 mm long. *Corolla white, tubular,* 6 mm long, with 5 free lobes at the tips that often arch backward with age. Styles 2, each stigma with a broad white club. The anthers are purple, the corolla sometimes tinged a similar purple. In flower May-Oct.

Easily overlooked, it has a tiny but beautiful flower. Found in coastal areas near the ocean but usually not in wet soils. Common in Seal Beach, Bolsa Chica Ecological Reserve, Newport Beach, upper Newport Bay, Laguna Beach, and atop the bluffs at San Clemente State Beach. Also found at some inland sites, where it grows in dry alkaline soils, such as in Irvine along roadsides and bicycle trails, along flood control channels in Tustin, and on the Rancho Mission Viejo. At the Nichols Road wetlands in the Elsinore Basin, it is parasitized by **salt dodder** (*Cuscuta salina*).

Named in 1818 by C.S. Kunth for the place of its original discovery near Truxillo, Peru.

Western Dichondra ★
Dichondra occidentalis House

A low-growing perennial, it spreads by brown stolons that creep above- or underground. *Leaves small, rounded or kidney-shaped,* 2 cm wide. The tiny short-lived flowers grow singly from the nodes. *Corolla* 3-3.5 mm long, *white or purplish, barely longer than the calyx. Fruit spherical to slightly 2-lobed, each half often 2-seeded, white,* densely covered with white hairs. In flower Mar-May.

Known only along the coast from Ventura to San Diego Counties and the Channel Islands. An uncommon plant, often overlooked and quickly being eliminated by coastal development. Grows on slopes and coastal bluffs, under shrubs in coastal sage scrub and nearby among *rocky-sandy soils.* Found in the James Dilley Reserve (Sycamore Hills), on Three Arches, Temple Hill, and Niguel Hill, and on the Dana Point Headlands. CRPR 4.2. People are often surprised to learn that dichondras are morning-glories and not grasses!

Named in 1906 by H.D. House for its western distribution (Latin, occidentalis = western).

Similar. Asian dichondra (*D. micrantha* Urb., not shown): *corolla white, shorter, only 2 mm long; fruit deeply 2-lobed, each half usually 1-seeded, brown;* requires much more water, cultivated as groundcover or lawn in homes; sometimes a yard weed, not often found in wildlands.

Dodders, Witch's Hairs: *Cuscuta*

Unusual plants, all are weak-stemmed annuals that grow as complete parasites on other plants (holoparasites). Their stems are yellowish to bright orange. Leaves are often absent but when present are only about 2 mm long and triangular or linear. Flowers are in a cluster or are solitary and arise from the nodes. Styles are 2-lobed from their base (like old-fashioned television "rabbit-ears" antennae), each topped with a spherical stigma. Fruit is a round capsule or is berry-like. Identification requires magnification to examine its stamens and "petal appendages" or "scales," small upright flaps that grow between the petals, just below the stamens. Three of our 4 species have tiny "fingers" on their appendages. This genus is sometimes placed into its own family, the Cuscutaceae.

	California *C. californica*	Field *C. campestris*	Canyon *C. subinclusa*	Goldenthread *C. pacifica*	Salt *C. salina*
Stems	Thin, orange to yellowish	Thick, deep orange to yellowish	Thick, deep orange	Thin, orange	Thin, orange
Corolla	A shallow bell; *tube broad, about as long as wide*	A shallow bell; *tube broad, wider than long*	Funnel-shaped; *the tube narrow and long*	Funnel-shaped; *the tube narrow and long*	Funnel-shaped; *the tube narrow and long*
Filaments	Long, at least as long as the anthers	Short, almost none	None or nearly so	Short, almost none	Short
Petal appendages	None (if present, then tiny, no "fingers")	Present and "finger-tipped," large	Present and "finger-tipped"	Present and "finger-tipped"	Present and "finger-tipped," too short to touch the stamens
Habitats	Coastal sage scrub and washes	Coastal, sandy soils	Canyons and washes	Coastal salt marshes	Inland salty soils
Hosts	California buckwheat, black sage, white sage, California lilacs, etc.	Sunflower and *pea* families	Most often laurel sumac, also mule fat	Pickleweeds, fleshy jaumea, saltbush, alkali plant, alkali heath	Pickleweeds, fleshy jaumea, saltbush, alkali plant, alkali heath
In flower	May-Aug	July-Nov (-Jan)	June-Oct	July-Oct	May-Oct

California Dodder

Cuscuta californica Hook. & Arn.

California Dodder

An orange to yellowish plant with slender hair-like stems. Corolla 3-5 mm long, shaped like a shallow bell, about as long as wide. The calyx and *corolla* each have 5 *lobes*, those on the corolla are very narrow and *often fold backward* late in the flowering period. Stamen filaments are as long as or longer than the anthers, so the *anthers are clearly held far above the corolla. Corolla appendages are absent or extremely tiny and without "fingers."* In flower May-Aug.

An abundant plant of coastal sage scrub, chaparral, and mixed coniferous forests, where it parasitizes many types of annuals and perennial plants such as California buckwheat and white sage. Found at higher elevations than our other dodder species, often on dry slopes and in the washes below. Quite widespread, found in Chino Hills, along Santa Ana Canyon, near Irvine Lake, on the Audubon Starr Ranch Sanctuary, and in Santa Ana Mountain canyons such as Santiago, Silverado, and San Juan. Abundant along the Holy Jim Trail, in Modjeska Canyon, Tucker Wildlife Sanctuary, in Trabuco Canyon to O'Neill Regional Park, and on Alberhill.

Named in 1838 by W.J. Hooker and G.A.W. Arnott in honor of the state of its discovery.

Similar. Canyon dodder (*C. subinclusa*): stems thicker, deeper orange; corolla tube narrow, very long. **Western field dodder** (*C. campestris*): stems thick or thin, orange to light yellow; corolla tube wider than long.

Western Field Dodder ✿
Cuscuta campestris Yuncker
[*C. pentagona* Engelm., in part]

Its main stems are thick and orange like canyon dodder, the side branches slender and lighter orange like California dodder. The corolla is 2-3 mm long, *shaped like a shallow bell; the tube is wider than long*. Petal edges broadly overlap at their base, giving the flower a 5-angled look, a feature best seen in bud. Anther filaments are short, about equal to the anthers. *The easily seen petal appendages are "finger-tipped," 1-2 mm long, able to touch the stamens, and often curved over the ovary*. In flower July-Nov (-Jan).

Grows at lower elevations than California and canyon dodder. Generally on shrubs, often those in the sunflower and pea families. Uncommon here, known from the San Joaquin Marsh in Irvine, where it lives on plants such as **wreath-plants** (*Stephanomeria* spp., Asteraceae) and **sweetclovers** (*Melilotus* spp., Fabaceae). More common and easily found on **beach bur-sage** (*Ambrosia chamissonis*, Asteraceae) at the Bolsa Chica Ecological Reserve and the southern end of Huntington State Beach.

Named in 1932 by T.G. Yuncker for its habitat (Latin campestris = growing in a field).

Similar. California dodder (*C. californica*): stems much thinner, yellower; corolla tube about as long as wide; generally farther inland. **Canyon dodder** (*C. subinclusa*): stems much thicker, deeper orange; corolla tube narrow and very long; almost exclusively on laurel sumac.

Canyon Dodder
Cuscuta subinclusa Durand & Hilg.
[*C. ceanothi* Behr]

A deep orange, thick-stemmed plant. Flowers are 5-6 mm long, funnel shaped, *the corolla tube is narrow and much longer than wide*. Its anthers are attached directly to the petals; it has no stamen filaments. The appendages are 1.6-2.1 mm long, spoon-shaped, and "finger-tipped." In flower June-Oct, sometimes all year.

Grows on woody shrubs in habitats similar to California dodder but mostly along wooded canyons and riparian communities. It is especially common on **laurel sumac** (*Malosma laurina*), less common on **poison oak** (*Toxicodendron diversilobum*, both Anacardiaceae), **mule fat** (*Baccharis salicifolia*, Asteraceae), and **blue elderberry** (*Sambucus nigra* ssp. *caerulea*, Adoxaceae) in riparian areas. In chaparral at higher elevations of the Santa Ana Mountains, it also lives on **California lilacs** (*Ceanothus* spp., Rhamnaceae) and **bush poppy** (*Dendromecon rigida*, Papaveraceae). Look for it along the Santa Ana River channel, Limestone Canyon, Santiago Canyon, Audubon Starr Ranch Sanctuary, Silverado Canyon, Trabuco Canyon, San Juan Canyon, and along the Holy Jim Trail.

Named in 1855 by E.M. Durand and T.C. Hilgard, in reference to the arrangement of its anthers near the free base of the petals, almost confined within the flower tube (Latin, sub- = under, below, near; inclusus = confined, shut up, included).

Similar. California dodder (*C. californica*): stems much thinner, yellow-orange; corolla tube about as long as wide; usually not along waterways. **Western field dodder** (*C. campestris*): stems thick or thin, orange to light yellow; corolla tube wider than long; coastal.

Above and previous page: California and western field dodder's flowers are short and broad with long lobes that often fold backward in age. They grow in asymmetric groups. Right and below: Canyon dodder's flowers are skinny and long with small lobes. They grow in spherical umbels along the stems.

Goldenthread Dodder

Cuscuta pacifica Costea & M. Wright **var. *pacifica***
  [*C. salina* Engelm. var. *major* Yunck.]

A medium-orange, *slender-stemmed* plant. Small flowers in umbels or heads. *Corolla 3.5-6 mm long, bell-shaped, the tube longer than wide like canyon dodder.* Stamen filaments are short, about equal length to the anthers or slightly longer. Petal appendages are "finger-tipped" but short, under 1.5 mm long, not long enough to touch the stamens. In flower July-Oct (-Nov), sometimes all year.

A plant of coastal salt marshes at sea level. Hosts are primarily **pickleweeds** (*Arthrocnemum* and *Salicornia* spp., Chenopodiaceae) and **fleshy jaumea** (*Jaumea carnosa*, Asteraceae). Known from Seal Beach, Newport Beach, and upper Newport Bay. *It is **not** known inland.*

Named in 2009 by M. Costea and M.A.R. Wright for its distribution along the Pacific Coast.

Goldenthread (far left) and salt dodder (left) are quite similar, though the flowers of goldenthread are generally larger (3.5-6 mm long versus 2.5-4.5 mm long). Salt dodder lives inland while goldenthread lives along the Pacific Coast.

Salt Dodder

Cuscuta salina Engelm.

A medium-orange, *slender-stemmed* plant. Small flowers in umbels. *Corolla 2.5-4.5 mm long, bell-shaped, the tube longer than wide like canyon dodder.* Stamen filaments are short, about equal length to the anthers or slightly shorter. Petal appendages are "finger-tipped" but short, under 1.5 mm long, not long enough to touch the stamens. In flower May-Oct (-Nov).

A plant of inland saline areas, where it lives on plants that grow in salty soils. Hosts include **saltbush** (*Atriplex* spp., Chenopodiaceae, not pictured), **Parish's pickleweed** (*Arthrocnemum subterminale*), **alkali plant** (*Cressa truxillensis*), and **alkali heath** (*Frankenia salina*, Frankeniaceae). Found in alkaline areas of the Elsinore Basin such as in Nichols Road Wetlands and those near Diamond Stadium. *It is **not** known from our coast.*

Named in 1876 by G. Engelmann for its salty habitat (Latin, salinus = salty).

CRASSULACEAE • STONECROP FAMILY

Annuals, perennials, and shrubs. *Ours are annuals or perennials,* rarely woody; *most are fleshy and hairless.* Their *succulent leaves* are undivided, often thick and rubbery, alternate, opposite, or whorled. Flowers are arranged in branched clusters among the nodes or on an upright stalk. Sepals 3-5, not fused. Petals 3-5, free or fused. *Stamens are equal or double in number to the sepals. Pistils also 3-5, each with a slender style.* Fruit is a collection of 3-5 teardrop-shaped follicles that spread apart and fall off at maturity. Seeds are tiny and numerous. A family of 1,500 species worldwide, 51 in California, 15 in our area. We have 2 genera, *Crassula* and *Dudleya.*

Several live-forevers (*Dudleya* spp.) are in cultivation, prized for their foliage as well as their flowers. They are easily transplanted and cared for, which unfortunately makes them targets for plant poachers. Many species are declining in the wild from habitat destruction and horticultural collecting, so some have been given legal protection. Fortunately, several species live in our wilderness parks, preserves, and national forests, where they are afforded extra protection. If you want them in your garden, please don't dig them up. Instead, purchase them from a reputable native plant nursery (see page 460).

Pygmy-stonecrops: *Crassula*

Tiny hairless annuals with one to many branches, *often growing in shallow standing water in their youth but left on drying ground as the water evaporates.* Leaves are both basal and on the stem; the stem leaves are opposite with fused bases. Their small size makes them easily mistaken for mosses, which sometimes grow with them. We have 2 native species.

Water Pygmy-stonecrop
Crassula aquatica (L.) Schönl.

A spreading plant with *multiple stems,* each 1-6 cm long, branched from the base. *Leaves* 3-6 mm long, linear, *with pointed tips. Sepals are rounded, much shorter than the whitish petals.* In flower Mar-July.

Known from vernal pools, ponds, and salt marshes. Recorded from Santa Ana, Laguna Canyon, and Santa Rosa Plateau. Certainly more common and overlooked.

It was named in 1753 by C. Linnaeus for its habit of growing in standing water (Latin, aquaticus = found in water).

Similar. Sand pygmy-stonecrop (*C. connata*): usually a single stem; leaves smaller, oval or rounded; sepals longer than petals; abundant.

Water Pygmy-stonecrop

Sand Pygmy-stonecrop

Sand Pygmy-stonecrop
Crassula connata (Ruiz & Pav.) A. Berger
[*Tillaea erecta* Hook. & Arn.]

An upright plant 2-7 cm tall, *generally not branched.* It often turns red as it ages. *Leaves* 1-3 mm long, oval or oblong, *usually round-tipped,* rarely pointed. *Sepals are pointed, usually longer than the whitish petals.* In flower Feb-May.

Widespread and abundant, found throughout our area in soil depressions, slightly moist spots, and sandy areas. Common along trails in most of our wilderness parks and natural areas.

Named in 1799 by H. Ruíz Lopez and J.A. Pavón for its fused leaf bases (Latin, connatus = fused).

Similar. Water pygmy-stonecrop (*C. aquatica*): usually many-stemmed; leaves longer, linear; sepals much shorter than petals; uncommon.

Live-forevers, Dudleyas: *Dudleya*

Fleshy perennials from a thick rootstock or underground tuber. Leaves are mostly in basal whorls, with fewer leaves arranged alternately on the flowering stalk. Flowers are in branched clusters on an upright stalk that withers after seed-set. Sepals and petals 5, stamens 10. Fruit of 5 follicles, held upright or outward at maturity.

Many species are well-suited to cultivation and make striking garden subjects in mixed plantings or alone as an attractive container plant. Along with the striking architectural form of the leaves, there exists a wide variety of flower colors from white to orange, yellow, and pink that put on a beautiful show from spring into mid-summer.

The genus was established in 1903 by N.L. Britton and J.N. Rose to honor W.R. Dudley.

There are about 45 species, all from southwestern North America. California hosts 25 species, 9 in our area. The genus *Dudleya* is divided into 3 groups, recognized by their leaves and flower posture. They are *Hasseanthus*, *Stylophyllum*, and *Dudleya*. Long ago, each group was considered a separate genus, but now each is treated as a subgenus.

Subgenus *Hasseanthus*: geophytes, flowers held wide open, fly- and bee-pollinated

Blochman's Live-forever ★

Dudleya blochmaniae (Eastw.) Moran **ssp. blochmaniae**

Its *small finger-like leaves*, 3-12 per rosette, appear during the winter-spring growing season. The *basal leaves* are spoon-shaped, narrow at the base, *broad and rounded at the tip.* Individual leaves are 7-25 mm in length and half as wide. The flowering stalk has several branches. *Its flowers are white, sometimes marked with red or purple.* In flower May-June.

Found from San Luis Obispo to northern Baja California, where it grows on dry rocky outcrops in serpentine or clay-dominated soils, in openings among grasslands and coastal sage scrub. Known from seaside bluffs in Laguna Niguel, Dana Point Headlands, and San Clemente. Once quite extensive in Dana Point, where it ranged inland of present-day Dana Hills High School and Ocean View, lost to rampant development in the 1970s-1980s. CRPR 1B.1. It is pollinated by flies and bees.

The epithet was given in 1896 by A. Eastwood in honor of I.M.T. Blochman (1854-1931).

Blochman's Live-forever

Young plant: Fred M. Roberts, Jr.

Many-stemmed Live-forever ★

Dudleya multicaulis (Rose) Moran

A small plant with a basal rosette of 6-15 grass-like fleshy leaves, it appears during the beginning of the rainy season. The *basal leaves* are longer than those of Blochman's live-forever, triangular at their base and *pointed at their tip.* Each hairless green leaf is 4-15 cm long and 2-6 mm wide, making it an easy plant to miss growing among grasses and shrubs. The flowering stalk is often multiple-branched with *lemon yellow flowers.* In flower Apr-June.

A plant of heavy clay and rocky soils in barren areas among coastal sage scrub and chaparral. Currently known from Newport Beach south into Camp Pendleton, Crystal Cove State Park, Laguna Canyon, Temple Hill, Rancho Mission Viejo, Caspers Wilderness Park, Audubon Starr Ranch Sanctuary, and Lucas Canyon and Oak Flats in the San Mateo Canyon Wilderness Area. Northward, known from Irvine Regional Park, the hills east of Orange, the Chino Hills, and Weir, Gypsum, and Coal Canyons. It also lives at the base of Horseshoe Canyon in the Temescal Valley and on Alberhill. A rare plant, threatened by development, roadways, grazing, and off-road recreation. CRPR 1B.2. Studies performed by Bob Allen showed that it is pollinated by flies and native bees.

The epithet was given in 1903 by J.N. Rose in reference to its many stems (Latin, multi- = many; caulis = stem).

Many-stemmed Live-forever

Subgenus *Stylophyllum:* perennial-leaved, flowers held wide open, fly- and bee-pollinated

Ladies'-fingers Live-forever
Dudleya edulis (Nutt.) Moran

An upright plant from a basal rosette of 10-20 leaves. Each leaf about 8-20 cm long, *light blue-green, finger-like, glaucous.* Inflorescence long, sometimes branched. *Flowers creamy white, often pink-tinged.* In flower May-June.

Typically a coastal species, it occurs in Orange and western Riverside Counties south into northern Baja California. Known locally from Newport Beach (extirpated?), Laguna Beach, San Clemente State Beach, cliffs on the south side of San Mateo estuary, and San Onofre State Beach. Also in a narrow strip of white sand from San Clemente, north into Cristianitos Canyon, through Richard and Donna O'Neill Conservancy, Rancho Mission Viejo, and sandstone bluffs in Caspers Wilderness Park. Pollinated by flies and bees.

The epithet was given in 1840 by T. Nuttall, who reported that native Americans ate its leaves (Latin, edulis = edible).

Ladies'-fingers

Sticky Live-forever ★
Dudleya viscida (S. Watson) Moran

An upright plant from a basal rosette of 15-35 leaves. Each leaf 6-15 cm long, *dark green, finger-like, covered with a thick sticky resin.* Inflorescence has 3 or more branches. Flowers *whitish, strongly marked with pink or red.* In flower May-June.

A plant of *rocky slopes and rock crevices near waterways,* in Orange, Riverside, and San Diego Counties. Known from upper San Juan Canyon and upper Hot Springs Canyon especially near its largest waterfall. Also known from San Mateo, Devil, Lucas, and Cold Spring Canyons in the San Mateo Canyon Wilderness Area. Abundant along Ortega Highway, where it grows on steep north-facing rocky cliffs among mixed chaparral. Pollinated by flies and bees. A rare plant listed as CRPR 1B.2, threatened by development and roadway construction.

The epithet was given in 1882 by S. Watson for its sticky-coated leaves (Latin, viscidus = sticky). It was first discovered in 1881 on the rocky walls along San Juan Creek just downstream of San Juan Hot Springs.

Sticky Live-forever

Subgenus *Dudleya:* perennial-leaved, flowers tubular, bumble bee- and/or hummingbird-pollinated

Canyon Live-forever ✿
Dudleya cymosa (Lem.) Britton & Rose

A small live-forever with a basal rosette of 1 to several leaves. Individual leaves are of various sizes, diamond- to spoon-shaped; the tip narrowed to a point. Leaves of high-elevation plants are *light grey-green and glaucous*; those in Santiago Canyon are green. *Petals bright yellow to gold, rarely red.* In flower Apr-July.

A plant of cliffs and rocky crevices among chaparral, it grows most often in full sun. Found on precarious cliffs in Santiago Canyon above the Modjeska Canyon community and in higher elevations of the Santa Ana Mountains such as in Fremont and Silverado Canyons and on Modjeska Peak. Pollinated by both bumble bees and hummingbirds.

The epithet was given in 1858 by C.A. Lemaire for its branched inflorescence (Latin, cymosa = full of shoots).

Canyon Live-forever

Lance-leaved Live-forever
Dudleya lanceolata (Nutt.) Britton & Rose

A plant with *variable growth habits,* always from a star-shaped basal rosette of leaves. *Each leaf is* 5-20 cm long, 1-3 cm wide, roughly flat, oblong to spear-shaped with a pointed tip, *light to medium green (to red) in color, sometimes lightly coated with whitish powder.* The *flower stalk is often multiple-branched mostly near its top and very tall,* sometimes up to 1 meter. *Corolla reddish, orangish, or yellowish,* usually darkest at the base and lighter near the tip. In flower May-July.

Widespread and abundant on rocky slopes and in sandy soils in coastal sage scrub and chaparral from Santa Barbara and Kern Counties south into northern Baja California. Known nearly throughout our entire area. A good plant for the garden, easily grown and tolerant of many soil, moisture, and light conditions. Pollinated mostly by bumble bees, sometimes also by hummingbirds.

The epithet was given in 1840 by T. Nuttall for its spear-shaped leaves (Latin, lanceolatus = lance-like, from lancea = a small light spear).

Lance-leaved Live-forever

Chalk Live-forever
Dudleya pulverulenta (Nutt.) Britton & Rose

A large, unusual live-forever that resembles a pale cabbage. It has a beautiful *large rosette* of 40-60 pale *bluish leaves covered with a delicate white powder (glaucous).* Each leaf is 8-25 cm long, 3-10 cm wide, and ends abruptly in a sharp point. The *branched flowering stalk* is 30-100 cm tall and *bears many red flowers that hang downward,* in position for the hummingbirds that pollinate it. In flower May-July.

It grows among coastal sage scrub and chaparral from San Luis Obispo County south into northern Baja California. In our area, found in nearly every wild area from coastline to foothill and mountain canyon. In a native garden, it makes a striking accent growing on a rocky slope that is lightly shaded during the summer months.

The epithet was given in 1840 by T. Nuttall for its chalky, dusty-looking covering (Latin, pulverulentus = dusty).

Chalk Live-forever

Laguna Beach Live-forever ★
Dudleya stolonifera Moran

A *small plant* with *medium to dark green broad leaves in a rosette, it spreads by slender stolons.* Each leaf is 3-7 cm long, 1.5-3 cm wide, the tip pointed. The *single flowering stalk* is 8-25 cm tall, rarely branched, and topped with *bright lemon yellow flowers* that are bee-pollinated. In flower May-July.

An Orange County endemic, it lives within Aliso and Laguna Canyons of Laguna Beach and a few tributaries. In Laguna Canyon, it grows among mosses on steep north-facing cliffs and boulders made of basaltic rock, rarely on sandstone. Hybrids between it and ladies' fingers are found near Aliso Canyon. Witnessing this stunning plant in full flower on these vertical sandstone walls and boulders is a journey every plant lover must take.

A very rare plant, listed by both state and federal agencies as Threatened. CRPR 1B.1. Impacts include development, roadways, recreation, competition from non-native plants, and suspected horticultural collecting. Easiest to see at Laguna Coast Wilderness Park, where rangers and docents lead hikes to see it in bloom. *Look but do not touch it!*

The epithet was given in 1950 by R.V. Moran for its underground stolons (Latin, stolonis = shoot or branch; fero = to bear).

Laguna Beach Live-forever

CUCURBITACEAE • CUCUMBER OR GOURD FAMILY

Our 2 species are *geophytes* that grow *from large fleshy, underground tubers*. Some species elsewhere are annuals. Their stems sprawl on the ground or climb over other plants. They usually have *1 tendril at each node. Leaves alternate, undivided, palmately lobed, or palmately compound*. Each inflorescence arises from a node. *Flowers are of only one sex, but both sexes are on each plant*. Male flowers are in clusters, rarely single, with 3-5 (or 1-3) stamens. Female flowers are most often solitary. The calyx is absent or has 5 free lobes. Corolla trumpet-shaped with 5 lobes. Ovary is inferior with 1-3 styles, the stigmas large and lobed. Fruit is a pepo, a berry with a leathery rind, usually round but sometimes elongate or snouted. To this family belong many cultivated fruits such as cucumbers, gourds, squash and pumpkins, and melons.

Coyote Melon, Calabazilla, Stinking Gourd
Cucurbita foetidissima Kunth

A vine that sprawls over the ground, 2-5 m long, it arises from a huge woody tuber. *Its foul odor is strong enough to detect from a distance.* It is entirely covered with stiff, bristly hairs. *Leaves* are 15-30 cm long, *gray-green, triangular or oval, folded up at the sides and stand up*. Male and female flowers arise from different nodes. Female flowers have 3 stigmas, each 2-lobed. Male flowers have 5 stamens, their filaments fused just above the base, their anthers twisted together. *The large bright orange corolla is a deep bell,* 9-12 cm long, the 5 tips curved back. Fruit is 7-8 cm across, spherical, green with irregular white stripes. They lie on the ground, looking like a group of striped tennis balls. The plant begins to grow new vegetation in mid- to late winter. In flower May-Aug.

Lives in sandy and gravelly soils at lower elevations, from the coast inland to the foothills, the Chino Hills, and Temescal Valley. Caspers Wilderness Park is a great place to see it. The common name "coyote melon" refers to its fruit, which is eaten by the **coyote** (*Canis latrans* Say, Canidae).

Named in 1817 by C.S. Kunth in reference to its odor (Latin, fetidus = ill-smelling; -issimus = very much, most).

Clockwise below: Vines on the ground, leaves point upward. Bottom: Female flower with ripe and broken fruit. Lower left: Male has no ovary; stamens are held to filament column, anthers twisted; female has an inferior ovary, lobed stigma.
Left: Fruit begins green, matures to yellow, often striped with white.

THE SQUASH GUILD

Striped Cucumber Beetle
Acalymma vittatum (Fabricius, Chrysomelidae)

Small beetles about 1 cm long with a black head, orange thorax, black abdomen, and whitish forewings that have 3 black stripes. Adults appear in early spring, mate, and lay eggs in the soil near the host plants as they start to resprout from their underground tuber. The young soon hatch from the egg and feed on the tuber and its roots. When mature after about 15 days, they pupate in the soil and emerge as adults about a week later. In our area, this generation mates, lays eggs, and produces another generation. *Adults feed on young plants, stems, leaves, and flowers of coyote melon.* The second-generation adults spend the winter hidden among plant debris away from the plants and appear the following spring. The beetles are commonly found together deep within the flowers. They become covered with pollen and may serve as a minor pollinator.

The epithet was given in 1775 by J.C. Fabricius for its black stripes (Latin, vittatus = striped).

Squash Bee
Peponapis pruinosa (Say, Apidae) [*P. angelica* Cockerell]

These bees specialize in gathering pollen from gourd, squash, and wild cucumber flowers. In fact, they are the most important pollinator of plants in the cucumber family. *Males* are 10-13 mm long, *gray with tan hair on their thorax and alternating bands of off-white and black on their abdomen. Females are slightly larger* (11-14 mm), stout, *and densely covered with orangish hairs.* The bees often sleep in the flowers, which close at night and open up in the early morning. Females fly into the base of the flower and crawl around rather comically in a circle around the style, gathering pollen on their body hairs as they go. Males enter the flowers in search of females. Both sexes unintentionally move pollen to the stigma and pollinate the plant. The female forms a pollen ball in her underground nest and lays one egg on it. The larva hatches, feeds on the pollen ball, pupates, and emerges as an adult 1-2 months later.

The epithet was given in 1837 by T. Say for colorful hairs on its body, as if frosted like a cake (Latin, pruinosus = frosted).

Wild Cucumber, Man-root
Marah macrocarpus (Greene) Greene

A climbing vine from an enormous tuber. Its many branches grow perhaps 2-5 meters long, right over other plants, and rarely lie on the ground. *Leaves* are 5-10 cm wide, *palmate, green, 5-7 lobed.* Male and female flowers arise from the same node. Males grow in a stalked cluster, their stamens fused and anthers twisted together. Female flowers are solitary. *The small corolla is whitish to yellowish, shaped like a shallow cup with 5 free lobes.* The fruit hangs downward from the vine, is 5-12 cm long, green, oblong, and covered with prickles that dry and stiffen with age. When dry, the far end opens and its large seeds fall to the ground. By summer, only the curious skeletons of the fruit remain. New plant growth appears in early winter. In flower Jan-Apr.

Widespread and abundant, especially in our wilderness parks. Found among coastal sage scrub, chaparral, and oak woodlands from coastline to mountains.

Named in 1885 by E.L. Greene for its large fruits (Greek, makros = long, large; karpos = fruit).

Wild cucumber, clockwise: Solitary female flower (note the spine-covered ovary), cluster of male flowers at right. Mature fruit with adult hand for scale. Top of exposed dormant tuber in Irvine Regional Park. A recent storm exposed this free-standing, 1.2 meter (4 feet) tall, still-living tuber in Coal Canyon.

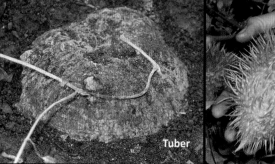

DATISCACEAE • DATISCA FAMILY

Upright hairless perennials with nitrogen-fixing root nodules. Leaves alternate, pinnately lobed or divided, tooth-edged. The small inflorescence appears in axillary clusters. Each plant has either all-male flowers or flowers that contain both sexes (in the Asian species, each plant has either all-male flowers or all-female flowers). *Male flowers have 4-9 unequal calyx lobes, no corolla, and 8-12 stamens with short filaments and long yellow anthers. Female and bisexual flowers have 3 calyx lobes; no corolla; 0 or 2-4 stamens; inferior ovary; and 3 pale styles, each thread-like and 2-forked.* Fruit is a capsule that opens at the top between the style bases. A tiny family of only 1 genus with 2 species, 1 in California, the other in Asia.

Durango Root ☠

Datisca glomerata (C. Presl) Baill.

An upright perennial herb, 1-2 meters tall. *Lower leaves often opposite or whorled, about 15 cm long, upper leaves alternate and smaller. Leaves oval to lance-shaped in outline, pinnate or pinnately lobed,* the lobes narrowed toward their tip, tooth-edged. Male flowers: calyx 2 mm long, anthers 4 mm long. Bisexual flowers: calyx 5-8 mm long, styles about 6 mm long. Fruit a capsule full of pitted oblong seeds. In flower May-July.

A common plant of shady streams and ephemeral waterways. Known from upper Santiago Canyon south into San Juan Canyon and San Mateo Canyon Wilderness Area to near the coast. Because of its similar growth habit and leaves, it is sometimes mistaken for marijuana and smoked by unwary individuals. *The entire Durango root plant is toxic.* Symptoms of poisoning include depression, diarrhea, increased respiration rate, and death.

The epithet was given in 1835 by C.B. Presl, perhaps for its ability to cause illness when ingested (Latin, glomeratus = wound).

Similar. Marijuana ✖ (*Cannabis sativa* L., Cannabaceae): odorous; leaves palmately compound, upper leaves sometimes narrower, lobed, or simple; flowers similar to nettles (Urticaceae); usually only one sex per plant (dioecious), sometimes both on same plant (monoecious); contains psychoactive chemicals; native to Asia, occasionally found in illicit plantings.

Durango root: long, skinny, pinnate leaves   *Marijuana: palmately compound leaves*

ERICACEAE • HEATH FAMILY

Perennials, shrubs, and trees. *Ours are handsome shrubs and trees.* Their bark is a lovely red, brown, or grey and is often self-shredding. *Leaves* are leathery, *simple and undivided,* usually alternate on the stem, and evergreen. The *calyx* generally has *4-5 free sepals* that often fall off while the flower is open. *The corolla is shaped like an urn and hangs upside-down with the opening toward the ground.* Its 4-5 white, pink, or red petals are mostly fused but often free at the tips. Stamens are 5, 8, or 10, often with 2 slender, curved appendages at their tips. The fruit is a capsule, drupe, or berry.

They are *buzz-pollinated* by native bees, although they are visited by many types of small native bees and flies capable of entering the small flower opening or possessing mouthparts long enough to reach the nectary at the ovary base. Butterflies alight on the flower clusters upside-down to take nectar. Hummingbirds take nectar and transfer pollen on the tip of their bill, a process called *bill-tip pollination.*

Some species form a tough, woody, fire-resistant region at the top of their root crown (burl) from which they can resprout after a burn.

Pacific Madrone, Madroño ✿
Arbutus menziesii Pursh

A stately, round-topped shrub or tree that grows from 5-40 meters tall but only to about 6 meters tall in our area. If fire-burned, it can regrow from its underground burl. The *thin red-brown bark self-shreds* to reveal a smooth honey-colored surface that fissures and darkens with age. *Leaves are ovate to oblong,* 5-12 cm long, entire to minutely toothed, bright green above and whitish or silvery below. *Corolla* 6-8 mm long, yellowish white to pink, and sweet-smelling. Each has 10 translucent "windows" near its base where internal nectaries are located. *Fruits are round,* 8-11 mm across, *covered with small bumps, arranged in grape-like clusters, and red-orange in color.* The fruits ripen Sep-Nov but remain on the plants until Jan if not eaten by wildlife. In flower Mar-May.

A rare shrub here, found sparingly at two locations in the Santa Ana Mountains. One grove is on the Riverside County side of Santiago Peak at 5000 feet elevation (though not found in many years). The better-known grove is in upper Trabuco Canyon on north-facing slopes at 3374 feet (1028 meters) elevation, southeast of Yaeger Mesa, where it grows among chaparral and mixed evergreen forest. To get there, hike up the Trabuco Trail east of Holy Jim Canyon. The plants are about one mile southeast of the trail's junction with West Horsethief Trail. The canyon narrows and the trail approaches the opposite canyon wall. There is a large **California sycamore** (*Platanus racemosa,* Platanaceae) in the creek channel immediately downslope from the trees. The madrones are visible from either side of the sycamore. You will need binoculars or a long camera lens to see them in detail. If you go in late fall or early winter, look for the eye-catching clusters of fruits hanging from the trees. It is possible but difficult to reach the base of the plants; besides, they are entwined with poison oak. Among and a few meters east of the madrones is a lush stand of **California bay laurel** (*Umbellularia californica,* Lauraceae), so thick that their odor fills the air, making the experience a truly sensory one.

It was named in 1814 by F.T. Pursh in honor of A. Menzies, who first collected the plant.

Pacific Madrone in upper Trabuco Canyon.
Top background: plants in flower, as viewed from Trabuco Trail with a wide-angle zoom lens at 105 mm, 15 April 2007.
Top insets: flowers of a cultivated specimen at Rancho Santa Ana Botanic Garden; red-brown outer bark self-shreds to reveal beautiful honey-colored inner bark.
Bottom background: plants in fruit, 24 October 2004, as viewed from Trabuco Trail with a 400 mm lens.
Bottom inset: fruit in the hand of Chris Barnhill (for scale).

Manzanita: *Arctostaphylos*

These are shrubs or small trees with reddish bark. Their *white to pink flowers* appear in winter. Four species live in our area: one in chaparral at lower elevations, two high up in the Santa Ana Mountains among high-elevation chaparral and conifers, and the fourth among chaparral in the San Mateo Canyon Wilderness Area and on the Santa Rosa Plateau. Identifying them in the field can be challenging.

Most manzanitas form the next year's flower buds during summer or fall. The buds grow to an observable size, stop growing well before winter starts, then resume development just prior to the plant's flowering period. The only exception is pink-bracted manzanita, which does not pre-form its buds.

The genus name is formed from the Greek words arktos (bear) and staphylos (cluster or bunch, as of grapes), because the flowers and fruits are borne in grape-like clusters and the fruits are eaten by bears. The common name, manzanita, hails from Spanish (manzana = apple, -ita = small) in reference to the small apple-like fruits.

Eastwood Manzanita

Arctostaphylos glandulosa Eastwood **ssp.** ***glandulosa***

A *rounded shrub* 1-2.5 meters tall, rarely to 4.6 meters. Twigs are slightly to very hairy, sometimes with sticky-tipped bristles. *Leaves* are *elliptical, oval, or round,* petiole 5-10 mm long, blade 2-4.5 cm long and 1-2.5 cm wide. *Upper and lower leaf surfaces are alike, dull gray-green, often covered with sticky-tipped hairs, and may be rough to the touch.* Inflorescence bracts are 5-15 mm long, green, and leaf-like. *The corolla is white,* 6-8 mm long, sometimes with a pink cast. Its *hairy, sticky fruits are flattened like a tangerine* (wider than tall), with well-developed pulp, red to red-brown, and 6-10 mm wide. Seeds are not fused into a single unit; when exposed, they separate into 3-5 individuals, *like a tangerine fruit.* In flower Jan-Apr.

An abundant shrub found from 300-1,800 meters (984-5,905 feet) throughout the Santa Ana Mountains, where it lives within chaparral and coniferous forest. After fire or other disturbance, it can regrow quickly from its *underground burl.* Good locations to find it include Bear Canyon Trail near the Ortega Candy Store, at Blue Jay Campground, Falcon Group Camp, Los Pinos Saddle, Trabuco Trail, West Horsethief Trail, and along most of North Main Divide Truck Trail, South Main Divide Road, south into the San Mateo Canyon Wilderness Area.

The epithet was named in 1897 by A. Eastwood in reference to its sticky leaves (Latin, glandulosus = glandular).

Eastwood Manzanita

Eastwood Manzanita

Top right: Typical bristly leaves and pink-tinged broad white flowers.
Bottom right: Hairy sticky fruits.
Below: Eastwood manzanita varies in its leaf color and texture. Of the two individuals shown, the one at left has greener leaves than the bluer-leaved individual at right. Both have bristly-haired leaves that are rough to the touch.

Eastwood Manzanita

Bigberry Manzanita
Arctostaphylos glauca Lindl.

A tree-like shrub, 2-8 meters tall. The bark is usually red-brown but is orange-brown in a few local individuals. Its *leaves are oval to roundish*, petiole 7-15 mm long, blade 2.5-5 cm long and 2-4 cm wide. *Leaf surfaces are alike, gray-green to white-glaucous, dull, and smooth.* The edges of young leaves may be toothed. *Leaves are neither hairy nor sticky.* Inflorescence bracts are green, short, 3-6 mm long, triangular, and held outward or curved back. *The corolla is white or pinkish*, 8-9 mm long. Its *large hairless fruits* are green and mature to red-brown, 12-15 mm wide, and very sticky. *Seeds are fused into a single rounded, ribbed stone with a sharp tip.* In flower Dec-Mar.

An *infrequent shrub* of chamise chaparral at elevations ranging from 450-1,400 meters (1,476-4,593 feet). Look for it in Coal Canyon, sparingly in Santiago Canyon within view of Santiago Canyon Road, in the community of El Cariso, but mostly along North and South Main Divide Roads within 6 miles either side of Ortega Highway. High in the Santa Ana Mountains are individuals of an uncommon form with slender pink flowers and orange-brown bark.

Lacking a burl, individuals are often killed by fire. The species survives fire not by resprouting but by its copious production of berries and seeds that germinate and produce many new individuals.

Its flowers are visited by hummingbirds, bees, flies, and butterflies. In mid-winter, watch wildlife come to this attractive shrub. South Main Divide Road is the easiest place to do so.

It was named in 1836 by J. Lindley for its leaf color (Greek, glaukos = silvery, gleaming, bluish-green or grey).

Bigberry Manzanita white form

Bigberry Manzanita pink form

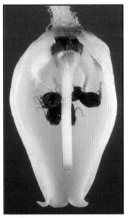

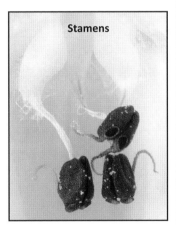

Stamens

Bigberry Manzanita

Bigberry Manzanita

Pink-bracted Manzanita ❀
Arctostaphylos pringlei Parry ssp. *drupacea* (Parry) P. V. Wells

A treelike shrub 2-5 meters tall, sometimes taller, with a sweet scent. *Leaves are elliptical, oval, or round,* petiole 5-10 mm long, blade 2-5 cm long and 1-4 cm wide, held upright. *Leaf surfaces alike, dull grey-green. Inflorescence bracts* are 6-10 mm long, *leaf-like, deep pink, and fall off during the flowering period.* The *corolla is rose-pink to white,* 7-8 mm long. *Mature fruits are red* and 6-23 mm wide. *Seeds are fused into a single rounded, ribbed stone.* Twigs, leaves, pedicels, and fruits are densely covered with sticky-tipped bristles. The pedicels and calyx lobes are all pink. It does not form a burl, nor does it pre-form its flower buds in summer or fall. In flower Apr-June.

This is an uncommon plant at higher elevations of the Santa Ana Mountains. It is much more common east of our area, near Idyllwild in the San Jacinto Mountains. In our area, it is quite rare; found only within manzanita chaparral around 1,219 meters (4,000 feet) elevation, on the southeast slope of Santiago Peak, 0.5 road miles down from the summit of the peak. Here it grows among Eastwood manzanita at the head of the Holy Jim/Coldwater Trail and possibly continues downslope into Mayhew Canyon. The ridgeline of the Santa Ana Mountains is lowered just above the head of Mayhew Canyon; this allows coastal fog to crest the ridge and spill down into Mayhew, providing it with fogdrip and cooler temperatures preferred by this manzanita.

The epithet was given in 1887 by C.C. Parry in honor of C.G. Pringle. Also in 1887, Parry named this form after the stone-like seeds within its fruit (Greek, drupa = a stone fruit; -aceus = of or pertaining to).

Pink-bracted Manzanita

Pink-bracted Manzanita

Rainbow Manzanita

Rainbow Manzanita ★
Arctostaphylos rainbowensis Keeley & Massihi
[*A. peninsularis* ssp. *keeleyi* Wells]

An upright shrub from 1-4 meters tall. *Leaves* are *elliptical or oval in shape,* petiole 6-12 mm long, blade 3.5-5 cm long and 2-3.5 cm wide. *Upper and lower leaf surfaces appear alike, gray-green to whitish, often tinged pinkish, hairless, covered with a bit of whitish powder,* margin occasionally toothed, rounded at the base. *Inflorescence bracts* are 2-4 mm long and *leaf-like.* The corolla is white, 6-8 mm long. *Mature fruit* 8-12 mm, *dark brown tinged with purple, covered with whitish bloom. Seeds fused into a single rounded stone that has a pointed tip and no ribs.* In flower Jan-Feb.

A rare burl-forming shrub, found only at elevations of 300-600 meters (984-1,969 feet) in upper San Mateo Canyon, south to Miller Mountain, and east to Santa Rosa Plateau, De Luz, Temecula, Rainbow, and foothills of the Agua Tibia Mountains. In some places, it lives with Eastwood manzanita (*A. glandulosa* ssp. *glandulosa*). CRPR 1B.1.

It was named in 1994 by J.E. Keeley and A. Massihi for the location in Rainbow, San Diego County, near the center of its distribution and where it was originally collected.

Rainbow Manzanita

Fruits of our two *common* manzanitas.

Upper row: Eastwood manzanita. Left to right: hairy, fleshy fruit, viewed from above; flesh removed to show ribs; side view, half of the seeds were easily removed; a single seed, shaped like a tangerine slice.

Lower row: Bigberry manzanita. Left to right: sticky but hairless fruit, side view; flesh removed to show the single stone with sharp point and shallow ribs. This stone is not easily opened.

Fruits of our two *uncommon* manzanitas.

Upper row: Rainbow manzanita. Left to right: hairless non-sticky fruit with whitish powder (glaucous); flesh removed to show the single stone with round shape, pointed tip, ribs and pits. This stone is not easily opened.

Lower row: Pink-bracted manzanita. Left to right: hairy and sticky fruit; flesh removed to show the single stone with elongate (or round) shape, very pointed ends, prominent ribs and sculpturing. This stone is not easily opened.

THE MANZANITA GUILD

Many species of butterflies, flies, and bees nectar from flowers of manzanitas. Two such butterflies are the **Red Admiral** (*Vanessa atalanta rubria* (Fruhstorfer, see also page 405) and the **Painted Lady** (*Vanessa cardui* (L.)), both Nymphalidae). To retrieve nectar from manzanitas, butterflies must rest head-down and insert their proboscis up into the flower. Larvae of the red admiral eat nettles (Urticaceae), while those of painted ladies feed on a wide variety of plants, primarily thistles, cudweeds, and everlastings (Asteraceae), mallows (Malvaceae), and borages (Boraginaceae).

Manzanita Leaf Gall Aphid

Tamalia coweni (Cockerell, Aphididae)

In early spring, odd-looking green to red "blisters" form along the leaf edges of bigberry manzanita. The leaf edges grow down and curve under the leaf, making what is called a roll gall. Within these galls are very small, dark-colored insects that feed on the leaves. Their mouthparts form a tube like a drinking straw with which they pierce the leaf and suck its fluids. By summer or fall, the gall turns brown and dries up but remains on the plant.

The epithet was given in 1905 by T.D.A. Cockerell in honor of J.H. Cowen.

Summer Holly ★
Comarostaphylis diversifolia (Parry) E. Greene var. *diversifolia*

Summer Holly

An unusual shrub up to 5 meters tall. *Its twigs are fuzzy and grey; the bark is rough and grey; its older branches shed their bark.* Leaves 3-8 cm long, reverse-oval, bright green above, paler and hairy below. Leaf edges generally rolled under, smooth or toothed. *Corolla white*, 3-5 mm long. *Fruits round, 5-6 mm wide, bright red, fleshy, bumpy-surfaced.* In flower May-June (-Aug). Fruits ripen about 1 month later.

A rare shrub in our area, found in two disjunct locations, both on north-facing slopes among chaparral. One population occurs in the southern end of the San Joaquin Hills in Upper Hobo Canyon (near Moulton Meadows Park) and on Niguel Hill (accessible from Sea View Park). The other is in the Santa Ana Mountains in an unnamed tributary to San Juan Canyon, south of and above Ortega Highway, 2.7 miles east of San Juan Fire Station. *The mountainside there is very steep, covered in poison oak, and dangerous to climb, so we do not recommend that you ascend it.* Since the plant is in cultivation, a wiser action is to buy it from a native plant nursery, grow it in your garden, and view it there. It is slow-growing and does not like summer watering. CRPR 1B.2.

The epithet was given in 1884 by C.C. Parry for the edges of its leaves, which are generally turned under (Latin, diversus = separated or turned; folium = leaf).

Summer Holly — Mission Manzanita

Young fruits of summer holly (right) and mission manzanita (far right). The leaf edges of both curl under. The flower stalk of summer holly generally stands out or up; that of mission manzanita hangs down. Also compare the fleshy, bumpy-surfaced fruits of summer holly with the dry, smooth fruits of mission manzanita.

Mission Manzanita
Xylococcus bicolor Nutt.

Mission Manzanita

The only member of its genus, this *burl-forming shrub* reaches 2-3 meters in height. *Twigs grey and fuzzy; bark smooth and red; older branches have shredding bark.* Leaves 2.4-4.5 cm long, alternate or opposite, *elliptic to oblong, with entire margins that are rolled under.* They are *dark green and hairless above, fuzzy and grey below. Calyx dark pink-red,* with 4-5 obvious lobes. *Corolla white or pink,* 8-9 mm long, with 5 (rarely 4) greenish lobes. Fruit round, red to red-black, 5-8 mm wide, with a solid stone-like seed cluster. In flower Dec-Feb.

At first sight, it looks like **California coffeeberry** (*Frangula californica* ssp. *californica,* Rhamnaceae), but the fuzzy twigs, shredding bark, rolled-under leaf edges, and manzanita-like flowers give it away. It is locally uncommon, found among chaparral, and difficult to access. One population is on a north-facing slope in upper Santiago Canyon, north-northwest of a northward loop in the Santiago Truck Trail, 0.34 air miles due east of its junction with The Luge Trail that runs south to Live Oak Canyon Road. It is more common to the southeast in Tenaja Canyon, along the Lucas Canyon Trail, in lower San Mateo Canyon, and on Miller Mountain. In cultivation, it makes a fine dense shrub with seasonal color.

In 1843, T. Nuttall named the genus for its woody fruit (Greek, xylon = wood; kokkos = berry) and the epithet for its two-colored leaves (Latin, bi- = two; color = color).

EUPHORBIACEAE • SPURGE FAMILY

A variable group, its members are *annuals or perennials* and assume just about all growth forms: herbs, vines, shrubs, and trees. *Many have milky-white or clear sap that contains highly toxic alkaloids and/or skin irritants.* Their stems are branching and fleshy, woody, or spiny. *Leaves are alternate or opposite,* with stipules and petioles, and undivided, toothed, or palmately lobed. Their flowers are often in groups variously arranged on the plant. Sepals 3-5. Petals usually none. *Each flower is either male or female, sometimes the entire plant has flowers of only one sex. Male flowers bear 1 to many stamens. Female flowers have 3 styles (rarely 2), each simple or divided.* The ovary is superior, spherical, often with 3 rounded lobes. It grows at the end of a stalk that lengthens and curves as the flower ages. Once fertilized, the 3-lobed ovary develops into a dry fruit that splits into 3 one-seeded pieces.

Familiar members include poinsettia, giant spiny cactus-like *Euphorbia* plants from Africa, and the plants from which we get tapioca, cassava, para rubber, and ceara rubber. This is a large family with over 6,000 species worldwide, 55 in California, and 20 in our area. Only a few are commonly seen here.

California Copperleaf
Acalypha californica Benth.

A *shrub* nearly 1.5 m tall covered with hairs that are somewhat sticky-tipped. *Its sap is clear. Leaves* 1-2 cm long, oval or triangular with scalloped edges, *green and copper-colored,* alternate on the stem. *Flowers are in spikes along the upper stems.* Male flowers have 4 sepals, no petals, and 4-8 spiraling stamens, in spikes 1.5-4 cm long. Female flowers have 3 to 5 sepals, no petals, and 3 reddish styles, in spikes up to 2 cm long. Fruit is spherical. In flower Jan-June.

Found in rocky soils of slopes, chaparral, and oak woodland. Known only from the area of Santa Ana Canyon in the Chino Hills, just east of Robbers Peak in Weir Canyon, and ridges east of Weir Canyon, to the head of Gypsum Canyon. It grows with **California wishbone bush** (*Mirabilis laevis* var. *crassifolius,* Nyctaginaceae). It makes a fine garden plant.

The genus was named in 1753 by C. Linnaeus for its resemblance to nettles (Greek, akalephes = a kind of nettle). The epithet was given in 1844 by G. Bentham for the place of its discovery.

Castor Bean ☠ ✖
Ricinus communis L.

A tree-like shrub, 1-3 meters tall. *Leaves large,* 10-50 cm long, *green to purple, palmate* with 7-11 lobes and toothed margins. Flowers of both sexes are on the same plant but grow separately. The inflorescence is 10-30 cm long, *female flowers above male.* Fruits are 1.2-2 cm across, spiny, green, mature reddish. *Seeds* 9-22 mm long, shiny, speckled with purple, *pretty to look at but fatal if eaten.* In flower nearly all year.

Its seeds contain 2 important compounds. *Castor oil* is used as a lubricant, to make soap, and in medicine. *Ricin* is one of the most toxic natural chemicals known. *Do not touch this plant, especially its seeds, leaves, and sap!*

Native to Europe or Africa. In the 1950s it was grown in California as a crop plant. It was also introduced into the ornamental plant trade but escaped from gardens into the wild. Now it is a noxious weed throughout the U.S., where it grows in disturbed areas, roadsides, and creekbeds. In addition to its toxic nature, it may cause contact dermatitis and pollen allergy in some people. *It should be eradicated wherever it occurs.* Managers of parks, playgrounds, and schools should be especially diligent in its safe removal.

Named in 1753 by C. Linnaeus either for its habit of growing in groups or its abundance (Latin, communis = growing in a society, common, general).

California Croton
Croton californicus Muell. Arg.

A *small shrubby perennial* less than 1 meter tall. *Its leaves and stems are covered with star-shaped scales, giving the plant a bristly feel.* It has clear sap. *Leaves 2-5.5 cm long, gray-green, elliptic to oblong, with a smooth margin. Plants have either male or female flowers, rarely both on the same plant.* Male flowers have 5 sepals, no petals, and 10-15 stamens with hairy filaments. Female flowers have 2 mm long sepals, and no petals. Style 2-lobed, each lobe is 2-forked. Seeds 3.5-5.5 mm long, smooth, round, and mottled. In flower Mar-Oct.

A common plant of *sandy soils and creekbeds* from coastline to canyons, including the Lomas de Santiago, Chino Hills, and Temescal Valley. Not found in the Santa Ana Mountains. Its seeds germinate readily after fire. In cultivation, it requires quick-draining soils such as sand and cannot tolerate clay soils or those with high organic content.

Named in 1866 by J. Mueller Argoviensis for the state of its discovery.

In the inset photograph at right, a cluster of male flowers was laid next to a cluster of female flowers. The two sexes do not grow on the same plant.

Doveweed, Turkey Mullein
Croton setiger Hooker
[*Eremocarpus setigerus* (Hook.) Benth.]

A low-growing soft-hairy annual less than 20 cm tall, up to 80 cm wide. It lies on the ground or stands just a bit above it. It has clear sap. *Leaves 1-6 cm long, oval or triangular, gray-green, with 3 obvious veins, covered with soft hairs. Flowers of both sexes grow on the same plant. Male flowers in clusters on top of the plant;* 5-6 sepals, no petals, 6-10 stamens. *Female flowers at nodes, below male flower clusters;* neither sepals nor petals. The ovary has only 1 chamber and produces only 1 seed, quite unusual for the family. Seeds 3-4 mm long and smooth. In flower May-Oct.

Occurs in *sunny dry places,* in disturbed soils and along trailsides throughout our area. Common in all of our wilderness parks, especially so at Caspers Wilderness Park.

The epithet was named in 1838 by W.J. Hooker, perhaps for its bristly flowers or fruits (Latin, seta = a bristle; gero = to bear or carry).

Linear-leaved Stillingia
Stillingia linearifolia S. Watson

An *upright shrub* to 70 cm tall. *Leaves 1-4 cm long, very narrow, linear,* undivided. *Flowers of both sexes grow on the same plant. Inflorescence long, narrow; male flowers above, female flowers below.* Male flowers crowded; cup-like, each with 2 stamens topped by broad anthers. Female flowers sparse; ovary hidden by 3 sepals; 3 long styles. Fruit broad, 3-lobed. In flower Mar-May.

A plant of dry soils in washes and exposed sites within coastal sage scrub. Known from Santa Ana Canyon in Yorba Linda, Temescal Valley, and lower Bedford Canyon. More common inland and in our deserts.

The epithet was given in 1879 by S. Watson for its narrow leaves (Latin, linearis = a line; folium = leaf).

Spurges: *Euphorbia*

Spurges bear their *flowers in clusters*, each with a single central female flower surrounded by several male flowers; each of those males has only 1 stamen. The clusters lie in a cup-shaped *involucre*. The involucre is often topped with red or yellow *nectar glands* that bear white *petal-like appendages*. The entire inflorescence is called a *cyathium*.

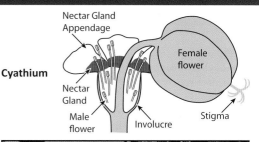

Rattlesnake Spurge

Euphorbia albomarginata Torr. & A. Grey
[*Chamaesyce albomarginata* (Torr. & A. Grey) Small]

A *hairless perennial* flat on the ground, *green stems* 5-25 cm long. It has milky-white sap. *Leaves* 3-8 mm long, round to oblong, *with a whitish margin. Stipules fused into a pale triangular scale with hairs along its margin.* Involucre 1.5-2 mm wide, bell-shaped; with 4 white petal-like appendages, triangular with a blunt tip, smooth or wavy-edged. Each appendage has a maroon oblong nectary at its base. *Fruit* 2-2.5 mm long, oval, sharply 3-sided, *hairless.* In flower Apr-Nov.

A common chaparral plant, most often found in full sun on slopes, barren cliff faces, and sandy creekbeds. Look for it in many of our wilderness parks and the Santa Ana Mountains.

Named in 1857 by J. Torrey and A. Gray for its white-edged leaves (Latin, albus = white; marginis = edge, border).

Golondrina, Small-seed Sandmat

Euphorbia polycarpa Benth. **var. *polycarpa***
[*Chamaesyce polycarpa* (Benth.) Millsp. var. *polycarpa*]

A *hairy (to hairless) perennial* flat on the ground or partly upright, *reddish stems* 5-25 cm long. It has milky-white sap. *Leaves* 1-10 mm long, round to oval. *Stipules linear or triangular, sometimes hairy at the margin. Stipules separate, 2-forked, slender, often hairy-edged, pale.* Flowers are solitary at the nodes. Involucre 1-1.5 mm wide, bell-shaped, hairless to hairy, with 4 petal-like appendages, white to red, triangular blunt-tipped, smooth or wavy edged. The nectary is maroon. *Fruit* 1-1.5 mm long, spherical but 3-lobed, *hairless to hairy.* In flower most of the year.

Widespread in our area, on sandy and disturbed soils among coastal sage scrub, chaparral, and roadcuts from coastline to mountain. Look for it on coastal bluffs and atop Temple Hill in Laguna Beach, Aliso and Wood Canyons Wilderness Park, Limestone Canyon, and throughout the Santa Ana Mountains.

Named in 1844 by G. Bentham for its small, plentiful seeds (Greek, poly = many; karpos = fruit).

Below: stipules near the stem tips are best viewed from the stem underside.

Rattlesnake spurge's stipules are fused into a hairy-edged triangle.

Golondrina's stipules are separate and slender.

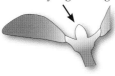

Cliff Spurge ★

Euphorbia misera Benth.

A *branching soft-wooded shrub* 0.5-1.5 m tall, hairy in youth but hairless with age. It has milky-white sap. *Young stems purplish. Leaves* 0.4-1.5 cm long, oval or round, hairy, folded upward. *Stipules divided into threads.* Cyathia are at the branch tips. Involucre 2-3 mm long, bell-shaped, hairy, with 5 white or greenish petal-like appendages, sculpted at the tip, each with 1 maroon nectary. Cyathia with 30-40 male flowers and 1 female. *Fruit* 4-5 mm long, spherical. In flower Jan-Aug.

Rare in California, found only on sea cliffs and some rocky canyon walls near the ocean from Corona del Mar to South Laguna, and Dana Point Headlands and bluffs bordering the Dana Point Marina. CRPR 2.2.

Named in 1844 by G. Bentham for its sometimes scraggly, unattractive appearance, especially when it drops its leaves in drought (Latin, misera = wretched, worthless, vile).

FABACEAE • PEA OR LEGUME FAMILY

A large diverse family of delicate annuals, perennial herbs, vines, shrubs, and stout trees. *Leaves of nearly all species are divided, most often pinnately or palmately so, arranged alternately on the stem.* Individual leaflets are usually entire, not divided or toothed. *Flowers bisexual. Sepals 5,* fused at the base with their tips free. *Petals 5,* free or the lower 2 fused. *Stamens 1 to many,* most often 10, in many species the lowermost 9 of them partly fused into a cylinder that surrounds the pistil, the uppermost stamen free. The *ovary is superior,* topped with a single style and stigma. Fruit is a dry or fleshy *legume,* usually a long pea pod, flat sided or inflated with air like a balloon. The pod is most often straight but in some it is curved or coiled. When ripe, the two sides of some fruit split open; in others it does not split open on its own.

In our species, the upper petal is largest, outermost, usually stands upright, and is 1-2 lobed; it is called the *banner petal* or standard. The petals on either side of the banner are identical to one another, usually held forward, and are called *wings.* The 2 lowest petals resemble each other and are fused together into a single boat-like structure called the *keel.* The flower's reproductive organs are held within the keel which itself is hidden between the wings. When a potential pollinating insect visits the flower, it alights on a platform made by the upper edges of the wings. The insects' weight pushes the wings and keel down, which forces the stamens and stigma out and up onto the belly of the insect. The anthers coat the insect with pollen, while the sticky-tipped stigma adheres to pollen already placed on the insect by a previous flower. In some species, after fertilization, the petals of that flower turn a different color and its nectar production stops, thought to be cues to insects that the flower is no longer accepting visitors.

Many species grow unusual nodules on their roots. These contain bacteria that capture nitrogen from the atmosphere and chemically alter it to a form useable by plants, a vital process called nitrogen fixation.

This family is the third largest in the world, with about 19,400 species worldwide. Many species are cultivated for food and fodder, such as peanuts, soybeans, beans, alfalfa, and clovers. Many contain toxic compounds, so peas from the wild should never be tasted or eaten.

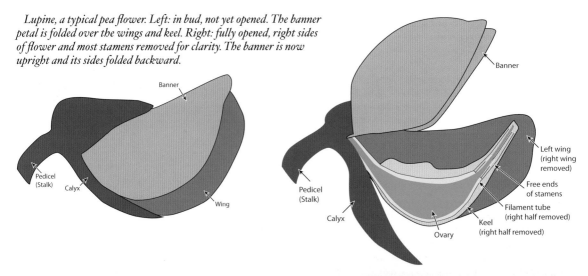

Lupine, a typical pea flower. Left: in bud, not yet opened. The banner petal is folded over the wings and keel. Right: fully opened, right sides of flower and most stamens removed for clarity. The banner is now upright and its sides folded backward.

False Indigo: *Amorpha*

Open deciduous shrubs with many upright spreading branches, the plants are dotted throughout with tiny yellow or purple glands; most plants have a *strong odor. Leaves are pinnate* with an odd number of leaflets though sometimes the leaflet at the tip is missing or deformed. The numerous flowers appear on long, slender, upcurved spikes at the branchtips. *Each flower is tiny, its corolla reduced to a single petal, the banner, which is purple and lies draped over the base of the stamens and pistil.* The 10 stamens and single pistil stick far out of the flower, allowing easy access for its pollinators, primarily bumble bees, other native bees, and butterflies. We have 2 similar species.

The genus was named in 1753 by C. Linnaeus in reference to its "deformed" (one-petaled) flowers (Greek, *amorphos* = without form, deformed).

Right: Crotch's bumble bee (Bombus crotchii *Cresson, Apidae) on flowers of California false indigo along upper Trabuco Trail.*

California False Indigo
Amorpha californica Nutt. **var.** ***californica***
The midrib of the leaf has tiny dark purple prickles on its underside. To check, gently run your hand under the midrib from leafbase to tip. *Each leaflet ends with a tiny dark gland that points down or out and may be flat or on a small bump (rarely elongated) but never a narrow bristle.* Examine several midribs and leaflet tips with a hand lens to be sure. Flowers are scattered on the spike. In flower May-July.

Widespread in our area, occurs in rocky or sandy soil among chaparral, dry riparian uplands, and oak woodlands, often in dense colonies. *Usually at higher elevations and drier locations* than Western false indigo. Known from Laurel Canyon and Sycamore Hills within Laguna Coast Wilderness Park in the San Joaquin Hills, Puente-Chino Hills, and Santa Ana Canyon. Also known from several sites in the Santa Ana Mountains such as Hall Canyon, along Maple Springs Truck Trail in upper Silverado Canyon, Maple Springs Saddle (a huge colony!), upper Trabuco Canyon, Trabuco Trail just below Los Pinos Saddle, upper Holy Jim Trail, upper Hot Springs Canyon, and Audubon Starr Ranch Sanctuary.

Named in 1838 by T. Nuttall for the place of its discovery.

Western False Indigo
Amorpha fruticosa L.
Nearly identical to California false indigo. Its leaf midrib has *no prickles* on the underside but may have small purple or green bumps. *Each leaflet ends in a single slender purple bristle that may have a gland on it.* Flowers are crowded on the spike. In flower May-July.

More often at lower elevations, mostly in sandy soils along streams and creeks. Known from the Santa Ana River channel, Blind Canyon, lower Silverado Canyon, San Juan Canyon, and San Juan Creek in San Juan Capistrano.

Named in 1753 by C. Linnaeus for its shrubby growth form (Latin, fruticis = a bush or shrub).

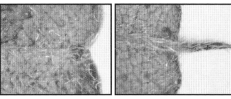

Leaflet tips. Left: California false indigo – tipped with a gland. Right: Western false indigo – tipped with a slender purple bristle.

THE FALSE INDIGO GUILD
California Dogface Butterfly
Zerene eurydice (Boisduval, Pieridae)
[*Colias eurydice* Boisduval]

A beautiful butterfly with *bright yellow pink-edged wings* and a wing span of 5-6.3 cm. Females are all yellow with a single spot on the forewing and smaller spots on the wing undersides. Males are also yellow but have large patches of black on the forewing and a dark spot in the center. This black pattern forms the profile of a dog's head on each forewing of the male. Upon emerging from the chrysalis, the yellow portion of the forewings of the male has a purplish iridescence. In flight spring-summer.

Larvae feed on leaves and flowers of California false indigo plants. There are no confirmed reports that they do or do not feed on western false indigo; a careful feeding study is needed. Larvae are green with stripes and/or crossbands of lighter colors. At maturity, they reach up to 2.5 cm long. The chrysalis is light green. Both males and females are easily found as they nectar on **California thistle** (*Cirsium occidentale* var. *californicum,* Asteraceae). *This butterfly was designated as California's State Insect in 1972.*

The epithet was given in 1855 by J.B.A.D. de Boisduval in reference to Eurydice, a female character from Greek mythology known for running through meadows with her lover, Orpheus. We find it to be a fair description of this lovely butterfly's flight.

Locoweeds, Rattleweeds, Milkvetches: *Astragalus*

A variable group of annuals and perennials. *Leaves are pinnate* with an odd number of leaflets. *Flowers arranged in stalked clusters* that arise from the branchtips and/or nodes. They can be solitary or in a round-headed group on the stalk. *Fruits in most are green, yellowish, or purplish, sometimes mottled with a darker color, and inflate with air as they ripen.* When dry, the seeds are loose and rattle inside the fruit. Many species contain toxic compounds. We have 9 types in our area, often difficult to identify. To do so, you need flower and fruit.

ANNUAL LOCOWEEDS
White Dwarf Locoweed
Astragalus didymocarpus Hook. & Arn.

White Dwarf Locoweed

A *small, slender, spreading or upright annual,* 25-45 cm tall. Leaves 0.8-7.5 cm long with 9-17 linear leaflets each notched at the tip. Inflorescence rounded, with 5-25 flowers. *Petals whitish tinged with purple.* Keel upwardly curved near its tip. *As developing pods inflate, they press against each other, giving the appearance of one large pod.* In flower Mar-May.

Mostly found in grasslands along the coast such as Pelican Point in Newport Beach, Sycamore Hills in Laguna Coast Wilderness Park, along San Juan Creek in San Juan Capistrano, Bee Canyon in the Chino Hills, Black Star Canyon, Temescal Valley, and the Elsinore Peak area.

The species was named in 1838 by W.J. Hooker and G.A.W. Arnott for its 2-lobed fruit (Greek, didymo = double, two-fold; karpos = fruit).

Gambel's Locoweed
Astragalus gambelianus E. Sheldon

Gambel's Locoweed

A *slender annual, spreading to upright,* 2-30 cm tall. Leaves 1-4 cm long with 7-15 linear leaflets, each 1-9 mm long, their tips blunt or notched. Inflorescence an upright spike with 4-15 flowers that stand up but later hang down. Petals whitish tinged with purple, the banner darker than the wings. *Fruits rounded, not crowded on the stalk.* In flower Mar-June.

In heavy clay soils among grasslands and coastal sage scrub near Gypsum Canyon, in Baker Canyon, Caspers Wilderness Park, Verdugo Canyon, San Mateo Canyon Wilderness Area, Temescal Valley, and Santa Rosa Plateau. Much more common in the years following a fire.

Named in 1894 by E.P. Sheldon in honor of W. Gambel.

PERENNIAL LOCOWEEDS
Braunton's Milkvetch ★
Astragalus brauntonii Parish

Braunton's Milkvetch

A upright, stout perennial, 70-150 cm tall, densely covered with white hairs that give it a grey appearance. Older plants have a thick taproot and a trunklike main stem. Leaves 3-16 cm long with 25-33 oval leaflets. Inflorescence a dense spike of 35-60 flowers. *Flowers dull lilac to magenta.* Fruit small, 6.5-9 mm long, 3-4 mm wide, slightly inflated. In flower Feb-June.

Found at fairly low elevations 15-450 meters (50-1,500 feet), in gravelly clay soils above granite or sandstone, among chaparral, and sometimes immediately downstream in riparian areas. *Seedlings sprout readily after physical soil disturbance or fire.* Occurs only in hills surrounding the Los Angeles basin. In our area, known from Coal and Gypsum Canyons and the ridge between them, and along the fire road that leads from Claymine Canyon to Sierra Peak. Recently found in lower Fremont Canyon, just north of the Santiago Dam. Pollinated by bumble bees.

Listed as Federally Endangered in 1997 by the US Fish and Wildlife Service because it has been eliminated from much of its range. CRPR 1B.1.

Named in 1903 by S.B. Parish in honor of E. Braunton.

Pomona Rattleweed
Astragalus pomonensis M.E. Jones

A hairless green or blue-green perennial with thick stems and many leaves, 25-80 cm tall. Leaves 5-20 cm long, divided into 25-41 elliptical leaflets. *Petals cream-colored or pale yellow-green,* arranged in a rounded or long cluster on the stalk. *Fruits very inflated and large,* 25-45 mm long, 10-20 mm wide, yellowish blushed with rose. In flower Mar-May.

In Orange County, known only from a single location in Aliso Viejo, just south of Aliso Viejo and Moulton Parkways, on west-facing slopes, east of Aliso Creek Riding and Hiking Trail, in sandy soil among limestone rock outcrops. Also in Bluewater and San Mateo Canyons within the San Mateo Canyon Wilderness Area. More common in Riverside County, such as along the Santa Ana River valley, through Temescal Valley, Elsinore Basin, Rancho California, and on the Santa Rosa Plateau.

Named in 1902 by M.E. Jones for the site of its original discovery (Pomona, a city in southern California; Latin, –ensis = place of origin, from).

Ocean Locoweed
Astragalus trichopodus (Nutt.) Gray **var. *lonchus*** (M.E. Jones) Barneby

A densely leaved and branched, upright perennial, 20-100 cm tall. *Leaves* 2.5-20 cm long, its 15-39 linear leaflets are 2-25 mm long, *dark green to grey-green,* sometimes margined with purple. *Petals greenish-white or cream-colored, sometimes with pale pink or purple veins, the keel rarely tipped with pink. Pod body broad,* 20-40 mm long, 8-21 mm wide, hairy in our area, upper and lower surfaces of the pod equally rounded. *The sparse fruits hang down and inflate with age.* In flower Feb-June.

Typically a coastal plant, but in our area it also occurs inland to our hills and mountains, mostly along canyons that eventually lead to the ocean. Near the coast, it is found at Aliso Canyon in South Laguna and Trafalgar Canyon in San Clemente. In the Chino Hills, it is in Brea, Carbon, and Tonner Canyons and above Horseshoe Bend near the Santa Ana River channel. Also at lower elevations of the Santa Ana Mountains such as Arroyo Trabuco in O'Neill Regional Park, Live Oak Canyon, on cliff faces lining the road up Trabuco Canyon, and on the former Robinson Ranch above Rancho Santa Margarita. Certainly also in Silverado Canyon.

The epithet was given in 1838 by T. Nuttall. The name refers to the thin hairlike stalk of the fruit (Greek, trichos = hair; podos = foot. The subepithet was given in 1923 by M.E. Jones in reference to the long stalk of the flower and fruit (Greek, lonche = spear).

Similar. Southern California locoweed (*Astragalus trichopodus* (Nutt.) A. Gray var. *trichopodus*): *petals greenish-white or cream-colored,* without pink or purple; *pod body narrow,* 15-35 mm long, 5-13 mm wide, hairless, upper surface somewhat flat, lower surface rounded, *ages red;* uncommon in coastal sage scrub in the Chino Hills (Carbon Canyon near La Vida Hot Springs, Soquel Canyon, and on slopes above the 57 Freeway), more abundant in the Puente Hills.

This plant is eaten by larvae of the **Western tailed blue butterfly** (*Everes amyntula* Boisduval, Lycaenidae). Eggs are laid on the flowers and the emerging caterpillars tunnel into the fruits, where they eat the developing seeds. If you kneel down to the level of the plant and align the sun or a flashlight behind a fruit, you can see the caterpillar inside. Please don't pick the fruit because doing so will kill both the pod and the caterpillar within it.

Teas: *Hoita, Rupertia*

A small group of upright or prostrate ephemeral perennials. *Leaves alternate, divided into 3 large leaflets covered with dot-like glands that release a strong odor.* Calyx hairy, its free lobes are long, the lowest often longer than the others. Corolla yellowish, pink, or purple. Fruit oval or elliptic, flattened side-to-side. We have 3 species in this group; all were originally placed together in the genus *Psoralea*, a word still often used as part of their common name.

The genus *Hoita* was established in 1919 by P.A. Rydberg. It is the name for these plants used by the Konkow Valley Band of the Maidu tribe from the Feather River region of California. Pronounce it "ho-I-tay" with short "i," long "a," and accent on the middle syllable (not "hoy-ta").

Leather Root Tea
Hoita macrostachya (DC.) Rydb.
[*Psoralea macrostachya* DC.]

Very tall, upright to 0.5-3 m tall. Leaflets 2-10 cm long, oval to broadly linear. *Flowers 9-10 mm long, pinkish-purple, in a long cluster.* Fruit 6-8 mm long. In flower May-Aug.

Found along mountain streams and moist canyons. Long ago also found along our coastline (such as Anaheim Marsh near Sunset Beach and Newport Bay) but not found there in many years, probably eliminated by development. Known from the Santa Ana River channel near Chino Hills, Holy Jim Canyon (easily found there), Trabuco Canyon, Audubon Starr Ranch Sanctuary, San Juan Canyon, Lion Canyon, Santa Rosa Plateau, and the San Mateo Canyon Wilderness Area. It is an easily grown garden plant, in moist soils under partial shade.

The epithet was given in 1825 by A.P. de Candolle in reference to the long spike of flowers (Greek, makros = long, large; stachyo = spike).

Leather Root Psoralea

Round-leaved Tea
Hoita orbicularis (Lindl.) Rydb.
[*Psoralea orbicularis* Lindl.]

A *low plant* that sprawls on the ground, the leaves and stemtips held upright to about 6.5 cm high. Leaflets 3-11 cm long, oval to round. *Flowers 12-23 mm long, pink, in an elongate cluster.* Fruit 6-9 mm long. In flower May-July.

Quite uncommon here, it grows in moist meadows, streams, and hillsides, often in shade. Known from the Santa Ana River channel near the Chino Hills, near Main Divide Truck Trail above Maple Springs, and San Juan Canyon, south into San Mateo Canyon Wilderness Area (Morgan Trail and near Miller Mountain).

Named in 1837 by J. Lindley for the shape of its leaves (Latin, orbicularis = circular).

Round-leaved Psoralea

California Tea
Rupertia physodes (Hooker) J. Grimes
[*Psoralea physodes* Hooker]

A *low perennial*, spreading or partially upright to about 0.5 m tall. *Leaflets* 2-6 cm long, triangular to linear, the *glands release a very strong odor.* Flowers in a dense rounded cluster on the end of a long stalk. *Petals light to medium yellow, sometimes purple-tipped.* Fruit 4-7 mm long, oval, flattened side-to-side, with a small point at the tip. In flower Apr-June.

Found in the Santa Ana Mountains on north-facing exposed areas among chaparral and within shady oak woodlands but not common. Known from the Maple Springs Truck Trail above Silverado Canyon, along Trabuco Canyon upstream of O'Neill Regional Park, San Juan Canyon, along Sitton Peak Truck Trail high above Ortega Highway, and in the San Mateo Canyon Wilderness Area.

The genus was established in 1990 by J.W. Grimes in honor of his colleague R.C. Barneby. This epithet was given in 1831 by W.J. Hooker, possibly for the shape of its fruit (Greek, physo = bladder; -odes = resemblance).

California Tea

Sweet Peas: *Lathyrus*

Familiar plants to most gardeners, these are hairy or hairless annuals or perennials. Their leaves are *pinnately divided into 2 or more linear leaflets. Each leaf ends in a branched and coiling tendril* that helps them climb over other plants. *Flowers are arranged on a stalk that arises from nodes along the stem.* Flowers are fairly large and showy, often shades of pink, purple, or red, sometimes white. Their pods are linear and flat. Seeds of most species are toxic and should not be eaten or tasted. Most species are pollinated only by bees.

The genus was named in 1753 by C. Linnaeus (Greek, lathyros = ancient name for a plant in the pea family).

Chaparral Sweet Pea
Lathyrus vestitus Nutt. var. *vestitus*

A hairless or slightly hairy perennial. In cross-section, its stem is angled or narrow-winged. Leaves divided into 8-12 leaflets, each 2-4.5 cm long, linear to ovate. Inflorescence with 8-15 flowers. *Corolla 14-18 mm long, pale lavender to purple* (white). *The banner is bent upward about 90 degrees and is often purple-veined.* In flower Apr-June.

Found in many habitats from coastline to mountaintop, in coastal sage scrub, chaparral, oak woodland, and coniferous forest. *In our area, this form is much more common, widespread, and at elevations higher than San Diego sweet pea.* Known from upper Newport Bay in Newport Beach, Top of the World in Laguna Beach, the Whittier Hills, Santa Ana Canyon, our wilderness parks, and throughout the Santa Ana Mountains.

The epithet was given in 1838 by T. Nuttall for its attractive flowers (Latin, vestitus = dress or attire).

Chaparral Sweet Pea

San Diego Sweet Pea
Lathyrus vestitus Nutt. var. *alefeldii* (T. White) D. Isely

A form of chaparral sweet pea. It is *completely hairless*. Corolla 16-20 mm long, *dark purple to wine red, the banner bent back more than 90 degrees.* In flower Apr-June.

Quite uncommon here. Known from Moro Canyon in Crystal Cove State Park, Aliso and Wood Canyons in Aliso and Wood Canyons Wilderness Park, Santa Ana Canyon near the Chino Hills, Santiago Dam and Loma Ridge in the Lomas de Santiago, many of the Santa Ana Mountain canyons such as Baker, Silverado, Trabuco, Leach, and San Juan, the Santa Rosa Plateau, and Mud Spring in upper Devil Canyon in the San Mateo Canyon Wilderness Area.

The plant was named in 1894 by T.G. White in honor of F.C.W. Alefeld.

Chaparral Sweet Pea leaf with leaflets and terminal tendrils

Chaparral Sweet Pea fruits

San Diego Sweet Pea

Chaparral Sweet Pea

Center: San Diego Sweet Pea, Fred M. Roberts, Jr.

The Lotus Clade: *Lotus, Hosackia, Acmispon*

Herbaceous annuals or perennials. *Leaves pinnately or palmately dissected with 3 or more leaflets.* Inflorescence a cluster of many flowers or only 1-2, on a stalk or attached directly to the stem. *Petals most often yellow,* though 2 of ours are whitish or pinkish, and age to orange, red, purple, or brown. Fruit is a typical pea-pod, often with a beak formed from the style.

The lotus group forms a clade within the pea family. Through the years a handful of species has been moved into and out of the genus *Lotus*. Recent work has moved the New World species out of *Lotus* and into the genera *Hosackia* and *Acmispon*. Locally, we have 1 *Lotus*, 1 *Hosackia*, and 10 species of *Acmispon*.

Don't confuse these lotus plants with those called waterlotus, popular among pond owners. While they are flowering plants, waterlotus are members of the family Nymphaeaceae within a very primitive clade called Nymphaeales (we have no waterlotus native to our area).

	Form	*Leaflets*	*Flower*	*Fruit*
Bird's-foot *Lotus corniculatus*	Mostly hairless perennial; low-growing	5; 2 basal, 3 palmate toward leaf tip	Long vertical stalk, about 3-7 flowered; bright yellow	Narrow-oblong, 1.5-3 cm long
Buck *Hosackia crassifolia* var. *crassifolia*	Mostly hairless ephemeral perennial; often tall	9-15, pinnate, large	*Hanging stalk,* 12-20-flowered; petals *yellow-green* marked with purplish-red on banner & wings	Oblong, 3.5-7 cm long
American *Acmispon americanus*	Delicate hairy annual, upright or mat-forming	3, *dark green*	*Solitary,* stalk about 15 mm; calyx shaggy-haired; *petals cream to pink,* often red-veined	Oblong, 1.5-3 cm long, beak curved
Silverleaf *Acmispon argophyllus*	Perennial covered with dense white silky hairs; flat	3-7, pinnate to palmate	Stalkless or short-stalked, 4-15(-20)-flowered; calyx white-hairy; corolla deep yellow	Roughly oval, curved; very short, same length as calyx; only 1-seeded
Hill *Acmispon brachycarpus*	Hairy, fleshy annual; flat or upright	Usually 4, pinnate to palmate	Solitary at nodes, stalkless; calyx hairy; corolla yellow	Oblong, 6-12 mm long, densely hairy, beak bent
Deer Broom *Acmispon glaber*	Mostly hairless upright perennial shrub; summer-deciduous	3-6; pinnate	*Stalkless, 2-7-flowered, numerous along long stems;* calyx hairless; corolla yellow	Linear, 1-1.5 cm long, curved upward, with a long beak
Woolly *Acmispon heermannii*	Hairy, perennial; flat	4-6, usually palmate	Very short-stalked; 3-5-flowered; calyx shaggy-haired; corolla deep yellow	Oblong, narrow, very short, barely longer than calyx, curves up and over
Alkali *Acmispon maritimus*	Mostly hairless fleshy annual, many-stemmed; flat to upright	3-7, pinnate	*Long-stalked, 5-15 mm, 2-4 flowered;* calyx sparsely-haired; corolla bright yellow	Narrow-oblong, 1.5-3 cm long, mostly hairless, curved, very narrow, beak hooked
San Diego *Acmispon micranthus*	Hairless or hairy annual; flat to upright	4-7, pinnate or palmate	*Stalkless or very short-stalked; 2-5-flowered;* calyx hairy; corolla yellow sometimes with red on banner and wings	Linear, 1-1.5 cm long, curved, end bent upward, tip hooked
Miniature *Acmispon parviflorus*	Mostly hairless delicate annual with slender stems, flat or upright	3-5, pinnate to palmate	*Solitary,* stalkless or very short-stalked; calyx hairless; *corolla only 5-6 mm long; whitish, pinkish, or pale salmon*	Linear, 1.5-2.5 cm long, hairy, straight with a short upcurved beak, narrowed between each seed, giving the pod a wavy margin
Strigose *Acmispon strigosus*	Hairless or slightly hairy fleshy annual, flat on the ground	4-9, pinnate, alternate	Stalked, 3-25 mm long, 1-2-flowered; calyx sparsely-haired; corolla yellow	Linear, 1-3.5 mm long, straight, curved near tip, with scattered hairs
California *Acmispon wrangelianus*	Sparsely haired annual, flat on the ground	4, alternate	*Solitary* at nodes, stalkless; calyx sparsely haired; corolla yellow	Linear, 10-18 mm long, reddish, tipped with an upward-hooked beak

Old World Lotus: *Lotus*
Bird's-foot Trefoil 💀 ✘
Lotus corniculatus L.

A *hairless or slightly hairy perennial, flat on the ground with inflorescences upright.* Leaves divided into 5 linear or oval leaflets, each 5-20 mm long. The lower 2 leaflets remain at the leaf base, the outer 3 leaflets are palmate and appear at the leaf tip. *Inflorescence on a long upright stalk,* 3-8 flowered, each 8-14 mm long. *Petals bright yellow,* the banner sometimes orange- or red-striped. Fruit narrow-oblong, 1.5-3 cm long. In flower late May-Sep.

Native to Eurasia, mostly found in disturbed *moist soils,* overwatered lawns, and run-off nearby. Known throughout our area, generally in or near lawns. Toxic if ingested.

The genus was named in 1753 by C. Linnaeus (Greek, Lotus = a name used for several different plants). The epithet was given in 1753 by C. Linnaeus, perhaps for the upcurved keel (Latin, corniculatus = with a small hornlike appendage, curved in the form of a horn).

Bird's-foot Trefoil

The 2 lowest leaflets are attached near the stipules

Perennial New World Lotus: *Hosackia*
Buck Lotus
Hosackia crassifolia Benth. **var. *crassifolia***
[*Lotus crassifolius* (Benth.) E. Greene var. *crassifolius*]

A large stout geophyte, hairless to slightly hairy, sprawling or upright to 1.5 m tall. Leaves divided into 9-15 elliptic or oval leaflets, each 20-30 mm long. *Inflorescence stalk 3-8 cm long, often hangs beneath the branches,* each with a cluster of 12-20 flowers that are 12-17 mm long. *Petals yellow-green marked with purplish-red on banner and wings.* Fruit oblong, 3.5-7 cm long. In flower May-Aug.

Lives in chaparral, oak woodlands, and coniferous forests, often in disturbed soils and along road cuts. Best known from Main Divide Truck Trail on Santiago Peak. Slightly resembles **false indigo** (*Amorpha* spp.); we know of no place where they co-occur.

The genus was named in 1829 by D. Douglas for D. Hosack. The epithet was given in 1837 by G. Bentham for its thick leaves (Latin, crassus = thick, heavy; -folius = leaved).

Buck Lotus

Buck Lotus fruits

Buck Lotus

Buck Lotus

New World Lotus: *Acmispon*

American Lotus

Acmispon americanus (Nutt.) Rydb. var. *americanus*
[*Lotus purshianus* (Benth.) Clements & E.G. Clements; *L. unifoliolatus* (Hooker) Benth.]

A delicate, hairy annual, sometimes mat-forming, upright or flat on the ground, with stems 15-80 cm long. Most of its *dark green leaves* are divided into 3 linear to elliptic leaflets, each 10-20 mm long. *Flowers 5-9 mm long, solitary* on a stalk up to 15 mm long, *a single leaf-like bract at the base of the flower*. Calyx with soft shaggy hairs. *Petals cream to pink, usually red-veined*. Fruit 1.5-3 cm long, oblong, beak curved, held outward or down. In flower May-Oct.

Widespread, along the coast, in foothill canyons, and along watercourses. Known from Dana Point, San Joaquin Hills, English Creek in Mission Viejo, and the Santa Ana Mountains including Santiago Canyon, Bell Canyon, San Juan Canyon, Alberhill, and Santa Rosa Plateau. An easily grown garden annual.

The genus was established in 1832 by C.S. Rafinesque for its pointed and often hook-tipped fruit (Greek, akme = a point, edge). The epithet was given in 1818 by T. Nuttall for its broad distribution in North America.

American Lotus

Silverleaf Lotus

Acmispon argophyllus (A. Gray) Brouillet
[*Lotus argophyllus* (A. Gray) E. Greene]

A perennial covered with dense white silky hairs that give it a silvery or gray color. Grows *flat on the ground* or with just the branchtips upright. Leaves pinnate to palmate, with 3-7 oval or linear leaflets, 6-12 mm long. Flowers attached to nodes or short-stalked, 4-8 flowers, each 6-10 mm long. *Calyx very hairy, engulfed in long white hairs. Petals deep yellow*, the banner ages to brown or purple. Fruit oval, curved, no longer than the calyx, only 1-seeded. In flower Apr-July.

Scarce here, found in canyons and on dry slopes in chaparral. There is a single record from San Juan Canyon. Old records for "Highway 71" inland of the Chino Hills and Temecula Canyon in Temecula. A recent record is from Cocklebur Canyon Beach on Camp Pendleton, not far from our area, so it might occur at San Onofre State Beach.

Named in 1854 by A. Gray for the white hairs that cover it (Greek, argos = bright, white; phyllus = -leaved).

Similar. San Diego lotus (*A. micranthus*): a lot less hairy; fruits much longer and less hairy; more common.

Silverleaf Lotus

Hill Lotus

Acmispon brachycarpus (Benth.) D.D. Sokoloff
[*Lotus humistratus* E. Greene]

A hairy, fleshy annual, it grows flat on the ground or upright. Leaves pinnate or palmate, generally with 4 elliptical or oval leaflets, each 4-12 mm long, arranged alternately on the midrib. *Flowers 5-9 mm long, solitary in the nodes, near the basal leaflet*. Calyx hairy. Petals yellow. Fruit 0.6-1.2 cm long, oblong, beak bent, densely hairy. In flower Mar-June.

Lives in grasslands and coastal sage scrub. Known from Lomas de Santiago near Irvine Lake, the Puente-Chino Hills, Temescal Valley, and Santa Rosa Plateau.

Named in 1848 by G. Bentham for its short fruit (Greek, brachys = short; karpos = fruit).

Hill Lotus

Coastal Deer Broom, California Broom, Coastal Deerweed
Acmispon glaber (Vogel) Brouillet **var. *glaber***
[*Lotus scoparius* (Nutt.) Ottley var. *scoparius*]

A hairless or sparsely-haired, shrubby perennial with straight wiry branches, 0.5-2 m tall, usually upright but sometimes low-growing near the coast. Leaves pinnate with 3-6 elliptic leaflets, each 6-15 mm long. The *inflorescence is attached directly to the stem* and has 2-7 flowers, each 7-12 mm long. Calyx hairless. *Petals yellow and age orange; the keel and wings are about the same length.* A series of inflorescences usually populates most of a stem; their flowers open at the same time and produce a shrubby sphere of color. Fruit 1-1.5 cm long, curved upward, with a long beak. In flower Mar-Aug.

Abundant throughout our area from coastline, inland to the foothills and the Santa Ana Mountains. In inland valleys, it occurs with desert deer broom. Primarily within coastal sage scrub but also in openings in chaparral, burned areas, and disturbed soils. Often summer deciduous, it drops leaves in summer and appears like a bunch of sticks or a broom buried upside-down in the ground (thus the common name *broom*). A hardy plant in the garden, it attracts butterflies and bees. If watered in summer, it will keep its leaves.

The epithet was given in 1835 by J.R.T. Vogel for it smooth hairless stems (Latin, glaber = smooth).

Coastal Deer Broom

Desert Deer Broom, Desert Deerweed
Acmispon glaber (Vogel) Brouillet **var. *brevialatus*** (Ottley) Brouillet
[*Lotus scoparius* (Nutt.) Ottley var. *brevialatus* Ottley]

A desert form of deer broom and very similar to it. The flower is 8-9 mm long. *The keel is clearly longer than the wings and looks like a banana hanging down below them.*

It ranges from the deserts westward to Temescal Valley and Elsinore Valley, where it co-occurs with coastal deer broom. It is occasionally found in the Santa Ana Mountains such as along North Main Divide Road "loop," in some parts of Caspers Wilderness Park, and on Loma Ridge near Irvine Lake.

Named in 1923 by A.M. Ottley for its short wings (Latin, brevi- = short; alatus = winged).

Right, upper: coastal deer broom, keel and wings the same length. Lower: desert deer broom, wings shorter than keel.

THE DEER BROOM GUILD
Bramble Green Hairstreak Butterfly
Callophrys perplexa B. & B. (Lycaenidae)

Adults have a wingspan of 2.5-3 cm. Their hindwings have no tails (related species have a tail on each hindwing). Wing uppersides usually drab gray to brown, sometimes orangish. *Forewing underside green* with a brown to orangish central patch. *Hindwing underside green, often with a few white spots.* Males perch on plants and/or hilltops and await females. After mating, females lay single eggs on flower buds of coastal deer broom. Larvae emerge, feed on flowers, fruits, and leaves. Pupation occurs on the plant in spring or summer and continues through winter. Adults emerge Feb-July.

The epithet was given in 1923 by W. Barnes and F.H. Benjamin because it was confused with other related species (Latin, perplexus = confused).

Bramble Hairstreak

Southern Woolly Lotus
Acmispon heermannii (Durand & Hilg.) Brouillet **var. heermannii** [*Lotus h.* (Durand & Hilg.) E. Greene var. *h.*]

A *perennial, it grows flat on the ground, forming a dense mat.* Leaves palmate, divided into 4-6 oval leaflets, 4-16 mm long. *Inflorescence very short-stalked, 3-8 flowered,* each 4-5 mm long. Calyx with shaggy hairs. Corolla bright yellow, sometimes darker tipped, aging red. Ovary covered with scattered hairs. *When mature, the fruit is short, barely longer than the calyx, and curves up and over.* In flower Mar-Oct.

A common creekside plant, known from the Santa Ana River bottom, Santiago Canyon, Audubon Starr Ranch Sanctuary, San Juan Canyon, Decker Canyon near El Cariso, Santa Rosa Plateau, and San Mateo Canyon Wilderness Area. An attractive garden plant placed near a pond or allowed to spill over the edge of a pot.

Named in 1855 by E.M. Durand and T.C. Hilgard to honor A.L. Heermann.

Similar. Northern woolly lotus (*A. heermannii* var. *orbicularis* (Gray) Isely, not pictured): leaflets rounded; corolla 5-6 mm long; ovary covered with soft spreading hairs. Uncommon in our area, known only from lower San Juan Creek and Trabuco Canyon.

Southern Woolly Lotus

Alkali or Coastal Lotus
Acmispon maritimus (Nutt.) D.D. Sokoloff **var. maritimus** [*Lotus salsuginosus* E. Greene ssp. *salsuginosus*]

A many-stemmed, fleshy annual, hairless or sparsely haired. *Grows flat on the ground* or upright to 30 cm tall, solitary or in dense mats. Leaves pinnate, divided into 3-7 oval or round leaflets, 5-15 mm long. *Inflorescence on a 1-4 cm long stalk and has 2-4 flowers,* each 3.5-10 mm long. Calyx sparsely haired. Petals yellow, the wings equal to or shorter than the keel. *Fruit 1.5-3 cm long, gently curved, narrow, oblong.* In flower Mar-June.

Primarily inland, less common near the coast. Known from Bolsa Chica Ecological Reserve, along trails in Crystal Cove State Park, bluffs in Laguna Beach, Puente-Chino Hills, ridge between Coal and Gypsum Canyons, Black Star Canyon, our wilderness parks, Verdugo Canyon on the Rancho Mission Viejo, San Juan Canyon, Temescal Valley, and San Mateo Canyon Wilderness Area. Abundant after fire.

Named in 1838 by T. Nuttall for its occurrence near the sea (Latin, maritimus = of or belonging to the sea).

Alkali Lotus

San Diego or Grab Lotus
Acmispon micranthus (Torr. & A. Gray) Brouillet [*Lotus hamatus* E. Greene]

A hairless or hairy annual. Its *stems lie flat on the ground* or grow upright. Leaves pinnately or palmately divided into 4-7 oval or elliptic leaflets, each 7-15 mm long. *Inflorescence attached to nodes or short-stalked, 2-5-flowered,* each 3-5 mm long. Calyx hairy. Petals yellow, sometimes with red on the banner and wings, aging orange. *Fruit 1-1.5 cm long, linear but bent upward with a hooked tip.* In flower Mar-June.

Lives in grasslands and coastal sage scrub of Sycamore Hills, Dana Point, bluffs in San Clemente, Loma Ridge, Whiting Ranch and Riley Wilderness Parks, Rancho Mission Viejo, Audubon Starr Ranch Sanctuary, Santa Ana Mountains from upper Lion Canyon south into San Mateo Canyon, and Santa Rosa Plateau. Often abundant after fire.

Named in 1838 by J. Torrey and A. Gray for its small flowers (Greek = mikros = small; anthos = flower).

Similar. Silverleaf lotus (*A. argophyllus*): much hairier; fruits much shorter and hairier.

San Diego Lotus

Miniature Lotus
Acmispon parviflorus (Benth.) D.D. Sokoloff
[*Lotus micranthus* Benth.]

A delicate often hairless annual with slender stems, it grows flat on the ground or upright to 30 cm tall. Leaves pinnate to palmate, divided into 3-5 elliptic or oval leaflets, each 4-12 mm long. *Flowers solitary, stalkless or very short-stalked, each 5-6 mm long. Calyx hairless. Petals whitish, pinkish, or pale salmon,* open briefly and fade quickly to red. *Fruits* hairless, 1.5-2.5 cm long, linear, straight with a short upcurved beak, *narrowed between seeds, giving the pod a wavy margin.* In flower Mar-July.

Grows in grasslands, coastal sage scrub, and chaparral. Uncommon, known from Coal Canyon, Santiago Peak, Yeager's Mesa, and Oak Flats in the Santa Ana Mountains.

Named in 1829 by G. Bentham for its small flowers (Latin, parvus = little, small; floris = flower).

Miniature Lotus

Strigose Lotus
Acmispon strigosus (Nutt.) Brouillet
[*Lotus strigosus* (Nutt.) E. Greene]

A hairless or slightly hairy fleshy annual that grows flat on the ground, branches to 30 cm long, sometimes with tips upright. Leaves pinnate with 4-9 linear to *oval, often skinny leaflets,* each 3-10 mm long, alternate on the midrib. *Inflorescence stalked, 3-25 mm long, 1-2 flowered, each 5-10 mm long.* Calyx sparsely haired. Petals yellow, age to orange or reddish. *Fruit* 1-3.5 mm *long,* straight, curved near the tip, with scattered hairs. In flower Mar-June.

Grows in coastal sage scrub, chaparral, and disturbed soils throughout our area. *It has both large-flowered and small-flowered forms;* the two grow together in places such as Limestone Canyon, sometimes along creekbeds.

Named in 1838 by T. Nuttall for its hairs (Latin, strigosus = covered with short, straight, rigid hairs that lie flat against the plant).

Strigose Lotus

California Lotus, Wrangel's Lotus
Acmispon wrangelianus (Fisch. & C.A. Mey.) D.D. Sokoloff
[*Lotus wrangelianus* Fischer & C.A. Meyer; *L. subpinnatus* Lag., misapplied]

A sparsely haired annual, it grows flat on the ground. Leaves pinnate with 4 elliptic or oval leaflets, each 4-15 mm long, alternate on the midrib. *The inflorescence is arranged like that of hill lotus: a single flower appears in the nodes near the basal leaflet.* Each flower 2.5-5 mm long. Calyx sparsely haired. Petals yellow, age to red-purple. *Fruit* 10-18 mm long, oblong, reddish, *straight,* with an upward-hooked slender beak. In flower Mar-June.

Grows on coastal bluffs and within chaparral, sometimes in disturbed soils. Known from Corona del Mar, San Joaquin Hills, Brea, Santa Ana River Canyon, Black Star Canyon, Crow Springs on the Audubon Starr Ranch Sanctuary, San Juan Canyon, Temescal Valley, and scattered sites within the San Mateo Canyon Wilderness Area.

Named in 1836 by F.E.L. von Fischer and C.A.A. von Meyer in honor of F.P. von Wrangel.

California Lotus

Lupines: *Lupinus*

Annuals, perennials, and shrubs; most grow upright. *Leaves palmately divided;* leaflets linear, smooth-edged. *Flowers alternate or whorled on a short or tall spike.* Each flower is a typical pea. The center base of the banner, *the banner patch,* often contrasts with the banner and turns to a different color after pollination. The *10 stamens* are fused into a tube at the flower base and free past it. *Five stamens have long filaments with short anthers, the other 5 have short filaments with long anthers. Fruit is a stout legume,* often hairy, flattened side-to-side, not inflated. As the pod dries, it splits open and the two sides twist in opposite directions, dropping the *rather large seeds.*

Lupines are cultivated worldwide as ornamentals, some as domestic animal food, though several species are toxic. The seeds of some species require treatment before they will germinate such as sand abrasion during a flood or heat from a fire. Many are quite common in the years after fire, diminishing in the years afterward.

The genus was formally named in 1753 by C. Linnaeus, but the name is an ancient one (Latin, lupinus = of or relating to a wolf), in reference to an old belief that these plants steal nutrients from the soil as a wolf might steal animals from shepherds and ranchers.

ANNUAL LUPINES
Chick Lupine
Lupinus microcarpus Sims **var. *microcarpus***

A *stout, upright* lupine, it grows 10-80 cm tall. It is *sparsely to densely hairy throughout.* The upper surface of each leaflet is hairless. Leaflets are 3-15 cm long. Inflorescence bracts and calyx have long shaggy hairs. Flowers are whorled, each flower is 8-18 mm long. Corolla pinkish to purple, rarely yellowish or white. *Fruits are held upright and appear on all sides of the stalk.* In flower Mar-June.

Widespread but seldom abundant, it grows in grasslands and on roadcuts. Found in Aliso and Wood Canyons Wilderness Park, in Tijeras and Chiquita Canyons on the Rancho Mission Viejo, along Santiago Canyon Road just west of Irvine Lake, Audubon Starr Ranch Sanctuary, and Santa Rosa Plateau.

Named in 1823 by J. Sims for its small fruit (Greek, mikros = small; karpos = fruit). The common name probably refers to the upright, white-fuzzy fruits which look like baby birds in the nest.

Similar. Dense-flowered chick lupine (*Lupinus microcarpus* var. *densiflorus* (Benth.) Jeps.): nearly identical; *corolla generally white to dark yellow, tinged pink or lavender,* rarely rose or purple; after flowers open, the stalks tilt or arch over and flowers grow on upper side; *all fruits thus appear on only one side of the stalk.* In flower Apr-June. Uncommon, known from Bee Canyon and Temple Hill, sometimes in hydroseed mixes. Named in 1835 by G. Bentham for its densely flowered inflorescence (Latin, densus = dense, compact; floris = flower).

Yellow-flowered form: Michael Charters

Miniature Lupine, Dove Lupine
Lupinus bicolor Lindl.

A *small hairy plant* that spreads out or stands upright, 10-40 cm tall. *Leaflets are hairy* above and below, often with longer hairs on the edges. Flowers 4-10 mm long, in whorls of 3-9 flowers at some or all nodes. *The wings are dark blue to purple, rarely pale lavender.* The banner is white to dark blue, flat-topped; *the banner patch is white sprinkled with dark blue dots, and ages to magenta.* In flower Mar-June.

Grows in grasslands and openings in coastal sage scrub and chaparral, also disturbed soils along trails and roadways. Commonly found throughout our area, from coastline to foothills, mountains, and inland valleys.

Named in 1827 by J. Lindley for its 2-colored flowers (Latin, bis = two; color = tint, hue).

Arroyo Lupine
Lupinus succulentus Douglas ex Koch.

An *upright annual* 20-100 cm tall with very *succulent stems.* Has short-flat or long-spreading hairs throughout. *Leaflets broad, large,* and dark green with rounded tips. *Flowers distinctly whorled, petals generally blue-purple, rarely white, lavender, or pink. Banner patch white with darker flecks, aging red-violet.* In flower Feb-May, longer in cultivation.

A common lupine found throughout our area, mostly in moist coastal areas and generally not at upper elevations. Often used in wildflower seed mixes and hydroseeded on slopes and along roadways.

Named in 1861 by D. Douglas in reference to its succulent stems (Latin, succulentus = juicy, succulent).

Arroyo Lupine

Agardh's Lupine
Lupinus agardhianus Heller

A small lupine, spreading and upright, with stems 5-20 cm long. *Overall green in color, it is covered with evenly spaced, short, straight hairs that stand outward or are angled upward.* Leaflets linear and noticeably broader at the far end; their upper surface is less hairy than their lower surface. Flower stalks are held above the leaves in taller plants, about even with the leaves in shorter individuals. Petals are bright blue or edged with rose-purple, the wings dark, the banner patch whitish with purple flecks, its tip pointed. Fruits are very hairy and held upward. In flower Mar-May.

Grows at medium to high elevations in decomposed granitic soils among coastal sage scrub, chaparral, and coniferous forests. Found on slopes near San Juan Fire Station, along South Main Divide Road, and on the slopes of Elsinore Peak.

Named in 1911 by A.A. Heller in honor of J.G. Agardh.

Some authors merge Agardh's lupine into bajada lupine; however, in our area they are recognizable and do not grow together, so we treat them separately.

Similar. Bajada lupine (*L. concinnus*): usually lies on the ground, sometimes stands upward; overall grey in color; hairs long, soft, shaggy; petals bluer; usually grows at lower elevations in dry, sandy, and disturbed soils.

Agardh's Lupine

Bajada Lupine, Elegant Lupine
Lupinus concinnus Agardh

A small lupine that usually lies on the ground, sometimes stands upright, with stems 10-30 cm long. *Overall gray in color, densely covered with long, soft, shaggy hairs.* Leaflets linear, only a bit broader at the far end, equally hairy on upper and lower surfaces. Flower clusters are hidden among the leaves. Petals pink, purple, or whitish; wings dark; the banner edged with reddish-purple, its patch whitish or yellowish with purple flecks, its tip rounded or notched inward. Fruits very hairy, held upward. In flower Mar-May.

Grows in dry, sandy, and disturbed soils among grasslands, coastal sage scrub, and chaparral, inland to the desert. Known from Gabino Canyon on the Rancho Mission Viejo, Hidden Ranch in Black Star Canyon, hillsides near Falcon Group Camp, near Oak Flats, the mouth of Bluewater Canyon, Temescal Valley, and Santa Rosa Plateau.

Named in 1835 by J.G. Agardh for its elegant appearance (Latin, concinnus = neat, pretty, elegant).

Similar. Agardh's lupine (*L. agardhianus*): more upright; overall green in color; hairs evenly spaced, short, straight, stand up or angled; petals blue to pinkish; usually at higher elevations in decomposed granitic soils.

Bajada Lupine

Coulter's Lupine
Lupinus sparsiflorus Benth.

A hairy annual with slender stems, upright 20-40 cm tall. The *7-11 leaflets are linear,* only 2-4 mm wide. *Flowers* 10-12 mm long and *spread out along the stalk. Petals dark blue, rarely pinkish, the banner with a yellow patch and red-purple flecks.* In age, the wings turn pale, the banner turns magenta. Most fruits appear on only one side of the stalk. In flower Mar-May.

Widespread, generally not common. A plant of coastal sage scrub and chaparral, sometimes in disturbed soils along trails. Known from Telegraph Canyon in Chino Hills, James Dilley Reserve in Laguna Beach, Whiting Ranch Wilderness Park, Silverado, Santiago, and Trabuco Canyons, along Harding Canyon Trail, Dana Point, San Juan Capistrano, Caspers Wilderness Park, Audubon Starr Ranch Sanctuary, San Juan Canyon, Temescal Valley, and Santa Rosa Plateau.

Named in 1848 by G. Bentham for its scattered flowers (Latin, sparsus = few, scattered; floris = flower).

Coulter's Lupine

Stinging Lupine
Lupinus hirsutissimus Benth.

Large, thick-stemmed, upright, 10-100 cm tall, covered with yellow, stiff, *stinging hairs* on all green parts. *Careful – those hairs are sharp and painful to the touch!* The 5-8 leaflets are very wide and rounded at their tips. Flowers are 12-18 mm long, all *petals red-violet to magenta.* The banner is narrowed at the top, sometimes with a whitish or yellowish patch, and has purple flecks. The banner patch ages red-violet. In flower Mar-May.

Very common in our local foothills, canyons, and mountains, long ago also right on the shore in Laguna Beach. Appears within coastal sage scrub and chaparral; *especially abundant in the years following a fire.* Easily found at Oak Canyon Nature Center, along Harding Canyon Truck Trail, Trabuco Canyon, Caspers Wilderness Park, Audubon Starr Ranch Sanctuary, along Ortega Highway in San Juan Canyon, and in the San Mateo Canyon Wilderness Area. Also known from the Whittier Hills and Temescal Valley.

Named in 1833 by G. Bentham for its dense covering of hairs (Latin, hirsutus = hairy; -issimus = very much, most).

Stinging Lupine

Collar Lupine
Lupinus truncatus Nutt. ex Hook. & Arn.

A dark green, nearly hairless annual, it grows upright with few branches, 20-50 cm tall. Its *5-8 leaflets are narrow, linear,* only 2-5 mm wide. *The leaflet tips are blunt* (truncated). The *base of each leaf and branch is swollen* (collared), often red. The inflorescence is sparsely flowered. *Flowers* 8-12 mm long, *disproportionately small for the plant's size.* Petals blue-purple to magenta with a white banner patch and dark blue specks, often with magenta surrounding the banner patch. The entire flower fades to magenta with age. In flower Mar-May.

Common throughout our area, from coastline to foothills, mountains, and inland valleys. Found among coastal sage scrub, chaparral, and disturbed soils along trails, especially in the years after a fire. Known from all of our wilderness parks and the Santa Ana Mountains. Look for it at Oak Canyon Nature Center, on Harding Canyon Trail, and near the Candy Store on Ortega Highway.

Named in 1840 by T. Nuttall for its truncated leaf tips (Latin, truncatus = cut off).

tips blunt • collar → • Collar Lupine

SHRUBBY PERENNIAL LUPINES

Our 4 species can be difficult to distinguish, even for experienced botanists. You must examine 4 important flower characteristics: (1) hairs on the back of the banner, best seen on buds; (2) keel-top hairs; (3) keel-bottom hairs, and (4) keel-top base lobes. Consult the keel drawings for help (base at left, tip at right).

Grape soda lupine (above and above right) brightens Main Divide Road between Santiago and Modjeska Peaks in June.

Silver lupine (right and below) is a woody shrub with very silvery foliage.

Silver Lupine

Lupinus albifrons Benth. **var.** ***albifrons***

A rounded, upright woody shrub with a trunk, generally 60-150 cm tall, covered throughout with soft, flattened hairs. *Leaves and stems appear silvery,* rarely greenish. *Leaflets* 6-10, each 10-45 mm long and *silvery on both sides,* the petiole 1-8 cm long. Flowers are usually in distinct whorls, though sometimes none are whorled. *Corolla violet to lavender,* 9-14 mm long. *The back of the banner is generally hairy,* a feature best seen when the flower is in bud. The banner patch is yellow or white, aging purple. Keel not lobed near base. Fruit 3-5 cm long and hairy. In flower Mar-June.

Uncommon but widespread in chaparral and foothill woodlands, coastline to the mountains. Known from Silverado Canyon and Audubon Starr Ranch Sanctuary. Very similar to grape soda lupine.

Named in 1835 by G. Bentham for its white leaves (Latin, albus = white; frons = leaf).

Grape Soda Lupine
Lupinus excubitus M.E. Jones **var. *hallii*** (Abrams) C.P. Smith

Grape Soda Lupine

A low-growing to upright, rounded shrub, generally 50-150 cm tall. Stems are woody at their base. *Leaves and stems greenish to silverish and hairy. Flowers 14-18 mm long, violet to lavender, with a distinctive sweet smell like grape juice when they are young.* The banner patch is bright yellow, aging purple. *The sides near the top of the banner petal are strongly folded backward. Banner-back hairy. Keel-top hairy from base to tip, densely so from middle to tip; keel bottom hairless; keel-top base lobed. The keel is sharply curved upward.* Fruit 3-5 cm long, covered with silky hairs. In flower Apr-June.

Found in coastal sage scrub, chaparral, and adjacent grasslands. Known primarily from moderate to high elevations in the Santa Ana Mountains as at Gypsum Canyon, Black Star Canyon, Silverado Canyon, Trabuco Canyon, San Juan Canyon, meadows near Blue Jay and Falcon Campgrounds, Temescal Valley, Santa Rosa Plateau, and scattered sites within the San Mateo Canyon Wilderness Area.

The species was named in 1898 by M.E. Jones because its upright flower spikes seem to stand guard within its habitat (Latin, excubitor = sentinel, watchman, guard). This variety was named in 1910 by L. Abrams in honor of H.M. Hall, who first collected it in 1901.

Grape soda lupine (right) and Southern mountain lupine (far right) are two forms of the same species (guard lupine, Lupinus excubitus *M.E. Jones). Their differences are often minor. Grape soda lupine is a much larger shrub, woody, and its leaves are more often green than silvery. Southern mountain lupine is a small non-woody shrub; its inflorescence stalks stand far above the rest of the plant. It lives at higher elevations.*

Southern Mountain Lupine

Southern Mountain Lupine
Lupinus excubitus M.E. Jones **var. *austromontanus*** (Heller) C.P. Smith

Similar to both grape soda and silver lupine. *A small low-growing shrub*, generally 20-50 cm tall. Stems are herbaceous, *not woody*. Leaves and stems are silverish and *covered with silvery hairs*. Banner-back is generally hairy, rarely hairless. *Keel-top hairy from base to tip, densely so from middle to tip; keel bottom hairless; keel-top base lobed. The keel is sharply curved upward.* In flower May-July.

Quite uncommon here, known from grasslands in the Elsinore Peak region. Possibly more widespread. Much more common among pines in larger mountain ranges.

Named in 1905 by A.A. Heller for its distribution in mountains of southern California (Latin, australis = southern; montanus = of mountains).

Parish's Stream Lupine
Lupinus latifolius Agardh **var. *parishii*** C.P. Smith

A bushy ephemeral perennial, 30-240 cm tall, *dark to bright green*. Stems hairless or nearly so. Upper leaf surfaces hairless to slightly hairy, lower surface somewhat hairy. *Leaflets large, 40-100 mm long*. Flowers generally whorled but not always. *Petals dark* blue to purple, rarely whitish. Banner patch white to yellowish, aging purple. *Banner back hairless. Keel top hairy from base to middle; keel bottom usually hairy; keel top base lobed.* In flower May-Aug.

Grows in very moist areas such as in seeps and along perennial streams, sometimes in oak woodlands, often in shade. Known from Silverado Canyon, below Yaeger Mesa in Trabuco Canyon, Viejo and San Juan Canyons, along the San Juan Trail in Chiquito Basin, Decker Canyon, and Bluewater Canyon within the San Mateo Canyon Wilderness Area. Easiest found along seeps near Upper San Juan Campground.

Named in 1835 by J.G. Agardh for its broad leaflets (Latin, latus = broad, wide; - folius = leaved). This variety was named in 1925 by C.P. Smith in honor of S.B. Parish.

Parish's stream lupine (left) has dark purple petals and deep green hairless foliage. It lives in moist soils along shady creeks and seeps. Pauma lupine (right) has pale blue-purple flowers on very long stalks and light green foliage covered in silvery hairs that make it look grey or silvery. It lives most often on north-facing slopes at lower elevations.

Pauma Lupine
Lupinus longifolius (S. Watson) Abrams

A large, rounded, silvery-green shrub, 50-150 cm tall, covered with soft short hairs. The 5-10 leaflets are 30-60 mm long, the petiole 4-10 cm long. Flowers more or less whorled or not. Flowers are large, 12-18 mm long, petals violet to blue, turning brown with age. Banner patch yellowish to white, sometimes not different from the rest of the banner, red-violet in age. *Banner back hairless. Keel top hairy from middle to tip; keel bottom hairless; keel top base lobed.* Fruit 4-6 cm long, dark and hairy. In flower Apr-June.

Often found on north-facing slopes among grasslands, coastal sage scrub, chaparral, and oak woodlands, from the coast inland to the foothills and mountains. Known from Aliso Canyon in the Chino Hills, Laguna Canyon and hillsides of Laguna Beach, Laurel Canyon in Laguna Coast Wilderness Park, Aliso Canyon in South Laguna, Chiquita Ridge on the Rancho Mission Viejo, Loma Ridge and Limestone Canyon Wilderness Park in the Lomas de Santiago, Hot Springs Canyon, and San Mateo Canyon in the San Mateo Canyon Wilderness Area.

Named in 1876 by S. Watson for its long leaves and petioles (Latin, longus = long; -folius = leaved).

Alfalfa ✖

Medicago sativa L.

An upright perennial, usually hairless, 20-80 cm tall. Stipules large and leafy. *Leaflets 3,* each narrow, linear. *Inflorescence an oval or rounded head* of 8-30 flowers. *Petals purple,* sometimes yellowish, yellow, or greenish. Stamens 10; filaments of 9 of them are fused most of their length, 1 filament is completely free. Fruit coiled, smooth. In flower Apr-Oct.

Native to Eurasia. An agricultural crop grown worldwide. Sometimes escapes into the wild, especially near active and former fields and along roadways. Recorded from places such as Telegraph Canyon in the Chino Hills, Santa Ana Canyon, Newport Beach, Mission Viejo, and Trabuco Canyon.

The epithet was given in 1753 by C. Linnaeus for its cultivation (Latin, sativus = that which is sown).

White Sweetclover ✖

Melilotus albus Medik.

A fast-growing upright annual with many sparsely-leaved branches, 0.5-2 meters tall. Stipules long, narrow, leaflike. *Leaves* odd-pinnate with *3 leaflets,* each 1-2.5 cm long, elliptical or oval, sometimes toothed at their edges. *Flowers white,* 4-5 mm long, hang down from a long stalk. Fruit 3-5 mm long, egg-shaped, etched with dark lines. In flower May-Sep.

A common weed of watercourses and parks throughout our area. Especially abundant along San Juan Creek in San Juan Capistrano. Mostly a plant of summertime.

Named in 1787 by F.C. Medikus for its white flowers (Latin, albus = white).

Yellow Sweetclover ✖

Melilotus indicus (L.) All.

An upright annual, 10-60 cm tall. Stipules long, narrow, leaflike. *Leaves* odd-pinnate, *3 leaflets,* each 1-2.5 cm long, elliptical or oval, *sharply toothed at their edges. Flowers yellow,* 2.5-3 mm long, on a tall stalk, often droop with age. Fruit 2-3 mm long, egg-shaped, bumpy or etched with faint lines. In flower Apr-Oct.

A plant of disturbed soils, often in gardens and adjacent wild lands, especially near water sources in wet soils. Known from Carbon Canyon, upper Newport Bay, San Joaquin Marsh, San Diego Creek, Rancho Mission Viejo, and the Whittier Hills. Native to the Mediterranean.

The epithet was given in 1753 by C. Linnaeus for its once-presumed place of origin, India (Latin, indicus = of India).

Chaparral Pea ★

Pickeringia montana Nutt. **var.** ***tomentosa*** (Abrams) Abrams

A bushy, many-branched, evergreen shrub 1-3 meters tall with smooth green bark and *thorns* among its nodes or tips. *Leaves small, without a petiole, palmately divided* into 3 elliptic or oval leaflets (rarely undivided), each 1-2 cm long. *Flowers solitary or in a small cluster at the branchtips.* Calyx bell-shaped with 5 short broad teeth. *Corolla pink-purple,* 1.5-1.8 cm long. *Unlike most members of this family, the two keel petals and 10 stamens of this one are free; none are fused to the others.* The pod is straight, not inflated, and a bit constricted between each of the 1-8 seeds. In flower May(-Aug).

Chaparral Pea

Grows sparingly among chaparral. In our area, known only from 2 locations in the Santa Ana Mountains: a ridge above Claymine Canyon (but not found there in many years) and along San Juan Trail, midway between Blue Jay Campground and Hot Springs Canyon. The population along the San Juan Trail is several miles from either end of the trail; there is no quick and easy way to get to it. This plant is breathtaking to see flowering in its natural habitat, and the hike to see it is worth every step, so plan a trip with some friends. Pack up *lots of water,* snacks, and lunch. Some parts of the trail are shady, many are not, so wear a comfortable hat. Leave a car at the Hot Springs Canyon trailhead, then drive a second car to Blue Jay Campground (there is a fee to park within the campground but with a Forest Adventure Pass you can park outside in marked parking areas). Hike downhill along the San Juan Trail. At an average pace, the plants are about 2.5 hours away, longer if you take time to watch wildflowers. Once you arrive at these plants, take time to enjoy them, then continue down the trail to Hot Springs Canyon. Use the car you left there to drive back up to get your second car. CRPR 4.3.

The genus was established in 1840 by T. Nuttall in honor of C. Pickering. Nuttall also assigned the epithet in 1840, for the mountainous habitat of the plant (Latin, montanus = of mountains).

Chaparral pea is pretty to look at but painful to touch because of the large thorns that are hidden among the leaves.

Spanish Broom ✖

Spartium junceum L.

A tall, nearly leafless shrub with straight, branches, it grows upright 1-3 m tall. *Leaves* 1-3 cm long, *linear, not divided,* and arranged alternately on the stem. The inflorescence is at the branchtips and bears many flowers. The calyx is split on top and has 5 small teeth. *Corolla* 2-2.5 cm long and *bright yellow.* The 10 stamens are all fused into a tube at their base. Fruit 5-10 cm long, flat, not inflated. In flower Apr-June.

Native to the Mediterranean. It has very tough roots that firmly hold the soil. For this reason, it has been used in hydroseed mixtures to stabilize cut roadways and loose mountain slopes. *Unfortunately, it often invades wildlands and crowds out native plants. Its use as a landscape plant should be discouraged and discontinued.* In Europe, its stems have been used for fiber in ropes and twine, its flowers for a yellow dye.

Found along Ortega Highway in San Juan Canyon all the way to Elsinore, Santiago Canyon, Silverado Canyon, and Trabuco Canyon, on Niguel Hill, and in Cold Spring and Lucas Canyons in the San Mateo Canyon Wilderness Area.

C. Linnaeus established both the genus and the epithet in 1753. The genus is from the Greek name for this plant (spartos); the epithet refers the similarity of its branches to those of rushes (Latin, junceus = rush-like).

Spanish Broom

Clovers: *Trifolium*

Annuals or perennials with leaves palmately divided into 3 leaflets (rarely 5-9), each leaflet usually toothed at its far end. Flowers arranged into a dense cluster at the stem tips. The inflorescence often has a flat or cup-like toothed disk below it called the *involucre*. Of the 10 stamens, 9 are fused at their base; the single topmost stamen is free. Fruit short and fat, usually shorter than the calyx; contains only 1-3 seeds. We have 16 species in our area, often difficult to identify, especially the non-native species. We present here 4 of our native species.

The genus was established in 1753 by C. Linnaeus for its 3 leaflets (Latin, tres = 3; folium = leaf).

Tree Clover

Trifolium ciliolatum Benth.

A nearly hairless, pale green annual, it grows upright, 20-50 cm tall. There is a lot of space between leaf nodes. *Leaflets* oblong, 1-3 cm long, *often with a pale spot in the middle. Inflorescence a round head* 1-2 cm across, *with no involucre below it.* Calyx 5-6 mm long, hairless but with bristles along its lobes. *Corolla* 6-7 mm long, *pinkish-purple. Flowers hang down with age.* Fruit small, 1-2 seeded. In flower Mar-June.

Found in grasslands, coastal sage scrub, and chaparral. Known uncommonly from the coast as at Laguna Beach and the Dana Point Headlands; more often found inland such as at Audubon Starr Ranch Sanctuary, Bear Canyon Trail (common), and the mountain canyons but not at higher elevations. Also known from Weir Canyon, Hicks Canyon, Alberhill, and Santa Rosa Plateau.

Named in 1848 by G. Bentham for the bristles along its calyx lobes (Latin, ciliolum = hair or hair-like process).

Tree clover (right) has no involucre. As its flowers age, they droop. Truncate sack clover (far right) has a tiny involucre. Its small flowers inflate in fruit.

Truncate Sack or Balloon Clover

Trifolium depauperatum Desv. **var. *truncatum*** (Greene) Isely [*T. truncatum* E. Greene]

A small hairless annual that lies on the ground or grows upright, its stems up to about 12 cm long. *Leaflets* oblong to oval, 0.5-2 cm long, *toothed at the tip, which is pointed or blunt. Inflorescence a round head* up to 1 cm wide, *its flowers held upward.* Involucre cup-like, with free or partly fused lobes, each 2-2.5 mm long. Calyx 2.5-4 mm long, hairless with no bristles. Corolla 4.5-7.5 mm long, whitish to purple, paler at the tips. *The banner inflates with age.* Fruit small, 1-5 seeded. In flower Mar-May.

Lives in clay soils among grasslands, coastal sage scrub, and oak woodlands, mostly along the coast. Known from UC Irvine Ecological Preserve, Bonita Canyon in the San Joaquin Hills, Dana Point, San Juan Capistrano, coastal bluffs in San Clemente, and Santa Rosa Plateau.

Named in 1814 by A.N. Desvaux for its small, slender stature (Latin, depauperatus = undeveloped, reduced, starved).

Creek or Clammy Clover
Trifolium obtusiflorum Hook. & Arn.

    A robust annual covered with sticky-tipped hairs. Leaflets 2-4 cm long, narrow-elliptic, edged with long sharp teeth. Inflorescence a round head 1.5-3 cm across, flowers held upward. Involucre flat and plate-like with many irregular teeth. Calyx covered with sticky-tipped hairs or bumpy-surfaced; with a split between the 2 uppermost lobes. *Corolla 14-18 mm long, pale to dull purple,* the tips lighter. Fruit small, 1-2 seeded. In flower Apr-July.

    A plant of marshy stream courses and wet disturbed soils. Known from Audubon Starr Ranch Sanctuary, Silverado Canyon, Santiago Canyon, San Juan Canyon, North Main Divide Road "loop," Elsinore Basin, Walker Canyon, and San Mateo Canyon in the San Mateo Canyon Wilderness Area.

    The epithet was given in 1838 by W.J. Hooker for its blunt-tipped banner petal (Latin, obtusus = blunt; floris = flower).

Creek clover (right) has a flat involucre with long teeth. Its flowers are pale, often with a darker banner patch. Its leaflets are broader, edged with long sharp teeth. Find it along ephemeral and perennial drainages.

Valley clover (far right) has a similar involucre. Its flowers are lavender to purple, the wings and keel often darker. Its leaflets are narrow, edged with short teeth.

Valley or Tomcat Clover
Trifolium willdenovii Spreng. [*T. tridentatum* Lindl.]

    A variable, hairless annual, can be very small and slender or large and stout, the range expressed even among individuals living closely together. It grows *upright or on the ground,* its stems 10-40 cm long. *Leaves and stems are sometimes burgundy.* Leaflets 1-5 cm long, *linear and narrow, with short teeth.* Inflorescence a round head 1.5-3 cm across, flowers held upward. Involucre flat and plate-like, 10-15 mm across, with many uneven sharp teeth along its edge. *Calyx hairless and smooth,* with a split between the 2 uppermost lobes. *Corolla* 8-15 mm long, *lavender to purple,* with wings and keel darker purple, all petals paler toward their tip, the wings and keel often white-tipped. In flower Mar-June.

    Prefers moist, heavy clay soils, usually among grasslands, coastal sage scrub, and disturbed sites. Widespread and very common from coastline to mountains but mostly found in the foothills.

    Named in 1826 by K.P.J. Sprengel in honor of C.L. Willdenow.

Vetch: *Vicia*

Annuals or perennial vines, these plants crawl over others like sweet peas (*Lathyrus* spp.), which they closely resemble. Leaves are even-pinnate, with 4 or more leaflets, each arranged alternately or opposite on the midrib, itself often tipped with a tendril. The inflorescence is at the branchtips or among the nodes.

Our vetches appear similar to sweet peas. Most vetches have flowers that are much smaller, narrower, and disproportionately longer than sweet peas. The American vetch, however, has large, wider flowers that are nearly identical to sweet peas. The best way to separate the two groups is to look at the style. In cross-section, the style of a vetch is round, that of a sweet pea is flattened top-to-bottom and has hairs along the top side.

The genus was named in 1753 by C. Linnaeus but the name is an ancient one for these plants (Latin, vicia = vetch). There are about 130 species, 15 in California, 9 in our area. Most are difficult to identify.

American Vetch
Vicia americana Muhl. ex Willd.
var. *americana*

A hairy or hairless perennial, the stems sprawl or climb, 20-120 cm long. Leaflets 8-16 elliptic or linear, each 1-3.5 cm long. *Inflorescence is on the top of an obvious stalk,* with 3-9 large flowers, uncrowded and arranged on more than one side of the stalk. Corolla 1.5-2.5 cm long, *blue-purple to lavender.* Fruit 2.5-3 cm long. In flower Apr-June.

Often mistaken for a sweet pea (*Lathyrus* spp.); study its style to be sure. Found along shady canyons at low elevations along the coast, the foothills, all the way up to the top of the Santa Ana Mountains. Known from Aliso and Wood Canyons Wilderness Park, Black Star Canyon, near Lower San Juan Picnic Ground, upper Harding Truck Trail, from Maple Springs Saddle up to Modjeska and Santiago Peaks, Verdugo Potrero and upper Bluewater and Devil Canyons within the San Mateo Canyon Wilderness Area.

Named in 1802 by H. Muhlenberg for the country of its discovery.

Southern Slender Vetch
Vicia ludoviciana Nutt. **var. *ludoviciana***
[*V. exigua* Nutt.]

A very slender annual with sprawling-climbing stems; often forms dense mats of growth. Leaves 4-10 cm long; leaflets very narrow, only 1-2.5 cm wide, 4-10 pairs per leaf. The long slender inflorescence is tipped with *1-3 small flowers close to each other. Corolla small, pale blue;* style hairy at the tip. Fruit 1.5-2.5 cm long, narrow, hairless. In flower Apr-June.

Generally uncommon, easily overlooked because of its small but pretty flowers and wispy growth. Found in grasslands, meadows, coastal sage scrub, oak woodlands, rarely in chaparral. Known from San Joaquin Hills, Lomas de Santiago, Caspers Wilderness Park, San Juan Canyon, Blue Jay Campground, Chiquito Basin, El Cariso, Temescal Valley, Santa Rosa Plateau, and San Mateo Canyon Wilderness Area.

The epithet was given in 1838 by T. Nuttall for the place of its initial discovery (Latin?, ludoviciana = from Louisiana).

Similar. Hasse's vetch (*V. hassei* S. Watson): *flowers 1(2), not close to each other; corolla 7-8 mm long, lavender to white; style hairs in a "unicorn" clump on the lower side;* one old record for Coal Canyon, perhaps more common and overlooked.

Spring Vetch, Common Vetch ✖
Vicia sativa L.

A hairy or hairless annual, the stems sprawl or climb, 30-80 cm long. Leaflets 8-14, 1.5-3.5 cm long, linear, often with a single tooth at the tip. *Flowers are nestled among the nodes, solitary or in a cluster of 2-3 flowers attached directly to the node or on a tiny stalk. Corolla pink-purple to whitish; the banner petal often remains straight and lies on the other petals.* Fruit 2.5-6 cm long, brown to black. In flower Apr-July.

Native to Europe, often found in the same areas as Bengal vetch but also at higher elevations of the Santa Ana Mountains such as at Chiquito Basin.

Named in 1753 by C. Linnaeus for its use in Europe as a cultivated crop for farm animals (Latin, sativus = sown, planted, cultivated).

Spring Vetch

Bengal Vetch, Purple Vetch ✖
Vicia benghalensis L.

A hairy annual, the stems sprawl or climb, 30-80 cm long. Leaflets 10-16, 1.5-3 cm long, elliptic to oblong. Like American vetch, its *inflorescence is on the top of an obvious stalk, with 3-12 flowers, crowded and arranged on only 1 side of the stalk. Corolla* 12-16 mm long, *rose-purple,* aging darker. Fruit hairy 2.5-3.5 cm long. In flower May-June.

A very common weed. Usually grows in disturbed grounds such as fallow and former pasturelands on the Irvine Ranch, Moulton Ranch, and Rancho Mission Viejo, in our wilderness parks, right along Ortega Highway, and in the Puente Hills.

Named in 1753 by C. Linnaeus, perhaps in reference to Bengal, India, where this plant may have originated.

Bengal Vetch

Winter Vetch, Hairy Vetch ✖
Vicia villosa Roth **ssp.** ***villosa***

An annual, its upper stems and leaves *covered sparsely or densely with hairs,* each 1-2 mm long. Its stems sprawl or climb. Leaflets 12-18, 1-2.5 cm long, narrowly oblong to elliptic, the tip rounded, each with a single tooth. *Inflorescence large and dense, usually with over 19 flowers.* Calyx lopsided and swollen at base, lower lobes 2-4 mm long, thread-like, often curved. *Corolla* 1.4-1.8 cm long, *violet-purple or lavender to white.* Fruit hairless, 1.5-4 cm long, 6-10 mm wide. In flower Apr-July.

Occasionally found here, most often in disturbed soils at low elevations, associated with ranches and old homesteads. Known from places such as the Puente Hills, Santa Ana River channel, Aliso and Wood Canyons Wilderness Park, Lomas de Santiago, Rancho Mission Viejo, Elsinore Peak, Lucas Canyon, Oak Flats, and San Mateo Canyon.

The epithet was given in 1793 by A.W. Roth for its hairy nature (Latin, villosus = hairy).

Similar. Variable winter vetch ✖ (*Vicia villosa* ssp. *varia* (Host) Corbière, not shown): *plant hairless to few-haired,* those hairs about 1 mm long; *inflorescence about 10-20-flowered, not as crowded;* lower calyx lobe 1-2.5 mm long; corolla 1-1.6 cm long; much less commonly found, known from Aliso and Wood Canyons Wilderness Park, Rancho Mission Viejo, Lucas Canyon, Wildomar region, and Santa Rosa Plateau.

Winter Vetch

FRANKENIACEAE • FRANKENIA FAMILY

A small family of annuals, perennials, and shrubs. Ours is a *low-growing shrub* that secretes salt from specialized glands. *Leaves opposite*, the edges rolled under. *Flowers long and tubular with 4-7 fused sepals. Petals, 4-7, free, white to pink or blue-purple, each with a small appendage at the throat.* Stamens variable, 3-12. Ovary superior, the style 1-4-branched. Fruit a capsule.

Alkali Heath
Frankenia salina (Molina) I.M. Johnst.
[*F. grandifolia* Cham. & Schldl.]

Alkali Heath

A low-growing shrub 10-60 cm tall from underground rhizomes, often grows as a dense mass. *Leaves opposite*, clustered, elliptical, leathery or fleshy, hairless to hairy, the edges rolled under. *Corolla pink*, sometimes white to blue-purple. *Stamens 6.* In flower May–Oct.

Common along the coastline and in upper reaches of salt marshes inland to deserts. Known from Seal Beach, Bolsa Chica Ecological Reserve, upper Newport Bay upstream to the San Joaquin Marsh, Corona del Mar, Laguna Beach inland to the Sycamore Hills, and along Horno Creek near San Juan Capistrano. Abundant at Nichols Road Wetlands in Elsinore Basin.

Similar. **Western sea-purslane** (*Sesuvium verrucosum*, Aizoaceae): water-warts; leaves linear to spoon-shaped, the edges sometimes rolled under; short flowers at nodes; calyx 5-lobed, purple-pink; petals absent; ovary green, short style is immediately 3-branched; *stamens numerous*, filaments purple-pink; grow together at Bolsa Chica Ecological Reserve, upper Newport Bay, and Nichols Road Wetlands.

Named in 1782 by J.I. Molina for its salty habitat (Latin, salinus = salty).

GARRYACEAE • SILK-TASSEL FAMILY

Evergreen shrubs or small trees, each plant is unisexual. Stems square in cross-section. *Leaves silvery to green, simple, leathery, opposite,* held close to the stem. *Inflorescence a long drooping multiflowered structure ("catkin"), covered with silky silvery hairs. Individual flowers unisexual, very small, and lacking petals.* Male flowers have 4 fuzzy-backed sepals often fused at their tips, plus 4 obvious stamens with yellow anthers. Female flowers have no sepals, an inferior ovary, and 2 styles (rarely 3). The 2 dark-tipped style branches look like a pair of antlers poking out of the female flower. Fruit is a round berry, at first green but maturing dark blue, black, or grayish, with dark purple to black pulp and only 2 seeds. They are all wind-pollinated.

Male catkins — Female catkin — Fruit

Pale Silk-Tassel
Garrya flavescens S. Watson

A rounded shrub 1.5-3 meters in height. Leaves 19-75 mm long, elliptic, flat or wavy-edged. *Leaf undersides hairless or with mostly straight hairs; those near the leaf tip are held tight against the leaf surface.* Male catkins generally 4-5 cm long, female catkins 2-3 cm long but lengthening to about 5 cm in fruit. In flower Jan-Apr.

Known from chaparral throughout the Santa Ana Mountains, especially along Main Divide Truck Trail.

The genus was named in 1834 by D. Douglas in honor of N. Garry. The epithet was given in 1873 by S. Watson for its tendency to turn pale yellow (Latin, flavescens = becoming yellow).

Similar. **Southern silk-tassel** (*G. veatchii* Kellogg): *leaf underside hairs wavy to curly, very dense, interwoven.* Male catkins generally 5-10 cm long, female catkins 2-3 cm, longer in fruit. In flower Feb-Apr. Serpentine soils among chaparral on Pleasants Peak in the Santa Ana Mountains. It was named in 1873 by A. Kellogg in honor of J.A. Veatch. **Manzanita** (*Arctostaphylos* spp., Ericaceae): leaves alternate; flowers bisexual, corolla urn-like and showy; sometimes co-occurs.

Pale Silk-Tassel — Male flower

GENTIANACEAE • GENTIAN FAMILY

Annuals or perennials, mostly upright though a few species lie on the ground and have upright branch tips. *Leaves opposite or whorled,* most without a petiole. *Flowers have 4-5 sepals, petals, and stamens but only 1 pistil.* Sepals and petals fused at the base, 4-5 free lobes above. Stamens alternate with corolla lobes. Ovary superior, fruit a capsule.

Alkali Chalice ✿
Eustoma exaltatum (L.) Don. **ssp.** *exaltatum*

A hairless, gray-green annual or short-lived perennial, its *stems are upright,* branched, 15-100 cm tall. Basal leaves 2-10 cm long, spoon-shaped to linear. Stem leaves 1.5-9 cm long and narrower. Sepals, petals, and stamens 5. *Calyx* 10-21 mm long, *each lobe narrow,* with a central ridge. *Corolla* 2-4.5 cm across; *blue to deep purple,* sometimes paler, rarely white; throat whitish with dark purple blotches. Stigma with 2 paddle-shaped yellowish lobes. In flower primarily June-Oct.

Very uncommon. Found in wet soils, seeps, and alkaline marshes, in full sun or partial shade. Known here only from upper Santa Ana River Canyon. In 2008, several plants were found about a half-mile downstream of Gypsum Canyon Road. On October 2, 2009, we found 6 plants at the edge of the river about one-eighth mile downstream of Gypsum Canyon Road, where it grows with **smooth bur-marigold** (*Bidens laevis*) and **cocklebur** (*Xanthium strumarium,* both Asteraceae). If you find it, please do not pick it.

The epithet was given in 1762 by C. Linnaeus for its height (Latin, exaltatus = very tall [most gentians are short]).

Alkali Chalice

Parry's Deer's Ears
Frasera parryi Torr. [*Swertia parryi* (Torr.) Kuntze]

Basal leaves 5-25 cm long, linear, *medium green with white wavy margins. Stem leaves* 4-10 cm long, *opposite,* linear or triangular. Inflorescence to 1.6 m tall with opposing branches. Flower parts 4 each, sepals and stamens alternate with the petals. Style short, stigma 2-branched. *Flowers face upward,* the calyx and corolla held flat out from the center as a landing pad for pollinators. *Corolla* lobes oval, pointed, whitish, greenish or yellowish, with many purple spots, and a *U-shaped, yellow-green nectary pit covered with long hairs.* In flower Apr-July.

A stunning 2-year perennial plant (biennial) with a stiff, heavy taproot and an unusual life cycle. It spends one season producing neither stalk nor flowers, just basal leaves that remain green for months and wither in late fall. The next season it sends up a few basal leaves, followed by a tall stalk with smaller stem leaves and numerous flowers. After the seeds set and mature, the plant dies. In some years of exceptional rainfall, it flowers both years.

Found in chaparral openings and dry uplands near meadows, often in the company of **white sage** (*Salvia apiana,* Lamiaceae). Known from the Bear Canyon trailhead near Ortega Candy Store, along South Main Divide Road, in Potrero Los Pinos near Falcon Group Camp, in Chiquito Basin along San Juan Trail south of Blue Jay Campground, along upper Long Canyon Road, and scattered sites in the San Mateo Canyon Wilderness Area.

The nectaries are often visited by small- to medium-sized native bees that appear to avoid the stamens and probably do not pollinate the plants. Larger insects such as bumble bees and carpenter bees are probably its most effective pollinators. Hummingbirds sometimes drink from its nectaries as well, possibly performing pollination.

The epithet was named in 1857 by J. Torrey for C.C. Parry.

Parry's Deer's Ears

Canchalagua, Charming Centaury
Zeltnera venusta (A. Gray) Mansion
[*Centaurium venustum* (A. Gray) Robinson]

An upright annual, 3-50 cm tall. Leaves 5-25 cm long, narrow, linear to oblong, broadly narrowed to a point. Flowers in clusters or solitary. Each flower 2-4 cm across, <1 cm long, its parts 5 in number. *Corolla throat whitish with red spots, free lobes brilliant magenta, rarely white,* held straight out away from the flower. *The bright orange anthers are straight when young but twist into a spiral as they shed their pollen and look like tiny corkscrews.* The lobes of the stigma are fan-shaped. Be sure to carry a hand lens so that you can observe these wild anthers and stigmas for yourself. In flower May-July.

Primarily a streamside plant but also found in grasslands and openings among chaparral; sometimes more common after fire. Known from low elevations in the Chino Hills near Bee Canyon, San Joaquin Hills on Temple Hill in Laguna Beach and Niguel Hill in Laguna Niguel. More common in the Santa Ana Mountains such as at Caspers Wilderness Park, along waterways in major canyons such as Santiago, Hot Springs, and San Juan, the Santa Rosa Plateau, and San Mateo Canyon Wilderness Area to San Clemente.

The genus was named in 2004 by G. Mansion in honor of husband and wife L. and N. Zeltner. The epithet was given in 1876 by A. Gray, no doubt for the character of the plant (Latin, venustus = charming, elegant). The Spanish common name canchalagua (can-chal-AG-wa; canchal = rocky place, agua = water) is in reference to the plant's tendency to grow in rocky soils, especially near water.

GERANIACEAE • GERANIUM FAMILY

Annuals or perennials, sometimes woody, usually hairy. Leaves rounded, kidney-shaped, and/or pinnate. Flowers with 5 sepals, free or fused at the base. Petals 5, all free, some with nectar glands between them. Stamens 5 or 10. Pistil is 5-lobed, 5 styles, ovary superior. In fruit the styles form an obvious beak (style column). When dry, each seed-filled lobe splits away from the beak and rolls or spirals upward, broadcasting the seeds. We have 7 species in our area; only 1 is native.

Large-leaved Filaree ★
California macrophylla (Hook. & Arn.) J.J. Aldasoro et al.
[*Erodium macrophyllum* Hook. & Arn.]

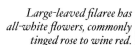

Large-leaved filaree has all-white flowers, commonly tinged rose to wine red.

An upright annual or biennial covered with short soft hairs, some hairs sticky-tipped. The main stem is very short. Petiole 3-12 cm long. *Leaf blade 2-5 cm wide, kidney-shaped, the edges toothed or shallowly lobed.* Sepals 8-10 mm long. *Petals 10-16 mm long; white, commonly tinged or entirely rose to wine red.* Stamens 5, *no staminodes* (compare with *Erodium* spp.). Fruit body 8-10 mm long, with 1 round pit on each side of beak segment, no ridges below the pits; fruit beak 3-5 cm long. In flower Mar-mid-Apr (-early May).

An uncommon plant of deep clay soils in grasslands and openings among coastal sage scrub. In our area, reported only from Temescal Valley, just south of DePalma Road, and on north-facing slopes between Indian Wash and Horsethief Wash. CRPR 1B.1.

The epithet was given in 1838 by W.J. Hooker and G.A.W. Arnott for its large leaves.

Similar. **Filarees** (*Erodium* spp.): leaves pinnately compound; corolla pink; stamens also 5, alternate with 5 scale-like staminodes. **Geraniums** (*Geranium* spp.): leaves palmately lobed or divided; stamens 10, no staminodes; grasslands, may co-occur. **Mallows** (Malvaceae): leaves often kidney-shaped; flowers somewhat similar; stamens numerous, filaments fused into a cylinder that surrounds the style; fruit flat, round, not beaked.

Filarees, Heronsbills, Storksbills: *Erodium*

Annuals or perennials, often from a deep taproot. *Leaves simple or pinnate,* basal leaves whorled or grouped, stem leaves opposite. Inflorescence is an umbel. *The 5 stamens alternate with 5 scale-like sterile stamens ("staminodes"). Petals are purple,* delicate, short-lived, and fall off easily. Sometimes they are marked with 3 long purple lines. Fruits narrow-billed; their length at maturity is important for identification. They are overly abundant invasive weeds thoughout the state, mostly in dry disturbed areas and grasslands. Native to Europe or Eurasia.

The genus was named in 1789 by C.L. L'Heritier de Brutelle for the resemblance of its beak to that of a heron, a type of shorebird (Greek, erodios = a heron).

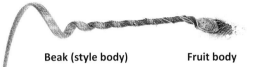

Beak (style body)   Fruit body

Filarees are well-known for their fruit. Upon fertilization, the style elongates into a style body or beak. The single seed resides within a spindle-shaped fruit body formed by the ovary. As the fruit matures and dries, it falls off of the plant and the beak twists spirally, driving the fruit body into the soil, planting the seed for next season.

Long-beaked (Short-fruited is identical)
White-stemmed
Red-stemmed

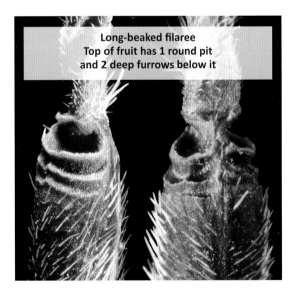

Long-beaked filaree
Top of fruit has 1 round pit and 2 deep furrows below it

Short-fruited filaree
Top of fruit has 1 narrow deep pit and 1 shallow furrow below it

Red-stemmed filaree
Top of fruit has 1 round pit and 1-2 furrows below it

White-stemmed filaree
Top of fruit has 1 round pit and 1 furrow below it; round glands in both

Long-beaked Filaree ✖
Erodium botrys (Cav.) Bertol.
Grows flat on the ground or upright, 10-90 cm tall, covered with short hairs. *Leaves are pinnate but the middle leaf vein is surrounded by leaf tissue; it is not a bare stem.* Leaf lobes number less than 9. Sometimes has upper leaves that are divided. *Petals are just barely longer than sepals, purple, veined with darker purple.* In fruit, the body is 8-11 mm long, the beak is 5-12 cm long, green or reddish. *The top of the fruit (just below the beak attachment) has 1 round pit and two deep furrows across it.* In flower Mar-May.

Widespread in our area, especially in grassy areas.

Named in 1819 by A.J. Cavanilles possibly for its cluster of purple flowers (Greek, botrys = a cluster or bunch of grapes).

Similar. Short-fruited filaree ✖ (*E. brachycarpum* (Godron) Thell.): leaf lobes <9; petals just barely longer than sepals; fruit body only 6-8 mm long, beak 5-8 cm long, striped with red. *The top of the fruit (just below the beak attachment) has 1 round deep pit and 1 shallow furrow below it.* In flower Apr-Aug. Widespread, common, often grows with other filarees.

Red-stemmed Filaree ✖
Erodium cicutarium (L.) L'Her.
Often a small plant, mostly grows flat on the ground, sometimes upward, 10-50 cm tall, covered with with sticky-tipped hairs. *Mature stems* and inflorescence stalk are *red* or reddish. *Leaves are 2-pinnate to the exposed mid-vein;* the 9-13 leaflets are slender and narrow-pointed. Flowers are in a sparse group atop a long stalk. *Sepals and petals about equal in length. Petals medium to deep pink.* Fruit body 4-7 mm long, the short beak only 2-5 cm long. *Top of fruit has 1 round pit and 1-2 furrows below it. The pit has no glands in it.* In flower Feb-May.

Widespread in grassy places.

Named in 1753 by C. Linnaeus for the similarity of its leaves to *Cicuta,* a genus of plants in the Carrot Family, Apiaceae (Latin, cicutarius = similar to *Cicuta*).

White-stemmed Filaree ✖
Erodium moschatum (L.) L'Her.
An annual, rarely a 2-year perennial, its *whitish stems* covered with short hairs. Grows partly to fully upright, 10-60 cm tall. Similar to red-stemmed filaree (originally named as a variety of it) but usually larger. *Leaves are pinnate to the mid-vein;* the 11-15 lobes are oval and unequal. Its leaflets are variable, sometimes similar to red-stemmed filaree. *Flowers are in a crowded group atop a long stalk. Petals much longer than the sepals, pale pink.* Fruit body 4-6 mm long, the short beak only 2-4 cm long. *Top of fruit has 1 round pit and 1 furrow below it. The pit and furrows have several round glands within them.* In flower Feb-May.

Widespread in grassy places throughout our area.

Named in 1753 by C. Linnaeus, supposedly in reference to is scent (Latin, moschatus = musky, musk-scented).

Cranesbills, Geraniums: *Geranium*

Ours are thin-stemmed hairy annuals. Stems commonly angled at the nodes. Leaves rounded or kidney-shaped in outline, often palmately lobed or divided. Stamens 10, arranged in two rings. Fruit beaked like filaree. We have 3 species: 1 native, 2 non-native.

The genus was named in 1753 by C. Linnaeus for the similarity of its fruit to the head of a crane (Greek, geranios = crane [a bird]).

Carolina Geranium
Geranium carolinianum L.

Carolina Geranium

A weak-stemmed upright annual, 10-70 cm tall. Leaf blades cut into wedge-shaped segments, each shallow-cut. Upper and lower surfaces often red- or dark-dotted. *Petals and sepals about equal in length, petals notched at the tip, white to pink. Stigmas yellow-green.* In flower Apr-June.

Found in the shade or open areas, usually among grasslands. Known from Sycamore Hills in Laguna Coast Wilderness Park, Riley Wilderness Park, Coal Canyon, Verdugo Canyon, Puente Hills, Walker Canyon, and scattered sites in the Santa Ana Mountains.

Named in 1753 by C. Linnaeus for Carolina in the eastern U.S., the state of its original discovery.

Carolina Geranium

Cutleaf Geranium

Roundleaf Geranium

Cutleaf Geranium ✖
Geranium dissectum L.

A slender-stemmed annual, upright, 20-80 cm tall. *Lower leaves divided 1-5 times, leaflets narrow and pointed. Petals and sepals about equal in length, petals notched at the tip, purple. Stigmas purple.* In flower Mar-May.

Widespread, known from places such as the San Joaquin Hills in Irvine, along Laguna Canyon Road in Laguna Beach, and on Rancho Mission Viejo such as in Verdugo Canyon. Common on the Santa Rosa Plateau. Sometimes a garden weed. Native to Europe.

Named in 1755 by C. Linnaeus for its deeply dissected leaves (Latin, dissectus = to cut up).

Roundleaf Geranium ✖
Geranium rotundifolium L.

A slender-stemmed sprawling annual, with stems 10-40 cm long. *Leaves round in outline; 5-7 major divisions;* multiple minor divisions, each roughly triangular-tipped. *Petals 6-7 mm long, pinkish. Stigmas pink.* In flower April-May.

A common weed in gardens, nurseries, recently spreading into wild lands. Native to Europe.

Named in 1753 by C. Linnaeus for its round leaves (Latin, rotundus = round; folium = leaf).

Cutleaf Geranium

GROSSULARIACEAE • GOOSEBERRY AND CURRANT FAMILY

Open shrubs to 2 or 3 meters tall. Those in our area are upright, but a few species found elsewhere in California trail on the ground. *Leaves are palmately veined* with 3-5 rounded lobes, *in bundles* alternately arranged on the stem. Most are summer deciduous; the new leaves appear in winter before or with the flowers. *Sepals and petals 5* (rarely 6, but only 4 in one local species) with an inferior ovary. The petals are held tight to the 5 stamens (rarely 4) while the *sepals are either held to the petals, curved outward, or folded back away from the flower tip*. Sepals are longer than the petals and often more colorful. Pollination is carried out by hummingbirds and small bees. The fruit is a berry, smooth-skinned or spine-covered. After flowering, the dried flower remains attached to the fruit.

The family contains 120 species, 30 in California, all in the genus *Ribes* (Arabic, ribas = a species of rhubarb, a word derived from a Persian word for acid-tasting, probably in reference to the taste of the fruits). They were once included within the saxifrage family, Saxifragaceae.

Two groups are easily recognized. **Gooseberries** have spines on their leaf nodes and fruits, and their long flowers hang downward loosely. Prior to opening, their long pistils grow far out of the bud, followed by the anthers. Only then do the sepals fold backward to reveal the petals. **Currants** lack spines, and their short flowers are in upright or arching clusters. Their pistils are short and do not pop out of the bud before the flowers open.

For identification, the flower tube (hypanthium) length is measured from its attachment with the ovary out to the point where sepals and petals separate. Its width is the average width of that tube. Look carefully at the flower and note placement and length of sepals and petals.

Ornithologists hypothesize that **Anna's hummingbird** (*Calypte anna* Lesson, Trochilidae) co-evolved with currants and gooseberries. The plants provide winter food that enables the birds to breed far earlier than all other hummingbirds; the bird successfully pollinates those flowers, contributing to the plants' reproductive success.

Many *Ribes* have horticultural value for their lovely flowers, unusual fruits, variable leaf textures, and growth forms. Flowering periods are generally short and mostly occur during winter-spring. They are best planted in partial shade beneath trees or near shrubs. Their roots and the soil above them should not be exposed to direct sun; for this reason they are often planted with another shrub that will provide ground-level shade. Because of their spines, gooseberries should not be planted near walkways or play areas.

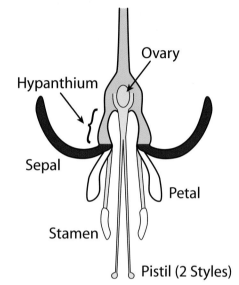

Slender-flowered Golden Currant

CURRANTS
Slender-flowered Golden Currant

Ribes aureum Pursh **var. *gracillimum*** (Cov. & Brit.) Jeps.
  Our only yellow-flowered Ribes. It may reach 3 meters in height and has flat, smooth, almost rubbery-textured, *light green leaves,* 15-50 mm long. Young leaves are sticky; older leaves are not. Flower clusters contain 5-15 flowers; *each flower has a very long tube.* Sepals 3-4 mm long, *yellow.* Petals 3-4 mm long, *yellow, turn red with age.* Fruit is hairless and red, orange, or black. In flower Feb-Apr.

In our area, known only from moist areas within Harding Canyon and in Temescal Valley (former Hunt Ranch just north of Glen Ivy Hot Springs).

The epithet was given in 1814 by F.T. Pursh, who described it from specimens collected by the Lewis and Clark Expedition, for its golden yellow flowers (Latin, aureolus = golden). This variety was named in 1908 by F.V. Coville and N.L. Britton in reference to the flower tube, which is more slender than the typical form of the species (Latin, gracillimum = most graceful, most slender).

Slender-flowered Golden Currant

White-flowered Currant
Ribes indecorum Eastwood

Our only white-flowered Ribes. It grows 2-3 meters tall and has *thick crinkled leaves,* 10-40 mm long. Leaf uppersides are medium green, hairy, and rough, while the undersides are whitish and fuzzy. *Its leaves are not sticky to the touch.* As it ages, the thin bark splits and shreds away. *Sepals and petals are white and very short,* only 1-2 mm long. Fruit purple, hairy, and sticky. In flower Nov-Mar.

A hardy, prolific bloomer, popular in the horticultural trade. Grows among coastal sage scrub and chaparral. Common in the Lomas de Santiago and Santa Ana Mountains: Santiago Canyon, Black Star Canyon, Caspers Wilderness Park, San Juan Hot Springs, Hot Springs Canyon, San Juan Canyon, near Blue Jay and Falcon Campgrounds, and up to Santiago Peak. Convenient places to see it are near the Ortega Candy Store and along San Juan Loop Trail.

Described in 1902 by A. Eastwood in reference to its plain white flowers (Latin, in- = not or without; decorus = elegant, decorative, or suitable).

White-flowered Currant

The fruit of white-flowered currant changes from green to yellowish, orange, red, and then purple at maturity.

White

Southern California

Southern California Currant
Ribes malvaceum* var. *viridifolium Abrams

A woody shrub, 1-2 meters tall. Its aging bark splits and shreds like white-flowered currant. *Leaves thick,* crinkled, large, 20-50 mm long, quite hairy, *very sticky,* and double-toothed. *Leaf color bright green above, light green (not white) below.* Flower clusters short and crowded, each flower all-pink. Sepals 4-6 mm long, twice as long as the petals. Fruit purple, covered with white powder and sticky-tipped white hairs. In flower Oct-Mar.

Widespread in chaparral but never abundant. Known from the Whittier and Chino Hills. In the Santa Ana Mountains, look for it in such places as Silverado Canyon, Santiago Canyon near Tucker Wildlife Sanctuary, along Harding Road, Black Star Canyon, along Main Divide Truck Trail west of Trabuco Peak, Trabuco Canyon, San Juan Hot Springs, along North Main Divide Road "Loop," the Santa Rosa Plateau, and the San Mateo Canyon Wilderness Area.

The epithet was given in 1815 by J.E. Smith, for the similiarity between its leaves and mallow (Latin, malva = mallow; -ceu = like).

Similar. Chaparral currant ✿ (*Ribes malvaceum* var. *malvaceum,* not pictured): leaves dull olive-green above, white below. Flower clusters long and open. Each flower is pinkish, flushed with white. Rare locally, known only among scrub oaks on Niguel Hill in Laguna Niguel.

Southern California Currant

GOOSEBERRIES

Fuchsia-flowered Gooseberry
Ribes speciosum Pursh

An arching shrub to 3 meters tall. Well-known for its numerous all-red flowers, 3 stout spines per node, and sharp prickles between the nodes. Leaves dark green, glossy, 10-35 mm long. Sepals and petals 4, red, 4-5 mm long, straight. Stamens 4, filaments red, anthers deep purple, pollen yellow. In flower Jan-Apr(-May).

This is our most common member of the genus. Found in moist natural habitats in the region, among coastal sage scrub and oak woodlands in canyons and on north-facing slopes. On the coast, look for it on the slopes surrounding upper Newport Bay and canyons in Dana Point, Laguna Beach, and Aliso Niguel. In Lomas de Santiago, it thrives in Limestone Canyon and Whiting Ranch Wilderness Park. In the Santa Ana Mountains, see it in Caspers Wilderness Park, San Juan Canyon, and Tucker Wildlife Sanctuary. Among the Chino Hills, you'll see it in Carbon and Tonner Canyons. Primarily a coastal plant, it occurs in Riverside County only at places with sea air influence such as portions of Corona and the San Mateo Canyon Wilderness Area.

Its flowers produce a copious amount of nectar, a favorite food of **Anna's hummingbird** (*Calypte anna* Lesson, Trochilidae). In winter or spring, find this plant and wait quietly for the birds to appear. A busy hummingbird will carry yellow pollen about halfway down its bill, easily seen with binoculars or camera with a long lens. It's simply amazing that the bird can hover below it and insert its beak into such a narrow flower!

The epithet was given in 1814 by F.T. Pursh in reference to its numerous showy flowers (Latin, speciosus = showy or brilliant).

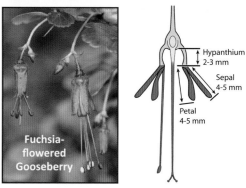

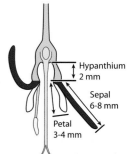

Southern California Hillside Gooseberry
Ribes californicum Hook. & Arn. **var. *hesperium*** (McClatchie) Jeps.

A shrub up to 2 meters tall with spreading branches. Stems have 1-3 spines per node. Leaves 10-30 mm long, glossy green above, lighter and hairy below but not sticky. The flower tube length is equal to its width. Sepals are 6-8 mm long, red-purple, and curve backwards away from the petals. *Petals are 3-4 mm long, white*, sometimes pinkish. Fruit red, covered with bristles that may be sticky-tipped. In flower Jan-Mar.

Its delightful flowers are a real treat to those who hike in winter. *Found in shady cold places* in San Juan, Santiago, Trabuco, and upper Silverado Canyons, south into western reaches of the San Mateo Canyon Wilderness Area. It is especially common beside the unpaved road through Arroyo Trabuco on the way to Holy Jim Canyon and the Trabuco Trail, along both the Maple Springs Truck Trail out of Silverado Canyon, and North Main Divide Road "Loop" between El Cariso and Long Canyon Road.

The epithet was given in 1838 by W.J. Hooker and G.A.W. Arnott in reference to the plant's occurrence in California. The variety was named in 1894 by A.J. McClatchie for its western distribution (Greek, hesperia = land of the west).

Bitter Gooseberry
Ribes amarum McClatchie

An upright, rounded shrub up to 2 meters tall. *It has 1-3 spines per leaf node, no spines between the nodes.* Leaves are hairy, sticky underneath, and 20-40 mm long. *Its flowers are the smallest of our gooseberries.* The flower tube is longer than wide. *Sepals 2-4 mm long, purple,* and curve backwards away from the petals. *Petals* 2 mm long, *white*. Fruit purple, covered with stiff bristles that may be sticky-tipped. In flower Mar-Apr.

Uncommon here, it grows among riparian, chaparral, and oak-conifer woodlands in the Santa Ana Mountains. Look for it in Silverado Canyon, Santiago Canyon, upper Trabuco and McVicker Canyons, the slopes of Santiago Peak, and along Bedford Canyon Road.

The epithet was given in 1894 by A.J. McClatchie in reference to the bitter taste of its berries (Latin, amarus = bitter). Early Western settlers mixed its berries with American bison meat to create a food called pemmican.

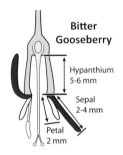
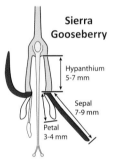

Sierra Gooseberry
Ribes roezlii Regel **var. *roezlii***

A sprawling shrub grows to about 1 meter tall. Stems have 1-3 spines per node. Leaves 12-25 mm long, sometimes covered with short hairs, light green above, lighter and hairy beneath but not sticky. The hairy flower tube is much longer than wide. *Sepals* 7-9 mm long, *purple, hairy,* and curve backwards. *Petals* 3-4 mm long, *white*. Fruit red or purple to black, covered with stout spines that are not sticky. Long hairs on the fruit's spines are sticky-tipped. In flower May-June.

Uncommon here, known only from riparian areas in mid-Santiago Canyon (Fleming Ranch), among conifers in upper Silverado Canyon, and among woodlands and chaparral. Best seen beneath oaks along North Main Divide Road on north-facing slopes of Santiago Peak.

Described in 1879 by E.A. von Regel in honor of B. Roezl.

THE CURRANT GUILD
Tailed Copper Butterfly
Lycaena arota (Boisduval, Lycaenidae)

A *small butterfly* with a wingspan of about 2 cm; the upper surface of the wings is orange-brown, underside banded with black and white. *Each hind wing has a small pointed tail* (most other coppers are tailless), outlined in black and filled with orange. Larvae feed on currants and gooseberries. They are found in major canyons of the Santa Ana Mountains such as Black Star, Silverado, Trabuco, and Holy Jim (near southern California hillside gooseberry), and on Santiago and Modjeska Peaks (near Sierra gooseberry).

HYPERICACEAE • ST. JOHN'S WORT FAMILY

Annuals, perennials, shrubs, and trees. Ours are perennials and shrubs. Leaves opposite or whorled, undivided, without petioles or stipules. Leaf edges dotted with black or clear glands. Flowers have 4-5 green sepals, mostly free or slightly fused at the base, sometimes falling off. Petals 5, *bright yellow and held straight out;* some species are dotted with black glands near their edges. Their *numerous long yellow stamens* are free or fused into 3-5 groups. They stand up above the petals or outward with them, some with a black gland on their anther. Ovary superior, the style 3-branched, yellow, sometimes with a red stigma tip. Fruit a capsule filled with numerous tiny seeds.

We have 3 species locally. The latter 2 mentioned below are perennials that green up and grow in spring, flower abundantly, set seed, and turn brown until the following year. The family is sometimes placed in the Clusiaceae.

Tinker's Penny
Hypericum anagalloides Cham. & Schltdl.

A *small herbaceous annual or perennial* that grows in low mats. Its stems are 3-25 cm long and lie on the ground with the tips upright. Leaves are 0.4-1.5 cm long, oval to round, the base somewhat clasps the stem. *Sepals and petals about equal, 2-4 mm long, petals golden to salmon-colored,* lacking black dots. Fruits are smooth and 3-lobed only at the tip. In flower June-Aug.

Uncommon, found only on a streamside rock outcrop in upper Devil Canyon within the San Mateo Canyon Wilderness Area. It sometimes grows with Scouler's St. John's wort (*H. scouleri*).

Named in 1828 by L.K.A. von Chamisso and D.F.L. von Schlechtendal in reference to the similarity of its leaves to a European plant, **scarlet pimpernel** ✖ (*Anagallis arvensis* L., Myrsinaceae, not shown) (Latin, -oides = like).

Tinker's Penny

Canary Islands St. John's Wort ✖
Hypericum canariense L.

A *medium to large shrub*, 2-5 m tall and as broad. Leaves 5-7 cm long, oblong to linear, narrow at their base, pointed at the tip. *Petals bright yellow*, 12-15 mm long. Fruits smooth, rounded, not lobed but break into 3 parts when ripe. In flower May-Apr.

An invasive weed, native to Spain's Canary Islands. We know of two places where it is established locally. One is in San Juan Capistrano near Valle Road at San Juan Creek Road, on road cuts among coastal sage scrub and up the nearby hills. The other site is among coastal sage scrub on the east side of the Dana Point Headlands, above Green Lantern Street and Pacific Coast Highway, now largely removed.

Named in 1753 by C. Linnaeus for its native location.

Scouler's St. John's Wort
Hypericum scouleri Hook.
[*H. formosum* HBK ssp. *scouleri* (Hook.) Thorne]

A *small herbaceous perennial* with upright stems that arise directly from the root. Inflorescence branched. *Leaves* 1-3 cm long, oval to oblong, *broad at the base which somewhat clasps the stem, broadly rounded at the tip, and with black dots along the edge. Petals yellow,* 7-12 mm long, *black-dotted near the edge.* Fruits are clearly divided into 3 lobes. In flower June-Aug.

Uncommon, found in a grassy meadow in the Chiquito Basin near the San Juan Trail, also in streambeds and rock pools in Lucas and Devil Canyons. One old record (1936) exists for Santa Ana Canyon, not found since. Another (1935) for North Main Divide Road "Loop," 1 mile from Ortega Highway.

Named in 1830 by W.J. Hooker in honor of his former student J. Scouler, who first collected the plant.

Scouler's St. John's Wort

LAMIACEAE • MINT FAMILY

A variable family of annuals, perennials, and shrubs, many with upright stems, hairy to hairless. Nearly all possess glands that release a *volatile oil* with a pleasant minty scent, a simple herbal smell, or a strong odor that isn't so pleasant. *Stems are square in cross-section.* Hold the stem between your thumb and forefinger then try to roll it; you will feel its 4-angled sides. *Leaves opposite,* the pairs often alternate on the stem at right angles to each other. *Flowers arranged in whorls around the stalk,* in most species the whorls are separated from each other by some distance. Calyx fused into a tube below, 5 free lobes above. *Corolla tubular, with 1-2 lips.* The upper lip is 1-2-lobed, the lower 3-lobed. *Stamens straight or curved, often protrude from the flower.* Most have 4 stamens of similar or different lengths. Some species have only 2 fertile stamens, the remaining 2 stamens lack anthers and are sterile. *Ovary superior,* the single style 2-lobed at its tip. *Fruit is a cluster of 4 nutlets* that ripen and fall out of the calyx. The nectary within the flower tube is popular with many types of birds, butterflies, bees, flies, and wasps.

About 7,200 species in the family, mostly from the Mediterranean region to central Asia Europe. Well-known cultivated members include sages, lavender, spear mint, pepper mint, oregano, rosemary, basil, and catnip. We present most of our mints in alphabetical order; some are placed next to similar species.

Heart-leaved Pitcher-sage ★
Lepechinia cardiophylla Epling

Heart-leaved Pitcher-sage

A fragrant, upright shrub nearly 2 m tall, densely or sparsely leaved, often quite hairy. *Large leaves, 4-12 cm long,* oblong to oval, lobed at their base, and *very broad,* the edges smooth or toothed. *Flowers hang like bells,* arranged in whorls on stalks near the branchtips; usually 2 flowers, sometimes only 1. *Corolla 3-3.5 cm long, large, tubular 2-lipped,* with irregular lobes, white to pinkish, often with pink veins inside the tube. Stamens, sometimes of different lengths. In flower Apr-July.

A rare plant, found only in the Santa Ana Mountains and the vicinity of Iron Mountain in San Diego County. It grows among chaparral, coniferous forests, and Tecate cypress woodland, often on north-facing slopes in decomposed granitic soils, usually in partial shade. Found in Coal Canyon, Claymine Canyon, near Sierra Peak, many places along Main Divide Truck Trail from the vicinity of Trabuco Peak north, and higher elevations of the Indian Truck Trail, where it was first collected. Pollinated by many species of bees, bumble bees, carpenter bees, and flies. Some bees bite open the flower base and steal nectar, which explains why some of its flowers are snipped open. CRPR 1B.2.

Named in 1948 by C.C. Epling in reference to the heart-shaped base of the leaves (Greek, cardio- = heart; phyllus = -leaved).

California Mountain Mint
Pycnanthemum californicum Torr.

California Mountain Mint

An upright perennial, 50-100 cm tall, from creeping rootstocks; its stems are hairy or hairless. *Leaves oval to somewhat triangular,* smooth-edged or toothed, hairless or covered in fine small hairs. *Flowers arranged in dense whorls,* each with a pair of bracts beneath. *Corolla 5-6.5 mm long, all white* or spotted with purple, 2-lipped, the upper lip 2-lobed, the lower 3-lobed. In flower June-Oct.

Uncommon here. It grows in seeps and moist creeks among other plant communities. Known from a creekside seep along San Juan Canyon, moist places along Holy Jim Trail, upper Santiago Canyon, the north fork of Cold Spring Canyon, San Mateo Canyon above Tenaja Falls, and Lucas Canyon. In our opinion, this plant and **San Miguel savory** (*Clinopodium chandleri*) are the best-smelling California plants. In cultivation, keep its roots moist and shaded.

The genus was named in 1803 by A. Michaux for its densely packed flowers (Greek, pycnos = close, dense, compact; anthos = flower). The epithet was given in 1855 by J. Torrey to honor the state of its discovery.

Common Henbit ✖

Lamium amplexicaule L.

An upright annual 10-40 cm tall. Leaves wavy-edged to lobed; the uppermost clasp the stem. Corolla tube very long, pink-purple, sometimes dark-spotted, middle lobe large and 2-lobed. Early- and late-season flowers do not open and instead undergo in-bud fertilization (cleistogamy). In flower Apr-Sep.

Native to Eurasia. Common in moist disturbed soils, especially citrus orchards and abandoned fields.

The epithet was given in 1753 by C. Linnaeus for the leaves that clasp its stem (Latin, amplexus = embracing, encircling; caulis = stem).

Common Henbit

Common Horehound ✖

Marrubium vulgare L.

A bushy perennial, 10-60 cm tall, covered with soft white woolly hairs. Leaf blade 1.5-5.5 cm long, oval to round, wrinkled. Flowers are in tight whorls. *Corolla white, 2-lipped,* the upper lip undivided or with 2 narrow lobes held upward like a rabbit's ears. Lower lip 3-lobed, central lobe broad and rounded. When the flower dries, the calyx becomes a bur that sticks in animal fur and hiker's socks, hitching a free ride to new locations. In flower spring-summer.

Native of Europe, now a worldwide weed. Commonly found in disturbed soils, along roadways, on former ranches, and along many trails in our wilderness parks, foothills, and mountains. Though the sap has a bitter taste, it is used to make candies and tea.

Both the genus and epithet were named in 1753 by C. Linnaeus for its bitter taste (Latinized from Hebrew, marrob = bitter juice, perhaps maror = bitter herbs eaten at Passover supper) and its abundance (Latin, vulgaris = common).

Common Horehound

Spear Mint ✖

Mentha spicata L. **var. *spicata***

A hairless perennial, 30-120 cm tall. No petiole; leaf blade 1-6 cm long, slender, oval to lance-shaped, the edges sharply toothed. *Flower whorls repeated in a long crowded spike, each whorl with a pair of very narrow bracts immediately beneath. Corolla white, pink, or lavender.* In flower July-Oct.

A garden escape, native of Europe. Currently known in the wild from Silverado Creek, mid Bell Canyon in Audubon Starr Ranch Sanctuary, San Mateo Creek, and Devil Canyon.

The epithet was given in 1753 by C. Linnaeus for its inflorescence which comes to a point, a spike of flowers (Latin, spicatus = spiked).

Spear Mint

Pepper Mint ✖

Mentha x piperita L.

A hairless perennial, 30-100 cm tall. Short-petioled; leaf blade 3-6(8) cm long, broad, oval to lance-shaped, edges sharply toothed. *Lowermost flower whorls separated by vertical spaces, those spaces reduced upward.* In some plants, the inflorescence is made up of a single dense head of whorls concentrated at the tip (there are no vertical spaces between whorls). Each whorl has a pair of narrow bracts immediately beneath. *Corolla white, pink, or violet.* In flower nearly year-round.

A garden escape, native of Europe. It is a hybrid between **water mint ✖** (*M. aquatica* L., not shown) and **spear mint**. It does not produce fertile seed; it only reproduces from rhizomes. Currently known in the wild only in San Juan Canyon in the creek downstream from San Juan Hot Springs.

The epithet was given in 1753 by C. Linnaeus (Latin, piper = pepper; Spanish, -ita = diminutive suffix for little or small).

Pepper Mint

Intermediate Thick-leaved Monardella ★
Monardella hypoleuca A. Gray **ssp. *intermedia***
A.C. Sanders & Elvin

An upright perennial from creeping rootstocks. Stems purple to greenish, hairless or sparsely hairy, 10-35 cm tall. *Leaves* 20-50 mm long, 5-9 mm wide, *linear to narrowly oval, stiff, very thick with obvious veins, dark green, sparsely haired above, white haired beneath, the edges strongly rolled under. Bracts purple,* hairy. *Corolla* 14-17 mm long, *white to pale lavender.* In flower Apr-Sep.

Found on dry slopes among chaparral, oak woodlands, and coniferous forests of the Santa Ana Mountains, eastward sparingly into the Palomar Mountains. Widespread but never abundant. Known from Baker Canyon, upper Silverado Canyon, Big Cone Springs, Harding Canyon Trail, Holy Jim Trail, Hot Springs Canyon, near Lower San Juan Picnic Ground, San Juan Trail south of Blue Jay Campground, and the Santa Ana Mountains ridgeline. CRPR 1B.3.

The epithet was given in 1878 by A. Gray for the white hair on its leaf undersides (Greek, hypo- = under, beneath; leuca = white). Our form was named in 2009 by A.C. Sanders and M.A. Elvin for characteristics that are intermediate between those of two other forms of *Monardella hypoleuca*.

Mustang Mint
Monardella breweri A. Gray **ssp. *lanceolata*** (A. Gray) A.C. Sanders & Elvin [*M. lanceolata* A. Gray]

An upright annual 20-50 cm tall. *Stems purple. Leaves* 30-40 mm long, *linear, soft to the touch. Bracts green* with purple tips and lots of net-like veins. *Corolla* 12-15 mm long, *rose-purple or paler.* In flower June-Sep.

Grows in dry soils and on disturbed sites such as road cuts, trailsides, and fireswept areas. Known from San Joaquin Hills, Laguna Coast Wilderness Park, Santa Ana Canyon, Maple Springs Truck Trail and Saddle, South Main Divide Road, Chiquito Basin Trail, Trabuco Trail, Modjeska Peak, Santa Rosa Plateau, and San Mateo Canyon Wilderness Area. More common after fire.

Named in 1876 by A. Gray for its linear leaves, which narrow to a point (Latin, lanceolatus = armed with a little lance or point).

Hall's Large-flowered Monardella ★
Monardella macrantha A. Gray **ssp. *hallii*** Abrams

A hairy perennial that creeps along the ground, it grows from a slender woody rhizome, the branches 10-50 cm long. *Leaves* 5-30 mm long, *triangular or oval,* hairy, smooth-edged or with tiny teeth. Flowers stand upright. *Corolla* 35-45 mm long, deep *red to red-orange,* rarely yellowish. In flower June-Aug.

Here, an understory plant *on talus slopes* among chaparral, oak woodland, and coniferous forests, often in the shade of oaks and other plants. Found only in the San Gabriel, San Bernardino, San Jacinto, and Cuyamaca Mountains and higher elevations of the Santa Ana Mountains. Known from Holy Jim Trail but more common in various places along Main Divide Truck Trail near the highest peaks such as west-facing talus slopes of Modjeska Peak and the summit of Santiago Peak.

A CRPR 1B plant, in need of proactive protection. Take care not to trample or pick it. Photographs cannot capture the beauty of this stunning plant. It is pollinated by hummingbirds and bee flies.

Named in 1876 by A. Gray for its large flowers (Greek, macranthus = large-flowered). Our form was named in 1912 by L. Abrams in honor of H.M. Hall.

Intermediate Thick-leaved Monardella

Mustang Mint

Hall's Large-flowered Monardella

Sages: *Salvia*

Annuals or perennials, soft herbs to woody shrubs. Leaf edges are smooth, lobed, or toothed, rarely spiny. *Flowers arranged into dense whorls far above most leaves.* The calyx is tubular, 2-lipped, with a variable number of free lobes. *Corolla is 2-lipped,* the upper lip not divided or 2-lobed, the lower with 3 lobes, the middle lobe by far the widest. *Only 2 stamens have anthers; the others are sterile or absent.* Stamens are free or attached to the corolla. *The style is forked at its tip.* A worldwide group with 17 species in California, 7 in our area. They are pollinated by bees, bumble bees, flies, butterflies, moths, and hummingbirds.

Sages are valued in the horticultural trade for their lovely flowers and pleasant scent. Their seeds are edible and were used extensively by Native Americans in food and drink.

The genus was named in 1753 by C. Linnaeus, who used the ancient Latin word for sage in reference to the healing properties of some species (Latin, salveus = safe, unhurt, well, sound).

Thistle Sage
Salvia carduacea Benth.

An annual, 10-100 cm tall, *covered in long white woolly hairs.* The *gray-green leaves* are 3-10 (rarely to 30) cm long, broadly linear, 1-pinnate, *with wavy, spine-tipped edges like a thistle* (Asteraceae), in a basal rosette. Flowers are arranged in rounded heads with spiny bracts. Calyx white-woolly, lobes spine-tipped. *Corolla lavender, blue, or white, 2-lipped, tips of the lower lips and sometimes also the upper lips are fringed.* The upper lip is held upright, 2-lobed, sometimes fringed at the tip. The lower lip is 3-lobed, fringed at the tip, and held outward. The 2 stamens are attached to the lower lip and stand up like insect antennae. In flower Mar-June.

Thistle Sage

Most often a desert plant but also found in coastal sage scrub and grasslands. Known long ago from Costa Mesa and Ortega Highway through Rancho Mission Viejo. More recently found at Alberhill in Temescal Valley and Elsinore-Temecula Valleys. Possibly still occurs along San Juan Creekbed as it flows through the north end of Caspers Wilderness Park, and undeveloped areas of Temescal Valley and Elsinore-Temecula Valleys.

Named in 1833 by G. Bentham for its thistle-like leaves (Latin, carduus = thistle; -acea = resemblance).

Chia
Salvia columbariae Benth.

An upright annual 10-50 cm tall. Its basal leaves are 2-10 cm long, oblong to oval, 1-2-pinnate with small rounded lobes, sometimes bristly but never spine-tipped. Flowers in 1-2 whorls per stalk, each whorl with many spine-tipped, green and purple bracts. Calyx is not woolly. *Corolla deep blue or paler, the upper lip 2-lobed and broadly rounded; the lower lip 3-lobed, its center lobe very large, 2-lobed, and white in its middle, sometimes with blue dots in the white patch.* The 2 blue stamens have long dark blue anthers, held straight out from the top of the flower tube with the single style. In flower Mar-June.

Chia

Widespread and common from coastline to foothills, mountains, and interior valleys, inland to the deserts. Often grows in dense groups, especially after soil disturbances such as fire or flood. Easily seen along the Harding Trail. A wonderful annual for the garden since it can grow practically anywhere.

Named in 1833 by G. Bentham for the resemblance of its flowers to the unrelated European pincushion plant (*Scabiosa columbaria* L., Dipsacaceae).

White Sage
Salvia apiana Jeps.

A *shrub* with multiple branches from the base. Non-flowering stems reach about 0.5 meters high. The slender inflorescence stalks begin to grow in winter and may reach 3 meters in height when mature. *Leaves white*, 4-8 cm long, broadly linear, edged with small rounded teeth. Flowers arranged in whorls on a long upright stalk. The stalk itself is pale whitish-purple, purple, or deep red. *Corolla white, speckled with lavender or all lavender;* upper lip tiny, only 2 mm long; lower lip large, 8-18 mm long, bent up at its base and cupped at the tip. *Stamens 2*, attached to the lower lip and standing up like insect antennae. The style also stands up or is held to one side of the flower; its 2-lobed tip is often purple. In flower Apr-July.

A common plant of coastal sage scrub, chaparral, and borders of oak woodland throughout our area, coastal but mostly inland on hot, dry, south-facing slopes in full sun. Found in all wilderness parks and many natural areas. A nearly pure stand occurs along the unpaved portion of Long Canyon Road below Blue Jay Campground. Readily hybridizes with black sage when they grow together.

Named in 1908 by W.L. Jepson for the bees that so commmonly visit its flowers (Latin, apis = bee; -ana = belonging to).

Similar. Desert brittlebush (*Encelia farinosa*, Asteraceae): leaves similar but lack minty scent, *alternate* on the stem; *yellow-rayed flowers in composite heads.*

POLLINATION IN WHITE SAGE

White sage flowers change in appearance as they mature. Fresh from the bud, the lower lip is folded up, which blocks the flower throat and forces the style to the side. In some flowers, the style is straight and stays to the side; in others, it curves around and extends toward the front of the flower. Its two stamens are held up and outward.

It is pollinated by **carpenter bees** (*Xylocopa* sp.) and **bumble bees** (*Bombus* spp., both Apidae). As the heavy bee alights on the first flower, it grasps the corolla and stops flying. The weight of the bee forces the corolla's lower lip downward; the bee can now use its tongue to access nectar at the inner base of the corolla. The stamens, which are attached to the base of the lower lip, are also forced downward. As they move, they strike and deposit pollen on the sides of the bee's body and wings. When the bee finishes feeding, it flies and releases the corolla, which snaps back upward. As the bee approaches a second flower, the style-tip contacts the bee. Pollen deposited on the bee by the first flower is captured by the stigmatic surface of the second flower and the flower is pollinated.

As the flower ages, the lower lip no longer remains folded upward. Instead, it relaxes and remains unfolded. This unfolding can also occur when bees roughly handle the flower and bite or cut the corolla base.

Above: Carpenter bee holding the lower lip down. The style and both stamens are toward us (the left stamen is usually on the bee's left side). As it approached and landed on the flower, the style struck the right side of its body, collecting pollen. As it opens the flower, the stamens brush pollen onto the sides and wings of the bee. More pollen is brushed onto the bee as it leaves the flower. Bottom: A single flower left open by a carpenter bee. After a visit by a heavy bee, the flower usually remains open.

WHITE SAGE IN PERIL?

Botanists have noticed a sharp decline in the distribution and abundance of white sage over the past 50 years. Habitat destruction certainly plays a role as humans remove plants to make way for buildings and roads. But many flowers on healthy plants in protected areas are not setting seed. It may be that pollinating insects have declined and white sage flowers are not being pollinated. For white sage to reproduce, carpenter and bumble bees need appropriate nesting habitat nearby. Carpenter bees need old wood in which to tunnel and nest. Bumble bees need undisturbed leaf litter on the ground for their nests. Another serious threat to the species is collection of its foliage for use in herbal remedies. It is being harvested on public lands and even from preserves. If you would like to use the plant for this purpose, please grow the plants on your property and do not take them from the wild.

Black Sage
Salvia mellifera E. Greene

Black Sage

A rounded shrub, 1-2 meters tall. *Leaves medium to dark green,* 2.5-7 cm long, linear or elliptic, hairless above, hairy below. *Corolla whitish to pale blue or lavender,* generally with no markings. Upper 2 lobes broad and rounded; *the central lobe of the lower lip has 2 very broad lobes.* The 2 stamens exit the top of the flower tube and arch over; anthers blue to purple. In flower Apr-July.

Widespread and common among coastal sage scrub from coastline to foothills of the mountains, absent from higher elevations. Capable of living in dry rocky soils but often found on north-facing slopes that retain soil moisture. In summer-fall, **lesser goldfinches** (*Spinus psaltria* (Say), Fringillidae) pull ripe seeds from the calyx and devour them.

Some debate exists about the origin of the common name; three explanations predominate. First, the dark green leaves sometimes appear darker at a distance. Second, the stems turn dark brown with age, so the plants appear blackish when the leaves are dropped. Third, the flower whorls turn dark with age and appear like dark spheres above the plant.

It was named in 1892 by E.L. Greene, probably because bees commonly visit its flowers and produce honey from its nectar (Latin, mellis = honey; fero = to bear).

Cleveland Sage
Salvia clevelandii (A. Gray) E. Greene

Cleveland Sage

A gray-green rounded shrub up to about 1 meter tall and wide. *Leaves 2-4 cm long, elliptic-oval, wrinkled, gray-green, with bases that narrow to the petiole. Corolla dark to medium blue or blue-violet with a long narrow tube.* Upper lip long, 2-lobed, and held upright; lower lip 3-lobed and held outward. *The central lobe of the lower lip is about the same length as the upper lip, the outer 2 lobes shorter than the upper lip.* The 2 stamens and single style are very long and are held outward. In flower July-Aug.

Found among chaparral in the San Mateo Wilderness from the south flank of Miller Mountain southward through San Diego County. A prolific bloomer and attractive shrub, very popular in the horticultural trade. Thought by many to have the most pleasant fragrance of the sages.

In young flowers, the stamens are held straight out and release their pollen. The style in those flowers is short, often held downward, out of the way, and is not receptive. As the flower ages and the anthers are emptied of their pollen, the anthers curl. The style elongates, matures, and becomes receptive to pollen brought to it by pollinators, primarily carpenter bees, bumble bees, and hummingbirds.

Named in 1874 by A. Gray in honor of D. Cleveland, who first collected the plant in San Diego County.

Purple Sage
Salvia leucophylla E. Greene

A *whitish-gray shrub* with many branches, upright or sprawling, just under 1.5 meters tall. *Leaves wrinkled, whitish,* 2-8 cm long, linear to oblong, with small rounded teeth, the edges sometimes rolled under. *The leaf base is blunt (truncate)* to heart-shaped. *Corolla large, lavender with a long narrow tube.* Upper lip long, 2-lobed, held upright; lower lip 3-lobed, held outward. *The upper lips are slightly shorter than the side lobes of the lower lips.* The central lobe of the lower lip is usually quite long. The 2 stamens curve upward; the single style is usually held outward. In flower May-July.

Grows among coastal sage scrub, often on north-facing slopes. Found in Carbon and Soquel Canyons in the Puente-Chino Hills, Black Star and Lower Blind Canyons in the Santa Ana Mountains, along Santiago Canyon Road near Irvine Lake in Santiago Canyon, and in Limestone Canyon. Its flowers mature in the same manner as Cleveland sage. It is also pollinated by carpenter bees, bumble bees, and hummingbirds. We know of no places where Cleveland and purple sage grow together.

Named in 1892 by E.L. Greene for its white leaves (Greek, leuco = white; -phyllus = -leaved).

Purple Sage

Purple sage flowers have long upper and lower lips. Compare this with Cleveland sage, which has long upper lips but short lower lips.

Hummingbird Sage
Salvia spathacea E. Greene

A perennial from creeping underground stems, 30-100 cm tall. *Leaves large,* 8-20 cm long, *oblong with a rounded arrowhead-shaped base,* very sticky, green and sparsely haired above, lighter and densely haired below. Inflorescence tall, upright, and covered with sticky-tipped hairs. *Flowers large, red, magenta, or salmon. The upper lip is short* and 2-lobed, lower lip much longer, 3-lobed; its central lobe is shallowly 2-lobed and nearly circular. The 2 stamens exit the top of the flower tube and arch over; the anthers are deep purple or red, the stigma is longer. In flower Mar-May.

Quite uncommon here, usually found beneath oaks in full or partial shade. We are at the southern end of its distribution, which extends from Orange County north to central California. Here known only from Laguna Canyon, Canyon Acres, and upper Mathis Canyon in Laguna Beach, Aliso Viejo (extirpated?), and Dripping Springs and Bee Flat Canyon in Limestone Canyon Wilderness Park. In gardens, often planted as an understory for oaks, to simulate its natural habitat.

Named in 1892 by E.L. Greene, probably for the spatula-shaped bracts and/or leaves (Latin, spatha = spatula; -acea = resemblance).

Hummingbird Sage

San Miguel Savory ★
Clinopodium chandleri (Brandegee) P.D. Cantino & Wagstaff [*Satureja chandleri* (Brandegee) Druce]

An upright or sprawling many-branched shrub under 0.5 m tall. *Leaves small*, 5-15 mm long, 4-16 mm wide, *triangular to oval*, hairy below, long-petioled. Flowers whorls of 1-6. *Corolla white*, 2-lipped, upper lip with 2 rounded lobes, lower with 3 rounded lobes. Stamens 4, in 2 pairs. The single style has a 2-lobed stigma; these parts exit the flower tube along the upper lip. In flower Mar-May, occasionally later.

Very rare, lives only from the Santa Ana Mountains, San Miguel Mountain, the Jamul Mountains in southwestern San Diego County, and northwestern Baja California. *Found in rocky soils among chaparral and oak woodlands, usually in shade, often on north-facing slopes.* Known from Hot Springs Canyon, San Juan Canyon north to Lion Canyon, and Chiquito Basin. Most easily found along the San Juan Trail below Blue Jay Campground. CRPR 1B.2.

In our opinion, this plant and **California mountain mint** (*Pycnanthemum californicum*) are the best-smelling California plants. Lightly rub a leaf and judge for yourself.

Named in 1905 by T.S. Brandegee to honor H.P. Chandler, who first collected it on San Miguel Mountain in 1904.

San Miguel Savory

Right: San Miguel savory has a white corolla with short round lobes.
Far right: Danny's skullcap has long tubular blue to purple corolla, with a cap-like upper lip that keeps the throat closed.

Danny's Skullcap
Scutellaria tuberosa Benth.

An ephemeral perennial that grows from creeping underground rhizomes and bulbous tubers, it stands upright to nearly 25 cm (but usually much shorter), covered with long sticky-tipped hairs. Leaf blades 1-2 cm long, wide and oval, tooth-edged. Its *large flowers* are solitary or in pairs, held upright. *Corolla blue to purple*, 13-20 mm long, tubular, with a small rounded cap-like upper lip and 3-lobed lower lip, which may be notched and have patches of white. In flower Apr-May.

Found mostly in *heavy clay soils,* in partial shade in openings among chaparral, more common after fire. Known from Carbon Canyon in the Chino Hills, Laguna Beach, Laguna Coast Wilderness Park, lower Hot Springs Canyon, along Bear Canyon Trail near the Ortega Candy Store, along South Main Divide Road, Santa Rosa Plateau, south into the San Mateo Canyon Wilderness Area.

Named in 1834 by G. Bentham for its fat tuber (Latin, tuber = a swelling; -osa = abundance).

Danny's Skullcap

Hedge-nettles: *Stachys*

Often strong-scented perennials, rarely annuals. Some have long hairs that look like those of nettles (*Urtica* spp., Urticaceae), but the hairs have no associated toxins. Flowers are in whorls on an upright inflorescence, each whorl subtended by a pair of leafy bracts. Calyx tubular, with 5 spine-tipped lobes. *Corolla white, pink, or purple; strongly 2-lipped.* The 4 stamens and single style are held against the upper lip. In all but one of our species, the underside of the corolla is constricted near its base and a backward-facing pouch formed.

Most of ours are found in moist soils, usually along perennial or ephemeral streams and seeps, often in full or partial shade. Pollinated by bumble bees and smaller bee species, sometimes visited by hummingbirds. They are home to many species of long-legged insects such as seed bugs (Lygaeidae) and stilt bugs (Berytidae) that can step over the plant's long hairs.

The genus was named in 1753 by C. Linnaeus in reference to the dense spike of flowers (inflorescence) that resembles an ear of corn (Greek, stachys = an ear of grain or spike).

	White-stem *Stachys albens*	**California** *Stachys bullata*	**Rigid** *Stachys rigida* var. *rigida*	**Hillside** *Stachys rigida* var. *quercetorum*	**Stebbins's** *Stachys stebbinsii*
Height	0.5-2.5 meters; upright	0.4-0.8 meters; upright	About 1 meter; upright	Under 1 meter; base usually spreading, tips upright	1.5 meter tall, robust; upright
Foliage hairs	White, cobwebby; soft, few prickly	Green; some stiff, some prickly	Green; soft-hairy, sometimes prickly	Green; soft-hairy (hairless), sometimes prickly	Green, soft hairy; very sticky
Leaves	Base roughly cordate; blade lance to oval; edges toothed or scalloped	Base roughly cordate; blade oval; edges toothed or scalloped	Base rounded to cordate; blade lance to oblong; edges smooth, small-toothed, or scalloped	Base cordate; blade oval; edges strongly scalloped	Base truncate to strongly cordate; blade oval; edges strongly scalloped
Flowers per whorl	10-12-flowered	6-flowered	6-10-flowered	6-10-flowered	6-flowered
Corolla	White	Pink to purplish	White to pink	White with purple	Pink
Corolla pouch	Pouched	Not pouched	Pouched	Pouched	Pouched
Internal corolla hairs	Angled	Perpendicular	Angled	Angled	Angled

California Hedge-nettle — No pouch

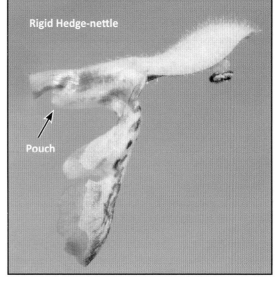

Rigid Hedge-nettle — Pouch

White-stem Hedge-nettle
Stachys albens A. Gray

A large upright perennial 0.5-2.5 meters tall. Densely covered with soft white cobwebby hairs. Leaf blades 1.5-2.5 cm long, oval to oblong, grayish, soft. Corolla white to pinkish, sometimes purple-streaked. Upper lip long, lower lip rounded. *Corolla base pouched.* In flower May-Oct.

A plant of very wet soils, swamps, and seeps. Known from San Mateo Canyon Wilderness Area, Santa Ana River channel at Chino Creek, upstream to Riverside, and into the San Bernardino Mountains. In cultivation, usually planted around water features.

The epithet was given in 1868 by A. Gray for its white appearance (Latin, albus = white).

Far left: White-stem hedge-nettle is covered with soft white hairs that give the plant a white cast. Left: California hedge-nettle flowers are pale to deep pink, sometimes mottled or striped with white.

California Hedge-nettle
Stachys bullata Benth.

An upright perennial 20-80 cm tall. Stems with surprisingly stiff and sharp hairs. Leaf blade 3-18 cm long, oval, dark green. *The deeply sunken leaf veins give the upper surface a bumpy or blistered appearance. Corolla pink, sometimes mottled or striped with white.* Upper lip short, *lower lip with a broad rounded central lobe. Corolla base not pouched.* In flower Apr-Sep.

Uncommon here, known from hillsides in upper Newport Bay (last reported in 1932, probably extirpated), Whittier Hills, Temescal Valley near Corona, and in a coastal canyon slightly south of San Onofre State Beach. Suspected to occur in San Mateo Estuary at the outfall of San Mateo Creek near Trestles. In cultivation, it is planted in moist shade and around water features.

The epithet was given in 1834 by G. Bentham for the blistered appearance of the leaves (Latin, bullatus = blistered).

Rigid Hedge-nettle
Stachys rigida Benth. **var. *rigida***

An upright perennial 60-100 cm tall. Stem hairs soft to stiff, sometimes sticky-tipped. Leaf blades 5-9 cm long, oval to lance-shaped, medium to light green, soft-hairy. *Corolla rose-purple or whitish with purple veins.* Lower corolla lip 4-6 mm wide. *Upper lip long, the central lobe of the lower lip generally lobed and angular. Corolla base pouched.* In flower July-Aug.

A common plant of active and ephemeral seeps and watercourses throughout our area. Find it at Crystal Cove State Park, Sycamore Hills, Dripping Springs in Limestone Canyon, Silverado Canyon, lower Holy Jim Trail, Bell Canyon in Caspers Wilderness Park, San Juan Canyon, and San Mateo Canyon Wilderness Area.

The epithet was given in 1848 by G. Bentham for its stiff stems or hairs (Latin, rigidus = stiff).

Similar. Stebbins's hedge-nettle (*S. stebbinsii* G.A. Mulligan & D.B. Munro, not pictured): *plant robust, densely glandular, resinous, musky-aromatic;* lower corolla lip 6–10 mm wide; in flower summer; uncommon, known from Silverado and San Juan Canyons, perhaps more widespread.

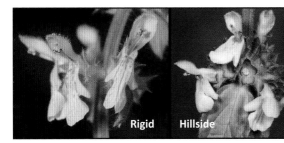

Far left: The lower lip of rigid hedge-nettle is often narrow and folded backward. Left: The lower lip of hillside hedge-nettle is often broad and held outward.

Hillside Hedge-nettle
Stachys rigida Benth. **var. *quercetorum*** (A.A. Heller) Epling

Often sprawling with only its tips upright, less often completely upright. Stems 15-40 cm long, hairs soft to stiff, sometimes sticky-tipped. Leaves oval, edges strongly scalloped, tip usually rounded, soft-hairy to soft. *Corolla white to pinkish; lower lobe wide, often held flat out.* In flower Mar-Oct.

Mostly at *lower elevations* from coast to foothills, typically *in heavy clay or sandy-clay soils on shaded north-facing slopes* among coastal sage scrub, grasslands, and oak woodlands. Formerly common and widespread but largely extirpated. Known from Crystal Cove State Park, San Joaquin Hills, Dana Point Headlands, Richard and Donna O'Neill Conservancy, Rancho Mission Viejo, Lomas de Santiago, and Lucas Canyon. Inland sites include Hagador Canyon, Temescal Canyon, Lee Lake, and the San Mateo Canyon Wilderness Area.

The name was given in 1907 by A.A. Heller for oak woodlands where it sometimes grows (Latin, quercetum = an oak wood).

Vinegar Bluecurls, Vinegar-weed
Trichostema lanceolatum Benth.

A hairy annual 10-60 cm tall with small glands that release an unforgettable *strong odor of vinegar*, especially when touched. Leaves 2-7 cm long, oval or triangular, with no petiole. Corolla tube strongly S-shaped. *Corolla light to medium blue with darker speckling on the lower lip.* Style and stamens are strongly arched. In flower Aug-Oct.

Widespread and common, often in dry disturbed soils. Known from the Puente-Chino Hills, San Joaquin Hills, lowlands near foothills and creekbeds, Lomas de Santiago, our wilderness parks, open flats in the Santa Ana Mountains such as at Chiquito Basin and Los Pinos Potrero near Falcon and Blue Jay Campgrounds, and Santa Rosa Plateau. Pollinated by native bees.

Named in 1835 by G. Bentham for its lance-shaped leaves (Latin, lanceolatus = lance-like).

Similar. San Jacinto bluecurls (*T. austromontanum* H. Lewis): annual; corolla tube curved gradually upward; stamens 2-6 mm long, no vinegar odor; in flower July-Oct; scarce, found in moist sand along Devil Creek about one mile downstream of Miller Creek.

Vinegar Bluecurls

Woolly Bluecurls
Trichostema lanatum Benth.

A shrub up to about 1.5 m tall. *Leaves 3.5-7.5 mm long, narrowly linear, dark green and hairless above, white-hairy below.* The inflorescence is covered with *white woolly hairs that hide the flower bases. Corolla tube straight. Corolla all-blue (rarely all-white).* Style and stamens long, lavender, nearly straight out or gently arched. In flower May-Aug.

Common and widespread mostly in dry soils among coastal sage scrub and chaparral. Known from Santa Ana Canyon, major canyons and higher elevations of the Santa Ana Mountains. Easiest to find in upper Silverado and Trabuco Canyons, on the San Juan Trail below Blue Jay Campground, and along North Main Divide Road. Pollinated by hummingbirds. In cultivation it prefers dry soils and bright sun.

Named in 1835 by G. Bentham for its woolly inflorescence (Latin, lanatus = woolly).

Woolly Bluecurls

Parish's Bluecurls
Trichostema parishii Vasey

A shrub to 1.2 m tall. *Leaves 2-6 cm long, linear, dark green and hairless above, gray-hairy below.* The inflorescence is fairly uncrowded, covered with *slender hairs* that *do not hide* the stems and pedicels. *Pedicel upcurved at tip, corolla tube straight or a bit curved. Corolla blue to blue-lavender,* its lower lip usually has one or more deep blue-purple patches surrounded by white. Style and stamens long, blue to blue-lavender, strongly arched. In flower May-Aug.

Long considered to be less common here than woolly bluecurls, but our fieldwork reveals that it is quite common. The two generally do not grow together. This species is typically found at higher elevations in the Santa Ana Mountains. Known from South Main Divide Road, North Main Divide Road from Los Pinos Potrero north to about Trabuco Peak, and Bedford Peak. Quite abundant along East Horsethief Trail. Pollinated by native bees.

Named in 1881 by G. Vasey in honor of S.B. Parish.

Parish's Bluecurls

LINACEAE • FLAX FAMILY

Upright annual and perennial herbs and shrubs. *Leaves* are mostly along the stem; *alternate*, opposite, or whorled in arrangement. Each leaf undivided, no petiole; linear or oval in shape; stipules tiny, with sticky glands, or absent. *Sepals and petals 4-5. Stamens 4-5, alternate with the petals.* Ovary superior, 4-6 chambered, styles 2-5, fruit usually a capsule. A small worldwide family of 300 species, 15 native to California, 1 in our area.

Small-flowered Dwarf Flax

Hesperolinon micranthum (A. Gray) Small

Small-flowered Dwarf Flax

An upright annual 5-20 cm tall, *with threadlike stems,* many-branched especially in the upper two-thirds of its height. Leaves 1-2.5 cm long, narrowly linear, alternate. *Petals small,* to 3.5 mm long, *white to pink,* sometimes with rose-colored lines. *Each petal has 3 small white scales at its inner base.* Styles 3(2). In flower May-July.

A delicate plant of heavy clay soils, widespread but not commonly found. Known from Blind Canyon (near Fremont Canyon) and Oak Flats, where it grows among chamise chaparral, Coal Canyon in openings between Tecate cypresses, Temescal Canyon on alluvial benches, and Santa Rosa Plateau. Self-pollinated.

The genus was established in 1865 by A. Gray for its distribution in the western U.S. (Greek, hesperos = western; linon, linum = flax). The epithet was given in 1868 by A. Gray for its tiny flowers (Greek, mikros = small; anthos = flower).

Similar. Small-flowered groundsmoke (*Gayophytum diffusum* ssp. *parviflorum*, Onagraceae): *sepals 4, usually folded backward*; petals 4, white; stamens 4 or 8; *ovary inferior; style 1, clubbed.*

LOASACEAE • STICK-LEAF OR LOASA FAMILY

Annuals or perennials covered with short barbed hairs. They stick to your clothes when you walk by! Leaves alternate, rarely opposite; pinnately lobed. Sepals and petals 5 each, petal bases free or fused. Corolla tubular, yellow to orange, lobes held outward. Stamens 5-to-many, threadlike or flat. Ovary inferior, style threadlike. When in fruit, petals wither and fall off but sepals remain. Fruit a dry capsule. A small family of about 200 species in the Americas, Africa, and Pacific Islands. We have 5 species in our area; 2 are presented here; the other 3 are very scarce.

Small-flowered Stick-leaf

Mentzelia micrantha (Hook. & Arn.) Torr. & A. Gray

Small-flowered Stick-leaf

An upright bushy annual 10-80 cm tall. Leaves 1-18 cm long, linear to oval, pinnately lobed, toothed, or wavy-edged. Corolla lobes completely free. Stamens about 20. Style tipped with a 3-lobed stigma. In flower Apr-May.

Our most common stick-leaf, it is found in the years *following fire* in chaparral, oak woodlands, and mixed coniferous forests. Known from Lomas de Santiago and our major canyons, mostly at higher elevations of the Santa Ana Mountains, plus scattered sites within the San Mateo Canyon Wilderness Area.

The genus was named in 1753 by C. Linnaeus in honor of C. Mentzel. This species was named in 1840 by W.J. Hooker and G.A.W. Arnott for its small flowers (Greek, mikros = small; anthos = flower).

Veatch's Stick-leaf

Mentzelia veatchiana Kellogg

Veatch's Stick-leaf

A stiff-stemmed upright annual, 3-45 cm tall. Leaves 1-18 cm long; basal lobed, stem leaves toothed to lobed. Petals 4-8 mm long, broad, orange or yellow-orange, often orange-throated, held straight out. The plant we photographed had notched petal-tips; most individuals of this species have pointed petal-tips. Stamens 20. In flower Apr-June.

Uncommon here, known only from Santiago and Modjeska Peaks in the first few years after fire.

The epithet was given in 1863 by A. Kellogg in honor of A.A. Veatch.

LYTHRACEAE • LOOSESTRIFE FAMILY

Annuals, perennials, shrubs, and trees. Ours are annuals or perennials. *Leaves linear, opposite,* rarely alternate. *Flowers tubular and held upward. Sepals mostly fused,* free only at their tips into 4-6 lobes that *alternate with tiny appendages* (epicalyx). *Petals fused into a tube* with 4-6 free lobes (or petals absent). *Stamens equal to or twice the number of petals.* Ovary superior; style single; stigma club-like. *Generally found in wet habitats* such as creeks and ponds.

Valley Red-stem
Ammannia coccinea Rottb.

A stiff-stemmed hairless annual, 10-100 cm tall, upright or sprawling. Leaves opposite, the pairs at 90 degrees to each other up the stem. Each leaf 2-8 cm long, linear to lance-shaped, "earlobes" at base; no petiole. Flowers usually 3-5 per leaf axil. *Longest peduncles 3-5(9) mm long.* Calyx lobes 4. *Petals 4, deep rose-purple.* Stamens 4(7). *Anthers deep yellow. Fruit 3-5 mm wide.* In flower June-Aug.

A plant of watercourse edges, lakes, ponds, and vernal pools. Common in the Santa Ana River channel. Uncommon in Peters Canyon, Laguna Lakes, and Rancho Mission Viejo.

The epithet was given in 1773 by C.F. Rottboll for its red flowers (Latin, coccineus = scarlet).

Similar. Robust red-stem (*A. robusta* Heer & Regel, not pictured): longest peduncles 0-3 mm long; *petals 4, pale lavender;* anthers pale yellow; *fruit 4-6 mm wide.* Laguna Lakes and Lambert Reservoir. The epithet was given in 1842 by O. von Heer and E.A. von Regel for its robust stems (Latin, robustus = strong, robust).

Valley Red-stem

California Loosestrife
Lythrum californicum Torr. & A. Gray

An upright hairless perennial, 20-60 cm tall. *Leaves* 1-7 cm long, narrowly linear, *flat,* those lower on the plant are opposite, those higher are alternate. Flowers 1-2 per upper leaf axil. Calyx tubular with triangular lobes. *Petals 4-8 mm long, purple. Stamens are of two lengths,* some hidden within the flower tube, some standing out of it. Style either long or short. In flower Apr-Oct.

Found at the edges of rivers, creeks, and ponds, such as the Santa Ana River bottom, Bolsa Chica, Newport Bay, Laguna Canyon, Sycamore Hills, Dripping Springs in the Lomas de Santiago, upper Santiago Canyon, lower Hot Springs Canyon, San Juan Canyon south into San Juan Capistrano, the Santa Rosa Plateau, and San Mateo Canyon.

It was named in 1840 by J. Torrey and A. Gray for the place of its discovery.

California Loosestrife

Hyssop Loosestrife, Grass Poly ✖
Lythrum hyssopifolia L.

A low-growing annual, sprawling with upcurved stem-tips. Leaves small, 5-30 mm long, *the sides curved under.* Flowers 1 per upper leaf axil. Petals 2-5 mm long, pink. Stamens all one length, hidden within the flower tube. Style barely pops out of the flower tube. In flower Apr-Oct.

Moist places, vernal pools, ponds; sometimes a landscape weed. Native to Europe.

The epithet was given in 1753 by C. Linnaeus, probably for its scented leaves (Greek, hyssopos = an aromatic plant; Latin folium = leaf).

Similar. Three-bract loosestrife ✖ (*L. tribracteatum* Spreng.): stamens and style hidden within flower; uncommon, recorded from Rancho Mission Viejo and Irvine.

Hyssop Loosestrife

MALVACEAE • MALLOW FAMILY

Annuals, perennials, shrubs, and trees *covered with star-shaped hairs that feel bristly to the touch*. Ours are annuals, perennials, and shrubs. *Leaves alternate, palmately veined and/or lobed,* with an obvious petiole. A whorl of small bracts (bractlets) sits right beneath the calyx of each flower. *Sepals and petals 5,* each oval or triangular. *The numerous filaments are fused into a tube that surrounds the style.* The sepals, petals, and filament tube are fused together at the base, so they fall off as a group when the flower ages. Ovary superior; the single style ends in multiple stigma lobes near its tip. The fruit is shaped like a wheel of cheese with 5 or more wedge-shaped pieces that fall out as it dries. The family has about 2,000 species worldwide, 50 native to California, 5 native species here.

Many-flowered Bush Mallow
Malacothamnus densiflorus (S. Watson) E. Greene

An upright shrub 1-2 meters tall , *with arching branches that often droop to the ground.* The leathery leaf blade is 3-6 cm long, oval in outline, sometimes with shallow rounded lobes. *Flowers are numerous, in tight dense clusters attached directly to the stalk, and widely spaced apart from each other.* Bracts beneath each flower are 5-15 mm long, generally about half as long as the calyx. Petals 10-15 mm long, rose-pink. In flower Apr-July.

An uncommon plant of coastal sage scrub and chaparral, more common in the years after fire. Known from the San Juan Trail near Sugarloaf Peak, the area south of Bedford Peak and along Bedford Truck Trail, and along South Main Divide Road near Elsinore Peak; more common in the San Mateo Canyon Wilderness Area. Perhaps more widespread but overlooked.

The genus was named in 1906 by E.L. Greene for its soft leaves (Greek, malakos = soft; thamnos = shrub). The epithet was given in 1882 by S. Watson for its dense clusters of flowers (Latin, densus = dense, compact; floris = flower).

Many-flowered Bush Mallow

Many-flowered

Chaparral

Chaparral Bush Mallow
Malacothamnus fasciculatus (Nutt. ex Torr. & A. Gray) E. Greene

A large shrub 1-5 meters tall *with upward arching branches, sometimes drooping to the ground, often covered with sticky-tipped hairs.* Leaf blade 2-11 cm long, thin and flexible, oval or round in outline, with angular or rounded lobes and toothed edges. *Flowers are in bundles that appear continuously all along the stalk, attached directly to it or on short side branches.* Bracts beneath each flower are 1-8 mm long, 1 mm wide or narrower, usually much shorter than the calyx. Petals 12-18 mm long, soft pink. In flower May-Aug.

Very common among coastal sage scrub and chaparral throughout our area, mostly in the years following fire. Known from all of our hills, canyons, and mountains, and most of our wilderness parks and wild areas. After the 1993 Laguna Beach fire, it became abundant in the hills there, especially near Big Bend along Laguna Canyon Road and adjacent Laguna Coast Wilderness Park.

The epithet was given in 1838 by T. Nuttall for its bundle of flowers (Latin, fasciculus = a bundle).

Chaparral Bush Mallow

Alkali Mallow
Malvella leprosa (Ortega) Krapovickas

Perennial from prostrate stems that turn upward at the tips, stems 10-40 cm long. Plant covered with dense white hairs that give it a gray or gray-green appearance. *Leaf blade* 1.5-4.5 cm wide, *kidney-shaped, wider than long.* Green bracts beneath the calyx are 1-3 in number, linear, smaller than the calyx lobes, and often fall off. Flowers appear at the nodes usually in groups of 1-3. Calyx lobes green and triangular. *Petals* 10-15 mm long, *creamy white to yellowish, often pinkish or pink-dotted on one side, making the buds appear pink.* In flower May-Oct.

A common plant of *alkaline soils* along the coast and elsewhere near marshes, creeks, streams, and soil depressions; sometimes a weed in orchards and gardens. A California native accidentally exported to other countries such as Australia, where it is considered a noxious weed. Often overlooked, it is an attractive plant pollinated by many insects such as bumble bees, bees, and butterflies.

The genus was named in 1853 by H.-F.C. de Jaubert and É. Spach for the plant's resemblance to smaller mallows (Latin, malva = mallow; -ella = small). The epithet was given in 1798 by C.G. Ortega, perhaps for its rough dense hairs or the pink spots on the petals (Latin, leprosus = scurfy, scaly deposits on a plant, or spotted like a leper.)

Alkali Mallow

Alkali Mallow

Coastal Checkerbloom

Right: coastal checkerbloom has delicate bowl-shaped flowers in shades of pink with white lines.

Coastal Checkerbloom
Sidalcea malviflora (DC.) Benth. **ssp.** *malviflora*

An upright or spreading hairy perennial, 15-60 cm tall, from heavy, spreading rootstocks. Leaves 2-6 cm wide, *toothed or lobed, the lobes cut deeper upward on the stalk,* mostly basal. Inflorescence a wand-like stalk. *Bractlets absent.* Petals 10-35 mm long, *lovely pink with white veins.* In flower Mar-July.

Mostly a plant of *grasslands, coastal sage scrub, and oak woodlands.* Found in our coastal parks, Limestone Canyon, Rancho Mission Viejo, meadows near Blue Jay and Falcon Campgrounds, south into the San Mateo Canyon Wilderness Area, and the Santa Rosa Plateau.

The genus was established in 1848 by A. Gray by combining the Greek names of two mallows it resembles; both names are also used as generic names: *Sida* and *Alcea.* The epithet was given in 1824 by A.P. de Candolle for its mallow-like flower (Latin, malva = mallow; florus = flower).

Similar. Salt spring checkerbloom ★ (*S. neomexicana* A. Gray, not shown): most leaves basal, blade 2-8 cm wide, fleshy, wavy-edged to shallowly lobed, upper leaves 5-lobed; *bractlets 2, fused at their base; flower smaller,* petals 0.6-1.8 cm long, pale pink with white veins. *Formerly widespread in wet areas, alkaline soils, and seeps,* mostly extirpated. Reports of it from Rancho Mission Viejo need confirmation. CRPR 2.2. *Very difficult to separate from coastal checkerbloom, expert determination is required.*

Coastal Checkerbloom

MONTIACEAE • MONTIA FAMILY

Annuals or perennials usually with fleshy, succulent hairless stems and leaves. Leaves undivided, opposite or alternate, *often basal*. They have a series of flower bracts that look like sepals, especially the cup-like innermost 2 (rarely as many as 9). Next is a series of petal-like lobes called tepals, most often 5 (sometimes 1-19), that easily fall off. Stamens 1-to-many, opposite the tepals and attached to them at the base. Styles (0, 1) 2-8, fused at their base. Ovary superior. Fruit usually a capsule. Most species were formerly in the purslane family, Portulacaceae.

Brewer's Red Maids ★

Calandrinia breweri S. Watson

A low-growing hairless annual, its *stem-tips often curve upward*. Its *flat leaves* are 2-8 cm long, oval to spoon-shaped. The 2 sepal-like flower bracts are 4-6 mm long, cupped, and overlap at their base, each hairless or with a few hairs along its edges. The 5 petal-like tepals are 3-5 mm long, have rounded tips, and are rose-red in color. *The mature fruit is much longer than the bracts.* In flower Mar-June.

A very uncommon plant of alluvial benches and hillsides, native grasslands, gravelly soils, and decomposed granitic soils amid chaparral. Known here only from upper Weir, Fremont, and Black Star Canyons and along Bear Canyon Trail. Sometimes locally abundant after fire. CRPR 4.2.

The epithet was given in 1876 by S. Watson in honor of W.H. Brewer.

Brewer's Red Maids

Brewer's Red Maids

Brewer's red maids and common red maids are similar. In Brewer's, the mature fruit body is longer than the two sepal-like bracts that surround it. In common, they are all about the same length.

Common Red Maids

Calandrinia ciliata (Ruíz Lopez and Pavón) DC.

A low-growing hairless annual with *spreading stems* up to 40 cm long. Its *flat leaves* are 1-10 cm long, linear or slightly spoon-shaped. The 2 sepal-like flower bracts are 2.5-8 mm long, cupped, and overlap at their base, each usually with hairs along its edges. The 5 petal-like tepals are 4-15 mm long, have rounded tips, and are rose-red in color, rarely white. *The mature fruit is as long as or barely longer than the bracts.* In flower Feb-May.

Widespread and common in grasslands, vernal pool margins, and moist openings in coastal sage scrub. Look for it in Telegraph Canyon in the Chino Hills, upper Newport Bay, the San Joaquin Hills, Caspers Wilderness Park, and San Mateo Canyon Wilderness Area. Quite common in upper elevations of Laguna Coast Wilderness Park.

Named in 1798 by H. Ruíz Lopez and J.A. Pavón for the hairs on its sepal-like flower bracts (Latin, ciliatus = furnished with hairs).

Common Red Maids

Seaside Pussypaws ★

Cistanthe maritima (Nutt.) Carolin ex Hershkovitz
[*Calandrinia maritima* S. Watson]

A low-growing hairless annual *covered with fine whitish powder* (glaucous). Its *flat, succulent leaves* are 1-6 cm long, *oval to spoon-shaped*. The 2 sepal-like flower bracts are 3-5 mm long, cupped, and overlap at their base, hairless, generally with purple veins. The 5 petal-like tepals are 3-6 mm long, have rounded tips, and are red-purple in color. *The mature fruit is barely longer than the bracts.* In flower Mar-Apr.

A plant of sandy soils near sea bluffs. Long ago (1926) known from sea bluffs in Laguna Beach, certainly extirpated by coastal development. A few plants were recently found by Fred Roberts on the Dana Point Headlands. CRPR 4.2.

The epithet was given in 1838 by T. Nuttall for its seaside habitat (Latin, maritimus = of or belonging to the sea).

Seaside Pussypaws

Common Pussypaws
Cistanthe monandra (Nutt.) Hershkovitz
[*Calyptridium monandrum* Nutt.]

A low-lying annual with spreading stems 1.5-18 cm long. Its stems start out green but soon turn red, followed by the remainder of the plant. *Basal leaves* 1-5 cm long, *narrow to spoon-shaped,* stem leaves generally much smaller. Inflorescence near stem-tip. Flowers small, often clustered. Sepals 1-2 mm. Petals usually 3, 1-3 mm long, pink to reddish. Stamen 1, anther pink, pollen yellow. In flower Jan-July.

An uncommon plant of *sandy soils,* more common after fire. Known from Lomas de Santiago, Coal and Gypsum Canyons, Lucas Canyon, San Juan Canyon, and the San Mateo Canyon Wilderness Area.

The epithet was given in 1838 by T. Nuttall for its single stamen (Greek, monos = single; andros = male). The common name refers to the inflorescence, which resembles the broad paw of a cat.

Miner's Lettuce: *Claytonia*

Succulent annuals with two types of leaves: those in a basal rosette and stem leaves fused into a cup that surrounds the stem. Flowers are on a 1-sided stalk atop the plant. The 2 sepal-like flower bracts are cupped and overlap at their base. The 5 petal-like white to pink tepals are slightly longer than the flower bracts. Both of our 2 species appear in many confusing named and unnamed forms, thought to be the result of extensive hybridization. Some are self-fertilizing while others require cross-pollination. They grow in shaded sites with moist soils, sometimes common in the years after fire. The genus was named in 1753 by C. Linnaeus in honor of J. Clayton.

Narrow-leaved Miner's Lettuce
Claytonia parviflora Hooker

An annual, 1-30 cm tall. *Its basal leaves are linear and taper gradually to a long petiole.* Tepals white to pink. Always self-pollinating. In flower Feb-May.

Lives in moist soils in openings among coastal sage scrub, chaparral, oak woodland, and mixed coniferous woodland communities. Known from Laguna Canyon and the hills of Mission Viejo, though more common in canyons of the Santa Ana Mountains such as upper Silverado, Trabuco, San Juan, and McVicker, Bear Canyon Trail, Puente Hills, Temescal Valley, and the Santa Rosa Plateau.

Named in 1832 by W.J. Hooker for its small flowers (Latin, parvus = little; floris = flower).

Common Miner's Lettuce
Claytonia perfoliata Willd.

An annual, 1-40 cm tall. *Its basal leaves are much broader than long, elliptical to kidney- or spoon-shaped, and taper abruptly to a long petiole.* A variable plant, usually light green but sometimes dark green, pink, red, or even green with whitish streaks. Tepals white to pink. Some forms are self-fertilizing, some require cross-pollination. In flower Feb-May.

Much more widespread and abundant than the preceding, it grows in moist and dryer soils among coastal sage scrub, chaparral, oak woodland, and mixed coniferous woodland communities. Known throughout our area, easily found in all of our wilderness parks and mountain canyons, and coastal bluffs, and the Puente-Chino Hills.

Named in 1798 by C.L. Willdenow for its main stem, which appears to pierce the fused cup-like stem leaves (Latin, perforatus = pierced through; folium = leaf).

NYCTAGINACEAE • FOUR O'CLOCK FAMILY

In our area, these are succulent ground-creeping perennials and small upright shrubs often covered with sticky-tipped hairs. Most perennial. Leaves opposite, of unequal size, neither divided nor lobed. Stems swollen at the nodes, often forked, very brittle, break apart easily. Inflorescence a spike, cluster, or rounded umbel. *Beneath each flower or group are bracts that are fused into a cup or remain free.* Flowers are tubular like a bell or trumpet. *The calyx is showy,* fused below, with 5 free lobes (rarely 3, 4, or 6 lobes), each sometimes notched at the tip. *Petals are absent.* Thus the bracts look like sepals, the sepals look like petals. Stamens 1 or many. Ovary is superior but appears inferior because the base of the calyx tightens and hardens around it as the fertilized flower ages. The single style and ovary produce only one seed per flower. The fruit is winged or has sticky glands that aid in wind or animal dispersal. A small family of 350 species, mostly in the New World; 29 in California, 3 in our area.

Sand-verbenas: *Abronia*

Red-stemmed annuals or perennials that sprawl over sandy soils on the coast and in our deserts. Leaves round or elliptical, often fleshy. *Flowers in rounded umbels* on stalks. *Calyx a colorful trumpet-shaped, long narrow tube with 5 small free lobes.* Stigma linear, style and stamens hidden within the flower. Fruit winged. Their flowers release a pleasant scent at night that attracts moths that pollinate them. During the day, they are pollinated by bumble bees.

The genus was named in 1789 by A.L. de Jussieu for its graceful flowers (Latinized from Greek, abros or habros = graceful).

Red Sand-verbena ★

Abronia maritima Nutt. ex S. Watson

A perennial *densely covered with sticky-tipped hairs,* its stem grows up to 2 meters long and has many branches. *Leaf blades* 5-7 cm long, oval to oblong, longer than wide, *thick and fleshy.* The *small flower tube* is 6-10 mm long, the face 7-10 mm wide. *Flowers are of a single color that ranges from deep crimson to red-purple.* In flower Feb-Oct.

A plant of shifting sands along the immediate coastline, mostly eliminated by development and trampling by beachgoers. In the past, reported from Bolsa Chica south to Capistrano Beach. Currently known only from the northern and southern ends of Crystal Cove State Park and at Trestles. CRPR 4.2.

The epithet was given in 1880 by T. Nuttall for its habitat near the ocean (Latin, maritimus = of the sea).

Beach Sand-verbena ✤

Abronia umbellata Lamarck **ssp.** *umbellata*

A nearly hairless perennial or annual, sometimes with sticky-tipped hairs. *The stems are only 1 meter long,* half that of red sand verbena. *Leaf blades* 1.5-7 cm long, oval to oblong, *thin, not fleshy.* Flower tube 9-13 mm long, the face 7-17 mm wide. *Flowers are two-colored: light pink to rose or magenta with a white center.* In flower most of the year.

Like red sand verbena, this plant has largely been eliminated from our coastline, but some of its populations remain. Also historically known from Bolsa Chica south to Capistrano Beach and Trestles. Some revegetation efforts have been undertaken at Bolsa Chica Ecological Reserve and Doheny State Beach. Still found sparingly on Balboa Peninsula.

This was the first plant collected from the coast of western North America and grown in Europe. Its seeds were gathered in 1786 at Monterey by French explorers with the Compte de La Pérouse Expedition to the Pacific coast in 1785-1788. In 1791, the French biologist J.B.P.A.M. de Lamarck studied them as they grew in France and established the epithet, in reference to the umbrella-like arrangement of the flowers (Latin, umbella, umbellula = sunshade).

Chaparral Sand-verbena ★
Abronia villosa S. Watson **var. *aurita*** (Abrams) Jeps.

An annual covered with sticky-tipped hairs, the stems nearly 1 meter long. *Leaf blade* 1-5 cm long, triangular, oval, or round, *thin, not fleshy.* The *large flower tube* is 20-35 mm long, the face over 15 mm wide. *Flowers are light purplish-rose to bright magenta, white in the center.* In flower Mar-Aug.

Very similar to beach sand verbena but *not found on the beach.* Instead, it lives on sandy benches along ephemeral watercourses among coastal sage scrub and chaparral. Historically known from Santa Ana Canyon nearly to the ocean. Also known from Temescal Valley near Alberhill, some of the lateral canyons south of Glen Ivy, and Temecula Creek. Many of these locations have been drastically altered by flood control activities, sand mining, and development. CRPR 1B.1.

The epithet was given in 1873 by S. Watson for the hairs that cover it (Latin, villosus = hairy). Our form was named in 1905 by L. Abrams for the wide ear-like wings on the fruits (Latin, auritus = eared).

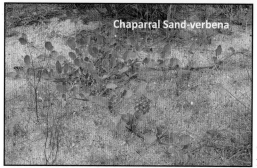
Chaparral Sand-verbena

Chaparral Sand-verbena

California Wishbone Bush
Mirabilis laevis (Benth.) Curran **var. *crassifolia*** (Choisy) Spellenberg
[*M. californica* A. Gray var. *californica*]

A small shrubby perennial, hairless or covered with sticky-tipped hairs. *The branches repeatedly fork like the wishbone of a chicken, the source of the common name.* Leaf blades 1-5 cm long, oval. Flowers are in clusters that have obvious leaf-like bracts at their base. The funnel-shaped calyx is 5-14 mm long and has a short tube with widely spreading lobes. *Its color ranges from purplish-red to pink, rarely white.* In flower Mar-May.

A common plant of sandy soils on coastal bluffs and among grasslands, coastal sage scrub, chaparral, and oak woodlands. When in need of water, it wilts and may drop its leaves. Known from coastline, valley, foothills, and throughout our mountains. Easily observed in our wilderness parks, near Tucker Wildlife Sanctuary, and along the Harding Canyon Trail. In Weir Canyon, it grows in sandstone with California copperleaf (*Acalypha californica,* Euphorbiaceae). A quaint garden plant, good for shady spots with no foot traffic.

The genus was established in 1753 by C. Linnaeus for its beauty (Latin, mirabilis = wonderful). The epithet was given in 1844 by G. Bentham for its smooth hairless leaves and stems (Latin, laevis = smooth). Our form was named in 1849 by J.D. Choisy for its thick leaves (Latin, crassus = thick; folium = leaf).

California Wishbone Bush

California Wishbone Bush

OLEACEAE • OLIVE FAMILY

The family is represented locally by trees and shrubs; elsewhere it also has vines and herbs. Worldwide the *leaves* are variable, but in ours they are *opposite and odd-pinnate. Their flowers are generally small, most often arranged in dense obvious clusters* of one or both sexes per plant. The calyx is tiny, 4-lobed (rarely more), and cup-shaped. *Petals absent or 2* (4 elsewhere). Stamens 2. A single superior ovary and style per flower, the stigma is 2-lobed. Fruit is fleshy and berry-like (like an olive) or dry with a single flat wing (called a *samara*). Familiar members of this family include European olive and valuable landscape plants such as ash, jasmine, privet, and true lilac.

California Ash
Fraxinus dipetala Hook. & Arn.

An upright slender shrub or medium-sized tree up to about 7 meters tall, though it usually only reaches half of that height. Mature trees in moist locations may become bushy and much wider than tall. Between the leaf nodes, its twigs are 4-angled in cross-section (rarely round). The *bright green leaves* are 5-19 cm long with 3-7 leaflets, each 1-7 cm long, oval to round, *usually broadest near the tip,* the edges toothed. *Flowers are of both sexes,* each with a tiny calyx, 2 petals, 2 stamens, and a single style. *Petals* are 6 mm long, *oval and white.* Stamens alternate with the petals and have large yellow anthers. The pale green stigma is scarcely 2-lobed. The body of the samara is flat in cross-section. In flower Mar-May.

When the word "wildflower" comes to mind, you might not think of an ash tree – that is, not until you witness the breathtaking display of this plant in spring! It is pollinated by bees, flies, and butterflies.

A very common plant in the Santa Ana Mountains, generally found on moist north-facing slopes and along canyon bottoms, often among chaparral and mixed coniferous forests. Best observed in Trabuco Canyon not far past the trailhead parking area (where these photos were taken), Harding Canyon Trail, upper Silverado Canyon, and along the paved section of North Main Divide Road a few miles from El Cariso and much of North Main Divide Road accessed from Bedford Road.

The epithet was given in 1840 by W.J. Hooker and G.A.W. Arnott for its 2 petals (Greek, di- = two; petalos = broad, flat, outspread [now taken to mean petal]).

Similar. Arizona ash (*F. velutina* Torr., not pictured): tall tree; twigs round in cross-section; leaves slightly darker green, larger; *flowers of only 1 sex per plant, petals absent; samara round in cross-section;* along watercourses.

California Ash

Its 2 petals alternate with its 2 stamens

Young fruits

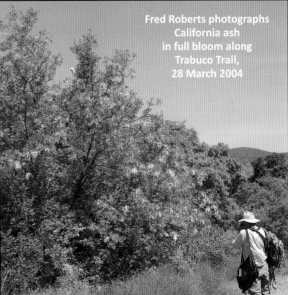
Fred Roberts photographs California ash in full bloom along Trabuco Trail, 28 March 2004

ONAGRACEAE • EVENING-PRIMROSE FAMILY

Ours are small annuals and shrubby perennials; some species elsewhere are shrubs or trees. Leaves often basal with smaller stem leaves; alternate, opposite, or whorled on the stem, toothed but undivided, rarely pinnate. Flower parts mostly in multiples of 2. *Sepals* usually 4, free but sometimes fused, *nearly always folded backward.* Petals 4, *large, showy*, usually yellow, white, or red, and age to a darker color. Stamens 4 or 8, large and obvious. The single style ends in a 4-lobed or clubbed stigma. *Ovary inferior, the flower tube often continues past it.* Fruit is an elongate capsule full of many tiny seeds. The flowers of some species open at dawn; others open at dusk, give off a scent, and remain open at night. Plants in this family are often eaten by larvae of sphinx moths (Sphingidae).

The shape of the stigma and its lobes are important for identification of evening-primroses. The stigma lobes of young flowers are closed like a fist and spread out with age, so study more than one flower to find mature stigmas.

	Camissonia, Camissoniopsis, Eulobus, Tetrapteron, Ludwigia, Gayophytum	*Epilobium* (but not *E. canum*)	*Clarkia* and *Epilobium canum*	*Oenothera*
Petals	Yellow	White to rosy	White, pink, or red	Yellow
Stigma	Spherical to hemispherical club	Oval or conical club	4 short lobes	4 long lobes

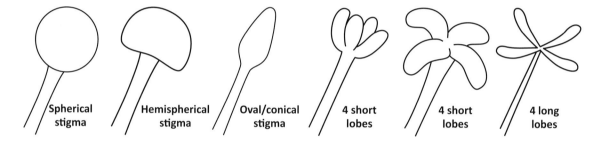

The Sun Cups Clade: *Camissonia, Camissoniopsis, Eulobus, Tetrapteron*

Annuals or perennials from fibrous or stout taproots. Most flowers open at dawn, face the sun, and move to face it all day. The 4 separate petals are often large and yellow, white, or lavender. The stigma is a round club. Fruit is straight or coiled, often ridged on the outside. When dry, it splits from the tip backward to release the seeds. We have 12 types in our area, most quite difficult to find and identify. We present the 7 most common here.

Sandysoil Sun Cup

Camissonia strigulosa (Fischer & C. Meyer) P.H. Raven

A slender upright annual with wiry stems that often arch over. Known to reach nearly 50 cm tall but usually only 15-20 cm tall. *Leaves* 8-35 mm long, *very narrow, the edge with tiny teeth.* The inflorescence nods before the flowers open. Once the flower opens, the sepals remain attached to each other in two groups of two. *Petals* 2.1-4.5 mm long, *bright yellow, age to reddish-orange, inside with 0-2 red dots.* Self-pollinated. In flower Mar-May.

A plant of sandy soils and decomposed granite. Not recently found in Orange County; last reported from Anaheim (1908), Garden Grove (1932), and San Juan Canyon ("5.6 miles NE from the mission in San Juan Capistrano," 1958). Recently found along the Bear Canyon Trail within sight of the Candy Store and near Elsinore Peak. Known from sandy washes in Temescal Valley, the Elsinore Basin, Temecula, the Santa Rosa Plateau, and Los Alamos Canyon in the San Mateo Canyon Wilderness Area. More common at inland sites east of our area. We suspect it may grow along the San Juan Trail, yet to be discovered.

The epithet was given in 1835 by F.E.L. von Fischer and C.A. von Meyer for the short stiff hairs that lie down on the plant's stems and leaves (Latin, strigosus = thin, lean).

Southern or California Sun Cup
Camissoniopsis bistorta (Nutt. ex Torr. & A. Gray) W.L. Wagner & Hoch
 [*Camissonia bistorta* (Nutt. ex Torr. & A. Gray) P.H. Raven]

An annual, rarely a short-lived perennial. Leaves are in a basal rosette. Stems sprawl on the ground, 50-80 cm long, the tips turned upright. The outer "skin" of the reddish stem peels away by itself with age. Leaves 1-12 cm long, elliptic to linear. *Petals are large and rounded,* 7-15 mm long (sometimes much smaller), bright yellow with 1-2 red dots at their base. *The pistil is longer than the anthers.* Fruits are straight or coiled 1-2 times and twisted. Cross-pollinated. In flower (Jan-)Mar-June.

Our most common and widespread sun cup, it grows along the coast and inland. It prefers sandy soils, often in hot, south-facing exposed sites. Known from the Santa Ana River bottom, upper Newport Bay, Corona del Mar, San Joaquin Hills, Dana Point Headlands, Capistrano Beach, Lomas de Santiago, Audubon Starr Ranch Sanctuary, San Juan Creek, major canyons of the Santa Ana Mountains, Santa Rosa Plateau, and Temescal Valley.

Named in 1840 by T. Nuttall for its coiled, twisted fruit (Latin, bis- = twice; tortus = twisted).

Southern is by far our most common sun cup. It can remain low on the ground or grow upright stalks that curve under at the tips.

Beach Evening-primrose
Camissoniopsis cheiranthifolia (Hornem. ex Spreng.) W.L. Wagner & Hoch **ssp. *suffruticosa*** (S. Watson) W.L. Wagner & Hoch
 [*Camissonia cheiranthifolia* (Spreng.) Raim.]

A mat-forming shrubby perennial densely covered with silvery hairs, its main stems grow over 1 meter long and lie on the ground. Leaves 5-50 cm long, oval with wavy edges and tiny teeth along the edges. Petals 6-20 mm long, bright yellow with red dots at the base, aging reddish. Fruits are coiled 1-2 times. Usually cross-pollinated. In flower Apr-Aug.

Found only in sandy beach soil along our immediate shoreline and slightly inland. Largely eliminated by development and trampling. Still found at Bolsa Chica Ecological Reserve, Balboa Peninsula, Corona del Mar, Laguna Niguel, Dana Point Headlands, San Clemente State Beach, and Trestles. A hardy plant and prolific bloomer in the garden, especially on a mound of sandy soil.

Named in 1825 by J.W. Hornemann for the similarity of its leaves to Mediterranean wallflowers (Brassicaceae) formerly in the genus *Cheiranthus,* now *Erysimum* (*Cheiranthus*; Latin, folium = leaf).

Petioled Evening-primrose
Camissoniopsis ignota (Jeps.) W.L. Wagner & Hoch
 [*Camissonia ignota* (Jeps.) P.H. Raven]

An annual from a basal rosette, nearly hairless or with short fine hairs, sticky-tipped in the inflorescence. *Few-branched; stems fleshy, red;* sprawling with tips upward or upright to 55 cm tall. Leaves <6 cm, narrow to oval, the upper on petioles up to 25 mm long. Petals 4-8 mm long, bright yellow, with or without basal red spots. Fruit coiled. Self-pollinated. In flower Mar-Aug.

An uncommon plant of rocky soils in the Santa Ana Mountains, known from Silverado and upper Holy Jim Canyons, Temescal Valley, and the Santa Rosa Plateau.

The epithet was given by W.L. Jepson, who felt it had been ignored (Latin, ignotus = unknown, ignored). It was first found in the Jurupa Hills near Riverside in 1905 but not described until 1925.

Small-flowered Evening-primrose

Camissoniopsis micrantha (Hornem. ex Spreng.) W.L. Wagner & Hoch [*Camissonia micrantha* (H. ex S.) P.H. Raven]

A hairy annual that grows from a basal rosette, its stems are under 60 cm long and often lean over or creep along the ground. *Leaves* are 1-12 cm long, *narrowly elliptic to linear, with wavy edges and tiny teeth.* Inflorescence often has sticky-tipped hairs. Petals small, only 1.5-4.5 mm long, bright yellow, sometimes with 1-2 red dots at the base, aging reddish. Fruits are straight or coiled. Self-pollinated. In flower Mar-May.

Found along the coast inland to our valleys, foothills, and in Temescal Valley. Abundant in the years after fire. Quite common in Laguna Coast and Caspers Wilderness Parks, Limestone Canyon, and Hot Springs Canyon.

Named in 1825 by J.W. Hornemann for its small flowers (Greek, mikros = small; anthos = flower).

Southern sun cup and small-flowered evening-primrose can be quite similar. The flowers of southern vary in size and can be as small as those of small-flowered. Look carefully – the pistil of southern is much longer than the anthers, while the pistil of small-flowered is about the same length (its anthers surround its stigma).

Small-flowered Evening-primrose

California False-mustard

Eulobus californicus Nutt. ex Torr. & A. Gray
[*Camissonia californica* (Nutt. ex T. & G.) P.H. Raven]

An upright stick-like hairless annual nearly 2 meters tall with few leaves, it looks like a slender mustard (Brassicaceae). *Leaves* 30 cm long or shorter, *narrowly elliptical, with sharp pinnate lobes.* Petals are 6-14 mm long, bright yellow or orange, often with red dots at their bases, aging pinkish-orange. *Fruits are straight and hang down.* Self-pollinated. In flower Apr-May.

Widespread but localized, it grows in disturbed soils in fire-burned areas, trails, and along roadways. Known mostly from the Puente-Chino Hills, Audubon Starr Ranch Sanctuary, Baker Canyon, Lomas de Santiago, throughout the Santa Ana Mountains, and the Santa Rosa Plateau. Very common in San Juan Canyon.

Named in 1840 by T. Nuttall for the place of its discovery.

From a distance, California false-mustard looks a lot like a mustard. Look for its large backward-folded sepals, huge club-tipped stigma, and inferior ovary. Mustards have smaller sepals that face forward or out, a slender round stigma, and superior ovary.

California False-mustard

Slender-flowered Evening-primrose

Tetrapteron graciliflorum (Hook. & Arn.) W.L. Wagner & Hoch [*Camissonia graciliflora* (Hook. & Arn.) P.H. Raven]

A stemless annual with long slender leaves, each 1-9.8 cm long. Petals are 5-18 mm long, bright yellow, aging red. The *fruit* is only 4-8 mm long, broad, *4-sided, held upright.* Usually self-pollinated. In flower Mar-May.

An uncommon plant of deep clay soils, it lies hidden among grasslands. Known from Tonner Canyon in the Chino Hills, Elsinore Peak, Oak Flats, and Miller Mountain.

The epithet was given in 1839 by W.J. Hooker and G.A.W. Arnott for its slender flowers (Latin, gracilis = slender; floris = flower).

Slender-flowered Evening-primrose

Farewell-to-springs, Fairyfans, Clarkias: *Clarkia*

Upright slender annuals mostly under 1 meter tall. Leaves linear, elliptic, or oval, sometimes toothed. When the flowers are in bud, their stalks are straight or curved downward, an important thing to notice. When the flower opens, the stalk straightens. *Sepals 4, fused at their tips* and usually remain so even after the flower opens; sometimes fused in pairs, rarely free. *Corolla often a colorful bowl. Stamens 8 or 4.* The ovary (and fruit) of some species has grooves running down its length. The stigma is divided into 4 lobes shaped like a Maltese cross or the letter "X"; the lobes are thick and linear or broad and rounded. The genus was named in 1814 by F.T. Pursh in honor of W.C. Clark. We have 7 species in our area. They are very popular in the horticultural trade, known for their beauty and ease of germination.

Punchbowl Clarkia

Clarkia bottae (Spach) Lewis & Lewis
[*C. deflexa* (Jeps.) Lewis & Lewis]

Usually 20-50 cm tall but may grow to about 1 meter in years with good rain. It has many branches and often many flowers. Leaf blades are 3-10 cm long and linear. *When in bud, the inflorescence stalk itself stands straight up but the buds droop downward; once the flowers open, they all stand straight up.* Sepals green to deep red. Once open, all 4 sepals stay fused together. *Each petal is 1.5-3 cm long, fan-shaped, pale to pinkish lavender, whitish toward the base, with red flecks.* The stigma is generally longer than the anthers. *The ovary has 4 grooves.* In flower Apr-June.

Quite common in mid to late spring among coastal sage scrub, chaparral, and oak woodlands. Widespread, known from Carbon Canyon and borders of Santa Ana Canyon in the Puente-Chino Hills, the San Joaquin Hills, Audubon Starr Ranch Sanctuary, and major canyons in the Santa Ana Mountains such as San Juan and Hot Springs Canyons.

Named in 1835 by É. Spach in honor of P.E. Botta.

Punchbowl's buds droop

Dudley's inflorescence droops

> *Punchbowl and Dudley's clarkia look very much alike. In bud, punchbowl's inflorescence remains upright (though the individual buds droop) while the entire inflorescence of Dudley's droops. The flowers of punchbowl are whitish toward the inner base and flecked with red; those of Dudley's are streaked with white and often have red flecks. When mature, the fruit of punchbowl has 4 low rounded ridges that alternate with 4 shallow grooves; those of Dudley's have 8 prominent ridges and grooves, easily seen (see inset at right). Punchbowl is common on the coastal side of the Santa Ana Mountains while Dudley's occurs on the inland side. Oddly, the situation is reversed in the Whittier-Puente-Chino Hills.*

Punchbowl

Dudley's

Punchbowl Clarkia

Dudley's Clarkia

Clarkia dudleyana (Abrams) J.F. Macbride

An upright plant to 70 cm tall. Leaf blade is 1.5-7 cm long, linear to lance-shaped. *In bud, the top of the inflorescence stalk is curved downward and the buds hang down; when the flowers open, both straighten up.* The 4 pink or purplish-red sepals remain fused together. *Petals 1-3 cm long, fan-shaped, lavender-pink and streaked with white, often dotted with red flecks.* The stigma is longer than the anthers and has 4 narrow lobes, longer than those of most clarkias but not nearly as long as those of *Oenothera. The ovary has 8 grooves.* In flower May-July.

Quite uncommon here, though perhaps thought to be punchbowl clarkia and not reported. Known from upper San Juan Canyon, Fisherman's Camp in the San Mateo Canyon Wilderness Area, the inland base of the Santa Ana Mountains such as the foot of East Horsethief Trail and Tin Mine Canyon, upper Gypsum Canyon, and mostly the coastal side of the Whittier-Puente-Chino Hills.

Named in 1904 by L. Abrams in honor of W.R. Dudley.

Dudley's Clarkia

Willow-herb Clarkia

Clarkia epilobioides (Nutt.) Nelson & J.F. Macbr.

A slender plant with few branches, it stands upright 20-70 cm tall. The leaf blade is 15-25 mm long, narrowly linear. In bud, the top of the inflorescence stalk is curved downward and the buds hang down; when the flowers open, both straighten up. The *4 red sepals* remain fused together or in 2 groups of 2. Petals are small, 0.5-1.0 cm long, oval, *white to pale cream*, and fade to pink with age. Unlike similar species, it has no purple or red flecks. The ovary has no obvious grooves. In flower Mar-May.

Fairly widespread, sometimes common but only in localized spots. Usually grows in shade among chaparral, oak woodland, riparian, and coniferous forest. Known from hillsides along the Santa Ana River channel, Sycamore Hills and Laguna Canyon in the San Joaquin Hills, Audubon Starr Ranch Sanctuary, and major canyons in the Santa Ana Mountains such as Santiago, Trabuco, San Juan, and Lucas. Also in Temescal Valley and the Santa Rosa Plateau.

Named in 1840 by T. Nuttall for its resemblance to willow-herbs in the genus *Epilobium* (Greek, -oides = resemblance).

Willow-herb Clarkia

Winecup Clarkia

Clarkia purpurea (Curtis) Nelson & J.F. Macbr.

A highly variable plant under 1 meter tall, upright or sprawling, few- or many-branched. The leaf blade is 1.5-7 cm long, linear, elliptic, or oval. *The inflorescence and its buds stand straight up* even before the flowers open. *Sepals remain fused* in 2 groups of 2 or all free. Petals 5-25 mm long and fan-shaped. Their color varies widely even within a single patch of the plants, from pale pink to purple to deep wine red, sometimes with a red or purple spot. The ovary has 8 grooves. In flower Apr-July.

Our most common clarkia, found most often among grasslands and openings among coastal sage scrub, chaparral, and oak woodlands, especially in the years following fire. Known throughout our area.

Named in 1796 by W. Curtis for its purplish flowers (Latin, purpureus = purple).

Winecup Clarkia

Winecup is our most variable clarkia. Its petals range from pale pink to purple to deep wine red. Those with a darker spot near the petal tip are sometimes called four-spot clarkia.

Winecup Clarkia

Canyon Clarkia
Clarkia similis H. Lewis & Ernst

Slender and upright, 30-90 cm tall. Leaf blade 2-4 cm long, linear to elliptic. As with willow-herb and tongue clarkia, *its inflorescence and buds hang down before the flowers open.* Once open, all four of its *red or greenish sepals stay fused at their tips. Petals small,* 0.6-1 cm long, diamond or reverse-oval-shaped, *pale pink, whitish near the base, and dotted with purple flecks;* they fade to pink with age. The ovary has 8 shallow grooves. In flower Apr-May.

Known from rocky slopes in San Juan Canyon, the north face of Sitton Peak, upper McVicker Canyon, Elsinore Peak, Temescal Valley, the Santa Rosa Plateau, and sparingly in the San Mateo Canyon Wilderness Area. Perhaps more widespread but easily mistaken for willow-herb clarkia (which never has purple or red flecks).

The epithet was given in 1953 by F.H. Lewis and W.R. Ernst for its similarity to willow-herb clarkia (*C. epilobioides*) (Latin, similis = like). They postulated that canyon clarkia is a stable hybrid between willow-herb clarkia and Waltham Creek clarkia (*C. modesta* Jeps., a species not found in our area).

Canyon Clarkia

Tongue Clarkia

Tongue or Diamond Clarkia
Clarkia rhomboidea Douglas

Upright to nearly 1 meter with few branches. Leaf blade 1-6 cm long linear, elliptic, or oval. Like willow-herb clarkia, *its inflorescence stalk and buds both hang down before the flowers open.* Sepals are all free. *Petals* 7-14 mm long, pinkish lavender to magenta with reddish flecks. *The lower half of each petal is quite narrow and forms a stalk (claw).* The base of the stalk has two lateral lobes (*wings*), the distal half of the petal is diamond- or spoon-shaped. Ovary 4-grooved. In flower May-July.

A lovely plant with unusual petals. In dry years, most seeds do not germinate and the plants are difficult to find. Even in a wet year, the plant is never abundant. Found on the upper slopes of Santiago and Modjeska Peaks and along trails in canyons not far below, and in Hagador Canyon.

Named in 1834 by D. Douglas for the shape of its petals (Greek, rhombos = diamond-shaped; -oides = resemblance).

Elegant Clarkia ✖
Clarkia unguiculata Lindl.

Upright to nearly 1 meter. Leaf blade 1-6 cm long, lance-shaped, elliptical, or oval. Prior to the flowers' opening, *the inflorescence stalk stands straight up but the buds hang down.* All 4 of its green to red sepals stay fused together. *Petals* 1-2.5 cm long, *lavender-pink, salmon, or dark reddish-purple. The lower half of each petal is quite narrow and forms a stalk (claw). There are no wings on the claw.* The distal half is heart- or triangle-shaped. Ovary 8 grooved. In flower May-June.

A California native but not from our area, it was introduced in seed mixes. Normally found in dry semi-shaded places, often on hillsides. Here, it generally persists only with irrigation. Cultivated varieties include those with frilly-edged and/or multiple petals. To date, found only in Santiago Oaks Regional Park and Laguna Coast Wilderness Park.

Named in 1840 by J. Lindley in reference to its long petal claw (Latin, unguiculus = nail or claw).

Elegant Clarkia

Willow-herbs and California Fuchsias: *Epilobium*

Upright annuals or shrubby perennials. Lower leaves opposite or in bunches along sides of the stem. Sepals 4, held straight or outward, not folded backward. *Petals 4, often notched inward at the tip.*

Summer Cottonweed
Epilobium brachycarpum C. Presl

Summer Cottonweed

An upright annual, 20 cm to 2 meters tall. Leaves narrow, linear, often curve backward, sides folded up. *Petals* 2-15(20) mm long, *white to pinkish, often streaked darker.* Stamens shorter than or equal to pistil length. Stigma clubbed. Fruits 15-32 mm long, very short for a tall plant in this genus. Seeds with a long hair tuft. Probably self-pollinated. In flower June-Oct.

Widespread but uncommon, it occurs in localized patches in hard-packed ground along channels, levees, and roadsides. Recorded from Aliso Creek, Laguna Lakes, Horseshoe Bend in the Santa Ana River Channel, Santiago Canyon inland of Lomas de Santiago, the Elsinore Basin, and the Santa Rosa Plateau. Osccasionally found along freeway offramps, such as the Costa Mesa Freeway (SR55) in Costa Mesa and the San Diego Freeway (I5) in Lake Forest.

The epithet was given in 1831 by C. Presl for its relatively short fruits (Greek, brachys = short; karpos = fruit).

Smooth Willow-herb
Epilobium campestre (Jeps.) Hoch & W.L. Wagner
[*E. pygmaeum* (Speg.) Hoch & P.H. Raven, illeg.]

A sprawling annual with stems 10-55 cm long. Leaves lance-shaped or triangular, distal leaves hairy. Inflorescence crowded among broad leaf-like bracts. *Petals* 1-3.2 mm long, *pink.* Stigma clubbed or with 4 short lobes. Fruit 3.5-8 mm long, seeds without a hair tuft. Self-pollinated, but most often they do not open and instead undergo in-bud fertilization (cleistogamy). In flower May-Sep.

Smooth Willow-herb

A uncommon plant of vernal pools and ponds in clay soils. Known from vernal pools on the Rancho Mission Viejo and in Fairview Park.

The epithet was given in 1901 by W.L. Jepson for its habitat (Latin, campestris = growing in a field).

Similar. Dense-flowered willow-herb (*E. densiflorum* (Lindl.) Hoch & P.H. Raven, not pictured): tall to 1.5 meters; leaves similar; inflorescence crowded among broad leaf-like bracts; petals 3-10 mm long, pinkish to white; stigma clubbed or with 4 short irregular lobes; fruit 4-11 mm long, seeds without a hair tuft; in flower May-Oct. Uncommon in wet soils and meadows. Known here only from Los Pinos Potrero near Blue Jay Campground, Caspers Wilderness Park, and the Santa Rosa Plateau.

Green Willow-herb
Epilobium ciliatum Raf. **ssp. *ciliatum***
[*E. adenocaulon* Haussknecht]

Green Willow-herb

An upright perennial 1-2 m tall, *from fleshy basal rosettes.* Leaves 1-5 cm long, linear to oval, *with lines of hairs.* Petals 2-14 mm long, white to rose-tinted. *Fruits very narrow,* 1.5-10 cm long, reddish. When ripe, they twist open to expose their seeds, each topped with a hair tuft. In flower (Mar-)May-Oct.

Quite abundant, especially along waterways, on streambanks, and in meadows. Occasionally appears in gardens, mostly in overwatered spots. Known throughout our area at all but the highest elevations.

Named in 1808 by C.S. Rafinesque-Schmaltz for the hair on its seeds (Latin, ciliatus = furnished with hair or a hairlike process).

Narrow-leaved California Fuchsia

Epilobium canum (E. Greene) P.H. Raven **ssp. *canum***
[*Zauschneria californica* ssp. *californica* E. Greene]

A hairy, shrubby perennial under 1 meter tall, usually not sticky. *Leaves linear, grayish, and arranged in bunches along the stems.* Flowers long, tubular, 2-3 cm long, red-orange to scarlet. Mature stigma 4-lobed, unusual for the genus. In flower Aug-Oct.

Widespread and common, especially near watercourses but not in wet soils. Known from the San Joaquin Hills, the Puente-Chino Hills, all elevations of the Santa Ana Mountains, and the Santa Rosa Plateau. Easily seen in upper Silverado Canyon and near Blue Jay Campground. Hummingbirds frequently visit this plant to take nectar and unwittingly transfer pollen with their bills. Males will often set up territories and chase away other birds that approach. Bumble bees and other bees take both nectar and pollen from it. Territorial hummingbirds also chase bees away from the plants. Many forms are widely grown as garden plants in this state and elsewhere. Its red flowers are quite distinct from the yellow flowers characteristic of most other fall-blooming wildflowers.

It was named in 1887 by E.L. Greene for the hairs that cover it (Latin, canus = white or grey). Many cultivars are used in horticulture.

Narrow-leaved California Fuchsia

Broad-leaved California Fuchsia

Epilobium canum (E. Greene) P.H. Raven **ssp. *latifolium*** (Hooker) Raven
[*Zauschneria californica* ssp. *latifolia* (Hook.) D.D. Keck]

Stands 10-50 cm tall, a bit shorter than narrow-leaved, usually covered with sticky-tipped hairs. *Leaves are broadly linear to oval, green in color, opposite on the stems.* Not as commonly found as narrow-leaved, mostly known from *upper elevations* of the Santa Ana Mountains, on Santiago and Modjeska Peaks.

Named in 1850 by W.J. Hooker for its broad leaves (Latin, latus = broad, wide; folium = leaf).

Broad-leaved California Fuchsia on Santiago Peak

Yellow Water-primrose

Ludwigia peploides (H.B.K.) Raven **ssp. *peploides***

A hairless creeping perennial. *Leaves alternate,* grouped, blade oblong to round, *tips rounded,* veins light and obvious. Sepals (4)5(6); *petals 5(6),* bright yellow. Stamens (8)10(12) in 2 unequal groups. In flower May-Oct.

A plant of wet habitats, it grows in slow-moving creeks and streams, rooted in mud with its upper parts resting on the water surface. When rooted out of water, it creeps along the ground. Known from the Santa Ana River, major creeks to the coast (San Diego and San Juan), and Temescal Valley.

Named in 1823 by F.W.H.A. von Humboldt, A.J.A. Bonpland, and C.S. Kunth, for its similarity to a European plant (*Peplis portula* L., Lythraceae; Greek, -oides = resemblance).

Similar. Red water-primrose ✘ (*L. repens* Forster, not shown): *leaves opposite,* blade narrow-elliptic to round; *sepals 4; petals 4,* yellow; *stamens 4.* An aquarium plant, released into the wild and now invasive. Uncommon in San Diego and San Juan Creeks. **Uruguayan water-primrose** ✘ (*L. hexapetala* (Hook. & Arn.) Zardini et al., not shown): creeping (to 4 meters long) or upright (to 2 meters tall) perennial; leaves alternate, blade elliptic, tip narrowly pointed, veins light; sepals 5(6); petals 5(6); stamens 10(12) in 2 unequal groups. Known here only from upper Newport Bay just below the Muth Environmental Center.

Yellow Water-primrose

Small-flowered Groundsmoke
Gayophytum diffusum Torr. & A. Gray **ssp.** ***parviflorum***
H. Lewis & J. Szweyk.

A many-branched, slender-stemmed, upright annual, generally 10-60 cm tall. *Leaves very narrow,* 1-6 cm long, uppermost often shorter. Petals tiny, only 1.2-3 mm long, pinkish when first opened, then white. Ovary 2-chambered. Style equal to or shorter than the anthers. Stigma rounded or club-shaped. Self-pollinated. In flower June-Aug.

It looks like a wisp of smoke, hence the common name. Abundant along the edges of dry trails and roadways in the upper reaches of the southeast side of Santiago Peak.

The genus was named in 1832 by A.L. de Jussieu in honor of C. Gay (Gay; Latin, phyton = plant). The epithet was given in 1840 by J. Torrey and A. Gray for its diffuse (not crowded) growth (Latin, diffusum = spread out, diffuse). This form was named in 1964 by F.H. Lewis and J. Szweykowski for its small flowers (Latin, parvus = little, small; floris = flower).

Similar. **Small-flowered dwarf flax** (*Hesperolinon micranthum*, Linaceae): *sepals 4-5, held forward;* petals 4-5, white; stamens 4-5; *ovary superior; styles (2)3, linear.*

Large Evening-primroses: *Oenothera*
Annuals, biennials, or perennials from a strong woody taproot. Leaves alternate or basal, toothed or lobed. *Sepals fold back when the flower opens.* Petals very large, yellow, white, or purplish, often age darker. *Stigma divided into 4 long slender linear lobes shaped like a Maltese cross or the letter "X".* Fruit is a straight or curved cylinder. Their flowers open just prior to or just after sunset. Many are self-pollinated; some are pollinated at night by sphinx or owlet moths.

California Evening-primrose
Oenothera californica (S. Watson) S. Watson
ssp. ***californica***

A low sprawling perennial with stems 10-80 cm long. When young, it grows from a basal rosette. Stem leaves 1-6 cm long, oblong to lance-shaped, shallowly to deeply lobed. *The flower buds hang down prior to opening.* Petals 15-35 long, white, fade to pink as they age. In flower Apr-June.

Found in sandy soil among open areas and ephemeral waterways, sometimes in coastal sage scrub, chaparral, and oak woodlands. Once common along the Santa Ana River channel from the San Bernardino Mountain foothills downstream to at least central Orange County. It may occur in Temescal Valley and Elsinore-Temecula valleys.

The epithet was given in 1873 by S. Watson to honor the state of its discovery.

Hooker's Evening-primrose
Oenothera elata Kunth **ssp.** ***hirsutissima***
(A. Gray ex S. Watson) W. Dietrich

A large, upright, hairy plant, with a stout main stem 0.5-2.5 m tall. Basal leaves large and broad, stem leaves 1-6 cm long, linear to triangular, smooth-edged or pinnate. *The buds hang down prior to opening.* Sepals fold back and remain attached to each other or break free. Petals 25-52 mm long, yellow, and fade to orange with age. In flower during summer months, generally June-Sep, longer with constant water.

Commonly found in moist soils along watercourses and ephemeral streams from the coast up into the foothills and mountains. Formerly common in Huntington Beach and Fountain Valley prior to creek channelizing. Look for it in upper Santa Ana Canyon, San Juan Creek, San Juan Canyon, and the Los Pinos Potrero, especially at seeps.

Named in 1823 by C.S. Kunth for its height (Latin, elatus = elevated). In 1873 A. Gray named our variety for its dense covering of hairs (Latin, hirsutus = hairy; -issimus = very much, most).

OROBANCHACEAE • BROOMRAPE FAMILY

All are parasites that use specialized structures (haustoria) on their roots to tap into the roots of adjacent plants to steal nutrients and water. Aboveground, some species are green in color and capable of making their own nutrients through photosynthesis. They are thus half or partial parasites (**hemiparasites**). The true broomrapes in the genus *Orobanche* are yellow or purple in color, incapable of photosynthesis, and are completely parasitic (**holoparasites**). Leaves reduced to small scales in the complete parasites, larger and often pinnate in the partial parasites. They all have 5 sepals and 5 petals, at least partially fused into a tube. The 2-lipped corolla has 5 free lobes with 2 facing up, 3 down. The 4 stamens are in 2 groups; a fifth stamen sometimes appears as a non-reproductive filament (staminode). In the paintbrushes (*Castilleja* spp.), colorful bracts surround the flowers to attract pollinators such as bees, wasps, flies, and birds. Seeds are small and dispersed by wind and water.

The genera *Castilleja, Chloropyron, Cordylanthus,* and *Pedicularis* used to be placed within the figwort family (Scrophulariaceae). Detailed studies of their parasitic lifestyle, floral characters, chemistry, and DNA revealed that they are closely related to the true broomrapes. As a result, these genera (and others not from our area) were reassigned to the broomrape family.

Paintbrushes: *Castilleja*

Perennials or slender annuals that punctuate their habitat with *bold, bright colors.* Leaves alternate, no petiole, undivided or lobed. *Flowers tubular,* slender, clustered near the stem tips, and *hidden among colorful bracts* (which are modified leaves). Flowers and bracts are yellow, red, orange, or pinkish-purple and white. Calyx unequally 4-lobed and usually of the same color as the bract tips. *The upper lip of the corolla looks like a bird's beak with the tip open; the lower lip is smaller and has 3 teeth or pouches.* The long pistil usually sticks out of the corolla, tipped with a clubbed stigma. Field identification is based on color, the shape of the bract that lies beneath each flower, calyx and corolla shape, and leaf texture. They are all partial parasites (holoparasites), able to manufacture their own food, but can supplement their diet by tapping into roots of adjacent plants. They are pollinated by hummingbirds and bumble bees.

The genus was established in 1781 by the botanical explorer J.C. Mutis in honor of D. Castillejo. The common name paintbrush refers to the appearance that the top of the plant had been dipped into a can of paint. Five species live in our area; all but one are common.

	Coastal C. *affinis* ssp. *affinis*	**Martin's** C. *applegatei* ssp. *martinii*	**Felt or Woolly** C. *foliolosa*	**Cal. Threadtorch** C. *minor* ssp. *spiralis*	**Purple Owl's-clover** C. *exserta*
Plant	Perennial; hairs bristly but not sticky	Perennial; short sticky-tipped hairs	Perennial to small shrub; short white to grey felt-like hairs	Annual; short sticky-tipped hairs	Annual; short hairs, often sticky-tipped
Leaves	Green, linear, lowest usually not divided, upper often 3-5-lobed; edges straight, not wavy	Green to gray-green, usually not divided (0-3-lobed); edges wavy	Grayish, linear, 0-3-lobed	Green, linear, not divided	Green, pinnately divided into 5-9 narrow thread-like lobes
Bracts	17-25 mm long, 3-5 lobes (sometimes more, rarely undivided), tips bright red to orange-red, rarely yellow	15-25 mm long, often 3-lobed (sometimes 0-7), tips muted red, tinged with orange or all yellow	15-25 mm long, 0-3-lobed, orange-red, rarely yellowish	20-50 mm long, leaf-like, undivided, with red or orangish tips	10-25 mm long, divided like the leaves, tipped with white, pale yellow, or purplish red
Calyx lobes	Long and narrow	Short and broad	Short and broad	Medium-long and narrow	Long and narrow
Corolla	25-40 mm long, beak 1 to 1.5 times the length of the tube; each of the 2 lobes of the lower lip has a tooth. Greenish, yellow, to red. Generally hidden among the long bracts	25-40 mm long, beak equal to or longer than the tube; no teeth on lower lip. Greenish, yellow, to red. Generally extends far beyond the bracts	18-25 mm long, beak about as long as the tube; no teeth on lower lip. Orange-red to yellowish. Generally somewhat hidden.	25-35 mm long, beak shorter than the tube; no teeth on lower lip. Greenish, yellow, to orange. Generally long and visible	12-30 mm long, beak much shorter than the tube, inflated; no teeth on lower lip. Generally red-purple (may be white, yellow, pale pink to deep red-purple). Inflated portion peeks above the bracts.

Coastal Paintbrush
Castilleja affinis Hook. & Arn. ssp. *affinis*

A perennial, sometimes from a woody base, 15-60 cm tall. Branches bristly, not sticky. Leaves linear, 0-5 lobes, each lobe broad or rounded at the tip. *Bract tips bright red* to orange-red, rarely yellow. *Each of the 2 lobes of the lower lip has a tooth.* In flower Mar-May.

A root parasite of many plants, it is very abundant in our region. Look for it in dry, wooded areas in coastal sage scrub, chaparral, and along roadsides, from the coast to the mountains, including the foothills on coastal and inland sides (more common on the coastal side), even on the very top of Santiago Peak. In Crystal Cove State Park, both yellow- and red-bracted forms grow right along the trails.

It was named in 1833 by W.J. Hooker and G.A.W. Arnott, perhaps because it is closely related to the other paintbrushes or because members of this species often grow quite near one another (Latin, affinis = related, adjacent).

Coastal Paintbrush

Bracts 0-5 lobed, often 1- to 2-pinnate

Leaves divided, mostly flat-edged

Martin's Paintbrush
Castilleja applegatei Fernald ssp. *martinii* (Abrams) Chuang & Heckard

Similar to coastal paintbrush but much less common. *Stems and leaves covered with short very sticky-tipped hairs. Leaves simple, usually not lobed (rarely up to 3 lobes), and wavy-edged. Bract tips muted red,* tinged with orange or all yellow. Calyx much shorter (15-25 mm) but corolla same size and color. *The 2 lobes of the lower lip have no teeth.* In flower Apr-July(-Oct).

Known from riparian woodlands in Trabuco Canyon and Bear Spring, dry chaparral-covered hillsides in Upper Hot Springs Canyon, along the Chiquito Basin Trail in Lion Canyon, and north-facing slopes along the Sitton Peak and Indian Truck Trails. Also appears in shady, moist road cuts along Main Divide Road in the northern portion of the Santa Ana Mountains. In the San Gabriel, San Bernardino, and San Jacinto Mountains, it also grows up into the pine forest, but it has not been found among pines in the Santa Ana Mountains. It parasitizes nearby shrubs.

First found in 1901 at Martin's Camp near Mt. Wilson in the San Gabriel Mountains, Los Angeles County, by L. Abrams and named by him the following year for that location.

Martin's Paintbrush

Bracts 0-3 lobed, wavy-edged

Leaves undivided, wavy-edged

Felt or Woolly Paintbrush
Castilleja foliolosa Hook. & Arn.

A perennial shrub, 30-60 cm tall. *Stems and leaves covered with short white or gray felt-like hairs.* Leaves linear, with 0-3 lobes. Bracts orange-red, rarely yellowish. In flower Jan-June.

Quite common and easily found here, it parasitizes many types of small associated shrubs such as **golden yarrow** (*Eriophyllum confertiflorum var. confertiflorum,* Asteraceae). It lives in dry rocky soils in coastal sage scrub and chaparral, from the immediate coast throughout the Santa Ana Mountains, the Chino Hills, and the inland foothills. In Crystal Cove State Park, the uncommon yellow-flowered form grows among the typical red-flowered form. It appears in nearly all of our wilderness parks, the Santa Ana Mountains, even atop Santiago Peak.

Named in 1833 by W.J. Hooker and G.A.W. Arnott in reference to the unusual felt-like leaves (Latin, foliolosus = full of leaves, leafy).

Yellow form of Felt Paintbrush

Felt Paintbrush

California Threadtorch
Castilleja minor (A. Gray) A. Gray **ssp. *spiralis*** (Jeps.) Chuang & Heckard

A slender, streamside annual with few branches, it stands upright, 30-150 cm tall. Foliage green to grayish, covered with short sticky-tipped hairs. Leaf linear, undivided. *Floral bracts leaf-like, undivided, with red or orangish tips that make the plant look like a thin orange torch.* Lower corollas and floral bracts are generally yellowish or greenish. Upper bracts and corollas are usually red or orangish. In flower May-Sep.

Find it in wet canyon watercourses and seeps in the Santa Ana Mountains, such as San Juan Canyon, Bear Springs, Holy Jim Canyon, and Silverado Canyon, and the Santa Rosa Plateau. After flowering, it dries up 1-2 months after ephemeral streams disappear. It parasitizes an assortment of nearby wetland plants.

Named in 1901 by W.L. Jepson in reference to its twisted floral bracts (Latin, spira = coil, twist).

California Threadtorch

Purple Owl's-clover
Castilleja exserta (A. Heller) Chuang & Heckard
[*Orthocarpus purpurascens* Benth.]

Upright annual, 10-45 cm tall, covered with short hairs, often sticky-tipped. Leaves 10-50 mm long, pinnately divided into 5-9 narrow thread-like lobes. Uppermost leaves may be tipped in pink or red. Floral bracts 10-25 mm long, divided like the leaves, and tipped with white, pale yellow, or purplish red. Corolla red-purple, two-lipped. Upper lip long, slender, covered with shaggy hairs (bearded), hooked at the tip. *Lower lip hairy with 3 inflated pouches that are white and/or yellowish on top, often with purplish dots.* Colors of the bracts, calyx, corolla and its markings can vary from pale pink to deep red-purple. In flower Mar-May.

The common name "owl's-clover" comes from the appearance of the corolla's inflated lower lip, which resembles a tiny owl that takes some imagination to see. When looking for the owl, ignore the upper lip (a feathery purple plume). Between the three inflated pouches are two small purple spots (eyes), sometimes with yellow. Some flowers have a dark purple band with a V shape on top, the white portion that enters the V represents the owl's beak. It isn't a clover, but it does grow in grasslands, as do many species of clover.

Stamens are hidden within the upper corolla lip. The stigma remains outside and looks like a fuzzy ball that sits just under the upper lip. A pollinating insect alights on the plant and uses its mouthparts and head to separate the corolla lips. Pollen is dusted onto its head as it feeds on the flower's nectar. When it visits the next flower and begins to open it, the pollen on its head contacts the stigma and is transferred to it. As it pries its head farther into the flower, it receives pollen that it transfers to the next flower as it feeds. The insect continues to feed and cross-pollinate in this manner.

Purple owl's clover often carpets our grasslands in spring. It also appears in grassy openings within coastal sage scrub and chaparral. You'll find it in nearly all of our wilderness parks, but the best display is usually in Limestone Canyon and along the Santiago Trail (the trailhead is at Modjeska Grade Road). If you'd prefer to drive up to it, the meadow at Los Pinos Potrero near Falcon Group Camp is beautiful and in flower later than in the lowlands.

Named in 1904 by A.A. Heller for its corolla, which sticks out above the bract (Latin, exsertus = protruding).

Purple Owl's-clover

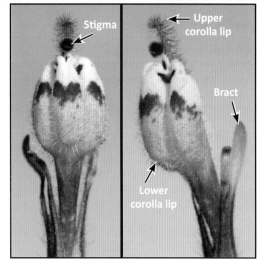
Stigma · Upper corolla lip · Bract · Lower corolla lip

Bird's-beaks: *Chloropyron, Cordylanthus*

Partial root parasites with yellow-colored roots. Leaf blade undivided or divided into several thread-like lobes. Each flower surrounded by two bracts and a calyx that looks like a bract. Corolla club-shaped, 2-lipped, lips about the same length. The tip of each lip curves toward the other, like the beak of a bird. Summer flowering. We have 2 local species, one rare, the other common. Both are annuals, pollinated by bumble bees and hummingbirds.

Salt Marsh Bird's-beak ★

Chloropyron maritimum (Benth.) A. Heller **ssp. maritimum** [*Cordylanthus maritimus* Benth. ssp. *maritimus*]

Sprawls on the ground or is loosely upright, 10-40 cm tall. Entirely gray-green, tinged purple or red, covered with short hairs, and often encrusted with salt. *Leaves linear, not divided, 5-25 mm long.* Flowers white to cream with lips that are pale to brownish or purplish red. In flower May-Oct.

Once found in *coastal salt marshes* as far north as San Francisco, today known only from Carpenteria south into Mexico. In our area found only at Seal Beach National Wildlife Refuge and upper Newport Bay. It is listed as Endangered by both state and federal agencies and is CRPR 1B.2. Habitat degradation and loss are given as causes for its decline. *You can help protect this plant by keeping out of its habitat. Do not walk off-trail in these areas! It is easily viewed from a distance with binoculars and/or camera with a long lens.*

Pollinated mostly by bumble bees, which are large enough for you to watch from a distance as they visit its flowers. It parasitizes nearby salt marsh and estuary plants, probably pickleweeds (*Salicornia* spp. and *Arthrocnemum* spp., Chenopodiaceae). The nature trail on Shellmaker Island and along Backbay Drive are good places to view it.

It was named in 1846 by G. Bentham for its habitat near the ocean (Latin, maritimus = of the sea).

Dark-tipped Bird's-beak

Cordylanthus rigidus (Benth.) Jeps. **ssp. *setigerus*** Chuang & Heckard

A slender, multiple-branched, upright annual, 30-150 cm tall. Foliage yellow-green, often tinged red, and covered with short stiff hairs. Leaves 10-40 mm long, not divided, often rolled up on the sides. Floral bracts and calyx are green, tipped with purple, and bristly. The outer bract is 3-lobed; the inner bract and the calyx are not divided. The corolla is 13-17 mm long, nearly hidden by the calyx and inner bract, and yellowish. The lower inside of the corolla is marked with a purplish letter U. In flower June-Aug.

Somewhat wispy and often overlooked even though it grows along many hiking trails in our area. Commonly found in openings in coastal sage scrub, oak woodlands, and chaparral where soil has been disturbed. Look for it in San Juan Canyon, Blue Jay Campground, the Santa Rosa Plateau, and most of our local wilderness parks (even in the Wilderness Glen Greenbelt in Mission Viejo). Easily grown from seed, it makes an unusual summer-blooming garden annual.

It is pollinated by bumble bees. Smaller bees are perhaps not strong enough to open its narrow-mouthed flowers. Hummingbirds are occasionally seen nectaring from it and may perform some pollination services as well. A generalist, it parasitizes the roots of almost any plant near it.

The epithet was given in 1836 by G. Bentham for its stiff stems (Latin, rigidus = rigid, stiff). Our local form was named in 1986 by T.-I.Chuang and L.R. Heckard for the bristles that cover it (Latin, seta = a bristle; -ger = bearing).

Warrior's Plume
Pedicularis densiflora Benth. ex. Hooker

Perennial herb, 6-55 cm tall, covered throughout with soft or coarse brown hairs. Leaves mostly basal, each long, pinnate, ruffle-edged, bronze at first then turning green with age. Inflorescence upright, 4-12 cm tall with dense clusters of flowers at its apex. *Corolla scarlet red, tubular,* flattened side-to-side with a conspicuous, straight upper lip. The lower lip is about half its length, with 3 white-tipped, rounded lobes. In flower Mar-May.

Uncommon in our area, it lives in dry chaparral and up into oak and pine forests, where it commonly parasitizes the roots of **chamise** (*Adenostoma fasciculatum,* Rosaceae) and **Eastwood manzanita** (*Arctostaphylos glandulosa* ssp. *glandulosa,* Ericaceae). Its seeds germinate the first few years following a fire. The plants thrive but then seem to die off after the dominant vegetation regrows and shades them out. Look for it in the Santa Ana Mountains such as above Coal Canyon, near Pleasants Peak, Main Divide Truck Trail near Black Star Canyon Road, upper Silverado Canyon, near Hot Springs Canyon, San Juan Canyon, Bear Canyon Trail, and into the San Mateo Canyon Wilderness Area. It is pollinated mostly by bumble bees and hummingbirds.

It was named in 1838 by G. Bentham for its dense flower clusters (Latin, densus = dense, compact; floris = flower).

True Broomrapes: *Orobanche*

These unusual plants are complete parasites that evolved the ability to steal nutrients from the roots of other plants and later lost the ability to photosynthesize; therefore they are non-green and cannot manufacture their own food. They are fleshy plants that have alternate scales instead of leaves. The corolla is tubular and 2-lipped, the upper lip has 2 lobes, the lower has 3. After flowering and setting seed, the plants shrivel and die, though some are geophytes that reappear the following spring. They are never abundant, and their small size makes them difficult to find. In dry years, they remain growing underground and do not appear above the soil surface.

The genus name *Orobanche* is both Latin and Greek for "vetch strangler" in reference to the plants' habit of parasitizing other plants (such as vetches, *Vicia* spp., Fabaceae). The common name broomrape probably refers to broom (basically, any plant that can be used as a broom) and Latin, rapax = seizing, grasping, greedy, again in reference to the plant's parasitism. We have 5 species in our area.

Chaparral Broomrape
Orobanche bulbosa G. Beck

Upright, 8-30 cm tall, from a large, thick, underground bulb-like stalk that is attached to the roots of its host plant. The entire plant is purple (rarely yellowish or brownish), including its sap. It is covered with whitish bumps and is not sticky. The inflorescence is very tight and pyramid- or mushroom-cap-shaped. *Flowers small, deep purple,* 10-18 mm long, on very short flower stalks. In flower Apr-July.

Most often a parasite of **chamise** (*Adenostoma fasciculatum,* Rosaceae). Chamise has wide-ranging roots, and the broomrape plant is sometimes found growing meters away from the host's main trunk. Found among chamise chaparral in rocky soils and decomposing shale and granite at higher elevations of the Santa Ana Mountains. Look for it along Long Canyon Road and Main Divide Truck Trail from Sierra Peak to Los Pinos Saddle.

It was named in 1890 by G. Beck-Mannagetta for its bulb-like underground stalk (Latin, bulbosus = full or bulbs, having bulbs).

Chaparral broomrape has a narrow tubular flower. You might overlook its beauty because of its small size, so be sure to examine it with a hand lens.

Clustered Broomrape
Orobanche fasciculata Nutt.

Upright, 5-20 cm tall. Flowers yellow or purple-tinged, 15-30 mm long, borne on long pedicels to form a loose cluster of flowers. It does not have a single stout central stalk like our other broom-rapes. In flower Apr-July.

Our most common broomrape, found in sandy soil on talus slopes near Modjeska Peak, in rocky soil on Santiago, Modjeska, Trabuco and Sierra Peaks, and in sandy washes such as Coal Canyon. Its primary hosts are **thickleaf yerba santa** (*Eriodictyon crassifolium* var. *crassifolium,* Boraginaceae), **rock buckwheat** (*Eriogonum saxatile,* Polygonaceae), and **California buckwheat** (*Eriogonum fasciculatum,* Polygonaceae), but it also parasitises **chamise** (*Adenostoma fasciculatum,* Rosaceae) and **manzanitas** (*Arctostaphylos* spp., Ericaceae). It grows very near the base of its host plant, where it may be partially hidden from view.

It was named in 1818 by T. Nuttall for the bundles of flowers on the ends of its stems (Latin, fasciculus = a bundle).

Clustered Broomrape

In our area, clustered broomrape exists in two forms. The more common form (top right) has short corolla lobes that are pale to medium yellow in color. It is known from Coal Canyon, Baker Canyon, and higher elevations in the Santa Ana Mountains.
The uncommon form (right) has much longer corolla lobes that are lemon yellow in color. It is currently known only from Modjeska Peak.
Parish's broomrape (bottom photos) has bold red veins on its long narrow corolla lobes.

Clustered Broomrape

Parish's Broomrape ✿
Orobanche parishii (Jeps.) Heckard ssp. ***parishii***

Another yellowish species, 15-26 cm in height. Flowers from a stout yellow-white stalk that is mostly underground. *Corolla 20-25 mm long, yellowish to pinkish, the upper lobes widely rounded.* In flower May-July.

Grows in heavy clay soils among coastal sage scrub and grasslands. Here known only from Ashbury Canyon near Arroyo Trabuco on the Rancho Mission Viejo and from Miller Mountain in the San Mateo Canyon Wilderness Area. In our area, its only recorded host is **sand aster** (*Corethrogyne filaginifolia,* Asteraceae).

It was named in 1925 by W.L. Jepson in honor of S.B. Parish.

Parish's Broomrape

Parish's Broomrape

Goldenrod Broomrape

Goldenrod Broomrape

Right: Goldenrod broomrape parasitizes the roots of California goldenrod.

Below, right: Valley broomrape parasitizes the roots of blue elderberry.

Goldenrod Broomrape ✿
***Orobanche* sp.**

A yellowish to pale-brownish plant, 10-35 cm tall, covered with short yellowy-brown hairs. Its *stout central stalk* is mostly underground. Each flower on a long pedicel. Corolla 25-40 mm long, *whitish or pale pinkish with pink or purple veins, and long narrow lobes, each* 10-14 *mm long. The top two lobes often lean toward each other; the cut between them is only one-third to one-half of their length.* In flower June-July.

Very rare, known here only from Whiting Ranch Wilderness Park where it parasitizes the roots of **California goldenrod** (*Solidago velutina* ssp. *californica*, Asteraceae), which grows among oak woodlands. It was discovered in July 2005 by Jennifer Naegele and Kathy Williams during preparation of this book. It may occur elsewhere in the Lomas de Santiago and the Santa Ana Mountains. Preliminary DNA sequence analysis of this plant has revealed that it is new to science.

Valley Broomrape ✿
Orobanche vallicola (Jeps.) Heckard

Similar to goldenrod broomrape and difficult to separate from it. Corolla similar, *but the top two lobes often stand straight up or lean away from each other; the cut between them is usually one-half of their length or more.* In flower Aug-Oct.

An uncommon plant throughout its range, known here only from lower Holy Jim Canyon in silty-clay soils, where it parasitizes **blue elderberry** (*Sambucus nigra* ssp. *caerulea*, Adoxaceae). This population was discovered in early 2006 by Bob Allen during preparation of this book. Though it resembles goldenrod broomrape, it utilizes a different host plant species and is active at a different time of year. DNA sequence analysis also revealed that they are different species.

Named in 1925 by W.L. Jepson for its occurrence in valleys (Latin, vallis = valley; -colo = to inhabit).

This is a cluster of several plants, not one large round plant. Valley Broomrape

Valley Broomrape

OXALIDACEAE • OXALIS OR WOOD-SORREL FAMILY

Small mound-shaped perennials, rarely geophytes. Leaves basal or scattered; clover-like, palmately divided into 3 leaflets, each heart-shaped. Flowers trumpet-shaped. Sepals and petals 5, stamens 10, sometimes of two lengths. Stigma 1, styles 5, ovary superior. Fruit a linear capsule that explodes when ripe, sending tiny seeds in all directions. The family takes its name from its largest genus, *Oxalis,* named for its sour sap (Greek, oxys = sour, acid).

California Wood-sorrel

Oxalis californica (Abrams) R. Knuth
[*Oxalis albicans* Kunth ssp. *californica* (Abrams) G. Eiten.]

A charming perennial from a deep woody taproot. Stems are 10-40 cm long, hairless or hairy, upright and sprawling, *never rooting at the leaf nodes.* Leaves are always green. Petals 8-12 mm long, medium to bright yellow, sometimes tinged with purple. In flower Feb-Apr(-June).

Our only native wood-sorrel. A common plant of coastal sage scrub and chaparral, often on slopes in moist rocky soils. Quite widespread from coastline to mountainside. Easily seen in places such as Silverado Canyon, Caspers Wilderness Park, and on the walls of San Juan Canyon.

Named in 1907 by L. Abrams for the state of its discovery.

Similar. Creeping wood-sorrel ✖ (*O. corniculata* L., not pictured): leaves green or purple; stems root at the leaf nodes; a common garden weed, sometimes in the wild.

California Wood-sorrel

Bermuda Buttercup, Sourgrass ✖

Oxalis pes-caprae L.

A large, *stout perennial* from a deep taproot and scaly bulbs. The stems do not sprawl, creep, or root. *Leaves* basal, green, *streaked or spotted with purple. Flowers* large, 1.5-2.5 cm long, *deep yellow.* In flower Nov-Mar or more.

A garden escape, native to South Africa. Abundant on some beaches and in urban areas, where it is an enemy of serious gardeners and orchard managers. In California, its flowers are sterile and produce neither fruits nor seeds; here it reproduces only by bulblets.

The epithet was established in 1753 by C. Linnaeus in reference to the shape of the leaf, like the foot of a goat (Latin, pes, pedis = foot; caper = goat or capra = female goat).

Bermuda Buttercup

PAEONIACEAE • PEONY FAMILY

Perennials or shrubs. *Leaves fleshy, alternate, divided into 3 broad lobes,* each of these usually divided 1-2 more times. Sepals 5-6, petals 5-10. *Many stamens, pistils 2-5, ovary superior.* The family takes its name from *Paeonia,* its only genus, named in honor of Paeon, mythical physician to the Greek gods.

California Peony

Paeonia californica Nutt. ex Torr. & A. Gray

A small bushy ephemeral spring perennial, 35-75 cm tall. *Leaves* 3-9 cm long, 1-6 cm wide, green above, paler beneath, and carrot-like; *divided into several narrow-tipped leaflets.* Sepals green. *Petals* 15-25 mm *long, elliptical, dark reddish-black.* Stamens numerous, bright yellow. In flower Jan-Mar.

When they first open, the flowers face out or toward the ground. As their fruits age, the flower stalk bends to the ground under the weight. The developing fruits look like 2-5 green lemons growing right out of the flower's center, even while the flower is still fresh. Clearly, pollination is accomplished (mostly by bees) early in its flowering period.

Grows in openings and beneath shrubs in coastal sage scrub and chaparral. Common throughout our area from coastline to mountaintop. Easily seen in Trabuco Canyon, Caspers Wilderness Park, San Juan Canyon, and woodlands surrounding the Los Pinos Potrero.

Named in 1838 by T. Nuttall for the region of its discovery.

California Peony

PAPAVERACEAE • POPPY FAMILY

Delicate annuals to bushy shrubs and small trees with clear or milky-yellow sap. The sap of some contains toxic alkaloids. Leaves basal and/or alternate, rarely opposite; undivided to finely divided. *Flowers usually bowl-shaped; flattened side-to-side in some. Sepals 2(3-4), fused in some, often drop when the flower opens.* Petals 4-6 or more, often showy. Stamens 4-to-many, to 250 in some. *Ovary superior, often resembles a bowling pin.* Style absent or short, stigma variable. *Fruit a capsule; the numerous tiny seeds are often shed through slits or pores like a saltshaker.* About 400 species, most from western North America. Pollinated primarily by beetles, flies, and bees.

Prickly Poppy, Chicalote

Argemone munita Durand & Hilgard

Prickly Poppy

A stout ephemeral perennial or annual, 60-150 cm tall, *covered with yellow spines.* Leaves 5-15 cm long, gray-green, broadly linear or oval with wavy edges; lower leaves pinnately lobed, upper shallowly lobed or not at all. *Leaf edges and surfaces armed with sharp prickles.* Flowers solitary on upper stalks; unopened buds stand straight up. Sepals 2 (3), prickly, fall off when the flower opens. *Petals 4(6) large, 25-50 mm long; crinkled; stark white.* Stamens 150-250. *Fruit 35-55 mm long, spine-covered.* In flower May-Aug.

Finding this plant is a memorable experience, especially if you touch it! Lives in dry rocky soils on slopes and in sandy soils of creekbeds and channels. Long ago found in Santa Ana River bed, apparently extirpated. Known from Santiago Creek north of Hangman's Canyon, Hagador Canyon, and Temescal Valley. Reliably found on Santiago and Modjeska Peaks, where they have far fewer prickles on their stems and leaves and their fruits have large, widely spaced spines all about the same size. It was named **robust prickly poppy** (*Argemone munita* ssp. *robusta* G. Ownbey) and is no longer formally recognized by some botanists.

The epithet was given in 1855 by E.M. Durand and T.C. Hilgard for its spines (Latin, munitus = armed, protected). Robust prickly poppy was named in 1958 by G.B. Ownbey for the large robust spines on its fruits (Latin, robustus = robust). The name chicalote is Spanish, derived from the Aztec word chicotl.

Similar. **Matilija poppies** (*Romneya coulteri* and *R. trichocalyx*): no prickles; flowers are larger.

Bush Poppy

Dendromecon rigida Benth.

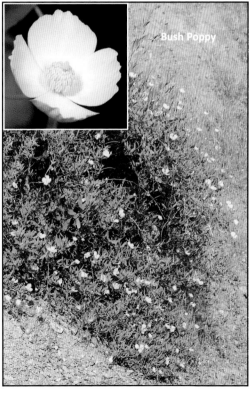

Bush Poppy

A hairless, evergreen shrub 1-3 meters tall. Its *stiff leaves* are 2.5-10 cm long, *yellow-, blue-, or gray-green* in color, elliptical to oblong, with tiny teeth along their edges. Unopened buds are held upright. The 2 sepals fall off just before the flower opens. The *4 petals* are 2-3 cm long, *bright yellow, and egg-shaped; they fall off just after pollination. Stigma flat, 2-lobed.* Fruit a narrow pod 5-10 cm long. In flower Apr-July.

A common shrub of dry rocky slopes and decomposed granitic soils, mostly among chaparral, sometimes in washes below it. More common in the years following fire. Known from the Santa Ana River channel near the base of the Chino Hills and the Santa Ana Mountains. Quite common along Main Divide Road, easily seen along the stretch from Ortega Highway to Falcon Group Camp.

Pollinated most often by beetles and bees. Its seeds are dispersed by native ants. A wonderful plant in cultivation, easy to grow and a prolific bloomer that produces flowers nearly year-round in the garden.

It was named in 1835 by G. Bentham for its stiff stems and leaves (Latin, rigidus = stiff).

Eardrops, Fire Hearts: *Ehrendorferia*

Unusual flowers flattened side-to-side. Outer 2 petals linear, curved. Inner 2 petals fused at their tips. Fire-followers.

Golden Eardrops

Ehrendorferia chrysantha (Hook. & Arn.) Rylander
  [*Dicentra chrysantha* (Hook. & Arn.) Walpers]

An upright, slender perennial, 0.5-1.6 meters tall. *Leaves pale gray, powder-covered, 15-20 cm long, 2-pinnately dissected. Petals golden yellow,* 12-16 mm long, the outer ones curving backward from about the middle. In flower Apr-Sep.

Occurs on dry hillsides and in disturbed areas among chaparral, most often in the years following fire. *More common in the central-to-southern portion of the Santa Ana Mountains.* Known from Baker Canyon up to Santiago Peak, upper San Juan Canyon, Los Pinos Potrero, upper Trabuco Canyon, inland slopes, and Temescal Valley.

Named in 1840 by W.J. Hooker and G.A.W. Arnott for its golden flowers (Greek, chrysos = gold; anthos = flower).

White Eardrops, Fire Hearts

Ehrendorferia ochroleuca (Engelm.) Fukuhara
  [*Dicentra ochroleuca* Engelmann]

Similar in many ways to golden eardrops but generally a bit shorter. *Leaves 15-30 cm long, powder-coated, but dissected 3 times instead of 2 times. Petals pale yellowish-white with purple tips,* 20-25 mm long, the outer petals curve backward only at their tips. In flower May-July.

Grows on dry, disturbed soils of hillsides and washes, usually in years after a fire. *Found more often at the north end of our area,* in the Chino Hills surrounding the Santa Ana River channel and the Santa Ana Mountains including Coal and Black Star Canyons. Generally only one species is observed in an area, but both have been found in Tin Mine Canyon.

Named in 1881 by G. Engelmann for its yellowish-white flowers (Greek, ochra = yellow-ochre; leukon = white).

Golden Eardrops

White Eardrops

Western Poppies: *Eschscholzia*

Upright or sprawling, somewhat succulent. Leaves basal or alternate, finely divided. Flowers usually solitary atop a long naked stalk. Sepals 2, fused into a conical "rocket ship" (*calyptra*) that falls off as the flower opens. Petals 4, rarely more. Stamens 12 to many. Style absent, stigma lobes 4-8. Fruit a long slender capsule, seeds numerous, thrown in all directions when it dries and explosively opens.

Tufted Poppy

Eschscholzia caespitosa Benth.

A hairless annual 5-30 cm tall, with leaves densely clumped (tufted) at the base and some leaves along the stem. Buds are held upright. At the flower's outside base, the receptacle gradually broadens, but there is no flat ring (torus). Sometimes a small torus is present, but it is no more than 0.3 mm wider than the receptacle edge. *Petals* 4, 10-25 mm long, *yellow, sometimes orange at the base.* Fruit 4-8 cm long. In flower Mar-June.

An uncommon plant of dry sparse chaparral slopes. Long ago found in the southern Chino Hills above Santa Ana Canyon. More recently found in Hagador Canyon and north of Walker Canyon. More common in the Gavilan Hills, the San Bernardino Valley, and the San Gabriel and Santa Monica Mountains. Most often found after fire.

The epithet was given in by G. Bentham for its tufted (densely clumped) leaves (Latin, caespitosus = tufted).

Tufted poppy (left) and California poppy (right) are nearly identical. Tufted has yellow petals and no or a tiny torus. California is orange to yellow and has a large obvious torus.

Left, top: A typical spring view in Caspers Wilderness Park. Left: Still in bud, the calyptra has not yet blasted off. Right: An open flower in Whiting Ranch Wilderness Park.

Right, top: A poppy party in Caspers Wilderness Park. Lower: A California poppy on a slope in the Santa Ana Mountains.

California Poppy
Eschscholzia californica Cham.

A hairless annual or short-lived perennial from a fleshy orange taproot, it stands upright 5-60 cm tall or sprawls on the ground. Sometimes tufted. Its *leaves and stems are often covered with a whitish powder and appear grayish.* Leaves mostly basal, with some stem leaves. Buds are held upright. *At the base of the flower is a flat ring called the torus, 0.5-5 mm wider than the receptacle edge. The torus is often pink or purple.* Petals 4, 20-60 cm long, *deep orange to light yellow, usually orange at the base.* Fruit 3-9 cm long. In flower Feb-Sep, sometimes longer.

Together with **coast redwood** (*Sequoia sempervirens* (D. Don) Endl., **giant redwood** (*Sequoiadendron giganteum* (Lindl.) J. Buchholz, both Cupressaceae), and **California lilacs** (*Ceanothus* spp., Rhamnaceae), California poppy is one of the most internationally recognized of all California plants. Grown in gardens worldwide. It is known from our entire area at most elevations and habitats.

The flowers generally open in bright sun. Their glossy petals strongly reflect glaring light, which makes photography a real challenge. In the evening and in daytime when shaded by heavy cloudcover, the flowers close back up. They are mostly pollinated by small soft-winged flower beetles (Melyridae) that spend the night within the closed-up flowers. Checkered beetles (Cleridae), long-horned beetles (Cerambycidae), and many types of native bees also frequent the flowers.

California poppies from the coast of central California have larger, deeper orange flowers than those in southern California. Introduced to our area by the horticulture trade, roadside plantings, and habitat restoration projects, the central coast poppies have escaped into the wild. These "foreign natives" generally require more water than those naturally occurring here and are more often found close to human settlements and roadsides that provide supplemental soil moisture. Seeds of our local poppy generally require fire treatment to germinate, while the central coast forms germinate on their own.

Following good winter rainfall, vast fields of poppies appear and put on tremendous displays of color. Once widespread, southern California's large poppy fields have nearly disappeared. Orange County's best display is in Caspers Wilderness Park.

California poppy is the State Flower of California. The California State Floral Society, a horticultural group based in San Francisco, voted on December 12, 1890, to select it as such and pushed for its adoption. On March 2, 1903, Governor George Pardee signed into law a bill making the designation official.

The genus was named in 1820 by A. von Chamisso (1781-1838) in honor of his good friend J.F. Eschscholtz (1793-1831). The epithet was also given in 1820 by von Chamisso for the region of its discovery. Both served as naturalists for the Romanzof expedition of 1815-1818 aboard the Russian ship *Rurik,* which traveled to the northern Pacific Ocean in search of an unfrozen passage from northern Russia to the Pacific. They collected the poppy in 1816 while visiting San Francisco.

Small-flowered Fairy Poppy
Meconella denticulata E. Greene

A small hairless annual, 5-30 cm tall. Leaves mostly basal, some up the stem, 1-4 cm long, linear, oval or spoon-shaped, smooth-edged or with small teeth. Sepals 3(2). *Petals 6(4), white, only 2-4 mm long, dropped after pollination.* Stamens 4-6, very short. Fruit slender, twisted, topped with 3 broad style lobes. In flower Mar-May.

A delicate plant of moist, shady spots on grassy slopes, often overlooked. More common after fire. Known from Silverado Canyon, Trabuco Canyon, upper Hot Springs Canyon, San Juan Canyon, the Santa Rosa Plateau, and the San Mateo Canyon Wilderness Area. Look for it along Bear Canyon Trail, San Juan Loop Trail, and Holy Jim Trail.

The genus was named in 1838 by T. Nuttall for its small size (Greek, mekon = poppy; -ella = small). The epithet was given in 1886 by E.L. Greene for the small teeth along its leaf edges (Greek, dentis = tooth; -ulata = minutely; Latin, denticulata = having small teeth).

Small-flowered Fairy Poppy

Fire Poppy
Papaver californicum A. Gray

A slender annual 30-60 cm tall, *hairless to quite hairy.* Leaves 3-9 cm long, pinnately divided. Flowers hang down when in bud, the 2 sepals fall off when the flower opens. *Petals 4, oval or triangular, brick red to orangeish with a green spot at the base.* Style absent. *The stigma sits atop the ovary like a flat disk with many small rounded lobes along its edge.* Fruit a flat-topped cylinder. When ripe, seeds fall out of pores that open on the sides just below the stigma. In flower Apr-May.

Uncommon, it grows in disturbed soils among chaparral and oak woodland, even along hiking and biking trails, most often in the years following fire. Known from the San Joaquin Hills, Claymine Canyon, Hot Springs Canyon, San Juan Canyon, and the San Mateo Canyon Wilderness Area.

The epithet was given in 1887 by A. Gray for the state of its discovery.

Similar. Wind poppy (*P. heterophyllum*): petals orange-red, a large purple spot at the base of each; "radio antenna" style topped with a club-like stigma.

Fire Poppy

Wind Poppy
Papaver heterophyllum (Benth.) Greene
[*Stylomecon heterophylla* (Benth.) G.C. Taylor]

A slender annual, 30-60 cm tall, *hairless to hairy.* Leaves 2-12 cm long, pinnately divided. Leaflets of 2 shapes: lower broad, upper narrowly linear. Flowers generally hang down when in bud, sepals fall off when the flower opens. *Petals 4, triangular, orange-red, each with a large purple spot at its base,* making the flower look dark-centered. *The single style is slender, topped with a club-like stigma, ringed with 4-11 small rounded lobes.* Fruit a flat-topped cylinder with the style still attached like a radio antenna. In flower Apr-May.

Found on grassy slopes and among chaparral, more often after fire. Quite uncommon here. Known from Silverado Canyon and Walker Canyon north of Lake Elsinore. Old records for "Laguna" and Olinda.

The epithet was given in 1835 by G. Bentham for the 2 shapes of its leaflets (Greek, heteros = other, different; phyllus = -leaved).

Similar. Fire poppy (*P. californicum*): petals brick red to orangeish, a green spot at the base of each; no style, stigma lies atop the ovary.

Wind Poppy

Cream Cups ✤
Platystemon californicus Benth.
  A bushy annual 10-30 cm tall, *covered with long shaggy hairs.* Leaves mostly basal, 2-8 cm long, usually linear, not toothed. Sepals 3, hairy on the outside. Petals 6, 8-16 mm long, whitish (yellow), base and/or tips may be yellow; not dropped after pollination, may turn pinkish. Stamens 13+, filaments flattened. Fruit cylindrical, bristly. In flower Mar-May.
  A plant of grasslands, more common after fire. Known from the Chino Hills, upper Newport Bay, South Laguna, the Dana Point Headlands, Rancho Mission Viejo, Caspers Wilderness Park, and San Juan Canyon. Greatly reduced in numbers; recent records only from Limestone Canyon.
  Named by G. Bentham, who established both the genus and the epithet in 1835. The genus was named for its flat stamen filaments (Greek, platys = flat; stemon = thread, stamon), the epithet for the place of its discovery.

Cream Cups

Matilija Poppies, Fried-Egg Flowers: *Romneya*
  Large, shrubby, woody-stemmed, 1-2.5 meters tall. Leaves gray-green, alternate, 3-5 deep lobes. Young leaves sometimes spiny. *Flowers large, face straight up.* Sepals 3, fall off as the flower opens. Petals 6, stark-white, wrinkled, shed after pollination. Stamens numerous, form a bright orange mound in the middle like a sunny-side-up fried egg. The genus was named in 1845 by W.H. Harvey to honor T. Romney Robinson.

Coulter's Matilija Poppy ★
Romneya coulteri Harvey
  Leaves 5-20 cm long. Sepals with a beak at the tip, the outer surface hairless. Petals 6-10 cm long. Fruit 3-4 cm long. Seeds bumpy-surfaced, dark brown. In flower Mar-July.
  Lives in washes, creekbeds, and on mountainsides only in the Santa Ana Mountains and in the foothills of northern Camp Pendleton in San Diego County. Known from the Chino Hills, the Santa Ana River channel, along major canyons such as Silverado, Santiago, Modjeska, Trabuco, and San Juan, and occasionally scattered at higher elevations of the Santa Ana Mountains. *This is the largest flower of any native California plant.* CRPR 4.2.
  The epithet was given in 1845 by W.H. Harvey in honor of T. Coulter, who collected it in California.
  Similar. Hairy matilija poppy (*R. trichocalyx*): smaller flowers; sepals not beaked, hairy. **Prickly poppy** (*Argemone munita*): plant and flowers smaller; plant prickly.

Coulter's Matilija Poppy

Hairy Matilija Poppy
Romneya trichocalyx Eastwood
  Leaves 3-10 cm long. Sepals have no beak at the tip; the outer surface is covered with flattened hairs. Petals 4-8 cm long. Fruit 2.5-3.5 cm long. Seeds smooth, brown. In flower Apr-June.
  Generally found in habitats similar to those of Coulter's, it has a broader range, from Ventura County south to Baja California. Very uncommon in our area, known only from the Santa Ana River channel, Silverado Canyon, and Caspers Wilderness Park, possibly elsewhere and assumed to be Coulter's matilija poppy.
  Named in 1898 by A. Eastwood for its hairy sepals (Greek, trichos = hair; kalyx = calyx of a flower).
  Similar. Coulter's matilija poppy (*R. coulteri*): larger flowers; sepals beaked, hairless. **Prickly poppy** (*Argemone munita*): plant and flowers smaller; plant prickly.

Coulter's   Hairy

Coulter's   Hairy

Since the sepals fall off as the flower opens, you must examine them while still in bud.

PHRYMACEAE • LOPSEED AND MONKEYFLOWER FAMILY

Annuals, perennials, and shrubs, many covered with sticky-tipped hairs. *Leaves opposite,* undivided. Flowers usually 2 per node. Calyx green, tubular, ribbed, with 5 free sharp-tipped lobes. *Corolla tubular, often 2-lipped: 2 lobes up, 3 down.* Stamens 4, all with fertile anthers. Anthers two-celled and fused at their tips; pollen is shed through a single common slit. *Ovary superior,* a single style topped with 2 oval-shaped stigma lobes that are sensitive to touch. Fruit is a capsule full of numerous tiny seeds.

Monkeyflowers are important in the horticultural trade as handsome shrubs and annuals with impressive flowers. For years, they were considered members of the figwort family (Scrophulariaceae). Detailed studies found them to be more closely related to members of the lopseed family (Phrymaceae), so in 1998 they were placed there. Our 9 species are easy to identify. Pollinated by hummingbirds, bees, and sphinx moths.

A recent paper (Barker, Nesom, Beardsley, and Fraga, 2012) proposes to split the monkeyflower genus *Mimulus* into multiple genera. Our species are now in the genera *Diplacus, Erythranthe,* and *Mimetanthe.*

PERENNIALS

Hairy Bush Monkeyflower
Diplacus longiflorus Nuttall
[*Mimulus aurantiacus* Curtis var. *pubescens* (Torr.) D.M. Thompson]

A medium to large shrub, 10-150 cm tall. *Its stems are hairy.* Leaves linear, 2-8 cm long, edges usually rolled under, covered in a very sticky substance; *hairless on top, slightly to very hairy below.* The lower leaf surface is much paler than the upper. Usually 2 flowers per node. *Calyx densely hairy. Corolla pale yellow to yellow-orange,* tubular, 2.5-6 cm long with five oddly shaped lobes: round-, square- or jagged-edged. The *anthers usually remain hidden within the corolla throat* (included). In flower Mar-July.

A common plant, it occurs from San Luis Obispo County and the Sierra Nevada range south through our area and into Baja California Norte. It grows among coastal sage scrub and other communities, usually in well-drained soils and between cracked rocks. It is widespread and common, found in most of our wilderness parks, in the Puente-Chino Hills, and at higher elevations. Pollinated by **Allen's hummingbird** (*Selasphorus sasin* (Lesson) Trochilidae).

A wonderful plant for the garden, tolerant of full sun or partial shade. If the soil is dry, its leaves fold farther under at the edges and may drop off in summer, but will leaf out again with rains of late fall and winter. With a bit of summer watering, its leaves remain healthy, and it often continues flowering year-round.

The epithet was given in 1838 by T. Nuttall for its long corolla (Latin, lonchus = long; floris = flower).

Hairy Bush Monkeyflower

Hairy stems and leaf undersides of Hairy Bush Monkeyflower

Hairy Bush Monkeyflower along North Main Divide Road "Loop"

Below: Hairy bush monkeyflower has hair on the calyx and leaf undersides.

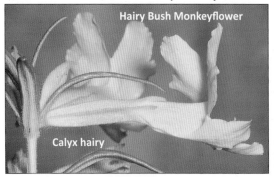
Hairy Bush Monkeyflower
Calyx hairy

Coastal Bush Monkeyflower
Diplacus puniceus Nutt.
[*Mimulus aurantiacus* Curtis var. *puniceus* (Nutt.) D.M. Thompson, *M. puniceus* (Nutt.) Steudel]

Stems and leaves hairless. Leaves are the same shade of green on top and bottom. It may have up to 4 flowers per node. *Calyx hairless. Corolla bright red-orange or red.* The *anthers usually extend out of the corolla throat* (excluded). In flower Mar-July.

This species lives in our area and on Santa Catalina Island, then continues south through San Diego County and into Baja California Norte. It grows among coastal sage scrub and chaparral in habitats that range from flat sandy soils to steep rocky hillsides. Find it in the San Joaquin Hills of Laguna Beach and Aliso and Wood Canyons Wilderness Park, south to the Dana Point Headlands. Also in the Santa Ana Mountains from Caspers Wilderness Park, San Juan Canyon, upper elevations, and inland near Glen Ivy Hot Springs. Not found in the Chino Hills.

It was named in 1838 by T. Nuttall for its red corolla (Latin, puniceus = reddish-colored).

Coastal Bush Monkeyflower along Bear Canyon Trail

Southern Bush Monkeyflower
Diplacus* x *australis (McMinn ex Munz) Tulig
[*Mimulus aurantiacus* Curtis ssp. *australis* McMinn ex Munz]

Hybrids between hairy bush and coastal bush monkeyflowers are given this name. Their *corolla* tends to be *light to dark orange, often with various shades of red,* especially near the coast, and with *few hairs on the calyx.* Like hairy bush monkeyflower, the *anthers are usually included within the corolla throat.*

This is the most commonly encountered bush monkeyflower in our area. It is primarily found in the San Joaquin Hills, Lomas de Santiago, and the coastal slope of the Santa Ana Mountains from Black Star Canyon south.

Studies of hairy, coastal, and southern bush monkeyflowers in San Diego County (Streisfeld and Kohn, 2005, 2006) found an absence of yellow-flowered plants in the coastal region and an absence of red-flowered plants in inland regions. In our area, we find both orange- and red-flowered plants along the coast, a predominance of orange-flowered plants inland to the Santa Ana Mountains (probably southern bush monkeyflower), and both orange- and red-flowered plants in the Santa Ana Mountains.

Right: southern bush monkeyflower, a progression of colors in the wild. All four photographs were taken in Hot Springs Canyon within 10 minutes of each other.

Southern Bush Monkeyflower — Calyx nearly hairless

Southern Bush Monkeyflower

Cleveland's Bush Monkeyflower ★
Diplacus clevelandii (Brandegee) Greene
[*Mimulus clevelandii* Brandegee]

An herbaceous perennial from a woody base. Covered with sticky-tipped hairs, it stands upright 30-90 cm tall, from few- to several-branched stems. *Leaves* 2-10 cm long, linear to oblong, *the edges rolled under. Flowers are golden yellow and very sticky.* The corolla tube is long and gradually widens toward the mouth. It has ridges in its throat like wide-throated monkeyflower. Sometimes mistaken for the 3 preceding species of bush monkeyflowers, but Cleveland *always has hairy leaf uppersides* (the others never do). In flower May-July.

Uncommon, found only among chaparral in the Santa Ana Mountains and Temescal Valley, southward to the Laguna and San Ysidro Mountains of San Diego County. Known from Holy Jim Canyon Trail, upper Trabuco Canyon, upper Hot Springs Canyon, and various places along Main Divide Truck Trail from Falcon Group Camp to Los Pinos Saddle and then to Santiago and Modjeska Peaks. In upper Silverado Canyon, it is known to hybridize with bush monkeyflowers. Pollinated by bees; it is our only perennial monkeyflower that is *not* pollinated by hummingbirds. CRPR 4.2.

Named in 1895 by T.S. Brandegee in honor of D. Cleveland.

Scarlet Monkeyflower
Erythranthe cardinalis (Douglas ex Benth.) Spach
[*Mimulus cardinalis* Douglas ex Benth.]

A hairy herbaceous perennial from creeping underground rhizomes, it stands 25-80 cm tall. *Leaves* 2-8 cm long, oblong to oval, *with 3-5 obvious main veins. Corolla red to red-orange, flattened side-to-side.* The upper lip is held forward with its lobes folded; the side lobes of the lower lip are folded back, its central lobe held down, folded or not. The stamens and pistil are held outside of the flower tube. In flower Apr-Oct.

A common plant of streams and seeps in most canyons of the Santa Ana Mountains. Look for it in the Arroyo Trabuco of O'Neill Regional Park, along Holy Jim Canyon Trail, and commonly in major canyons such as Silverado, Harding, and San Juan. Pollinated mostly by hummingbirds, carpenter bees, bumble bees, and butterflies such as skippers and the California dogface.

Named in 1835 by D. Douglas for the red color of its flowers (Latin, cardinalis = chief, principal, or red). The term was first applied to church cardinals but was later used to refer to the red color of their garments.

ANNUALS

Palomar Monkeyflower ★
Erythranthe diffusa (A.L. Grant) N.S. Fraga
  [*Mimulus diffusus* A.L. Grant]

A wiry-stemmed delicate annual 1-28 cm tall. Leaves 3-28 mm long, linear or oval. Flower face rounded in outline, not strongly 2-lipped like most monkeyflowers. Throat held open, not closed. *Corolla 1.2-1.5 cm long, purple to rose-violet, marked with yellow and purple,* free lobes usually notched. *Throat floor lined with clubbed hairs.* Anthers hairless. In flower Apr-June.

A plant of sandy soils, full sun, or partially shaded grassy areas within chaparral and oak woodlands. It occurs only on the coastal side of the Peninsular mountain ranges (Santa Ana, San Jacinto, and Agua Tibia Mountains) south into Baja California. Locally known only from the vicinity of Trabuco Peak, Upper San Juan Campground (moist soils along seeps), Elsinore Peak, San Mateo Canyon, and the south flank of Miller Mountain. More common in the years after fire. Easily overlooked and trampled; roads and trail systems should be carefully routed around its populations. CRPR 4.3.

The epithet was established in 1925 by A.L. Grant. The name refers to the outspread stems that give the plant a sparse appearance (Latin, diffusus = spread out, extended, or dispersed). First collected on Palomar Mountain in 1901.

Fremont's Monkeyflower ✿
Diplacus fremontii (Benth.) G.L. Nesom
  [*Mimulus fremontii* (Benth.) A. Gray]

A small annual 4-20 cm tall *with reddish stems and leaves.* Only 1 flower per node (most monkeyflowers have 2 or more flowers per node). *Corolla 2-2.5 cm long, lobes smooth- or wavy-edged,* magenta to red-purple with yellow on the hairless throat floor. Stigma lobes about equal. In flower Mar-June.

Found in sandy soils and dry areas, especially during the first spring after fire. Uncommon here; known from San Juan Canyon, South Main Divide Road, and scattered sites in Temescal Valley such as Indian Creek wash.

Named in 1846 by G. Bentham in honor of J.C. Fremont, who first collected it in California in 1843-1844.

Similar. Palomar monkeyflower (*Erethranthe diffusa*): long wiry naked peduncles; corolla lobes notched; throat and lower lip lined with clubbed hairs; much smaller. **Rattan's monkeyflower** (*D. rattanii*): 1 flower per node; flowers much smaller; most of corolla tube remains within the calyx, lower stigma lobe much (5-7 times) longer than upper lobe.

Rattan's Monkeyflower ✿
Diplacus rattanii (A. Gray) G.L. Nesom
[*Mimulus rattanii* A. Gray]

A diminutive annual, 1–18 cm tall, *densely covered in straight sticky-tipped hairs*. Leaves 3-46 mm long, narrow; lowest leaves hairless. Only 1 flower per node, though the uppermost flowers may appear as though sharing a node. *Calyx inflated, with tall ridges. Most of the corolla tube remains within the calyx.* Corolla pink to magenta, tube throat 7-10 mm long, face 4-7 mm wide. Lower stigma lobe 5-7 times longer than the upper lobe. In flower Apr-July.

A small uncommon plant of sandy soils or sandstones, most often found the first spring after fire. Known here from only one documented occurrence on 30 May 1931, along the Glen Ivy (Coldwater) Trail, eastern slope of Santiago Peak.

The epithet was given in 1885 by A. Gray in honor of V. Rattan who collected the plant in 1884 in Colusa County.

Similar. Fremont's monkeyflower (*Diplacus fremontii*): flowers are larger, most of the corolla tube extends out of the calyx, stigma lobes about equal in length.

Rattan's Monkeyflower

Aaron Schusteff

Wide-throated Monkeyflower
Diplacus brevipes (Benth.) G.L. Nesom
[*Mimulus brevipes* Benth.]

A slender, sticky, hairy annual 5-80 cm tall with stiff upright stems and few branches. Leaves about 1-9 cm long, lower long and oval, upper shorter and linear. *Flowers large, bright yellow*. Corolla tube short and narrow within the calyx but quickly widens after emerging from it. *Throat floor and back have ridges spotted with maroon*. In flower Apr-June.

Common on dry slopes in the San Joaquin Hills and Santa Ana Mountains, more abundant after fire and winter or spring rains. In some years, it puts on a show visible for miles. Fairly regular along Main Divide Truck Trail from Falcon Group Camp up to Los Pinos Saddle and inland along upper Indian Truck Trail.

Named in 1835 by G. Bentham for its short corolla tube (Latin, brevis = short).

Wide-throated Monkeyflower

Seep Monkeyflower
Erythranthe guttata (Fisch. ex DC.) G.L. Nesom
[*Mimulus guttatus* Fischer ex DC.]

A hairy or hairless annual, sometimes a perennial from above- and belowground rhizomes. Stems 2-150 cm long, upright or sprawling, take root at the nodes. *Leaves 4-125 mm long, oval or round, often toothed; upper pairs fused.* Calyx green, red-spotted, or red; vase-shaped with unequal lobes, upper clearly longer than lower. *Corolla bright yellow, lower lip variously red-spotted*. Base of lower lip swollen, often closing the flower. The calyx inflates in fruit. In flower Mar-Aug(+).

Abundant in moist or wet soils, along creeks and seeps. Found throughout our area in such places as the Santa Ana River channel, Laguna Lakes, San Juan Creek, San Juan Canyon, and our wilderness parks. A popular garden plant, easily grown from seeds or cuttings of its rhizomes. Plant it in partial shade and keep its roots fairly moist. Does fine in a container or in the ground. It sometimes undergoes in-bud fertilization.

It was named in 1813 by F.E.L. von Fischer for the red spots on its flowers (Latin, guttatus = with drop-like spots, spotted, speckled).

Seep Monkeyflower

Slimy Monkeyflower

Erythranthe floribunda (Douglas ex Lindl.) G.L. Nesom
[*Mimulus floribundus* Douglas ex Lindl.

A small hairy annual often covered in a slimy film. It grows 3-50 cm long, sprawling, climbing on other plants, or upright. Stems and leaves green to yellowish, often purplish. *Leaves 5-45 mm long, linear to oval, sometimes toothed, the base round or heart-shaped.* The petiole of lower leaves is almost as long as the blade. Calyx tubular with 5 pointed lobes of equal length and shape. *Corolla tiny, bright yellow,* the inside of the lower lip with 2 yellow ridges spotted with red. The flower throat remains open. Unlike that of the similar but much larger seep monkeyflower, the lower lip is not swollen and does not close the throat. In fruit, the calyx does not inflate. Fruits are hairless. In flower Apr-Aug.

Found in rock crevices and moist sand along streams and seeps. Uncommon, known from scattered localities in the San Joaquin Hills and Santa Ana Mountains, such as Sycamore Hills, Laguna Lakes, upper Santiago Canyon, Arroyo Trabuco, San Juan Hot Springs, San Juan Canyon, along upper San Juan Trail, and the San Mateo Canyon Wilderness Area.

It was named in 1828 by by D. Douglas in reference to its many flowers (Latin, floris = flower; abundo = to overflow, abound with plants, or grow up with luxuriance).

Similar. Downy monkeyflower (*Mimetanthe pilosa*): upright; never slimy; leaves linear to oblong, no petiole; fruit hairy; sometimes together. **Seep monkeyflower** (*E. guttata*): upright, usually much larger; upper leaves fused; flower larger, lower lip red-spotted; calyx inflates in fruit; abundant.

Middle right: Slimy and Downy monkeyflowers. Note the broad long-petioled leaves of slimy, versus the narrow, petiole-less leaves of downy.

Downy Monkeyflower

Mimetanthe pilosa (Benth.) Greene
[*Mimulus pilosus* (Benth.) S. Watson]

A small upright annual that stands 2-35 cm tall, *covered with long soft white hairs* (only the corolla and fruit are hairless). Stems green to yellowish, often purplish. *Leaves 1-3 cm long, linear or oblong and have no petiole.* Calyx vase-shaped, upper lobes longer than lower. *Corolla tiny, bright yellow,* the inside of the lower lip with 2 maroon dots, sometimes more dots extending into the throat (rarely there are no dots at all). In fruit, the calyx does not inflate. The hairy oval fruit is surprisingly stout for such a small flower. In flower Apr-Sep.

Grows in moist sandy or gravelly places such as creekbeds and trails but is generally not abundant and is easily overlooked. Look for it along Borrego Creek Trail in Whiting Ranch Wilderness Park, Aliso Creek in Mission Viejo, Holy Jim Canyon, San Juan Creek, in moist spots among chaparral along South Main Divide Road, and the Santa Rosa Plateau.

It was named in 1836 by G. Bentham for its long white hairs (Latin, pilosus = hairy).

Similar. Slimy monkeyflower (*Erethranthe floribunda*): often sprawling; slimy film; hairy; leaves often toothed, broad based, long petiole; hairless fruit; sometimes together. **Seep monkeyflower** (*E. guttatus*): upright, usually much larger; upper leaves fused; flower larger, lower lip red-spotted; calyx inflates in fruit; abundant.

PLANTAGINACEAE • PLANTAIN AND SNAPDRAGON FAMILY

Annual herbs to shrubs. Leaf arrangement alternate and spiral, opposite or whorled in some. Leaves undivided with smooth edges or small teeth. *Flowers tubular and 2-lipped (snapdragon-type) or bowl-shaped (plantain-type).* Sepals fused at least near their bases; snapdragon-type with 5 free lobes, plantain-type with 4 free lobes. *Petals 5,* though in plantains the upper 2 are fused and the flower appears 4-petaled. Stamens 2-4, snapdragon-type with a sterile fifth, their filaments partially fused to the corolla. Each anther has 2 elliptical parts; each part has a pollen slit parallel to the stamen filament or is shaped like an upside-down letter U or V. *Ovary superior, style single with a clubbed or 2-lobed stigma.* Snapdragon-types have a nectary near the ovary and are pollinated by bees, flies, and/or hummingbirds. Plantains have no nectary and are wind-pollinated.

Fruit is an oval or spherical capsule. When ripe, the capsules release their seeds by different means. In the true plantains, the hat-like top half of the capsule falls off, allowing the seeds to spill out. In the snapdragons, the capsule develops 2-3 holes near its tip and the seeds spill out, but in some the end bursts opens and spreads the seeds. In 1998, snapdragons were removed from the figwort family (Scrophulariaceae) and placed into the plantain family. Our local snapdragons and plantains don't seem to share many characteristics, but worldwide species of the two are actually quite similar. One European species of plantain introduced to our area resembles a snapdragon, and its identity is a mystery until you study its flowers.

Snapdragons: *Antirrhinum*

Corolla 2-lipped, with a large bulbous platform or "palate" on the lower lip. The palate keeps the throat closed; it opens only when pollinators, primarily small native bees, force it open to collect nectar and pollen. The upper lip is 2-lobed, the lower 3-lobed. The common name comes from the flowers' similarity to a dragon's head and their habit of snapping shut when forced opened and released. The genus was named in 1753 by C. Linnaeus in reference their nose- or snoutlike flowers (Greek, anti- = resemblance; rhinos = a nose).

	Coulter's *Antirrhinum coulterianum*	Kellogg's *Antirrhinum kelloggii*	Chaparral *Antirrhinum multiflorum*	Nuttall's *Antirrhinum nuttallianum*
Growth	Slender annual, upright	Slender annual, vine-like	Stout perennial (annual), upright	Bushy annual to biennial, many branches
Hairs	Hairy (hairless), usually not sticky	Hairless, not sticky	Densely hairy, very sticky	Short hairs, sticky
Flowers	White to pale lavender, palate pale-yellowish	Lavender to deep blue; bulbous palate whitish with dark purple veins	Pink to red; palate white, compressed side-to-side, puckered at the tip	Lavender to deep blue-purple; palate white-blotched with blue veins

Coulter's Snapdragon

Coulter's Snapdragon

Antirrhinum coulterianum Benth. in DC.

A large upright annual 30-120 cm tall. *Leaves linear to oval, often in a basal rosette,* unique among snapdragons. Stems generally hairy, more so near the flowers than below. *Corolla is 9-12 mm long, white to pale lavender with a pale yellowish palate.* The upper 2 lobes are narrow and held upright like jackrabbit ears. In flower Feb-June.

Not commonly found near the coast but fairly common in the foothills and upper elevations of the Santa Ana Mountains, especially following fires. After good wet winters, it is quite abundant on the top and inland side, as along Indian Truck Trail, South Main Divide Road, and atop Santiago Peak. Also found in Coal Canyon, Limestone Canyon, Holy Jim Canyon, Hot Springs Canyon, along San Juan Trail, and on the Santa Rosa Plateau. Those at Oak Canyon Nature Center in the Anaheim Hills are unusual (plant and lower inset); they are all white-flowered, less hairy than elsewhere, up to 2 meters tall, and have few branches. Those at most other locations are pale lavender-flowered, quite hairy, much shorter, and many-branched.

It was named in 1846 by G. Bentham for T. Coulter.

Kellogg's Snapdragon
Antirrhinum kelloggii E. Greene

Often overlooked because of its *vine-like growth* through other plants. A hairless annual, it reaches 30-100 cm in length. *Its stems and pedicels are thin, wiry, and seemingly invisible.* It climbs by twisting its pedicels around objects, using them for support and continuing to grow. Its lower leaves are oval, upper leaves narrower. *Corolla* 10-14 mm long, *lavender to deep blue-purple with dark veins.* The upper 2 lobes are short, broad, and rounded. The bulbous lower lip is usually whitish with obvious dark purple veins. In flower Mar-May.

It comes up more commonly after chaparral areas have been cleared by fire or physical means. Known from major canyons such as Fremont, Weir, Hicks, upper Santiago, and San Juan. Also known from Harding Canyon Trail, Temescal Valley, the Santa Rosa Plateau, the Whittier Hills, and the San Mateo Canyon Wilderness Area.

It was named in 1883 by E.L. Greene for A. Kellogg, who first collected the plant in Alameda in 1870.

Kellogg's Snapdragon

Chaparral Snapdragon
Antirrhinum multiflorum Pennell

A stout, upright, shrubby perennial (annual) with long wandlike branches, 60-150 cm tall, *densely covered with sticky-tipped hairs.* The abundant leaves are 1-6 cm long, lance-shaped, and alternate on the stem, though seedlings have opposite basal leaves. *The tall inflorescence is crowded with flowers;* adjacent flowers open together, making a fine floral display. Upper calyx lobes are longer than the lower lobes. *Corolla* 13-18 mm long, *the lobes pink to red, hairs in the throat are white.* The white palate is compressed side-to-side and puckered inward at the tip, which ages yellow to brown. In flower May-July.

A plant of chaparral and coniferous forests from the San Bernardino, San Gabriel, and Santa Monica Mountains north to coastal central California. First discovered in our area during study of the post-fire recovery of Fremont Canyon in spring 2006. Pollinated by both bees and hummingbirds.

The epithet was given in 1951 by F.W. Pennell for its numerous flowers (Latin, multus = many; floris = flower).

Chaparral Snapdragon

Nuttall's Snapdragon
Antirrhinum nuttallianum Benth. in DC.

An upright often shrubby-looking annual 10-100 cm tall, *covered with short sticky-tipped hairs.* Leaves oval, the lower leaves opposite, alternate above. Flowers are somewhat solitary; adjacent flowers are generally not open at the same time. Calyx lobes are all about the same length. *Corolla* 7-12 mm long, *lavender to blue-purple with a large white blotch at the base of the lower lip, criss-crossed with blue veins.* Like Coulter's snapdragon, this one has upper lobes that are narrow and held upright. Hairs in the throat are golden. In flower Mar-May.

A widespread snapdragon, found in many places and habitats from coastline to mountains, more common after fire. Grows in open sandy soils on the Dana Point Headlands, among coastal sage scrub in the San Joaquin Hills of Irvine and Laguna Beach, and chaparral in the Santa Ana Mountains such as in Fremont Canyon, along the Harding Canyon Truck Trail, and in San Juan Canyon. Quite common near the Willow Canyon parking area in Laguna Coast Wilderness Park.

It was named in 1846 by G. Bentham in honor of T. Nuttall.

Nuttall's Snapdragon

Chinese Houses and Blue-eyed Mary: *Collinsia*

Flowers of these stunning annuals are two-lipped; the upper lip stands straight up and the lower lip straight out or angled down. The common name comes from the angle between the two lips, which resembles the angled roof structure of a Chinese pagoda. Their leaves are opposite; the uppermost ones have no petiole. The genus was named in 1817 by T. Nuttall in honor of Z. Collins.

Purple Chinese Houses
Collinsia heterophylla Graham

A delicate upright annual that grows 10-50 cm tall. *Leaves soft,* 1-7 cm long, *quite variable: linear to triangular; hairless, hairy, and/or sticky and smooth- or tooth-edged.* Flowers strongly 2-lipped. Upper lip stands straight up and is pale lavender (rarely white) with wine-colored spots in the banner. Banner patch usually has a horizontal purplish stripe or strong band of spots. Lower lip projects outward from the stem and is purple or whitish. Fruit many-seeded. In flower Mar-June.

Our largest and most abundant *Collinsia*. It grows in moist, shady spots, often beneath oaks, on north-facing hillsides, and in seeps. Find it in the San Joaquin Hills, Laguna Coast Wilderness Park, Santiago and Silverado Canyons, Whiting Ranch Wilderness Park, Limestone Canyon, Caspers Wilderness Park, Hot Springs Canyon, Bear Canyon Trail, Main Divide Truck Trail, and Indian Truck Trail, and the San Mateo Canyon Wilderness Area. A popular garden plant. Sow its seeds in partially shady areas, water occasionally, and enjoy the bloom.

Named in 1838 by R. Graham for its variable leaves (Greek, heteros = other or different; phyllon = leaf).

Similar. Southern Chinese houses (*C. concolor*): often smaller; calyx lobes blunt-tipped; each banner lobe is notched inward; banner patch markings are arranged into a triangular shape; never has a horizontal stripe in the banner patch; no backward spur at base of the two uppermost filaments (if spur is present, then it is no more than 0.5 mm long); uncommon in our area.

Southern Chinese Houses
Collinsia concolor E. Greene

    A small annual, 15-40 cm tall. Leaves narrowly oblong to lance-shaped. Uppermost flowers crowded into whorls that often hide the stem. *Lower pedicels shorter than calyx.* Sometimes has solitary flowers at the end of long pedicels, below the crowded whorls. Corolla 11-15 mm long, upper and lower lips about equal in length, generally blue to bluish-purple. *The base of the upper lip is whitish and generally dotted with purple in a triangular-shaped patch.* Fruit many-seeded. In flower Apr-June.

    Uncommon in Orange County. Known from lower Silverado Canyon, the vicinity of Limestone Canyon, San Juan Canyon, and North Main Divide Road "Loop." Also within Temescal Valley such as at Horsethief Canyon and lower Indian Truck Trail. Generally in the shade of chaparral shrubs or conifers. Most local records for this species turn out to be misidentified individuals of purple Chinese houses.

    The epithet was given in 1895 by E.L. Greene for the single-colored lower corolla lobes (Latin, concolor = of the same color, one-colored).

    Similar. Purple Chinese houses (*C. heterophylla*): usually larger; calyx lobes sharp-pointed; banner lobes rounded or small-toothed, usually not notched inward; banner patch markings are variable but are almost never triangular; banner patch usually has a horizontal purplish stripe; base of the two uppermost filaments has a backward-pointing spur (1-2 mm long) that extends into the pouch at the base of the flower; abundant.

Southern Chinese Houses

Southern Chinese Houses

Parry's Blue-eyed Mary

Parry's Blue-eyed Mary
Collinsia parryi A. Gray

    A small annual, 10-40 cm tall. Leaves generally lance-shaped, uppermost broad at the base and clasp the stem. Flowers solitary or whorled, not very crowded. *Lower pedicels longer than calyx. Calyx lobes blunt-tipped.* Corolla 4-10 mm long, upper and lower lips about equal in length, blue-violet to lavender (white). *Fruit 8-12-seeded.* In flower Mar-June.

    Uncommon to locally abundant, in moist soils among grasslands and decomposed granite in chaparral in the Santa Ana Mountains. Rarely found, easily overlooked because of its small size. Known from San Juan Trail in Lower Hot Springs Canyon, lower Bear Canyon Trail, various spots along North and South Main Divide Road, San Mateo Creek, and north of Walker Canyon. More common in the years after fire.

    The epithet was given in 1878 by A. Gray for C.C. Parry.

    Similar. Blue-eyed Mary (*C. parviflora* Lindl., not pictured): calyx lobes sharp-tipped; fruit 4-seeded; uncommon.

Parry's Blue-eyed Mary

Yellow False Pimpernel
Lindernia dubia (L.) Pennell

Annual (rarely perennial), *sprawling or upright, slender-stemmed*, under 27 cm tall. Leaves 1-37 mm long, lance-shaped to oval, opposite. Calyx lobes 5, slender. *Corolla 7-10 mm long, tubular, 2-lipped, white, blue, or lavender.* Filaments fused to corolla; fertile stamens 2; staminodes 2, forming ridges in the throat. Pistil 1, stigmas 2. In flower June-Aug.

An uncommon plant of very wet soils and pond edges. Known from Laguna Lakes, where it is sometimes locally abundant. Recently placed by some botanists into a separate family, the Linderniaceae.

The epithet was given in 1753 by C. Linnaeus perhaps for its uncertain place of origin (Latin, dubius = uncertain).

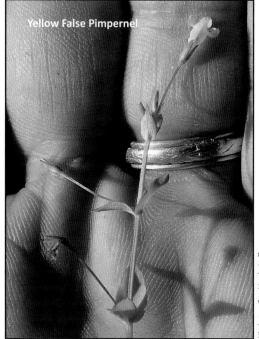
Yellow False Pimpernel

Beardtongues: *Keckiella* and *Penstemon*

Spectacular tubular or bell-shaped flowers admired by wildflower lovers and gardeners around the world. Each flower has 5 stamens, of which 4 are fertile and grouped into 2 pairs. Fertile stamens arch upward and are positioned in the roof and sides within the flower. The fifth stamen is sterile (staminode; it has no anther), has a hairy upper surface in some species, and lies in the floor of the flower throat like a bearded tongue that hangs outside or remains inside the tube (thus the name "beardtongue"). The shape of the anthers is unique in a few species, easily seen with a hand lens. Some species are more common in the years following a fire. Most are cultivated and make unusual garden plants that provide splashes of red, orange, yellow, blue, and purple in spring. Among horticulturalists, an affinity for these plants is known as Penstemania. It is a contagious affliction; both of us grow them in our home gardens.

Two genera of beardtongues live in our area: *Keckiella* and *Penstemon*. The genus *Keckiella* was named in 1967 by R.M. Straw in honor of D.D. Keck. The genus *Penstemon* was established in 1748 by J. Mitchell in reference to their five stamens (Greek, pente = five; stemon = stamen).

Above right: Grinnell's beardtongue has a long hairy (bearded) tongue. Right: Royal beardtongue has a long hairless (beardless) tongue. Below: Heart-leaved bush beardtongue has a short hairy (bearded) tongue.

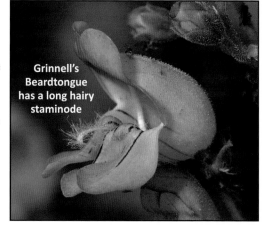
Grinnell's Beardtongue has a long hairy staminode

Royal Beardtongue has a medium-length hairless staminode

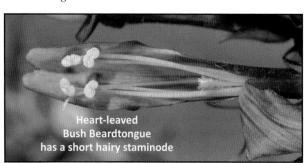
Heart-leaved Bush Beardtongue has a short hairy staminode

Yellow Bush Beardtongue
Keckiella antirrhinoides (Benth.) Straw
ssp. *antirrhinoides*

A very bushy shrub with several branches, up to 2.5 m tall. *Leaves 5-20 mm long, linear with smooth edges.* The outside of the flower is covered with short sticky-tipped hairs. *Corolla 15-23 mm long, bright yellow, with a broad throat held wide open.* The upper lip is tall and arching, with 2 small lobes. The lower lip has 2 large linear lobes held downward. Its 4 fertile stamens are held inside the top of the flower. *The sterile staminode is densely bearded with yellow hairs and sticks out of the lower lip like a fuzzy tongue.* In flower Apr-June.

An abundant shrub of hot interior canyons. Common in many of our major canyons such as Santa Ana, Santiago, Black Star, Silverado, and San Juan and the northern reaches of the San Mateo Canyon Wilderness Area. Widespread through the inland side of the Santa Ana Mountains, easily seen along Bedford Road and Indian Truck Trail. After flowering, the leaves often turn yellow or brown and fall off, leaving the plant leafless in summer and fall. In some areas, such as about 2 miles up from the USFS gate in Silverado Canyon, some south- and east-facing slopes are abundantly covered by this plant, which turn bright yellow with its blooms in April.

Named in 1846 by G. Bentham for the similarity of its flowers to those of true snapdragons in the genus *Antirrhinum* (Latin, -oides = likeness).

Yellow Bush Beardtongue

Yellow Bush Beardtongue

Chalcedona Checkerspot

Heart-leaved Bush Beardtongue

Heart-leaved Bush Beardtongue
Keckiella cordifolia (Benth.) Straw

A climbing shrub with stems that reach up to 3 m in length, clambering through or over other plants. *Leaves* flat, oval to triangular *with a heart-shaped base,* blade 20-65 mm long, edges with small teeth. *Corolla tubular,* 18-25 mm long, *scarlet red or red-orange, upper lip long and straight or bent upward, tipped with 2 small free lobes.* Lower lip with 3 long, narrow, twisted lobes held down or backward. Pistil and 4 fertile stamens are held just under the upper lip, easily seen. The *sterile staminode is covered with yellow-brown hairs, short,* and held within the flower. In flower (Mar-)May-Aug.

A common component of coastal sage scrub and chaparral, more common in moist areas that receive coastal fog in June. Widespread throughout our area, in all of our wilderness parks and major canyons. Popular in gardens, best planted near other plants on which to climb. A very important larval host plant of the **chalcedona checkerspot butterfly** (*Euphydryas chalcedona* E. Doubleday, Nymphalidae).

It was named in 1835 by G. Bentham for its heart-shaped leaves (Latin, cordis = heart; folium = leaf).

Heart-leaved Bush Beardtongue

Blue-stemmed Bush Beardtongue
Keckiella ternata (Torr.) Straw **ssp. *ternata***

A twiggy-looking shrub with upright stems to about 2.5 m tall. Upper leaves whorled in groups of 3(-4), lower ones are opposite. Leaf blade linear, 15-60 mm long, edges toothed, often folded up at the sides. *Corolla tubular, 21-21 mm long, red to red-orange.* Upper lip short, made up mostly of 2 linear free lobes. Lower lip has 3 flat linear lobes held downward, often curled backward. Pistil and 4 fertile stamens are held just under the upper lip. The *sterile staminode is densely covered with yellow hairs* and held in the flower. In flower June-Sep.

A plant of higher elevations in the Santa Ana Mountains, primarily on Santiago and Modjeska Peaks. Easily seen along Main Divide Road in spring and early summer.

Named in 1859 by J. Torrey for its 3-whorled leaves (Latin, ternatus = three each).

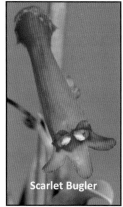

Far left: Blue-stemmed bush beardtongue has a red to red orange corolla with long lobes; its staminode is hairy and hidden in the throat. Left: Scarlet bugler has a bright red corolla with short lobes; its staminode is hairless and hidden in the throat. In our area, they are not known to grow near each other.

Scarlet Bugler
Penstemon centranthifolius Benth.

A hairless upright plant with few wand-like stems, 30-120 cm tall. Stems and leaves covered with white powder, giving the plant a gray-green appearance. Lowest leaves 3-7 cm long and spatula-shaped, middle leaves linear and longer, upper leaves linear, shorter, and sometimes fused to each other at their bases. *Corolla tubular, 20-30 mm long, bright red, held outward or downward in bud but drooping down loosely when open.* The free lobes are all short and rounded, held straight forward or nearly so, not folded up or back. Pistil and 4 fertile stamens are all held within the flower. The *short sterile staminode is yellow, hairless,* much shorter, and hidden well within the flower tube. In flower April-July.

Quite uncommon here, usually found growing with **coast live oak** (*Quercus agricola,* Fagaceae). Known from only a few locations in the Santa Ana Mountains such as near Upper San Juan Campground, Bear Canyon Trail, along North Main Divide Road between Santiago and Modjeska Peaks, and along South Main Divide Road en route to Elsinore Peak. Much more common farther north and in higher mountains east of our area. Pollinated by hummingbirds, probably also by sphinx moths.

Named in 1835 by G. Bentham for the resemblance of its leaves to those of valerian (Valerianaceae), a European plant in the genus *Centranthus* (Latin, folium = leaf).

Grinnell's Beardtongue
Penstemon grinnellii **ssp. *grinnellii*** Eastwood

A low, rounded, hairless shrub with many branches, it grows 10-60 cm tall and much wider. *Leaves thick, stiff,* 50-90 mm long, light- to yellow-green, linear to oval, toothed, *and narrow at their base.* Corolla rounded, 22-30 mm long, *pale lavender or pinkish,* covered inside and outside with short sticky-tipped hairs. The flower throat is quite narrow near the ovary, but once past the calyx it expands abruptly to give the flower a swollen look, as if it could swallow a bumble bee or a marble. The upper lip is 2-lobed, each lobe held upward or forward. The lower lip is 3-lobed, each lobe wavy-edged and held down or backward. *The floor and lower lip are heavily striped with dark purple "nectar guides" that act as roadways to help pollinators find their way into the flower.* The single white pistil and 4 fertile stamens are held inside the flower, just beneath the upper lip. *The very long sterile staminode is densely covered with long white hairs and held outside the flower like a tongue.* In flower May-Aug.

Normally a plant of higher elevations and more extensive coniferous forests, in the Santa Ana Mountains it grows on Santiago and Modjeska Peaks, even right along North Main Divide Truck Trail, and sparingly in upper Trabuco Canyon. The flowers give off a sweet, intoxicating scent that fills the air around them. Once pollinated, the flower ceases to produce the scent. It is pollinated primarily by carpenter bees but is also visited by many other species of local bees and flies. Often used in gardens.

In 1905, A. Eastwood named the plant in honor of F. Grinnell, an entomologist who collected the plant in 1903 on Mount Wilson in the San Gabriel Mountains, probably during an insect-collecting trip.

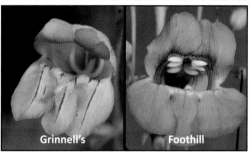

Right: Grinnell's beardtongue has a rounded corolla; its staminode is hairy and long. Far right, Foothill beardtongue has a long corolla; its staminode is hairless and medium length. In our area, they sometimes grow near each other.

Foothill Beardtongue
Penstemon heterophyllus **Lindl. var. *heterophyllus***

A hairless woody perennial, <60(25-150) cm tall. *Leaves* 2-4.5 cm long, *somewhat linear, narrow.* Corolla hairless, 23-40 mm long; it looks too large to be produced by such a slender plant. *Corolla color varies from magenta to blue, the free lobes blue or lilac.* The single purple pistil and 4 fertile stamens curve out toward the sides and up toward the roof of the flower. *The short sterile staminode is totally hairless, white, tipped with lavender, and lies on the floor of the flower tube.* In flower Apr-July.

Locally uncommon. Look for it on Sitton Peak and a few places on Santiago Peak along Main Divide Road.

Southern Foothill Beardtongue
Penstemon heterophyllus **Lindl. var. *australis*** Munz & I.M. Johnston (not pictured)

Similar to foothill beard-tongue but covered with short hairs and with narrower leaves. In flower May-June. This is the common form widely found in openings among chaparral in the Santa Ana Mountains such as along North Main Divide Road above Falcon Group Camp and south into the San Mateo Canyon Wilderness Area.

The epithet was given in 1836 by J. Lindley for its variable leaf widths (Greek, heteros = other, different; phyllon = leaf). The southern variety was named in 1924 by P.A. Munz and I.M. Johnston for its southern California distribution (Latin, australis = southern).

Royal Beardtongue
Penstemon spectabilis Thurber **ssp.** ***spectabilis***

An upright, hairless, many-stemmed perennial 80-120 cm tall. Stems and leaves green or blue-green, sometimes covered with a fine white powder and appearing tinged with gray. Leaves 35-100 mm long, linear to oval, toothed. *Most upper leaves are widely fused at their base. Corolla* 24-34 mm long and tubular, *blue to purplish in color.* The inner roof of the flower is white and has several short broad hairs near its mouth that glisten when lit. The pistil and 4 fertile stamens curve out toward the sides and up toward the roof of the flower. The pistil is purple, tipped with a white stigma. Stamens are white and lavender-tipped, their anthers all-white. *The sterile staminode is hairless, white, sometimes lavender-tipped, and lies on the floor of the flower tube.* In flower Apr-June.

A common plant among coastal sage scrub, chaparral, and oak woodlands mostly at higher elevations of the Santa Ana Mountains. Find it along Trabuco and Horsethief Trails, near Los Pinos Saddle, amid higher peaks in the range, Caspers Wilderness Park, and scattered sites within the San Mateo Canyon Wilderness Area. It is pollinated by many types of bees and wasps. A popular plant in the horticultural trade, it is easily cultivated and self-seeds into other parts of the garden.

It was named in by G. Thurber for its beautiful flowers (Latin, spectabilis = visible, remarkable).

Royal Beardtongue

Right: Royal beardtongue has long tubular flowers of blues and purples; its staminode is hairless and medium length. Far right: Blue toadflax has blue to blue-violet flowers with darker veins; it has no staminode.

Blue Toadflax
Nuttallanthus texanus (Scheele) D.A. Sutton
[*Linaria canadensis* (L.) Dum-Cours. var. *texana* (Scheele) Pennell; *Linaria texana* Scheele]

A slender, hairless, weak-stemmed annual, rarely a biennial, 10-60 cm tall, *with several non-flowering branches that trail at the base, perhaps for stability.* Its narrow linear leaves are 5-25 mm long, those on the non-flowering branches are oval to linear. *Corolla blue-lavender to blue-violet marked with purple veins, strongly 2-lipped, all 5 lobes are rounded.* It has a large whitish bulbous platform (palate) on the lower lip like a snapdragon. The bottom of the flower throat has a long backward spur as long as the flower itself. The 4 stamens are in 2 groups of different lengths. In flower Mar-May.

Widespread and common along the coast among grasslands and grassy openings in coastal sage scrub, though often overlooked because of its narrow stems and tiny flowers. Known from the slopes that surround upper Newport Bay, the San Joaquin Hills, and the Dana Point Headlands, inland to Lomas de Santiago, Mission Viejo, and Caspers Wilderness Park. Also in the San Mateo Canyon Wilderness Area, as at Oak Flats, Bluewater Canyon, and the north fork of Bear Canyon.

The epithet was given in 1848 by G.H.A. Scheele in reference to the state of Texas, where it was first collected.

Blue Toadflax

Speedwells: *Veronica*

Upright or sprawling annuals or perennials that grow from taproots or rhizomes *in soil inundated with shallow freshwater or very moist soils.* Leaves opposite and linear to oval. Their small flowers appear in the upper parts of the plant, solitary or in dense clusters. Calyx with 4-5 free lobes. *Corolla with 5 lobes, but the uppermost 2 lobes are fused together into a single broad lobe so it appears to have only 4 lobes.* The 2 stamens and single style extend outside of the flower. We have 4 species in our area; only 1 is native; 2 are found in or near waterways, and the last (*V. arvensis* L., not pictured) is a weed of overwatered lawns.

Mexican Speedwell

Mexican Speedwell
Veronica peregrina Pennell **ssp. *xalapensis*** (Kunth) Pennell

A taprooted annual, often reddish, covered with sticky-tipped hairs. Leaves linear to oval, 5-25 mm long, *very fleshy,* sometimes tipped with red. *Flowers appear singly among nodes,* with a leaf-like bract beneath each. *Corolla* 2-3 mm wide, *whitish,* all 4 lobes about equal in size and shape. In flower Apr-Aug.

A plant of moist ground and vernal pools. Known from vernal pools near Chiquita Ridge on the Rancho Mission Viejo, Fairview Park in Costa Mesa, and muddy soils near Quail Hill in Irvine, Laguna Lakes, and Yorba Linda. Also in Temescal Valley, Lake Elsinore, and the Santa Rosa Plateau.

Named in 1753 by C. Linnaeus probably for its white-flowers, which are unusual among speedwells (Latin, peregrinus = strange, foreign). Our form was named in 1818 by C.S. Kunth for the location of its original discovery in Jalapa, Veracruz, Mexico (Jalapa or Xalapa; Latin, -ensis = belonging to).

Great Water Speedwell ✖
Veronica anagallis-aquatica L.

A large succulent hairless perennial, it generally grows 10-60 cm tall. Leaves 20-80 mm long, broad and oval. Flowers are in dense clusters on opposing stalks that arise from the nodes. *Corolla* 5-10 mm wide, *pale lavender with violet lines,* upper petal much broader than others. In flower May-Sep.

A native of Europe, it has become a common and widespread plant of ponds and moist creekbeds, mostly at lower elevations. Look for it in such places as the Santa Ana River channel, upper Newport Bay, Audubon Starr Ranch Sanctuary, and creeks such as San Diego, San Juan, and Aliso.

The epithet was given in 1753 by C. Linnaeus for its flowers, which close at night and open up again when the sunlight reappears and strikes them, and for its preference for growing in water (Greek, ana- = back, again; agallein = to delight in; Latin, aquaticus = found in the water).

Great Water Speedwell

Persian Speedwell ✖
Veronica persica Poiret

A low-growing, hairy annual with branches up to 60 cm long, usually much shorter. *Leaves* 5-25 mm long, *oval or triangular,* often wavy-edged and toothed. *Corolla* 8-12 mm wide, *blue lined with purple, whitish in the center.* In flower Feb-May.

A native of Persia and the surrounding region, here it occurs as a weed in moist fields, creek edges, and overwatered lawns.

Named in 1808 by J.L.M. Poiret for the place of its discovery (Latin, persicus = belonging to Persia).

Persian Speedwell

Plantains: *Plantago*

Soft, often hairy annuals and perennials, usually without branches. *Most of our species have basal leaves with parallel veins and ribs that run down the length of the leaf; because of this they are sometimes mistaken for monocots such as grasses, rushes, or lilies.* One of our plantains, however, has branched stems and opposite leaves on the stems, making it look more like a snapdragon than a plantain. Flowers bowl-shaped, crowded in an oval head or long cylindrical spike. Each flower has beneath it a single bract. Sepals fused at the base and have 4 free lobes. Petals 5, but the upper 2 are fused so the flower appears to have only 4 petals, each colorless or green. Stamens 4 or 2, on long filaments that hold the anthers way out past the petals. Ovary superior. The single style has a very hairy clubbed stigma. Their flowers have no nectary; they are wind-pollinated and do not need to attract pollinating insects, though bees and flies collect and eat its pollen. The fruit is a capsule that when ripe loses its hat-like top half and spills out its seeds (*circumscissile capsule*).

The genus was named in 1753 by C. Linnaeus, probably for its distribution along footpaths (Latin, planta = sole of the foot or footprint). In the table, native species are listed first; non-natives are indicated by an asterisk (*).

 **Long spike** **Oval head**

	Growth	Leaf	Inflorescence	Flowers
Alkali *P. elongata*	Annual, spreading to upright	3-10 cm long; linear to thread-like; held flat or upright; *mostly hairless*	2-18 cm tall; *oval head or short spike on a long peduncle*	Petals narrow: 3 out or back, 1 upward with style; *only 2 stamens; fruit red,* 4-9-seeded, seeds dark
California *P. erecta* ssp. *erecta*	Annual, upright; *mostly coastal and inland scrub*	3-13 cm long; narrowly linear to thread-like but wider near tip; held upright; *sparse long silky hairs*	3-13 cm tall; *oval head on a short peduncle*	Petals round to oval, held outward or folded backward; 4 stamens; fruit green to tan, 2-seeded
Woolly *P. ovata*	Annual, upright; mostly desert and its edges	2-17 cm long; linear to oblong; *dense silky hairs*	2-27 cm tall; *oval head on a short peduncle, very woolly*	Petals round to oval, held outward; 4 stamens; bract beneath each flower oval to round, about equal in size to sepal; fruit 2-seeded
Mexican *P. subnuda*	Perennial, upright; *separate sexes*	12-40 cm long; elliptical to linear; *mostly hairless*	9-50 cm tall; *very long narrow spike on a long peduncle*	Petals linear with a pointed tip; males: petals upright, 4 stamens; females: petals outward, fruit 3-seeded
Red-seeded *P. rhodosperma*	Annual or biennial; *separate sexes*	5.5-30 cm long; linear to oval, broad near tip; narrow long hairs	3-30 cm tall; long narrow spike on a long peduncle	Petals linear with a pointed tip; males: petals curve upward, 4 stamens; females: petals outward, fruit 2-seeded, *seeds red*
Sand *P. arenaria**	Annual, upright, *many branches*	2-4.5 cm long; string-like or linear, opposite on stem	1-2 cm; *oval heads on a short peduncle, paired, from nodes*	Petals narrowly oval, held outward; 4 stamens; fruit 2-seeded
Cutleaf *P. coronopus**	Annual, biennial, upright	4-25 cm long; narrow lance; *pinnately lobed,* often lie flat	5-50 cm; *long spike on a medium peduncle; nodding in flower, upright in fruit*	Petals lance-ovate, 3 outward, 1 up; fruit 3-seeded
English *P. lanceolata**	Perennial, upright, *large*	5-25 cm long; linear to oblong; short-hairy	2-27 cm tall; *oval head on a long peduncle*	Petals oval, pointed tip, held outward; 4 stamens; fruit 1-2-seeded
Common *P. major**	Perennial, upright, low, short	5-18 cm long; elliptical or heart-shaped; *hairless*	5-60 cm tall; very long narrow spike on a very long peduncle	Petals tiny, oval to linear, held outward; 4 stamens; fruit 5-16-seeded
Dwarf *P. pusilla**	Annual, upright	1-10 cm long; thread-like; mostly hairless	2-19 cm tall; long narrow spike on a long peduncle	Petals lance, upright; *stamens 2; fruit 4-seeded*
Virginia *P. virginica**	Annual or biennial; *separate sexes*	2-12 cm long; linear to oval, broad near tip; narrow long hairs	3-33 cm tall; long narrow spike on a long peduncle	Petals linear with a pointed tip; males: petals curve upward, 4 stamens; females: petals outward, fruit 2-seeded

Alkali Plantain ✿
Plantago elongata Pursh

Spreading to upright annual, *hairless* or sparsely haired. Leaves very slender, linear, gently tapered to base, held upright. Inflorescence 2-18 cm tall; head oval to cylindric, stalk short to medium long. Petals narrow. *Stamens only 2. Fruit red,* usually 4-9-seeded (3-14), seeds dark brown to black. In flower Apr-June.

A plant of *salty alkaline soils, beaches, and vernal pools,* largely eliminated by development. Known from beaches in Corona del Mar, coastal bluffs in San Clemente, (former) vernal pools in Dana Point and Laguna Beach, and salty soils near Lake Elsinore.

Named in 1814 by F.T. Pursh for its long leaves (Latin, elongatus = elongated).

California or Dotseed Plantain
Plantago erecta Morris **ssp.** *erecta*

Upright annual, long-haired. *Leaves very slender, linear, gently tapered to base, held upright.* Inflorescence 10(3-30) cm tall; head short round to oval (short cylindric); stalk short; *hairs straight, usually point upward.* Petals outward or folded backward. Stamens 4. Fruit 2-seeded, *seeds dull brown, not shiny.* In flower Mar-May.

A common plant of many different situations and soils such as sand, clay, and serpentine. Found in openings among grasslands, coastal sage scrub, chaparral, and oak woodland throughout our area. Known from bluffs and canyons in Corona del Mar and Newport Beach, the San Joaquin Hills, Sycamore Hills, the Dana Point Headlands, bluffs in San Clemente, Mission Viejo, the Chino Hills, Lomas de Santiago, hills north of Irvine Lake, Audubon Starr Ranch Sanctuary, Trabuco Canyon, up to Sierra and Bedford Peaks, the Santa Rosa Plateau, the San Mateo Canyon Wilderness Area, and throughout Temescal Valley. It is the primary foodplant for larvae of the **quino checkerspot butterfly** (*Euphydryas editha quino* Behr, Nymphalidae), which is federally listed as Endangered.

Named in 1900 by E.L. Morris for its upright leaves and stalks (Latin, erectus = erect).

Similar. **Woolly plantain** (*P. ovata*): nearly identical; densely covered with silky hairs; leaves wider; petals held outward; seeds reddish-yellow, shiny; in dry soils; uncommon.

Woolly Plantain
Plantago ovata Forsskål

Upright annual, covered with silky hairs. *Leaves very slender, linear, gently tapered to base, held upright.* Inflorescence 2-27 cm tall; head short oval to cylindric, covered with woolly hairs; stalk short; hairs *wavy, usually point in all directions* (upward). Petals held outward. Stamens 4. Fruit 2-seeded, *seeds reddish-yellow and shiny.* In flower Feb-Apr.

Known from the Chino Hills, Dana Point, and Trabuco Canyon. Perhaps it was once more common and widespread, but its frequent use in restoration projects has obscured its native range.

Named in 1775 by P. Forsskål, probably for the shape of its petals or bracts (Latin, ovatus = egg-shaped).

Similar. **California plantain** (*P. erecta*): *nearly identical;* less hairy; leaves similar; petals held outward or backward; seeds dull brown, not shiny; in moist or dry soils; widespread, common.

Red-seeded Plantain
Plantago rhodosperma Decne.
   Annual to biennial. *Leaves long, broad, oval, gently tapered to base, usually tooth-edged;* leaves low, sometimes flat on the ground. *Inflorescence 3-30 cm tall; a long hairy cylinder. Flowers of only one sex.* Stamens 4 in male flower. Fruit 2-seeded, *seeds deep red to red-brown, visible through fruit wall.* In flower May.
   Found in rocky to sandy soils along the coast, at Pelican Hill in 1988 and Corona del Mar in 1908, possibly extirpated.
   Named in 1852 by J. Decaisne for its red seeds (Greek, rhodon = a rose, red; sperma = seed).
   Similar. Virginia plantain ✖ (*P. virginica* L., not pictured): nearly identical; *seeds pale brown;* native to central and eastern U.S.

Red-seeded Plantain

Sand Plantain ✖
Plantago arenaria Waldst. & Kit. [*P. indica* L.]
   Annual. *Many-branched,* quite unusual for a plantain. *Leaves linear, short, opposite on stem. Inflorescence an oval head on a short stalk, opposite on stem.* Stamens 4. Fruit 2-seeded. In flower July-Nov.
   Grows in sandy soils near the coast, as at Bolsa Chica Ecological Reserve, the lower Santa Ana River channel, and Newport Beach. Native to Eurasia, sometimes used in commercial bird seed.
   Named in 1801 by F.P.A. von Waldstein and P. Kitaibel for its sandy habitat (Latin, arenarius = pertaining to sand).

Sand Plantain

English Plantain ✖
Plantago lanceolata L.
   Large perennial. *Leaves very long,* slender, linear, gradually tapered to base. *Inflorescence an oval head on a very long stalk,* 2-27 cm tall. Stamens 4. Fruit 1-2-seeded. In flower Apr-Aug.
   Native to Europe, a common weed of watercourses in wild areas.
   Named in 1753 by C. Linnaeus for the shape of its leaves (Latin, lanceolatus = lance-like, armed with a pointed weapon).

Common Plantain ✖
Plantago major L.
   Annual to perennial. *Leaf blade wide, oval, abruptly narrowed to the petiole;* leaves usually low, sometimes flat on the ground. *Inflorescence 5-60 cm tall; a long cylinder on a very long stalk.* Stamens 4. Fruit 5-16-seeded. In flower Apr-Sep.
   Native to Europe, a common weed of disturbed places and watercourses in wild areas.
   Named in 1753 by C. Linnaeus, probably for its large wide leaves (Latin, major = greater, larger).
   Similar. Mexican plantain (*P. subnuda* Pilger, not shown): perennial native; *leaves elliptic, gradually narrowed to a wide-winged petiole;* often lie flat on the ground; inflorescence a very long narrow spike on a long stalk; *flowers of only one sex,* stamens 4 in male; fruit 2-seeded; moist soils near the coast; in flower May-Sep. **Cutleaf plantain ✖** (*P. coronopus* L., not shown): *Leaves pinnately lobed,* often lie flat; *inflorescence similar, nodding in flower, upright and/or curved in fruit;* stamens 4; fruit 3 (2-4)-seeded; in flower Apr-July; coastal, salty soils; Seal Beach National Wildlife Refuge; native to Europe.

English Plantain

Dwarf Plantain ✖
Plantago pusilla Nutt. (not shown)
   Annual, *mostly hairless.* Leaves very slender. Inflorescence 2-19 cm tall; a long cylinder. *Stamens 2. Fruit 4-seeded.* In flower Mar-May.
   Uncommon, known only from depressions in sandy soils in Baker Canyon. Native to eastern U.S.
   The epithet was established in 1818 by T. Nuttall for its small size (Latin, pusillus = very small, weak).

Common Plantain

PLATANACEAE • PLANE TREE OR SYCAMORE FAMILY

Large, winter-deciduous trees with watery sap and thin bark. *Leaves, large, palmate,* 3-5-lobed, arranged alternately on the stem. Young leaves have a large leafy stipule at their base that may later detach but often remains on the stem like a collar. The axillary bud is enclosed by the petiole base. *The unisexual flowers are minute and grow in unisexual spherical inflorescences that hang down on a stalk.* Male flowers have 3-6 yellow stamens. Female flowers have 9 superior ovaries, each tipped with a single dark red fuzzy stigma. Males release their yellow pollen to the wind, which carries it to the female flowers. After pollen release, male flowers and spheres dry up and fall off. Fertilized female flowers produce bristly fruits that ripen, turn brown, and fall apart during winter storms as many dry 1-seeded achenes. One genus, with 8 species.

California Sycamore, Aliso
Platanus racemosa Nutt.

A tall tree, 10-35 meters in height, *from a single or multiple trunk that often leans or curves.* Its multicolored white, tan, and grey bark peels away by itself in thin, puzzle-like pieces. *Leaves alternate; large, palmate, hairy, especially below, with white to tan star-shaped hairs that fall off with age.* The petiole is swollen and hollow at its base, which hides the developing bud. In flower Feb-Apr.

It grows throughout our area along stream and river courses with deep sandy soils, mostly at low elevations but also in seeps at higher elevations. Look for it among riparian systems in all of our wilderness parks, canyons, hills, and mountains.

The genus was named in 1753 by C. Linnaeus, who used the ancient name for plane tree (Latin, platanus; Greek platanos = the Oriental plane tree). The epithet was given in 1842 by T. Nuttall in reference to its fruits (Latin, racemulus = the stalk of a cluster or bunch of berries or grapes).

Similar. Big-leaf maple (*Acer macrophyllum,* Sapindaceae): *leaves opposite* with 5 large lobes, each with secondary lobes, slightly hairy below; fruits are winged samaras.

Right: a very large, very old, multitrunked tree at Tree of Life Nursery. (The dark green shrub at lower left foreground is a toyon, Heteromeles arbutifolia, Rosaceae).

THE SYCAMORE GUILD

Many living things rely on sycamore for food and shelter. Here are just a few (see also the mistletoe family, Viscaceae).

Sycamore Anthracnose Fungus
Apiognomonia veneta (Sacc. & Speg.) Höhnel (Valsaceae)

This fungus feeds on the young leaves and stem tissue in spring, killing some leaves and twigs and forcing the development of side shoots, thus leading to the plants' angular growth form. In years of good rainfall and/or high humidity, the fungus often thrives and kills off the first crop of leaves and twigs. As the air dries out in late spring, the fungus enters dormancy and the tree produces a healthy second crop of leaves and twigs. The epithet was given in 1878 by P.A. Saccardo for its blue color (Latin, venetus = sea-colored, bluish).

Sycamore anthracnose fungus often kills the first crop of leaves and twigs.

Pseudoscorpions
(Arachnida: Pseudoscorpiones)

These harmless relatives of true scorpions are *small, generally under 5 mm long, have large claws but no tail.* They live under loose bark of sycamores and other trees, where they feed on mites and other small arthropods.

Pseudoscorpions are small and harmless

Western Sycamore Lace Bug
Corythucha confraterna Gibson (Tingidae)

A small flat true bug, 3-5 mm long when adult. The thorax and forewings are modified into glasslike structures with lacy designs. The thorax also has a hood-like extension that covers the head. On these lacy structures, the ridges are white, the open areas mostly clear; a few gray or brown patches provide the only color. Nymphs are black and white. All life stages live together on the underside of leaves, where they use their piercing-sucking mouthparts to feed on plant sap. Small round black dots visible in the photograph are fecal piles. They spend the winter as adults, hiding in bark cracks and fallen leaves.

Named in 1918 by E.H. Gibson for its close association with others of its species, crowded together on the leaves (Latin, con- = with; frater = a brother).

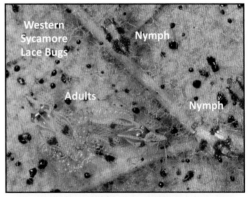
Western Sycamore Lace Bugs, Nymph, Adults, Nymph

20-spotted Ladybird Beetle
Psyllobora vigintimaculata Say (Coccinellidae)

A small ladybird beetle, 1.75-3 mm long; pale, marked with dark brown spots. On some individuals, a number of the spots are lighter brown than others. A common insect found on sycamores throughout our area mostly in spring and summer. Long thought to feed on **sycamore scales** (*Stomacoccus platani* Ferris, Steingeliidae), it actually feeds on mildew fungi. It is not known if it also feeds on sycamore anthracnose fungi, but we have observed it on leaves infected with the fungus.

The epithet was given in 1824 by T. Say for its 20 spots (Latin, viginti = 20; maculatus = spotted).

Similar. Northern 20-spotted ladybird (*P. borealis* Casey, not pictured): a bit larger, 2.4-3.1 mm long, broader, huskier; most forewing spots larger, two spots at tip of forewings farther away from the other spots; also found on sycamore but less common.

20-spotted Ladybird

Sycamore Borer Moth
Synanthedon resplendens (Hy. Edwards, Sesiidae)
The adult moth resembles a paper wasp. Larvae tunnel through the bark of sycamore and oaks, causing it to darken, thicken, roughen, and crumble. Tell-tale signs of the larvae are the many holes they leave in the bark. Larvae pupate in bark with their head-end facing outward and emerge as adults in spring-summer.

The epithet was given in 1881 by H. Edwards for the glitter-like scales on the adult (Latin, resplendens = glittering, shining).

Sycamore borer moth.
Left: California sycamore bark roughened by tunneling and feeding by larvae.
Right: Adult nectaring on narrow-leaved milkweed in the daytime.

Western Tiger Swallowtail Butterfly
Papilio rutulus rutulus Lucas (Papilionidae)
Its green larvae have two large false eyes ("eyespots") near the head-end. They feed on sycamore and willow leaves, pupate on their branches, and emerge as adults in spring-summer. The late-season generation spends the winter in the pupal stage and emerges in late winter-early spring. *Adults have a wingspan of 7-10 cm, bright yellow striped with black; the back edge of the hindwing has red and blue patches within the black marginal band.* They are commonly seen throughout our area as they sail effortlessly through the air.

The epithet was given in 1852 by H. Lucas in reference to the red patches on the hindwing (Latin, rutilis = ruddy).

Right, upper: An adult western tiger swallowtail rests on coast live oak (Quercus agrifolia, Fagaceae). They often bask with wings open, presenting more surface area to the sun and collecting heat. Near right: The larva is green, sometimes brownish-orange, always with blue eyespots surrounded by yellow. Far right: The pupa (chrysalis) sits upright, is medium brown, and is held in place by a silk "belt." You can see an outline of a wing facing the twig.

Anna's Hummingbird
Calypte anna (Lesson, Trochilidae)
An average-sized hummingbird with a body 9-10 cm long. Its belly is medium to pale gray, the wings dark. Upper parts of *adult male* are bright green to bluish; *his gorget* (throat-patch) *and crown* (top of head) *are rose red to coppery red. Adult female bright green; her gorget is a small central patch of red to copper.* Nests any time of year in urban and suburban areas, mostly Nov-Apr in rural and wild lands.

A year-round resident, common and widespread throughout our area. A mated female constructs her nest from hairs pulled from the underside of sycamore leaves and other plant parts and animal hair, bound together with spider silk collected from webs. The resulting nest is a soft-textured cup, usually nestled among the branches of trees and shrubs, occasionally on man-made objects such as signs.

The epithet was given in 1829 by R.P. Lesson in honor of Anna Masséna, Duchess of Rivoli (France).

PLUMBAGINACEAE • LEADWORT FAMILY

Annuals, perennials, shrubs, or vines. *Ours are perennials with a basal rosette of oval or linear leaves.* The inflorescence is a head or a branched cluster of small flowers above the leaves, atop a leafless stalk. *Calyx bell-shaped or tubular,* often tipped with 5 free lobes, each with a ridge on the outside; has the feel of dry straw. *Corolla tubular with 5 free tips, usually blue, purple, pink, or white.* Stamens 5, opposite the petals. Ovary superior with 5 linear styles, sometimes fused together. Fruit a capsule or dry achene, enclosed by the calyx. The corolla closes up at night and reopens the following morning.

Some European species are cultivated here as landscape ornamentals; many have escaped and invaded natural habitats. The floral industry uses them for fresh and dried flower arrangements. We have 7 in our area, 1 native, 6 introduced.

California Sea-lavender, Western Marsh-rosemary
Limonium californicum (Boissier) Heller

A perennial, the main stem red, thick, and woody at its base. Its *long leaf blades* are 10-30 cm long, 1-6 cm wide, *oblong to oval, flat or wavy, sometimes thick and leathery, tapered at their base. The petiole is shorter than the leaf blade and has narrow "winged" edges.* The leaf and petiole have 3 obvious long veins (best seen from the underside), each with many forked side branches. *Inflorescence 20-50 cm tall, hairless, not winged.* Flowers small, about 5-6 mm long, 2 mm wide. *Calyx white and hairy. Corolla pale violet to blue, only slightly longer than the calyx.* In flower July-Dec.

Widespread and common in salt marshes along our coast, most often rooted in very wet sandy soil, sometimes within the high tide region. Look for it at Seal Beach, Bolsa Chica Ecological Reserve, upper Newport Bay (easily found at North Star Beach), and Little Corona Beach.

The epithet was given in 1848 by E.P. Boissier in honor of the state of its discovery.

Perez's Sea-lavender ✖
Limonium perezii (Stapf) F.T. Hubbard

A stout perennial, woody at its base. *Leaf blade 4-15 cm long, round to oval, edges wavy, hairy, base abruptly widened (not tapered to the petiole). Leaf tip with a long point, 4 mm or longer.* Inflorescence many-branched, 15-45 cm tall, uppermost branchlets winged, hairless to short-haired. Flowers about 10 mm long, 4-5 mm wide, arranged in pairs. *Calyx blue-purple. Corolla white (pale yellow).* In flower Mar-Sep.

Native to the Canary Islands. Cultivated, sometimes escapes. In our area, known from sandy soils along the coast such as at upper Newport Bay and Dana Point.

The epithet was given in 1908 by O. Stapf in honor of J.V. Perez.

Similar. Despréaux's sea-lavender ✖ (*L. preauxii* (Webb & Berthel.) Kuntze, not pictured): Leaves pinnately lobed at base, blade base broad, heart-shaped, edges hairless to few-haired; named in 1891 for J.M. Despréaux. **Sventenius' sea-lavender ✖** (*L. sventenii* A. Santos & M. Fernandez, not pictured): Leaves pinnately lobed at base, blade base tapered, edges hairless to few-haired; named in 1983 for E.R. Sventenius. Both are uncommon garden escapes, found in Lake Forest. Native to the Canary Islands.

European Sea-lavender ✖

Limonium duriusculum (Girard) Fourr.

An upright or spreading perennial to 30 cm tall, basal tuft not crowded. *Basal leaves* 1-4 cm long, 5-9 mm wide, *tips very broadly rounded.* Inflorescence few-branched, open. Flowers slender, in clusters of 1-3 . Calyx 5-6 mm long, green to red. *Corolla* about 8 mm long, *pale pink, held widely open.* In flower Sep-June.

Occurs in coastal salt marsh, riparian, and disturbed areas. Known from Yorba Linda, upper Newport Bay and blufftops. Native to the Mediterranean region.

The epithet was named in 1844 by F. de Girard for its woody stem (Latin, durusculus = hard or woody).

Similar. Rock sea-lavender ✖ (*L. binervosum* (G.E. Sm.) C.E. Salmon, not pictured): basal tuft dense; basal leaves 3-10 cm long, 0.7-1.4 cm wide, reverse-oval, base tapered, tip broadly rounded, sometimes with a single spine; *leaf and petiole surrounded by 2 thin branchless veins that are parallel to the main vein;* inflorescence many-branched, hairless; flowers small and narrow; calyx red with red central vein and whitish tips, generally hairless; corolla lavender to pink; in flower June-July; upper Newport Bay and Irvine; named in 1831 by G.E. Smith for the 2 thin veins in its leaves (Latin, bi- = 2; nervus = a sinew, tendon, nerve).

Algerian Sea-lavender ✖

Limonium ramosissimum (Poiret) Maire

Upright perennial *from a very dense basal tuft. Basal leaves* 3-10 cm long, 7-20 mm wide; *often spatula-shaped, tips pointed to rounded.* Inflorescence many-branched, not winged. *Flowers in clusters of 2-5, crowded at inflorescence tips.* Calyx white. Corolla pale pink. In flower June-July.

A native of the Mediterranean region, used in the horticultural trade, from which it has escaped. Recorded from upper Newport Bay, Irvine, Laguna Niguel, Lake Forest, Rancho Santa Margarita, Corona, and Temescal Valley, seemingly anywhere near planted landscapes.

The epithet was given in 1789 by J.L.M. Poiret for its many-branched inflorescence (Latin, ramulosus = full of branches; -issimus = very much, most).

Similar. Rock sea-lavender ✖ (*L. binervosum,* not pictured): see note under European sea-lavender. **European sea-lavender ✖** (*L. duriusculum*): Upright to sprawling; leaves smaller, shorter, tips much rounder; inflorescence less crowded at the tips; flowers more broadly opened; in flower Sep-June; Yorba Linda, upper Newport Bay.

Winged Sea-lavender ✖

Limonium sinuatum (L.) Mill.

A rough-haired perennial or biennial. Leaf blade 3-12 cm long, oval in outline, pinnately lobed, tapered at the base. *Both the petiole and the flowering stalk are winged.* Along the inflorescence, the wings produce linear leaf-like appendages. Flowers about 1 cm long, in groups of 3-5. *Calyx blue to white, funnel-shaped, ruffled at its tip, without obvious free lobes. Corolla pale yellowish-white with rounded lobes, just a bit longer than the calyx.* Cultivated forms include those with yellow, white, purple, and pink calyces and corollas. In flower June-Oct.

A native of the Mediterranean, escaped from cultivation and the cut-flower industry. Known from Santiago Creek in Orange, Serrano Creek, and upper Newport Bay.

The epithet was given in 1753 by C. Linnaeus for its sinuous leaf edges (Latin, sinuatus = bent, curved).

POLEMONIACEAE • PHLOX FAMILY

Delicate annuals, herbaceous perennials, or shrubs. Leaves simple and undivided in some, pinnate in many. *Calyx and corolla funnel-shaped or tubular,* both with 5 free lobes. *Corolla tube often very long.* Stamens 5, connected to the corolla. Anthers and pollen often brightly colored and helpful for identification. *Ovary superior; the single slender style is often longer than the stamens and ends in 3 narrow branches.* Fruit a capsule. Pollinated mostly by butterflies, moths, and long-tongued bees (common, Apidae, and leafcutter, Megachilidae), bee flies (Bombyliidae), and various beetles. About 320 species worldwide, 170 in California, 23 in our area.

Blue False-gilia

Allophyllum glutinosum (Benth.) A.D. Grant & V. Grant

A delicate annual 10-60 cm tall, reported to have a skunk-like odor (but neither Bob nor Fred can smell it). Stems green to red, covered with long hairs, usually sticky-tipped. Leaves light to dark green with many pinnate lobes, alternate. Lowest leaves in a basal cluster, 3-10 pairs of narrow lobes. Upper leaves with fewer but wider pinnate lobes, uppermost often palmate. Flowers in groups of 2-8. *Corolla 6-11 mm long, often 2-lipped, pale blue to lavender.* Style and stamens lie on the lower lip and curve upward. In flower Apr-June(-Sep).

A common plant of *sandy or rocky soils* in openings among coastal sage scrub, chaparral, oak woodlands, and trail edges, often in shade and after fire. Known from canyons and major peaks of the Santa Ana Mountains, including the summits of Santiago and Modjeska Peaks, and near Lake Elsinore.

The epithet was given in 1833 by G. Bentham for the sticky substance along its stems (Latin, gluta = glue; -osum = full of).

Blue False-gilia

Santa Ana River Woolly-star ★

Eriastrum densifolium (Benth.) H. Mason
ssp. *sanctorum* (Milliken) H. Mason

An upright perennial to about 1 meter, *covered with white-woolly hairs;* the main stem becomes woody with age. *Leaves 10-50 mm long, pinnate, with 2-6 narrow lobes, each spine-tipped;* alternate. Inflorescence crowded. Calyx lobes unequal in length, very woolly. *Corolla 25-32 mm long, funnel-shaped with a long tube, medium blue with darker lines,* rarely white or pinkish. In flower May-Aug.

The sandy gravelly soils in which the plant is found are occasionally scoured by floods. Known only from the Santa Ana River channel, historically from the base of the San Bernardino Mountains downstream to Anaheim. Much of it has been eliminated by habitat destruction and flood control such as the construction of Prado Dam. Now found mostly near Mentone and Redlands. Though it was last seen in our area in 1929, the Santa Ana River channel should be routinely surveyed for its presence. Listed as Endangered by federal and state agencies. CRPR 1B.1.

The epithet was given in 1833 by G. Bentham for its many leaves (Latin, densus = dense, compact; folium = leaf). Our form was named in 1904 by J. Milliken for the plant's habitat, along the Santa Ana ("Saint Anne") River (Latin, sanctus = sacred, saintly).

Santa Ana River Woolly-star

THE WOOLLY-STAR GUILD
Acton Giant Flower-loving Fly

Rhaphiomidas acton Coquillett (Apioceridae)

A large stout fly about 4 cm long with a very long proboscis held directly in front of the body. Little is known about its life cycle. The larvae spend 1-2 years (or more) in loose sand, where they probably feed on other insects. After pupation, adults emerge as adults in summer and nectar from Santa Ana River woolly-star (perhaps its sole pollinator). Eggs are laid on plants or directly in sand.

Acton Giant Flower-loving Fly

Sapphire Woolly-star
Eriastrum sapphirinum (Eastwood) H. Mason

An upright annual, 5-40 cm tall, often many-branched, covered with tiny sticky-tipped hairs. Leaves mostly 1-3 cm long, linear, undivided, the upper sometimes 3-lobed. *Corolla* funnel-shaped with a short tube and long narrow lobes. *Color variable, bright to pale blue (rarely white), with a white or yellow throat, sometimes also with a yellow blotch at the base of each lobe.* Anthers yellow to pale white. In flower May-Sep.

A delightful plant that brightens our trails with its brilliant flowers. It grows in sandy or rocky soil among openings in coastal sage scrub and chaparral, often in hot full sun. Commonly found throughout the Puente-Chino Hills, Lomas de Santiago, Santa Ana Mountains, and Temescal Valley. Less abundant in the San Joaquin Hills of Laguna Beach and foothills down to Mission Viejo.

The epithet was given in 1904 by A. Eastwood for the color of its corolla (Latin, sapphirinus = of sapphire).

Sapphire Woolly-star

Los Angeles or Grassland Gilia
Gilia angelensis V. Grant

An upright annual, 7 -70 cm tall, with many branches. Leaves hairy, 2-7 cm long, 1-3 pinnate, in a basal rosette. Inflorescence 1-10-flowered. *Corolla* about 5-12 mm across, *bell-shaped, lavender, the lobes round, sometimes pointed. The throat is short, yellowish, and has no spots.* Stamens and style only slightly longer than the throat. Anthers blue-lavender to whitish. The corolla falls off as the flower ages. In flower Mar-May.

A common widely distributed plant found in sandy or rocky soils, often in grasslands. Known from many places in our area, generally not at high elevations. Recorded from such places as upper Newport Bay, UCI Ecological Preserve, Laguna Niguel, the Dana Point Headlands, coastal bluffs in San Clemente, the Puente Hills, Lomas de Santiago, Limestone Canyon, Arroyo Trabuco in O'Neill Regional Park, Caspers Wilderness Park, the Santa Rosa Plateau, the San Mateo Canyon Wilderness Area, and Temescal Valley.

The epithet was given in 1952 by V.E. Grant for Los Angeles County, where it was first found (Latin, -ensis = belonging to).

Los Angeles Gilia

Ball Gilia
Gilia capitata Sims **ssp.** ***abrotanifolia*** (E. Greene) V. Grant

An upright annual 20-80 cm tall, hairless or covered with sticky-tipped hairs. *Lower leaves often in a basal rosette, 1-2-pinnate;* upper leaves 1-pinnate or undivided. Inflorescence stalk long, often leafless, topped by a round sphere of 25-50 flowers. *Corolla* about 1 cm across, *light to medium blue or blue-violet, rarely white.* Stamens and style are clearly longer than the throat, as long as or longer than the corolla lobes. Anthers blue-lavender to white. In flower Apr-May.

Known from sandy soils among coastal sage scrub and chaparral, more common after fires. Widely known from the Chino Hills (Telegraph Canyon) and Santa Ana Mountains, as at Silverado Canyon, Maple Springs Saddle, Bear Springs, Holy Jim Canyon, Modjeska Canyon, Modjeska Peak, and Indian Truck Trail. In cultivation as a popular garden annual.

The epithet was given in 1826 by John Sims for its head-like inflorescence (Latin, capitatus = having a head). The subepithet was named in 1895 by E.L. Greene for its leaves (Latin, abrotes = delicacy, splendor; folium = leaf).

Ball Gilia

Slender Desert Gilia

Gilia ochroleuca M.E. Jones **ssp. *exilis*** (A. Gray) A.D. Grant & V. Grant

An unusual annual, its central stem is 6-15 cm tall. It usually has several branches from the base that are much taller, to 30 cm tall (unless stunted by drought). *Leaves with cobweb-like hairs. Lowest leaves 2-3 cm long, pinnate, in a basal rosette.* Upper leaves usually not divided but sharp-toothed. Inflorescence a loose cluster of flowers. *Corolla funnel-shaped, lobes rounded, pink to white. The throat shallow and broad, blue in upper portion, yellow in lower.* Stamens and style only slightly longer than the throat, clearly shorter than the corolla lobes. In flower Mar-June.

Occurs in sandy soils of the Temescal Valley, from Corona to Lake Elsinore and Temecula. Common in Rice and Indian Canyons (even near the I-15, Temescal Canyon Road, and south of Glen Ivy). Diminishing because of development.

The epithet was given in 1898 by M.E. Jones for its yellowish-white corolla throat (Greek, ochra = yellow-ochre; leukon = white). Our local form was named in 1886 by A. Gray probably for its slender calyx lobes or its thin stems (Latin exile = small, thin, slender).

Southern Gilia

Saltugilia australis (H. Mason & A.D. Grant) L.A. Johnson
[*Gilia australis* (H. Mason & A.D. Grant) V. Grant]

An upright annual, 20-45 cm tall (rarely to 70 cm). *Basal leaves 2-7 cm long, 2-pinnate (1-3-pinnate), hairy; stem leaves linear, may be lobed at their base.* The inflorescence is atop a long stalk. Calyx hairless, green, sometimes purple or purple-marked, the lobes long and narrow. *Corolla 5-10 mm long, funnel-shaped, the tube and throat broad, pale to yellow with yellow spots, much longer than calyx. The corolla lobes are lavender to white, commonly streaked with purple on the outside.* Stamens and style are held just outside the throat, anthers and pollen blue. In flower Mar-June.

An uncommon plant of sandy soils and openings among grasslands and chaparral, more common after fire. Known from the hills above San Clemente, at the summit of Ortega Highway near El Cariso, sparingly in the San Mateo Canyon Wilderness Area, and in the hills in northern Lake Elsinore.

The epithet was given in 1948 by H.Mason and A.D. Grant for its southern distribution (Latin, australis = southern).

Whisker Brush

Leptosiphon ciliatus (Benth.) Jeps.
[*Linanthus ciliatus* (Benth.) E. Greene]

Upright annual, 2-30 cm tall, *branched from the base, covered with stiff hairs. Leaves 5-20 mm long, palmately lobed into narrow segments, each tipped with a tiny spine.* Leaves opposite, placed far apart, exposing the reddish or greenish stem. Inflorescence a crowded sphere at the branchtips. *Corolla tube 10-25 mm long, hairy on the outside, white or pink. Corolla throat yellow, sometimes white above, each pink to white lobe with a red or pink patch near the base.* Stamens and style are held just barely out of the throat; anthers and pollen orange or yellow. In flower Apr-July.

Uncommon here, more abundant in taller, wetter mountain ranges elsewhere. Found in dry openings in coastal sage scrub and chaparral. Known from northwestern Mission Viejo and Maple Springs.

The epithet was given in 1849 by G. Bentham for the little hairs that cover the outside of its corolla tube (Latin, ciliatus = with little hairs).

Many-flowered Linanthus
Leptosiphon floribundus (A. Gray) J.M. Porter & L.A. Johnson **ssp. *floribundus***
[*Linanthus floribundus* (A. Gray) Milliken; *L. nuttallii* (A. Gray) E. Greene ssp. *floribundus* (A. Gray) Munz]

A low rounded perennial 10-40 cm tall *with dense foliage and hairy stems, leaves, and calyx.* Leaves 8-20 mm long, palmately lobed into 3-5 narrow segments, each tipped with a tiny spine; opposite on the stem. Inflorescence is a bit crowded at the branchtips. *Corolla funnel-shaped with a short tube, only 5-9 mm long, white with a yellow throat.* The anthers and pollen are orange or yellow, held just inside the throat. In flower May-Aug.

Occurs in openings and road cuts among chaparral and coniferous forests, often in decomposing granite soils or cracked rocks, frequently in shaded sites. Common and widespread in the Santa Ana Mountains such as its major canyons (Black Star, Silverado, upper Trabuco, San Juan), Audubon Starr Ranch Sanctuary, many places along North Main Divide Truck Trail, south into the San Mateo Canyon Wilderness Area. Easiest seen along North Main Divide Road "Loop."

Many of its flowers open up simultaneously, covering the plant with a blanket of white. Oddly, the scent of its flowers is often compared to that of horse urine, but you should judge it for yourself. Reported to be largely self-pollinated but likely pollinated by flies.

The epithet was given in 1870 by A. Gray for its abundant flowers (Latin, floris = flower; abundo = abundant).

Similar. Smooth many-flowered linanthus (*L. floribundus* ssp. *glaber* (R. Patterson) J.M. Porter & L.A. Johnson, not pictured): *stem, leaves, and calyx hairless;* known from the Modjeska (Fleming) Ranch and many places along Main Divide Truck Trail, never growing together.

Many-flowered Linanthus

Many-flowered Linanthus

Flax-flowered Linanthus
Leptosiphon liniflorus (Benth.) J.M. Porter & L.A. Johnson
[*Linanthus liniflorus* (Benth.) E. Greene]

A slender delicate annual 10-50 cm tall *with thread-like stems,* generally branched above the base. Leaves 10-30 mm long, palmately lobed into 3-9 narrow segments, each tipped with a tiny spine. Leaves opposite, placed far apart on the reddish or greenish stem. Inflorescence very loose, few-flowered, or flowers solitary. *Corolla funnel-shaped with a short, relatively broad tube; lilac, pink, or whitish, often with purplish veins, the throat yellowish.* Corolla lobes 8-10 mm long, rounded or tapered. Stamens and style are held far above the throat; anthers and pollen orange or yellow. Stamens hairy at their base. In flower Apr-July.

An unusual-looking plant, easily overlooked. Its slender stems are seemingly invisible and make the flowers appear to float unattached. Every spring we receive many requests to identify this wildflower. Common in grasslands, meadows, and openings among coastal sage scrub and chaparral. Known from the Chino Hills, Lomas de Santiago, and Santa Ana Mountains, in places such as Rancho Mission Viejo, Chiquito Basin, and many sites along Main Divide Truck Trail. Easiest found at Whiting Ranch Wilderness Park and Limestone Canyon.

The epithet was given in 1833 by G. Bentham for its flowers, which resemble flax (the genus *Linum*; Latin, floris = flower).

Flax-flowered Linanthus

Coast Baby-star
Leptosiphon parviflorus Benth.
[*Linanthus parviflorus* (Benth.) E. Greene]

An upright hairy annual, 5-25 cm tall. Leaves 8-25 mm cm long, *palmately lobed into 5-9 narrow segments, each tipped with a tiny spine*. Leaves opposite, placed far apart, exposing the reddish or greenish stem. Inflorescence is a crowded sphere at the branchtips. The *narrow corolla tube* is 20-35 mm long, hairy on the outside, often red at the base. *The corolla throat is yellow, the lobes white, yellow, or pink (rarely blue or purple)*. Stamens and style are held very far out of the throat; anthers and pollen orange or yellow. In flower Apr-May.

One of the most charming wildflowers in our area, *it often grows in dense masses*. Their small size makes you sit down and take notice. In cultivation, available in native wildflower seed packs. Pollinated by moths and butterflies, which use their long proboscis to get at the nectar held near the ovary at the base of the long tube.

Known throughout our area, mostly in openings among grasslands, coastal sage scrub, and chaparral. Often quite abundant after fire. Find it in the San Joaquin Hills of Laguna Beach, the Chino Hills, upper Fremont Canyon (white form), Limestone Canyon (white), Hot Springs Canyon, San Juan Trail below Blue Jay Campground, Temescal Valley at Indian Canyon (yellow), Elsinore Peak (yellow), the Santa Rosa Plateau (white), and San Mateo Creek (white).

The epithet was given in 1833 by G. Bentham for its small flowers (Latin, parvus = little, small; floris = flower).

Above: Yellow and white forms of coast baby-star. See page 474 for a spring bloom near Elsinore Peak.

Mainland Pygmy Linanthus
Leptosiphon pygmaeus (Brand) J.M. Porter & L.A. Johnson **ssp. *continentalis*** (P.H. Raven) J.M. Porter & L.A. Johnson
[*Linanthus pygmaeus* Brand]

A slender-stemmed upright annual, 2-10 cm tall, usually branched from the base. *Leaves palmately lobed 3-5 times*, each lobe 2-6 mm long, very slender, spine-tipped. *Flowers solitary*. Calyx 3-5 mm long, fused for about 2/3 of its length. *Corolla funnel-shaped with a long narrow yellow tube, lobes 1-2 mm long, rounded, white*. Stamens and style are held far above the throat; anthers and pollen orange or yellow. Stamens hairless. In flower Mar-June.

A plant of dry soils in openings among chaparral and oak woodlands. Known sparingly from the Temescal Valley near Glen Ivy and Alberhill, the Santa Rosa Plateau, and scattered sites in the San Mateo Canyon Wilderness Area.

The epithet was given in 1907 by A. Brand for its small size (Greek, pygmaios = dwarfish). The subepithet was given in 1965 by P.H. Raven for its distribution on the mainland (continent); the typical form has a blue corolla and is only known from San Clemente and Guadalupe Islands.

Similar. Flax-flowered linanthus (*L. liniflorus*): much taller, 10-50 cm tall, branched above the base; leaves much longer, 10-30 mm long; stamens hairy at base; often abundant.

Prickly Phlox

Linanthus californicus (Hook. & Arn.) J.M. Porter & L.A. Johnson [*Leptodactylon californicum* Hook. & Arn.]

    A rounded or upright shrubby perennial, 30 cm to 1 meter tall, often wider. *Leaves 3-12 mm long, palmately lobed, sharply spine-tipped, alternate on stem, with a bundle of additional smaller leaves at their base. Corolla funnel-shaped with a short tube* only 10-15 mm long, *the throat yellow, white above.* Flower face about 3 cm across, 5-lobed though commonly 4-7, *pink, magenta, or white.* Stamens and style are held well inside the throat; anthers and pollen orange or yellow. In flower Mar-June.

    Grows in decomposed granite, sandstone, and rock cracks. Uncommon in the Chino Hills and Lomas de Santiago, common in the Santa Ana Mountains. Easiest found along Ortega Highway such as at Bear Canyon and San Juan Loop Trails and the stretch from The Lookout to Lake Elsinore. Its flowers open during the day, mostly when hit directly with bright sunshine. *The spiny leaves can inflict a painful sting.*

    The epithet was given in 1838 by W.J. Hooker and G.A.W. Arnott for the place of its discovery.

Prickly Phlox

Ground Pink

Linanthus dianthiflorus (Benth.) E. Greene

    A hairy annual, 5-12 cm tall, with few stems, rarely several. *Leaves* 5-20 mm long, *hairless, thread-like or linear, not divided. Corolla* funnel-shaped with a short tube. Its *large lobes are pink to white with a purple patch near the base, the edges toothed.* Stamens and style are held well inside the yellow throat; anthers and pollen orange or yellow. In flower Feb-Apr.

    A delicate wildflower of openings among grasslands and coastal sage scrub, generally in fast-draining soils, more abundant after fire. Known throughout our area at relatively low elevations, from coastal bluffs and foothills to scattered sites in the San Mateo Canyon Wilderness Area, the Santa Rosa Plateau, and Temescal Valley. Best seen in Limestone Canyon and Caspers Wilderness Park after good winter rainfall.

    The epithet was given in 1833 by G. Bentham for the similarity of its flowers to carnations in the genus *Dianthus* (Latin, floris = flower).

Ground Pink

Slender Phlox

Microsteris gracilis (Hook.) E. Greene
  [*Phlox gracilis* (Hooker) E. Greene]

    An upright or sprawling annual; often many-branched, stems to 20 cm long. *Upper parts usually covered with sticky-tipped hairs.* Leaves 10-30 mm long, *lance-shaped, not divided, opposite (alternate above).* Corolla funnel-shaped, the yellowish tube is just slightly longer than the calyx lobes. *Corolla lobes pink to white, blunt at the tip or notched inward.* Stamens and style are held well inside the throat; anthers and pollen are orange or yellow. In flower Mar-Aug.

    Occasionally found in openings among chaparral and oak woodlands, mostly after fire. Known only from Maple Springs Saddle, Caspers Wilderness Park near the Ortega Highway overcrossing of San Juan Creek, Bedford Canyon near the forest gate, and Tenaja Canyon.

    The epithet was given in 1829 by W.J. Hooker for its slender stems (Latin, gracilis = slender).

    Similar. Small-flowered morning-glory (*Convolvulus simulans,* Convolvulaceae): no sticky-tipped hairs; leaves oblong or linear, alternate; corolla lobes pinkish to bluish, their tips pointed or rounded; only 2 style branches; fruits nod (held downward).

Slender Phlox

Holly-leaved Skunkweed

Navarretia atractyloides (Benth.) Hook. & Arn.

An upright annual, 5-29 cm tall with 1-several stems. Leaves variable, often in tight clusters, generally pinnate with long spines. Inflorescence is a tight sphere of green to red, spiny, curved bracts. *Corolla 8-9 mm long, funnel-shaped, the tube white to yellowish, slightly longer than the calyx and bracts. Corolla lobes white to pale lavender, often with a white ring atop the throat.* Stamens and style are held above the throat; anthers and pollen blue. The stigma is 3-lobed. In flower May-July.

A small plant of sandy or rocky soils in openings among coastal sage scrub, in meadows, and in vernal pools; sometimes along hard-packed unpaved roads. Known from Santa Ana Canyon, San Joaquin Hills (vernal pools along Nyes Place), Whiting Ranch Wilderness Park (on the former Glenn Ranch), Chiquito Basin, throughout Temescal Valley and the San Mateo Canyon Wilderness Area. In defiance of its common name, this "skunk" has no odor.

The epithet was given in 1838 by G. Bentham for its similarity to a thistle (Greek, atraktylus = a thistle-like plant).

Similar. Holly-leaved skunkweed (*N. hamata* ssp. *leptantha*): skunk-like odor; spines of inflorescence bracts are curved and each tipped with a hook; corolla pink, tube purple.

Holly-leaved Skunkweed

Southern Hooked Navarretia

Navarretia hamata E. Greene **ssp. *leptantha*** (E. Greene) H. Mason

An upright or sprawling annual, 8-30 cm tall, small-haired and sticky throughout. The entire plant has a skunk-like odor. *Leaves pinnate with long spines; those at the tip are 3-lobed, their spines usually hooked.* Inflorescence crowded with green to red bracts, each pinnate and edged with spines that are hooked at their tips. *Corolla small, funnel-shaped, the narrow tube purple, much longer than the calyx and bracts. Corolla lobes 2-3 mm long, pink, often white or yellowish atop the throat.* Stamens and style are held just outside the throat; anthers and pollen blue. The stigma is 3-lobed. In flower Apr-June.

Typically a coastal plant of sandy or rocky openings among coastal sage scrub and chaparral, sometimes vernal pools, more common after fire. Known historically from Corona del Mar, more recently found in Fullerton Nature Park, San Juan Canyon, and at higher elevations. Look for it among chamise and manzanitas along South Main Divide Road.

The epithet was given in 1887 by E.L. Greene for its hook-tipped bracts (Latin, hamatus = barbed, hooked at the tip). In 1889 he named this form for its narrow corolla tube (Greek, leptos = slender, thin; anthos = flower).

Similar. Holly-leaved skunkweed (*N. atractyloides*): no skunk-like odor; spines of inflorescence bracts are each curved but not tipped with a hook; *corolla white to pale lavender,* tube white to yellowish.

Southern Hooked Navarretia

Hooked Navarretia

Navarretia hamata E. Greene **ssp. *hamata***

Similar to southern hooked navarretia. *Corolla tube short, broad, same length as the calyx and bracts. Corolla lobes 3-5 mm long, bright pink or purple.* Usually in hard-packed soils, uncommon. Known here only from the San Mateo Canyon Wilderness Area.

Hooked Navarretia

Spreading Navarretia ★
Navarretia fossalis Moran

A short annual with a single central stem, 1-15 cm tall, and long side stems. *Rarely does it lie flat on the ground.* Leaves hairless, 1(-2) pinnately lobed. Each inflorescence is spherical. *Corolla small, all-white, funnel-shaped, the tube about the same length as the calyx and bracts.* Stamens and style are held slightly outside the throat; anthers and pollen pale yellow to white. Stigma 2-lobed. In flower Apr-June.

A plant of vernal pools and alkali wetlands. Known in our area only from Lake Elsinore, more common eastward to Hemet and at Camp Pendleton. Nearby, old records exist for Murrieta Hot Springs, both old and new records for the Santa Rosa Plateau. Ranked Federally Threatened and CRPR 1B.1, it is endangered by a multitude of human activities and by competition/crowding from non-native plants.

The epithet was given in 1977 by R. Moran for its tendency to live in soil depressions and ditches (Latin, fossa = ditch).

Similar. **Prostrate navarretia** (*N. prostrata*): almost no central stem, *lies flat on the ground; corolla pale blue to white;* stamens and style held far outside the throat; stigma 3-lobed; vernal pools and marshes along the coast; rare.

Spreading Navarretia

Prostrate Navarretia

Prostrate Navarretia ★
Navarretia prostrata A. Gray

An annual that has little or no central stem; it lies completely flat on the ground. Leaves hairless, 1-2-pinnately lobed. At the center of the plant is a spherical inflorescence, surrounded by spreading branches, each tipped with its own spherical head. *Corolla small, funnel-shaped, the tube pale blue to white, about the same length as the calyx and bracts. Corolla lobes blue to white.* Stamens and style are held far outside the throat; anthers and pollen pale yellow to white. Stigma 3-lobed. In flower Apr-May.

Historically known from a marsh in Laguna Beach. Recent records are from vernal pools at San Clemente State Beach and in Fairview Park in Costa Mesa. Not known from Temescal Valley, though it is found on the Santa Rosa Plateau. Ranked CRPR 1B.1, it is endangered by development and non-native plants.

The epithet was given in 1881 by A. Gray for its low-growing habit (Latin, prostratus = prostrate).

Similar. **Spreading navarretia** ★ (*N. fossalis*): *central stem present, usually not flat on the ground; corolla all white;* stamens and style slightly outside the throat; stigma 2-lobed; found in vernal pools of Lake Elsinore to Hemet and coastal pools on Camp Pendleton; rare.

Prostrate Navarretia

POLYGALACEAE • MILKWORT FAMILY

Annuals, perennials, or shrubs, rarely vines or trees. Leaves simple, generally alternate. *Flowers 2-lipped and look like that of a pea* (Fabaceae, its sister family). Sepals 5, the lateral pair enlarged and petal-like, forming lateral wings. Petals 3 or 5, the lower 1 formed into a keel, the upper 2 strap-like, the remaining 2 formed as small lateral petals. Fruit is a flat capsule, the seed often with a long snout called an aril. In our single species, the keel petal has an obvious beak. This family has few species in North America north of Mexico; most are tropical.

Fish's Milkwort ★
Polygala cornuta Kellogg **var.** *fishiae* (C. Parry) Munz

An open-branched shrub with deciduous leaves, 0.6-2.5 meters tall, that forms thickets in openings and beneath other shrubs and trees. Leaves linear, elliptical, or oval, 10-65 mm long. *Flowers dark pink to purple*, 7-11 mm long; the keel petal is yellow with an obvious beak. In flower May-Aug.

Grows among chaparral and oak woodlands, often as a sparse understory shrub. Known from the Chino Hills, Coal Canyon, the San Joaquin Hills, Laguna Beach, Santiago Canyon, Hot Springs Canyon, San Juan Canyon, along the San Juan Trail near Blue Jay Campground, near Pleasants Peak in the Santa Ana Mountains, Santa Rosa Plateau, and the San Mateo Canyon Wilderness Area. A common sight along the Trabuco and Holy Jim Trails. Considered rare (CRPR 4.3) because of its limited distribution.

The species was named in 1855 by A. Kellogg in reference to the obvious beak on the keel petal (Latin, cornutus = horned). Our local form was named in 1886 by C.C. Parry in honor of F.E. Fish, who first discovered it near Sauzal on Todos Santos Bay, Lower California.

Fish's Milkwort

Fish's Milkwort

Fruits

Fish's Milkwort

Fish's Milkwort

POLYGONACEAE • BUCKWHEAT FAMILY

Annuals, perennials, and shrubs, rarely trees. *Ours are annuals, perennials, or shrubs. Leaf nodes often swollen* along the thin, brittle, branching stems. Leaves basal and/or along the stem, alternate or in bundles, rarely opposite or whorled. Some have leaf-like bracts on the flowering stems. Flowers are very small and quite numerous, packed in tight clusters. Each flower may be enclosed by bracts that are round- or spine-tipped. *The calyx has 4-6 free lobes that resemble petals, usually green, white, yellow, pink, or red in color. Petals are absent.* Stamens 6 or 9 (rarely 3), often large and obvious. *Ovary superior,* topped with a 2- or 3-lobed style. Fruit is a dry 3-sided, 1-seeded achene, sometimes with papery wings. A moderately sized family of about 1,100 species, 35 in our area. They are quite important as wildlife shelter and food (pollen, nectar, and leaves are eaten). They are pollinated by bees, ants, flies, and butterflies. Many are cultivated as ornamentals, a few others for food.

Spineflowers: *Chorizanthe, Dodecahema, Lastarriaea, Sidotheca*

Attractive small-flowered annuals with stiff stems that grow upright or lie flat on the ground. Stems are often angular and brittle, breaking off in segments with age. Leaves all or mostly basal. Leaf-like bracts appear on the flowering stems. Cup-like involucral bracts (collectively, the *involucre*) enclose the small flowers, each bract with multiple teeth, tipped with sharp spines that catch on fur and skin, making them easily carried and dispersed by mammals. *Calyx lobes 6, each white, pink, or yellow.* Species in the genera *Chorizanthe, Dodecahema,* and *Sidotheca* have 9 stamens, *Lastarriaea* has only 3 stamens. Some spineflowers release a sweet scent; all are pollinated by ants, bees, and flies. We have 11 types in our area. Some are rare, so field surveys for them are needed and correct identification is critical. You will need a hand lens or microscope to examine their flower structure and fruit. Use this book as a guide but consult a technical manual and/or a professional botanist to confirm identification when doing so is critical.

	Form	Bract Teeth	Calyx	Habitat
Fringed *Chorizanthe fimbriata*	Upright or sprawling; stems hairy, sticky	6, of 2 alternating lengths; spines straight	Large; pink, *fringed*	Chaparral slopes; Santa Ana Mountains
Turkish Rugging *Ch. staticoides* ssp. *staticoides*	Upright; often single-stemmed	6, of 2 alternating lengths; spines curved	Small, pale rose, pink, or red, hairy on the outside; lobes about equal in size	Coastal sage scrub and chaparral slopes
OC Turkish Rugging *Ch. staticoides* ssp. *chrysacantha*	Upright; often single-stemmed; stem short and thick	6, of 2 alternating lengths; spines curved	Small, pale rose, pink, or red, hairy on the outside; lobes unequal in size	Seashores and slopes above
Peninsular *Chorizanthe leptotheca*	Spreading to upright; stems thinly haired	6, of 2 alternating lengths; spines curved	Medium; rose to red, hairy	Chaparral; sandy or gravelly soils in Santa Ana Mountains and washes
Parry's *Chorizanthe parryi* var. *parryi*	Sprawling; many-branched	6, of 2 alternating lengths; spines hooked	Tiny; white; lobes of 2 sizes	Sandy channels, openings in chaparral nearby
San Fernando Vly *Chorizanthe parryi* var. *fernandina*	Sprawling; many-branched	6, of 2 alternating lengths; spines straight	Tiny; white; lobes of about equal size	Sandy channels, openings in chaparral nearby
Long-spined *Ch. polygonoides* var. *longispina*	Sprawl flat on the ground; reddish stems	6, but appear to have only 3; spines long, hooked at the tip	Tiny; white to rose; lobes notched	Clay soils; inland
Prostrate *Chorizanthe procumbens*	Sprawl flat on the ground; yellowish-green stems	6, of 2 alternating lengths; spines curved at tips	Small; *yellowish* to whitish	Sandy soils, coastal, mountains; Temecula
Leather *Lastarriaea coriacea*	Sprawls to upright; yellow-green to red-brown	None	Tiny; light green to whitish; lobes spined, hooked at tip	Openings in coastal sage scrub and chaparral
Slender-horned *Dodecahema leptoceros*	Sprawls or slightly upright; red stems	6, all long, straight; bottom with 6, all curved	Tiny; pink to whitish; hairy	Sandy to gravelly soils of major washes; Indian Canyon wash
Puncturebract *Sidotheca trilobata*	Upright, single-stemmed; stems purple	5-6, spine-tipped, beneath groups of 3-10 flowers	Small; white to pink; each lobe 3-lobed, pointed	Decomposing granite; Santa Ana Mountains

Fringed Spineflower
Chorizanthe fimbriata Nutt. **var.** *fimbriata*

Stems hairy, sticky, green to red, upright, 5-30 cm tall, or sprawling to 50 cm long. Leaf blades 5-35 mm long, oval, basal, hairless above, sparsely haired beneath. *Bract teeth 6, the inner 3 much smaller than the outer, all tipped with straight spines. Calyx large, 6-10 mm long, free lobes pink to whitish, with fringed edges.* In flower Feb-June.

A common annual of the Santa Ana Mountains. Look for it in our major canyons such as Silverado, Hot Springs, San Juan, and Lucas, also in the Los Pinos Potrero, Long Canyon Road, and Elsinore Peak. In good wet winters, south-facing hillsides along Ortega Highway are ablaze with vibrant patches of its flowers during April.

The genus was named in 1836 by R. Brown for its divided calyx (Greek, chorizo = to divide; anthos = flower). The epithet was named in 1848 by T. Nuttall for its fringed calyx lobes (Latin, fimbriatus = fringed).

Turkish Rugging
Chorizanthe staticoides Benth. **ssp.** *staticoides*

An upright annual, 5-60 cm tall, the stems greenish to reddish, *often from a single main stem that branches above.* Leaf blade 5-80 mm long, oblong to oval, nearly hairless above, very hairy below, mostly basal. The first leaves are indented at the tip. *Bract teeth 6, of 2 lengths, the spines curved. Calyx small, 4-5 mm long, lobes about equal in size, pale rose, pink, or red, hairy on the outside.* In flower Apr-May.

A common widespread spring annual of coastal sage scrub and chaparral. Grows in sandy soils, often on hot, exposed, south-facing slopes. Known from the San Joaquin Hills south to San Clemente, Lomas de Santiago, Mission Viejo, Rancho Mission Viejo, and Richard and Donna O'Neill Conservancy. Less commonly found at places such as Hot Springs and San Juan Canyons, the Santa Rosa Plateau, the San Mateo Canyon Wilderness Area, and coastal to inland slopes of the Santa Ana Mountains.

Named in 1836 by G. Bentham for its resemblance to plants in the genus *Statice* (Plumbaginaceae) (Greek, -oides = resemblance).

Orange County Turkish Rugging ✿
Chorizanthe staticoides Benth. **ssp.** *chrysacantha* (Goodman) Munz

A coastal form of Turkish rugging, no longer recognized by some botanists, but we still consider it distinct. Best identified by comparing it with typical Turkish rugging. *It usually stands shorter and has thicker stems. Its flower clusters are much more densely-packed.* Involucral bracts are longer (5-7 mm long instead of 3-4 mm), lower ones leaf-like. *The calyx is longer* (5-5.6 mm long instead of 4-5 mm long) *and its lobes clearly unequal* (instead of about equal). In flower Apr-May.

Known from openings in coastal sage scrub along the coast, such as the hills surrounding Newport Bay, ocean bluffs in Corona del Mar, Laguna Beach, Crystal Cove State Park (even right on the beach), and Dana Point.

Named in 1934 by G.J. Goodman for the gold- or amber-yellow-colored spines on its flower bracts (Greek, chrysos = gold; akantha = a spine).

Peninsular Spineflower ★
Chorizanthe leptotheca Goodman
A thinly haired spreading annual, 5–3.5 cm tall, 5–3(5) cm across, its green stems age to red. Leaf blade 0.5–2(3) cm long, 0.3–0.5(0.7) cm wide, usually hairy above. *Bract teeth 6, of 2 lengths, the spines curved. Calyx* 4.5–6 mm, *rose to red, hairy.* In flower May–Aug.

A very uncommon plant of sandy or gravelly soils. Known here from Long Canyon Road near Ortega Highway in the Santa Ana Mountains and on alluvial benches in Indian Wash along Temescal Valley. A bit more common east and south of our area. CRPR 4.2; much of its habitat is already lost to development projects; also threatened by competition from non-native grasses.

It was named in 1934 by G.J. Goodman for its slender flower bracts (Greek, leptos = slender; theke = a box).

Parry's Spineflower ★
Chorizanthe parryi S. Watson **var.** ***parryi***
An annual, its *many-branched stems sprawl over the ground* and reach 2–30 cm long, variously covered with straggly hairs. Leaf blades 5–40 mm long, linear to oblong, mostly basal; the first leaves are spatula-shaped with a rounded tip. *Bract teeth 6, of 2 alternating lengths, each tipped with a hooked spine. Calyx* 2.5–3 mm long, the tube yellow-green, *free lobes white.* When open, the flowers are 2–3 mm wide, the free calyx lobes are of 2 alternating shapes unequal in length: the 3 outer lobes oval and longer, the 3 inner lobes linear and shorter. In flower Apr–June.

Primarily a plant of the Perris-Aguanga Basin of western Riverside County and the Santa Ana River valley. Perhaps found in the Temescal Valley. More common along the western edge of the Mojave Desert. CRPR 1B.1; it is endangered by altered flood regime, development, mining, non-native plants, and off-road vehicles.

The epithet was given in 1877 by S. Watson in honor of C.C. Parry.

San Fernando Valley Spineflower ★
Chorizanthe parryi S. Watson **var.** ***fernandina***
(S. Watson) Jeps.
A variant of Parry's spineflower (above). *Its 6 bract teeth are each tipped with a straight spine. The 6 calyx lobes are about the same length.* In flower Apr–June.

Grows in sunny openings among grasslands and coastal sage scrub. Known in our area from specimens collected in 1908 in "hills near Santa Ana," which may refer to Red Hill, Lemon Heights, the Orange Hills, Lomas de Santiago, or the San Joaquin Hills. Possibly eliminated from this area by development, but we keep looking for it. Last seen in 1940 and long thought to be extinct, it was rediscovered in Ventura County by local botanist Tony Bomkamp in 1999. Now known from Ahmanson Ranch in Ventura County and Newhall Ranch in Los Angeles County. The flowers release a mild sweet scent. Pollinated primarily by native ants. CRPR 1B.1 and State Endangered. Known from only 3 populations, it is endangered by development and non-native plants.

The variety was named in 1923 by S. Watson in reference to the San Fernando Valley of Los Angeles County, where the plant was then common and widespread.

Long-spined Spineflower ★
Chorizanthe polygonoides Torr. and A. Gray
var. ***longispina*** (Goodman) Munz

Reddish stems sprawl flat on the ground, 1-15 cm long, covered with soft hairs. Leaf blades 3-10 mm long, linear to elliptical, basal. Bract teeth 6, but 3 are tiny or absent, *so they appear to have only 3 bracts, the spines long and hooked at the tip.* Calyx 1.5-2 mm long, white to rose, densely hairy. In flower Mar-June.

Uncommon and easily overlooked. Usually found in clay soils. In our area, known only from Gypsum Canyon, Elsinore Peak, Temescal Valley, and near Temecula. CRPR 1B.2; it is threatened by development, non-native grasses, recreational activities, vehicles, and grazing.

Named in 1870 by J. Torrey and A. Gray for its resemblance to knotweeds in the genus *Polygonum* (Greek, -oides = resemblance). The variety was named in 1955 by G.J. Goodman for its long spines (Latin, longus = long; spina = full of spines).

Prostrate Spineflower ✿
Chorizanthe procumbens Nutt.

Sparsely haired yellowish-green stems sprawl flat on the ground, 2-25 cm long. Leaf blades 5-40 mm long, oblong, mostly basal. The first leaves are spatula-shaped and rounded at the tip. *Bract teeth 6, of 2 lengths, held straight out, the spines curved at the tips. Calyx yellowish to whitish,* 2-2.5 mm long, the lobes equal in size. In flower Apr-June.

Known from sandy soils mostly along our coastline. Locally diminished by development. Look for it in the San Joaquin Hills, Laguna Coast Wilderness Park, Sycamore Hills near Laguna Beach, near Salt Creek, on the Dana Point Headlands, Serrano Creek in Lake Forest, Lomas de Santiago, near Irvine Lake, and the Temecula Valley. Also in the Santa Ana Mountains such as North Main Divide Road near El Cariso, South Main Divide Road, San Mateo Canyon, Los Alamos Canyon, and Tenaja Guard Station, south to Rainbow.

Named in 1848 by T. Nuttall for its prostrate growth (Latin, procumbens = falling forward, prostrate).

Leather Spineflower
Lastarriaea coriacea (Goodman) Hoover
[*Chorizanthe coriacea* Goodman]

A low-growing annual with brittle reddish stems that grow 2-15 cm long. Its *narrow linear leaves* are 5-30 cm long, hairy, and have no petiole. They are mostly basal, those on the stem in whorls of 4-5. Leaf-like bracts are 5-10 mm long and narrow; their spines curve outward and end in a hooked tip. *There are no involucral bracts. Calyx 4 mm long, greenish, leathery, and divided into 6 narrow lobes.* Like the leaf-like bracts, *each calyx lobe is spined and ends in a hooked tip.* It has only 3 stamens. In flower Apr-June.

Usually not abundant, small populations are found among sandy openings in coastal sage scrub and chaparral. Known from Sycamore Hills in Laguna Canyon, Lomas de Santiago, both Riley and Caspers Wilderness Parks, and Los Alamos Creek near Wildhorse Creek. Fairly common in Temescal Valley.

The genus was established in 1851 by E.J. Rémy in honor of J.V. Lastarria Santander. The epithet was given in 1943 by G.J. Goodman for its leathery calyx (Latin, coriacea = made of leather).

Slender-horned Spineflower ★
Dodecahema leptoceras (A. Gray) Reveal & Hardham
[*Centrostegia leptoceras* A. Gray, *Chorizanthe leptoceras* (A. Gray) S. Watson]

An annual that sprawls over the ground or stands somewhat upright, with reddish branches 3-10 cm long. Leaves 10-60 mm long, linear, hairless, and all basal. Its leaf-like bracts are stiff, 3-4 lobed, tipped with red spines, and held on one side of the stem. *Involucral bracts 2-4 mm long, the top 6 teeth tipped with large straight red spines of 2 lengths, the bottom with 6 slender curved red spines. No other local spineflower has spines at the bottom of the involucre.* Calyx 1.2-2 mm long, *pink to whitish, hairy,* with 6 spatula-shaped lobes. In flower Apr-June.

Very rare, it occurs in sandy to gravelly soils of major washes in western Riverside, San Bernardino, and eastern Los Angeles Counties. In our area, best known from Indian Canyon and Glen Eden Sun Club (the latter is private land, not open for casual visits). It often grows with **leather spineflower** (*Lastarriaea coriacea*). Listed as an Endangered Species by both Federal and State agencies. CRPR 1B.1. It is threatened by development and off-road vehicles.

Its 12 hooks help it to hang on tightly to mammal fur. It is thought to have been transported mostly by **American antelope** (*Antilocapra americana* Ord, Antilocapridae), which once ranged widely in southern California. Both the antelope and the plant declined sharply with human settling and development of southern California.

The genus was named in 1989 by J.L. Reveal and C.B. Hardham for the 12 hooks on its involucral bracts (Greek, dodeca- = twelve; Latin, hamus = hook). The epithet was given in 1870 by A. Gray for its slender spines (Greek, leptos = slender; keras = horn).

Slender-horned Spineflower

Slender-horned Spineflower

Three-lobed Starry Puncturebract ✤
***Sidotheca trilobata* (A. Gray) Reveal**
[*Oxytheca trilobata* A. Gray]

A red or purple-stemmed annual usually from a single main stem, its brittle upper branches are zigzagged and forked, often held parallel to the ground, each 7-50 cm long. Leaf blade 1-9 cm long, linear to weakly spoon-shaped, no petiole. Leaves are all basal, green or reddish, aging red, drying up when flowers open. Every fork of the stem has a stiff, 3-lobed leaf-like bract held on one side of the stem, each lobe tipped with a long sharp yellow spine. *Flowers are in groups of 3-10 with 5-6 broad spine-tipped bracts close beneath them. Calyx white to pink, darker pink at the base. The 6 calyx lobes are divided into 3 linear pointed lobes each.* In flower Apr-Sep.

Generally uncommon, it grows in soils of *loose decomposing granite,* mostly on south-facing barrens in openings among chaparral and coniferous forests that receive fog. Its cup-like spine-tipped bracts capture condensed fog and direct it to the soil, providing soil moisture. Found along upper Holy Jim Trail and the southeast side of Santiago Peak mostly within 1-5 miles south of the summit, right along Main Divide Truck Trail. Pollinated by small flies and bees.

The genus was named in 1848 by T. Nuttall for the sharp spines on its bracts (Greek, oxys = sharp; theke = case). The epithet was given in 1876 by A. Gray for its three-lobed bracts (Latin, tri- = 3; lobatus = lobed).

Three-lobed Starry Puncturebract

Wild Buckwheats: *Eriogonum*

A variable group of annuals and perennials. *Leaves are basal or along the stem, alternate, sometimes in bundles, usually covered with fuzzy hairs beneath.* Individual flowers are small, arranged singly or in clusters, surrounded by a 3-10-lobed funnel-like *involucre that lacks spines.* These are in sparse or dense groups on stems generally near the top of the plant, often quite hairy. *Calyx lobes 6, white, yellow, or red.* Stamens 9, each topped with red or pink anthers. About 250 species of *Eriogonum* are known, most from North America. This is the largest eudicot genus in the state, with 112 species and numerous named forms. We have 10 species in our area. Many wild buckwheat species are in cultivation, easily grown from seed and potted nursery stock. No southern California garden should be without them. The genus was named in 1803 by A. Michaux for its hairy nodes, a condition found on the first of its species described (Greek, erion = wool; gonu = knee).

	Form	Leaves	Inflorescence	Habitat
Perennial Wild Buckwheats				
Gray Coast *E. cinereum*	Perennial; mounded shrub; *covered in ashy gray hairs*	Stem; oval, edges wavy, folded under; covered in gray hairs	Tight clusters along and atop stalks; calyx white to pinkish	Coastal; uncommon
Long-stemmed *E. elongatum*	Perennial; upright, *long wand-like stems,* narrowly forked	Stem; oval to elliptical, hairy beneath; *short petiole*	Knob-like clusters along and atop long stalks; calyx white	Coastal, foothills, mountains
Coastal California *E. fasciculatum* ssp. *fasciculatum*	Perennial; rounded shrub	Stem, in bundles; *dark green or gray-green,* hairless above, edges rolled under	*Pom-poms, branched;* calyx white to pinkish	Coastal to coastal slopes of Santa Ana Mountains
Mojave Desert California *E. fasciculatum* ssp. *polifolium*	Perennial; rounded shrub	Stem, in bundles; *dense white hairs above, appear grayish;* edges not rolled under	Pom-poms, solitary or few in number; calyx white to pinkish	Foothills, mountains, and inland; usually in hot dry sites
Leafy California *E. fasciculatum* ssp. *foliolosum*	Perennial; rounded shrub	Stem, in bundles; abundant, *dark to gray-green, thinly haired above, tightly rolled under*	*Pom-poms on branched stalks;* calyx white to pinkish	Hot situations inland, away from coastal influence
Naked-stemmed *E. nudum* var. *pauc.*	Perennial; *long widely-forked branches*	Basal; linear, oval, to heart-shaped; *long petiole*	Rounded clusters, at tips of *long naked stalks;* calyx white	Santiago and Modjeska Peaks
Bluff *E. parvifolium*	Perennial; *low mounded or mat-like shrub*	Stem; linear to round or *triangular;* hairless above, white-fuzzy below	Pom-poms on simple or branched stalks; calyx white	Bluffs, cliffs, and dunes along the immediate coast
Rock *E. saxatile*	Perennial; upright to spreading	Basal; blade elliptic to round, gray-hairy, light gray	Atop branched stalks; calyx white, pinkish, or yellowish	*Talus slopes* on Modjeska Peak
Short-stemmed *E. wrightii* ssp. *sub.*	Perennial; *low, mat-like shrub*	Stem; linear to elliptic; hairy above, densely beneath; gray-green	*Atop and along short stalks;* calyx white to pinkish	Santiago Peak
Annual Wild Buckwheats				
Davidson's *E. davidsonii*	Annual; *small, hairless, upright*	Basal; round, wavy-edged, fuzzy beneath	Atop and along stalks; calyx small, white to pink	High elevations of Santa Ana Mountains
Thurber's *E. thurberi*	Annual; often 3-branched; *woolly, especially below*	*Basal, abundant;* elliptic to round; very hairy, especially below	Atop stalks; many-flowered; calyx white to pink, age reddish	Santa Ana River channel
Slender *E. gracile* var. *gracile*	Annual; small, upright; *main stem(s) hairy*	Basal and stem; very fuzzy below; basal oblong, stem elliptic	Atop and along stalks; calyx white to pale yellow	Coastal: plains, hills, mountains
Smooth-stemmed *E. gracile* var. *incultum*	*Main stem hairless or sparsely hairy*	Basal and stem; very fuzzy below; basal oblong, stem elliptic	Atop and along stalks; calyx white to pale yellow	Inland: mountains and Temescal Valley

PERENNIAL WILD BUCKWHEATS

Small or medium-sized *shrubs or subshrubs* with woody main stems. Some are mounded in overall shape when mature and have leaves along the stems. Others are stick-like and upright with mostly basal leaves.

Gray Coast or Ash-leaved Buckwheat
Eriogonum cinereum Benth.

A beautiful mounded shrub about 0.5-2 meters high, often a bit broader. *Densely covered with gray hairs that give the plant a gray appearance. Leaf blade* 15-30 mm long, *oval, the edges wavy and often folded under, hairy above, even hairier below.* Leaves appear along the stem. Calyx white to pinkish, in tight dense head-like clusters along and atop the angled stalks. In flower June-Dec.

Sparingly found in sandy soils along beaches and coastal bluffs in our area, mostly near upper Newport Bay and canyons upstream (though perhaps introduced there). Possibly more widespread along our coast but eliminated by development and activity. A popular shrub in cultivation, easily grown in coastal areas.

Named in 1844 by G. Bentham for its ash-gray color (Latin, cinereus = ash-colored).

Gray Coast Buckwheat

Long-stemmed Buckwheat
Eriogonum elongatum Benth. **var.** ***elongatum***

An upright perennial with a woody base and long wand-like stems, 60-180 cm tall, *covered with whitish hairs.* The branches fork narrowly and remain mostly parallel to each other. *Leaf blade* 10-30 mm long, *oval to elliptical, with wavy-edges, sparsely-haired above but quite hairy beneath. The petiole is noticeably shorter than the leaf.* Most leaves are basal, though there is *often a cluster of leaves above the base,* as shown in the photo. Nearly all leaves are dried up when the plant is in flower. Flowers arranged in loose knob-like clusters, distanced from each other, along and atop the long straight, leafless stems. Calyx white. In flower Aug-Nov.

A common plant of dry, rocky, gravelly, or sandy soils among coastal sage scrub and chaparral in coastal, foothill, and mountain sites. Look for it in upper Newport Bay, Limestone Canyon, Silverado Canyon, Hot Springs Canyon, San Juan Canyon, and openings in the Santa Ana Mountains.

Named in 1844 by G. Bentham for the distance between flower clusters on its long stems (Latin, elongatum = removed).

Long-stemmed Buckwheat

California Buckwheat
Eriogonum fasciculatum (Benth.) Torr. and A. Gray

A low-lying or mounding shrub up to 0.5-1(2) meters tall and 3 meters wide. *Leaves* 6-18 mm long, *narrow and linear, arranged in bundles that are arranged alternately on the stem.* Each leaf is hairless to hairy above and densely white hairy below. Flowers are arranged in dense rounded heads (like pom-poms) at the top of single stalks or grouped in umbels. Calyx white to pinkish, held firmly in the cluster. In flower most of the year but less commonly during cold winters.

A common plant throughout our area and most of southern California. Grows in all of our wilderness parks and forest lands. Used for food and shelter by a plethora of wildlife species. An excellent garden plant.

It was named in 1837 by G. Bentham for its bundled leaves (Latin, fasciculus = a bundle). Three subspecies (sometimes treated as varieties) occur in our area, arranged here from coast to inland types.

Meadow at Los Pinos Potrero, almost completely filled with Leafy California Buckwheat

Coastal California Buckwheat

Eriogonum fasciculatum (Benth.) Torr. & A. Gray
ssp. *fasciculatum*
[*E. fasciculatum* var. *fasciculatum*]

Noted for its *leaves, which are dark green or gray-green and hairless above, their edges rolled under. Inflorescence branched, often umbel-like.* It lives on *coastal bluffs* in Newport Beach, upper Newport Bay, Corona del Mar, Laguna Beach, Dana Point, San Clemente, San Joaquin Hills, *and inland to areas with coastal climatic influence* such as southern Mission Viejo, Rancho Mission Viejo, and lower San Juan Canyon.

Leaf cross-section of coastal California buckwheat, showing hairs and leaf edges.

Hairless above

Leaf edges rolled under

Mojave Desert California Buckwheat

Eriogonum fasciculatum* ssp. *polifolium (Benth.) Stokes
[*E. fasciculatum* var. *polifolium* (A. DC.) Torr. and A. Gray]

Its *leaves are much broader than those of the other forms, densely covered with white hairs above and appear grayish. The edges are not rolled under or are only slightly rolled.* The entire inflorescence is very hairy. *Each flower head is atop a solitary stalk* or shares a short stalk with few other flower heads. Mostly a foothill, mountain, inland, and desert form. Found in O'Neill Regional Park, and Silverado and Santiago Canyons, on Santiago Peak, and in many places in the Santa Ana Mountains. Uncommon in scattered sites in the San Mateo Canyon Wilderness Area. On the inland side of the Santa Ana Mountains, it is much less common than leafy California buckwheat and lives on drier, rockier slopes.

Named in 1870 by G. Bentham for its gray-colored leaves (Greek, polios = gray; folium = leaf)

Leaf cross-section of Mojave Desert (left) and leafy (right) California buckwheat, showing hairs and leaf edges.

Densely hairy above  **Thinly hairy above**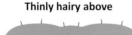

Leaf edges not rolled under or slightly rolled   **Leaf edges tightly rolled under**

Leafy California Buckwheat

Eriogonum fasciculatum* ssp. *foliolosum (Nutt.) Stokes
[*E. fasciculatum* var. *foliolosum* (Nutt.) Abrams]

Its abundant leaves are dark to gray-green, thinly hairy above, and tightly rolled under. It lives in hot situations generally *away* from coastal influence. Found on dry slopes in places such as the Chino Hills, Mission Viejo, and major canyons: Santa Ana, Silverado, Limestone, Santiago, Trabuco, and San Juan. Also in interior canyons and slopes of the Santa Ana Mountains, the San Mateo Canyon Wilderness Area, and throughout Temescal Valley, Elsinore Basin, and Temecula Valley.

Named in 1848 by T. Nuttall for its numerous leaves (Latin, foliolum = leaflet; -osum = abundance).

THE CALIFORNIA BUCKWHEAT GUILD

California buckwheat serves as host for many wildlife species. Its nectar is eaten by beetles, butterflies, moths, flies, ants, bees, and wasps. Some bees collect its pollen to provision their nests. Grasshoppers and the larvae of beetles, butterflies, and moths feed on its roots, stems, leaves, and flowers. Other forms of wildlife such as lizards, snakes, birds, and beetles use it indirectly as a form of shelter. Here are just a few members of this very large guild.

Six-spotted Longhorn Beetle

Six-spotted Longhorn Beetle
Judolia sexspilota (LeConte) (Cerambycidae)

Commonly found on buckwheat flowers during the day. *Adults are about 12 mm long, black-bodied with brown legs, forewings yellow with 6 black spots, and long antennae.* Larvae tunnel through woody shrubs, pupate in the tunnels, and emerge as adults in spring.

The epithet was given in 1859 by J.L. LeConte for the 6 black spots on its forewings (Latin, sex- = six; spilos = spot).

Black-winged Blister Beetle
Nemognatha nigripennis LeConte (Meloidae)

A slender beetle, about 1 cm long, with orange head, thorax, and leg bases. The rest of the body is black. Adults feed on nectar of buckwheat flowers in spring and summer. *Beetles in this genus are unusual in that some of their mouthparts are formed into a long straw-like tube with which they drink nectar like butterflies and moths.* Larvae feed on the larvae of solitary ground-nesting native bees in their nest. They feed, mature, and pupate in the bee nest, then emerge as adults and crawl out of the nest to find flowers and a mate.

Black-winged Blister Beetle

The epithet was given in 1853 by J.L. LeConte for its black forewings (Latin, nigra = black; penna = wing). The **orange blister beetle** (*N. lurida* LeConte) has orange to yellow head, thorax, and forewings. It was also named in 1853 by J.L. LeConte for its color (Latin, luridus = pale yellow).

Tumbling Flower Beetles
(Mordellidae)

Tiny humpbacked beetles, about 3-10 mm long, *usually all-black, quite narrow-bodied, and tapered to a point at the hind end.* When disturbed they fold up their legs and drop to the ground. They drink nectar from buckwheat and other wildflowers.

Tumbling Flower Beetle

Bee Flies
(Bombyliidae)

As their name implies, *most bee flies look a bit like bees.* They are *quite hairy, emit a loud buzz as they fly, and feed on nectar.* As flies, they have only two wings (the front wings); the hind wings are reduced to lollipop-shaped halteres that act as stabilizing gyroscopes while in flight. This explains how flies can hover with such precision. *Bee fly adults have a long straw-like proboscis through which they imbibe nectar.* The larvae of most bee flies feed on the larvae of other insects that parasitize other insects, a feeding mode called *hyperparasitism.* Their usual hosts are flies, ants, bees, and wasps.

Bee Fly

Unlike bees, bee flies can neither sting nor bite. They are often seen nectaring on buckwheats and other plants. They will also alight on the ground with their wings spread in a characteristic V shape. Some have interesting patterns on their wings. Many species are known in our area.

Electric Buck Moth ✿

Hemileuca electra W.G. Wright (Saturniidae)

A day-flying moth of fall. Wingspan 2.3-3.1 cm. *Forewings whitish with black markings and heavier black edges. Hindwings red, marked with black.* Body hairy, banded red-and-black. In fall, mated females lay eggs on twigs of California buckwheat. Larvae emerge in Jan or Feb. They are all-black, covered in urticating spines, and feed together. Pupation occurs on the plant in spring or early summer. *Adults emerge mid-Sep through early Dec but most commonly in Oct. Mating occurs on or near California buckwheat.* Once quite common here wherever California buckwheat grew, severely devastated by development, they are now rarely found. The epithet was given in 1884 by W.G. Wright for its bright colors (Greek, elektron = amber or something bright). He first collected it in October 1883 along the Santa Ana River Valley in San Bernardino, not far from his home.

Electric Buck Moth

Buckwheat Root Borer Moth

Synanthedon polygoni (Hy. Edwards, Sesiidae)

A day-flying moth of spring and summer. Wingspan about 1.5-2.2 cm. *Adults dark blue-black, marked with red and orange.* Forewings generally dark, hind wings largely transparent. Color markings and amount of wing transparency are quite variable. Males and females are similar. Larvae tunnel within roots and lowest stems of California, slender, bluff, and short-stemmed buckwheats. Known locally from Laguna Beach and mountain meadows on Santiago and Elsinore Peaks. The epithet was given in 1881 by H. Edwards in reference to the feeding habits of the larvae, on buckwheat plants (Polygonaceae).

Buckwheat Root Borer Moth

Bernardino Blue Butterfly

Euphilotes bernardino bernardino (W. Barnes & McDunnough, Lycaenidae)

Very small, wingspan 1.7-2.1 cm. Males bright blue above with a broad black border; *no orange band on the upperside.* Females black above, often with blue scales, edges narrowly black-bordered; *an orange band on the hindwing upperside. Undersides light gray, heavily spotted with black.* On the forewing underside, *the black mark closest to the body is a line* (not a dot). *The hindwing orange spots are usually separated from the outer black spots. Wing fringes black-and-white checkered.* All stages feed on California buckwheat. Adults emerge late May-early July. The epithet was named in 1916 by W. Barnes and J.H. McDunnough for the place of its discovery, near San Bernardino.

Bernardino Blue — Line

Acmon Blue Butterfly

Plebejus acmon acmon (Westwood & Hewitson)
[*Icaricia acmon acmon* (Wwd. & Hewits.)] (Lycaenidae)

Small, wingspan 2-2.9 cm. Males bright blue above. Females brownish above, often with blue scales. *Hindwing of both sexes with pink to orangish spots near hind edge. Undersides dull white to gray, spotted with black, each edged with a halo of white.* On the forewing underside, *the black mark closest to the body is dot-like* (not a line). *The hindwing orange spots contain black spots that are ringed with irridescent green or blue,* best seen in strong sunlight. *Wing fringes white, not checkered.* Our most common blue. Adults emerge Mar-Oct. Larvae feed primarily on **coastal deer broom** (*Acmispon glaber* var. *glaber,* Fabaceae) and California buckwheat. The epithet was given in 1852 by J.O. Westwood and W.C. Hewitson to honor the Greek god Acmon.

Acmon Blue — Irridescent scales — Dot-like mark

Naked-stemmed Buckwheat
Eriogonum nudum Douglas ex Benth. **var. *pauciflorum*** S. Watson

An upright perennial similar to long-stemmed buckwheat but with forked branches that spread apart farther; the flowers are borne in rounded clusters on spreading stems at the top of the plant. It stands 10-200 cm tall. Leaves are 15-30 mm long, linear, oval, or heart-shaped, (nearly) hairless above, white hairy below, the edges usually flat, not wavy. The petiole is about the same length as the leaf or longer. In flower Aug-Oct.

Found on dry slopes near oaks and coniferous forests, never abundant here. In the Santa Ana Mountains, known from Maple Springs, the saddle above it, and the upper reaches of Santiago Peak.

Named in 1836 by D. Douglas for its leafless stems (Latin, nudus = naked). Our variety was named in 1877 by S. Watson for its few flowers (Latin, paucus = few; florum = flowers).

Bluff Buckwheat
Eriogonum parvifolium Smith

An open shrub when mature, 30-100 cm tall, 50-200 cm wide, it grows flat on the ground in windy areas. Leaf blade 5-30 mm long (though often on the small side), *linear to round, sometimes with the sides folded under and appearing triangular, the petioles shorter. Leaves dark green, sometimes reddish,* sometimes hairless above, *densely white fuzzy below, arranged in bundles along the stem.* Flowers in dense rounded heads 2-4 cm across, along and atop the stalk. The calyx whitish to pale pinkish. In flower year-round but mostly during summer.

A plant of bluffs, cliffs, and dunes along the immediate coast. Known from upper Newport Bay, Crystal Cove State Park, Laguna Beach, Dana Point Headlands, and San Clemente.

Named in 1809 by J.E. Smith for its small leaves (Latin, parvus = little; folium = leaf).

Rock Buckwheat ✿
Eriogonum saxatile S. Watson

An unusual upright or spreading plant, 10-20 cm tall, about as broad. *Leaves are crowded at the base,* the blade is 3-25 mm long, the petiole as long or longer. *Each leaf is elliptical to round, covered with whitish hairs, and appears light grey.* Their white, pinkish, or yellowish flowers are borne in loose rounded clusters along branched stalks. In flower May-July.

Grows on mountain slopes among loose, broken rocks. Known only from Modjeska Peak, easily seen along Main Divide Truck Trail. It also occurs in the San Gabriel, San Bernardino, and San Jacinto Mountains, in addition to many desert mountain ranges. In cultivation, best used in dry gardens with plenty of loose rocks. On talus slopes, it is sometimes parasitized by **clustered broomrape** (*Orobanche fasciculata,* Orobanchaceae).

Named in 1877 by S. Watson for its association with rocks (Latin, saxatile = growing among rocks).

Short-stemmed Buckwheat ✿
Eriogonum wrightii Benth. **ssp.** ***subscaposum*** (S. Watson) Stokes

A small mat-like shrub, 5-20 cm tall and twice as wide. *Leaf blade flat,* 5-12 mm long, *linear to elliptic, hairy above but densely so beneath.* Leaves gray-green, crowded on the stems. The *short, slender, leafless flower stalks are few-branched and stand straight up.* The whitish to pinkish flowers are in loose rounded heads. In flower July-Oct.

Found in *gravelly or rocky soils* in openings among chaparral at high elevations. Once known from the very top of Santiago Peak but mostly eliminated with the construction of additional communication towers and overzealous brush-clearing. Still found sparingly on its north-facing slopes below the summit, near Main Divide Road.

The species was named in 1856 by G. Bentham for its original collector, C. Wright. Our form was named in 1880 by S. Watson for its short flower stalks (Latin, sub- = under or below; scapus = the stalk of a plant).

Short-stemmed Buckwheat

ANNUAL WILD BUCKWHEATS
Small, delicate, upright annuals with basal leaves, usually with one or few main flowering stalks. We have 3 species.

Davidson's Buckwheat ✿
Eriogonum davidsonii E. Greene

A hairless upright annual, 5-40 cm tall. *Leaf blade round,* 10-20 mm long and wide, has wavy edges, and is *very fuzzy beneath.* The petiole is longer than the blade. *The flowering stalk branches about halfway up its length and is often red; the small white to pink flowers are arranged in small loose groups with lots of space between them.* In flower June-Sep.

A lovely delicate buckwheat, worth the bumpy ride up to Santiago Peak to behold. Found in loose, well-draining soils such as decomposed granite in open areas atop Santiago Peak (extirpated in the 1980s by tower construction and clearing) and occasionally along Main Divide Road near the Coldwater Trail but never common. When found, it forms a localized group of individuals. Perhaps its seeds do not disperse far from the parent plant, so a colony maintains itself in the same location every year. Pollinated by small native bees and blue butterflies (Lycaenidae).

Named in 1892 by E.L. Greene in honor of A. Davidson.

Davidson's Buckwheat

Thurber's Buckwheat
Eriogonum thurberi Torr.

An upright annual, 5-40 cm tall. Its sticky stalk often has three branches at each fork. *The lower branches are wool-covered, upper branches less so. Leaf blade* 8-45 mm long, *elliptic to round, very hairy, more so below.* Leaves plentiful, basal, on petioles about the same length. *Calyx white to pink, ages to deeper pink or red.* The outermost calyx lobes are broad and fan-shaped, their bases very narrow. In flower Apr-July.

Uncommon in our area but sometimes found in dense patches at a few localities. Old records exist from open sandy areas near Horseshoe Bend along the Santa Ana River bed, upstream in Riverside County, the mouth of Brea Canyon, Chino Creek, the Elsinore Basin, and near Murrieta. Modern sightings are only from the Santa Ana River bed through Corona-Riverside and near Murrieta.

Named in 1859 by J. Torrey in honor of G. Thurber.

Thurber's Buckwheat

Slender Buckwheat
Eriogonum gracile Benth. **var.** *gracile*

A very slender upright annual, 20-50 cm tall, with a single main stem or with multiple stems that arise from the very base of the plant. *It is densely to sparsely woolly throughout.* Flowering stalks branch unevenly at the top of the plant. Most leaves are basal with a few on the lower stem, often dried up by the time the flowers open. Both the *leaf blade and petiole* are about 10-60 mm long, *wavy-edged, very fuzzy below.* Basal leaves oblong, stem leaves elliptic. Calyx small, yellowish, in clusters along upper parts of the long arching stems. The stems turn a lovely burnt red with age. In flower spring to fall but mostly June-Oct.

Common in many places in our area but often overlooked because of its slender stems, habit of growing among other plants, and flowering period during hot summer and fall months, when many naturalists don't venture afar. Occurs most often in *dry sandy soils* in creekbeds, at the edges of grasslands, and in openings among coastal sage scrub, chaparral, and oak woodland. Found in the San Joaquin Hills, Sheep Hills, Santa Ana River bed, Chino Hills, Lomas de Santiago, San Juan Canyon, and nearly all elevations in the Santa Ana Mountains. Easiest seen in Riley Wilderness Park and in Limestone and Hot Springs Canyons.

The epithet was given in 1844 by G. Bentham for its slender stems (Latin, gracilis = slender).

Slender buckwheat is a very important nectar source for small butterflies, flies, and bees. Here a late-season acmon blue (Plebejus acmon acmon, Lycaenidae) enjoys a drink in late August at Riley Wilderness Park.

Smooth-stemmed Slender Buckwheat
Eriogonum gracile Benth. **var.** *incultum* Reveal

An uncommon *inland form* of slender buckwheat. *Its main stem is hairless or sparsely hairy. Calyx white to pale yellow.* In flower July-Oct.

Similar to Davidson's buckwheat, with which it grows in some areas. *Its basal leaves are oblong, stem leaves elliptic.* Davidson's buckwheat has round leaves. Uncommon, found in sandy soils of southwestern Riverside County. Known from Temescal Valley, the San Mateo Canyon Wilderness Area (Tenaja and San Mateo Canyons), and Temecula, eastward along Highway 79.

The variety was named in 1989 by J.L. Reveal for its plain, mostly hairless stems (Latin, incultus = unadorned).

ASSORTED BUCKWHEATS

Coastal Woollyheads ★
Nemacaulis denudata Nutt. var. ***denudata***

A delicate coastal annual that grows upright or flat on the ground, with slender threadlike wool-covered stems 4-40 cm long. *Leaf blade 5-80 mm long, has no petiole. Leaves basal,* linear or spoon-shaped, *ruffled, covered with white woolly hairs. Calyx tiny, greenish, yellow, pink, or dark red, arranged in tight round clusters of 5-30 flowers.* In flower Apr-Sep.

Found in loose sandy soils along the coast. Known from Seal Beach, Bolsa Chica Ecological Reserve, Newport Beach, and upper Newport Bay. Possibly at other places such as Crystal Cove State Park but easily overlooked because of its slender stems and short stature. CRPR 1B.2; it has been wiped out by coastal development. Easily trampled by beachgoers and maintenance crews, it needs strict protection and monitoring.

Named in 1848 by T. Nuttall, the genus refers to its slender thread-like stems (Greek, nema = thread; kaulos = stem), the epithet for its bare, leafless stems (Latin, denudatus = stripped, made bare).

Coastal Woollyheads
Long flower stalks seem to hover over the ground
Flower clusters face downward

Pale Smartweed ✖?
Persicaria lapathifolia (L.) Gray
[*Polygonum lapathifolium* L.]

An upright annual to about 1 meter tall. *Papery sheath at the base of the leaf (ocrea) 4-25 cm long, cylindric. Leaf blade 4-12(22) cm long; lance-shaped;* upper surface sometimes with dark blotches. *Inflorescence droops, each with 4-13 flowers. Flower 3-4 mm long, bell-shaped, whitish to pink.* In flower June-Oct.

Abundant in wet soils, mostly along creeks and rivers. Found throughout our area. Disagreement exists whether it is native or not. Five additional species occur here, rather similar.

The epithet was given in 1753 by C. Linnaeus for its leaves, which often have dark blotches like some species of wood-sorrel (*Oxalis* spp., Oxalidaceae) (Latin, lapathium = sorrel; folium = a leaf).

Pale Smartweed

Granny's Hairnet, Woodland Threadstem
Pterostegia drymarioides Fischer & C. Meyer

A weak-stemmed trailing annual that grows flat on the ground, sometimes crawling over other plants. Stems are generally 10-40 cm long, rarely to 1 meter. *In shade it is usually green but often turns red when growing in full sun.* Leaf blade 3-20 mm long, broader, with a short slender petiole. *Each leaf is fan-shaped, undivided or notched at the tip into 2 lobes, opposite on the stem.* One to two tiny flowers appear at the nodes, each with a yellow to pink 6-lobed calyx. Flowers are unisexual. Female flowers are surrounded by 2-lobed bracts that swell after fertilization to loosely surround the developing ovary. In flower Mar-July.

Very common but easily overlooked. Found mostly beneath plants and among rocks in coastal sage scrub and chaparral communities. Known from coastline to mountaintop, inland canyons, and valleys. Look for it on the Dana Point Headlands, in all of our wilderness parks, and along trails such as Harding Canyon and Holy Jim.

The genus was named in 1835 by F.E.L. von Fischer and C.A. von Meyer for the bracts that surround its fruit (Greek, pteron = wing; stege = covering, roof). They chose the epithet for the plant's similarity to South American chickweed plants in the genus *Drymaria* (Latin, -oides = likeness).

Granny's Hairnet

PRIMULACEAE • PRIMROSE FAMILY

Annuals or perennials, often hairless. *Leaves opposite, whorled, or alternate, many with basal leaves and upright flower stalks. Calyx deeply divided into 4 or 5 small sepals*, still connected at their base. *Corolla 5-lobed* (rarely 4-lobed), the *petals held outward or folded backward*. Stamens equal to the petals in number. *Ovary superior,* rarely half-inferior. A single style with a simple head-like stigma at the tip. Fruit is a capsule. A small family of 600 species, 21 in California, 1 in our area. Many are cultivated as ornamentals.

Padre's Shooting Star
Dodecatheon clevelandii E. Greene **ssp. *clevelandii***

An ephemeral spring perennial from white fibrous roots and a short ground-level stem. *Leaves 5-11 cm long, in a dense basal rosette.* The single flower stalk is upright, 18-40 cm tall, covered with short sticky-tipped hairs, and topped with an umbel of 1-16 flowers. Calyx 5-lobed, folded back when in flower, folded forward when in fruit. *The corolla has a shallow tube ringed with dark maroon followed by a ring of yellow. The 5 long, linear petals fold straight back. In coastal areas, the petals are white, sometimes tinged with pink. Inland they are dark pink to magenta.* The 5 stamens are arranged like a cone around the style; their dark maroon filaments are short and fused at the base, the anthers a deep maroon. The outside of each anther has a swollen tissue that is maroon and smooth at its base, turning creamy yellow and horizontally wrinkled as it tapers toward the tip. The single style is straight, longer than the stamens, and tipped with a small round stigma. In flower Jan-Apr.

In bud, the flowers face straight *up*. As the individual stalk grows, it arches over, the flower opens and faces *down*. After fertilization, the stalk straightens out again and faces straight *up*. Petals and stamens fall off, but the sepals remain with the developing fruit, an egg-shaped capsule that when ripe loses its cap and spills out numerous tiny seeds.

Bumble bees pollinate it by a process called *buzz pollination*. A bee lands on a flower, grasps the petals, and vibrates its wing muscles but not its wings. The anthers release pollen onto the bee's belly. As the bee lands on the next flower, pollen is transferred from its belly onto the flower's stigma.

It grows in *deep clay soils* among grasslands, often on slopes. On the coast side of the Santa Ana Mountains, it is found at low elevations below about 2,000 feet, up to nearly 4,000 feet on the inland side. Unfortunately, this plant thrives on the same flats and gentle slopes upon which we build our homes. Many extensive fields of shooting stars have given way to development projects.

Known from Crystal Cove State Park, the San Joaquin Hills, Temple Hill, the Dana Point Headlands, and inland through the hills surrounding Dana Hills High School (once vast fields, now eliminated). Uncommon in San Juan Capistrano and Hot Springs Canyon, quite common throughout Rancho Mission Viejo, Richard and Donna O'Neill Conservancy, Audubon Starr Ranch Sanctuary, and Caspers Wilderness Park. Uncommon in the San Mateo Canyon Wilderness Area as at Indian Potrero and Lucas Canyon Trails. Inland, it is known from Temescal Valley, south through the Elsinore-Temecula Valleys. The dark magenta form is abundant along Bedford Truck Trail and on the Santa Rosa Plateau.

The common name refers to its similarity to a shooting star rocketing through space. The genus was named in 1753 by C. Linnaeus, who used a name coined by Pliny the Elder for an unknown plant, thought to refer to the number of flowers in an inflorescence, comparing it to the number of Olympian gods (Greek, dodeca = twelve; theos = god). Our species was named in 1888 by E.L. Greene in honor of D. Cleveland.

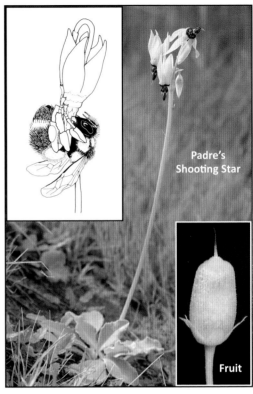

Padre's Shooting Star

Fruit

Bumble bee illustration by Heather C. Proctor, from: Harder and Barclay (1994). Used with permission of Lawrence Harder, University of Calgary.

Padre's Shooting Star

RANUNCULACEAE • CROWFOOT AND BUTTERCUP FAMILY

Annuals, perennials, and geophytes, some are woody vines. Leaves alternate, opposite, or basal, undivided or divided. *The base of each petiole is flat and broad, often wrapped around the stem (sheathed) like the carrot family (Apiaceae).* Their flowers are quite variable, arranged in a branched inflorescence or solitary on a single stalk. *Flower parts are generally free, not fused.* Sepals usually 5 but range from 3 to 15. Petals 0-15. *When petals are lacking, the sepals are often colorful and petal-like.* Stamens 10 or many more. Pistils 1-many, ovary superior with a single style that often forms a beak in fruit. *The fruits vary as well, usually a cluster of achenes or multiparted pod, rarely a berry.* A primitive eudicot family. These are beautiful, fascinating flowers that assume intricate forms and colors and are sure to be on your list of favorites. Many species are cultivated worldwide; some are highly toxic if ingested. A worldwide family of 1,700 species, 88 in California, 12 in our area.

Virgin's-bowers: *Clematis*

Woody vines that grasp with their long petioles and climb over other plants and objects. *Leaves opposite, 1-2-pinnate,* the leaflets often lobed and toothed. *Flowers stand upright* and are in a branched cluster or on an individual stalk. Each flower contains only one sex; usually both sexes are on the same plant (*monoecious*), but sometimes a plant has flowers of only one sex (dioecious). *Sepals 4, all alike, linear and petal-like, cream-colored in our area* (many Asian species are brightly colored). *Petals are absent.* Stamens and pistils numerous. *Fruit a cluster of achenes, each with a long fuzzy style,* nicknamed *weasel tail* by some. Pollinated by many types of insects and hummingbirds. We have 3 species in our area. They are in cultivation and make great garden plants. The genus was named in 1753 by C. Linnaeus for its long twiggy stems (Greek, klematis = branch, twig, vine-twig; a climbing plant).

	Chaparral *Clematis lasiantha*	Western *Clematis ligusticifolia*	Southern California *Clematis pauciflora*
Vine	4-5 meters long	4-6 meters long	2-4 meters long
Leaflets	3-5 in number, each generally 3-lobed, the largest 1.5-6 cm long	5-15 in number, lobed or toothed, the largest 2-9 cm long	3-5(-9) in number, each generally 3-lobed and toothed, the largest 1-3 cm long
Inflorescence	Each flower appears on its own stalk	In a many-flowered cluster	Flower stalks 2-3 cm long, topped with 1-3(-12) flowers
Sepals	10-21 mm long, hairy on the upper and lower sides	6-10 mm long, hairy on the upper and lower sides	Sepals 7-12 mm long, hairless upper side, hairy lower side
Stamens	Much shorter than sepals	About equal in length	About equal in length
Achene	Hairy	Hairy	Hairless and smooth
In flower	Jan-June	June-Sep	Jan-June

Chaparral Virgin's-bower, Pipestem
Clematis lasiantha Nutt.

A vine 4-5 meters long, the *branches very hairy. Leaflets 3-5 in number, each generally 3-lobed,* the largest 1.5-6 cm long. *Each flower appears on its own stalk* 4-12 cm long that arises from a node on the main stem. *Sepals 10-21 mm long, hairy on the upper and lower sides.* Stamens much shorter than sepals. The achene is hairy. In flower Jan-June.

Our most common and widespread virgin's-bower. Found among coastal sage scrub, chaparral, and woodlands in coastal canyons, foothills, and throughout the Santa Ana Mountains. Easily seen in Caspers Wilderness Park and upper Silverado and Trabuco Canyons.

Named in 1838 by T. Nuttall for its hairy flowers (Greek, lasios = hairy, woolly, shaggy; anthos = flower).

Western Virgin's-bower
Clematis ligusticifolia Nutt.

A vine 4-6 meters long, sometimes longer, *most branches nearly hairless. Leaflets 5-15 in number, lobed or toothed,* the largest 2-9 cm long. *Flowers are arranged in a many-flowered cluster,* a spectacular sight in full bloom. *Sepals 6-10 mm long, hairy on the upper and lower sides. Stamens and sepals about equal in length. The achene is hairy, like chaparral virgin's bower. In flower June-Sep,* much later than its relatives.

A common plant of variable habitats, mostly in moist soils along waterways but also common in dry regions of the Santa Ana Mountains, even right atop Santiago Peak. Known from the southern Chino Hills, Silverado Canyon, Limestone Canyon, San Juan Hot Springs, and San Juan Canyon. Long ago, even found in Anaheim Marsh near Sunset Beach. Those in Moro Canyon of Crystal Cove State Park have pink tails on their fruits, yellowish elsewhere.

Named in 1838 by T. Nuttall for the resemblance of its leaves to species of licorice-root (*Ligusticum* spp., Apiaceae) (Latin, Ligusticum; folium = leaf).

Flowers in dense clusters

Western Virgin's-bower

Southern California Virgin's-bower, Ropevine
Clematis pauciflora Nutt.

A vine 2-4 meters long, *the branches sparsely hairy. Leaflets 3-5(-9) in number, each generally 3-lobed and toothed,* the largest 1-3 cm long. *Flower stalks 2-3 cm long, topped with 1-3 flowers. Sepals 7-12 mm long, hairless on the upper side, hairy on the lower side. Stamens and sepals about equal in length. The achene is hairless and smooth. In flower Jan-June.*

Uncommon though perhaps overlooked. Grows in dry chaparral on rocky slopes and in canyons. Known from Santa Ana River Canyon, Silverado Canyon, Caspers Wilderness Park, San Juan Hot Springs, San Juan Canyon, cliff edges at the Dana Point Headlands, Temescal Valley, and the Santa Rosa Plateau.

Named in 1838 by T. Nuttall for its relatively few flowers (Latin, paucus = few; floris = flower).

Southern California Virgin's-bower

Hairless achenes

Flowers in clusters of 1-3

Larkspurs: *Delphinium*

Ephemeral spring perennials from fibrous to fleshy tubers and a single upright main stem. *Leaves palmately lobed or divided, lower and upper leaves sometimes quite different.* Their numerous unusual flowers appear in the upper portion of the main stalk. *Sepals 5, petal-like, colorful and highly modified, the uppermost sepal has a long backward-pointing spur. Petals 4, much smaller and 2-lipped.* The uppermost 2 petals are spurred, contain nectar, and are tucked into the uppermost sepal. The lower 2 petals are hairy, notched at their tips, and folded over the stamens. There are 3 pistils (rarely 4 or 5) each maturing into an upright pod with a pointed tip. The genus was established in 1753 by C. Linnaeus and based on the Greek name for larkspur, delphinion, in reference to the similarity of its flower buds to dolphins (Greek and Latin, delphis = dolphin; -ium = resemblance).

	Scarlet Delphinium cardinale	Parry's Delphinium parryi ssp. parryi	Liver-leaved Delphinium patens ssp. hepaticoideum
Height	1-2 meters	0.5-1 meter	0.25-0.8 meter
Leaves	Large and highly divided	Lower leaves divided into broad rounded lobes, middle and upper leaves variously divided into narrow linear lobes	Much broader than long, divided into 3-5 broad rounded lobes
Sepals	Scarlet red, sometimes yellow	Deep purplish-blue	Medium to dark sapphire blue
Petals	Yellow with scarlet tips	Upper petals whitish; lower petals purplish-blue and sparsely covered with long white hairs	Upper petals whitish with blue lines; lower petals medium to dark blue, sometimes whitish

Scarlet Larkspur
Delphinium cardinale Hooker

Our largest larkspur, it grows 1-2 meters tall. In our area, those in the mountains are taller than those in the lower-elevation washes. *Leaves are large and highly divided. Sepals are scarlet red, sometimes yellow.* Petals yellow with scarlet tips. In flower May-July.

Widespread and localized. Known from Carbon Canyon and the edge of Santa Ana Canyon in the Chino Hills. More common in the Santa Ana Mountains such as along Santiago Truck Trail, lower Trabuco Canyon, Long Canyon Road en route to Blue Jay Campground, North Main Divide Truck Trail near Falcon Group Camp and southwest of Trabuco Peak. In Temescal Valley, you'll find it in washes of Indian Canyon and at the Glen Eden Sun Club, where it grows with **slender-horned spineflower** (*Dodecahema leptoceros*, Polygonaceae). Often super-abundant after chaparral fire, along with **golden eardrops** (*Ehrendorferia chrysantha*, Papaveraceae); the play of scarlet and golden flowers is spectacular.

A very popular nectar source for hummingbirds. Bumble bees and other native bees gather pollen from it, but most cannot reach its nectaries.

After wet winters, spectacular pockets occur in northeast-facing drainages southwest of Upper San Juan Campground. To see it, start at the parking lot for San Juan Loop Trail and head south on Ortega Highway. Stop at the first or second turnout on the west (right) side of the road, above the campground. Look southwest. When in peak bloom, you can easily see masses of bright red flow through the chaparral in the drainages, best watched with binoculars, spotting scope, or long photographic lens.

Named in 1855 by W.J. Hooker for its red flower color (Latin, cardinalis = chief, principal, or red).

Scarlet larkspur's leaves are divided many times. Early in the season, you might see its stout stalks and a few leaves.

Parry's Larkspur
Delphinium parryi A. Gray ssp. *parryi*

Reaches about 1 meter but is usually only half that height. *Lower leaves are divided into broad rounded lobes;* middle and upper leaves are variously divided into narrower linear lobes. *Sepals are deep purplish-blue.* Upper petals whitish, lower petals purplish-blue and sparsely covered with long white hairs. In flower all spring.

Widespread and common, it grows among grasslands, coastal sage scrub, chaparral, and oak woodlands at many elevations. Known from the coast at upper Newport Bay, on bluffs in Corona del Mar and Laguna Beach, and on the Dana Point Headlands. Also found in the Whittier Hills, Chino Hills just above the Santa Ana River channel, Limestone Canyon in the Lomas de Santiago, Rancho Mission Viejo, Bell Canyon in Caspers Wilderness Park, Audubon Starr Ranch Sanctuary, and major canyons such as Sierra, Baker, Black Star, Silverado, Santiago, Hot Springs, and San Juan. It carpets a few mountain meadows such as the one near the entrance to Falcon Group Camp. Reported to hybridize with scarlet larkspur elsewhere, but we have not observed such hybridization in this area over our many years of field work.

Named in 1887 by A. Gray in honor of C.C. Parry.

Liver-leaved Larkspur

Delphinium patens Benth. **ssp. *hepaticoideum*** Ewan

Typically short, it reaches 25-80 cm tall. Its *leaves* are much broader than long and *divided into 3-5 broad rounded lobes. Sepals are medium to dark sapphire blue.* Upper petals whitish with blue lines, lower petals medium to dark blue, sometimes whitish. In flower Apr-May.

Grows in *shaded canyons,* grasslands, and the edges of oak woodlands. Known from upper elevations of the Santa Ana Mountains and major canyons such as Hot Springs, San Juan, Fremont, Leach, Black Star, Pine, Silverado, Trabuco, and McVicker. Also known from shady north-facing road cuts along Main Divide Truck Trail southeast of Trabuco Peak and on *north-facing talus slopes* of Modjeska Peak.

The epithet was given in 1849 by G. Bentham for its uncrowded inflorescence (Latin, patens = standing open or spreading). Our form was named in 1945 by J.A. Ewan for the similarity of its leaves to our 3-lobed liver (Greek, hepatikos = pertaining to the liver; -oideum = resemblance).

Little Mouse-tail ✿

Myosurus minimus L.

An annual from fibrous roots, it stands 2-9 cm tall and looks a little like plantain (*Plantago* spp., Plantaginaceae). *Its basal leaves are narrowly linear, almost thread-like. Flowers appear on stout naked stalks.* The numerous pistils are arranged on a long spike that extends way above the flower base but is not longer than the leaves. The 5-10 sepals are white, green, or pinkish and have a long spur at their base. The 5 white to greenish petals have a nectary at their base and often fall off early. Stamens 10, held upright or outward. In flower Mar-June.

An uncommon annual, found in vernal pools and marshes with very wet soils. In Orange County, only from the vernal pools in Costa Mesa's Fairview Park. Inland, from a terrace in Temescal Valley and a pool south of Lake Elsinore.

Named in 1753 by C. Linnaeus for its small size (Latin, minimus = least or smallest).

Common Meadow-rue

Thalictrum fendleri A. Gray **var. *polycarpum*** Torr.
[*T. polycarpum* (Torr.) S. Watson]

An upright ephemeral spring perennial to 2 meters tall. *Leaves 7-46 cm long, triangular in outline, divided 1-4 times into fan-shaped leaflets with rounded or pointed teeth.* The inflorescence is a branched, upright structure. *Each plant has flowers of only one sex (dioecious). Sepals 4-5, small, greenish-white to purplish. Petals absent.* Male flowers have 15-28 narrow stamens with yellow anthers that hang down. Female flowers have 2-20 yellowish to red pistils that usually stand up. In flower Mar-June.

Widespread and common in shady woods and sunny scrub in moist soils, among coastal sage scrub, chaparral, and oak woodlands. Found in all of our major mountain canyons. It thrives in a variety of habitats along many parts of the Trabuco Trail. In cultivation, it does best in shady places with regular watering. The leaves give off an unusual, spicy scent.

Named in 1849 by A. Gray for A. Fendler, who first collected it in New Mexico in 1847. Our form was named in 1856 by J. Torrey for its many fruits (Greek, poly- = many; karpos = fruit).

Aquatic Buttercup ✿
Ranunculus aquatilis L. **var. *diffusus*** With.

A perennial with stems 20-80 cm long, *it lives submerged in freshwater and grows on mud. Leaves submerged and floating, dissected into slender thread-like segments.* Sepals 5, held outward or backward, fall off early. *Petals 5, white, sometimes yellow at the base.* Fruit a cluster of 15 or more circular, flattened achenes, each topped with a thread-like beak. In flower Mar-Sep.

A plant of wetlands such as ponds, creeks, rivers, and lakes. Widespread but uncommon. Known from a small branch of Chino Creek in the Prado Basin and from vernal pools on the Santa Rosa Plateau.

The epithet was given in 1753 by C. Linnaeus for its aquatic habitat (Latin, aquatilis = living in or near water). The subepithet was given in 1796 by W. Withering for its openly branching growth form (Latin, diffusus = diffuse or loosely branched).

California Buttercup
Ranunculus californicus Benth.

An upright ephemeral spring perennial, long stems to 70 cm. Basal leaves round or heart-shaped in outline, divided into 3-5 leaflets, upper leaves deeply cut into thinner leaflets, each with sharp-pointed teeth. Sepals green to yellow, often folded backward, and shorter than most petals. *Petals 5-13 mm long, narrow, 7-22 in number, extremely shiny and bright yellow.* Fruit is a cluster of 5 or more circular, flattened, mostly hairless achenes, each topped with a curved beak-like style. In flower Feb-May.

Widespread and common in *heavy clay soils* that are wet in late winter and spring, among meadows, grasslands, and oak woodlands. Common in most of our wilderness parks, also the San Joaquin Hills, Chino Hills, Lomas de Santiago, Santa Ana Mountains, and Santa Rosa Plateau.

Named in 1849 by G. Bentham for the place of its discovery.

Similar. Western buttercup (*R. occidentalis* Nutt.): *petals 5-6, round*; Santa Rosa Plateau. **Desert buttercup** (*R. cymbalaria* Pursh, not pictured): *very short*, from underground runners (*stolons*); *leaves round, undivided*, the tips usually lobed; sepals and petals 5, narrow; grows in very wet areas. Old records for the Santa Ana River channel in Yorba Linda, Chino Creek, the Santa Ana River near Corona. Last reported in 1966 from upper Newport Bay. In flower May-Aug.

Delicate Buttercup
Ranunculus hebecarpus Hook. & Arn.

A delicate upright annual, 2-30 cm tall, covered in soft hairs. Leaves deeply 3-5 lobed, each lobe often toothed or lobed again. Lowest leaf blades 0.5-2 cm long, upper smaller. *Tiny flowers, much smaller than its fruits.* Sepals 1-2 mm long, yellowish. Petals 0-3(5), 1-2 mm long, whitish-yellow, fall off early and easily. Fruit a cluster of 3-11 circular, flattened, hairy achenes, each topped with a curved beak. In flower Mar-May.

A small, slender-stemmed wildflower, often unnoticed. The sepals and petals are short-lived; the flower sets fruit rather quickly. It typically grows in moist shaded spots within scrub, chaparral, and oak woodlands. Fairly common in the Santa Ana Mountains, such as in Silverado, Lost Woman, and Trabuco Canyons, and right along Holy Jim Canyon, Bear Canyon, and San Juan Loop Trails.

The epithet was given in 1828 by W.J. Hooker and G.A.W. Arnott for its hairy fruit (Greek, hebe = pubescence; karpos = fruit).

RHAMNACEAE • BUCKTHORN FAMILY

Many-branched shrubs or trees, some with thorns. Leaves undivided; alternate, rarely opposite; usually with obvious stipules. Flowers tiny, in clusters among the outermost branches. *Sepals 4-5,* tubular at their base, free above. *Petals absent or 4-5, often colorful.* Stamens 4-5 alternate with the sepals. *Ovary superior or partly inferior, 2-3-lobed.* The single style is 1-4-lobed at the top. Fruit a capsule or drupe. A family of 900 species in temperate and tropical regions, 57 members in California, 14 in our area. They are important as sources of wildlife cover, food, pollen, and nectar. Many species are prized in cultivation.

California Lilacs, Ceanothus: *Ceanothus*

Handsome plants in a wide variety of sizes and forms, from low-lying ground covers to rounded or erect shrubs and small trees. Flowers small, densely arranged in elongate or rounded clusters. Each flower is no larger than 5 mm long and has 5 sepals, petals, and stamens. Sepal and petal bases fused into an urn-like hypanthium. *Sepals petal-like, triangular,* often curved toward the flower's center (**incurved**), and remain on the flower. Petals skinny at the base (**clawed**) with a broad scoop-shaped (**hooded**) distal region (**blade**); they often fall off of the flower during bloom. Stamens long, opposite the petals, and stand straight up or outward when mature. The single style is 3-lobed at the top. The *colorful nectary* surrounds the ovary and secretes nectar so plentiful that the flower center glistens with it. *Fruit is a 3-lobed capsule* that separates into 3 parts when mature.

Their flowering period is rather short, with adjacent members of a species all in flower at once, so it is easy to visit them too early or too late to witness peak bloom. From January through May, a hike or drive along Main Divide Road will treat you to spectacular views of the various species in prolific bloom. Many species are adapted to fire and will *resprout from burned stumps.* The *long-lived seeds* of most germinate more readily after being heated or scarred by fire. They are highly prized worldwide in the horticultural trade; many hybrids have been produced and are sold by native plant nurseries. We recommend a visit to Rancho Santa Ana Botanic Garden, where several species are grown, in order to familiarize yourself with the growth habits of these plants before seeking them in the field.

Of the 55 species of California lilac, 46 are native to California, 10 of those in our area. Some are difficult to identify, so we arranged them into 4 logical groups based on twig, leaf, flower, and fruit characteristics. Consult the table on the following page. When you find an unknown species, determine to which group it belongs and then study the species in that group to identify your plant.

The genus was named in 1753 by C. Linnaeus, who reused an ancient Greek name (Greek, Keanothus = a name used by Theophrastus for a thorny plant).

In addition to *The Jepson Manual* (Baldwin et al. 2012), important references for California lilacs include Fross and Wilken (2006) and McMinn (1939, 1964).

Below: Our most widespread species, thick-leaved lilac (Ceanothus crassifolius*), covers the slopes along South Main Divide Road. California lilacs paint our hills and mountains with abundant flowers and fill the air with characteristic scents.*

Thick-leaved Lilac

	Twigs	Stipules	Leaves	Flowers	Fruit	Location
Horned Lilacs: stipules thick and corky; flowers white; fruit with 3 large, rounded horns						
Thick-leaved *C. crassifolius* var. *crassifolius*	Rigid	Thick and corky	Edges toothed, rolled under; *opposite;* petiole 2-5 mm; evergreen	White	3-horned	Southern to north-central Santa Ana Mountains
Bigpod *C. megacarpus* ssp. *megacarpus*	Rigid	Thick and corky	Edges smooth, flat; *mostly alternate;* petiole 1-4 mm; evergreen	White	3-horned	San Joaquin Hills, northern Santa Ana Mountains, Chino Hills
Thorny Lilacs: stipules leaf-like; twigs rigid and thorny; flowers white to blue; fruit smooth or 3-crested						
Ch. Whitethorn *C. leucodermis*	Rigid and thorny; *bark white*	Thin and leaf-like, fall off	3 main veins, edges smooth; alternate	White to blue	Smooth, sticky, tip sunken in	Santa Ana Mountains
Greenbark *C. spinosus*	Rigid and thorny; *bark green*	Thin and leaf-like, fall off	1 main vein, edges smooth; alternate; evergreen	Pale blue to whitish	Smooth or with 3 low fin-like crests	San Joaquin Hills, Santa Ana Mountains, Chino Hills
Oval-leaf Lilacs: stipules leaf-like; leaves broadly oval, edges toothed; flowers blue; fruit 3-crested						
Hairy *C. oliganthus* var. *oliganthus*	Flexible; red-brown, smooth or warty, very hairy	Thin and leaf-like, fall off easily	3 main veins; edges toothed, not sticky; upper long-haired, *only lower veins are hairy;* alternate; petiole 3-7 mm	Deep blue or purplish; ovary smooth.	3 fin-like crests	Santa Ana Mountains
San Diego *C. oliganthus* var. *orcuttii*	Flexible; red-brown, smooth or warty, very hairy	Thin and leaf-like, fall off easily	3 main veins; edges toothed, not sticky; upper long-haired, *only lower veins are hairy;* alternate; petiole 3-7 mm	Pale blue to blue; ovary short-hairy.	3 fin-like crests; wrinkled, short-hairy, especially when young.	Santa Ana Mountains, more common on east slopes.
Jim Brush *C. oliganthus* var. *sorediatus*	Rigid; gray-green or purplish; hairless (small hairs)	Thin and leaf-like, fall off easily	3 main veins; edges toothed, not sticky; upper shiny, mostly hairless; *lower paler, small hairs on veins;* alternate; petiole 3-7 mm	Pale to deep blue; ovary smooth.	Smooth, or with 3 low crests	Santa Ana Mountains and their foothills
Woollyleaf *C. tomentosus*	*Rust-red, covered with black sticky warts and rusty hairs*	Thin and leaf-like, fall off easily	3(1) main veins, edges toothed and sticky; *lower surface white-hairy with veins;* alternate; petiole 1-3 mm	Deep blue or purplish (whitish)	3 fin-like crests	Santa Ana Mountains and their foothills.
Mountain Lilacs: stipules leaf-like; leaves much longer than wide; flowers white or blue; fruit with 3 low crests						
So. Deer Brush *C. integerrimus* var. *macrothyrsus*	Flexible; greenish	Thin and leaf-like, fall off easily	3 main veins; edges smooth; some soft hairs above; winter deciduous; alternate	White	3 fin-like crests, tip sunken in	Santa Ana Mountains, high elevations
Palmer *C. palmeri*	Flexible; gray-green to brownish	Thin and leaf-like, fall off	1 main vein; edges smooth; hairless above; evergreen; alternate	White	3 sticky fin-like crests	Santa Ana Mountains, high elevations
Wartleaf *C. papillosus*	Green or pink, covered with dense woolly hairs	Thin and leaf-like, fall off easily	1 main vein; *wart-covered;* edges toothed, *strongly rolled under;* very hairy beneath; alternate	Dark blue to purplish	3 narrow fin-like crests	Santa Ana Mountains, high elevations

HORNED LILACS

Twigs are not thorny. Their leaf undersides have obvious sunken pits. *Stipules are thick and corky, easily seen, and do not fall off by themselves.* Flowers are always white. *Each fruit has 3 distinct horns at its tip.*

Thick-leaved Lilac
Ceanothus crassifolius Torr. **var. *crassifolius***

A stout, upright shrub, 2-3.5 meters tall. Stipules thick and corky. Its *thick leathery leaves* have 1 main vein and are 1.5-3 cm long, the *blade broadly elliptical; evergreen, opposite on the stem.* Leaf edges usually toothed and folded under. Upper leaf surface smooth, hairless, and olive-green. *Lower leaf surface densely covered with white hairs. Flowers are white and release a heavy perfume.* Fruits have 3 horns at their tip. In flower Jan-Apr.

An abundant shrub on chaparral slopes and in washes below. After fire, its seeds germinate and mature shrubs stump-sprout. In some places, such as the hills east of San Juan Canyon, it forms dense thickets difficult to hike through. Known from the Santa Ana River wash near the Chino Hills and major canyons of the Santa Ana Mountains but not on the higher peaks. Easily seen along Ortega Highway in Hot Springs and San Juan Canyons, also along Long Canyon Road.

Named in 1857 by J. Torrey for its thick leaves (Latin, crassus = thick; folium = leaf).

Similar. Flat-leaved lilac (*C. crassifolius* Torr. var. *planus* Abrams, not pictured): leaf blade underside hairless to sparsely haired; leaves flat, edge not rolled under. Crow and Bell Canyons, perhaps elsewhere. Uncommon.

Thick-leaved Lilac

Thick-leaved

Thick-leaved | Bigpod

Bigpod

Thick-leaved | Bigpod

Bigpod Lilac
Ceanothus megacarpus Nutt. **ssp. *megacarpus***

A many-branched, dense shrub with rounded outline, 1-4 meters tall, often wider. Stipules thick and corky. *Its leathery leaves have 1 main vein,* the blade 1-2.5 cm long, elliptical to reverse-oval, blunt or notched at the tip; evergreen, *mostly alternate on the stem* and crowded on the twigs. *Leaf edges smooth and flat,* neither toothed nor folded under. Upper leaf surface smooth, hairless, and dull-green. *Lower leaf surface hairless or sparsely haired between the veins. Flowers white.* Fruits have 3 horns at their tip. In flower Jan-Apr.

In coastal and mountain chaparral. Scarce in the Santa Ana River wash, more common at the north end of the Santa Ana Mountains such as in Coal and Claymine Canyons and on Sierra Peak. Abundant in the San Joaquin Hills such as Los Trancos Canyon, Temple Hill, and Niguel Hill.

Named in 1846 by T. Nuttall for its large fruits (mega = large, carpus = fruit).

Bigpod Lilac

THORNY LILACS

Twigs are rigid and thorny. Leaf undersides do not have sunken pits. *Stipules are thin and leaf-like, sometimes missing.* Flowers white to blue. Fruits are smooth or with 3 thin fin-like crests up their sides.

Chaparral Whitethorn
Ceanothus leucodermis E. Greene

A stiff shrub 2-4 meters tall with spreading branches and rigid, thorny twigs. *Its bark is smooth and pale green, its twigs are covered with whitish powder and so appear white.* Its mostly hairless leaves have 3 main veins and are 1-2.5 cm long, the *blade oval to long-elliptic, evergreen, alternate on the stem.* Leaf edges are smooth but may be toothed on stump-sprouts and new growth. Both surfaces are covered with a whitish powder, easily rubbed off. *Flowers are white to blue.* Fruits sticky and smooth, the top-center sunken in. In flower April-June.

Grows on dry rocky and sandy soils among chaparral and mixed coniferous forests. Known sparingly from the Santa Ana River wash near the Chino Hills but mostly at middle to high elevations of the Santa Ana Mountains such as Santiago and Modjeska Peaks, along Main Divide, Harding, and Santiago Truck Trails. Also at Los Pinos Saddle, sporadically along the Main Divide Truck Trail to Horsethief Trail, near Bear Spring, and in Tenaja Canyon.

Named in 1895 by E.L. Greene for the white bark of its twigs (Latin, leuco = white; dermis = skin).

Chaparral Whitethorn

Chaparral Whitethorn | Greenbark Lilac

Greenbark Lilac
Ceanothus spinosus Torr. & A. Gray
[*C. spinosus* Torr. & A. Gray var. *spinosus*]

A large tree-like shrub 2-6 meters tall *with green bark, many rigid branches, and stiff thorns.* Its *mostly hairless leaves* have 1 main vein and are 1.2-3 cm long, the blade elliptic to oblong; evergreen; alternate on the stem. The edges are smooth, the tip often notched inward. Upper surface bright to medium green, paler below. *Flowers pale blue to whitish.* Fruits smooth or with low crests. In flower Feb-May.

A plant of *coastal hills* from San Luis Obispo County southward, it grows among coastal sage scrub and chaparral in the San Joaquin Hills, where it is found on Niguel Hill and in major canyons of the Santa Ana Mountains at low and middle (but not high) elevations. Known from Silverado, upper Santiago, and Trabuco Canyons. Also at Caspers Wilderness Park and Audubon Starr Ranch Sanctuary but most abundantly throughout San Juan Canyon and the western portion of the San Mateo Canyon Wilderness Area. Its seeds germinate and burned shrubs stump-sprout vigorously after fire.

Named in 1838 by J. Torrey and A. Gray for its long thorns (Latin, spinosus = spiny).

Greenbark Lilac

OVAL-LEAF LILACS

Twigs are not thorny. Leaves are broadly oval or heart-shaped, dark green above, the edges toothed. Stipules are thin, sometimes missing. Flowers blue to whitish. Fruits crested. *The lilacs in this group are difficult to separate.* Apparent hybrids between woollyleaf and hairy lilac are known from Bedford and Trabuco Canyons.

Hairy Lilac
Ceanothus oliganthus* Nutt. var. *oliganthus

A handsome shrub 1-3 meters tall with short, stiff branches. *Young branches and twigs are red-brown, smooth or wart-covered, and very hairy. Petiole 3-7 mm long, hairy.* Leaves have 3 obvious main veins from the base; blade 1.5-4 cm long, oval, long-oval, or elliptic; evergreen; alternate. Leaf edges toothed, sticky or not. *Upper leaf surface covered with long soft hairs.* Lower surface hairy, mostly on the veins. *Ovary hairless. Fruit with 3 obvious fin-like crests up the sides.* Flowers deep blue to purplish. In flower Mar-Apr.

A plant of dry brushy chaparral slopes in the upper elevations of the Santa Ana Mountains. Known from lower Pine Canyon near Silverado, along upper Holy Jim Trail, Upper Trabuco Canyon, upper Hot Springs Canyon near Falcon Group Camp, Los Pinos Spring, Los Pinos Saddle and northward along Main Divide Road, Modjeska Peak, Bedford Canyon, and the San Mateo Canyon Wilderness Area. Easiest seen along North Main Divide Road "Loop."

Named in 1838 by T. Nuttall in reference to its loose, open, uncrowded inflorescence, which appears to have few flowers (oligos = few; anthos = flower).

Similar. Jim brush (*C.o.* var. *sorediatus*): petiole 3-7 mm long, hairless to slightly hairy; leaf upper surface shiny and totally or nearly hairless, lower surface paler and with small hairs; flowers identical; fruits usually smooth, hairless, no fin-like crests. **San Diego hairy lilac** (*C.o.* var. *orcuttii* (Parry) Jeps., not pictured): flowers paler blue; *ovary with short hairs; fruit hairy and wrinkled;* Santa Ana Mountains (Coldwater Canyon Trail, Los Pinos Saddle), southward; uncommon. Named in 1889 by C.C. Parry for C.R. Orcutt. **Woollyleaf lilac** (*C. tomentosus*): shrub-like; *young twigs rust-red; petiole 1-3 mm long;* leaf edges toothed *and* sticky; leaf upperside with short ashy-grey hairs, underside with short white hairs mostly on veins; flowers generally lighter; local hybrids occur.

Jim Brush
Ceanothus oliganthus* var. *sorediatus (Hook. & Arn.) Hoover [*C. sorediatus* Hook. & Arn.]

Twigs gray-green or purplish, *hairless or with small hairs. Petiole 3-7 mm long, hairless to slightly hairy. Upper leaf surface shiny and totally or nearly hairless.* Lower surface paler and with small hairs. Inflorescence crowded. Flowers pale to deep blue. *Ovary hairless,* no fin-like crests. In flower Feb-May.

Grows in canyons and on chaparral slopes, generally at lower elevations than hairy lilac. In the Santa Ana Mountains, it occurs at low to medium elevations, sparingly at higher elevations. Known from Silverado and Trabuco Canyons, San Juan and Lucas Canyons in Caspers Wilderness Park, Santiago Peak, Main Divide Truck Trail near Trabuco Peak, and along North Main Divide Road on the El Cariso-Long Canyon loop.

Named in 1840 by W.J. Hooker and G.A.W. Arnott for its tightly packed inflorescence (Latin, sored, soros = a heap).

Similar. Woollyleaf lilac and **San Diego hairy lilac**: described under entry for hairy lilac, above. **Hairy lilac** (*C.o.* var. *oliganthus*): tree-like; *petiole 3-7 mm long, hairy;* leaf upperside with long hairs, underside with hairs mostly on veins; flowers identical; fruits usually bumpy (smooth), lobed, crested, hairy or not.

Jim Brush

Jim Brush fruit

Hairy Lilac fruit

Woollyleaf Lilac

Ceanothus tomentosus C. Parry
[*C.t.* C. Parry ssp. *olivaceus* (Jeps.) Munz]

An upright shrub 1-3 meters tall with long slender branches. *Its young branches and twigs are often rust-red, covered with conspicuous black glandular warts and rust-colored hairs.* The bark ages to gray-brown or reddish. Leaves have 3 main veins (rarely 1) from the base, the blade 1-2.5 cm long, oval to elliptic, evergreen, alternate on the stem. *Leaf edges have tiny gland-tipped, sticky teeth.* Upper leaf surface is covered with short ashy-grey hairs. Lower surface is gray-green and covered with short white hairs mostly on the veins. Flowers deep blue or purplish to almost white. Fruits with 3 obvious fin-like crests up the sides. In flower Mar-May.

Found among chaparral at low to medium elevations of the Santa Ana Mountains, uncommon in the San Mateo Canyon Wilderness Area. Known from Coal, Leach, and Silverado Canyons, Sierra Peak, and the junction of Trabuco and Horsethief Trails.

The epithet was given in 1899 by C.C. Parry for the white woolly hairs on its leaf undersides (Latin, tomentum = woolly hairs).

Similar. Hairy lilac (*C. oliganthus* var. *oliganthus*) and **Jim brush** (*C.o.* var. *sorediatus*): young twigs red-brown; *petiole 3-7 mm long, hairy;* leaf edges toothed but *not sticky;* leaf upperside with long hairs, underside with hairs mostly on veins; flowers generally darker; hybrids are known.

Woollyleaf Lilac

Teeth and glands along leaf edges

MOUNTAIN LILACS

Twigs are not thorny. *Leaves linear or elliptical, clearly much longer than wide, the edges smooth or toothed.* Stipules thin, sometimes missing. Flowers white to dark blue. *Fruits have low crests.* Mostly in higher elevations of the Santa Ana Mountains.

Southern Deer Brush

Ceanothus integerrimus Hook. & Arn. **var. *macrothyrsus*** (Torr.) G.T. Benson
[*C. integerrimus* Hook. & Arn. var. *puberulus* (Greene) Abrams]

A loosely branched upright shrub, 1-4 meters tall. *Its thin leaves have (1-)3 main veins from the base and are 2.5-7 cm long, broadly oval, alternate on the stem, deciduous.* Leaf edges smooth, sometimes minutely toothed toward the tip. Upper leaf surface is smooth, sparsely hairy, and light green. Lower leaf surface is paler and hairier. Stipules are thin and fall off by themselves. *In our area, the flowers are white* (elsewhere to dark blue, rarely pink). *Fruits have a vertical crest on the side of each lobe; the top-center is sunken in.* In flower May-June.

A plant of dry chaparral slopes, ridges, and canyons below. Known mostly from higher elevations of the Santa Ana Mountains such as Santiago and Modjeska Peaks, upper Harding and Santiago Truck Trails, and Santiago Canyon near the Fleming Ranch. Also various sites along the Main Divide Truck Trail, especially on north-facing slopes southeast of Santiago Peak. Known to stump-sprout after fire.

It was named in 1840 by W.J. Hooker and G.A.W. Arnott for its smooth-edged leaves (Latin, integerrimus = very perfect, complete). Our form was named in 1874 by J. Torrey for its long inflorescence (Greek, makros = long, large; thrysus = stalk, wand).

Similar. Palmer lilac (*C. palmeri*): leaves thicker, partly winter deciduous, 1-veined; white flowers; high elevations; uncommon.

Southern Deer Brush

Southern Deer Brush

Palmer Lilac
Ceanothus palmeri Trel.
[*C. spinosus* Torr. & A. Gray var. *palmeri* (Trel.) K. Brandegee]

A large upright shrub 1-3.5 meters tall. *Its thick, firm leaves have 1 main vein from the base, the blade 1.5-3.5 cm long, oblong, tips sometimes notched inward; alternate; mostly evergreen.* Leaf edges smooth and flat. Upper surface hairless and light green, underside even lighter and sometimes a little hairy on the veins. Leaves are partly winter deciduous (some fall off, some remain on the plant). Flowers are white, rarely pale blue. Fruits usually have 3 thin, sticky, fin-like crests up their sides. In flower May-June.

Found on dry slopes in chaparral and coniferous forests though uncommon here. Known primarily from high elevations (over 4,000 feet) such as Santiago Peak and above Silverado Canyon.

Named in 1888 by W. Trelease in honor of E. Palmer, who first collected this species in the Cuyamaca Mountains of San Diego County in 1875.

Similar. Greenbark lilac (*C. spinosus*): spiny branchlets; flowers blue, pale blue, or whitish; fruits circular in cross-section, smooth, rarely with low ridges; usually at lower elevations; San Joaquin Hills, Chino Hills, Santa Ana Mountains; common. **Southern deer brush** (*C. integerrimus* var. *macrothyrsus*), leaves thinner, completely winter deciduous, (1- or) 3-veined, the lateral pair sometimes quite long; white flowers; Santa Ana Mountains; much more common.

Palmer Lilac

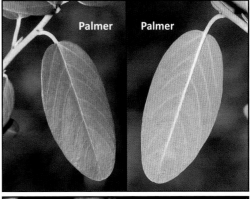
Palmer Palmer

Palmer Lilac

Wartleaf Lilac ✿
Ceanothus papillosus Torr. & A. Gray

An upright plant in nature but often low-growing in gardens, it stands 1-5 meters tall. *Young twigs are green or pink and covered with dense woolly hairs. Its unusual leaves have one obvious main vein; the slender linear blade is 1.5-5 cm long, less than 1 cm wide, the tip blunt or notched inward; alternate; evergreen.* The edges are toothed and rolled under. Upper surface is dark green and covered with hairs and numerous sticky-tipped "warts." Lower surface is paler and much hairier. Each inflorescence is a small upright umbel-like cluster of flowers. Flowers are deep dark blue, almost purple (rarely whitish). Fruits have low, narrow vertical crests up their sides. In flower Feb-June, generally the last week of Apr to the first week of May in our area.

Grows among dense chaparral on rocky slopes between 2,000-4,000 feet in elevation. Known from a few scattered locations in our area: among knobcone pines and chaparral on Pleasants Peak and northwest of it (mostly the south side, adjacent to Main Divide Road), at the head of Hagador Canyon, on the slopes of Santiago Peak along upper Holy Jim Trail (the easiest place to see it, 2.5 miles up the trail), in upper Trabuco Canyon, near Los Pinos Peak along Los Pinos Trail, and on "old" Sugarloaf Peak along San Juan Trail. On Los Pinos Peak, it grows among Eastwood manzanita (*Arctostaphylos glandulosa* ssp. *glandulosa*, Ericaceae), where it often pops up amongst a dense sea of manzanita.

The epithet was given in 1838 by J. Torrey and A. Gray in reference to the nipple-like or wart-like bumps on the leaves (Latin, papilla = nipple).

Wartleaf Lilac

THE CALIFORNIA LILAC GUILD

California lilacs provide shelter and food for invertebrate and vertebrate species alike. Many insects such as flies, butterflies, beetles, and bees take nectar and pollen from the flowers. Deer often browse its young shoots and leaves. We present here three insect members of this large guild.

Ceanothus Silk Moth
Hyalophora euryalus Boisduval (Saturniidae)

Ceanothus Silk Moth

This lovely moth utilizes California lilacs (*Ceanothus* spp.), coffeeberries (*Frangula* spp.), and laurel sumac (*Malosma laurina,* Anacardiaceae) as larval foodplants. Their *large light green caterpillars* sport thorny yellow warts on their backs and often blue bumps on their sides. They pupate right on the twigs in large cocoons made from medium-brown silk. The cocoon is unusual in that it is double-layered, *a cocoon within a cocoon.*

Cocoon of Ceanothus Silk Moth

Adults emerge in spring, mostly during April. From midnight almost to daybreak, females release pheromones that attract males. After mating, females deposit eggs on the larval foodplants. The most commonly used species here are thick-leaved and greenbark lilacs. Once they hatch, larvae feed and grow until late fall when they pupate. Adults have no functioning mouthparts and cannot feed; they live off stored fat reserves for about 1-2 weeks.

Named in 1855 by J.B.A.D. de Boisduval for its broad wings (Greek, euryalos = broad).

Ceanothus Rain Beetle
Pleocoma puncticollis puncticollis Rivers (Pleocomidae)

An unusual beetle completely dependent upon ceanothus. *Their larvae develop slowly and feed underground on the roots for 8-13 years.* When mature, they pupate in summer or fall. During and just after the first rains arrive in September-November, adults emerge from their pupal case and dig upward to the soil surface. The *shiny black fuzzy-bellied males fly* and seek out the *brownish flightless females, which remain in their burrows and release pheromones into the air.* Mating takes place in an underground burrow; the male dies soon thereafter. The female deposits her eggs near roots and dies, perhaps within a month. Eggs hatch, the young larvae feed on ceanothus roots, and the cycle begins again. Adults have an incomplete digestive system, so they cannot feed; like the ceanothus silk moth, they burn stored fat for energy. This beetle lives near the coast at Del Mar and north through the Santa Ana and Santa Monica Mountains.

Ceanothus Rain Beetle

Named in 1889 by J.J. Rivers for the numerous tiny pinpricks on its thorax (Latin, puncti = small hole, puncture; collis = neck).

Ceanothus Longhorn Beetle
Callimoxys sanguinicollis fuscipennis (LeConte, Cerambycidae)

Slender beetles, 1-1.5 cm long, black above, paler below. *Females larger than males and have a red thorax. Hind legs yellowish; toward the "knee" the femur is black and swollen like a club.* Larvae are wood-borers that tunnel within the stems of ceanothus. They pupate and winter in the plant, emerge as adults in spring, and feed on nectar and pollen. The species is found throughout North America, this form from British Columbia to southern California.

Ceanothus Longhorn Beetles

Named in 1861 by J.L. LeConte for its dark wings (Latin, fuscus = brown, dark, dusky; penn, pinna = wing).

Coffeeberries, Buckthorns, Redberries: *Frangula, Rhamnus*

Upright woody shrubs, sometimes tree-like. Branches are alternate, twig-tips sometimes thorny (buckthorns and redberries, *Rhamnus*). Leaves deciduous or evergreen, smooth-edged or toothed. *Flowers tiny; solitary or arranged in rounded umbels.* Fruit is berry-like drupe, round, with 2-4 separate stones inside. None have large showy flowers; their beauty lies in clusters of tiny flowers visited by a variety of insects, followed by attractive fruits that are often eaten by birds. They all make wonderful garden plants, easily grown and free from most plant pests and diseases. Numerous cultivated varieties exist in the horticultural trade. We have 4 species, sometimes combined within the genus *Rhamnus*. The genus *Frangula* was named in 1754 by P. Miller, presumably for its fruits; the stones break apart when the fruit is opened (Latin, frangere = to break). The genus *Rhamnus* was named in 1753 by C. Linnaeus, who used the ancient Greek name for buckthorn (Greek, rhamnos = buckthorn plant).

	Coffeeberries: *Frangula*	**Redberries:** *Rhamnus*
Winter bud scales	Absent	Present
Sepals	Fleshy, erect (held upward), keeled on their upper (inner) side	Thin, spreading (held outward), not keeled on their upper (inner) side
Petals	5	Absent or 4-5
Fruit	Hard, red-brown to black, opaque, never semi-transparent	Soft, succulent, red, semi-transparent when mature

California Coffeeberry
Frangula californica (Eschsch.) A. Gray **ssp. *californica***
[*Rhamnus californica* Eschsch. ssp. *californica*]

An upright, densely or sparsely branched shrub, 1-4 meters tall. Its bark is bright gray or brown, the youngest twigs reddish, hairless or covered with fine hairs. *Leaves are evergreen,* the blade 20-80 mm long, oval, oblong, or elliptical. *Leaf upper surface is dark green, the lower bright green or yellow, both surfaces mostly hairless. Leaf edge is smooth, rarely toothed, sometimes rolled under.* The tiny flowers are arranged in umbels, each flower with 5 (rarely 4) yellowish petals hidden among the stamens, each 2-lobed and notched inward in the center. Its hard fruits are red, age to brown then black; each contains 2 seeds. In flower Mar-May.

Grows at low to middle elevations of the Santa Ana Mountains, among coastal sage scrub, chaparral, oak woodlands, and riparian communities, *commonly in shady canyons.* Especially abundant in Santiago, Silverado, Trabuco, and San Juan Canyons.

An excellent garden plant, tolerant of cold and heat, generally not browsed by deer. Caterpillars of the **pale swallowtail butterfly** (*Papilio eurymedon*, Papilionidae) feed on its leaves.

Named in 1823 by J.F. Eschscholtz for the place of its discovery; he was also its first collector.

Similar. Chaparral coffeeberry (*F. californica* ssp. *tomentella* [(*Rhamnus tomentella* ssp. *tomentella*]): twigs gray, woolly; leaf upper surface dull green and mostly hairless or white-woolly; *leaf lower surface soft velvety to silvery,* edges smooth or toothed; chaparral and mixed woodland, Fremont Canyon; in flower Jan-Apr. **Desert coffeeberry** (*F.c.* ssp. *cuspidata* [(*Rhamnus tomentella* ssp. *cuspidata*], not shown): twigs red, hairs of 2 lengths; leaves mostly hairless above, very white-hairy below, edges toothed; chaparral and woodlands, Black Star Canyon; in flower Apr-July.

California Coffeeberry

California (left, green, hairless) / Chaparral (right, white-woolly)
California (left, green, mostly hairless) / Chaparral (right, white-woolly)

Above: Leaf upper surfaces (left side) and lower surfaces (right side)

Spiny Redberry
Rhamnus crocea Nutt.

A dense shrub with many rigid, spiny branches, it stands 1-2 meters tall. The youngest growth is often yellow-green. *Leaves evergreen, thick and small, the blade 10-15 mm long, reverse-oval in shape.* The upper surface is dark to pale green, the lower brown or saffron-yellow. The edges are variable: flat or wavy, smooth or toothed, sometimes folded under. Flowers are in small groups of 1-6, each very tiny, without petals, and of only one sex; both sexes are on the same plant. The fruit is bright red, clear, and succulent, often fed upon by birds. In flower Mar-Apr.

Usually a shrub of dry inland chaparral, but in Orange County it is uncommon and known only from maritime chaparral along the immediate coast. Found on Temple Hill in Laguna Beach, Niguel Hill in Laguna Niguel, and other scattered sites in the San Joaquin Hills. Easily seen along trails on the inland side of Aliso Beach County Park in Laguna Beach. In western Riverside County, it occurs in the Temescal Valley at the foot of the Santa Ana Mountains such as along Indian Truck Trail.

Named in 1838 by T. Nuttall for its saffron-yellow leaf undersides (Latin, croceus = saffron-colored).

Similar. Holly-leaved redberry (*R. ilicifolia*): taller and broader; leaves larger and rounder; abundant.

Spiny Redberry

The fruits of redberries are soft, succulent, red, and semi-transparent when mature. Not only are they beautiful, but they are used as food by many types of insects, birds, and mammals.

Holly-leaved Redberry

Holly-leaved Redberry
Rhamnus ilicifolia Kellogg
[*Rhamnus crocea* ssp. *ilicifolia* C.B. Wolf]

A dense upright shrub, often tree-like, 1.5-4 meters tall. *Leaves evergreen, thick, the blade 20-40 mm long, oval to round in shape.* The upper surface is dark green, the lower paler and brownish, both surfaces hairless or hairy. The edges are spiny, rarely smooth; wavy or flat; the sides usually curved over. *The leaf tip is rounded though sometimes notched inward.* Like those of spiny redberry, the flowers are in small groups of 1-6, each very tiny, without petals, and of only one sex. The fruit is bright red, clear, and succulent, often fed upon by birds. In flower Mar-June.

A very common shrub of moist coastal sage scrub, chaparral, oak woodland, and mixed coniferous forest communities. Known from the San Joaquin Hills, the Puente-Chino Hills, Lomas de Santiago, and throughout the Santa Ana Mountains. Easily seen in the vicinity of Tucker Wildlife Sanctuary, the forest gate in Silverado Canyon, Borrego Trail in Whiting Ranch Wilderness Park, and along the San Juan Trail in lower Hot Springs Canyon.

The epithet was given in 1863 by A. Kellogg for the similarity of its leaves to **English holly** (*Ilex aquifolium* L., Aquifoliaceae) (Latin, -folius = leaved).

Similar. Spiny redberry (*R. crocea*): shorter and narrower; leaves smaller and narrower; uncommon and localized.
Holly-leaved cherry (*Prunus ilicifolia,* Rosaceae): leaves oval to rounded, the edges always crinkled or very wavy with strong spine-tipped teeth, the tip pointed; generally at higher elevations, less common.

Holly-leaved Redberry

ROSACEAE • ROSE FAMILY

A variable family of annuals, perennials, vines, shrubs, and trees, both evergreen and deciduous forms. *Leaves of some species are simple, those of others are divided into leaflets.* The bases of the calyx, corolla, and stamens are all fused together to form a *funnel-shaped hypanthium* at the base of the flower. The free lobes of those parts radiate out from it like spokes of a bicycle wheel. Calyx with 5 free lobes. Corolla with 5 free lobes, 0 in some species. *Stamens numerous, often more than 10, rarely 5 or absent. Ovary superior, inferior, or half-inferior; styles 1-5.* Fruits range widely from a dry achene (chamise) to a plump pome (apple) or a cluster of fleshy spheres (blackberries).

An important plant family with many species vital to all sorts of wildlife, primarily for foliage, cover, and fruit. Humans value their horticultural uses as groundcovers, woodland shrubs, and showy flowers, and have cultivated some types for thousands of years. Familiar crops include strawberries, blackberries, peaches, nectarines, plums, cherries, pears, apples, and roses. Worldwide, there are about 3,000 species, 151 in California, 18 in our area.

Ribbonshanks, Red Shanks
Adenostoma sparsifolium Torr.

Ribbonshanks

An upright shrub 2-6 meters tall, often tree-like. Young twigs are bright green and hairless, sometimes covered in sticky resin. The bark ages to grey-brown. On older, larger branches, it shreds away to reveal beautiful smooth red-brown bark beneath. Its *light green sticky leaves* are 4-15 mm long, *linear, most of them arranged alternately on the stem, spaced far apart and not in bundles.* Flowers are in loose open clusters. Unlike common chamise, the flowers have individual stalks and are not held as close to the inflorescence stalk. Flowers tiny, about 3 mm long, with no glands in the throat of its hypanthium (funnel). *Petals white or pinkish.* Fruit is a dry achene encased by the funnel. In flower July-Aug.

Found in chaparral and conifer forests, usually at higher elevations than common chamise, but the two often grow together. When viewed from a distance, ribbonshanks looks yellow-green, easily seen against common chamise, which is medium to dark green. Ribbonshanks is also much taller than common chamise. It has a open, airy appearance because the leaf-bearing twigs are only on the outermost branchtips; there are no leafy twigs in the middle of the plant.

Ribbonshanks is quite rare in our area. Our only population is a group of about 20 plants near Bear Canyon. To see them, take Bear Canyon Trail around, up, and over the ridge notable for its giant granite boulders. Not far after you overtop the ridge but before you intersect the Morgan Trail, face south. The plants are in a low spot on the hillside to the southwest, a few hundred meters away, at an elevation of 668 meters (2,192 feet). They are surrounded by a dense mass of common chamise and thick-leaved lilac and are best observed during bloom with binoculars or long camera lens. At present, no trail exists to reach them; we had to crawl through dense chaparral to get these photos. It's much easier to see it at Tree of Life Nursery and on the William Harding Nature Trail in Irvine Regional Park.

Ribbonshanks

This population is slightly east of the Orange/Riverside County line, barely within Riverside County. It still has not been found in Orange County. Other populations may exist in our area. We hope our readers will make those discoveries and notify us through the herbaria at Rancho Santa Ana Botanic Garden and UC Riverside. The nearest populations to this one are to the south in the Cuyamaca Mountains of San Diego County, southeast in the Agua Tibia Mountains of Riverside County, to the east near Hemet in Riverside County, and to the north in the Santa Monica Mountains of Los Angeles County.

The common name refers to the ribbon-like pieces of bark that self-shred from the branches ("shanks"). The epithet was given in 1848 by J. Torrey for its few leaves (Latin, sparsus = few, scattered; folium = leaf).

Common Chamise, Greasewood
Adenostoma fasciculatum Hook. & Arn.
var. *fasciculatum*

An upright woody shrub 0.5-3.5 meters tall. Trunk bark gray-brown, its outer layer may shred away with age to reveal gray-brown or reddish bark beneath. Twigs usually hairless. Its *dark green stiff leaves* are 4-10 mm long, linear, *tapered to the end, arranged in bundles* that alternate along the stem. Inflorescence a dense, compact cluster. *Flowers have no individual stalks and are held closely to the main stalk.* Flowers tiny, about 3 mm long; each has 5 glands in the throat of its hypanthium. Petals are creamy white. Fruit is a dry achene encased by the funnel. In flower May-June.

Chamise contains resinous sap that is presumably flammable. It has an obvious basal burl from which it resprouts after fire. Stump-sprouts and seedlings have feathery pinnate leaves that look nothing like its mature leaves. It is the host of a root parasite, **chaparral broomrape** (*Orobanche bulbosa*, Orobanchaceae).

An abundant component of chaparral, often grows in dense forests made up only of chamise plants, higher in elevation than coastal sage scrub and lower than conifer woodlands. It is considered an *indicator species;* its dominant presence in a plant community designates it as chaparral. When fruits mature and dry they turn deep red-brown, giving the entire community a red-brown look. Later, the fruits simply fall to the ground. They crumble in your hand when touched.

Known from the San Joaquin Hills, Chino Hills, Lomas de Santiago, and throughout the Santa Ana Mountains. It is easily observed at Tucker Wildlife Sanctuary and Caspers Wilderness Park.

The genus was established in 1832 by W.J. Hooker and G.A.W. Arnott for the glands in the flower funnel (Greek, adenos = gland; stoma = mouth). In the same paper, they named this species for its bundled leaves (Latin, fasciculus = a bundle).

San Diego Chamise
Adenostoma fasciculatum Hook. & Arn.
var. *obtusifolium* S. Watson

Plant under 2 meters tall, appears stunted when compared with common chamise. Twigs covered with short to long wavy hairs. *Leaves 2-6.5 mm long, fatter distally, blunt at the end, each tipped with a tiny spine.* In flower May-June.

A coastal plant, found here *only* on San Onofre Breccia soils on Temple and Niguel Hills. Easiest to find near Moulton Meadows Park. More common in San Diego County.

Named in by S. Watson for its blunt-ended leaves (Latin, obtusus = blunt; folium = leaf).

Common Chamise

Common Chamise

San Diego Chamise

Common Chamise

San Diego Chamise

Top: Common chamise in upper Fremont Canyon. As the dominant species in chamise chaparral, it often forms a dense, impenetrable thicket. When in flower during spring, the hills appear snow-covered.

Bottom: Chamise chaparral in spring 2006, a few months after the Sierra Peak fire. Chamise is a stump-sprouter. In a few short years, new growth will again cover this hillside and hide evidence of the burn.

Western Lady's Mantle

Aphanes occidentalis (Nutt.) Rydb.
[*Alchemilla occidentalis* Nutt. ex Torr. & A. Grey]

  A weak-stemmed annual, 2-10 cm tall, upright or sprawled out. Leaves hairy, 3-12 mm long, *deeply, palmately 3-7-lobed, often hair-tipped; alternate. Flowers numerous, tiny,* often hidden by the large cup-like stipules. Sepals 4, petals 0, stamens 1, pistils 1-2, ovary superior. In flower Feb-June.

  Usually found in canyons under the shade of coast live oaks or among moist grasslands. Known from the San Joaquin Hills above Laguna Beach, canyons on the Rancho Mission Viejo, Caspers Wilderness Park, the Chiquito Basin, and the San Mateo Canyon Wilderness Area. Certainly more common and widespread but overlooked.

  The epithet was given in 1840 by T. Nuttall for its distribution in western North America (Latin, occidentalis = western). The common name "mantle" refers to a woman's cloak; the ruffled leaf edges of a European species reminded some people of a cloak's pleated edges.

  Similar. American bowlesia (*Bowlesia incana*, Apiaceae): covered with star-shaped hairs; *leaves opposite,* similar but broader and *not as deeply lobed*; petals yellowish-green; ovary inferior; fruits 2-lobed.

California Mountain-mahogany

Cercocarpus betuloides Torr. & A. Gray **var. *betuloides***

  A large shrub to small tree 2-8 meters tall, its long branches stand upright or spread outward. Bark is smooth and light grey-brown. *Leaf blade 1-4 cm long, reverse-oval, elliptical to round, with very straight veins, arranged alternately on short twigs.* Leaf edge smooth from base to middle, toothed from middle to tip. *Leaf undersides are whitish and somewhat hairy.* Flowers yellowish-white, 6-8 mm wide at the top of the funnel, in groups of 1-3. *Calyx with 5 short free lobes. Petals absent.* The 10-45 stamens are attached to the funnel in 2-3 rows. The single style is hairy, the stigma narrow and hairless. After fertilization, the style grows much longer and hairier, often twisting. In flower Mar-Apr.

  A common plant of chaparral on slopes and nearby washes. Known from our foothill and mountain areas but oddly rare in the Lomas de Santiago. Look for it in Aliso and Wood Canyons, Laguna Coast, and Caspers Wilderness Parks. Less common in the San Mateo Canyon Wilderness Area. A beautiful shrub in flower and fruit. The view of morning light behind it when in fruit is especially striking.

  The genus was named in 1823 by F.W.A. von Humboldt, A.J.A. Bonpland, and C.S. Kunth, for the twisted tail of its fruit (Greek, kerkis = tail; karpos = fruit). The epithet was given in 1840 by J. Torrey and A. Gray for the similarity of the leaves to birch trees, in the genus *Betula* (Latin, betula = birch; -oides = similar).

  Similar. San Diego mountain-mahogany ✿ (*Cercocarpus minutiflorus* Abrams): *leaves shorter,* 1-2.5 cm long, light green above, yellow-green, *nearly hairless below; flowers much smaller,* 2-5 mm wide at top of the funnel; *solitary,* rarely groups of 2-3; in flower Mar-May. Uncommon in Orange County: San Joaquin Hills, Niguel Hill, Aliso and Wood Canyons Wilderness Park, Richard and Donna O'Neill Conservancy; common in southwestern portions of the San Mateo Canyon Wilderness Area.

Toyon, Christmas Berry
Heteromeles arbutifolia (Lindl.) M. Roemer

A large evergreen shrub or tree to 10 meters tall, the canopy often wider. Bark gray and hairless, that of young twigs is slightly hairy. *Leaf blade 4-11 cm long, leathery, elliptical or oblong, the edge with many (rarely few) sharp teeth; dark green and shiny above, yellow-green below.* Dense clusters of numerous flowers appear at the branchtips. The flower funnel is shaped like a shallow bowl. Calyx with 5 free sepals. Petals 5, white, large, and rounded. Stamens 10, arranged in pairs, held opposite the sepals. There are 2-3 styles above the inferior ovary. Fruits are pulpy and bright red, rarely yellow. In flower June-July, sometimes all year. Fruits ripen Nov-Jan.

Abundant in chaparral, oak woodlands, and coniferous forests, more common on moist north-facing slopes and canyon bottoms. In summer, we get to enjoy its beautiful flowers. *The scent given off by its flowers is unpleasant to many people; some liken it to that of a dead animal.* The scent doesn't travel far from the plant, so most people don't notice it. Its flowers are visited and pollinated by all sorts of insects, mostly flies and bees. In winter, its cheerful berries brighten up wild lands and gardens, as do the birds that feed on them.

In the 1800s and early 1900s, it was popular to make holiday wreaths from sprigs of fruiting toyon. To make it resemble **European holly** (*Ilex aquifolium* L., Aquifoliaceae), sprigs of native oaks were added.

The common name toyon was anglicized from Spanish tollon, itself from Greek tolon, both with meanings unknown. The genus was named in 1847 by M.J. Roemer for the similarity of its fruit to an apple (Greek, heteros- = other; *Malus* = the genus of apple). The epithet was given in 1820 by J. Lindley for the similarity of its leaves to plants in the genus *Arbutus* (Ericaceae) such as Pacific madrone and strawberry tree (*Arbutus;* -folium = leaf).

Along the coast, the day-flying brown ctenucha moth (Ctenucha brunnea Stretch, Ctenuchidae) nectars from toyon in summer.

Toyon

Toyon

Toyon

Oceanspray

Holodiscus discolor (Pursh) Maxim. **var. *discolor***

An upright or spreading hairy-stemmed shrub, 1.5-6 meters tall or wide. *Petiole obvious*, 2-15 mm long. *Leaves* 1.5-8 cm long, oval to elliptic, *large-toothed from near base to tip, each tooth tipped with smaller teeth. Sepals and petals 5, white to pink, held straight out.* Inside the flower funnel is a complete ring-like nectar disk to which 15-20 stamens are attached. Ovary superior, pistils 5. Fruits small, 1-seeded, hairy achenes. In flower May-Aug.

A delightful plant of oak and conifer woodlands and moist, rocky north-facing chaparral slopes. Widespread in our area, known sparingly from the Chino Hills, more common in sites such as Dripping Springs Canyon in the San Joaquin Hills and the Santa Ana Mountains such as Los Pinos Peak and in Coal, Silverado, upper Santiago, and Horsethief Canyons. Also in the Tenaja/San Mateo Creek confluence.

The genus was named in 1879 by C.J. Maximowicz for the disk that completely surrounds the ovary (Greek, holos = whole, entire; diskos = disk). The epithet was given in 1814 by F.T. Pursh for its white (colorless) flowers (Latin, dis- = negative; color = color).

Similar. Mountain spiraea (*H. discolor* var. *microphyllus* (Rydb.) Jepson [*H. microphyllus* Rydb.]): shorter, to 1 meter tall; *leaves almost lack a petiole; edges toothed only from middle to tip, the teeth of only 1 type.* In flower June-Aug. *Higher elevations:* Fremont Canyon, Santiago and Modjeska Peaks.

Oceanspray

Oceanspray

Oceanspray | Mountain Spiraea

THE OCEANSPRAY GUILD
Oceanspray Longhorn Moth

Adela septentrionella (Walsingham, Adelidae)

Adults are shiny dark brown with 2 white bands on the forewings, white dots near the wingtips, body covered with brown hairs. *Antennae very long*, white with dark rings at the base, all-white farther up. Females have a yellow-brown tuft of hair on their heads. Mated females insert an egg into a flower bud. The caterpillar hatches and feeds on developing seeds. Weeks later, it exits the flower and drops to the ground. It builds a case from fallen leaves, sews it up with silk, and uses it as shelter from which it feeds on fallen oceanspray leaves. It pupates within the shelter, spends the winter, and emerges as an adult in May-June. Adults sip nectar from nearby blooming plants. Known from the Santa Ana Mountains north to British Columbia.

The epithet was given in 1880 by T. de G. Walsingham for its northern distribution (Latin, septentrionalis = belonging to the north).

Oceanspray Longhorn Moth female laying eggs into buds

Native "Strawberries": *Horkelia*, *Drymocallis*, *Potentilla*

While not true strawberries (*Fragaria* spp.), plants in these genera are similar-looking and related to them. They are also often confused with each other.

Top: Mesa horkelia. Petiole usually green; leaf pinnate, the leaflets narrowly oval, the edges toothed; lighter green.
Bottom: Sticky cinquefoil. Petiole green, often red; leaf pinnate, the leaflets round or oval, the edges toothed; darker green.

Mesa Horkelia ★
Horkelia cuneata* ssp. *puberula (E. Greene) Keck

A green to gray perennial that grows in low-lying mats, covered with sticky-tipped hairs. Leaves 10-30 cm long; *pinnate, 5-12 leaflets per side, each 10-25 mm long, tooth-edged. Leaflets all about the same size, usually rounded in outline at the tip and tooth-edged.* Petals narrow, 4-8 mm long, 1.5-4 mm wide. Inner rim of the hypanthium mostly hairless. In flower Feb-Sep.

Found in dry to moist sandy soils among chaparral and oak woodland, sparingly in the San Joaquin Hills, more common in foothills of the Santa Ana Mountains. Look for it along the San Juan Loop Trail, Trabuco Trail, and San Juan and Chiquito Basin Trails in the Chiquito Basin and Lion Canyon. Considered a rare plant (CRPR 1B.1) because much of its habitat has been destroyed by development.

The genus was named in 1827 by L.K.A. von Chamisso and D.F.L. von Schlechtendal in honor of J. Horkel. The epithet was given in 1837 by J. Lindley for the wedge-shaped leaflet base (Latin, cuneatus = wedge-shaped). This form was named in 1887 by E.L. Greene for its hairy leaves (Latin, pubescentis = with downy hairs).

Similar. Wedge-leaved horkelia (*H.c.* ssp. *cuneata*): most hairs on the plant are not sticky-tipped, inner rim of the hypanthium hairy, in flower Apr-Sep; formerly found sparingly in sandy soils and old sand dunes among coastal sage scrub along the coast at Pelican Hill, probably eliminated from our area by development.

Mesa Horkelia

Mesa Horkelia — Leaflet tips rounded

Ramona Horkelia — Leaflet tips blunt

Ramona Horkelia ★
Horkelia truncata Rydb.

An upright green perennial that grows in tufts. Main stem 20-60 cm long. Leaves 4-13 cm long; pinnate; 1-3 leaflets per side, each 10-30 mm long, tooth-edged. *The leaflet at the tip is clearly longer than the others, squarely cut off (truncate) at the end and tooth-edged.* Inflorescence not crowded, 5-20 flowers each. Petals broad and rounded, 5-7 mm long, at most about 5 mm wide. In flower May-June.

An inland plant of scrub and chaparral. Known here from the southeastern portion of the San Mateo Canyon Wilderness Area, around Santa Margarita Peak and upper Devil Canyon. Listed as CRPR 1B.3, threatened by recreational activities, possibly also by chaparral "management," road maintenance, mining, and grazing.

The epithet was given in 1908 by P.A. Rydberg for its truncate central leaflet (Latin, truncatus = cut off).

Sticky Cinquefoil
Drymocallis glandulosa (Lindl.) Rydb. **ssp. *glandulosa***
[*Potentilla glandulosa* Lindl. ssp. *glandulosa*]

An upright perennial from branching stems, 20-80 cm tall, *densely covered with sticky-tipped hairs. Leaves pinnate,* those closest to the ground are 10-25 cm long. Leaflets only 3-5 per side, reverse-oval to round, the edges toothed. Inflorescence branched; contains leafy bracts and bears 2-30 flowers. The *broad, oval petals are pale yellow to creamy white,* 4.5-7 mm long, about equal in size to the sepals. In flower May-July.

A common plant of foothills and mountains, often in shade on north-facing slopes and near seeps. Known from the San Joaquin Hills such as in Sycamore Hills and Laguna Coast Wilderness Park. In the Santa Ana Mountains, found at Black Star Canyon, Audubon Starr Ranch Sanctuary, San Juan Canyon, Falcon Group Camp, upper Silverado and Trabuco Canyons, Trabuco Trail, and Santiago Peak.

In peak flower, it releases a sweet scent that fills the air. Many insects such as beetles and flies visit its flowers, take nectar, and transfer pollen.

The genus was named in 1753 by C. Linnaeus in reference to its supposed powerful medicinal properties (Latin, potens = powerful). The epithet was given in 1838 by J. Lindley for its sticky-tipped "glandular" hairs (Latin, glandulosus = glandular).

Sticky Cinquefoil

Greene's Cinquefoil
Drymocallis glandulosa* ssp. *reflexa (Greene) Ertter
[*Potentilla glandulosa* ssp. *reflexa* (E. Greene) Keck]

A smaller form of sticky cinquefoil. Its basal leaves are 7-15 cm long with leaflets 15-40 mm long. *Petals yellow,* narrowly oblong, widest toward the tip, shorter than the sepals. *The petals fold backward when in full flower.* In flower May-July.

Found in moist soils, meadows, and shady sites in the Santa Ana Mountains such as Los Pinos Potrero, the Chiquito Basin along the San Juan Trail south of Blue Jay Campground, and sparingly along San Juan Canyon.

This form was named in 1891 by E.L. Greene for its folded petals (Latin, reflexus = turned back).

Stem Sawfly wasp on Greene's Cinquefoil

Pacific Silverweed ✿
Potentilla anserina L. **ssp. *pacifica*** (T.J. Howell) Rousi

A low-growing non-sticky perennial with long horizontal runners that root along their length. It has no upright main stem; the leaves, runners, and individual flowers grow directly from a central knot-like main stem at ground level. *Leaves 3-50 cm long, pinnate, usually with a silvery sheen on the underside.* Each solitary flower is on a stalk 5-30 cm long. Sepals and petals are oval to round in shape, the petals yellow, larger than the sepals. In flower May-Oct.

A plant of low-elevation wetlands, mostly along the immediate coast in springs and freshwater marshes, especially those with saltwater intrusion. In our area, known from the lower reaches of the eastern Wintersburg Channel just below the mesa in Bolsa Chica Ecological Reserve (recorded there in 1924, 1932, and 1970). A search of that area in 2009 failed to rediscover the plant. A small population was recently found in the freshwater marsh along San Mateo Creek near Trestles, between Interstate 5 and the ocean.

The epithet was given in 1753 by C. Linnaeus, perhaps for its growing location (Latin, anserinus = pertaining to geese or growing on land grazed by geese). This form was named in 1898 by T.J. Howell for its Pacific coast distribution.

Pacific Silverweed. Leaves green above, silvery below

Holly-leaved Cherry
Prunus ilicifolia (Nutt.) Walpers **ssp.** *ilicifolia*
An upright woody shrub, 1-8 meters tall, *densely covered with evergreen foliage that usually hides the trunk and stems.* Leaf blade 1.6-12 cm long, *glossy, widely oval to round, the tip pointed, the edges wavy and spiny-toothed.* Inflorescence long, many-flowered. Petals 1-3 mm long, bright white. Fruit spherical, 12-18 mm across, bright red or yellow. In flower Apr-May.

A plant of chaparral on slopes and in canyons. Widespread in the Santa Ana Mountains. Known from Fremont Canyon and above it along Main Divide Road, south to Pleasants Peak, along Bedford Canyon Road, middle to upper Silverado Canyon, and in Hot Springs and San Juan Canyons. When in bloom, the slopes above Silverado Canyon are ablaze with its flowers. An important larval food plant of the **pale swallowtail butterfly** (*Papilio eurymedon,* Papilionidae).

Studies have shown that its seeds have a low germination rate. However, seeds that have passed through the digestive tract of a coyote have a much higher germination rate. Thus, areas with dense populations of this plant likely have (or had) healthy populations of coyotes that distributed the seeds.

The epithet was given in 1832 by T. Nuttall for the similarity of its leaves to **English holly** (*Ilex aquifolium* L., Aquifoliaceae) (Latin, -folius = leaved).

Similar. **Holly-leaved redberry** (*Rhamnus ilicifolia,* Rhamnaceae): leaves more pliable, generally small-toothed, somewhat flat, tip often indented; flowers tiny, yellow-green.

Holly-leaved Cherry

California Rose
Rosa californica Cham. & Schltdl.
An upright shrub 1-3 meters tall, *its stems armed with sharp prickles that are flattened side-to-side and curved backward.* Leaves pinnate, the central leaf stalk with small prickles beneath, leaflets elliptical or oval, their edges toothed. The 5 triangular sepals have a long narrow tip. The 5 large pink petals are oval or rounded, sometimes curved inward at the tip. Stamens are held outward, 10-20 or more in number. In flower May-Aug.

A widespread, common shrub of oak and riparian woodlands, often in shady canyons, frequently near California sycamore and coast live oak trees. Wonderful thickets are found at Caspers Wilderness Park (Dick Loskorn Trail at Bell Canyon), Whiting Ranch Wilderness Park (Borrego Trail), Limestone Canyon (Dripping Springs East Road), and Laguna Coast Wilderness Park (Laurel Canyon).

Pollinated by a variety of insects, mostly bees and flies, sometimes butterflies. After fertilization, the petals fall off but the stamens remain, turn brown, and shrivel as the fruit develops and turns red (*rose hips*).

The genus was given in 1753 by C. Linnaeus who used the ancient Latin name for rose. The epithet was given in 1827 by L.K.A. von Chamisso and D.F.L. von Schlechtendal for the place of its discovery.

California Rose
Fruit

THE ROSE GUILD

Spiny Leaf Gall Wasp
Diplolepis polita (Ashmead, Cynipidae)
These galls appear on the upper surface of native rose leaves. *Each green to yellowish or red gall grows to 5 mm across, with many spines on top and sides.* Adult wasps emerge in spring. Very common on roses in our area.

The wasp was named in 1890 by W.H. Ashmead for its smooth body (Latin, politus = smooth).

Spiny Leaf Gall Wasp

California Blackberry
Rubus ursinus Cham. & Schltdl.

A long vine that mostly remains on the ground and grows in mounds, generally does not climb shrubs and trees. Stems *covered in fine white power;* armed with slender prickles, straight or curved. In cross-section, the main stems are round. *Leaves 3-lobed or pinnately divided into 3(5) leaflets.* The midrib and leaf undersides are covered with prickles. *Leaflets are tooth-edged,* elliptical to oval, the end leaflet often lobed. The 5 sepals are broadly triangular and pointed, hairy on the outside. The 5 *white petals* are oval to rounded, sometimes with a pointed tip. *Each plant usually produces flowers of only one sex (*dioecious*).* All-male flowers have smaller petals than all-female flowers. Fruit an aggregate of flesh-covered achenes. In flower Mar-July. Fruits ripen about 1 month later.

A plant of moist ground in shady canyons and near seeps. Widespread and common in our area. Look for it near Sleepy Hollow in the Chino Hills, canyons in Crystal Cove State Park, along Borrego Trail in Whiting Ranch Wilderness Park, and upper Silverado, Trabuco, Hot Springs, and San Juan Canyons.

The flowers release a sweet fragrance that most people don't notice, probably because they keep their noses away from those prickles! Commonly pollinated by beetles, flies, and bees.

Each female flower has a cluster of 10 or more pistils. When fertilized, each develops into a round fleshy fruit; together this cluster makes a egg-shaped aggregate that turns black when ripe. They are edible and sweet-tasting, a real treat while hiking. Carefully identify the plant before you eat its fruit. In many places, it grows entangled with poison oak.

The genus was named in 1753 by C. Linnaeus for its growth habit (Latin, rubus = bramble, derived from ruber or rubeo = red, for the reddish stems and prickles). The epithet was given in 1827 by L.K.A. von Chamisso and D.F.L. von Schlechtendal, in reference to the fruits eaten by black and grizzly bears that were abundant throughout California (Latin, ursus = bear).

Similar. Western poison oak (*Toxicodendron diversilobum,* Anacardiaceae): no stem prickles; flowers small, creamy yellowish; fruit is a single berry. Abundant.

Above: California blackberry and western poison oak often grow together, sometimes intertwined.

Himalayan Blackberry ✖
Rubus armeniacus Focke
[*Rubus discolor* Weihe & Nees, misapplied]

A large, stout vine to 3 meters long. Prickles numerous, very large and heavy. *Stems hairless, green, not covered in fine white power; angled in cross-section.* Most leaves are very large, somewhat circular in outline, palmately compound, (3 or) 5 leaflets; *white-woolly beneath.* Uppermost leaves often very different, smaller, sometimes linear and not divided. Petals 10-15 mm long, reverse oval, white to pink. Fruit black. In flower Mar-June.

Native to Eurasia. A garden escape, probably also intentionally planted in or near wild lands. Known from wet riparian woodlands and similar situations. Uncommon in most of Orange County, known from places such as Huntington Beach and upper Newport Bay. Far too abundant in the Chino Hills, north through the Puente and Whittier Hills where it strangles out native riparian vegetation.

The epithet was given in 1874 by W.O. Focke for Armenia, its original place of discovery.

RUBIACEAE • MADDER FAMILY

Quite a variable family, its members run the gamut of life forms as annuals and perennials, shrubs, vines, and trees. *Ours are annuals, perennials, and small shrubs. Leaves are entire and usually opposite, but in some species they are in whorls. Flowers are generally in a rounded cluster with 4 or no sepals and 4 petals. Ovary is inferior* in most. The family includes cultivated plants such as coffee and many handsome ornamentals.

Bedstraws: *Galium*

Annuals or perennials, often shrubby, vine-like, and clamber over other plants, aided by small bristles on their stems. The stems are square in cross-section, especially when young. *The stipules appear like leaves and grow alternately to them.* For simplicity, these leaf-like stipules are counted as leaves. *Leaves (true leaves and leaf-like stipules) are in whorls of 4 or more (rarely 3).* Flowers are quite tiny, somewhat bell-shaped, and have no sepals. *Petals are 4 (or 3) in number, yellow, white, or green in color.* Ovary inferior; styles 2, fused at their base. Fruit is a pair of nutlets or a single berry.

We have 6 species, all in the genus *Galium*. Their flowers are quite similar. To identify them, you need to make note of their growth (weak-stemmed or woody-stemmed); number of leaves per whorl; leaf shape and tip (pointed or not); sex (one sex or both sexes in the same flower); number of corolla lobes; fruits (a single nutlet or two; hairs straight, curved, or absent); and habitat (soil moisture and plant community).

The common name bedstraw refers to their growth habit as an intertwined mass, formerly used as bedding. The genus was named in 1753 by C. Linnaeus from the use of lady's bedstraw (*Galium verum* L.), an English species, as an aid to curdling milk for cheesemaking (Greek, gala = milk).

	Weak-stemmed Bedstraws			Woody-stemmed Bedstraws		
	Common *G. aparine*	**Parisian** *G. parisiense*	**Pacific** *G. trifidum*	**Chaparral** *G. angustifol.*	**San Diego** *G. nuttallii*	**Climbing** *G. porrigens*
Growth	Weak-stemmed annual; covered with small hooked prickles	Brittle-stemmed annual; covered with short hairs	Weak-stemmed perennial, rarely an annual; covered with minute prickles	Woody-based perennial; hairless to hairy	Woody-based perennial; covered with tiny, sharp curved hairs; ages dark red	Woody-based perennial; covered with tiny prickles
Leaves	Linear; tip broad to narrow, often pointed; in whorls of 6-8	Linear; in whorls of (4-)6; tip broad and pointed; fold backward with age	Linear to oval; rounded tip; in whorls of 4-6	Linear, narrow; tip pointed or not; in whorls of 4	Linear to triangular-ovate; tip narrow and pointed, sharp to the touch	Widely ovate to oblong; tip broad, may be pointed but is not sharp to the touch
Sex	Male and female organs in same flowers on same plant	Male and female organs in same flowers on same plant	Male and female organs in same flowers on same plant	Male and female organs in separate flowers on separate plants	Male and female organs in separate flowers on separate plants	Male and female organs in separate flowers on separate plants
Flowers	4 corolla lobes (rarely 3)	4 corolla lobes; whitish or purplish	3 corolla lobes (rarely 4)	4 corolla lobes	4 corolla lobes	4 corolla lobes
Fruits	Pair of nutlets; covered with short hooked hairs	Pair of nutlets; covered with short hooked hairs (rarely hairless)	Pair of nutlets; smooth and hairless; hard and black when dry	Pair of nutlets; densely covered with long, stiff, straight hairs	Single, smooth, hairless berry	Single, smooth, hairless berry
Habitat	Moist soils in disturbed sites	Rocky soils in Santa Ana Mountains	Wetlands and other moist soils	Coastal sagebrush scrub and chaparral	Chaparral	Chaparral

WEAK-STEMMED BEDSTRAWS

Annuals, a few perennials; all have rather weak non-woody stems. *All flowers contain both sexes (the plants are monoecious).*

Common Bedstraw, Goosegrass
Galium aparine L.

A weak-stemmed annual covered with tiny hooked prickles. Stems 30-90 cm long. Upper leaves 13-31 mm long, narrow, and in whorls of 6-8. Leaves held up or out, generally not folded back. Flowers held close to the nodes. Petals 4, whitish. Fruit is a pair of 2 nutlets covered with short hooked hairs. In flower Mar-July.

Often grows in moist or shaded soils, beneath other plants or climbing through them. Lives in grasslands and as understory in many plant communities but mostly in disturbed soils and in gardens. Found on both coasts of North America and in Europe. Some botanists suspect it was introduced into North America from Europe.

Named in 1753 by C. Linnaeus, who used the existing Greek word for the plant (Greek, aparine = bedstraw).

Common Bedstraw, thicker

Parisian Bedstraw, very slender

Parisian Bedstraw ✖
Galium parisiense L.

A small, very slender, brittle-stemmed upright annual, covered with short hairs. Stems 15-25 cm long. Leaves linear, in whorls of (4-)6, tip broad and pointed, fold backward with age. Flowers on long thread-like stalks. Petals 4, whitish or purplish. Fruit is a pair of elongated nutlets covered with short hooked hairs (rarely hairless). The entire plant often turns red with age. In flower during spring.

Formerly uncommon in our area but now spreading. Found in warm exposed sites with rocky soils in the Lomas de Santiago and Santa Ana Mountains. Known from Limestone Canyon, Whiting Ranch Wilderness Park, Trampas Canyon on the Rancho Mission Viejo, Audubon Starr Ranch Sanctuary, Lucas Canyon Trail, and the confluence of San Juan and Lion Canyons. Native to Europe.

Named in 1753 by C. Linnaeus for the location of its original discovery in Paris (Latin, -ensis, = belonging to). A mistranslation of the epithet (Latin, paries = a wall) is probably responsible for an alternate common name, wall bedstraw.

Parisian Bedstraw

Pacific Three-petal Bedstraw ✿
Galium trifidum L. var. *pacificum* Wiegand

A weak, straggling perennial (rarely an annual), covered with very tiny prickles. Stems 10-50 cm long. Leaves 4-19 mm long, linear or oval with a rounded tip, in whorls of 4-6. Petals 3 (rarely 4), white to pinkish. Fruit is a pair of hairless nutlets that are hard and black when dry. In flower during spring, into summer with more water.

Quite uncommon and difficult to find, it grows in wetlands and similar moist soils. Known in our area only from upper Newport Bay (in 1932) and Hagador Canyon (in 1966). We've been unable to relocate it but are hopeful that it will turn up, quite possibly in the wetlands along Laguna Canyon Road in Laguna Beach. If you find it, please notify the herbaria at Rancho Santa Ana Botanic Garden or UC Riverside.

The name was given in 1753 by C. Linnaeus in reference to its 3 petals (Latin, tri- = three; -fidus = divided). The variety was named in 1897 by K.M. Wiegand for its distribution, only along the Pacific Coast from Alaska to Mexico.

Pacific Three-petal Bedstraw

WOODY-STEMMED BEDSTRAWS

We have 3 species of woody-stemmed bedstraws. Chaparral bedstraw is easily identified by its long narrow leaves. The other two are quite similar and harder to distinguish. All three grow together in places such as Temple Hill in Laguna Beach, Niguel Hill in Laguna Niguel, and Caspers Wilderness Park. All have leaves arranged in whorls of 4. *Each plant has only all-male or only all-female flowers (dioecious).* Male flowers have a reduced non-functional (sterile) pistil. Female flowers have 4 reduced non-functional (sterile, empty or missing anthers) stamens. Since the fruits are major characteristics and male plants do not produce fruits, male-flowered plants usually cannot be identified with certainty. Their identity is often inferred by identification of female-flowered plants that grow nearby. Additionally, the species differ in minute characteristics of their leaves, petals, and fruits.

Above: single-sex flowers of Galium angustifolium. Left: male flower (4 functional stamens, a non-functional pistil in center). Right: female flowers (2 functional styles, 4 non-functional stamens spread out), the ovaries with long white bristles. Not to scale - the flowers of males and females are the same size.

Chaparral or Narrow-leaved Bedstraw
Galium angustifolium Nutt. **ssp. *angustifolium***

Woody especially at the base, *It ranges from hairless to hairy.* Stems 15-100 cm long with many ridges along them. *Leaves narrowly linear, 5-27 mm long when mature, in whorls of 4.* Flowers are in many small side-branches, each with 4 yellowish petals. *Fruit is a pair of nutlets densely covered with long, stiff, straight white hairs.* In flower all spring.

This is our most often encountered woody-stemmed bedstraw. It typically grows alone, not clambering through and over other plants. Common on cliffs, canyons, hills, and mountains.

Named in 1841 by T. Nuttall in reference to its narrow leaves (Latin, angustus = narrow; folium = leaf).

San Diego Bedstraw
Galium nuttallii A. Gray **ssp. *nuttallii***

Stems and leaves covered with tiny, sharp, curved hairs that help it climb high into other plants. The plant turns dark red as it ages. Leaves linear to triangular-oval, 3-8 mm long, leathery, in whorls of 4; *their tips are sharp to the touch.* Male flowers are in clusters along the stems, female flowers alone at the base of the leaf whorls. Each flower has 4 yellow-reddish petals. *Fruit is a single, smooth, hairless berry.* In flower Mar-June.

A common *chaparral plant,* found from coastline to canyons and mountains. Often grows with or near climbing bedstraw. It clambers through and over other plants.

Named in 1852 by A. Gray in honor of T. Nuttall.

Climbing or Graceful Bedstraw
Galium porrigens Dempster **var. *porrigens***

Another climbing, woody perennial, its stems are slender and covered with tiny prickles. Leaves are widely oval to oblong, 2-18 mm long, in whorls of 4. *The leaf end is rounded and tipped with a pointed hair that is not sharp to the touch.* Like those of San Diego bedstraw, male flowers are in clusters along the stems, female flowers alone at the base of the leaf whorls. Its 4 petals are yellowish or reddish. *Fruit is a single, smooth, hairless berry.* In flower during spring.

Commonly found in chaparral, it often occurs with or near San Diego bedstraw, but is more commonly found. It typically clambers through and over other plants.

Named in 1974 by L. Dempster for its habit of growing out over other plants (Latin, porrectus = projected or stretched outward and forward).

RUTACEAE • RUE FAMILY

A family of *strong-scented* perennials, shrubs, and trees. *Ours are perennials and shrubs. Leaves alternate* (rarely opposite), simple or pinnate. *Flowers have 4 sepals and 4-5 petals. Ovary superior, 4-5 lobed.* Leaves and fruits are *covered with small aromatic oil glands* that are easily seen when held up to light. The familiar citrus trees (oranges, lemons, grapefruits, etc.) are members of this family. Three species are native to California, only 1 in our area, plus 2 garden escapes.

Bushrue, Coastal Spice Bush
Cneoridium dumosum (Nutt.) Baillon

A low-growing *evergreen woody shrub* under 1.5 m tall. *Leaves* are 1-2.5 cm long, *linear, undivided, hairless,* and arranged in opposite pairs or in bundles. Flowers are about 1.2 cm across. Sepals 4, tiny. *Petals* 4(5-7) *oval, white.* Stamens 8, of 2 lengths: 4 short, 4 long. *Fruits spherical,* 5-6 mm across, *green or reddish,* and turn red-brown with age. In flower Nov-Mar.

Lives among coastal sage scrub and chaparral in rocky and sandstone soils on coastal hills, sea bluffs, and steep canyon walls in Orange and San Diego Counties. Here restricted to the southern coastal portions of the San Joaquin Hills. Found on Temple Hill, in Aliso Canyon, and on Niguel Hill. Its flowers release a wonderful citrus scent.

The plant's sap contains (a) chemical(s) that, upon contact with human skin and exposed to light, produces a severe skin rash called *phytophotodermatitis*. People with sensitive skin should avoid contact with it. A lovely garden plant but should not be planted in school yards and public parks.

The genus was established by J.D. Hooker, who felt it resembled European plants in the somewhat related genus *Cneorum* (Latin, Cneoridium = resembling *Cneorum*), a name established in 1735 by C. Linnaeus. The epithet was given in 1838 by T. Nuttall to describe the plant's bushy appearance (Latin, dumosus = bushy).

Fringed Rue ✖
Ruta chalepensis L.

An upright shrub to 80 cm tall, hairless, lightly covered in whitish powder (*glaucous*). Leaves 2-3-pinnate. Flowers nearly 2 cm across, *petals bright yellow, cupped, the edges long-fringed.* Fruit 4-5-lobed. In flower Mar-Aug.

Native to the Mediterranean, it sometimes persists from abandoned homesteads. Uncommon, known here only from the Santa River wash near Coal Canyon. *The entire plant gives off a strong scent that persists on your hands if you touch it.*

The epithet was given in 1767 by C. Linnaeus for its place of collection in Africa, presumably a place called Chalep (Latin, -ensis,= belonging to).

Similar. Common rue ✖ (*R. graveolens* L.): *petals yellow, edges wavy but not long-fringed;* a garden escape, native to southern Europe, uncommonly found in upper Big Canyon above Newport Bay. Both are used medicinally and for cooking. Both species can cause phytophotodermatitis.

SAPINDACEAE • SOAPBERRY AND MAPLE FAMILY

Vines, shrubs, or large trees with watery, often sugary sap. In our species, the leaves are opposite, palmately lobed, though other species are rarely pinnate. The leaves of most maples turn bright red or yellow in fall and drop off in winter, to reappear in late winter or spring. *Flowers are arranged in hanging clusters.* In some species each plant has only male or female flowers, while in other species each flower has both male and female flowers. *Flowers are tiny, yellow-green, usually with 4 sepals, 5 petals, and 8 stamens.* The ovary is superior. The fruit of maples consists of a pair of dry, flat-winged pods together called a *samara*. The family has 1,560 species worldwide, 5 in California but only 1 native in our area. Ours is a member of the genus *Acer*, the maples, formerly placed in their own family, the Aceraceae.

Big-leaf Maple
Acer macrophyllum Pursh

A medium to large stately tree with a rounded canopy, 5-30 meters tall. Twigs and petioles smooth and hairless. *Leaves opposite, 10-25 cm wide with large-toothed edges, hairless above, slightly hairy below. Flowers are in a long drooping cluster* of 30 or more flowers and appear before or with the leaves. Each cluster has a mixture bisexual flowers together with male-only flowers. The mature fruit is brown-hairy near the seeds and has 2 large paper-thin wings. The two halves often separate before they fall from the tree and twirl to the ground. In flower late Mar-May.

A plant of rocky soils on north-facing slopes and in canyon bottoms, often with big-cone Douglas-fir and canyon live oak. Found in upper Bell Canyon on the Audubon Starr Ranch Sanctuary, Claymine Canyon, Old Woman Canyon, Silverado Canyon, Holy Jim Canyon Trail, and along Main Divide Truck Trail southeast of Trabuco Peak. Easy to find in Trabuco Canyon upstream of O'Neill Regional Park where it grows right along Trabuco Creek Road, in the creek channel, and on north-facing slopes throughout upper Trabuco Canyon visible from Trabuco Trail. Apparently absent from the San Mateo Canyon Wilderness Area.

The genus was named in 1753 by C. Linnaeus, who used the ancient name (Latin, acer = maple tree). The epithet was given in 1814 by F.T. Pursh in reference to its large leaves (Greek, macros = long, large; phyllon = leaf).

Similar. California sycamore (*Platanus racemosa*, Platanaceae): *leaves alternate* with 5 large lobes, *no secondary lobes, hairy above, very hairy below;* plants monoecious but flowers unisexual; *fruits are small dry 1-seeded achenes.*

Big-leaf Maple

Big-leaf Maple

Fruit

Big-leaf Maple

SAXIFRAGACEAE • SAXIFRAGE FAMILY

Hairy herbs and small shrubs; *most of ours are geophytes* that grow from an underground bulb-like tuber. *Leaves are mostly or all basal, lobed or toothed, often with palmate veins.* Sepals and petals 5, the petals usually white and narrow at their bases. Stamens 5 or 10. Ovary superior to inferior, with 2 simple pistils or 1 compound pistil. A small family of 600 species mostly in cold northern regions, 64 in California, 4 in our area.

Round-leaved Boykinia ✤
Boykinia rotundifolia C. Parry

An upright perennial, 30-80 cm tall, densely covered with long sticky-tipped hairs. *Leaves round to oval, edges round-lobed, toothed,* lowest leaf blades nearly 20 cm long and wide, upper leaves smaller. Flowers small, petals white, barely longer than the sepals. Stamens 5. Styles 2. In flower June-July.

A plant of wet canyons, shaded seeps, and springs, often among riparian and oak woodlands. Known from Lost Woman Canyon near Silverado Canyon, upper Santiago Canyon, Holy Jim Falls, Bear Spring near Main Divide Truck Trail, Upper McVicker Canyon, and Mayhew Canyon. Probably in other moist sites as well. Uncommon throughout its range, not abundant when found. In cultivation it is quite successful in shaded spots, both potted and in the ground.

The genus was named in 1834 by T. Nuttall in honor of S. Boykin. The epithet was given in 1878 by C.C. Parry for its round leaves (Latin, rotundus = round; folium = leaf).

Round-leaved Boykinia

Coast Jepsonia
Jepsonia parryi (Torr.) Small

A delicate, slender geophyte 10-30 cm tall *from a bulb-like tuber. Leaves 2-6 cm across, round to heart-shaped, basal, often flat on the ground.* Its *single leaf per year* (rarely 2-3) appears in Dec-Jan and remains through spring, then withers. In fall, its flowers appear atop a naked stalk, generally 1-5 per stalk. *Petals 5 white, spoon-shaped, often with light purplish veins.* Stamens 10. Styles 2. In flower Oct-Dec.

Within each population of this plant exist individuals that produce only one of two different flower types. About half of the plants have flowers with long styles, large stigmas, short stamens, and small pollen grains. The rest have flowers with short styles, small stigmas, long stamens, and large pollen grains. *Each flower type can only be fertilized by pollen from the other flower type.* Coast jepsonia flowers are thus self-incompatible; their flowers must receive pollen from another plant, a system that maintains genetic diversity. Flower flies (family Syrphidae) and metallic sweat bees (family Halictidae) perform the required cross-pollination services. Carefully observe the plants to identify the different flower types and watch for their pollinators.

A plant of deep clay and rocky soils on gentle to steep slopes, north-facing and in partial shade. Often overlooked, locally common in the San Joaquin Hills and low to mid elevations of the Santa Ana Mountains. Known from Aliso and Wood Canyons Wilderness Park, Sycamore Hills, Aliso Canyon, Temple Hill, Niguel Hill, Lomas de Santiago, upper Trabuco Canyon, Richard and Donna O'Neill Conservancy, San Juan Canyon, lower San Juan Trail, along San Juan Loop Trail, and the Santa Rosa Plateau. Recently found at San Clemente State Beach.

The genus was named in 1896 by J.K. Small in honor of W.L. Jepson. The epithet was given in 1859 by J. Torrey to honor C.C. Parry.

Coast Jepsonia

Woodland Star
Lithophragma affine A. Gray

An upright slender ephemeral spring perennial 10-60 cm tall. Leaves mostly basal, a few up the stem; round to heart-shaped, 3-lobed, each lobe further lobed or broad-toothed. Inflorescence 3-15-flowered. Petals white, 5-13 mm long, *generally 3-lobed at the tip, the lobes sometimes further lobed.* Stamens 10, short, hidden within the flower, as are the 3 styles. In flower Feb-June.

A plant of *moist soils on slopes, in full to partial shade,* among chaparral, oak, and riparian woodlands. Usually found in clusters of individuals. Known from the San Joaquin Hills such as Temple Hill and Laguna Coast Wilderness Park. Widespread in the Santa Ana Mountains, such as Sierra, Silverado, Trabuco, Hot Springs, and San Juan Canyons and the Santa Rosa Plateau. Common on shady north-facing road edges along Main Divide Truck Trail between Trabuco and Santiago Peaks.

The genus was named in 1834 by T. Nuttall for its common habitat among rocks (Greek, lithos = stone; phragma = hedge, fence). The epithet was given in 1865 by A. Gray for its close relation to Hill Star (*L. heterophyllum* (Hook. & Arn.) A. Gray & Torr.) (Latin, affinis = related, adjacent).

Woodland Star

THE WOODLAND STAR GUILD
Woodland Star Moth
Greya politella (Walsingham) (Prodoxidae)

Larvae of this moth feed only on woodland star. *Wingspan of the adult is 11-20 mm; its wings are brownish-gray with a bronzy sheen.* During the day, females alight on the face of the flower, bend their abdomen downward, and use their long spear-like ovipositor to pierce the ovary and deposit eggs directly into it. In the process, pollen sticks to their abdomen, and they accidentally transfer it to other flowers, pollinating them. The young feed on one developing seed, tunnel out of the ovary, find other shelter, and enter summer dormancy when the flower stalks fall. In winter as new leaves emerge, the larvae waken, fold a basal leaf and sew it in place, then feed on and pupate within it. Adults emerge in spring. Since woodland star produces several seeds and leaves and the larvae eat very few of those seeds, the value of the adult moth as a pollinator outweighs the minimal damage done by larvae.

Woodland star moth, right after laying an egg in the flower's ovary. Soon, the tiny bee fly on the top petal will walk to the center of the flower and drink sap from the damaged ovary wall.

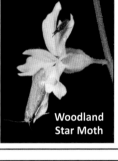
Woodland Star Moth

California Saxifrage ✿
Micranthes californica (Greene) Small
[*Saxifraga californica* Greene]

A small upright herbaceous perennial, 15-35 cm tall. *Leaves basal; blade 4-10 cm long, oval to elliptical; edges shallow-toothed, tips rounded or pointed; petiole broad.* Sepals green to purplish, folded backward in age. *Petals white, oval,* held upward or outward. Stamens 10, styles 2. In flower Feb-June.

Lives in moist soils in shady places, more often on north-facing slopes, in grassy openings within coastal sage scrub and chaparral. Known sparingly from Temple Hill in Laguna Beach and Limestone, Silverado, San Juan, and Hot Springs Canyons. In the San Mateo Canyon Wilderness Area, it is known from the bottom of Los Alamos Canyon one-half mile east of San Mateo Canyon and other scattered sites.

The epithet was given in 1889 by E.L. Greene for the state of its discovery.

California Saxifrage

SCROPHULARIACEAE • FIGWORT FAMILY

Upright herbs or shrubs that grow as annuals, biennials, and perennials. Leaves alternate, spirally arranged, or opposite. Undivided in most species, smooth-edged or toothed. Flowers commonly 2-lipped, *radial* (radiate like spokes of a bicycle wheel). Calyx with 5(3-4) free lobes. Petals fused at their base, often tubular and 2-lipped, with 5(4) free lobes. Stamens 5, fused to the corolla at their base. Each anther is elliptical; its base is not shaped like an arrowhead. Pollen is released through slits that are perpendicular to the stamen filament. Ovary superior. Style 1, stigma club-like. Fruit a round or oval capsule. The true figworts (*Scrophularia*) have 4 fertile stamens and 1 sterile one (it has no anther). The mulleins (*Verbascum*) have 5 fertile stamens.

California Figwort, Bee-plant
Scrophularia californica Cham. & Schltdl.

An upright perennial covered with short sticky-tipped hairs, *its stems are square in cross-section*, 80-120 cm tall. *Leaf blade 7-19 cm long, oval to triangular, the edges with large teeth, sometimes lobed; opposite on the stem.* Corolla urn-shaped, 2-lipped, with 5 small free lobes, red-brown to maroon above, paler below. Its 4 fertile stamens are easily seen in the flower throat, the sterile fifth stamen is held close to the upper 2 lobes. The single style sticks out of the flower, often on the lower lobe, and curves backward. In flower Mar-May.

A common plant mostly along the coast and in moist foothill canyons, often in shade and from boulder crevices, among coastal sage scrub, chaparral, oak and riparian woodlands. Look for it in most of our wilderness parks and forests such as Temple Hill in Laguna Beach, Niguel Hill in South Laguna, the Oak Canyon Nature Center, and Silverado, Hot Springs, San Juan, and Trabuco Canyons.

The nectary is at the base of the ovary within its small flowers. It is pollinated by small native bees and wasps.

The common name refers to the resemblance of its root nodules to neck nodules caused by a tuberculosus lymph node infection of the neck called scrofula. The epithet was given in 1827 by L.K.A. von Chamisso and D.F.L. von Schlechtendal for the place of its discovery.

California Figwort

Woolly Mullein ✖
Verbascum thapsus L.

A stout biennial densely covered with woolly hairs, its thick flowering stalk to 2 meters tall. Flowers are crowded like a corn-cob. Filaments of the upper 3 stamens are covered with white or yellow hairs, the lower 2 hairless or nearly so. Fruit woolly. In flower June-Sep.

Native to Eurasia. An uncommon weed, known from the Santa Ana River wash and Silverado Canyon.

The genus was named in 1753 by C. Linnaeus, who used an ancient name for "bearded"; verbascum is a variant of barbascum (Latin, barba = a beard; cum = with). Linnaeus also gave the epithet in 1753, for Thapsus, a North African seaport where the plant was first collected.

Woolly Mullein

Wand Mullein ✖
Verbascum virgatum Stokes

Similar to woolly mullein, but shorter and skinnier, to 1.2 meters tall, sometimes branched. Its basal leaves are 10-30 cm long, reverse-oval and often wavy-edged. *Flowers are spaced widely apart, not tight like a corn-cob. All stamen filaments are covered with purple hairs.* Fruit mostly hairless. In flower May-Sep.

Native to Eurasia. Common in Corona, Weir Canyon, O'Neill Regional Park, San Juan Creek, Rancho Mission Viejo, and Whiting Ranch and Caspers Wilderness Parks.

The epithet was given in 1787 by J.S. Stokes for its wand-like twigs (Latin, virgatus = twiggy, made of twigs).

Wand Mullein

SOLANACEAE • NIGHTSHADE FAMILY

A family of small to medium-sized herbs or shrubs. *Leaves alternate,* usually not divided but deeply lobed in some species. *Calyx bowl- or funnel-shaped with 5 free lobes often spread out like a 5-pointed star.* Corolla mostly fused at the base, sometimes long and trumpet-shaped, with 5 free lobes. *Stamens 5, often of the same length, the anthers sometimes connected into a cone around the style.* Ovary superior, style, stigma clubbed. Fruit is a berry or capsule.

Mostly found in the tropics and subtropics, especially diverse in Australia and Central and South America. The family is known for ornamentals such as petunias and angel's trumpet but is better known for food plants such as tomatoes, potatoes, eggplant, and chili peppers. Many are highly poisonous, including tobacco and nightshades. Some have hallucinogenic properties, among them belladonna and jimsonweed which are accompanied by side effects that are sometimes fatal. There are about 3,000 species worldwide, 58 in California, 32 in our area (many are introduced and weedy).

Jimsonweeds, Thorn-apples: *Datura*

Foul-smelling hairy annuals or perennials that often grow in a mound shape, up to about 1 meter tall, usually wider. Leaves large, oval or roughly triangular, sometimes lobed. Calyx green, tubular, with 5 straight free lobes. Corolla large, white to purplish, funnel-shaped, folded spirally in bud, generally with 5 lobes, sometimes as many as 10. Fruit is a round capsule covered with prickles. We have 2 species in our area, both highly toxic. The genus was named in 1753 by C. Linnaeus, who used the Hindustani name for an Asian species (Hindustani, Dhatura).

Jimsonweed ☠ ✖
Datura stramonium L.

An annual, mostly hairless. Leaf blade 5-15 cm long, 4-10 cm wide, wavy-edged, with large teeth, sometimes lobed. *Corolla 6-9 cm long, white to pale blue-purple, much smaller than that of western jimsonweed. Fruits are held upright;* covered with prickles, those facing upward are longer than those facing downward. In flower mostly during summer months.

Native to Mexico and presumably the eastern United States, now a worldwide weed. *Grows near flowing water and seasonal wetlands.* Found occasionally along the Santa Ana River wash, flood control channels, and in Peters Canyon Regional Park.

The epithet was given in 1753 by C. Linnaeus, who used the ancient name for a concoction of this plant long used in the treatment of asthma.

Similar. Western jimsonweed (*D. wrightii*): hairy stems; corolla much larger; fruits held downward; found near water and in dry places.

Jimsonweed

Western Jimsonweed ☠
Datura wrightii Regel [*Datura meteloides* A.DC.)

An annual or perennial covered with whitish hairs. Leaf blade 7-20 cm long, oval in outline, often toothed. *Corolla 15-20 cm long, white, much larger than that of jimsonweed. Fruits are held downward;* covered with prickles of about the same length. In flower mostly during summer months.

Native to the American Southwest, south through Mexico and into South America, though it may have been introduced into North America by native Americans or early Spanish settlers. *Widespread and common in our area, often in sandy or gravelly soils.*

Both jimsonweeds are open at night, presumably to attract night-time pollinators such as sphinx moths and beetles. It is not uncommon to find June beetles within the flower tube in the morning.

The epithet was given in 1859 by E.A. von Regel in honor of C. Wright, who collected the plant in Texas just prior to working for the Mexican Boundary Survey of 1851.

Similar. Jimsonweed (*D. stramonium*): (nearly) hairless stems; corolla much smaller; fruits held upward; found only near water.

Western Jimsonweed

THE JIMSONWEED GUILD

Dark-headed June Beetle
Cyclocephala melanocephala (Fabricius, Scarabaeidae)
    A stout beetle about 15 mm long, *pale with a very dark red head and thorax.* Adults are commonly found inside jimsonweed flowers in the morning. In the larval stage, it lives underground and feeds on plant roots, probably jimsonweed.

    The epithet was given in 1775 by J.C. Fabricius for its dark head (Greek, melas = black; kephale = head).

Three-lined Datura Beetle
Lema dataraphila Kogan & Goeden (Chrysomelidae) [*L. trilineata* (Olivier)]
    A small, soft-bodied beetle, 5-8 mm long. *Elytra yellow and white, striped with black.* Thorax yellow, reddish, often spotted, sometimes all-black. Adults and the slug-like, yellow to gray larvae feed on jimsonweeds in spring-summer.

    The epithet was given in 1970 M. Kogan and R.D. Goeden for its love of jimsonweeds (Datura; Greek, philos = loving).

Small-flowered Wild Petunia ✿
Petunia parviflora Juss. [*Calibrachoa p.* (Juss.) D'Arcy]
    A low-growing annual, often a dense mat flat on the ground, the stems 10-40 cm long. Entirely covered with sticky-tipped hairs. Leaves fleshy, 5-14 mm long, somewhat oval. Calyx with 5 very long free lobes. *Corolla 4-6 mm long, pink, purplish, to white.* In flower Apr-Aug (-winter).

    Uncommon, it occurs in perennially wet or seasonally moist soils such as sandy to silty creekbeds and major water channels. Known from lower San Juan Creek in Dana Point, the lower Santa Ana River, Laguna Lakes, and Mile Square Park, inland from Santa Ana River in Corona and the Temecula-Murrieta region.

    The epithet refers to the flower's small size (Latin, parvus= little, small, pretty; floris = flower).

California Box-thorn ★
Lycium californicum Nutt.
    A dense shrub 1-2 meters tall *with stiff angular branches and spine-tipped twigs.* Leaves tiny, 3-10 mm long, *linear, fleshy, more or less round in cross-section.* Corolla tube short, 4-6 mm, *white to purplish.* Fruits red, 3-6 mm, *only 2 seeds each.* In flower Mar-July.

    A plant of coastal bluffs among coastal bluff scrub. Known from Crystal Cove State Park, Newport Bay, Laguna Beach, the Dana Point Headlands, San Clemente State Beach, and near San Mateo Creek. CRPR 4.2.

    Named in 1876 by T. Nuttall for the state of its discovery.

    Similar. Anderson's desert-thorn (*L. andersoni* A. Gray, not pictured): *leaves somewhat flat in cross-section; corolla tube long, 5-10 mm, white to violet; fruit several-seeded;* a desert species, it approaches the coast on dry slopes among coastal sage scrub in Lomas de Santiago (evidently eliminated), Temescal Valley near Corona, and near Lake Elsinore. **Santa Catalina Island desert-thorn ★** (*L. brevipes* Benth. var. *hassei* (Greene) Hitch.): *leaves somewhat flat in cross-section; corolla tube long, 6-10 mm, lavender to whitish; fruit several-seeded;* in coastal bluff scrub at San Clemente and Corona del Mar State Beaches; CRPR 1B.1.

Tobacco: *Nicotiana* ☠

Ill-scented slender-stemmed annuals, except for one semi-woody, tree-like perennial. Leaves undivided, elliptical to oval. Calyx bowl-shaped or tubular, 5-lobed. Corolla with a long tube, tipped with 5 lobes that flare out. Fruit is a capsule.

These plants contain highly toxic alkaloids, mostly nicotine or anabasine. Smoking or otherwise ingesting them results in nicotine poisoning with symptoms such as salivation, nausea, vomiting, diarrhea, clammy skin, labored breathing, weak irregular pulse, and paralysis of the respiratory system. Nicotine overdose produces death in a few minutes.

There are about 60 species of tobacco, most native to the Americas. We have 4 species in our area, though only 2 are common. They are more common after fire and other soil disturbances.

The genus was named in 1753 by C. Linnaeus in honor of J. Nicot, who introduced tobacco into France about 1560.

Tree Tobacco ☠ ✖

Nicotiana glauca Graham

Upright or arching, woody, tree-like, 2-6 meters tall, hairless, covered with a whitish powder (glaucous). Leaves blue-green, oval, 5-21 cm long, feel like rubber. Corolla yellow, 30-35 cm long, tipped with 5 very small lobes. In flower Apr-Nov, sometimes year-round.

Native to South America and introduced to North America, Africa, and the Mediterranean region. It grows in disturbed soils, especially common in the years after fire. *An aggressive weed, known for crowding out native species of plants. It should be eradicated wherever it occurs.* Ingestion of its parts has caused numerous poisonings and several human deaths in California, even recently. We've observed docents and park naturalists sucking nectar from the base of the flower and suggesting that their visitors do likewise. Because of the plant's toxins, this practice should be discouraged.

The epithet was given in 1828 by R. Graham for its leaves (Greek, *glaukos* = bluish-green or grey).

Tree Tobacco

Wallace's Tobacco ☠

Nicotiana quadrivalvis Pursh
[*N. bigelovii* (Torr.) S. Watson var. *wallacei* A. Gray]

A slender-stemmed annual, 30-120 cm tall, covered with sticky-tipped hairs. Leaves 4-15 cm long, elliptical to oval; stem leaves lack a petiole. Corolla tube 25-50 mm long, 5-lobed, white, sometimes tinged green or purple. In flower Apr-Oct.

Our most common native *Nicotiana*. A plant of creekbeds and washes, sometimes among coastal sage scrub and chaparral. More common after fire. Known from Corona del Mar, lower San Diego Creek, the San Joaquin Hills, Lomas de Santiago (especially Hicks Canyon), Caspers Wilderness Park, lower San Juan Creek, and Tenaja Trail.

The epithet was given in 1814 by F.T. Pursh for its 4-chambered fruit (Latin, *quadris* = fourfold; *valvus* = leaf of a folding door). When you open the fruit, it appears as if the sides were folded inward like folding doors. The plant was described from an unusually large 4-chambered form (2-chambered is typical).

Similar. Coyote tobacco ☠ (*N. attenuata* Torr., not pictured): *all leaves petioled; corolla tube 20-27 mm long, 5-lobed, greenish, pink-tinged;* last documented from Laguna Beach in 1931. **Cleveland tobacco** ☠ (*N. clevelandii* A. Gray, not pictured): *stem leaves lack petiole; corolla tube only 15-20 mm long, 4-lobed; green-white;* upper Newport Bay below Mesa Drive and Laguna Beach.

Wallace's tobacco (below). Left: stem leaves have no petiole. Right: flower opens at night, closes in daytime.

Wallace's Tobacco

Thick-leaved Ground-cherry ☠
Physalis crassifolia Bentham
[includes *Physalis greenei* Vasey & Rose]

An upright hairy perennial, 20-80 cm tall; *its stems have longitudinal ridges and grow in a zigzag pattern. Leaves* 1-3 cm long, *oval, thick,* wavy-edged or toothed. *Calyx tubular* with 5 short lobes. *Corolla yellow,* broadly bell-shaped, tubular at its base, the 5 lobes fused and pointed. In flower Mar-June.

A lovely flower from which develops a most unusual fruit. After the flower is fertilized, it gradually rotates and faces downward, and the corolla withers and falls off. *The calyx lengthens and expands like an air-filled balloon, and its 5 lobes tightly close at the tip.* The developing fruit remains within this "bag." Its unripe fruit is toxic.

Primarily a desert plant, in our area it occurs uncommonly on dry rocky slopes, sometimes following fire. Known from the Sycamore Hills of Laguna Beach and San Juan Canyon within Caspers Wilderness Park. Some of our plants are a coastal form that occurs in Orange and San Diego Counties and south into Baja California. It was once considered to be a separate species, **Greene's ground-cherry** (*Physalis greenei* Vasey & Rose).

The genus was named in 1753 by C. Linnaeus for its bladder-like calyx (Greek, physalis = bladder or bubble). The epithet was given in 1844 by G. Bentham for its thick leaves (Latin, crassus = thick; folium = leaf).

Right: Thick-leaved ground-cherries from the Mojave Desert, where it is more upright, less hairy, thinner-stemmed, and has a deeper-yellow flower than those from our coastline.

Pampas Lily-of-the-Valley ✖
Salpichroa origanifolia (Lam.) Baillon

A hairy crawling and climbing plant with stems to 2 meters long. It spreads from a creeping rhizome. *Leaves* 1-3.5 cm *long, oval, fleshy.* Calyx lobes long, narrow. *Corolla* 5-7 mm long, *urn-shaped, white to greenish.* Fruit 1-1.5 cm long, a white to yellowish fleshy berry. In flower July-Oct.

Native to the Andes Mountains of South America. An aggressive weed of hillsides, ravines, and canyons. Known from places such as lower Moro Canyon in Crystal Cove State Park, along 7th Avenue near Mission Hospital Laguna Beach, and in some parts of Dana Point.

The epithet was given in 1794 by J.B. Lamarck for its mountain habitat and beautiful leaves (Greek, oros = mountain; ganos = beauty; Latin, folium = leaf).

Nightshades: *Solanum* ☠

Small shrubs and herbs, hairless or hairy, sometimes spiny. Their leaves are undivided; some are toothed or wavy-edged, rarely lobed. The calyx is bell-shaped, 5-lobed, relatively small. The corolla is shaped like a wide bell, the 5 lobes free or fused. The 5 stamens are held tightly in a cone around the style. Fruit is round and green like a small tomato. Most are highly toxic, especially the fruits. The toxins affect children more severely than adults. Symptoms appear hours after ingestion and include gastric pain and fever; severe poisonings may lead to coma and death.

Their flowers have no nectar to attract and reward pollinators; insects (primarily bees) visit only to gather pollen. Bees alight on a flower, hold tightly, and buzz their wings to shake pollen from the anthers like a salt shaker (*buzz pollination*). Bumble bees are the primary pollinators of this genus.

Worldwide, there are over 1,000 species in this genus, 19 in California more than half of which are introduced weeds. We have 12 in our area, only 4 of them native: 1 white-flowered, 3 purple-flowered. The 8 non-native species are urban weeds generally not found in wild areas and are not discussed here. The genus was named in 1753 by C. Linnaeus in reference to the narcotic properties, used in ancient medications (Latin, solamen = quieting).

	Stems	Leaves	Corolla size	Corolla color	Fruit
Native					
Douglas's *S. douglasii*	Green; curved white hairs	Oval; smooth to wavy-toothed	*Small,* about 10 mm wide	*White* to lavender	6-9 mm, black
Parish's *S. parishii* (see entry under Blue Witch)	Green; mostly hairless	Lance to elliptic; smooth, sometimes wavy; tapered to base	Large, 17-22 mm wide	Blue-purple; 2 greenish glands at base of each lobe	7-10 mm; green (to purple?)
Blue Witch *S. umbelliferum*	Green; hairs dense, branched	Elliptic to oval; usually smooth	Large, 16-25 mm wide	Lavender to blue-purple; 2 greenish glands at base of each lobe	12-14 mm; green (to purple?)
Chaparral *S. xanti* (see entry under Blue Witch)	Green; mostly hairy, some sticky	Lance to oval; smooth or 1-2-lobed at base	Large, 15-30 mm wide	Dark blue to lavender	10-15 mm, greenish
Non-native					
White *S. americanum*	Green, not sticky; mostly hairless	Oval; smooth to wavy-toothed	*Tiny,* 3-6 mm wide	White to white tinged with purple	5-8 mm, black to greenish
Silverleaf Horse-nettle *S. eleagnifolium*	Silvery, not sticky; prickly; hairs star-shaped	Oblong; smooth to somewhat lobed, yellowish	Large, 20-30 mm wide	Purple or blue	8-15 mm, *orange*
Lance-leaf *S. lanceolatum*	Sparsely prickly; hairs star-shaped, some sticky	Lance; smooth to lobed	Large, 25-30 mm wide	Blue-purple	7-15 mm, *yellow-orange*
Black *S. nigrum*	Green; short-haired; some sticky, some not	Oval; smooth to wavy-toothed	*Small,* 10 mm wide	White	6-8 mm, black
Hairy *S. physalifolium*	Green, sticky; short-haired; some sticky	Oval; toothed to shallowly lobed	*Tiny,* 3-5 mm wide	White	6-7 mm, yellowish to green
Buffalo-bur *S. rostratum*	Green, not sticky; densely prickly; hairs star-shaped	*Pinnate; prickly*	Large, 25-35 mm wide	Yellow	9-15 mm, green, in a tan spiny case

Douglas's Nightshade ☠☠
Solanum douglasii Dunal

An upright perennial 1-2 meters tall, its stems covered with rough hairs. *Leaves* 1-9 cm long, *oval in outline, often wavy-edged or toothed.* Calyx small, lobes long. *Corolla bell-shaped,* about 1 cm wide, greenish at the base, *with 5 long lobes, bright white to pale purple-tinged.* The fruit turns dark purplish-black when ripe. In flower all year.

A common plant throughout our area, among many plant communities, often in moist soils along watercourses and disturbed ground. Common in upper Newport Bay, most of our wilderness parks, and in major canyons of the Santa Ana Mountains.

Named in 1852 by M.F. Dunal to honor D. Douglas.

Douglas's Nightshade

Blue Witch ☠☠
Solanum umbelliferum Eschscholtz

An upright shrubby perennial 0.6-1 meter tall, often shorter near the coast. *Leaves* 1-4 cm long, *elliptical to oval or reverse-oval.* Inflorescence is a dense umbrella-shaped, arched cluster from a stalk at the top or sides of the plant. Calyx with small lobes. *Corolla lobes mostly fused, lavender to dark blue-purple. The base of each lobe has two green spots with a white halo.* Its bright yellow anthers contrast sharply with the corolla. In flower Jan-June.

Widespread and common, in grasslands and among openings in coastal sage scrub, chaparral, and oak woodlands. Known from the San Joaquin Hills including Moro Canyon, Temple Hill, Niguel Hill, and Salt Creek. Found in major canyons of the Santa Ana Mountains, such as Silverado, Santiago, Trabuco, and San Juan, and in Rancho Mission Viejo, O'Neill Regional Park, and Caspers Wilderness Park.

Technically, we have three native purple-flowered nightshades in our area. The other two are **chaparral nightshade** (*S. xanti* A. Gray) and **Parish's nightshade** (*S. parishii* (A.A. Heller). However, they are nearly identical and differ only in minor physical features that vary widely, overlap, and are inconsistent. Their characteristics in the *Solanum* table (above) are from published literature but are not reliable. These three nightshades are probably best considered members of a single species. If so, the rules of botanical naming (International Code of Botanical Nomenclature, ICBN) require they be referred to by the earliest published name, that of *Solanum umbelliferum*.

The epithet was given in 1826 by J.F. Eschscholtz in reference to the shape of its inflorescence (Latin, umbellula = sunshade).

Blue Witch

Buffalo-bur ✖
Solanum rostratum Dunal

An upright annual to about 0.5 meter tall, *densely covered in long sharp prickles. Leaves* 5-15 cm long, *pinnate,* very similar to those of watermelon. Calyx lobes very prickly. *Corolla* 25-35 mm wide, *yellow.* Anthers unusual; 4 straight, short; 1 curved, much longer. Style curved as well. Fruit dry, tightly surrounded by prickly calyx. In flower May-Sep.

A plant of disturbed soils, often those with seasonal moisture such as creeks, seeps, roadsides, and vernal pools. Known from Fairview Park, Silverado Canyon, Modjeska Grade Road, and Dana Point Harbor. Native to the American Great Plains.

The epithet was given in 1813 by M.F. Dunal for its curved long anther and style (Latin, rostratus = beaked, hooked).

Buffalo-bur

STYRACACEAE • STORAX FAMILY

Shrubs or small trees with winter-deciduous leaves and showy flowers. Calyx a short tube with (0)5(-9) short teeth. Corolla lobes (4)5(10), fused toward the base, free toward the tips. Stamens are 2(-4) times the number of petals. Ovary superior, rarely inferior. Fruit is a capsule, rarely a drupe. A small family of about 120 species worldwide and only a single species in California. Some members are cultivated as ornamentals; a few have medicinal uses. The resin called storax is obtained from the bark and is used in perfumes, incense, and medicine.

Southern California Storax, Southern Snowdrop Bush
Styrax redivivus (Torr.) L.C. Wheeler

Southern California Storax

An upright shrub, 1-4 meters tall, often with multiple slender trunks and an open branching structure. Leaves usually partly or completely winter deciduous. *Leaf blade* 2-8 cm long, *oval or rounded, with wavy margins and often an indented heart-shaped base*. Upper leaf surface hairless, lower surface paler with star-shaped hairs. Inflorescence droops and bears 1-6 flowers that have a strong sweet fragrance. Calyx short, tubular, 6-9 lobed. *Corolla* 12-26 mm long, *5-10-lobed, bell-shaped, bright white*. Pistil pale green, longer than the stamens. Stamens 10-16 with white filaments and bright orange anthers that contrast with the white petals. Fruit a semi-dry capsule. In flower Mar-June.

A real treat to find in the wild. Once the flowers wither and fall off, the shrub blends in with its neighbors and is difficult to locate. Grows among chaparral on slopes and in canyon bottoms, most often in shade. In our area, known from southern slopes and ridges of Fremont Canyon, Trabuco Canyon, along the San Juan Trail, and scattered sites along San Juan Canyon, but is more common on the north-facing slopes below Sitton Peak Truck Trail. Oddly, it is also found in great numbers on the south-facing slopes of Los Pinos Peak, among **Eastwood manzanita** (*Arctostaphylos glandulosa* ssp. *glandulosa*, Ericaceae) at higher elevations and chamise at lower elevations. A small pocket of it also occurs on north-facing slopes right along the Lucas Canyon Trail in the San Mateo Canyon Wilderness Area.

The genus was established in 1753 by C. Linnaeus, who used its ancient name (Greek, styrax or styrakos = a European tree from whose bark is produced a fragrant gummy resin called storax). The epithet was given in 1851 by J. Torrey, possibly for its sweet "reviving" scent (Latin, redivivus = revived).

Fruit

Southern California Storax along the Lucas Canyon Trail

THEOPHRASTACEAE • THEOPHRASTA FAMILY

Perennial herbs, shrubs, and trees, usually dotted with glands or dotted or streaked with resin canals. *Leaves undivided; smooth or tooth-edged; clustered along the stem or at the tips or alternate;* stipules absent. Inflorescence usually at stem tips, sometimes axillary. *Sepals 5,* free or fused at base. *Petals 5,* fused at base, lobes held straight or outward. *Stamens 5,* each in front of a petal; some species also with staminodes. *Ovary superior or half-inferior,* style and stigma 1. Fruit berrylike or a capsule. Pollinated by bees and flies such as midges. A small family of about 100 species, all New World, mostly tropical. All or just some of its species are sometimes included within the primrose family (Primulaceae). Some botanists place the genus *Samolus* within its own family, the Samolaceae.

Water-pimpernel, Seaside Brookweed
Samolus parviflorus Raf.

An upright, arching, or sprawling perennial, stems 10-40 cm long. Leaves 2-5 cm long, mostly oval; petiole winged or nearly absent. *Inflorescence usually long and spread out,* pedicels 1-2 cm long with a tiny bract about halfway up. *Flower only about 1.5 mm wide.* Calyx lobes 5, each 1-2 mm long, triangular. *Corolla lobes 5, 3-5 mm long, oval, white, straight or spread out.* Stamens 5, held against the pistil when young, spreading out in age. *Staminodes 5(10), alternate to the stamens, sometimes petal-like and white.* Ovary half-inferior. Fruit spherical, about 2.5 mm across. In flower spring-summer.

A small plant of very moist sites, especially along shaded streams and seeps, sometimes along moist outfalls near the beach. Known from Lower Cañada Chiquita on Rancho Mission Viejo, Los Trancos Canyon in Crystal Cove State Park, Peters Canyon Regional Park, and Santa Rosa Plateau. Old records exist from Bolsa Chica (1932), Wintersberg (1927), Santa Ana River in Yorba Linda (1927) and near Corona (1917), and Chino Creek (1932-33).

The epithet was given in 1818 by C.S. Rafinesque for its tiny flowers (Latin, parvus = small; floris = flower).

Left: Water-pimpernel has tiny flowers, each usually with 5 (or 10) white staminodes that alternate with the petals.

TROPAEOLACEAE • NASTURTIUM FAMILY

Low-growing annuals or perennials *with succulent stems. Leaves* alternate; often rounded, *umbrella-like (peltate), with palmate veins.* Flowers solitary on a long pedicel. *Sepals 5, uppermost with a long spur.* Corolla 2-lipped. *Petals 5, the base of each narrowly-clawed,* the distal end broad and rounded. Stamens 8, in 2 unequal groups. Ovary superior; single style, 3-branched near tip. Fruit nut-like. A small family of about 90 species, none native to North America.

Garden Nasturtium ✖
Tropaeolum majus L.

A long-stemmed trailing annual or perennial. Leaf blade 3-12 cm wide, round. *Flowers large and showy,* 2.5-6 cm across. *Upper sepal with a very long spur.* Petals orange or yellow, often with red or orange patches and/or stripes. *The 3 lower petals have eyelash-like fringes.* In flower Mar-June or longer.

A garden escape, spreading into riparian and other moist habitats, mostly near our coastline. Known from upper Newport Bay and lower Aliso Canyon.

The name was given in 1753 by C. Linnaeus for its large flowers (Latin, majus = great, large).

URTICACEAE • NETTLE FAMILY

Annuals, perennials, shrubs, and soft-wooded trees. *Ours are annuals or perennials with watery sap, some hairless but many with stinging hairs. Leaves alternate or opposite,* often toothed, with an obvious petiole. Embedded in the leaves are crystals of calcium carbonate that appear like raised dots, visible with a hand lens. The inflorescence, a dense cluster of many small flowers, appears among the nodes. *Flowers are often of only 1 sex, small and greenish.* Calyx fused or 4-5-lobed (rarely 2-3). *Corolla absent.* Male-only flowers have 4-5 stamens. Pollen is released explosively from the anthers. Female-only flowers have a superior ovary, a single style (sometimes 2 or none), and a fuzzy-tipped stigma. Fruit is a tiny achene surrounded by the calyx.

Nettle flowers have no nectary, so their insect visitors arrive *only* to collect pollen from male flowers. The plants rely on the wind to carry pollen from male to female flowers.

A family of only 700 species worldwide, 7 in California, 5 in our area. All but 2 are small and often overlooked.

Dwarf Nettle

Nettles often go unnoticed, that is, until you brush against them with bare skin. Their stiff hairs inject a chemical cocktail that causes immediate pain, itching, and swelling. Nettles have a way of getting your attention.
Stinging nettles (western, hoary, and dwarf) have specialized hollow liquid-filled brittle hairs with a swollen base and slender tip. If you contact the hair, its tip is broken and easily penetrates your skin. Existing pressure from the base forces liquid out and into your skin. It contains histamines, acetocholines, and formic acid (the latter is an active ingredient in ant venom; thus nettle stings feel like ant stings). Initial symptoms are red, itchy skin, advancing to swelling and burning. Some people feel only mild pain, others feel quite intense pain. The effects are not permanent, but the memories are!

Western Nettle
Hesperocnide tenella Torr.

An upright annual, 5-40 cm tall, its *weak stems* are often pink. *Leaves* opposite, 4-40 mm long, oval, the *edges with round teeth, upper surface with bold black dots.* Each spherical inflorescence contains both male and female flowers. In flower Feb-June.

Uncommon in shaded, *moist soils* within coastal sage scrub, chaparral, and riparian woodlands, often under shrubs and rock edges. Known primarily from near the coast, such as at the Dana Point Headlands and the San Joaquin Hills; also in Lucas Canyon.

The epithet was given in 1857 by J. Torrey in reference to its slender, delicate stems (Latin, tenellus = quite delicate).

Similar. Dwarf nettle ✘ (*Urtica urens*): usually taller; stem green; leaves with pointed teeth, no black dots on upper surface, opposite; in flower Jan-Apr. An abundant, widespread weed.

Western Nettle

Western Pellitory
Parietaria hespera Hinton **var. *hespera***

A sprawling annual, branched from the base. *It has no stinging hairs.* Leaves alternate, blade 0.5-2 cm long, oval to lance-shaped, longer than wide. Inflorescence few-flowered, polygamous. Calyx lobe tips slender or abruptly narrowed below the tip; apex with a hairlike tip. In flower Mar-June.

Occasionally found on *moist shaded slopes* among coastal sage scrub and chaparral; San Joaquin Hills, Laguna Canyon, Aliso Creek, Chino Hills, Fremont Canyon in the Santa Ana Mountains.

The epithet was named in 1969 by B.D. Hinton for its western distribution (Greek, hesperos = western).

Similar. California pellitory (*P.h.* var. *californica* Hinton, not shown): *leaf blade round to oval, width about equal to length; calyx lobes long-tapered to the tip;* in flower Feb-May. Scarce, similar habitats; moist soils among coastal sage scrub in the San Joaquin Hills and Dana Point Headlands.

Western Pellitory has alternate leaves

Hoary Nettle ☠
Urtica dioica L. **ssp. *holosericea*** (Nutt.) Thorne

A large upright perennial 1-3 meters tall covered with stinging hairs. Stems square with few or no branches. Leaves 5-12 mm long, linear or oval, often folded, the edges toothed; opposite. Both stems and leaf undersides are covered in many long, soft, non-stinging hairs. Of course, its stems and leaves *also* contain many stinging hairs. Each inflorescence contains flowers of only one sex, in loose clusters nearly as long as the leaves. In flower June-Sep.

A common plant of watercourses, seeps, and wet soils. Known from many places such as upper Newport Bay, Crystal Cove State Park, San Juan Creek, Santiago Canyon, Whiting Ranch Wilderness Park, and Holy Jim Canyon.

The genus was named in 1753 by C. Linnaeus who used the ancient name for nettle, in reference to their burning hairs (Latin, uro = to burn). He named the epithet the same year for the European form of this species in which each plant bears flowers of only one sex (Greek, di- = 2; oikos = house). Our form was named in 1848 by T. Nuttall in reference to the soft non-stinging hairs on its stems and leaf undersides (Greek, holos = whole, entire; Latin, sericeus = pertaining to silk).

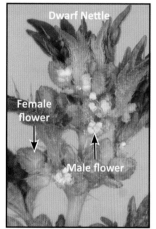

Far left: The inflorescence of dwarf nettle has clusters of male and female flowers. Females are surrounded by the green calyx, like a clamshell. Males have 4-5 yellowish stamens that unfurl and spread out. Left: Male flowers, the middle one is half open, the lower one just starting to open.

Dwarf Nettle ☠ ✖
Urtica urens L.

An annual 10-60 cm tall with a single stem or branched from the base, covered with stinging hairs. Leaf blade 1.5-3 mm long, elliptical to oval, the edge with large slender teeth. Each inflorescence contains both male and female flowers, in clusters shorter than the petiole. In flower Jan-Apr.

Native to Europe. A common weed in moist soils within orchards, gardens, parks, and on some trails.

The epithet was given in 1753 by C. Linnaeus for its stinging hairs (Latin, urens = stinging).

Similar. Western nettle (*Hesperocnide tenella* Torr.): shorter; *leaves with rounded teeth, upper surface with black dots;* in flower Feb-June; uncommon in moist soils near the coast, such as at the Dana Point Headlands.

THE NETTLE GUILD

The larvae of two local butterflies eat nettles. Both are true brushfoots (family Nymphalidae).

Satyr Anglewing
Polygonia satyrus (Edwards)

Adults have a wingspan range of 4.5-7.6 cm. Wings are rusty orange-brown above, dotted with brown and black, the wing edges are sharply angular. *Larvae feed only on hoary nettles and always sew the leaf edges downward. They have large branched spines on the head and body.* Found along riparian waterways away from urban areas. We've seen them along San Juan Creek in San Juan Capistrano, upstream through San Juan Canyon, Whiting Ranch Wilderness Park, and in upper Silverado Canyon. Also reported from Featherly Park, Irvine Regional Park, and upper Trabuco Canyon.

The epithet was given in 1869 by W.H. Edwards as a tribute to the mischievous Greek deity, Satyros.

Larvae and their shelters. Upper right: The satyr anglewing sews leaf edges downward. Right: Larvae have large branched spines on the head and body. Lower right: The red admiral sews leaf edges upward. Right: Larvae have stout spines on the body, smaller ones toward the head. Both larvae have recently shed their skin and look pale. Soon, they will darken in color.

Red Admiral
Vanessa atalanta rubria (Fruhstorfer)

Adults have a wingspan range of 4.5-6.4 cm. Wings are dark brownish-black above. Forewings have a bold, bright red-orange stripe that runs *across* the wing, the tips have white patches and spots. Hindwings have red-orange along the *back edge* of the wing and blue eyespots at the back angle. *Larvae feed on both hoary and dwarf nettles and sew the leaves upward to make a shelter. They have stout spines on the body, smaller spines toward the head.* In urban areas, they will also feed on **baby's tears** (*Soleirolia soleirolii* (Req.) Dandy), a non-native, low-growing garden nettle that has no stinging hairs. Adults and larvae are commonly found in urban and suburban areas, up into the foothills, but not at higher elevations. See the entry for The Manzanita Guild (Ericaceae) for a photo of the adult butterfly nectaring from flowers of **bigberry manzanita** (*Arctostaphylos glauca*, Ericaceae).

The epithet was given in 1758 by C. Linnaeus for Atalanta, a female character from Greek mythology known as a fast runner and fierce fighter. This butterfly flies *very* fast. The subepithet was given in 1909 by H. Fruhstorfer for its red-orange bands which can be redder than the European form of the species (Latin, rubra = red).

VALERIANACEAE • VALERIAN FAMILY

Annuals or perennials, some species ill-scented. *Basal leaves usually whorled, stem leaves opposite. Inflorescence* is a *dense cluster* of numerous small flowers. Calyx fused to the top of the inferior ovary. *Corolla tubular, lobes usually 5, the base with an obvious spur.* Stamens 1-3. Fruit is a cypsela (some botanists call its fruit an achene). A small family of 300 species, 8 native to California, 1 of them in our area, and 1 garden escape.

Long-spurred Plectritis
Plectritis ciliosa (E. Greene) Jeps.

An upright hairless annual, 10-50 cm tall. Leaves about 1-3 cm long; lower ones with a petiole, upper ones without; opposite on the stem. *Corolla* 1.5-3.5 mm long, *medium pink to dark pink, the lower lip with 2 red dots*. The spur is less than one-half the total length of the corolla, usually shorter than the ovary. The fruit has winged edges, often with 2 rows of hairs. In flower Mar-May.

Generally not abundant, but in years of good winter rainfall it can be quite common in dense, localized patches. In drought years, it may not come up at all. Grows in meadows, grasslands, and oak woodlands, often in shade, sometimes on north-facing slopes. Known from Maple Springs Truck Trail and Saddle, Los Pinos Potrero near Falcon Group Camp, along San Juan Trail in the Chiquito Basin, and along South Main Divide Truck Trail between El Cariso and San Mateo Canyon. In the San Mateo Canyon Wilderness Area, known from lower Los Alamos Canyon, south through San Mateo Canyon to Tenaja Canyon, along Tenaja Trail near Pigeon Spring, the Oak Flats area, and Indian Potrero Trail. Also on the Santa Rosa Plateau.

The genus was named in 1830 by J. Lindley for its spurred corolla (Greek, plektron = a rooster's spur). The epithet was given in 1895 by E.L. Greene for its hairy fruit (Latin, ciliosus = fringed, full of hairs).

Long-spurred Plectritis

Long-spurred Plectritis

Red Valerian

Red Valerian ✘
Centranthus ruber (L.) DC.

An upright hairless perennial, 0.3-1 meter tall, covered with whitish powder. Leaves 5-8 cm long; lower ones with a petiole, upper ones without; broadly oblong, elliptic, to lance-shaped. *Corolla* 14-18 mm long, *purplish red or pinkish, rarely lavender or white*. Each flower has only 1 stamen. The spur is twice as long as the ovary. In flower Apr-Aug.

A common garden plant, used more often in years past than today. Mostly found in older settlements in canyons from where it has escaped and now commonly grows along creeks and moist rock walls nearby. Known from Newport Beach, Corona del Mar, Silverado Canyon near the Maple Springs Visitor Center, lower Hot Springs Canyon, and along Ortega Highway above Lake Elsinore. Native to the Mediterranean region of Europe.

The genus was established in 1805 by A.P. de Candolle for its spurred flowers (Greek, kentron = a point, spine, center of a circle; anthos = flower). The epithet was given in 1753 by C. Linnaeus for its red flowers (Latin, ruber = red).

Red Valerian

VERBENACEAE • VERVAIN FAMILY

Annuals, perennials, shrubs, and rarely trees. *Ours are annual or perennial herbs, or shrubs with square stems*, often scented, and *covered with bristly hairs or prickles. Leaves opposite or whorled*, usually toothed. *Inflorescence a dense cluster shaped like a round-topped mushroom or a long spike. Calyx tubular, 4-5* (rarely 2-4) *teeth or free lobes. Corolla tubular, trumpet-shaped*, longer than the calyx tube, with 5 (rarely 4) *spreading lobes, sometimes 2-lipped. Stamens 4-5*, unequal in length. *Ovary superior with a single forked style, only 1 fork bearing the stigma. Fruit is a cluster of 2-4 nutlets.* A family of 1,900 species, mostly in the American tropics, 17 in California, 8 in our area, only 2 of them are common.

Garden Frog-fruit
Phyla nodiflora (L.) E. Greene [*Lippia nodiflora* Michx.]
  A mat-like creeping perennial. Leaf blade 5-30 mm long, oval to wedge-shaped, tooth-edged; usually opposite. Inflorescence a short, dense, head-like spike. Calyx 2-4-toothed. *Corolla weakly 2-lipped, white, pink, or reddish, often with a yellow throat patch.* Fruit is a pair of 2 nutlets. In flower May-Nov.

  An uncommon plant of moist soils in drainages and waterways, also at the moist base of some cliffs. Known from upper Newport Bay, Trampas Canyon, Dana Point Harbor, Santiago Creek in Irvine Regional Park, Irvine Lake, and Silverado Canyon. In cultivation, used as a groundcover plant. Possibly introduced from South America.

  The epithet was given in 1753 by C. Linnaeus for its knob-like cluster of flowers (Latin, nodus, nodulus = knotty, nobby; flor = flower).

Bracted Vervain
Verbena bracteata Lag. & Rodr.
  An annual or short-lived perennial. Its stems spread along the ground, the tips sometimes upturned, 8-30 cm long. Leaves 1-3(6) cm long, tooth-edged or lobed, covered with rough hairs. *Inflorescence a crowded long spike of small flowers.* Each flower has an obvious bract beneath it, as long as or longer than the flower. *Corolla 4-5 mm long, white, lavender, or blue.* Fruit is a cluster of 4 nutlets. In flower Apr-June(-Oct).

  Generally found in drying mud of vernal pools and margins of reservoirs. Occurs among vernal pools in Costa Mesa's Fairview Park and those on Rancho Mission Viejo, and at Bonita Canyon Reservoir, Laguna Lakes, and Irvine Lake. Known long ago from the Chino Hills (near Round Top) and Placentia.

  The epithet was given in 1801 by M. Lagasca y Segura and J.D. Rodriguez for the numerous bracts in its inflorescence (Latin, bractea = a thin plate of metal).

Western Verbena
Verbena lasiostachys Link
  An upright perennial 35-80 cm tall *covered with stiff, bristly hairs. Leaf blade 4-10 cm long, oval, toothed, with 1-2 lobes near the base*, the edges tapered and continued down the petiole. Inflorescence a long dense hairy spike of flowers, often 2-3-branched at its base. Calyx 5-toothed, very hairy. Corolla weakly 2-lipped, blue to purple, the tube darker. Fruit is a cluster of 4 nutlets. In flower May-Sep.

  Found in dry grasslands, moist meadows, creeks, and oak woodlands. Widespread but generally not abundant. Known from many places such as Crystal Cove State Park, Sycamore Hills, Dana Point, San Juan Creek, Whiting Ranch Wilderness Park, Audubon Starr Ranch Sanctuary, Corona, Temescal Valley, and Tenaja Canyon.

  The epithet was named in 1753 by J.H.F. Link in reference to the densely hairy spike of flowers that resembles an ear of corn (Greek, lasio = hairy; stachys = an ear of grain or spike).

VIOLACEAE • VIOLET FAMILY

Annuals and perennials, rarely shrubs, trees, or vines. *Ours are small perennials or annuals. Leaves basal and up the stem, alternate,* not divided in most species. Inflorescence a loose cluster or flowers are solitary. *Flowers 2-lipped, petals of unequal size and shape.* Sepals and petals 5. The lowest petal is broad and has a backward-pointing spur or pouch at its base. The others are rounded or oval, the top pair folded back and often backed with purple or brown. Stamens 5, alternate with the petals. Ovary superior with a single style. Some species have unusual flowers that don't fully open; instead they remain closed and self-fertilize (in-bud fertilization or *cleistogamy*). *Fruit is a dry 3-lobed capsule that explodes open and shoots out the seeds.* The seeds of some species have an appendage on them; ants pick up the seeds by the appendage and carry them off to their nest, eat the appendage, and discard the seed, thus distributing the seeds to new habitats. We have 4 violets, found in moist meadows, among coastal sage scrub, and on shaded slopes of our foothills and mountains.

Johnny Jump-ups
Viola pedunculata Torr. & A. Grey

An upright perennial, 5-39 cm tall, from a deep-rooted rhizome. Leaves undivided, 10-55 mm long, triangular, oval, or heart-shaped. *Stem leaves only, no basal leaves.* Petals bright orange-yellow, the lower 3 have dark-brown veins. The side pair has thick paddle-like yellow hairs at the throat. *The upper pair has red-brown on the outside.* In flower Feb-Apr.

Our most common and widespread violet, found both in the San Joaquin Hills and throughout the Santa Ana Mountains. It grows in grasslands and coastal sage scrub at low to middle elevations and in meadows. Formerly quite abundant on the hillsides now occupied by the communities of Las Flores and Ladera Ranch, only scattered patches of it remain on the lower slopes of Chiquita Ridge to its east. Easy to find in portions of Caspers Wilderness Park and in Los Pinos Potrero near Falcon Group Camp.

Named in 1838 by J. Torrey and A. Gray in reference to its long peduncles (Latin, pedunculus = foot).

Oak Yellow Violet
Viola purpurea Kellogg **ssp. quercetorum** (M. Baker & J. Clausen) R.J. Little

A variable, low-growing to upright perennial, 4-25 cm tall, from a hairy woody taproot. Leaves undivided, sometimes toothed, basal and stem leaves. *Fresh leaves light grey-green above, sometimes purple-tinted below. Petals deep lemon-yellow. Purplish markings on the outside of the upper 2 petals.* In flower Feb-July.

Found at moderate and high elevations, in drier habitats, among grasslands, shrubs, chaparral, and oak woodlands, generally not with conifers. Known from Coyote Canyon in the San Joaquin Hills, Holy Jim Trail, a grassy north-facing area on Modjeska Peak, along Main Divide Truck Trail above Maple Springs and near Trabuco Peak, and shaded sites on Santiago Peak. Uncommon in the northern region of the San Mateo Canyon Wilderness Area.

The epithet was given in 1855 by A. Kellogg for its purplish leaves (Latin, purpureus = reddish, violet, or purple). The subepithet was given in 1948 by M.S. Baker and J.C. Clausen. The name is in reference to its occurrence among oaks (Latin, Quercus = oak; -orum = of).

Similar. **Mountain yellow violet** (*Viola purpurea* ssp. *purpurea*, not shown): fresh leaves dull to bright green above, tinted purplish; often strongly purple on undersides; uncommon, higher elevations, usually with conifers, upper Holy Jim Canyon Trail, Main Divide Truck Trail, atop Santiago Peak; in flower Mar-July. **Johnny jump-ups** (*Viola pedunculata*): stem leaves only, no basal leaves, leaves green (not purple); purplish markings on the outside of the upper 2 petals (instead of dark brown); common and widespread.

Shelton's Violet ✿
Viola sheltonii Torr.

An unusual perennial from an upright rhizome. *Its stems remain underground,* 3-27 cm long. *Leaves dark green, dissected; most lie flat on the ground,* the blade is under 7 cm long. Petals deep lemon yellow. In our area, most plants have self-fertilizing flowers that never open. In flower Apr-July.

A true gem, quite rare in the Santa Ana Mountains. We are at the extreme southern end of its distribution, which extends northward to Washington. It grows in loose decomposing soils and talus slopes on only a few north- and northwest-facing regions of the Santa Ana Mountains on the shady flanks of Santiago and Modjeska Peaks, often in the shade of **canyon live oaks** (*Quercus chrysolepis,* Fagaceae).

Named in 1856 by J. Torrey, in honor of C.A. Shelton.

Shelton's Violet

Fruit

THE VIOLET GUILD
The larvae of two local butterfly species, both true fritillaries or "silverspots" (*Speyeria* spp., Nymphalidae), feed only on violets.

Comstock's Fritillary
Speyeria callippe comstocki (Gunder)

A medium-sized butterfly with wingspan about 4-5.5 cm. *Wing upper surface muted or pale orange with heavy black markings and brown near the wing bases.* Formerly abundant throughout our area, even along the coast and in low-elevation cities such as San Juan Capistrano and Orange, but development projects have seriously destroyed populations of its larval foodplants and thus the butterfly. It is still fairly common in medium to high elevations within the Cleveland National Forest such as at Los Pinos Potrero, Blue Jay Campground, and atop Santiago Peak. In years of good rainfall, both the violets and the butterflies can be quite common. One of the many delights of late spring is watching the females flit about meadows laying eggs on violets. *Larvae black and spiny with whitish spines.* They feed primarily on *Viola pedunculata.* In flight May-July.

The epithet was given in 1852 by J.B.A.D. de Boisduval, probably in honor of Callippus of Cyzicus. The subepithet was given in 1925 by J.D. Gunder in honor of J.A. Comstock.

Comstock's Fritillary

Comstock's   Semiramis

Semiramis Fritillary
Speyeria coronis semiramis (W.H. Edwards)

A medium-large butterfly, its wingspan is about 5-7 cm. *Wing upper surface is brighter orange and has much less black than Comstock's fritillary, and the wing bases are clearly orange.* Fairly uncommon here, it is most often seen in rapid flight across meadows and conifer forests at middle-to-higher elevations in the Santa Ana Mountains. It stops to take nectar from purple flowers such as California thistle (*Cirsium occidentale* var. *californicum,* Asteraceae). Known from Live Oak Canyon, Trabuco Canyon, Los Pinos Potrero, and atop Santiago Peak. *Larvae black, their spines black with orange bristles.* They feed on one or both forms of *Viola purpurea.* In flight June-Sep, its peak flight is July.

The epithet was given in 1864 by H.H. Behr, probably for the silver spots on the underside of its hind wings; those nearest the wing edge are capped ("crowned") with green or brown toward the wing base (Latin, corona = crown). The subepithet was given in 1886 by W.H. Edwards in honor of Semiramis, a legendary Queen of Assyria.

Semiramis Fritillary

VISCACEAE • MISTLETOE FAMILY

Perennials or shrubby plants that live as partial parasites (hemiparasites) on aboveground stems of woody plants. Each plant has flowers of both sexes or of only one sex (only one sex in our area). Their *jointed stems are brittle* and often break off and fall during winds and rains. *Leaves linear to oval and opposite,* though in some species found elsewhere they are small and scale-like. *Flowers tiny, in clusters, sunken into the stem.* Fruit is a shiny spherical berry.

This is yet another family of plants with tiny unspectacular flowers, included here because of their unusual growth habit. All parts of mistletoe plants are toxic to humans and other mammals. At low doses, symptoms include abdominal pain and diarrhea; higher doses produce more intense pain and sometimes death.

Their attractive fruits are fed upon by birds such as the phainopepla (see next page). The seeds are not digested and simply pass through birds unharmed. When birds defecate onto plant stems, they deposit seeds that germinate and grow in place; thus mistletoes are bird-distributed.

About 450 species worldwide, 20 in California, 3 in our area. They are easiest identified by determining the identity of the host plant.

Dense Mistletoe

Phoradendron bolleanum (Seem.) Eichler
[*P.b.* ssp. *densum* (Trel.) Wiens; *P. densum* Trel.]

An upright plant, stems hairless, 30-60 cm long, green to olive-green. Leaves (5-)10-25 mm long, very narrow, only 2-8 mm wide. In general, its leaves are 3 times (or more) longer than wide. Fruit hairless, about 4 mm across, white, straw-colored, or pinkish. In flower June-Aug.

Commonly found on **California juniper** (*Juniperus californica* Carrierre, Cupressaceae). This juniper is uncommon in our area, known from Richard and Donna O'Neill Conservancy, Rancho Mission Viejo, San Juan Canyon within Caspers Wilderness Park, San Juan Loop Trail, Long Canyon, and sparingly along North Main Divide Road "Loop". The mistletoe has not been recorded from junipers those areas. Both California juniper and dense mistletoe are found in the Temescal Valley and on the Gavilan Hills just east of our area. It is also known to parasitize cypresses and may occur on **Tecate cypress** (*Hesperocyparis forbesii* (Jeps.) Bartel, Cupressaceae), which lives in Coal and upper Fremont Canyons. Thus, this mistletoe may occur in the Santa Ana Mountains and is yet undetected.

The genus was named in 1848 by T. Nuttall for its habit of stealing from trees (Greek, phor = thief; dendron = tree). The epithet was given in 1856 by B.C. Seemann, who first collected the plant in the Sierra Madre of Durango, Mexico, in 1853 and named it in honor of his friend, C.A. Bolle.

Dense Mistletoe on California Juniper

Large-leaved Mistletoe

Phoradendron serotinum (Raf.) M.C. Johnston
ssp. ***macrophyllum*** (Engelmann) Kuijt
[*P. macrophyllum* (Engelmann) Cockerell]

Upright, spreading, or hanging, up to about 1 meter long, it often grows in long heavy masses. The youngest green stems are covered in short hairs. Leaves broad, yellow-green, shiny 30-43 mm long, 15-23 mm wide, reverse-oval to rounded, hairless or slightly hairy. Fruit 4-5 mm wide, white, sometimes pink-tinged. In flower Dec-Mar.

Commonly found along watercourses, primarily on **California sycamore** (*Platanus racemosa,* Platanaceae). Also occurs on other woody trees and shrubs including alder, ash, walnut, cottonwood, and willows. *Never found on oaks.*

The epithet was given in 1820 by C.S. Rafinesque-Schmaltz for its cold-season flowering period, in winter-early spring (Latin, serotinus = late ripe, backward). The subepithet was given in 1878 by G. Engelmann for its large leaves (Greek, makros = long, large; phyllon = leaf).

Large-leaved Mistletoe on California Sycamore

Hairy Mistletoe or Oak Mistletoe ☠

Phoradendron serotinum (Raf.) M.C. Johnston
ssp. *tomentosum* (DC.) Kuijt [*P. villosum* (Nutt.) Nutt.]

An upright gray-green to yellow plant up to about 80 cm long, the leaves and youngest stems densely covered in short hairs. Leaves gray-green, dull (not shiny), 15-47 mm long, 10-25 mm wide, reverse-oval to elliptical. Fruit 3-4 mm wide, pinkish-white, with short hairs at its tip. In flower July-Sep.

Relatively uncommon here. Found in oak woodlands, where it lives almost exclusively on oaks, sometimes on nearby plants such as chamise, manzanita, sugar bush, and California bay laurel. In our area, it lives primarily on **coast live oak** (*Quercus agrifolia*) and **California scrub oak** (*Quercus berberidifolia,* both Fagaceae).

The subepithet was given in 1830 by A.P. de Candolle for its hairy leaves and young stems (Latin, tomentosus = densely covered with wool or short hair).

Hairy Mistletoe on California Scrub Oak

THE MISTLETOE GUILD
Western Great Purple Hairstreak
Atlides halesus estesi Clench (Lycaenidae)

A stunning butterfly whose beauty cannot adequately be captured by a photograph; you must see it in natural sunlight! *Its basic colors are black and gray, overlain with metallic blue, dotted with red beneath.* On the upper surface of the wings, males have more extensive blue scaling and a black patch (stigma) on each forewing. Females are slightly larger (wingspan 20-22 mm) than males (18-20 mm).

Larvae feed primarily on large-leaved mistletoe in California sycamores along watercourses, less commonly on dense mistletoe. When mature, larvae crawl down the trunk to pupate on and beneath bark. Early-season individuals pupate for a few weeks then emerge as adults. Late-season individuals spend the winter in the pupal stage and emerge the following spring. In flight late Mar-early Oct.

Look for adults nectaring on flowers in areas where mistletoe is common, such as along the Santa Ana River, parts of Fullerton, Orange, and Villa Park, Trabuco Canyon through Arroyo Trabuco, and San Juan Creek. Favorite flowers include scale-broom (*Lepidospartum squamatum,* Asteraceae) and woolly-fruited milkweed (*Asclepias eriocarpa,* Apocynaceae).

The subepithet was given in 1940 by H.K. Clench in honor of F. Estes, who collected the original specimens in his home town of Riverside.

Male

Female

Phainopepla
Phainopepla nitens (Swainson, Ptilogonatidae)

This lovely bird feeds on mistletoe fruit, recorded at up to 1,100 of them a day! *Males are glossy black with white wing patches,* easiest seen in flight. *Females are similar but are grey instead of black. Both sexes have red eyes.* Their call is a sweet whistle, though they have several other calls and are known to mimic many other birds. Look for them anywhere you find any species of mistletoe in fruit. We've seen them in many places such as Blackstar Canyon, Limestone Canyon, Silverado Canyon, Tucker Wildlife Sanctuary, and near Blue Jay Campground.

The genus was established in 1858 by S.F. Baird for its attractive appearance (Greek, phaino = to show; peplos = a robe or coat). The epithet was given in 1838 by W.J. Swainson for its shiny feathers (Latin, nitens = shining).

Phainopepla

VITACEAE • GRAPE FAMILY

Woody vines that climb with the help of modified flowerless inflorescences (*tendrils*) arranged opposite the leaves. *Stems are swollen at the nodes. Leaves alternate, simple or palmate, often large.* Calyx is a simple doughnut-like rim or has 4-6 lobes. *Corolla 4-6 lobed,* with separate petals or fused together at the far end to form a cap. *Fruit is a round fleshy berry with 4 hard seeds,* each with a round bump or pit on its side called a *chalazal knot*. Many types of grapes are cultivated and used for table grapes, raisins, and wines. There are about 800 species worldwide, 4 in California (2 native, 2 exotic), and only 1 in our area.

Desert Wild Grape
Vitis girdiana Munson

Long climbing vines, the older woody stems with self-shredding brown bark. *Stems and leaf undersides are densely covered with white fuzzy hairs. Leaf blade 5-16 cm wide, oval to heart-shaped with toothed edges, sometimes 3-lobed.* Calyx is a simple green rim with no lobes. *The cap-like corolla has 6 yellowish linear lobes, fused together at their tip; the entire cap falls away when the flower opens. Flowers of this grape are of only one sex; both types of flowers appear on the same plant.* Male flowers have 6 stamens, female flowers have a single short style and clublike stigma. The nectary forms a ring around the base of the ovary. The black edible fruit is 3-6 mm wide, usually covered with a thin layer of whitish powder. The chalazal knot on the pear-shaped seed is a sunken pit. In flower May-June. Fruits ripen during summer and early fall; they taste tart to sweet when ripe.

Very common along watercourses, springs, and seeps from our deserts to the southern California coastline, where it climbs over shrubs and up into trees, often forming dense blankets. Look for it along the Santa Ana River channel and washes such as at Green River and Featherly Regional Park. Also abundant at Tucker Wildlife Sanctuary and Audubon Starr Ranch Sanctuary and in our major canyons such as Silverado, Trabuco, Holy Jim, San Juan, and Hot Springs.

A lovely sight in fall, when its leaves turn yellow-gold to red and drift to the ground, leaving its bare stems draped over trees and shrubs.

The genus was established in 1753 by C. Linnaeus for its growth form (Latin, vitis = vine). The epithet was given in 1887 by T.V. Munson in honor of H.H. Gird, a rancher in Fallbrook who first brought the plant to his attention.

*Desert wild grape often covers waterways and hillside seeps, as it does here in Trabuco Canyon (right and below). In the photograph at right, the grape has climbed and covered a California sycamore (*Platanus racemosa, *Platanaceae).*

MONOCOTS

Terrestrial or aquatic flowering plants with a variety of growth forms. Cotyledons 1 per seed; after seed germination, it usually serves as an early stem, a conduit to a food reserve, or remains within the seed (but is never a storage organ). In most species, the *leaf veins are parallel* or penni-parallel. *Leaf base usually with a sheath that wraps around the stem. Flower parts generally in multiples of 3.* Pollen aperture 1 per pollen grain.

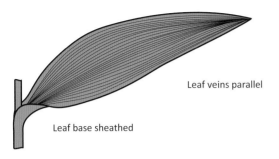

Leaf veins parallel

Leaf base sheathed

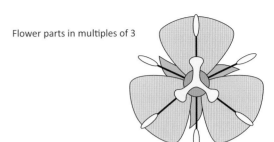

Flower parts in multiples of 3

AGAVACEAE • AGAVE FAMILY

Small to gigantic *perennials* with *dense spiraling rosettes of leaves at base or at branchtips.* Underground, some grow from a tough central crown, others from a bulb or rhizome. *Leaves are long-lance or triangle-shaped;* pliable, fibrous, and/or fleshy. *Sepals and petals similar* (tepals), with many easily seen veins. *Stamens 6,* slender or broad and fleshy. *Ovary superior* (or half-inferior), with 3 internal chambers. Nectaries are embedded exteriorly, within the tissue separating the ovary chambers. *Single style, 3-branched at the tip.* Fruit a capsule; seeds globular or flat, seed coat strongly bonded to the seed. A small family of 637 species; 18 in California, 3 in our area.

Soap Plants: *Chlorogalum*

Geophytes from an underground bulb. *Leaves are pliable and not spine-tipped;* their edges are wavy. Flowers thin and pliable. Stamens 6, filaments long, *anthers* large and *face inward* toward the center of the flower. Pollen is released from each of 2 anther sections through 2 separate slits. The single style is long, tipped with 3 slender lobes. *Seeds globular* (not flat), black. Of the 5 described species, 2 occur in our area. **California mule deer** (*Odocoileus hemionus californicus* (Caton), Cervidae, not shown) often eat the inflorescence, leaving only the leaves and a cut-off stalk

Below: Wavy-leaved soap plant with bulb and roots exposed. Each bulb is up to the size of a baseball and is clothed in a shaggy tunic of dark brown fibers.

Wavy-leaved Soap Plant, bulb exposed

Below: Wavy-leaved soap plant with young inflorescence and buds. Deer commonly eat them in this stage.

Wavy-leaved Soap Plant in bud

Small-flowered Soap Plant

Chlorogalum parviflorum S. Watson

Leaves 10-20 cm long, 3-9 mm wide. Inflorescence stalk 30-90 cm tall. *Two or more flowers per node. Tepals 7-8 mm long, strongly curved backward, white or pinkish, often with a green center stripe. The style is longer than the tepals.* In flower May-June.

A day-bloomer. Its flowers open early in the morning (5-6 a.m.) and close in late afternoon (4-5 p.m.). Fairly uncommon in our area, generally found among openings in coastal sage scrub and chaparral. Known along the coast at lower Shady Canyon and Niguel Hill in the San Joaquin Hills, the Dana Point Headlands, upper Aliso Creek, the Santa Ana Mountains such as in Blind and Baker Canyons, along Harding Truck Trail, ridges above Fremont Canyon, Caspers Wilderness Park; Alberhill, the San Mateo Canyon Wilderness Area, and the Santa Rosa Plateau.

The epithet was given in 1879 by S. Watson for its small flowers (Latin, parvus = little, small; floris = flower).

Similar. Wavy-leaved soap plant (*C. pomeridianum*): taller; flowers open mostly at night; one flower per node; tepals long and narrow; style shorter than or equal to tepals; common and widespread.

Wavy-leaved Soap Plant, Amole

Chlorogalum pomeridianum (DC.) Kunth
var. *pomeridianum*

Leaves 20-70 cm long, 6-25 mm wide. Inflorescence stalk 60 cm-2.5 meters tall. *One flower per node. Tepals 15-25 mm long, white, often with a green or pink center stripe, held open and gently curved backward. The style is shorter than or equal to the tepals.* In flower May-July.

Mostly blooms at night or in the evening. Its flowers open in late afternoon (about 3 p.m.) or evening and generally close soon after daybreak but may remain open longer in fog or under overcast skies. Once very common here; much of its coastal habitat lost to development. Widespread, known from such places as the Dana Point Headlands, San Juan Canyon, Santiago Canyon, Harding Canyon Truck Trail, and the San Mateo Canyon Wilderness Area.

The epithet was given in 1813 by A.P. de Candolle for the time of its flower opening (Latin, pomeridianus = in the afternoon).

Similar. Small-flowered soap plant (*C. parviflorum*): shorter; flowers open in daytime; two or more flowers per mode; tepals short and broad; style longer than tepals; uncommon.

Above: Small-flowered soap plant blooms from morning to late afternoon. These plants along the Harding Truck Trail were still open at 3 p.m.

Below: Wavy-leaved soap plant is abundant at the same location; its flowers are just starting to open. These photographs were made just minutes and meters apart. We find the same situation at Caspers Wilderness Park.

Western Yucca: *Hesperoyucca*

Persistent perennials from a tough central crown. *Leaves are stiff, fibrous, and tipped with a stout spine.* Tepals thick and waxy. *Filaments thick, club-shaped,* about 1.3 cm long, as long or slightly longer than the pistil. Each anther is forked and tipped with two separate pollinia. The style thick and fleshy, 3-branched at the tip. Ripe fruit splits open *between* the ovary chamber edges. In the similar genus *Yucca,* fruit splits open *at* the ovary chamber edges. Seeds flat (not globular), black, quite numerous. Three species, 1 in our area.

Chaparral Yucca

Hesperoyucca whipplei (Torr.) Trel. [*Yucca whipplei* Torr.]

Plants form a *densely leaved rosette* with no aboveground stems. *Leaves* 40 cm to 1 meter long, *stiff, gray-green, spine-tipped;* the edges have tiny sharp teeth that can easily cut your skin. The *inflorescence stalk grows rapidly*, often 2-4 meters tall. Flowers are 3 cm across and face outward or downward. The *broad tepals are white or yellowish, often purple-tinged*. Filaments thick and white. The two club-like anther sections are bright orange; each section bears a single pollinium. The 3-lobed green stigma has several clear fringes around its edges. In flower Apr-May.

The plants grow for about 5-8 years before they reach maturity. Then in winter they send up a single flower stalk that grows an amazing 10-15 cm a day. When the flowers open, the stalks brightly dot the hills and from afar look like lit candles. Each plant flowers only once in its life, then completely dies off after its fruits mature in summer. It occurs among coastal sage scrub and chaparral throughout our area, generally in sandy or rocky soils. Flower buds and young stalks of chaparral yucca are roasted and eaten. Tepals can be eaten raw.

The epithet was given in 1859 by J. Torrey in honor of A.W. Whipple. The plant was first collected by J.M. Bigelow along Cajon Pass near San Bernardino, California.

Chaparral Yucca

Similar. Chaparral beargrass (*Nolina cismontana*, Ruscaceae): leaves narrower, more flexible, pointed but not spine-tipped; flowers smaller; plants with flowers of only one functioning sex; wind-pollinated; fruit dry and 3-winged, each with 1-3 brown seeds; in flower Mar-June; sometimes grow together.

Chaparral Yucca

Chaparral Yucca

THE YUCCA GUILD
Yucca Moths: *Tegeticula, Prodoxus*

We have two types of yucca moths: true (*Tegeticula* spp.) and bogus (*Prodoxus* spp.; both are members of the family Prodoxidae). Both types lay their eggs in yucca plants, but only true yucca moths pollinate the yucca flowers afterward.

Southern California Yucca Moth
Tegeticula maculata extranea (H. Edwards)

A small moth with a wingspread of 14-19 mm. *Its body and front wings are completely black. The hind wings are grey, thinly scaled, and transparent.* Its *mouthparts are orange*, easily seen against the black body. It lives in Los Angeles, Orange, Riverside, and San Diego Counties. This species is active mostly during the day, while species in other areas are active at night. They are responsible for pollinating chaparral yucca. Without them, the plants could not set seed. Unable to reproduce, chaparral yucca would surely face extinction.

Adult yucca moths do not feed. Each of the maxillary palps (two finger-like projections near the mouth) of the female has a stout elongated extra appendage covered with bristles (easily seen with a hand lens). She uses these "tentacles" to pick up and hold a pollinium (pollen package) beneath her head, packing it in tightly with her mouthparts and forelegs. When the pollen is secured, she flies to another plant and alights on and enters a flower. She first investigates the condition of the ovary; it must be of the proper age (not too young or old) and must not already have had yucca moth eggs laid in it. If satisfied, she crawls up a stamen or the ovary, then walks backward until she sits on the ovary, positioned between two stamens. She uses a saw-like structure at the tip of her abdomen to pierce the ovary and deposit one egg into the nearest of the 3 ovary chambers. Then she withdraws the "saw," crawls up the ovary, places the pollinium on the stigma, and packs it down tightly with her mouthparts; this ensures fertilization and seed development. She then gathers another pollinium and repeats the process.

Larvae hatch in about one week and feed on a few of the seeds; most seeds remain uneaten and left to mature, which allows the plant to reproduce. As the fruits mature in summer or fall, the larvae tunnel out, crawl or fall to the ground, dig a few centimeters into the soil, and make a loose silken cocoon surrounded by a dense outer soil lining. The soil lining prevents the insect from drying out. Adults emerge from the cocoon the following spring, although if winter-spring rainfall is low they may remain in the pupal stage for years, a phenomenon called diapause. After emergence from the cocoon, the adults fly up to nearby yucca flowers, mate, and continue their life cycle.

The epithet was given in 1881 by C.V. Riley for the dark spots on the typical form of the moth (not found in our area), which is white with dark spots (Latin, maculatus = spotted). Our form was described in 1888 by H. Edwards for its unusual totally dark appearance (Latin, extraneus = strange, outside).

When their wings are closed, southern California yucca moths are about 9-10 mm long from head to wingtip. Bogus yucca moths are only about 4.5-7.5 mm long.

These two female moths are laying eggs inside different chambers of the same ovary. Note the orange pollinium held beneath the tentacles of each moth. After laying an egg, she will place the pollinium onto the stigma, ensuring pollination and fertilization of the ovules within the ovary.

Ash-colored Bogus Yucca Moth
Prodoxus cinereus Riley

Generally smaller than the Southern California yucca moth and often found with it. *Its wingspread is 9-15 mm. The entire body and front wings are light brown with no markings. Hind wings are paler, sparsely scaled, and partially transparent.* It is known from Los Angeles, Orange, Riverside, and San Diego Counties. Quite common on chaparral yucca throughout our area.

Bogus yucca moths are found in yucca flowers often in large numbers, though they do not pollinate the flowers. Females lack the mouth "tentacles" of the true yucca moths but do possess the same saw-like egg-laying device. They lay their eggs in the base of the ovary, not in the ovary itself. Their young tunnel within the plant and feed on the stalk, leaves, or fruit but not the seeds. *Bogus yucca moth larvae pupate within the yucca, while those of true yucca moths pupate within a cocoon in soil at the base of the plant.* When mature, the bogus yucca moth larva excavates a narrow chamber that faces outward and leaves in place a thin outer layer of yucca tissue, as a sort of lid. It pupates in that chamber, with its head facing the "lid." The top of the pupa has a ramrod-shape. When the adult is ready to emerge, it wriggles and pushes outward, using its "ramrod" to pop off the lid. It continues to wriggle until part of the pupal case protrudes from the plant; then the adult crawls out of the case, dries its wings, and flies off, leaving the empty pupal case sticking out of the plant. You can easily find these cases on dead yucca stalks.

The epithet was given in 1881 by C.V. Riley in reference to its color (Latin, cinereus = ash-colored).

Ash-colored Bogus Yucca Moth resting on the base of a stamen, shown for scale

Adults of bogus yucca moths (Prodoxus spp.) are much smaller than the filaments of chaparral yucca flowers, while adults of true yucca moths (Tegeticula spp.) are about as long as the filaments. Bogus and true yucca moths are often found on the same plant and even within the same flowers.

Bordered Bogus Yucca Moth
Prodoxus marginatus Riley
[*P. pulverulentus* Riley]

Typically a bit smaller than the ash-colored bogus yucca moth, adults have a wingspread of 8-12 mm. *The head, thorax, and front wings are covered with white scales. Females have a patch of dark scales at the tips of their front wings; these are scattered or absent in males.* A second color form occurs here as well, formerly considered a separate species (*Prodoxus pulverulentus*). Those females have several dark scales scattered throughout the forewing (mostly on the outer half), giving it a gray shading. Males often have a gray streak parallel with the outer edge of the front wing. In both color forms, the hindwings are light to medium gray.

Known from central California south into Mexico. The white color form is known in the northern portion of the range; the darker color form increases in abundance southward. Here it is known in the Santa Ana Mountains, mostly at higher elevations, where it sometimes co-occurs with the ash-colored bogus yucca moth.

The epithet was given in 1881 by C.V. Riley for the dark markings at the edge of the forewings (Latin, marginis = edge, border; -atus = provided with).

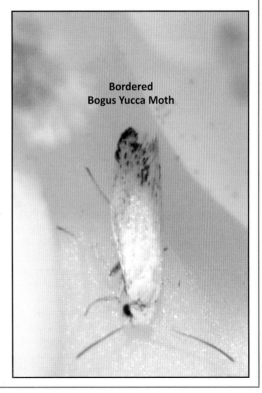

Bordered Bogus Yucca Moth

Yucca Weevil
Scyphophorus yuccae Horn (Curculionidae)

Weevils are unusual slow-moving beetles with part of their face modified to form a snout tipped with a pair of tiny mandibles. Their antennae are attached to the snout, in front of the eyes, made up of a single long segment and an "elbow joint" followed by smaller swiveling segments that end in a club. Ants and some tiny wasps also have these "elbowed" antennae.

One of the largest weevils in the state, the yucca weevil is black, 12-20 mm long, flattened front to back, with tiny pits on the thorax and grooves that run down the length of the front wings. It eats *only* chaparral yucca. Adults begin to appear on the plants in late March (January) through July or so. They mate on the plant, and females lay eggs at the leaf bases. Larvae tunnel through the green flowering stalk, mostly near the base, and feed on the tissue within. Before the stalk dies off, they bore out, crawl or fall to the ground, dig into the soil, and pupate. Adults emerge the following year during spring and summer. To feed, adults bite into the plant and ingest its sap. They are often quite abundant among the leaf bases.

To see them, gently press the leaves down with a stick (not your hands) and watch them scurry about. They are harmless and can be held in your hand; however, their legs are strong and can tightly grasp your skin with their hook-tipped feet. For this reason, very young children probably should not hold them.

The epithet was given in 1873 by G. Horn in reference to the host plant.

Yucca Weevil

Spotted Leaf Beetle
Pseudoluperus maculicollis (LeConte, Chrysomelidae)

A *small beetle*, about 7 mm in length, with striking coloration. *Its body and legs are black, the face red-orange, the thorax red-orange with a central black stripe. Front wings are a brilliant metallic blue-green that glistens in sunlight.* Commonly found in great numbers on blooming chaparral yucca, especially at higher elevations of the Santa Ana Mountains, such as near the junction of the Coldwater and Bear Canyon Trails.

A member of the leaf beetle family, known for chewing on all plant parts. Most leaf beetle larvae feed on leaves, while those of some species feed internally within a plant's roots, stems, or leaves. The larvae of this species feed on the tissue between the leaf veins; they leave the veins intact. Eaten leaves look like skeletons of their former selves. Insects that feed in this manner are said to be leaf skeletonizers. We have not observed the larvae to eat chaparral yucca; we doubt they would eat its very tough leaves. Elsewhere larvae are known to feed on a member of the pea family (Fabaceae) that does not live in our area. Associated plants at the locality above include three likely candidates: **chaparral sweet pea** (*Lathyrus vestitus* var. *vestitus,* Fabaceae), **California lilacs** (*Ceanothus* spp., Rhamnaceae), and **slender sunflower** (*Helianthus gracilentus,* Asteraceae). Adults of this species seem to collect nectar from the nectaries on the ovary of chaparral yucca but may be eating sap secreted by the plant after yucca moths oviposit.

The epithet was given in 1884 by J.L. LeConte for the spots on its thorax (Latin, macula = spot; collum = neck).

Spotted Leaf Beetle

ALISMATACEAE • WATER-PLANTAIN FAMILY

Most are perennials, a few annuals. *Aquatic or semi-aquatic, from thickened roots or shoots below or above ground, in standing water or in mud.* Roots shallow, fibrous. Stem very short. *Leaves simple, basal,* often long petioled; *palmately veined.* Submerged leaves often different from emergent (held out of the water) and terrestrial (land) leaves. Plants monoecious with bisexual and unisexual flowers or dioecious with unisexual flowers. *Inflorescence often branched, flowers whorled along the branches.* Sepals 3, green. *Petals 3, white or pink.* Stamens and pistils 6 to many. Fruit heads often spherical, fruit a beaked achene. A family of about 100 species, mostly tropical, 11 in California and 4 in our area.

Northern Water-plantain
Alisma triviale Pursh [*A. plantago-aquatica* L., misapplied])

A *perennial* herb. Leaves 7-45 cm long, blade 5.5-15 cm long, the base truncate or somewhat lobed. Inflorescence usually much longer than the leaves. Flowers bisexual. *Petals 3-5 mm long, white to pinkish,* often cut or jagged-edged. Stamens 6. Pistils many; *styles upright, tips curved outward. Fruit head doughnut-shaped,* fruit body 1.5-3 mm long, with a tiny beak on its outer side. In flower spring-fall.

Last reported from Los Alamitos on July 21, 1908; probably extirpated. A small population was discovered in the San Mateo Canyon Wilderness Area, along the San Mateo Trail about halfway between Bluewater and Clark Trails, by Ron Vanderhoff on September 5, 2011.

The epithet was given in 1814 by F.C. Pursh for its abundance or wide distribution (Latin, trivialis = common, ordinary, found everywhere).

Similar. Tule-potato (*Sagittaria latifolia* Willd.): perennial; leaf blades arrowhead-shaped, basal lobes large. River Road bridge over the Santa Ana River in Corona. **Montevideo arrowhead** (*Sagittaria montevidensis* Cham. & Schltdl. ssp. *calycina* (Engelm.) Bogin): Like tule-potato but annual. Found once in Twin Lakes Park in Garden Grove in 1978.

Upright Burhead
Echinodorus berteroi (Spreng.) Fassett

An *annual* or biennial herb. *Leaves 8-30 cm long; submerged blades linear, very long, frail; emergent and terrestrial blades 6-14 cm long, broadly elliptic to heart-shaped.* Inflorescence upright, many-branched, usually longer than the leaves. Flowers bisexual, 1-3(4) per whorl. Petals 6-9 mm long, white, ruffled or smooth-edged. Stamens 9-15. Pistils many. *Fruit heads spherical; fruit body* 1.5-3 mm long, *tipped with a long beak, the fruit head spiny and bur-like.* In flower midsummer-fall.

A plant of wet ponds and lake margins, mostly at low elevations. Once widespread and common, extirpated from many of its former locations. Old records for Laguna Lakes (1918, 1924), Wintersburg Channel (1975), and Lake Elsinore (1891). Recent records only for Delhi Channel just above upper Newport Bay, Talbert Regional Park in Costa Mesa, and Irvine Lake, where it is abundant.

The epithet was given in 1825 by C.P. Sprengel for C.L.G. Bertero.

Similar. See above.

ALLIACEAE • ONION FAMILY

These are true onions with a characteristic *onion odor* due to sulfur compounds within them. They are all perennials from an underground bulb, surrounded by a sheath made from dry leaf bases of past years. Leaves are basal, long but narrow, and V-shaped, cylindrical, or flat in cross-section. *The inflorescence is a leafless stalk topped by an umbel with bracts at its base.* The pedicel of some species is jointed where it meets the flower. Individual flowers have no bracts. *Each flower has 6 tepals* (3 sepals and 3 petals), generally white (some striped), magenta, or purple. Nectaries are born on the sepals. Stamens 6, their filaments free or fused to the tepals. All stamens are fertile; their anthers are all filled with pollen. *Ovary superior,* the tip of the style clubbed or 3-branched. Fruit is a capsule filled with black crescent-shaped seeds.

The family has about 600 species worldwide, 51 in California, and 6 in our area. Onions are closely related to the brodiaea (Themidaceae) and amaryllis (Amaryllidaceae) families. Our native species are all in the genus *Allium,* named in 1753 by C. Linnaeus, who used the ancient Latin name for garlic (Latin, allium = garlic). Cultivated members of the family include onion, garlic, leek, and chive.

Onions can be a challenge to identify, as many species cannot be easily separated without close examination of the ovary and microscopic examination of the bulb skin. Fortunately, our local species are fairly simple to identify from easily seen characters. They are all geophytes.

	Red-skinned *Allium haematochiton*	**Pitted** *Allium lacunosum*	**Munz's** *Allium munzii*	**Early** *Allium praecox*	**San Bdno Mtn** *Allium monticola*	**Red-flowered** *Allium peninsulare*
Leaves	3-6, narrow, round	2-3, slightly cylindrical	1, narrow, cylindric	2-3, broadly channeled or keeled	1, cylindric	2-3, channeled to nearly round
Tepals	Broadly oval, tips rounded or sharp; mostly white to pale rose with a dark reddish midvein	Narrowly to broadly oval, tips mostly rounded; white to pale pink with a dark midvein	Broadly oval; white, often tinged with rose or red in age, with a red or green midvein	Narrowly oval, tips pointed; white to pale rose or lavender, midvein purple	Tubular, long and narrow, tip curved backward; pink to dark rose-purple	Tubular, long and narrow, tip curved backward; rich magenta or red-purple, inner tepals shorter
Stamens	Nearly as long as tepals	Shorter than tepals	About equal to tepals	Nearly as long as tepals	Much shorter than tepals	Much shorter than tepals
Ovary	6 short rounded pink crests densely covered with tiny "warts"	3 small, mostly rounded pinkish crests, each 2-lobed, often with minute dense covering of "warts"	6 prominently, triangular green crests, each with fine teeth	3 minute rounded and smooth-surfaced pink crests	6 prominently linear to triangular crests; ovary and crests are either white or green	3 minute low "bumps," each 2-lobed and smooth-surfaced; ovary and crests are either white or green

Allium haematochiton

Allium lacunosum

Allium munzii

Allium praecox

Allium monticola

Allium peninsulare

Red-skinned Onion
Allium haematochiton S. Watson

A *bunching onion* that grows from a cluster of oblong bulbs with reddish papery bulb skins. There are *3-6 narrow leaves,* round in cross-section. Flowers are arranged in an umbel at the top of a leafless stalk. Each flower 6-8 mm long, shaped like an open bell. *Tepals* broadly oval, tips rounded or sharp, *mostly white to pale rose with a dark reddish mid-vein.* Stamens and style nearly as long as the tepals. Unopened anthers are pink or pale brown. The stigma is undivided. *The ovary has six short rounded pink crests densely covered with tiny "warts," each 6-8 mm long.* In flower Mar-May.

Found in heavy clay soils on the grassy hills near UC Irvine and in foothills of the Santa Ana Mountains at Caspers Wilderness Park. Also found in the Santa Ana Mountains, especially in the north, where it is known from Gypsum, Coal, and Fremont Canyons, and Sierra Peak where it is patchy but often fairly abundant when found. Common on Maple Springs Saddle, from Elsinore Peak southward, on the Santa Rosa Plateau, in the San Mateo Canyon Wilderness Area, and in the Temescal Valley on the east slope of the mountains. In cultivation as an easily grown garden accent that prefers moist soils.

The epithet was given in 1885 by S. Watson in reference to the color of the bulb skin (Latin, haematos = blood; Greek, chiton = an outer covering).

Similar. The number of leaves and clustering habit generally distinguish this onion from other species in this area. **Munz's onion** (*A. munzii*): has only *a single leaf* and a prominently crested ovary with minute teeth. Its single leaf is not always immediately evident because it sometimes grows in dense clusters and is thus easily mistaken for red-skinned onion in a casual look. The ovary crests on Munz's onion have fine teeth.

Pitted Onion
Allium lacunosum S. Watson **var. *lacunosum***

A *slender plant* that grows from a brown-coated ovoid bulb. It has *2-3 slightly cylindrical leaves with sheathing bases.* Flowers are in dense umbels at the end of a leafless stalk. Pedicels 5-15 mm long. *Flowers 6-9 mm long, often bell-shaped and white to pale pink with dark midvein.* Tepals narrowly to broadly oval, most often with rounded tips. Stamens shorter than the tepals; unopened anthers yellow or purple. *The ovary has 3 small, mostly rounded pinkish crests, each 2-lobed, these often with minute dense covering of "warts."* In flower Apr-May.

In our area, known from the rocky, heavy clay soils of Elsinore Peak and the north slope of Miller Mountain, where it is local but fairly common in native grasslands that border chaparral.

The epithet was given in 1879 by S. Watson for the squarish pits on its bulb skin (Latin, lacuna = ditch or pit).

Similar. Munz's onion (*A. munzii*): *a single cylindrical leaf* and prominent ovary crests with minute teeth; often grows together. **Red-skinned onion** (*A. haematochiton*): *typically found in dense clusters; its 3-6 leaves are round in cross-section.* **Early onion** (*A. praecox*): very similar but a more open umbel with pedicels over 15 mm long and broader, mostly broadly channeled or keeled leaves. The ovary crests of early onion are smooth while those of pitted onion have minute "warts," but the warts can be very hard to see and do not appear to be present on all plants.

Munz's Onion ★
Allium munzii (F. Ownbey & Aase ex Traub) D. McNeal

A *very slender plant* that grows from a reddish bulb and *produces a single stem*. It has only *a single narrow, cylindric leaf per stem, which can be its only aboveground evidence in dry years*. The flowers are in *dense umbels* at the end of a leafless stalk. *Each flower* is 6-8 mm long and *bell-shaped with broadly oval white (rarely pink) tepals, which often become tinged with rose or red in age. The tepals have a red or green midvein*. Stamens are about equal in length to the tepals; unopened anthers are yellow. The stigma has three linear lobes, which are often long and recurved. *The ovary has 6 prominent triangular green crests, each with fine teeth*. In flower Apr-May.

Found in heavy, often rocky, *clay soils* of grasslands and openings of coastal sage scrub on Elsinore Peak and in native grasslands and openings of chaparral in the Temescal Valley as at Indian Canyon and Alberhill near Lake Elsinore. The type locality at the mouth of Indian Canyon has been greatly reduced or eliminated by residential development and construction of the I-15 Freeway. It is listed by the State of California as Threatened and by the U.S. Fish and Wildlife Service as Endangered, primarily because of habitat loss. CRPR 1B.1. It is endemic to Riverside County.

The epithet was given in 1972 by F.M. Ownbey and H.C. Aase in honor of P.A. Munz.

Similar. This is the only onion in our area with distinctive fine teeth at the crest of the ovaries. In overall appearance it can be mistaken for **Pitted onion** (*A. lacunosum*), **Red-skinned onion** (*A. haematochiton*), and **Early onion** (*A. praecox*), all of which have two or more leaves.

Munz's Onion

Munz's Onion

Early Onion
Allium praecox Brandegee

A *upright, single-stemmed onion* from a spherical bulb. *The 2-3 leaves are broadly channeled or somewhat keeled*. Flowers are in *open umbels* at the end of a leafless stalk. The pedicels are 15-40 mm long. *Each flower* is 8-12 mm long, *tepals white to pale rose or lavender, spreading and narrowly oval with somewhat pointed tips. The tepal midveins are usually purple*. Stamens are nearly as long as the tepals; unopened anthers are yellow, pink, red, or purple. The stigma is undivided or slightly lobed. *The ovaries have 3 minute rounded and smooth-surfaced pink crests*. In flower Mar-May.

Found in moist draws and grassy slopes in native grassland and coastal sage scrub in the Chino Hills, San Joaquin Hills, mountain foothills, and Santa Ana Mountains. See it in Brea Canyon, Santa Ana Canyon, UC Irvine Ecological Preserve, Coyote Canyon, just west of Gypsum Canyon, Lomas de Santiago, Richard and Donna O'Neill Conservancy, and Coal, Silverado, San Juan, and Lucas Canyons. In spring it is quite numerous in Laguna Coast Wilderness Park.

The epithet was given in 1906 by T.S. Brandegee for its early flowering period (Latin, praecox or praecocis = early, premature).

Similar. **Red-skinned onion** (*A. haematochiton*): frequently has more than three cylindric leaves; it also grows in clumps of plants. **Munz's onion** (*A. munzii*): has a single leaf and prominent ovary crests with fine teeth. **Pitted onion** (*A. lacunosum*): has a denser umbel with pedicels shorter than 15 mm, cylindric leaves, and often minute "warts" on the ovary crest.

Early Onion

San Bernardino Mountains Onion
Allium monticola A. Davidson

A short onion mostly less than 20 cm tall, from a gray-brown bulb. The *single leaf* is cylindric in cross-section. Flowers are in dense umbels at the end of 1-3 leafless stalks. *Each flower is 12-19 mm long, tubular, pink to dark rose-purple.* Tepals are long and narrow, often slightly curved backward at the tip. Stamens are much shorter than the tepals; unopened anthers are purple. The stigma is slightly head-like or smoothly narrowed to the tip. The ovary has 6 prominently linear to triangular white or green crests. In flower late May-July.

It lives in openings of chaparral on loose talus slopes of the upper Santa Ana Mountains, especially on the western arm of Modjeska Peak.

The epithet was given in 1921 by A. Davidson for its habit of living in mountains (Latin, monticulus = mountains).

Similar. Red-flowered onion (*A. peninsulare*): taller plant; magenta to red-purple flowers; clay to rocky soils at lower elevations.

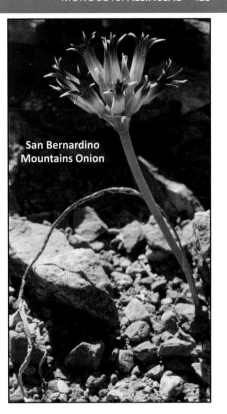

San Bernardino Mountains Onion

Red-flowered Onion
Allium peninsulare Lemmon **var.** ***peninsulare***

This onion grows from brown to gray-brown bulbs. The *2-3 leaves* are channeled to nearly round in cross-section. The flowers, in dense umbels at the end of a leafless stalk, are 10-15 mm long, bell-shaped with tepals held upright. *Flowers are a rich magenta or red-purple color.* The tepals are unequal: inner tepals are narrow with pointed, backward-curved tips, outer tepals are broadly triangular with rounded or short-pointed tips. Stamens are much shorter than the tepals; unopened anthers are pinkish to yellowish. The ovary has 3 minute, low, white or green crests, each 2-lobed and smooth-surfaced. The stigma is 3-lobed, sometimes head-like. In flower Mar-June.

Found in clay soils among openings in chaparral, coastal sage scrub, and grassland, often on rocky slopes. Rarely seen in Orange County, where it is found along the crest of the northern Santa Ana Mountains above Silverado Canyon and in Santa Ana Canyon. Fairly widespread and common on Elsinore Peak and along the eastern slopes of the Santa Ana Mountains in the Temescal Valley. Formerly abundant in grasslands up to the forest gate along Bedford Canyon Road, but that population was apparently destroyed in 2004-2005 by construction of a large housing development and alteration of the forest entrance.

The epithet was given in 1888 by J.G. Lemmon for the location of its discovery, on the peninsula of Baja California (Latin, peninsularis = growing on a peninsula).

Similar. San Bernardino Mountains onion (*A. monticola*): shorter plant; pink to purplish flowers; talus slopes at higher elevations.

Red-flowered Onion

Below, left: San Bernardino Mountains onion; right, Red-flowered onion. The former lives in deep rocky talus slopes, the latter in clay soils. We know of no sites in our area where they grow together.

San Bernardino Mountains Onion

Red-flowered Onion

ASPHODELACEAE • ASPHODEL FAMILY

A small family of *herbs*, though *some are woody shrubs and trees. Leaves basal in most species,* succulent and aggregated at the tips in woody species. Inflorescence is upright, single-stalked or branched, often many-flowered. *Flowers flat or tubular; sepals and petals similar,* all free (*Asphodelus*) or fused into tubes (*Aloë*). Stamens 6, all similar or of 2 lengths. Ovary superior, style long, fruit a capsule, each seed has an aril.

No California natives; the species here have escaped from cultivation or are remnants from old homesteads.

Soap Aloe ✖

Aloë maculata All. [*A. saponaria* (Ait.) Haw.]

An upright, succulent perennial to 0.2 meters tall, unbranched or branched. Sap yellow. *Leaves* 10-30 cm long, thick, *triangular, broad-based; upper surface green to purplish with whitish spots; marginal teeth brown to red.* Inflorescence upright, 10-30 cm long, densely flowered; lowest flowers open first, droop at opening. *Flower* 2.5-4 cm long; *tubular, narrow just past the ovary, the tip with 6 free tepals; orangish to pinkish, sometimes yellow.* Stamens 6, of 2 lengths. In flower Mar-Apr.

Native to South Africa. Known from the Dana Point Headlands and old mining homesteads in Lucas Canyon, probably elsewhere along the coast. Pollinated here by hummingbirds, in its native lands by sunbirds.

The epithet was given in about 1800 by C. Allioni for spots on its leaves (Latin, maculatus = spotted).

Similar. Candelabra aloe ✖ (*A. arborescens* Mill.): *tree-like shrub,* 2-2.5 meters tall, with main stem or trunk; *leaves* to 60 cm long, thick, grey-green to bluish, marginal teeth white to reddish; *flower deep orange with yellow* and/or green throat. Known from lower Big Canyon in upper Newport Bay and old homesteads in Lucas Canyon. In flower May-July. **Chabaud's aloe ✖** (*A. chabaudii* Schonl.): *perennial* up to 0.5 tall; sap clear; little or no visible stem; *leaves relatively thin, very broad-based, grey-green to bluish, sometimes tinged or entirely pink, faintly spotted or spotless; marginal teeth white; flower pink, peach, or salmon.* Known from old homesteads in Lucas Canyon. In flower Feb-Mar.

Hollow-stem Asphodel, Onion Weed ✖

Asphodelus fistulosus L.

An onion-like plant but with no onion odor. Annual or short-lived perennial, 15-70 cm tall, from a shallow stem base (not a bulb) and numerous tough roots. *Leaves cylindrical in cross-section, hollow, numerous.* Flowering stalk upright (not an umbel), many-flowered. Each flower faces out and downward, closes up at night, opens in bright sunlight. *Tepals free, all similar, white to pinkish, held straight out. Each tepal with a central stripe that is reddish or red-brown, rarely white.* Stamens 6, of two lengths; mostly held toward lower side of flower; filaments broad, their bases covered with whitish hairs; anthers and pollen orange. Ovary hidden by the hairy filament bases, dull green, smooth-topped; style long. Stigma pink, 3-branched. In flower Mar-June.

An invasive plant of roadsides and grasslands. Native to the Mediterranean region. Known from Yorba Linda, Santiago Canyon, South Main Divide Road, and along roadsides from San Clemente to Camp Pendleton and San Onofre State Beach. Land managers should carefully confirm its identification before eradicating it.

The epithet was given in 1753 by C. Linnaeus for its hollow stems and leaves (Latin, fistulosus = full of pipes).

Similar. Onions (*Allium* spp., Alliaceae): *onion odor; flowers in a round-topped umbel.* **Soap plants** (*Chlorogalum* spp., Agavaceae): leaves flat, not hollow; stamens of equal length, filaments slender and hairless, anthers yellow or purple, pollen yellow or cream.

Soap Aloe
Fred M. Roberts, Jr.

Candelabra Aloe • Chabaud's Aloe

Hollow-stem Asphodel

IRIDACEAE • IRIS FAMILY

Perennials from rhizomes, or geophytes from bulbs and corms. Leaves grasslike, most basal, few along the stem, overlapping. *Each leaf is completely folded along the midrib; all those on a stem are flattened in the same plane.* One or more leaf-like bracts lie beneath each inflorescence. Flowers with 3 sepals and 3 petals, fused at their base into a narrow tube right above the ovary. The 3 stamens are attached to the sepals. In some, the filaments are fused into a tube that surrounds the pistil. *Ovary inferior with 3 chambers.* The single style is tipped with 3 lobes. In some irises, each of the 3 style lobes is itself 2-lobed and the stigma located on the underside of those lobes. In other species, the 3 styles are simple (not further divided). Fruit is a capsule.

About 1,500 species worldwide, mostly in Africa; 20 native to California, 1 in our area. Familiar cultivated garden plants include *Iris*, *Gladiolus*, *Crocus*, and *Freesia*.

Western Blue-eyed-grass
Sisyrinchium bellum S. Watson

A perennial from a compact rhizome, it stands about 10-60 cm tall. *Most stems have 1 or more leaf-bearing nodes.* Inflorescence of 2 similar bracts that hide its base, usually with (1-)2 or more peduncles. *Sepals and petals similar, oval, tipped with a slender point; deep blue, purplish, or white, often with darker veins and a yellowish blotch at the base.* Stamen filaments completely fused into a long tube, the anthers held tightly just below the stigma lobes; the entire structure looks like a club. In flower Mar-May.

A common plant of seasonally wet clay soils among grasslands, coastal sage scrub, and oak woodlands. More abundant in coastal areas, especially on north-facing slopes, sometimes in shade. Found throughout our area, including all of our wilderness parks; much less common on the inland side of the Santa Ana Mountains. Look for it in places such as Laguna Coast Wilderness Park, Caspers Wilderness Park, and the Chiquito Basin. In cultivation as a spring bloomer, it needs adequate water and often does best in partial shade.

The epithet was given in 1877 by S. Watson for its attractive appearance (Latin, bellus = neat, charming, handsome).

Most flowers are blue or purple; only rarely do we find white flowers.

LILIACEAE • LILY FAMILY

True lilies that generally lack an odor. All are perennials from bulbs that have contractile roots, some from bulb-like rhizomes. Leaves are generally narrow or grasslike, sometimes broad; V-shaped or flat in cross-section. Leaf arrangement is variable: some basal, others alternate, spiraled, or whorled along the stem. Stipules are absent. The inflorescence appears near the top of the plant, few to many branched, one to many flowered. The flowers are large and conspicuous. There are 3 sepals and 3 petals; each petal has a basal nectary that faces the interior of the flower. Stamens 6, each anther attached to a filament at its base or middle. Ovary superior with a single style that is undivided or 3-lobed. Fruit is a 3-chambered capsule. Seeds flat, pale-colored.

The large showy flowers of lilies have been valued for centuries as favorite garden plants and cut flowers. They are among the most beautiful of our local wildflowers.

A worldwide family of about 635 species, 90 species in California, 8 locally.

Mariposa Lilies, Butterfly Tulips: *Calochortus*

Geophytes from a fibrous- or membranous-coated bulb with a single basal leaf that often withers by flowering time. There are zero-to-several leaves on the stem, sometimes appearing basal. Bulblets are sometimes formed in the axil of the basal leaves. Inflorescence few-branched or with only a single flower. Flowers are large and showy, globose to broadly bell-shaped. The 3 sepals are narrower and smaller than the 3 petals, the former with a hint of color similar to the petals or greenish. Petals white, yellow, rose, or purple, often with splotches of darker color. Each petal has a conspicuous nectary gland at the base, usually surrounded by hairs; it often sits in a sunken pit that appears as a bump on the outside of the petal. Fruit is an elongate capsule, often narrowing to a point at each end but sometimes rounded; some have 3 ribs (wings) down their length. We have 6 species here.

The genus was established in 1814 by F.T. Pursh for its lovely flowers and grasslike leaves (Greek, kalos = beautiful; chortus = grass).

White Fairy Lantern
Calochortus albus

Catalina Mariposa Lily
Calochortus catalinae

Shy Mariposa Lily
Calochortus invenustus

Plummer's Mariposa Lily
Calochortus plummerae

Intermediate Mariposa Lily
Calochortus weedii var. *intermedius*

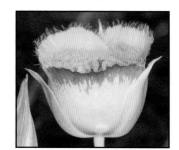

Weed's Mariposa Lily
Calochortus weedii var. *weedii*

Splendid Mariposa Lily
Calochortus splendens

California Chocolate Lily
Fritillaria biflora

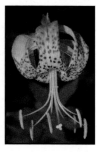

Ocellated Humboldt Lily
Lilium humboldtii ssp. *ocellatum*

White Fairy Lantern
Calochortus albus (Benth.) Douglas ex Benth.

This plant grows from a bulb with a membranous bulb coat. Stems 20-80 cm tall. The persistent basal leaf is strap-shaped, stem leaves short. *Inflorescence few-branched with 2-many nodding, delicate, ball-shaped closed flowers.* Each flower is associated with a long leaf-like bract. Sepals 10-15 mm long, green or white, held against the petals. *Petals elliptical,* 20-25 mm long, *white,* and often tinged green or reddish toward the base (all pink in central coastal California). The nectary, found at the base of the petal, is sunken and has several fringed membranes; the lower membrane covers 1/3-2/3 of the petal. *Inner face of the petal is hairy,* the hairs over the nectary long, white, and slender. Anthers are cream-colored. Fruit nodding and 3-winged, 20-40 mm long, about a third as broad. Seeds are irregular in shape. In flower Apr-June.

Found mostly on shaded, sometimes rocky, slopes in openings of chaparral and oak woodland. Known here only in the Santa Ana Mountains, where it is fairly widespread but seldom seen except following fire. It has been found in Mabey Canyon on Skyline Drive near Sierra Peak, Baker Canyon, upper Falls Canyon, along Holy Jim and Trabuco Trails, in San Juan Canyon, and on the Santa Rosa Plateau.

The epithet was given in 1835 by G. Bentham for its white flowers (Latin, albus = white).

White Fairy Lantern

Right: broad fruit of Catalina mariposa lily

Catalina Mariposa Lily

Catalina Mariposa Lily ★
Calochortus catalinae S. Watson

A geophyte from a bulb with a membranous sheath. Basal leaf narrowly strap-shaped; usually withers before the flowers open. Inflorescence few-branched with 1-4 flowers on stems 20-60 cm tall. Bulblets often form at the stem base. *The flowers are open, face upward, broadly bell-shaped with narrow bracts.* Sepals 20-30 mm long, greenish with purple spots near the base. *Petals white, sometimes tinged pink,* 2-50 mm long with a rounded or abruptly pointed tip. Base of the petal often dark maroon. Nectary oblong, densely covered with branched hairs and not sunken. *Petal face without hair.* Anthers pinkish. Fruit upright, oblong, broad, rounded in cross-section, 2-5 cm long, seeds flat and lens-shaped. In flower Feb-May, peaking in late Mar and early Apr.

Relatively uncommon but fairly widespread in our area. It occurs mostly in native grasslands and openings among coastal sage scrub and chaparral at low and moderate elevations. Found in the Chino Hills, Santa Ana Canyon, Crystal Cove State Park, Aliso and Wood Canyons Wilderness Park, Hicks and Bee Canyons of the Lomas de Santiago, Audubon Starr Ranch Sanctuary, Rancho Mission Viejo, Caspers Wilderness Park, and major canyons such as Gypsum, Fremont, Black Star, Trabuco, and San Juan. It is especially abundant following fire. A rare plant, it is ranked as CRPR 4.2.

The epithet was given in 1879 by S. Watson for the site of its original discovery, on Santa Catalina Island.

Catalina Mariposa Lily

Shy Mariposa Lily
Calochortus invenustus E. Greene

Geophyte from a membranous-coated bulb. Stem slender, twisted, 20-50 cm tall, bulblets form at its base. Single basal leaf linear, 10-20 cm long, withered when the flowers open. Inflorescence umbel-like, with 1-6 open, bell-shaped flowers. Sepals 2-3 cm long, greenish. Petals 2-4 cm long, white to pale pinkish-purple, with a distinctive green central stripe on the outer surface and a few hairs near the base of the inner surface. Nectary round, covered with branching hairs and surrounded by a fringed membrane. Anthers purplish or yellowish. Fruit upright, linear, 5-7 cm long. Seeds flat and lens-shaped. In flower May-June.

Mostly a foothill and mountain species, found here on sunny rocky openings in montane chaparral. Known from our area only from near the summits of Modjeska and Santiago Peaks. While not uncommon at these localities, it is rarely seen because few people visit these peaks while it is in bloom and its light-colored flowers are easily overlooked.

The epithet was given in 1890 by E.L. Greene for its flower, which to him was rather plain-looking when compared with others in its genus (Latin, in = not; venustus = beautiful). From the original description, "A homely species of the higher mountains to the westward of the Mojave Desert, where it was collected by the writer, June 25, 1889."

Splendid Mariposa Lily
Calochortus splendens Benth.

A geophyte from a bulb with a membranous coat. Basal leaves narrowly strap-shaped, usually withered before the flower opens. Stems 20-60 cm tall. Inflorescence with a distinct central axis, flowers 1-4, erect, bell-shaped. Bulblets generally lacking. Sepals 20-30 mm long, lilac, often spotted. Petals 30-50 mm long, pale to rich pink, or pinkish-lilac, often paler toward lower third, sometimes purplish red at base. Petal margins entire or shallowly toothed. Face of petal with long white hairs over the lower third. Nectary is round to squarish, mostly densely covered with branched hairs. Anthers purple and short to lilac and narrow, mostly shorter than the filaments. Fruit upright, linear, angle-edged, 5-6 cm long. Seeds flat and lens-shaped. In flower Apr-early June.

Those in our area appear to be a form that has been called **Davidson's mariposa lily** (*Calochortus davidsonianus* Abrams), which has short purple anthers that are shorter than the filaments and hairs only over the lower third of the petal face. Some botanists consider it to be a minor form of *Calochortus splendens*. While currently not officially recognized as distinct, *Calochortus davidsonianus* is tetraploid (4n = 28) and *Calochortus splendens* is diploid (2n = 14). Recent gene sequencing suggests that these two lilies may not be closely related. Further study is clearly needed to sort this out.

Our most common mariposa, found in a range of habitats from grasslands, coastal sage scrub, and openings in oak woodland to pine forest. See it in the Chino Hills, Crystal Cove State Park, Laguna Coast and Aliso and Wood Canyons Wilderness Parks, Limestone Canyon, Richard and Donna O'Neill Conservancy, Audubon Starr Ranch Sanctuary, Caspers Wilderness Park, and along most trails in the Santa Ana Mountains.

The epithet was given in 1835 by G. Bentham for its splendid flowers (Latin, splendidus = splendid).

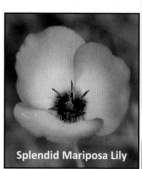

Plummer's Mariposa Lily ★
Calochortus plummerae E. Greene

Geophytes from a robust bulb with a thick, fibrous coat. Basal leaf one, strap-shaped, mostly withered before the flower opens. Edges of the upper stem leaves are rolled inward. Stems lack bulblets at the base. Inflorescence usually 2-6-branched with open bell-shaped flowers. Sepals 3-5 cm long with a few hairs near the base, mostly pink to rose. Petals 3-4 cm long, *pink to rose-purple, sometimes pale yellow toward the base, with long conspicuous yellow-orange hairs along the lower inner face, upper half to one-third mostly without hairs. Petal margin mostly smooth, sometimes toothed.* Nectary round, hairless, slightly sunken, and conspicuous as a "wart" on the outer surface of the petal. Anthers pinkish-orange. *Fruit narrow,* 4-8 cm long. In flower May-July.

A rare plant, it just barely enters our area in the northern Santa Ana Mountains, where it occurs sparingly in Fremont Canyon and near Oak Flat south of Sierra Peak. It is more frequently seen in the Santa Monica Mountains and along the foothills of the San Gabriel, San Bernardino, and northern San Jacinto Mountains. It grows on rocky soils from chaparral to pine forest, most commonly following fire. CRPR 4.2.

The epithet was given in 1890 by E.L. Greene to honor S.A. Plummer.

Similar. Intermediate mariposa lily (*C. weedii* var. *intermedius*): conspicuous yellow hairs with a dark spot at the base along the entire inner petal face; petal margins fringed with long, conspicuous hairs; pale lemon yellow flower with dark purplish or maroon spots, blotches, or bands; however, in places such as the Chino Hills, it can have a pinkish-purple flower quite similar to Plummer's.

Above: Plummer's mariposa lily.

Weed's Mariposa Lily
Calochortus weedii Alph. Wood

Geophyte from a robust, fibrous-coated bulb. Basal leaves 20-40 cm long, withering while flowers open. Stem slender, branched, 30-90 cm tall, inflorescence 2-6-flowered. Bracts resemble leaves but are shorter. Flowers open, bell-shaped. Sepals 20-30 mm long, oval to lance-shaped, tapered to a pointed tip, and yellowish to purple-brown in color. *Petals 20-30 mm long, yellowish and square- or diamond-shaped. All or most of the petal face is covered with long dark yellow hairs, each hair often with dark spots at the base. The petal margins are fringed with dark hairs.* Nectary round, slightly sunken, and surrounded with long hairs. The nectary itself lacks hair. Anthers large, orange or purple, mostly as long as or longer than the filaments. *Fruit* 40-50 cm *linear, erect, and angled.*

Found on rocky soils in chaparral and dense coastal sage scrub from eastern Los Angeles, Orange, and Riverside Counties, south into Baja California. It is one of the last *Calochortus* to flower; the first flowers appear at the end of May and the last are seen about mid-July.

Three varieties of this species are recognized; two of them occur here.

The epithet was given in 1868 by A. Wood to honor A. Weed.

Below: Weed's mariposa lily, nominate variety on left, intermediate on right.

Weed's Mariposa Lily (nominate variety)
Calochortus weedii Alph. Wood **var. *weedii***

Recognized by its bright, bold, orange-yellow petals. These often have maroon-brown spots or blotches, and occasionally a maroon-brown band across the tip of the petal. The sepals are often narrower and more symmetric than those of the intermediate variety (see below).

Known from the central Santa Ana Mountains and southern Riverside County south through San Diego County, where they are quite abundant, to the border region of the United States and Mexico (in Baja California this plant is mostly replaced by **peninsular mariposa lily**, *C. weedii* var. *peninsularis*). In our area it is found on the eastern slope of the Santa Ana Mountains, Temescal Valley, San Juan Canyon, and southward. Many plants on Rancho Mission Viejo are a mixture of intermediate and Weed's mariposa lilies, although at the extreme southern boundary of Rancho Mission Viejo and Camp Pendleton many have the characteristic bright yellow flowers of this largely San Diego County plant.

Weed's Mariposa Lily

Intermediate Mariposa Lily ★
Calochortus weedii **var. *intermedius*** F. Ownbey

Its petals are pale lemon-yellow to pale yellow-orange in color, with dark purple brown blotches, spots, and bands across the petals. Some flowers are purple with pinkish-purple near the base. The sepals are often broad and asymmetric.

This rare plant is found almost entirely in Orange County, with a few small populations in Los Angeles and western Riverside County. It is found throughout the Chino and San Joaquin Hills, foothills of the Santa Ana Mountains from Rancho Mission Viejo north, and western slopes of the Santa Ana Mountains, where it is primarily found from Trabuco Canyon north. On Rancho Mission Viejo, it co-occurs with Weed's mariposa lily, and the populations gradually become more consistent with the nominate variety as one moves south. CRPR 1B.2.

The epithet was given in 1940 by F.M. Ownbey for its appearance, which lies between Weed's and Plummer's mariposa lilies (Latin, intermedius = middle, between).

Similar. Plummer's mariposa lily (*C. plummerae*): mostly pinkish-purple flowers with *hairs only on the lower half* to two-thirds of the petal face and *no* (or very few short) *hairs along the petal margin*.

Intermediate Mariposa Lily

Below: Intermediate mariposa lily comes in a wide variety of colors and hair densities. Even within small populations, the variation can be surprising.

California Chocolate Lily, Mission Bells ✣
Fritillaria biflora Lindl. var. ***biflora***

Erect geophyte from bulb, stems 1-4.5 cm tall. Leaves often crowded at base of plant and fewer above, each 5-19 cm long, oblong to narrowly oval, alternate on stem. Flower nodding, 18-40 mm long, odorless. Tepals 6, all similar, dark brown or greenish-purple. Nectary large, forming a narrow line 2/3 of flower length, purplish to greenish. Style branches obvious, divided over half length. Fruit an angled capsule, rounded or flattened at both ends. In flower Mar-Apr.

Found on heavy clay soils in grasslands and openings among chaparral and coastal sage scrub, especially on north-facing slopes. Scattered and difficult to find in our area, although it can be fairly common on Elsinore Peak and on benches within the Temescal Valley. In Orange County it has been seen in Limestone Canyon, Cristianitos Canyon, Audubon Starr Ranch Sanctuary, Caspers Wilderness Park, the eastern boundary of Coto de Caza, and Weir, Fremont, Baker Canyons and near Mabey Canyon in the northern Santa Ana Mountains. Easily found and viewed on the Santa Rosa Plateau.

The common name refers to the chocolate brown color of the flowers. The epithet was named in 1834 by J. Lindley, probably in reference to an early specimen that had only two flowers per stalk (Latin, bi- = two; floris = flower).

California Chocolate Lily

Right: Tepal removed to show pendulous stamens and 3 recurved style branches. Far right: Young fruit; note that though flowers hang, when in fruit the unit turns upward, matures, dries, and later releases seeds. Below, right: Population in Limestone Canyon, recovering after the Santiago Fire of October 2007. Below: Numerous plants (each dark spot is a flower) on a clay slope near Elsinore Peak.

California Chocolate Lilies

California Chocolate Lilies

Ocellated Humboldt Lily ★

Lilium humboldtii Roezl & Leichtlin **ssp. *ocellatum*** (Kellogg) Thorne

Tall geophytes from bulb-like scaly rhizomes. Stems erect, to 3.5 meters tall, with 4-9 whorls of leaves. Leaves attached directly to the stem, 4-15 cm long, widest above the middle and often wavy-edged. Inflorescence with 1-25 flowers, which hang downward. Flower very large, showy, with strongly recurved tepals, each 5-10 cm long, yellow-orange or light orange with red to red-purple spots and occasionally blotches. The spots have lighter edges (like a halo) and most are larger toward the tip of the tepal. Pistil 4.5-7 cm long. Filaments much longer than the flower. Anthers 11-19 mm long, pinkish-tan or peach, becoming yellow, attached at mid-length to the filament. Fruit a ribbed capsule, 2-6 cm long. In flower late May-July.

A striking and unmistakable plant found along creeks in shaded canyon bottoms within oak and riparian forests throughout the Santa Ana Mountains. Known from Gypsum, Fremont, Black Star, Silverado, Ladd, Holy Jim, Trabuco, and San Juan Canyons in Orange County and on Skyline Drive between Tin Mine Canyon and Santiago Peak, Horsethief Canyon, and Fisherman's Camp at Tenaja Canyon in Riverside County. Outside the Cleveland National Forest it is known only from Bell Canyon on Audubon Starr Ranch Sanctuary and Caspers Wilderness Park below San Juan Hot Springs. CRPR 4.2.

The epithet was given in 1871 by B. Roezl and M. Leichtlin in honor of F.W.H.A. von Humboldt. Our form was named in 1873 by A. Kellogg for the spots on the tepals (Latin, ocellatus = spotted as with little eyes).

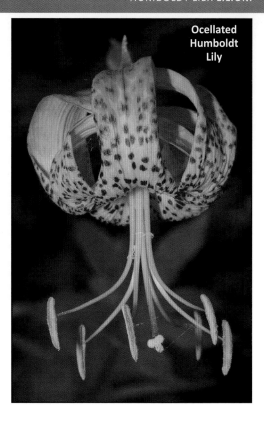

Ocellated Humboldt Lily

Ocellated Humboldt Lily

Below: This plant was roughly 3 meters (9.8 feet) tall and bent over with the weight of its abundant flowers. Others, visible in the background, were of similar size

Humboldt lilies are pollinated primarily by large butterflies, especially **western tiger swallowtails** (*Papilio rutulus rutulus* Lucas, Papilionidae) and **pale swallowtails** (*Papilio eurymedon* Lucas, Papilionidae), and to a lesser extent by **monarchs** (*Danaus plexippus* (L.), Nymphalidae). We've watched both swallowtails take nectar from and pollinate flowers in Santiago, Holy Jim, and Trabuco Canyons.

In some years, the bloom of ocellated Humboldt lily is spectacular. It is most often found along waterways, in the shade of **coast live oak** (*Quercus agrifolia* Nee, Fagaceae) or **white alder** (*Alnus rhombifolia* Nutt., Betulaceae), rarely in full sun. Its preference for shade and dappled sunlight makes it a difficult subject to photograph.

Plants along roads and popular trails often have their blossoms cut off and taken by flower-lovers. These plants put a lot of their hard-earned energy and growth into those structures in order to reproduce. Removing the flowers eliminates their chance of setting seed and continuing their species, removes valuable nectar and pollen sources from visiting animals, and reduces the number of beautiful wildflowers that enthusiasts get to enjoy. Please do not pick these flowers!

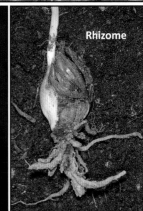

Upper right: Nancy and Chris Barnhill examine a lily that towers over their heads in Trabuco Canyon. Center, left: Dried fruit; note that though flowers hang, when in fruit the unit turns upward, matures, dries, and later releases seeds. Right: Bulb-like rhizome with long reddish scales. Below, left: Pale swallowtail butterfly nectaring in upper Santiago Canyon. Center: Large whorls of linear leaves which wither when the flowers open. Right: Entire plant, pre-bloom.

MELANTHIACEAE • FALSE-HELLEBORE FAMILY

Geophytes from a short thick rhizome or bulb, some from a corm. *Leaves alternate, large and basal in most species,* along the stem in a few. *Inflorescence upright, long,* generally branched. Flowers funnel- or cup-shaped or held flat open. Tepals 6, with a nectary at the base of each or just at the base of each sepal. Stamens 6. Ovary superior to half-inferior, with 3 separate sections or internal chambers. *Style of 3 separate sections or a single 3-branched structure; slender.* Fruit a capsule, seeds flattened, rounded, often winged, pale in color (not black). A small family of about 130 species, mostly in the northern hemisphere, 27 in California, 2 in our area.

Fremont's Death Camas
Toxicoscordion fremontii (Torr.) Rydb.
[*Zigadenus fremontii* (Torr.) S. Watson]

An upright geophyte, its stems 40-90 cm tall. Leaves 20-50 cm long, 8-30 mm wide, curved from base to tip, folded up at the sides, the edges straight (not wavy). Inflorescence stalk 5-40+ cm tall. Flowers generally in small groups (sometimes solitary) on slender lateral stalks. Tepals 5-15 mm long, oval, yellowish-white with a broad yellow-green nectary that is notched inward toward its tip. Style with 3 slender lobes. Fruit an elongate capsule. In flower Feb-June.

Generally quite abundant but only within the first 5-10 years after fire or other soil disturbance. Found in grasslands, coastal sage scrub, and chaparral. Known from the San Joaquin Hills (Laurel Canyon, Aliso and Wood Canyons Wilderness Park) and the Santa Ana Mountains (Coal Canyon, Fremont Canyon, Audubon Starr Ranch Sanctuary, Hot Springs Canyon) south through the San Mateo Canyon Wilderness Area and on the Santa Rosa Plateau. Expected in most of our wild places that experience fire. Toxic if ingested.

The epithet was given in 1857 by J. Torrey in honor of J.C. Fremont.

Similar. Meadow death camas (*T. venenosum* var. *venenosum*): slender, shorter plant; flowers solitary on unbranched lateral stalk from the inflorescence; nectary not notched inward near tip; clay soils at higher elevations.

Meadow Death Camas
Toxicoscordion venenosum (S. Watson) Rydb.
var. *venenosum*
[*Zigadenus venenosus* S. Watson var. *venenosus*]

Also an upright geophyte but more diminutive than Fremont's, its slender stems 15-70 cm tall. Leaves 10-40 cm long, 4-10 mm wide, curved, the edges straight. Inflorescence stalk 5-25 cm tall. Most flowers solitary on their own slender lateral stalks. Tepals 4-6 mm long, oval, yellowish-white with a yellow-green nectary. Style with 3 slender lobes. Fruit an elongate capsule. In flower May-July.

Widespread in the Pacific Northwest, high plains, sparingly in southern California. Here known only from clay soils among grasslands and chaparral on the southeast side of Elsinore Peak. Toxic if ingested.

The epithet was named in 1879 by S. Watson for its toxicity (Latin, venenosus = full of poison, very poisonous).

Similar. Fremont's death camas (*T. fremontii*): stouter, taller plant; flowers usually on branched lateral stalk from the inflorescence; nectary large, notched inward near tip; rocky soils at many elevations.

ORCHIDACEAE • ORCHID FAMILY

Perennials, some persistent, others ephemeral from rhizomes or tuber-like roots. *Leaves linear, oval, or rounded, their bases sheathed.* Inflorescence branched or simple, with a bract beneath each flower. *Flower 2-lipped.* Before each flower opens, the ovary twists 180 degrees, positioning the flower upside-down. *Sepals 3 (1 up, central; 2 down, lateral), usually petal-like. Petals 3 (2 up, lateral; 1 down, central, forming a distinct "lip").* The lip of some species has a *hollow backward spur* that contains nectar. Stamen 1, fused with the style and stigma into a column. Pollen held in 2 sacks (pollinia) that stick to pollinating insects. Stigma 3-lobed, hidden beneath the column. *Ovary inferior,* with a single chamber. Seeds tiny and numerous.

The largest plant family, with 25,000 species, mostly in the tropics; 34 in California, 4 in our area. An extract from the large fruits of the vanilla orchid is used to flavor food and drink.

Stream
Epipactis gigantea
Large flowers,
very colorful,
lobes broad,
no spur

Cooper's Rein
Piperia cooperi
Small flowers,
yellow-green,
lobes rounded,
short spur

Wood Rein
Piperia leptopetala
Small flowers,
yellow-green,
lobes narrow,
medium spur

Thin Wood Rein
Piperia elongata
Small flowers,
yellow-green,
lobes narrow,
very long spur

Stream Orchid

Epipactis gigantea Hooker

A stout perennial from tuber-like roots, 30 cm to 1 meter tall. Leaves 5-15 cm long, lance-shaped to elliptical, with long ribs you can feel, basal and/or alternate on the stem. The floral bracts look like leaves. Flowers appear mostly along one side of the stalk. The 3 sepals are similar, green, with purple veins. The 2 upper petals are similar, often pinkish, with purple veins. The lip 14-20 mm long, broad and round at the sides near the base, pinkish or greenish. The outer half of the lip is much narrower; white, yellow, orange, and/or pink (rarely greenish). The lip has no spur. In flower May-Aug.

A large orchid of creeks, meadows, seeps, and seasonal water sources, not commonly found here. Known from San Juan Canyon, Lomas de Santiago, a seep along North Main Divide Road "Loop," Hagador Canyon, and San Mateo Creek in the San Mateo Canyon Wilderness Area. Easiest seen on the face of Dripping Springs in Limestone Canyon. Often pollinated by flies, most commonly flower flies (Syrphidae). A fine plant for shady moist soils and near garden ponds.

The genus was named in 1800 by O.P. Swartz for a related European orchid that, when added to milk, curdles it (Greek, epipaktis or epipegnuo = hellebore orchid). The epithet was given in 1839 by W.J. Hooker for its large flowers (Greek, gigantos = giant).

Stream Orchid

THE STREAM ORCHID GUILD

Stream orchids are pollinated by flower and hover flies (Syrphidae). Right: This **pied hover fly** (*Scaeva pyrastri* (L.)) visited stream orchids near Los Pinos Potrero. When it flew away with pollinia on its back, Chris Barnhill ran after it, saw it alight on a coast live oak leaf, and captured this image. Far right: Bob Allen photographed this female **swift flower fly** (*Eupeodes volucris* Osten-Sacken) as it exited a flower and methodically wiped off the pollinia.

Rein Orchids: *Piperia*

Slender upright geophytes. Most leaves are basal, 2-6 in number, linear or lance-shaped, *often dried up at flowering time.* Stem leaves are much smaller and narrower. Flower bracts very slender, not leaf-like. *Flowers small, yellow-green or whitish.* The lip has a *hollow backward spur* at its base that contains nectar. Each plant remains in flower for 4-6 weeks. As the fruits mature, they turn upright and remain held tightly against the stem.

Rein orchids are a joy to find in the field, but the experience comes at a price. The mountain-dwelling species bloom when biting flies (mostly deer flies, horse flies, and snipe flies) are at their peak abundance in June-July. We found it best to have one person shoo away the flies while another photographs the plants.

To attract their pollinators, rein orchid flowers release a pleasant scent at night. Some species found elsewhere are known to be pollinated at night by moths. This is probably true for our species as well, but, to date, no one has observed and identified the pollinators in our area.

Unusual for orchids, they grow on dry hillsides in bright sun or shade. Most of our populations occur on protected lands such as wilderness parks, preserves, and the Cleveland National Forest. All 10 species in the genus occur in California, 3 of them in our area.

The genus was named in 1901 by P.A. Rydberg in honor of his friend C.V. Piper.

Cooper's Rein Orchid, broad lower lip, very short spur

Cooper's Rein Orchid ★

Piperia cooperi (S. Watson) Rydb.

The stalk is 15-90 cm tall. Its 2-4 narrow basal leaves are 8-20 cm long, 8-30 mm wide. Flowers yellow-green. *The lip 2-4 mm long, broadly oval to triangular. Spur 2.5-6 mm long, pointed back or down. The short, stubby spur is about the same length as the lip.* Seeds are blackish-brown. In flower Mar-June.

Its first leaves appear in late fall and mature as the flower stalks grow. When the flowers open up in March, the lowest leaves are usually brown. It gives off a honey-like scent, mostly at night. Peak flowering is in late Apr-early May. It often grows right up through other plants and remains unnoticed until the adjacent plants are removed by fire or brushing (mechanical removal).

Though it is uncommon, it is by far the most often-encountered rein orchid in our area. It is found in dry soils on hillsides among coastal sage scrub, chaparral, and oak woodlands. Widely distributed, known from canyons such as Limestone, Santiago, Silverado, Trabuco, and San Juan and scattered sites in the San Mateo Canyon Wilderness Area. Also found along San Juan Loop Trail and both ends of the San Juan Trail. A population on Niguel Hill, the only one near our coast, was destroyed by a housing development in the 1980s. It may still occur elsewhere in the San Joaquin Hills, yet to be rediscovered. CRPR 4.2.

The epithet was given in 1876 by S. Watson to honor J.G. Cooper, who first collected it near San Diego.

Similar. Wood rein orchid (*P. elongata*): lip also triangular but much longer; spur much longer; mixed coniferous forest.
Thin wood rein orchid (*P. leptopetala*): lip narrow and very long; spur much longer; oak woodland.

Cooper's Rein Orchid, leaves withered at flowering time

Cooper's Rein Orchid, basal leaves appear between fall and spring

Wood Rein Orchid ❧
Piperia elongata Rydb.

A tall stalk, 20-100(-130) cm tall. Its 2-4 basal leaves are 8-30 cm long, 10-65 mm wide, broader than other local rein orchids. Flowers greenish. *The lip 2-5.5 mm long, triangular, folded backward. Spur (0.8-)6.5-15 mm long, pointed down. The very long, slender spur is much longer than the triangular lip, often longer than the ovary as well.* Seeds are light brown. In flower May-July.

The first leaves appear in March and mature through spring and early summer as the flower stalks elongate. The earliest flowers open in late May with peak bloom in mid-June, sometimes mid-July. It gives off a honey-like scent at night. The flowering stalks are very tall and after drying in summer, remain in place through the fall and winter. They commonly stand next to the following year's flower stalks.

In the Santa Ana Mountains, it occurs in mixed coniferous forest, under moderate shade of **big-cone Douglas-fir** (*Pseudotsuga macrocarpa,* Pinaceae) and **California bay laurel** (*Umbellularia californica,* Lauraceae). Elsewhere in southern California it is found in chaparral, sometimes with **Cooper's rein orchid**. Known here from upper Trabuco Canyon and sparingly in Silverado Canyon. Both populations were discovered by us during preparation of this book.

The epithet was given in 1901 by P.A. Rydberg for its long spur (Latin, elongatus = removed [long]).

Similar. Cooper's rein orchid (*P. cooperi*): lip short-triangular; lip and spur much shorter; chaparral and oak woodlands. **Thin wood rein orchid** (*P. leptopetala*): lip much narrower and longer; spur shorter; oak woodlands.

Wood Rein Orchid

Thin Wood Rein Orchid ★
Piperia leptopetala Rydb.

The stalk is 15-70 cm tall. Basal leaves 7-15 cm long, 15-30 mm wide, narrow. *Flowers pale greenish, sepals and petals long and narrow. The lip 2.5-5 mm long, narrow and lance-shaped, pointed forward or downward. Spur 4-7 mm long, but often variable in length, curved and pointed down. The long, slender spur is usually just a bit longer than the lance-shaped lip.* Seeds are cinnamon brown. In flower May-July.

At night it gives off a faint lemon scent. Known only from California and difficult to find, this is a very rare orchid. It was found only once in our area, on a steep slope among chaparral in Holy Jim Canyon, reported at 2,000 feet elevation, on 19 June 1948. Repeated searches of the area for many years have so far failed to rediscover it. If you are fortunate enough to find it, please notify us through the Herbarium at either Rancho Santa Ana Botanic Garden or UC Riverside. CRPR 4.3.

The epithet was given in 1901 by P.A. Rydberg for its slender petals (Greek, leptos = slender, thin; petalon = flower leaf [petal]).

Similar. Cooper's rein orchid (*P. cooperi*): lip short-triangular; lip and spur much shorter; chaparral and oak woodland. **Wood rein orchid** (*P. elongata*): lip long-triangular but shorter; spur much longer; mixed coniferous forest.

Thin Wood Rein Orchid

Robert Lauri

RUSCACEAE • BUTCHER'S BROOM FAMILY

Upright herbs, shrubs, or trees. Herbaceous forms often with creeping rhizomes, tree forms with a woody trunk. Leaves with parallel venation, usually alternate on stem, sometimes in a basal or terminal rosette. Stem leaves clasp the stem or are sheathed at base. Individual flowers small, arranged in a sparse or dense, branched inflorescence. Tepals 6, all similar, generally white or cream, free or fused into a bell or urn. Most species have flowers of both sexes. Stamens 6. Ovary superior, (2)3-chambered, the single stigma is clubbed or 3-lobed. Nectaries present, embedded in the external surface of the walls that separate the internal chambers of the ovary. Fruit a berry or dry capsule, seeds oval, never flat.

A worldwide family of 500 species with 3 genera in California, 2 genera and 2 species in our area.

Western False Solomon's Seal

Maianthemum racemosum (L.) Link
[*Smilacina racemosa* (L.) Desf., Convallariaceae]

False Solomon's Seal

An herbaceous perennial 30-90 cm tall, from a thick creeping rhizome. The stem is usually zig-zagged from leaf to leaf. Leaves 7-20 cm long, oval, hairless or finely hairy; the base clasps the stem. The inflorescence is very showy and dainty. Each tiny flower has 6 short, clean white tepals, 6 stamens, and a single style topped by a 3-lobed stigma. Fruit a round red or pink berry, often spotted with purple. In flower Mar-July.

An understory plant in moist soils of riparian, oak, and conifer forests, generally not found in full sun. Quite uncommon here, found only once, in late May 1968. The locality is rather vague: along Main Divide Truck Trail, between El Cariso and Santiago Peak, at an elevation of 4,000 feet. Several places in the Santa Ana Mountains may host the plant, such as Bear Springs, Maple Springs, Bigcone Springs, Coldwater Canyon, Mayhew Canyon, and generally moist, shady spots. Its flowers are pollinated by bees and wasps.

The epithet was given in 1753 by C. Linnaeus for the structure of its inflorescence (Latin, racemulus = the stalk of a cluster).

Chaparral Beargrass ★

Nolina cismontana Dice

Chaparral Beargrass

An upright shrub, 0.5-1.5 meters tall, with a very obvious central unbranched or many-branched woody stem. Leaves 0.5-1.4 meters long when mature, lance-shaped, the edges finely toothed, the tip pointed but without a stout spine, green sometimes covered with a thin white powder and appear grayish. Each terminal rosette is made up of 30-90 leaves. Inflorescence arises from the stem tip, 1-3 meters long. Each dry papery flower has 3 sepals and 3 petals, usually whitish or yellowish. Flowers are mostly of only one functional sex. Male flowers have 6 stamens with thin filaments; each anther is a single structure (not forked); the vestigial pistil exists as a mound in the center. Female flowers have a superior ovary; the style is short and 3-branched at the tip; vestigial stamens appear as short stalks surrounding the ovary. Fruit is a dry, papery, 3-winged capsule, inflated with air. Seeds 4-5 mm long, 3-4 mm diameter, reddish-brown, oval, not flattened, dispersed by winds. In flower Mar-June.

Nolina and its close relatives are sometimes placed into the nolina family, Nolinaceae. An unusual plant with unique flower structures. The inflorescences of male-flowered and female-flowered plants are different enough that their sex can be determined from a distance. Found mostly on dry rocky slopes, in sandstone and shale-derived soils within chaparral. Healthy plants can survive most fires and resprout from stems. They don't all flower each year. We notice that every 2-4 years, nearly every plant comes into flower and makes for an impressive display.

Though a rare species, it is quite abundant in some parts of our area. Known here from the coastal slope of the Santa Ana Mountains, in the Lomas de Santiago, Arroyo Trabuco, and southeast to the ridges above Hot Springs Canyon along the San Juan Trail (about 3-4 miles above the trailhead). Common in parts of Coal, Gypsum, Fremont, Weir, Black Star, Santiago, Silverado, Trabuco, and San Juan Canyons. Easily observed in Whiting Ranch Wilderness Park (due west of the Glenn Ranch parking area), O'Neill Regional Park, and above the Trabuco Canyon Post Office. Also known from the greenbelt adjacent to Vista del Lago to the east of Los Alisos Boulevard in Mission Viejo. CRPR 1B.2.

The genus was established in 1803 by A. Michaux in honor of P.C. Nolin. The epithet was given in 1995 by J.C. Dice for its distribution in cismontane southern California (Latin, cis = on this side; montanus = mountains).

Similar. Chaparral yucca (*Hesperoyucca whipplei*, Agavaceae): no obvious main stem, leaves tipped with a stout spine; flowers larger, each with both sexes; moth-pollinated; filaments thin, anthers forked; fruit a succulent capsule, each with several black, flat seeds.

Top right: upper Fremont Canyon.
Right: male flowers at left, female flowers at right.
Below: Close-up of male flowers.

THE BEARGRASS GUILD

Beargrass Moth
Mesepiola specca Davis (Prodoxidae)

A small moth with a wingspan of 8-12 mm. Adults of both sexes are similar, each with narrow forewings speckled with white, dusky, and rust-colored scales. Hindwings brownish, their trailing edge with long hair-like fringes. Underside of the abdomen is whitish. Females have a hooked appendage on the upper (dorsal) side near the tip of their abdomen; they hold it against the calyx as they use the tip of their abdomen to cut into an ovary and deposit an egg. Adults are active at dusk as they fly around the flowers, drink nectar, mate, and lay eggs. Larvae feed inside one or more seeds, then exit the seed and ovary, burrow into the soil, and pupate. Adults emerge in Mar-June when the plants are in flower.

This is the only insect recorded to feed on a related plant, **Parry's beargrass** (*Nolina parryi* S. Watson). The moth and Parry's beargrass are both known from Arizona and parts of southern California. On 17 May 1938, one adult female beargrass moth was collected in El Toro (now Lake Forest) by C.M. Dammers. Since Parry's beargrass is not in our area, the moths here most certainly feed on chaparral beargrass.

The epithet was given in 1967 by D.R. Davis for its speckled forewings (Anglo-Saxon, specca = speckled).

THEMIDACEAE • BRODIAEA FAMILY

These are onion-like plants that lack the characteristic onion odor. They are all perennials from an underground corm. Every year, new corms grow from the old one, surrounded by a tunic made from dry leaf bases of past years. Leaves are long and generally flat or channeled with an expanded base; they are often dried up when the flower is at peak bloom. The inflorescence is atop a long naked stalk, the flowers of our species arranged in a loose umbel with bracts at its base but no bracts at the bases of individual flowers. *Each pedicel is often jointed at its base or where it meets the flower.* Tepals 6, in various shades of white, yellow, blue, or purple. Stamens 6, their filaments free or fused to the tepals. In some species, all 6 stamens are fertile, but in the genus *Brodiaea*, 3 of those stamens are sterile (staminodes). Ovary superior, the tip of the style clubbed or 3-branched. Fruit is a capsule filled with stout triangular seeds.

A small family of 62 species found from western North America to Central America, 37 in California, 6 in our area. This family is closely related to the hyacinth family (Hyacinthaceae) but in the past has been grouped with the lily (Liliaceae), onion (Alliaceae), and amaryllis (Amaryllidaceae) families. Recent studies based on DNA and detailed analyses of their structures split them off into their own family. Interestingly, recognition of the family was first proposed in 1866.

Common Goldenstar

Bloomeria crocea (Torr.) Coville

A perennial that grows from a fibrous-coated corm. It has only a single leaf, which is flat or slightly V shaped in cross-section, often has a keel, and is bright green in color. The inflorescence is an open spherical umbel atop a leafless stalk. Each pedicel is jointed; it has a thickened area near its tip, just below its attachment with the flower. *Tepals 8-12 mm long, yellow, widely spread apart. Each tepal has a brownish or greenish mid-vein stripe, often giving the outer surface a yellow-orange color, especially in bud.* The stamens are longer than the style, attached at the base of the tepals. Each stamen has basal appendages tipped with two shallow points; in combination, the appendages form a nectar-filled cup-like base around the ovary. Anthers are blue or whitish and attached near their base to a filament. The ovary is 3-parted, rounded and smooth. In flower Apr-June.

Common to locally abundant on gentle slopes in native grassland and openings of coastal sage scrub; especially common in the Chino Hills, the San Joaquin Hills, and the foothills on either slope of the Santa Ana Mountains. In the mountains it is often found in meadows and grassy openings of oak woodland. It can be seen along many trails in the county. Look for this plant at Crystal Cove State Park, Laguna Coast Wilderness Park, Aliso and Woods Canyon Wilderness Park, and Caspers Wilderness Park.

In spring you may notice odd swellings near the upper portion of the stalk or on the pedicel (see photo, lower right). These are stem galls; we suspect they are caused by a species of **lily stem gall midge** (perhaps *Lasioptera* sp., Cecidomyiidae). Several orange-colored larvae live within the gall, where they feed on its tissue. They pupate in the gall during summer and emerge as adults the following spring. Though we haven't yet studied them in great detail, the plants seem unaffected by the galls.

The genus was established in 1863 by A. Kellogg in honor of H.G. Bloomer. The epithet was given in 1859 by J. Torrey for its saffron (yellow-orange) flowers (Latin, croceus = saffron-colored).

Common Goldenstar

Corm

Lily stem gall midge

Cluster-lilies, Brodiaeas: *Brodiaea*

Perennials that grow from *fibrous corms*. Leaves 1-6, often narrow, channeled, or slightly flattened, dark green in color. The inflorescence is an open skyward-facing *umbel* atop a leafless stalk. *Tepals bluish or purple-blue,* upright with the tips spreading outward, sometimes strongly held outward or curved backward in larger flowers. The outer tepals are narrower than the inner tepals and have a nectary at their base. At the flower base, the tepals are united to form a short basal tube. Past the tube, the tepals are free (not united). This makes the flower's overall form tubular or bell-shaped. They are pollinated by native beetles and bees.

Plants in this genus have *only 3 fertile stamens,* each one attached at its base to an inner tepal. *The other 3 stamens have no anther; they are reduced to sterile (non-reproductive) structures called staminodes.* Each staminode is attached at its base to an outer tepal. They come in a variety of distinctive forms that are often key to identification. Some are narrow like a typical filament, others are broadly flattened and often shaped like the blade of a small handheld shovel. One species has no staminodes at all.

The ovary is three-lobed, green or rarely purplish. After fertilization, the tepals dry up but remain on the flower. As the developing seeds grow, the ovary expands, often pushing against the basal tube. When this happens, the tube may become totally or partially transparent and may split apart. In some species, the basal tube isn't changed at all.

The genus was established in 1810 by J.E. Smith in honor of J.J. Brodie.

There are about 18 species in the genus, with several others in need of being described. We have at least 3 species locally, although the actual number has been in debate in recent years. One of them is considered by some botanists to be a hybrid between two existing species; other botanists consider it a valid species. We present it here as a valid species.

Santa Rosa Basalt Brodiaea

Cluster-lilies are popular with small native bees such as this **sweat bee** (*Lasioglossum* sp., Halictidae). Bob watched it for several minutes as it carefully removed pollen from the anthers.

Many populations and entire species of cluster-lilies face the threat of extinction. Much of their habitat has been cut, filled, leveled, and built upon, a practice that continues today.

Cluster-lilies can be difficult to identify. Normally, one can use the staminodes as a defining feature, but some beetles like to eat them, making the task more challenging. The table below should help.

Santa Rosa Basalt Brodiaea with *Lasioglossum* bee

	Thread-leaved *Brodiaea filifolia*	**Santa Rosa Basalt** *Brodiaea santarosae*	**Earth** *Brodiaea terrestris*
Inflorescence	20-30 cm	9-36 cm	2-20 cm
Flower	Small, only 14-20 mm long; basal tube 6-8 mm; tepals free for 10-14 mm	Very large, 24-36 mm long; basal tube 6-11 mm; tepals free for 10-14 mm	Very large, 26-33 mm long; basal tube 6-13 mm; tepals free for 15-30 mm
Filaments	0.5-1.5 mm long	2.4-8.2 mm long	2-4.5 mm long
Anthers	3-6 mm long, tip with a wide notch between chambers	5.4-8.9 mm long, tip with a V-shaped notch between chambers	(2-)4.5-7 mm long; sometimes with a "tooth" in notch between chambers
Staminodes	1-4 mm long; linear to thread-like, base triangular, tapered to a pointed tip; pale purple	Up to 8 mm long (sometimes absent), linear, tapered to a pointed tip; pale purple	5.5-7 mm long; broad, edges often rolled in, tip often rolled over like a hood, 2-notched; violet (to whitish)

Thread-leaved	Santa Rosa	Earth	Thread-leaved	Santa Rosa	Earth
Brodiaea filifolia	*Brodiaea santarosae*	*Brodiaea terrestris*	*Brodiaea filifolia*	*Brodiaea santarosae*	*Brodiaea terrestris*

A corm is a short, thick, food-storage, underground stem. In cluster-lilies, the corm is surrounded by fibrous bases of non-photosynthetic leaves, often called a coat or a tunic. Note the yellowish growth plate from which adventitious roots emerge. Corms can produce small cormlets from the plate, as seen in the left side of the photograph on the right. New plants grow from both cormlets and seeds.

Thread-leaved Brodiaea ★

Brodiaea filifolia S. Watson

Tepals violet-reddish-purple or bluish, their lobes spread strongly outward. *The inconspicuous staminodes are whitish to pale purple or pale blue, triangular at their base, linear or thread-like, and taper to a point.* They often subtly lie against the tepals. Ovary green. In flower Apr-June.

Uncommon and patchy in native grasslands on *heavy clay soils* within southern Orange County, mostly in the vicinity of Cañada Chiquita, Cristianitos, and Cañada Gobernadora on Rancho Mission Viejo and the hills behind San Clemente. It is also found sparingly in Aliso and Wood Canyons Wilderness Park, the hills north of Lake Forest, Caspers Wilderness Park, and lower Devil Canyon in the San Mateo Canyon Wilderness Area. A very rare plant, it has been listed by the State of California as Endangered and by the US Fish and Wildlife Service as Threatened because of loss of habitat to urbanization, flood control, and highway projects. CRPR 1B.1. Disturbing or picking this rare beauty will contribute to its decline, so please just look and do not touch.

The epithet was named in 1882 by S. Watson in reference to its narrow, threadlike leaves (Latin, filium = a thread; folium = a leaf).

Similar. **Orcutt's brodiaea ★** (*B. orcuttii* (Greene) Baker, not pictured): no staminodes; San Diego County only. **Santa Rosa Basalt brodiaea ★** (*B. santarosae*): flowers much larger; filaments 2.4-8.2 mm long; staminodes absent or up to 8 mm long, pointed.

Santa Rosa Basalt Brodiaea ★

Brodiaea santarosae T.J. Chester, W.P. Armstrong, & Madore

Tepals blue-purple, their lobes spread outward like a bell. *Staminodes narrow, variable, sometimes absent; purplish or whitish; triangular at the base, tapered to the tip. Staminode length varies between and within flowers.* Ovary green. In flower May-June.

Found in native grasslands, sometimes near vernal pools, on heavy clay soils derived from the Santa Rosa Basalt and Santiago Peak Metavolcanic Basalt, volcanic rock formations. In our area it is known only from grasslands of Elsinore Peak, grasslands and drainages of Miller Mountain, and the Santa Rosa Plateau where it is much more common. CRPR 3.

The epithet was given in 2007 by T.J. Chester, W.P. Armstrong, and L.K. Madore for Santa Rosa Basalt.

For years, botanists have considered this plant to be a hybrid between **thread-leaved brodiaea** (*B. filifolia*) and **Orcutt's brodiaea** (*B. orcuttii*); some still do. For discussions, see Chester, Armstrong, and Madore (2007) and Roberts et al. (2004, 2007).

Similar. Thread-leaved brodiaea ★ (*B. filifolia*): filaments short; staminodes short, pointed. Individuals of *B. santarosae* that lack staminodes may be confused with **Orcutt's brodiaea** ★ (*B. orcuttii*): basal tube 3-5 mm long, free tepals 12-19 mm long, filaments 4-8 mm long, San Diego County only.

Earth Brodiaea

Brodiaea terrestris Kellogg

Tepals pinkish-purple to violet purple, their lobes spread outward somewhat like a bell, often curved backward near the tip. In some plants, the filament tip continues past the anthers as a "tooth" between the anther chambers. *The broad staminodes are pale purple to almost white, somewhat rectangular in outline, the edges rolled toward the center (sometimes flat). The tip of each staminode is flat or rolled over like a hood and notched inward.* The staminodes often lean into the center of the flower, partially hiding the style. Ovary green. In flower Apr-June.

Found in native grasslands and grassy openings in oak woodlands on sunny slopes in the southern Santa Ana Mountains near Blue Jay Campground, at Elsinore Peak, on the north flank of Miller Mountain, and on clay benches in Temescal Valley. At lower elevations west of the mountains it is relatively scarce, found only in a small patch on the UC Irvine Ecological Preserve, in San Clemente, and on a grassy mesa near Cristianitos Canyon. Fairly common on the Santa Rosa Plateau.

The epithet was given in 1859 by A. Kellogg in reference to its typical growth habit, close to the ground (Latin, *terrestris* = of or belonging to the earth).

Similar. Some low-elevation plants with hooded staminodes have been historically treated as **mesa brodiaea** (*B. jolonensis* Eastw.), which is found only in Monterey County. Ours may be an undescribed species or a subspecies of earth brodiaea. For now, we call it **"Mystery" brodiaea**.

Above: "Mystery" Brodiaea. Similar to both earth and mesa brodiaeas, usually with white staminodes, is known from the San Joaquin Hills (UC Irvine campus), portions of southeastern Orange County, the Santa Ana Mountains above Corona, in Temescal Valley (such as Indian Canyon at I-15), and on the Santa Rosa Plateau. It is currently under study by W.P. Armstrong and T.J. Chester.

School Bells
Dichelostemma capitatum (Benth.) Alph. Wood ssp. *capitatum*
[*D. pulchellum* (Salisbury) A. Heller var. *pulchellum*]

A perennial that grows from a fibrous-coated corm. Its 2-3 leaves are mostly dark green, triangular in cross-section when young, becoming flattened in age, and can be quite wide, especially after fire. Sometimes the leaves have a keel. Some plants that sprout after a fire have unusually broad leaves (see bottom photo). *Flowers are arranged in a fairly compact umbel atop a leafless stalk. The tepals are bell-shaped, bluish-purple to reddish-purple, united at the base into a short round tube.* The lobes are upright to spreading, 7-12 mm long. There are 6 stamens, all fertile; those attached to the outer tepals have longer, broader filaments and shorter anthers than those attached to the inner tepals. Filaments of the 3 inner stamens have almost no free central stalk but are expanded at the sides into a pair of white, deeply notched, crown segments, 4-6 mm long, that tilt inward to hide the stamens and curve outward at their tips. In flower Jan-May.

Commonly seen on sunny slopes in grasslands and grassy openings in coastal sage scrub, chaparral, and woodlands throughout our area, especially during wet years and following fire. It can be seen at Chino Hills State Park, upper Newport Bay, Crystal Cove State Park, Laguna Coast Wilderness Park, Aliso and Woods Canyons Wilderness Park, the Dana Point Headlands, Audubon Starr Ranch Sanctuary, Richard and Donna O'Neill Conservancy, Caspers Wilderness Park, the Temescal Valley, and trails throughout the Santa Ana Mountains. It is in cultivation as an easily grown garden accent. It reproduces by both seed and bulblets. Plants from bulblets often result in a cluster of plants, probably a source of the alternate common name "cluster-lily."

Confusion exists with the oft-used common name "blue-dicks," a never-ending source of giggles among novice wildflower watchers. It is unclear where the name originated. Some botanists have speculated that the word "dicks" is simply a shortened form of the genus name, *Dichelostemma* (so it should be spelled "dichs"). The flowers are bluish in color, thus the name "blue-dicks." However, "blue-dicks" was in use before *Dichelostemma* was published, so that is unlikely. Additional common names include cluster-lily and wild hyacinth, but it's neither a true lily (which are in the lily family, Liliaceae) nor a true hyacinth (which are in the hyacinth family, Hyacinthaceae). We prefer the older common name "school bells," which probably refers to schoolchildren hearing the school bells ring as they walked through fields of this common wildflower, or it may describe the flower since each is shaped like a small bell. Suggested pronunciation of the genus is dih-keh-low-STEM-mah (the dih syllable contains a short "i" as in "sit").

The genus was named in 1843 by K.S. Kunth in reference to its crown segments (Greek, dicha = bifid; stemma = garland). The epithet was given in 1857 by G. Bentham for its tight head-like cluster of flowers (Latin, capitatus = having a head). Once known as *Dichelostemma pulchellum* based on *Hookeria pulchella*, a name given in 1808 by R.A. Salisbury for its beauty (Latin, pulchellus = beautiful). However, the specimen upon which *Hookeria pulchella* was based includes two specimens on the sheet and an uncertain locality, invalidating its use.

Right: After fire, school bells sometimes produces very wide leaves.

School Bells

Above: School bells is one of our most abundant and popular wildflowers. Fields like this are easily found in Caspers Wilderness Park, Richard and Donna O'Neill Conservancy, and Limestone Canyon, and in many other places that have pockets of clay soils.

Common Muilla ✿
Muilla maritima (Torr.) S. Watson

A *delicate geophyte* from a small fibrous-coated corm. These plants have 3-10 narrow, somewhat flattened leaves. The flowers are in open umbels at the top of leafless stalks. The pedicels are not jointed at the flower. The stalks are mostly quite short in our area, seldom more than 15 cm tall, although they can be considerably taller elsewhere. *Flowers are greenish white with 6 spreading tepals, united only slightly at the base, otherwise free.* The lobes are 3-6 mm long, the inner set slightly broader, and have a greenish or brownish central stripe on the outer surface. *The filaments are broadly triangular at the base.* Anthers are blue, green, or purple and attached to the filaments near their middle. In flower mostly Mar-Apr in our area.

Ironically, common muilla is now uncommon on the coastal side of the Santa Ana Mountains. It generally occurs on hillsides and coastal mesas, in native grasslands associated with clay soils. It was once fairly often seen in Dana Point before much of the area was urbanized. It is now known here only from San Clemente State Park, southern portions of Rancho Mission Viejo, and Miller Mountain. It is still common inland on clay benches in the Temescal Valley, but that area is rapidly being developed. It is easiest to find on the Santa Rosa Plateau.

As the new leaves break through the soil in mid-winter, they lie flat and twisted on the ground, like thin green strings. Mature leaves are more stout and arch back toward the ground.

The genus was established in 1883 by S. Watson as a clever anagram (letter reversal) of *Allium,* probably in reference to its similarity to onions in the genus *Allium*. The epithet was given in 1857 by J. Torrey in reference to its close proximity to the sea (Latin, maritimus = of the sea).

Similar. In general appearance, similar to some white-flowered **onions** (*Allium* spp., Alliaceae), but the flowers of common muilla are smaller, appear more rigid, and lack an onion odor.

Common Muilla

Common Muilla

TYPHACEAE • CATTAIL FAMILY

Upright hairless perennials that grow from long submerged rhizomes, they generally form dense colonies in and around clear slow-moving or murky standing water. Their long leaves are linear, flat, and ribbon-like. The leaf base has a sheath that wraps around the stem. The female flowers are tightly packed into an inflorescence that looks like a brown sausage on a stick. Above them are the male flowers in a cluster that looks like a slender fuzzy bottlebrush. Each tiny flower has sepals and petals that are reduced to long slender threads with a single bract at its base. Male flowers with 2-7 stamens and many papery scales. Female flowers with long hairs, fertile and sterile flowers mixed together. The fertile females have one stalked ovary and a single style; the hairs originate from the base of the flower. Sterile female flowers look like a hairy matchstick with a non-functional ovary at the tip of a central stalk, the hairs arranged in multiple whorls below it. Fruit is an achene or nutlet.

They are wind-pollinated; male flowers release lots of pollen that blows in the wind and alights on receptive female flowers. After pollen is expended, male flowers wither and fall away, leaving a bare stick above the females.

Cattails are important members of pond and marsh communities, where they serve as food for species of grasshoppers, moths, and beetles. They also form habitat and cover for larger forms of wildlife such as **red-wing blackbirds** (*Agelaius phoeniceus* L., Icteridae), **marsh wrens** (*Cistothorus palustris* Wilson, Troglodytidae), and **common yellowthroats** (*Geothlypis trichas* L., Parulidae). Some human communities use the rhizomes and pollen for food and weave the leaves into baskets and bedding.

There are about 12 species worldwide, 3 in California and our area. The genus was established in 1753 by C. Linnaeus, who used the ancient Greek name for them.

Cattails can be difficult to identify. Not only are their defining characteristics a bit vague, but our species hybridize. Two features illustrated here are leaf sheath tops and inflorescences. In narrow-leaved and broad-tailed cattails, the top of the leaf sheath is broad, shaped like an upside-down ear lobe, and continues as a broad, tapering sheath. Southern cattail lacks this ear lobe but does have a narrow sheath along the edge. In narrow-leaved and southern cattails, there is an obvious space between the clusters of male (upper) and female (lower) flowers. In broad-leaved cattail, the male and female flower clusters meet.

	Narrow-leaved cattail *Typha angustifolia*	**Southern cattail** *Typha domingensis*	**Broad-leaved cattail** *Typha latifolia*
Leaves	U-shaped in cross-section at the base	V-shaped in cross-section at the base; base of upper surface gland-dotted	Flat in cross-section at the base
Sheath	Upside-down "ear lobe" at top of leaf sheath	No upside-down "ear lobe" at top of leaf sheath	Upside-down "ear lobe" at top of leaf sheath
Inflorescence	An obvious space between male and female flowers	An obvious space between male and female flowers	No space between male and female flowers
Female	Dark brown to reddish brown in flower and fruit	Bright yellow-brown ("buff") to orange-brown in flower and fruit	Green in flower, brown in fruit
Abundance	Scarce, only in very wet coastal region of lower San Mateo Creek	Most common cattail in Orange County; less common in western Riverside County	Uncommon in Orange County; most common cattail in western Riverside County

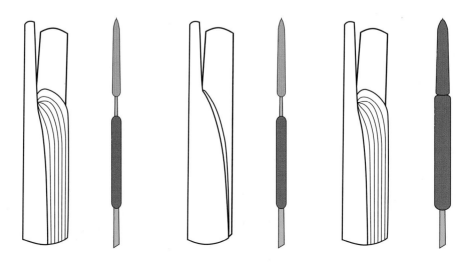

Narrow-leaved Cattail ✖?
Typha angustifolia L. (not pictured)

An upright perennial, 1.5-3 meters tall. The deep green *leaves and stems are very slender, taller than the flower stalk. Leaves are V-shaped in cross-section at the base.* The top of the leaf sheath looks like an *upside-down ear lobe* on each side. The male inflorescence sits far above the female, leaving a clear space between them. The female inflorescence is medium to dark brown when in flower and fruit. In flower May-July.

A plant of freshwater marshes and salty soils at low elevations, mostly in wet northern regions of the U.S., quite scarce in southern California. Known from San Mateo Creek, 0.5 miles upstream from I-5. Also reported from alkali wetlands off Avenida Presidio near North La Esperanza in San Clemente but not yet confirmed. Known from Eurasia and the Americas. Recent studies suggest that it is native to Europe and may have been introduced to North America in colonial times.

It was named in 1753 by C. Linnaeus for its narrow leaves (Latin, angustus = narrow; folius = leaf).

Southern Cattail

Southern or Slender Cattail
Typha domingensis Pers.

An upright perennial, 1.5-4 meters tall. The pale green *leaves and stems are very slender, generally shorter than the flower stalk. Leaves are U-shaped in cross-section at the base. The upper surface of the leaf is gland-dotted near the base.* The top of the leaf sheath tapers to the leaf; *it does not have "ear lobes."* As in narrow-leaved cattail, the male inflorescence sits far above the female, leaving a clear space between the two. The female inflorescence is bright yellow-brown to orange-brown when in flower and fruit. In flower June-July.

Found in wet soils, ponds, and marshes at many elevations. This is the most common cattail in Orange County.

It was named in 1806 by C.H. Persoon in reference to Santo Domingo, one of the Bahama Islands, where the first specimens were probably collected.

Narrow-leaved and southern cattail hybridize; their offspring are highly fertile and exhibit intermediate characteristics, making identification uncertain. Hybrids between broad-leaved cattail and either of the other species are sterile and less common.

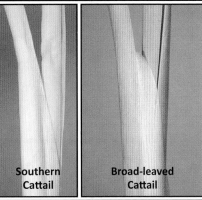

Southern Cattail | Broad-leaved Cattail

Broad-leaved Cattail
Typha latifolia L.

An upright perennial, 1.5-3 meters tall. Our largest cattail, its *leaves and stems are thicker than those of the other two species. Leaves are flat in cross-section at the base.* The top of the leaf sheath looks like an *upside-down ear lobe* on each side. *The male inflorescence usually sits right on top of the female, leaving no space between them.* The female inflorescence is green when in flower and brown in fruit. In flower June-July.

Grows in in wet soils, ponds, and marshes throughout our area. Less common in Orange County than narrow-leaved cattail, but the two are about equally common in western Riverside County.

It was named in 1753 by C. Linnaeus for its broad leaves (Latin, latus = broad; folius = leaf).

Broad-leaved Cattail

THE CATTAIL GUILD

Numerous species of wild animals and plants live in, on, and near cattails. Here are three of them.

Red-winged Blackbird
Agelaius phoeniceus (L.) (Icteridae)

A medium-sized songbird about 22 cm long from head to tail tip. *Adult males are glossy black and develop red shoulder patches, buffy yellow distally.* Females are brown with a light eyebrow stripe and a light breast and belly that is boldly streaked with brown; they sometimes have some red on the shoulders or pink on the chin and throat. First-year males are streaked like females but have the red shoulder patch.

A common resident of our area. They nest among cattails, often in large groups. Males are known to have a harem of up to 15 females and are extremely territorial. They have numerous different songs and calls, some melodic, some harsh.

The epithet was assigned in 1766 by C. Linnaeus for the male's red shoulder patch (Greek, phoenikos = crimson).

Red-winged Blackbird

Western Cattail Toothpick Grasshopper
Leptysma marginicollis hebardi Rehn & Eades (Acrididae)

A slender grasshopper, 28-45 mm long, with broad, flattened antennae, and a slanted face. Between the antennae bases, the head continues in a cone shape. It is usually brown, sometimes tinged with red or pink, and has a whitish, yellowish, or brownish stripe from the back edge of the eye to the base of the legs. Between the base of the front legs, it has a single small spine.

A common year-round inhabitant of cattails on which it feeds. When it sees you or perceives a threat, it walks to the other side of the leaf or stem. If disturbed, it flies off and lands on another plant; rarely does it land on the ground. We find it on southern cattail, though it probably feeds on broad-leaved cattail as well.

The subepithet was given in 1961 by J.A.G. Rehn and D.C. Eades in honor of M. Hebard.

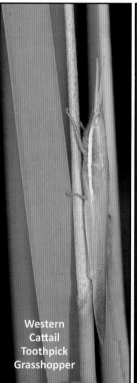

Western Cattail Toothpick Grasshopper

Cardinal Meadowhawk Dragonfly
Sympetrum illotum (Hagen) (Libellulidae)

A small dragonfly, its body is only 38-40 mm long. The body and wings are *red in color*, not orange as in some other local species. Each side of the thorax has two faint lines. Its red legs bear black spines and claws. Adults are in flight in late Mar-Nov.

Generally uncommon here but likely overlooked because of its small size. Like all adult dragonflies, they feed on flying insects and often alight on cattails. The small, broad, short-bodied nymphs live in the bottom of slow-moving ponds, where they feed on any aquatic animal they encounter.

The epithet was given in 1861 by H.A. Hagen. The beautiful red color of this dragonfly fades to dusky brown after its death. Since Hagen described it from dried-out, faded specimens, the name probably refers to the "dirty" appearance of those specimens (Latin, illotus = dirty, unwashed).

Cardinal Meadowhawk

Where to Go Wildflower-Watching

While the plains and lower foothills of Orange County and the adjacent regions of Los Angeles, San Bernardino, and Riverside counties have been developed to a large extent, there still remain wonderful natural areas in which to see wildflowers. A good number of our wildflowers appear during spring, but in our Mediterranean climate (cool and wet in winter, moist amd sunny in spring, hot and dry in summer and fall), something is always in bloom. Most annuals and perennials bloom in winter or spring, tougher annuals in early summer, and many perennials, especially those in the sunflower family, during fall. Even in the heat of summer, creekside annuals such as Rothrock's lobelia, canchalagua, and stream orchid are still putting on a show.

On county, state, and federal lands, there may be a fee to enter or to park your car. Most city lands are free. For frequent visitors to Orange County-owned parks, the Orange County Parks Annual Pass is an economical option. The pass is available at most parks, bicycle shops, and some sporting goods stores.

State Park daily and annual passes are available at the parks, possibly also at some sporting goods stores. Entry into the Cleveland National Forest is free, but in order to park your car within it (even along roads) you'll need to purchase and display a daily or annual National Forest Adventure Pass, available from federal ranger stations, some businesses along Ortega Highway, and some sporting goods stores.

Following is a brief list of good places to watch wildflowers, listed alphabetically within general place type. Within the book, each wildflower entry includes suggested places to see that species. For brevity, the following abbreviations are used: OC (Orange County), LAC (Los Angeles County), SB (San Bernardino County), RC (Riverside County), SD (San Diego County), CDFG (California Department of Fish and Game), CNF (Cleveland National Forest), CSP (California State Parks).

Coastal

These locations provide access to the coast and its wildflowers: Bolsa Chica Ecological Reserve, Bolsa Chica State Beach, Huntington city and state beaches, Balboa Peninsula, upper Newport Bay, Corona del Mar State Beach, Crystal Cove State Park (coastal side), Aliso Beach Park, Salt Creek Beach Park, Dana Point Headlands, Dana Point Harbor, Doheny State Beach, San Clemente State Beach, and San Onofre State Beach.

Dana Point Headlands

Foothills

There are several good foothill locations to find wildflowers: Crystal Cove State Park (inland side), Laguna Coast Wilderness Park, James Dilley Preserve, Moulton Meadows Park, Aliso and Wood Canyons Wilderness Park, Laguna Niguel Regional Park, Arroyo Pescadero Park, Schabarum Regional Park, Chino Hills State Park, Carbon Canyon Regional Park, Coal Canyon, Featherly Canyon RV Park, Oak Canyon Nature Center, Santiago Oaks Regional Park, Irvine Regional Park, Peters Canyon Regional Park, Limestone Canyon and Whiting Ranch Wilderness Park, O'Neill Regional Park, Riley Wilderness Park, Richard and Donna O'Neill Conservancy, and Caspers Wilderness Park.

O'Neill Conservancy

Mountains

Vehicle-accessible mountain spots include Silverado Canyon (past the residential community), Tucker Wildlife Sanctuary in upper Santiago Canyon, North Main Divide Road "Loop", Blue Jay Campground, El Cariso Campground, and The Lookout. The Santa Rosa Plateau is accessed from Clinton Keith Road, a paved road from Murrieta. Access to the higher elevations of the Santa Ana Mountains can be rather tricky. At this writing, the only paved road through them is Ortega Highway. The other access routes are unpaved, rugged, and steep; most require a high-clearance vehicle. These routes are Skyline Drive, Bedford Canyon Truck Trail, Indian Truck Trail, Main Divide Truck Trail, and Maple Springs Truck Trail out of Silverado Canyon. Non-motorized routes into the mountains include trails such as Coal Canyon, Coldwater Canyon, Harding Truck Trail, Holy Jim, Horsethief, Joplin, Santiago, and Trabuco.

Santa Ana Mountains

Favorite Places on Public Lands

Here are some of our publicly accessible favorite places to watch wildflowers. Some places host several wildflower species, while others have a few choice species generally not found elsewhere. Many of the photographs in this book were taken at the locations in this list. For discussions of local trails, see the wonderful hiking guides by Jerry Schad and Karin Klein, available in most bookstores and sporting goods stores.

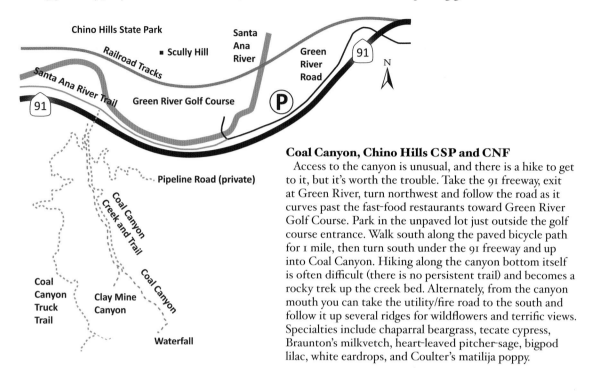

Coal Canyon, Chino Hills CSP and CNF

Access to the canyon is unusual, and there is a hike to get to it, but it's worth the trouble. Take the 91 freeway, exit at Green River, turn northwest and follow the road as it curves past the fast-food restaurants toward Green River Golf Course. Park in the unpaved lot just outside the golf course entrance. Walk south along the paved bicycle path for 1 mile, then turn south under the 91 freeway and up into Coal Canyon. Hiking along the canyon bottom itself is often difficult (there is no persistent trail) and becomes a rocky trek up the creek bed. Alternately, from the canyon mouth you can take the utility/fire road to the south and follow it up several ridges for wildflowers and terrific views. Specialties include chaparral beargrass, tecate cypress, Braunton's milkvetch, heart-leaved pitcher-sage, bigpod lilac, white eardrops, and Coulter's matilija poppy.

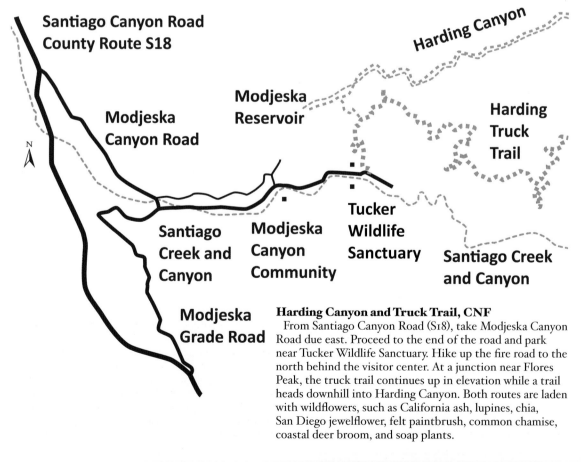

Harding Canyon and Truck Trail, CNF

From Santiago Canyon Road (S18), take Modjeska Canyon Road due east. Proceed to the end of the road and park near Tucker Wildlife Sanctuary. Hike up the fire road to the north behind the visitor center. At a junction near Flores Peak, the truck trail continues up in elevation while a trail heads downhill into Harding Canyon. Both routes are laden with wildflowers, such as California ash, lupines, chia, San Diego jewelflower, felt paintbrush, common chamise, coastal deer broom, and soap plants.

Holy Jim Trail, CNF

Suggested only for vehicles with higher than average clearance. From Live Oak Canyon Road / Trabuco Canyon Road near O'Neill Regional Park, drive up the rocky dirt road through Arroyo Trabuco (careful, the road washes out most winters). It passes through coastal sage scrub, chaparral, and a variety of riparian communities. Be respectful of the private homes and their property surrounding the road. Just past the Holy Jim fire station is a parking lot at Alder Spring. Park here (remember your National Forest Adventure Pass) and hike up Holy Jim Canyon Trail (northern fork from the parking lot) to see oak woodland, riparian forests, and chaparral. Marvel at ocellated Humboldt lilies in July. About 1.5 miles up from the forest gate is a small trail spur to Holy Jim Falls, a good place to rest and snack. The rock face is host to maidenhair fern and fall-blooming Rothrock's lobelia. This is the easiest place in our area to see round-leaved boykinia. The giant chain ferns here are spectacular. From the spur, Holy Jim Trail continues up toward Main Divide Truck Trail. Along it are California ash and the elusive wartleaf lilac, and striking views of the canyon below. As it nears Main Divide Truck Trail, there is another population of round-leaved boykinia, plus Lemmon's campion. Optionally, you can continue along Main Divide Truck Trail to Santiago Peak or return on the same trail.

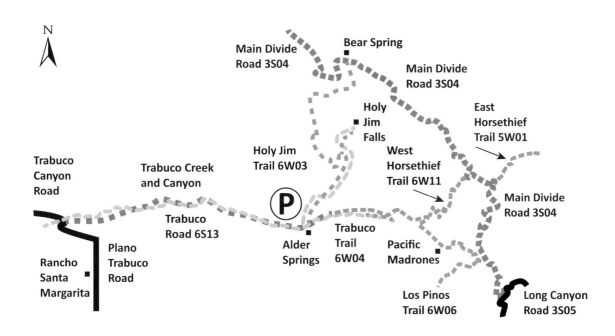

Trabuco Trail/Trabuco Canyon, CNF

Follow the directions for Holy Jim Trail to Alder Spring. Continue up the main road another mile or so to the trailhead for Trabuco Trail. There are very few parking spots there, so good luck finding a spot. If not, head back and park along a safe pull-out that doesn't block the road or a driveway or park in the lot at Alder Spring and hike to the trailhead. Long ago this was a truck trail that intersected Main Divide Truck Trail at Los Pinos Saddle, but the road was wiped out many storms ago. Plant communities along this trail include coastal sage scrub, chaparral, oak woodland, riparian forests, and an extensive mixed conifer forest with bigcone Douglas-fir and bigleaf maple. Trailside wildflowers are numerous and include chaparral yucca, Weed's mariposa lily, California ash, Martin's paintbrush, scarlet larkspur, chaparral whitethorn, and toyon. About 2.5 miles up is our only accessible stand of Pacific madrone. Afterward, you can return to the trail head or continue up to Los Pinos Saddle.

Ronald W. Caspers Wilderness Park, OC

Along Ortega Highway, nearly 7.5 miles northeast of the Santa Ana (I-5) Freeway in San Juan Capistrano (entrance fee). An 8,000-acre wilderness, its extensive hiking trails travel through vast habitat and scenery. The area is so large and rugged that many parts remain to be explored. While all of our park areas are especially nice places to see wildflowers, hike, and photograph, Caspers holds a special place in our hearts. We have both been visiting the park since it opened in 1974.

Most wildflowers in this book are found in Caspers. Specialty wildflowers here include yellow pincushion, scale-broom, valley cholla, small Venus's looking-glass, ladies's fingers, small-flowered morning glory, lupines, Engelmann oak, felt paintbrush, poppies, purple Chinese houses, padre's shooting star, several species of everlasting, phacelia, and monkeyflower, and nice stands of California sycamore, coffeeberry, and rose. It is one of few places that has three species of mariposa lilies: Catalina, splendid, and Weed's.

To get started, park at the north end of Bell Canyon Road and head north on Bell Canyon Trail. Along the way are padre's shooting stars (in early spring), mariposa lilies (spring), and Coulter's matilija poppy (spring-summer), among others. Either return on the same trail, or loop back on Star Rise and Oak trails. For a more adventurous hike, head to the Dick Loskorn Trail where coast live oaks and a California trio of sycamore, coffeeberry, and rose forms a shady woodland. The trail leads up a steep hill to the West Ridge Trail with Catalina mariposa lily along the way. In years of good rainfall, the East Ridge Trail explodes with wildflowers such as California poppies, ladies's fingers, padre's shooting star, false rosinweed, canchalagua, and wide-throated monkeyflower. The sandy bench between San Juan Creek and Ortega Flats Campground has an extensive formation of arroyo scrub with yellow pincushion (spring), valley cholla (spring), scale-broom (fall), and Sonora everlasting (fall). Shady oak woodlands often host purple Chinese houses.

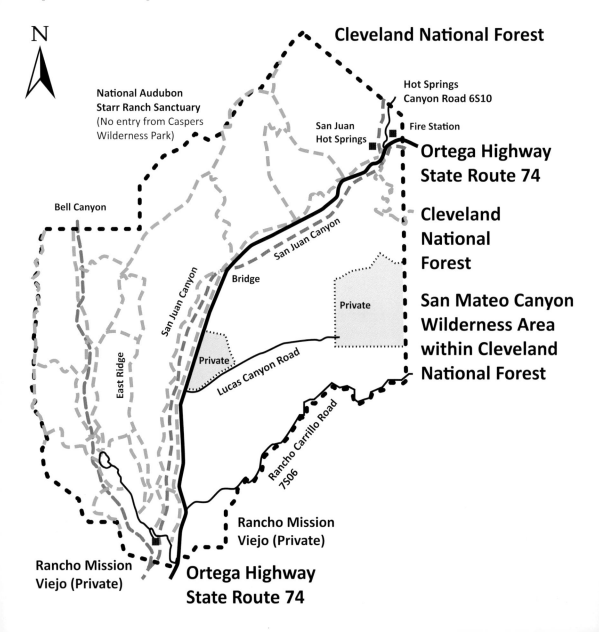

San Juan Loop Trail, CNF

Follow Ortega Highway to the loop trail parking lot, opposite the Ortega Candy Store. Park in the loop trail's lot (Forest Adventure Pass required); do not park in the candy store's lot. San Juan Loop Trail is most often hiked counterclockwise from the north end of the parking lot, though you can head clockwise from the south end of the parking lot. Going north, the trail meanders over rocky ground through oak woodlands, chaparral, and a moist north-facing hillside, then to an overlook above Decker Canyon Creek. A short optional spur takes you down to water's edge and its riparian community. There are a few gnarled California junipers growing out of cracks in bare rock by the creek, fine subjects for photography. A scramble back up to the main trail takes you westward around the hillside, which is strewn with common chamise, sugar bush, ferns, and other assorted wildflowers. Switchbacks and rollercoaster segments bring you to San Juan Creek where large coast live oaks provide welcome shade. The Chiquito Trail crosses the creek and joins the loop trial, which then passes through Upper San Juan Campground. A short uphill hike brings you back to the parking lot; be sure to look at the ferns and wildflowers as you progress to the south end of the parking lot. Specialties along this trail include fringepods, popcorn flowers, coast jepsonia, California peony, Palomar monkeyflower, Parry's blue-eyed Mary, sticky cinquefoil, and mesa horkelia.

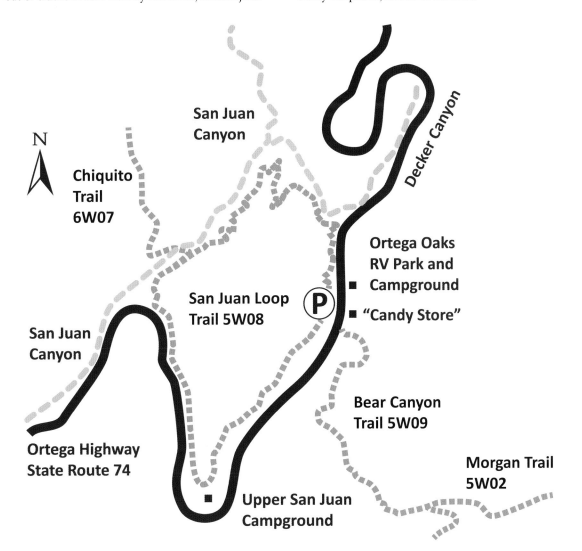

Bear Canyon Trail, CNF

Same parking as San Juan Loop Trail (above). Carefully cross Ortega Highway on foot. The trailhead is slightly south of the candy store. Just beyond a billboard kiosk is a sign-in station; be sure to sign in and out. The trail heads toward giant granitic boulders to the south and passes over a ridge into Morrell Canyon. In the chaparral right at the trailhead is an amazing abundance of wildflowers such as slender sunflower, canyon dodder, canchalagua, deer's ears, white-flowered currant, California peony, and Parry's blue-eyed Mary. Up and over the ridge, at the edge of Morrell Canyon is the only population of ribbonshanks known in the Santa Ana Mountains (about 0.5 mile south of the trail, binoculars needed). In trailside meadows farther along are wildflowers such as California tea, intermediate thick-leaved monardella, and purple Chinese houses.

Bluejay Campground, Los Pinos Potrero, Chiquito Basin, San Juan Trail, Ortega Highway, CNF

From Ortega Highway, head west up Long Canyon Road (more direct but steep) or Tenaja Road (also called North Main Divide Road, Forest Service Route 3S04, a little longer drive but gentler) and follow it to Bluejay Campground. There are day-use and camping fees. Park within the campground or outside of its entrance, where there are a few parking spots and a Forest Adventure Pass is required.

Plant communities in Blue Jay Campground, Falcon Group Camp, and in neighboring Los Pinos Potrero include oak woodland, coastal sage scrub, chaparral, and grasslands. Coulter pines in and around the potrero (pasture) were planted by the US Forest Service. Specialty wildflowers include woolly-fruited wing-fruit, mountain dandelions, common hareleaf, San Bernardino aster, southern mule's ears, American winter cress, bluecups, Eastwood manzanita, lupines, Parry's deer's ears, San Miguel savory, and Fish's milkwort.

A day trip to the area will be pleasantly sufficient for most, but more adventurous wildflower-watchers will want to wander farther. A stroll down Falcon Trail, especially as it turns toward Hot Springs Falls, will reveal several wildflowers and specialties such as California barberry, California saxifrage, long-spurred plectritis, and white fairy lantern. From Blue Jay Campground, San Juan Trail takes off to the south, passes though the Chiquito Basin, and eventually ends up in Hot Springs Canyon, an 11-mile hike (of course you can hike part-way and return). Special treats there include California milkweed, Douglas's stitchwort, Parry's deer's ears, Scouler's St. John's Wort, and Greene's cinquefoil. About midway along San Juan Trail, near Sugarloaf and Old Sugarloaf Peaks, is a population of chaparral pea.

If you plan to take the entire trail, leave a car at the Hot Springs Canyon trailhead (remember your Forest Adventure Pass) and carpool up to Blue Jay Campground. The northern portion of the trail is moderately oak-shaded, the middle and lower portions have little to no shade at all. Take lots of water, bring a map, and wear a hat.

Another way to do the San Juan Trail is to start at the Hot Springs Canyon trailhead and hike up. From Ortega Highway, turn north onto Hot Springs Canyon Road (Forest Service Route 6S10, often called Lazy W Ranch Road) and continue about 1 mile to the well-marked trailhead. Park your car (Forest Adventure Pass required) and head up the trail. It's quite steep and full of switchbacks, but in years of good winter/spring rainfall the wildflowers are spectacular. Specialties along the first half-mile include Weed's mariposa lily, Cooper's rein orchid, fire poppy, sapphire woolly-star, Nuttall's snapdragon, and coast jepsonia.

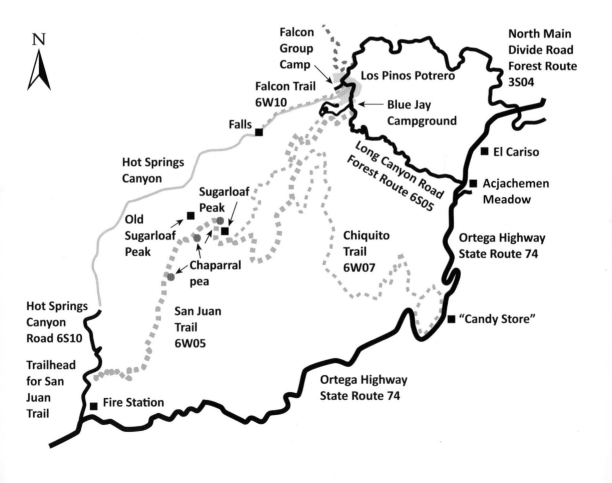

North Main Divide Road "Loop," CNF

A fairly easy counterclockwise drive that begins and ends on Ortega Highway. Take Ortega Highway to El Cariso and turn northward onto Tenaja Road (also called North Main Divide Road, Forest Service Route 3S04). The first one-third of the loop passes through terrain that treats you to California junipers, bigberry manzanitas, various California lilacs, southern California currant, southern California hillside gooseberry, and Parish's goldenbush. At 4 miles past Ortega Highway, you'll come to a junction with Long Canyon Road (though unlabeled, it is Forest Service Route 6S05). At the junction is Eastwood manzanita, Cleveland's bush monkeyflower, pale silk-tassel bush, bush poppy, and a roadside seep with stream orchid and California threadtorch. To proceed with the loop, turn onto the left fork (6S05, Long Canyon Road). The road goes downhill and passes by Falcon Group Camp and Los Pinos Potrero; parking is allowed just before Falcon Group Camp and again just after the potrero. Along this stretch are Parry's larkspur, purple owl's-clover, Parry's deer's ears, and southern mule's ears. Next comes Blue Jay Campground (see above) followed by a moderate downhill grade. Along this stretch of the road is an unusually dense, nearly pure stand of white sage. After it passes under a shady grove of coast live oaks, turn left. Soon it becomes quite a steep downhill grade. Dotting its southern side is a spectacular population of scarlet larkspur, some of which reach 7 feet tall, a favorite hangout for thirsty hummingbirds in late spring. The road continues directly to Ortega Highway.

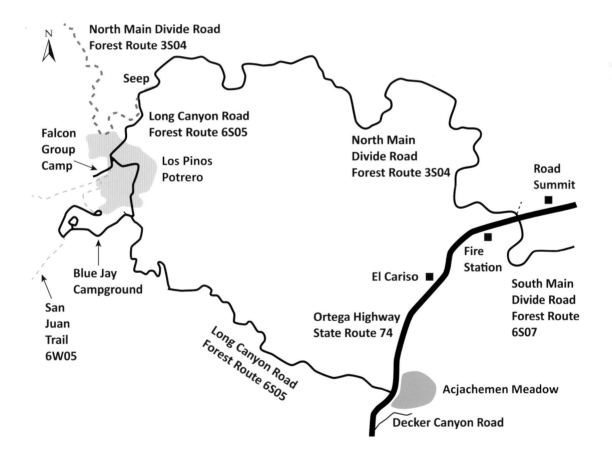

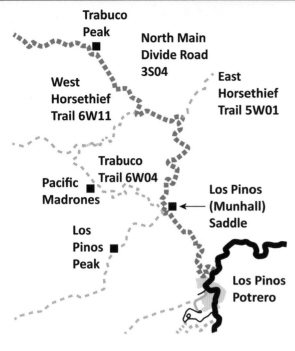

Los Pinos Saddle, CNF

A steep, loose dirt road, not for underpowered or low-slung vehicles. From Ortega Highway (SR74) near the community of El Cariso, take Tenaja Road (3S04, North Main Divide Road) northwest. At 4 miles past Ortega Highway, the road forks. Stay to the right on North Main Divide Road (3S04) and continue through the forest gate, which is usually open (call the US Forest Service at 951-736-1811 to find out). The road goes uphill for about 2 miles and takes you to a flat parking area at Los Pinos Saddle, also called Munhall Saddle (Forest Adventure Pass required). The Forest Service gate there is usually open, but you can stop here at the saddle. On the way up, stop along turnouts in the road when you see wildflowers appear. The variety here is remarkable and includes lupines, felt paintbrush, woolly bluecurls, Eastwood manzanita, California lilacs, Cleveland's bush monkeyflower, intermediate thick-leaved monardella, and slender sunflower. Right in the parking area is a nice population of poodle-dog bush with its trumpet-like purple flowers in late spring, a favorite of local hummingbirds and native bees. You can remain here at the saddle or head up Los Pinos Trail or down Trabuco Trail. From here, the first mile down the Trabuco Trail is densely shaded by a forest of canyon live oak.

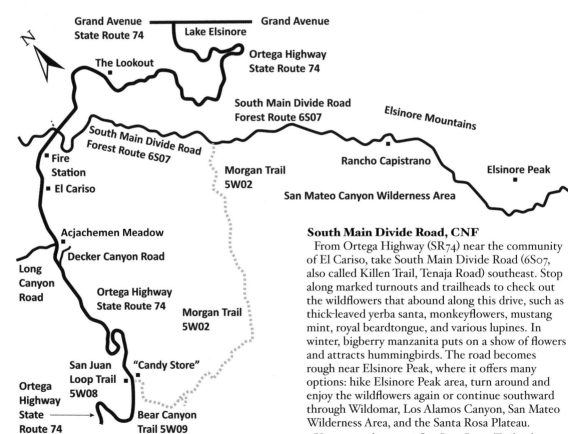

South Main Divide Road, CNF

From Ortega Highway (SR74) near the community of El Cariso, take South Main Divide Road (6S07, also called Killen Trail, Tenaja Road) southeast. Stop along marked turnouts and trailheads to check out the wildflowers that abound along this drive, such as thick-leaved yerba santa, monkeyflowers, mustang mint, royal beardtongue, and various lupines. In winter, bigberry manzanita puts on a show of flowers and attracts hummingbirds. The road becomes rough near Elsinore Peak, where it offers many options: hike Elsinore Peak area, turn around and enjoy the wildflowers again or continue southward through Wildomar, Los Alamos Canyon, San Mateo Wilderness Area, and the Santa Rosa Plateau.

You can park a car at San Juan Loop Trail, take another car up to South Main Divide Road, park at Morgan Trail (Forest Adventure Pass needed for both parking places), and hike south along the Morgan Trail to the Bear Canyon Trail. It offers oak woodlands, chaparral, riparian, and scenic views.

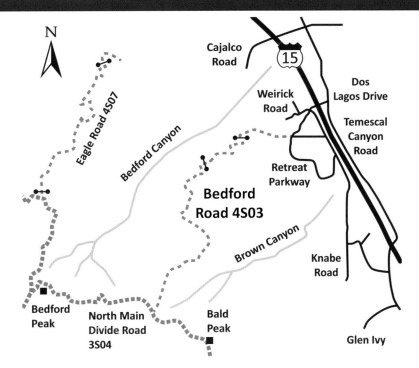

Bedford Canyon Road, Main Divide Road, CNF

From Interstate 15, take Wyrick Road offramp, then go up Wyrick Road and head southwest toward the Santa Ana Mountains. The road meanders uphill through an older neighborhood and then a newer one before reaching the Forest Service gate, which is closed at unpredictable times. From the forest gate up to Main Divide Road, keep on the lookout for wildflowers and stop along the way. Treasures include California lilacs, yellow bush beardtongue, tidy tips, red-flowered onion, school bells, Catalina mariposa lily, padre's shooting stars, and California barberry.

Indian Truck Trail to Main Divide Road, CNF

A rough, pitted dirt road, not for low-slung vehicles. From Interstate 15, exit at Indian Truck Trail and head southwest toward the Santa Ana Mountains. With many private dirt roads appearing along the way, it's a little tricky to keep on the right one, but stick to the most-worn road and you'll be fine. Highlights include Coulter's snapdragon (by the acre in a good year), southern pink, Lemmon's campion, canyon live-forever, Martin's paintbrush, heart-leaved pitcher-sage, sand-wash butterweed, wide-throated monkeyflower, and California lilacs.

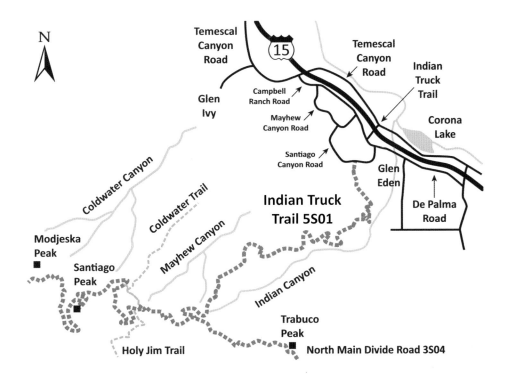

Parks and Nature Centers

There are many local places at which to see wildflowers. Many have hiking guidelines and maps, available in a kiosk or at a visitor center.

Orange County Parks. In the OC Park system, their names tell you something about their purpose and hint about their appearance and function; a general description follows. For more information about parks operated by the County of Orange, see www.ocparks.com.

- A **Regional Park** is a property developed for use by people. They often have maintained lawns, trees, picnic benches, and barbeques. Some of them include natural areas and trails that allow hiking and biking. Some also offer equestrian use and camping.
- A **Wilderness Park** is an undeveloped property set aside for wildlife and for the enjoyment of people. They have established trails and allow hiking, biking, and often equestrian use. Some include a visitor center, a building with educational displays where visitors can learn about wildlife in the area.
- A **Nature Center** is a property that includes a visitor center and is often staffed with employees or volunteers who give nature walks and talks. Most include a surrounding natural area with a system of hiking trails.
- A **Reserve** is a property left undeveloped and set aside for a special purpose, often as "open space" or future transformation into another type of use such as a wilderness park. A **Preserve** is an area left undeveloped and is specifically for the protection of wildlife or other natural resource. Reserves and preserves are not established for the purpose of human recreation and typically do not promote mass visitation by people.

Local Parks

Aliso and Wood Canyons Wilderness Park, OC
28373 Alicia Parkway, Laguna Niguel, CA 92677, 949-923-2200. A 4,000-acre wilderness park, open for hiking, mountain biking, and equestrian use.

Aliso Beach Park, OC
31131 S. Pacific Coast Highway, Laguna Beach, CA 92652, 949-923-2280. Good coastal plants and geology. Canyon sunflower is found on its inland portion.

Arroyo Trabuco, *see O'Neill Regional Park.*

Bolsa Chica State Beach, CSP
17851 Pacific Coast Highway, Huntington Beach, CA 92648, 714-846-3460, 714-377-2481, www.parks.ca.gov/?page_id=642. A three-mile-long beach from Warner Avenue south to Seapoint Avenue. Our best place to see bather's delight, beach morning-glory, and beach evening primrose.

Bolsa Chica State Ecological Reserve, CDFG and the Bolsa Chica Conservancy
3842 Warner Avenue, Huntington Beach, CA 92469-4263, 714-846-1114, www.bolsachica.org. A 900-acre coastal wetland, nationally famous among birders. Its sandy banks host various coastal plants and dune wildflowers.

Carbon Canyon Regional Park, OC
4442 Carbon Canyon Road, Brea, CA 92823, 714-973-3160. A 124-acre developed park with access to hiking into the Chino Hills.

Ronald W. Caspers Wilderness Park, OC
33401 Ortega Highway, San Juan Capistrano, CA 92675, 949-923-2210. The jewel of the OC Park system. Its holdings are immense (8,000 acres) and provide access to a wide range of habitats.

Chino Hills State Park, CSP
4721 Sapphire Road, Chino Hills, CA 91709, 760-389-2281. Largely undeveloped, 13,000 acres in the expansive Chino Hills, host of the southernmost large population of southern California walnut trees.

Cleveland National Forest, U.S. Forest Service
1147 East 6th Street, Corona, CA 92879, 909-736-1811. Spectacular 135,000-acre national forest. Parking requires a Forest Adventure Pass.

Crystal Cove State Park, CSP
8471 North Pacific Coast Highway, Laguna Beach, CA 92651, 949-497-7647, 949-494-3539. A 2,800-acre park with habitats that begin underwater and progress upland onto beaches, ancient marine terraces, and rugged backcountry hills above Laguna Beach.

Dana Point Harbor, OC
949-496-1094. Take Dana Point Harbor Drive to its west end and park. Walk into the nearby marine preserve, but do not climb the cliffs.

Dana Point Nature Interpretive Center, Dana Point Headlands, City of Dana Point
34558 Scenic Drive, Dana Point, CA 92629, 949-542-4755. Access on Cove Road from Dana Point Harbor or on Street of the Green Lantern from Pacific Coast Highway. Park near the interpretive center and hike the well-marked trails. Incredible ocean and coastline view; coastal sage scrub and coastal bluff scrub.

El Dorado Nature Center, City of Long Beach
7550 E. Spring Street, Long Beach, CA 90815, 562-570-1773, www.longbeach.gov/naturecenter. A museum and 105-acre nature preserve provide sanctuary for animals and plant life.

James Dilley Preserve
See Laguna Coast Wilderness Park.

Fairview Park, City of Costa Mesa
2525 Placentia Avenue, Costa Mesa, CA 92627, www.cmfairviewpark.org. Our largest vernal pool complex, known to biologists through about 1940, later "forgotten," then "rediscovered" in the 1990s.

Featherly Regional Park, OC
24001 Santa Ana Canyon Road, Anaheim, CA 92808, 714-771-6731, 714-637-0210. Situated along the Santa Ana River, the park contains both natural riparian woodland and maintained picnic areas.

City of Irvine Open Space Preserve
Managed by Irvine Ranch Conservancy, 714-508-4757, www.irconservancy.org. Owned by City of Irvine, 949-724-6738, www.cityofirvine.org/cityhall/cs/openspace/default.asp. Includes Bommer Canyon, Shady Canyon, Quail Hill, and portions of Loma Ridge. Some sections are open for 7-day access, others are by advance permission. Visit www.irlandmarks.org for more information on these and other wildlands.

PARKS AND NATURE CENTERS

Irvine Ranch Open Space, OC
Managed by Irvine Ranch Conservancy, 714-508-4757, www.irconservancy.org. A 20,000-acre reserve on former ranchland. Properties include Limestone, Weir, Gypsum, Fremont, and Agua Chinon canyons and portions of Blackstar Canyon. Entry is by advance permission. Visit www.irlandmarks.org for more information on these and other wildlands.

Irvine Regional Park, OC
1 Irvine Park Road, Orange, CA 92862, 714-973-6835. At 477 acres, this is Orange County's oldest park, established in 1897. The body of the park is developed for family use, but Santiago Creek and its surrounding foothills provide good habitat for wildflowers. Includes a nature center and Orange County Zoo.

Laguna Coast Wilderness Park, OC
20101 Laguna Canyon Road, Laguna Beach, CA 92651, 949-923-2235. Includes the 76-acre Laguna Laurel Ecological Reserve (California Department of Fish and Game), Jim Dilley Greenbelt Preserve (City of Laguna Beach), and Nix Nature Center (18751 Laguna Canyon Road, 949-923-2235). A 6,200-acre wilderness in the rugged hills above Laguna Beach. Spectacular spring wildflowers include Allen's daisy, coastal bush monkeyflower, and California buttercups.

Laguna Niguel Regional Park, OC
28241 La Paz Road, Laguna Niguel, CA 92677, 949-923-2240. A developed regional park. Most native plants appear around the lake.

Limestone Canyon Nature Preserve
A large preserve accessed from Santiago Canyon Road. Entry is only with Docent-led hikes by Irvine Ranch Conservancy. Abundant wildflowers.

William R. Mason Regional Park, OC
18712 University Drive, Irvine, CA 92612-2601, 949-923-2220. At 345 acres, the northern portion is a typical regional park, while the southern part is undeveloped and supports native vegetation and wildlife. Coastal goldenbush is plentiful here.

Nix Nature Center
See Laguna Coast Wilderness Park.

Oak Canyon Nature Center, City of Anaheim
6700 East Walnut Canyon Road, Anaheim, CA 92807, 714-998-8380. This 58-acre wilderness park packs in a surprising variety of habitats and wildflowers. In spring, the trails are colored with baby blue eyes, lupines, and white snapdragon.

O'Neill Regional Park, OC
30892 Trabuco Canyon Road, Trabuco Canyon, CA 92679, 949-923-2260. One of Orange County's oldest parks, encompassing 3,100 acres. The north end is developed mostly for camping and day use. Park at the southern end and enjoy the network of trails: a ridgetop to the east and Arroyo Trabuco due south.

Peters Canyon Regional Park, OC
8548 East Canyon View Avenue, Orange, CA 92869, 714-973-6611. A 359-acre park, centered around Peters Canyon Creek and Reservoir. Offers good riparian and coastal sage scrub habitats.

Richard and Donna O'Neill Conservancy, part of The Reserve at Rancho Mission Viejo
P.O. Box 802, San Juan Capistrano, CA 92693, 949-489-9778, theconservancy.org. A 1,200-acre wilderness reserve open only for scheduled nature walks and other events. Located 5 miles east of San Juan Capistrano, just south of Ortega Hwy. On private land and not open for drop-in visits; reservations are required. Great displays of padre shooting stars, San Diego tarplant, many-stemmed live-forever.

Thomas F. Riley Wilderness Park, OC
Formerly Wagon Wheel Wilderness Park. 30952 Oso Parkway (at Coto de Caza Drive), Coto de Caza, CA 92679, 949-923-2265. For many years a landlocked preserve adjacent to Caspers Wilderness Park, development of Coto de Caza brought with it roads for easy access into the 1,200-acre park. From Interstate 5, head east along Oso Parkway for 6 miles.

Salt Creek Beach Park, OC
33333 S. Pacific Coast Highway, Dana Point, CA 92629, 949-923-2280. Provides access to 18 acres of sandy beaches, rock outcrops, and cliffs. Coastal wildflowers can be viewed along the cliff bases.

San Clemente State Beach, CSP
225 Avenida Calafia, San Clemente, CA 92672, 949-492-0802, 949-366-8594. Primarily a 110-acre camping and surfing beach, but coastal wildflowers still remain along the bluffs. From Interstate 5, exit at Avenida Calafia, turn southwest, follow it to the beach and into the park.

San Joaquin Wildlife Sanctuary, Irvine Ranch Water District
P.O. Box 5447, Irvine, CA 92616, 949-261-7963. A joint venture between the Irvine Ranch Water District and Sea and Sage Chapter of the National Audubon Society. A 300-acre sanctuary. mostly a series of man-made ponds formed in San Joaquin Marsh near San Diego Creek. Wildflowers are most common along nature trails through the adjacent floodplain to the west. Accessed from Riparian View via Campus Drive.

San Mateo Wilderness, *See Cleveland National Forest.*

San Onofre State Beach, CSP
www.parks.ca.gov. A 3,000-acre park along San Mateo Creek to its ocean outfall at Trestles Beach, south to San Onofre Surf Beach and coastal bluffs. Specialty wildflowers include California box-thorn and beach evening primrose. For easy coastal access to Trestles, take the San Diego Freeway (I-5) to San Clemente, exit at Cristianitos Road, turn inland over the bridge, and park at the public lot nearby. San Mateo Campground is farther northeast along Cristianitos. San Onofre Bluffs and San Onofre Surf Beach are farther south, exit I-5 at Basilone Road.

Santiago Oaks Regional Park, OC
2145 North Windes Drive, Orange, CA 92669, 714-973-6620. A 350-acre park nestled along Santiago Creek in the Santa Ana Mountain foothills. Its trails provide access to alluvial scrub and coastal sage scrub habitat, home to numerous wildflowers.

Shipley Nature Center, City of Huntington Beach
17829 Goldenwest Street, Huntington Beach, CA 92647, 714-842-4772, www.shipleynature.org. An 18-acre fenced nature center within Huntington Beach Central Park. Access from Goldenwest Street, about 3 miles south of I-405.

Starr Ranch Sanctuary, National Audubon Society
100 Bell Canyon Road, Trabuco Canyon, CA 92679, 949-858-0209, 949-858-3537, www.starrranch.org. Occupying upper Bell Canyon, north of Caspers Wilderness Park (both were once owned by the Eugene Starr family), host to many natural habitats and wildflowers. Access restricted (no drop-in visits).

Talbert Regional Park (formerly called Talbert Nature Preserve), OC
1298 Victoria Avenue, Costa Mesa, CA 92627, 949-923-2250. Located along the Santa Ana River not far above its outfall at the Pacific Ocean. Features plants native to the river channel.

Tucker Wildlife Sanctuary, California State University, Fullerton
29322 Modjeska Canyon Road, Modjeska Canyon, CA 92676-9801, 714-649-2760, www.tuckerwildlife.org. A 12-acre preserve in the community of Modjeska Canyon, nestled in upper Santiago Canyon. Offers a nice visitor center, observation deck, and nature trail. The adjacent Harding Canyon Truck Trail leads up into the Santa Ana Mountains.

Turtle Rock Nature Center
See Irvine Open Space Reserve.

Upper Newport Bay Nature Preserve, OC and CDFG
2301 University Drive, Newport Beach, CA 92660, 949-923-2290, www.ocparks.com/uppernewportbay. Features both a 752-acre reserve and a 140-acre nature preserve. The bay is ringed by coastal sage scrub with riparian and marsh habitats in the lowlands near the water. Both play host to a wonderful array of wildflowers. This is the only place in our area to see saltmarsh bird's beak. Includes the Peter and Mary Muth Interpretive Center.

Harriett M. Wieder Regional Park, OC
19251 Seapoint Avenue, Huntington Beach, CA 92648, 949-923-2250. A link between Huntington Beach Central Park and Bolsa Chica State Beach, the park hosts developed areas and native plant gardens. Of particular value is its view of the nearby Bolsa Chica Ecological Reserve.

Whiting Ranch Wilderness Park, OC
P.O. Box 156, Trabuco Canyon, CA 92678, 949-923-2245. Until recently, managed as a unit with Limestone Canyon Nature Reserve. A 4,000-acre park that occupies 2 disjunct properties, the largest and most popular accessed from Borrego Trail just off Portola Parkway at Market Street in Foothill Ranch. Wildflowers are abundant.

Native Plant Gardens, Arboreta, and Local Retail Sources
Orange County
Fullerton Arboretum
1900 Associated Road, Fullerton, CA 92831, 714-278-3579, arboretum.fullerton.edu.

Goldenwest College, California Native Garden
15744 Goldenwest Street, Huntington Beach, CA 92647, 714-892-7711 x52181, www.goldenwestcollege.edu/garden.

Tree of Life Nursery
33201 Ortega Highway, P.O. Box 635, San Juan Capistrano, CA 92693, 949-728-0685, www.californianativeplants.com.

University of California, Irvine, Arboretum
19172F Jamboree Boulevard (at Campus Dr.), Irvine, CA 92697, 949-824-5833, arboretum.bio.uci.edu.

Los Angeles County
Rancho Santa Ana Botanic Garden
1500 North College Avenue, Claremont, CA 91711-3157, 909-625-8767, www.rsabg.org.

Theodore Payne Foundation
10459 Tuxford Street, Sun Valley, CA 91352-2116, 818-768-1802. Recorded wildflower hot line for all of southern California: 818-768-3533, www.theodorepayne.org.

Riverside County
University of California, Riverside, Botanic Gardens
900 University Avenue, Riverside, CA 92521-0124, 909-784-6962, botanicgardens.ucr.edu.

San Diego County
Las Pilitas Nursery
8331 Nelson Way, Escondido, CA 92026, 760-749-5930, www.laspilitas.com.

San Diego Botanic Garden (formerly Quail Botanic Gardens)
230 Quail Gardens Drive, Encinitas CA 92024, 760-436-3036, www.SDBGarden.org.

Organizations
California Native Plant Society (CNPS)
2707 K Street, Suite 1, Sacramento, CA 95816-5113, 916-447-2677, www.cnps.org. Founded in 1965, dedicated to the study and conservation of California native plants. Publishes *Fremontia,* a quarterly journal about native plants. Has several chapters.

Laguna Canyon Foundation
384 Legion Street, Laguna Beach CA 92651, 949-497-8324, www.lagunacanyon.org. Preserves, protects, enhances, and promotes the South Coast Wilderness Area in Orange County. Provides educational/recreational programs and activities, coordinates land acquisitions, habitat restoration, and funding.

Laguna Greenbelt, Inc.
P.O. Box 860, Laguna Beach, CA 92652, www.lagunagreenbelt.org. Founded in 1968; promotes the preservation of Orange County's wilderness and open space, primarily near Laguna Beach.

National Audubon Society (NAS)
www.audubon.org. Founded in 1886; dedicated to conservation and restoration of natural ecosystems; sponsors scientific and educational programs. Historically a bird-centered organization, it has broadened its focus to include native plants and other forms of wildlife. Has several chapters.

Sierra Club
85 Second Street, 2nd Floor, San Francisco, CA 94105, 415-977-5500, www.sierraclub.org. Founded by John Muir and friends in 1892 and dedicated to all aspects of environmental protection.

Southern California Botanists (SCB)
Southern California Botanists, 1500 North College Avenue, Claremont, CA 91711, www.socalbot.org. Founded in 1927, devoted to the study, preservation, and conservation of the native plants and plant communities of southern California. Publishes *Crossosoma*, a quarterly journal about native plants, and holds an annual botanical conference.

The Nature Conservancy, California Chapter
201 Mission Street, 4th Floor, San Francisco, CA 94105-1832, 415-777-0487, nature.org. Founded in 1951. Dedicated to habitat preservation by forming partnerships and purchasing lands for preservation.

Local Chapters and Organizations

Orange County
Orange County Chapter, CNPS
P.O. Box 54891, Irvine CA 92619-4891. www.occnps.org.

Orange County Wild
www.orangecountywild.org.

Laguna Canyon Foundation
www.lagunacanyon.org.

Laguna Hills Chapter, NAS
P.O. Box 2652, Laguna Hills, CA 92654

Sea and Sage Chapter, NAS
P.O. Box 5447, Irvine, CA 92616, 949-261-7963. www.seaandsageaudubon.org.

Orange County Group, Sierra Club
www.angeles.sierraclub.org/orange.

Sierra Sage Group, Sierra Club
www.angeles.sierraclub.org/sage.

Los Angeles County
Los Angeles-Santa Monica Mountains, CNPS
(818) 881-3706, www.lacnps.org.

San Gabriel Mountains Chapter, CNPS
c/o Eaton Canyon Nature Center, 1750 North Altadena Drive, Pasadena, California 91107, www.cnps-sgm.org.

South Coast Chapter, CNPS
www.sccnps.org.

El Dorado Chapter, NAS
P.O. Box 90713, Long Beach, CA 90809-0713, 562-961-5711, www.geocities.com/eldoraudubon.

Los Angeles Chapter, NAS
7377 Santa Monica Boulevard, West Hollywood, CA 90046, 323-876-0202, losangelesaudubon.org.

Palos Verdes/South Bay Chapter, NAS
P.O. Box 2582, Palos Verde, CA 90274, 310-722-7777, www.lmconsult.com/pvaudubon.

Pasadena Chapter, NAS
1750 North Altadena Drive, Pasadena, CA 91107, 626-355-9412, www.pasadenaaudubon.org.

San Fernando Valley Chapter, NAS
P.O. Box 7769, Van Nuys, CA 91409, 818-347-3205

Whittier Chapter, NAS
P.O. Box 548, Whittier, CA 90608, www.whittieraudubon.org.

Chino Hills State Park Interpretive Association
4717 Sapphire Road, Chino Hills, CA 91709, www.chinohillsstatepark.org.

Santa Ana Mountains Natural History Association
www.freewebs.com/santaanamountains.

Riverside County
Riverside-San Bernardino Chapter, CNPS
www.enceliaCNPS.org.

Pomona Valley Chapter, NAS
www.pomonavalleyaudubon.org.

San Bernardino County
Riverside-San Bernardino Chapter, CNPS
See entry under Riverside County.

San Bernardino Valley Chapter, NAS
www.sbvas.org.

San Diego County
San Diego Chapter, CNPS
c/o San Diego Natural History Museum, P.O. Box 121390, San Diego, CA 92112-1390, 619-685-7321, www.cnpssd.org.

Buena Vista Chapter, NAS
2202 South Coast Highway, Oceanside, CA 92056, 760-439-2473, www.bvaudubon.org.

Palomar Chapter, NAS
P.O. Box 2483, Escondido, CA 92033, 858-487-4831, palomaraudubon.org.

San Diego Chapter, NAS
4891 Pacific Highway #112, San Diego, CA 92110, 619-275-0557, www.sandiegoaudubon.org.

Registered Herbaria

A herbarium is a repository of preserved plant specimens. Plants are collected, pressed, dried, and mounted on a sheet of archival herbarium paper and a data label added. A label is affixed which includes the plant's name, where and when it was collected, a brief description of the plant, associated plants, the date, and the name of the person who collected it. These specimens are known as *vouchers* since they vouch (confirm) that the plant specimen is what the collector claims it is. Voucher specimens collected from a specific area serve to represent the species of plants present at a study site. Years later, plants at the same site can be re-studied and compared with what was recorded there in the past. Vouchers and the registered herbaria that house them are vital tools for scientists. We have 4 primary herbaria in our area. After each name is its Index Herbariorum identifier code in parentheses.

Fay A. McFadden Herbarium (MACF)
California State University, Dept of Biological Science, 800 North State College Blvd., Fullerton, CA 92831-3599, 657-278-3614, biology.fullerton.edu.

Rancho Santa Ana Botanic Garden (RSA)
1500 North College Avenue, Claremont, CA 91711-3157, 909-625-8767, www.rsabg.org.

University of California, Irvine (IRVC)
Dr. Peter A. Bowler, Arboretum Director, University of California, Irvine Arboretum, Irvine, CA 92697-1450, 949-824-0157, 949-824-5833, arboretum.bio.uci.edu/herbarium.cfm.

University of California, Riverside (UCR)
University of California, Department of Botany and Plant Sciences, Riverside, CA 92521, (951) 827-4619, www.herbarium.ucr.edu.

References

Abrams, L. 1904. Flora of Los Angeles and Vicinity. First edition 1904 and "Supplemented edition" 1911, both by Stanford University Press, Stanford, CA; pocket size edition 1917, by The New Era Printing Company, Lancaster, PA.

Abrams, L. 1940. Illustrated Flora of the Pacific States: Washington, Oregon, and California. Volume I. Stanford University Press, Stanford, CA.

Abrams, L. 1944. Illustrated Flora of the Pacific States: Washington, Oregon, and California. Volume II. Stanford University Press, Stanford, CA.

Abrams, L. 1951. Illustrated Flora of the Pacific States: Washington, Oregon, and California. Volume III. Stanford University Press, Stanford, CA.

Abrams, L. and R.S. Ferris. 1960. Illustrated Flora of the Pacific States: Washington, Oregon, and California. Volume IV. Stanford University Press, Stanford, CA.

Allen, R.L. 1983. Natural Vegetation of Orange County, California. pp. 1-10. in: Stadum, C., editor, The Natural Sciences of Orange County, California, Volume 1. Natural History Foundation of Orange County, Newport Beach, CA.

Allen, R.L. 1999. Stalking the Wild Arthropod: The Lorquin Entomological Society's Guide to Photographing Arthropods. Lorquin Entomological Society, Los Angeles, CA.

Angel, H. 1998. How to Photograph Flowers. Stackpole Books, Mechanicsburg, PA.

Angiosperm Phylogeny Group. 1998. An Ordinal Classification for the Families of Flowering Plants. Annals of the Missouri Botanical Garden 85(4):531-553.

Baldwin, B.G., D.H. Goldman, D.J. Keil, R. Patterson, T.J. Rosatti, and D.H. Wilken, editors. 2012. The Jepson Manual: Higher Plants of California, second edition. UC Press, Berkeley, CA.

Balls, E.K. 1962. Early Uses of California Plants. California Natural History Guides: 10. UC Press, Berkeley, CA.

Barker, W.R., G.L. Nesom, P.M. Beardsley, and N.S. Fraga. 2012. A taxonomic conspectus of Phrymaceae: A narrowed circumscriptions for Mimulus, new and resurrected genera, and new names and combinations. Phytoneuron 2012-39: 1-60.

Beauchamp, R.M. 1986. A Flora of San Diego County, California. Sweetwater River Press, San Diego, CA.

Belzer, T.J. 1984. Roadside Plants of Southern California. Mountain Press Publishing Co., Missoula, MT.

Bell, A.D. and A. Bryan. 2008. Plant Form: An Illustrated Guide to Flowering Plant Morphology, second edition. Timber Press, Portland, OR.

Bolli, R. 1994. Revision of the genus *Sambucus*. Dissertationes Botanicae 223:1-227-(256).

Bornstein, C., D. Fross, and B. O'Brien. 2005. California Native Plants for the Garden. Cachuma Press, Los Olivos, CA.

Bornstein, C., D. Fross, and B. O'Brien. 2011. Reimagining the California Lawn: Water-conserving Plants, Practices, and Designs. Cachuma Press, Los Olivos, CA.

Boughey, A.S. 1968. A Checklist of Orange County Flowering Plants, Museum of Systematic Biology: Research Series No. 1, UC Irvine, CA.

Bowler, P.A. and M.E. Elvin. 2003. The vascular plant checklist for the University of California Natural Reserve System's San Joaquin Freshwater Marsh Reserve. Crossosoma 29(2): 45-66.

Bowler, P.A. and D. Bramlet. 2002. Vascular plants of the University of California, Irvine Ecological Preserve. Crossosoma 28(2): 27-49.

Boyd, S.D. 2001. New records for the vascular flora of the Santa Ana Mountains, California. Aliso 20: 43-44.

Boyd, S.D., T.S. Ross, O. Mistretta, and D. Bramlet. 1995. Vascular Flora of the San Mateo Canyon Wilderness Area, Cleveland National Forest, California. Aliso 14(2): 109-139.

Boyd, S.D., T.S. Ross, and F.M. Roberts, Jr. 1995. Additions to the Vascular Flora of the Santa Ana Mountains, California. Aliso 14(2): 105-108.

Brandenburg, D.M. 2010. Field Guide to Wildflowers of North America. Sterling Publishing, Inc., New York.

Brown, B.V., J.N. Hogue, and F.C. Thompson. 2011. Flower Flies of Los Angeles County. Natural History Museum of Los Angeles County, CA.

Brown, L.R. and C.O. Eads. 1965a. A Technical Study of Insects Affecting the Oak Tree in Southern California. California Agricultural Experiment Station Bull. 810. Berkeley, CA.

Brown, L.R. and C.O. Eads. 1965b. A Technical Study of Insects Affecting the Sycamore Tree in Southern California. California Agricultural Experiment Station Bulletin 818.

Capinera, J. 2010. Insects and Wildlife. Arthropods and their relationships with wild vertebrate animals. Wiley-Blackwell, Hoboken, NJ.

Castner, J.L. 2006. Photographic Atlas of Botany and Guide to Plant Identification. Feline Press, Gainesville, FL.

Chester, T., W. Armstrong, and K. Madore. *Brodiaea santarosae* (Themidaceae), a New Rare Species from the Santa Rosa Basalt Area of the Santa Ana Mountains of Southern California. Madroño 54(2): 187–198.

Clarke, C.B. 1977. Edible and Useful Plants of California. California Natural History Guides: 41. UC Press, Berkeley, CA.

Clarke, O.F., D. Svehla, G. Ballmer, and A. Montalvo. 2007. Flora of the Santa River and Environs, with References to World Botany. Heyday Books, Berkeley, CA.

CNPS. 2011. Inventory of Rare, Threatened, and Endangered Plants of California. California Native Plant Society, Sacramento, CA. Online at: www.rareplants.cnps.org.

Coleman, R.A. 1995. The Wild Orchids of California. Comstock Publishing Associates, Cornell University Press, Ithaca, NY.

Collins, B.J. 1974. Key to Trees and Wildflowers of the Mountains of Southern California. California Lutheran College, Thousand Oaks, CA.

Collins, B.J. 2000. Key to Coastal and Chaparral Flowering Plants of Southern California, third edition. Kendall/Hunt, Dubuque, IA.

Crampton, B. 1974. Grasses in California. California Natural History Guides: 33. UC Press, Berkeley, CA.

Crosby, D.G. 2004. Poisoned Weed: Plants Toxic to Skin. Oxford University Press, New York.

Cruden, R.W. 1972. Pollination Biology of *Nemophila menziesii* (Hydrophyllaceae) with Comments on the Evolution of Oligolectic Bees. Evolution 26(3): 373-389.

Dale, N. 2000. Flowering Plants: The Santa Monica Mountains, Coastal and Chaparral Regions of Southern California, revised second edition. California Native Plant Society, Sacramento, CA.

Davidson, A. and G.L. Moxley. 1923. Flora of Southern California. Times-Mirror Press, Los Angeles, CA.

Davis, D.R., O. Pellmyr, and J.N. Thompson. 1992. Biology and Systematics of *Greya* Busck and *Tetragma* n. gen. (Lepidoptera: Prodoxidae). Smithsonian Contributions to Zoology 524:1-88.

Dawson, Y. and M.S. Foster. 1982. Seashore Plants of California. California Natural History Guides: 47. UC Press, Berkeley, CA.

Dawson, Y. 1966. The Cacti of California. California Natural History Guides: 18. UC Press, Berkeley, CA.

Eckenwalder, J.E. 2009. Conifers of the World: the Complete Reference. Timber Press, Portland, OR.

Eisner, T. 2006. For Love of Insects. The Belknap Press of Harvard University Press, Cambridge, MA.

Evans, A.V. 2007. National Wildlife Federation Field Guide to Insects and Spiders of North America. Sterling Publishing Company, Inc., NY.

Evans, A.V. and J.N. Hogue. 2006. Field Guide to Beetles of California. California Natural History Field Guides: 88. UC Press, Berkeley, CA.

Flora of North America Editorial Committee (FNA). 1997 et seq. Flora of North America. NY. www.fna.org.

Fross, D. and D. Wilken. 2006. Ceanothus. Timber Press, Portland, OR.

Fuller, T.C. and E. McClintock. 1986. Poisonous Plants of California. California Natural History Guides: 53. UC Press, Berkeley, CA.

Greenfield, A.B. 2005. A Perfect Red: Empire, Espionage, and the Quest for the Color of Desire. HarperCollins Publishers, New York.

Griffin, J.R., and W.B. Critchfield. 1972. The Distribution of Forest Trees in California. USDA Forest Service. Pacific Southwest Forest and Range Experiment Station Research Paper PSW-82. Berkeley, CA.

Grillos, S.F. 1966. Ferns and Fern Allies. California Natural History Guides: 16. UC Press, Berkeley, CA.

Harder, L.D. and R.M.R. Barclay. 1994. The Functional Significance of Porocidal Anthers and Buzz Pollination: Controlled Pollen Removal from *Dodecatheon*. Functional Ecology 8(4): 509-517.

Harrington, H.D. and L.W. Durrell. 1985. How to Identify Plants. Swallow Press Books, Ohio University Press, Athens, OH.

Harris, J.G. and M.W. Harris. 2001. Plant Identification Terminology: An Illustrated Glossary, second edition. Spring Lake Publishing, Payson, UT.

Head, W.S. 1972. California Chaparral: An Elfin Forest. Naturegraph Publishers, Happy Camp, CA.

Heide-Jorgensen, H.S. 2008. Parasitic Flowering Plants. Brill Academic Publishers, Leiden, The Netherlands.

Hess, W.J. and J.C. Dice. 1995. *Nolina cismontana* (Nolinaceae), a New Species Name for an Old Taxon. Novon 5: 162-164.

Hickman, J.C., editor. 1993. The Jepson Manual: Higher Plants of California. UC Press, Berkeley, CA.

Hogue, C.L. 1993. Insects of the Los Angeles Basin, second edition. Natural History Museum of Los Angeles County, CA.

Holland, R.F. 1986. Preliminary Descriptions of the Terrestrial Natural Communities of California. State of California, The Resources Agency, Department of Fish and Game, Sacramento, CA.

Holland, V.L. and D.J. Keil. 1996. California Vegetation. Kendall/Hunt Publishing Company, Dubuque, IA.

Jaeger, E.C. and A.C. Smith. 1966. Introduction to the Natural History of Southern California. California Natural History Guides: 13. UC Press, Berkeley, CA.

Jayne, S.B. 1990. Plants of the Crystal Cove Backcountry. Crossosoma 16(3):1-8

Jepson Flora Project. 2012 (v. 1.0). Jepson eFlora. ucjeps.berkeley.edu/IJM.html.

Johnston, V.R. 1994. California Forests and Woodlands: A Natural History. California Natural History Guides: 58. UC Press, Berkeley, CA.

Jones, C.E. and R.J. Little, editors. 1983. Handbook of Experimental Pollination Biology. Scientific and Academic Editions, Division of Van Nostrand Reinhold Co., New York.

Judd, W.S., C.S. Campbell, E.A. Kellogg, P.F. Stevens, and M.I. Donoghue. 2002. Plant Systematics: A Phylogenetic Approach, second edition. Sinauer Associates, Inc., Sunderland, MA.

Keator, G. and A. Middlebrook. 2007. Designing California Native Gardens: The Plant Community Approach to Artful, Ecological Gardens. UC Press, Berkeley, CA.

Keil, D.J. 2007 [2008]. *Pentachaeta aurea* subsp. *allenii* (Asteraceae), a New Subspecies from Orange County, California. Madroño 54(4): 343-344.

Klein, K. 2010. 50 Hikes in Orange County. The Countryman Press, Woodstock, VT.

Kruckeberg, A.R. 2006. An Introduction to California Soils and Plants: Serpentine, Vernal Pools, and Other Geobotanical Wonders. California Natural History Field Guides: 86. UC Press, Berkeley, CA.

Kuijt, J. 1969. The Biology of Parasitic Flowering Plants. UC Press, Berkeley, CA.

Lanner, R.M. 1999. Conifers of California. Cachuma Press, Los Olivos, CA.

Lathrop, E.W. and R.F. Thorne. 1978. A Flora of the Santa Ana Mountains, California: An Annotated List of the Vascular Plants and the Plant Communities of the Santa Rosa Plateau, Santa Ana Mountains. Aliso 9(2):197-278.

Lathrop, E.W. and R.F. Thorne. 1985. A Flora of the Santa Rosa Plateau: An Annotated List of the Vascular Plants and the Plant Communities of the Santa Rosa Plateau, Santa Ana Mountains. Southern California Botanists, Special Publication No. 1. Southern California Botanists, Fullerton, CA.

Lightner, J. 2011. San Diego County Native Plants, third edition. San Diego Flora, San Diego, CA.

Manhoff, D. and S. Vogel. 1996. Mosby's Outdoor Emergency Medical Guide. What to Do in an Emergency When Help May Take Some Time to Arrive. Mosby-Year Book, Inc., St. Louis, MO. www.beechwoodhealthbooks.com/outdoor.html.

Mason, H.L. 1957. Flora of the Marshes of California. UC Press, Berkeley, CA.

McMinn, H.E. 1964. An Illustrated Manual of California Shrubs. UC Press, Berkeley, CA.

Minnich, R.A. 2008. California's Fading Wildflowers. Lost Legacy and Biological Invasions. UC Press, Berkeley, CA.

Morhardt, S. and E. Morhardt. 2004. California Desert Flowers: An Introduction to Families, Genera, and Species. UC Press, Berkeley, CA.

Munz, P.A. 1935. A Manual of Southern California Botany. Claremont Colleges, Claremont, CA.

Munz, P.A. 1961. California Spring Wildflowers. UC Press, Berkeley, CA.

Munz, P.A. 1962. California Desert Wildflowers. UC Press, Berkeley, CA.

Munz, P.A. 1963. California Mountain Wildflowers. UC Press, Berkeley, CA.

Munz, P.A. 1964. Shore Wildflowers of California, Oregon, and Washington. UC Press, Berkeley, CA.

Munz, P.A. 1974. A Flora of Southern California. UC Press, Berkeley, CA.

Munz, P.A. and D.D. Keck. 1973. A California Flora and Supplement (combined edition). UC Press, Berkeley, CA.

Munz, P.A., D. Lake, and P.M. Faber. 2003a. California Mountain Wildflowers, revised edition. California Natural History Guides: 68. UC Press, Berkeley, CA.

Munz, P.A., D. Lake, and P.M. Faber. 2003b. Shore Wildflowers of California, Oregon, and Washington, revised edition. California Natural History Guides: 67. UC Press, Berkeley, CA.

Munz, P.A., D. Lake, and P.M. Faber. 2004. California Spring Wildflowers, revised edition. California Natural History Guides: 75. UC Press, Berkeley, CA.

Munz, P.A., D.L. Renshaw, and P.M. Faber. 2004. California Desert Wildflowers, revised edition. California Natural History Guides: 74. UC Press, Berkeley, CA.

Niehaus, T.F. and C.L. Ripper. 1976. A Field Guide to Pacific States Wildflowers. The Peterson Field Guide Series: 22. Houghton Mifflin Co., Boston, MA.

Ornduff, R., P.M. Faber, and T. Keeler-Wolf. 2003. Introduction to California Plant Life, second edition. California Natural History Guides: 69. UC Press, Berkeley, CA.

Papp, C.S. and L.A. Swan. 1983. A Guide to Biting and Stinging Insects and Other Arthropods, second enlarged edition. Entomography Publications, Sacramento, CA.

Pavlik, B.M., P.C. Muick, S. Johnson, and M. Popper. 1991. Oaks of California. Cachuma Press and the California Oak Foundation, Los Olivos, CA.

Pequegnat, W.E. 1951. The Biota of the Santa Ana Mountains. Pomona College Journal of Entomology and Zoology 42(3and4): 1-84.

Pellmyr, O. 1996. *Mesepiola specca* Davis. Version 01 January 1996 (under construction). tolweb.org/Mesepiola_specca/12416/1996.01.01 in The Tree of Life Web Project, tolweb.org.

Peterson, P.V. 1970. Native Trees of Southern California. California Natural History Guides: 14. UC Press, Berkeley, CA.

Popper, H. 2012. California Native Gardening: A Month-by-Month Guide. UC Press, Berkeley, CA.

Powell, J.A. and C.L. Hogue. 1979. California Insects. California Natural History Guides: 44. UC Press, Berkeley, CA.

Powell, J.A. and P.A. Opler. 2009. Moths of Western North America. UC Press, Berkeley, CA.

Pritchard, A.E. 1953. The Gall Midges of California; Diptera: Itonididae (Cecidomyiidae). Bulletin of the California Insect Survey, Vol. 2, No. 2. UC Press, Berkeley, CA.

Proctor, M., P. Yeo, and A. Lack. 1996. The Natural History of Pollination. Timber Press, Portland, OR.

Raven, P.H. 1966. Native Shrubs of Southern California. California Natural History Guides: 15. UC Press, Berkeley, CA.

Rea, B. K. Oberhauser, and M.A. Quinn. 2003. Milkweed, Monarchs, and More. A Field Guide to the Invertebrate Community in the Milkweed Patch. Bas Relief Publishing Group, Glenshaw, PA.

Rebman, J.P. and N.C. Roberts. 2012. Baja California Plant Field Guide, third edition. Sunbelt Publications, El Cajon, CA.

Rebman, J.P. and M.G. Simpson. 2007. Checklist of the Vascular Plants of San Diego County, fourth edition. San Diego Natural History Museum, San Diego, CA.

Reed, F.M. 1916. A Catalog of the Plants of Riverside and Vicinity. Unpublished manuscript. Available in the Science Library, UC Riverside.

Roberts, F.M., Jr. 1990. Rare and Endangered Plants of Orange County. Crossosoma 16(2): 3-12.

Roberts, F.M., Jr. 1995. Illustrated Guide to the Oaks of the Southern Californian Floristic Province: The Oaks of Coastal Southern California and

Northwestern Baja California, Mexico. F.M. Roberts Publications, Encinitas, CA.

Roberts, F.M., Jr. 2008. The Vascular Plants of Orange County, California, an Annotated Checklist. F.M. Roberts Publications, San Luis Rey, CA.

Roberts, F.M., Jr. and D.E. Bramlet. 2007. Vascular Plants of the Donna O'Neill Land Conservancy, Rancho Mission Viejo, Orange Co., California. Crossosoma 33: 2-40.

Roberts, F.M., Jr., S.D. White, A.C. Sanders, D.E. Bramlet, and S. Boyd. 2004. The Vascular Plants of Western Riverside County, California: An Annotated Checklist. F.M. Roberts Publications, San Luis Rey, CA.

Roberts, F.M., Jr., S.D. White, A.C. Sanders, D.E. Bramlet, and S. Boyd. 2007. Additions to the Flora of Western Riverside County, California. Crossosoma 33(2): 55-69.

Roberts, F.M., Jr. 2007. An Inventory of Orange County Herbarium Specimens. F.M. Roberts Publications, San Luis Rey, CA.

Roque, N., D.J. Keil, and A. Susanna. 2009. Illustrated Glossary of Compositae. Pp. 781-806 in Funk, V.A., A. Susanna, T.F. Stuessy, and R.J. Bayer, (editors). Systematics, Evolution, and Biogeography of Compositae. Vienna, Austria: International Association for Plant Taxonomy.

Rotenberg, N. and M. Lustbader. 1999. How to Photograph Close-ups in Nature. Stackpole Books, Mechanicsburg, PA.

Rundel, P.W. and R. Gustafson. 2005. Introduction to the Plant Life of Southern California. California Natural History Guides: 85. UC Press, Berkeley, CA.

Russo, R.A. 2006. Plant Galls of California and Other Western States. California Natural History Field Guides: 91. UC Press, Berkeley, CA.

Sawyer, J.O., T. Keeler-Wolf, and J. Evens. 2009. A Manual of California Vegetation, second edition. California Native Plant Society, Sacramento, CA.

Schad, J. 2006. Afoot and Afield in Orange County, third edition. Wilderness Press, Berkeley, CA.

Schiffman, P.M. 2005. The Los Angeles Prairie. in Deverell, W. and G. Hise, editors. Land of Sunshine: An Environmental History of Metropolitan Los Angeles. University of Pittsburgh Press, Pittsburgh, PA.

Schmidt, M.G., K.L. Greenberg, and B. Merrick 2012. Growing California Natives, second edition. UC Press, Berkeley, CA.

Schneider-Ljubenkov, J.A. and T.S. Ross. 2001. An Annotated Checklist of the Vascular Plants of the Whittier Hills, Los Angeles County, California. Crossosoma 27: 1-23.

Schoellhamer, J. E. 1981. Geology of the Northern Santa Ana Mountains, California. Geology of the Eastern Los Angeles Basin, Southern California. Geological Survey Professional Paper 420-D. U.S. Geological Survey, Department of the Interior, Washington, DC.

Schoenherr, A.A. 1992. A Natural History of California. California Natural History Guides: 56. UC Press, Berkeley, CA.

Schoenherr, A.A. 2011. Wild and Beautiful: A Natural History of Open Spaces in Orange County. Laguna Wilderness Press, Laguna Beach, CA.

Simpson, M.G. 2010. Plant Systematics, second edition. Elsevier Academic Press, Amsterdam, The Netherlands.

Smith, J.P., Jr. 1977. Vascular Plant Families. Mad River Press, Eureka, CA.

Streisfeld, M.A. and J.R. Kohn. 2005. Contrasting patterns of floral and molecular variation across a cline in *Mimulus aurantiacus*. Evolution 59: 2548–2559.

Streisfeld, M.A. and J.R. Kohn. 2007. Environment and pollinator-mediated selection on parapatric floral races of *Mimulus aurantiacus*. J. Evol. Biol. 20: 122–132.

Stuart, J.D. and J.O. Sawyer. 2001. Trees and Shrubs of California. California Natural History Guides: 62. UC Press, Berkeley, CA.

Thompson, D.M. 2005. Systematics of *Mimulus* Subgenus *Schizoplacus* (Scrophulariaceae). Systematic Botany Monographs, Volume 75. The American Society of Plant Taxonomists, Ann Arbor, MI.

Tracy, H.H. 1936. Brief Notes on Southern California Pteridophytes. American Fern Journal, 26(3): 87-91.

Tulag, M.C. & G.L. Nesom. 2012. Taxonomic overview of *Diplacus* Sect. *Diplacus* (Phrymaceae). Phytoneuron 2012-45: 1-20.

Tunget, C.L., S.G. Turchen, A.S. Manoguerra, R.F. Clark, and D.E. Pudoff. 1994. Sunlight and the Plant: A Toxic Combination: Severe Phytophotodermatitis from *Cneoridium dumosum*. Cutis 54:400-402.

Vasek, F.C. 1980. A Vegetative Guide to Perennial Plants of Southern California. San Bernardino County Museum Association, Redlands, CA.

Vessel, F.F. and H.H. Wong. 1989. Natural History of Vacant Lots. California Natural History Guides: 50. UC Press, Berkeley, CA.

Walawender, M.J. 2000. The Peninsular Ranges: A Geological Guide to San Diego's Back Country. Kendall/Hunt Publishing Co., Dubuque, IA.

Walters, D.R., D.J. Keil, B. Walters, and Z.E. Murrell. 2006. Vascular Plant Taxonomy, fifth edition. Kendall/Hunt Publishing Co., Dubuque, IA.

Wilson, E.O. 1986. Biophilia. Harvard University Press, Cambridge, MA.

Wilson, R.C. and R.J. Vogl. 1965. Manzanita Chaparral in the Santa Ana Mountains, California. Madroño 18:47-62.

Yerkes, R. F., T.H. McCulloh, J.E. Schoellhamer, and J.G. Vedder. 1965. Geology of the Los Angeles Basin, California – an introduction. Geology of the Eastern Los Angeles Basin, Southern California. Geological Survey Professional Paper 420-A. U.S. Geological Survey, Department of the Interior, Washington, DC.

Zomlefer, W.B. 1994. Guide to Flowering Plant Families. University of North Carolina Press, Chapel Hill, NC.

Glossary

Many technical terms are used by botanists to describe plants in order to communicate with other people. We have tried to keep our use of such terms to a minimum. Even so, there are many that we could not avoid. There is no need to memorize all of these terms; just place a bookmark here and refer to this glossary when needed.

Our definitions are based on our own working definitions and those in Baldwin et al. (2012), Bell and Bryan (2008), Bidlack and Jansky (2008), Harrington and Durrell (1985), Harris and Harris (2001), and Simpson (2010).

[ ] **Square brackets**. These are used to enclose a synonym of a scientific name, usually a name no longer in use.

( ) **Parentheses**. These usually enclose information that is only rarely true, such as an atypical number of items or an atypical maximum size of a flower. In the statement, "petals (0)4-5," the plant usually has 4-5 petals, rarely none. Sometimes, parentheses are used to surround incidental information.

★ **Sensitive**. A native plant or animal that has federal and/or state status as Endangered or Threatened and/or a native plant that has a California Rare Plant Rank (CRPR). See full definition below.

✩ **Locally sensitive**. Few or dwindling in number, and/or in danger of extirpation within our area. See full definition below.

✖ **Non-native.** A weedy non-native plant or a non-native animal. See full definition below.

☠ **Toxic**. Poisonous, relating to or caused by poison. The toxin can be acquired through contact, inhalation, and/or ingestion.

0 **Zero, none, absent**. In the example "petals 0" the flower has no petals. In the statement "petals (0)4-5" the flower usually has 4-5 petals, rarely none.

1-Pinnate. A pinnate leaf that has one set of (primary) leaflets (elderberry). Compare with 2-Pinnate.

2-Pinnate. A pinnate leaf that has two sets of leaflets (Parish's tauschia). The primary leaflets are themselves divided into secondary leaflets. Compare with 1-Pinnate. Leaves that vary from one to two sets of leaflets are 1-2 pinnate (purple sanicle).

Achene. A dry one-seeded fruit that develops from a one-chambered superior ovary. This term is sometimes used for fruit of the aster family. Found in the grass family. Compare with **cypsela**.

Adventitious root. A root that grows from an organ other than a root, usually from a stem (cluster-lilies, Western poison oak).

Alternate. Having only one leaf per node (lemonade berry).

Annual. A plant that lives for only one year or growing season.

Anther. A male reproductive structure in which pollen is formed. When mature, pollen is released from it through a pore or a slit.

Apex. Tip, such as the portion of a leaf blade or the petal farthest from the base.

Apical. Toward the apex (tip) of an object.

Areole. A region of specialized tissue on a cactus stem, which may bear leaves, spines, and/or hairs.

Aril. An appendage on a seed, rarely covering the entire seed. Sometimes nutritious and eaten by insects.

Armature. Protective covering on a plant or an animal. See **prickle**, **spine**, and **thorn**.

Arrowhead-shaped. A leaf or leaf base shaped like the head of an arrow with basal lobes that point backward, sometimes with multiple lobes. Technically called **sagittate**.

Asexual. A flower that contains no fertile (working) reproductive parts. Also said of an organism that reproduces without the union of gametes (sperm and egg).

Authority. The name of the person who published a scientific name or category of name.

Axil. Stem tissue at the upper base of a leaf.

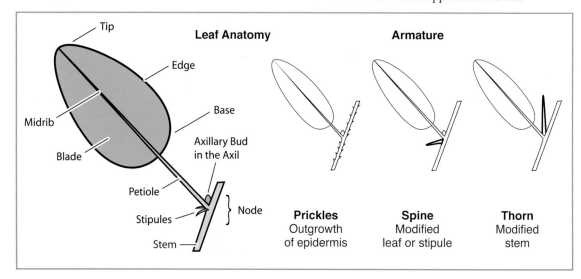

Leaf Anatomy: Tip, Edge, Base, Axillary Bud in the Axil, Node, Stem, Stipules, Petiole, Blade, Midrib

Armature: **Prickles** Outgrowth of epidermis; **Spine** Modified leaf or stipule; **Thorn** Modified stem

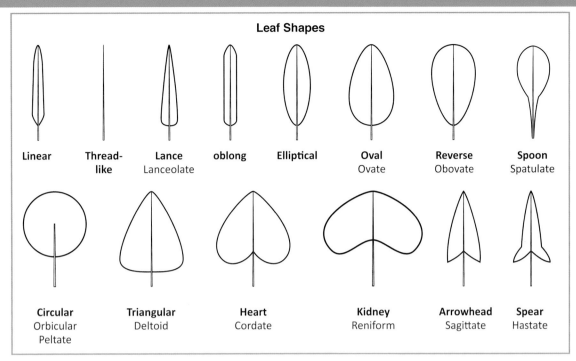

Axillary bud. A bud in the axil that may grow into a stem (shoot), leaf, or inflorescence.

Basal leaves. Leaves that arise from the base of the plant (chaparral yucca, popcorn flowers).

Basal rosette. Basal leaves arranged in a tight cluster, sometimes whorled or nearly so (Padre's shooting star).

Base. Area of a leaf blade closest to the petiole or the area of a petal closest to the receptacle.

Biennial. A plant that lives for only two years or growing seasons.

Bilabiate floret. A floret in a flowerhead of a plant that is a member of the aster family. The corolla is distinctly two-lipped; one lip has two slender lobes, the other has a single broad three-lobed ray (sacapellote).

Bilateral symmetry. Symmetry whereby an object, such as a flower or fruit, can be divided into equal halves only if cut along a lengthwise plane that includes the object's center. In plants, also called **irregular** or **zygomorphic**.

Bisexual. Containing fertile reproductive parts of both sexes (at least one pistil and one stamen). In botany, bisexual flowers are also called **perfect**.

Blade. The expanded part of a leaf.

Bract. A small structure in an inflorescence or flower, usually below it. Looks like a leaf or scale. Not all inflorescences and flowers have bracts.

Bulb. A belowground structure composed of overlapping fleshy scale leaves atop a very small stem. The outermost scale leaves (tunic) may be membranous or fibrous (onions).

Bundled. Growing in a group from the same short shoot. Technically called **fasciculate**. The short shoots are almost always alternate on the stem (California buckwheat, chamise).

California Endangered Species Act (CESA). A state law to provide for the conservation of endangered and threatened species of fish, wildlife, and plants, and their habitats. Codified in California Fish and Game Code, Sections 2050 et seq.

California Rare Plant Rank (CRPR). A numerical ranking system for sensitive species of plants. In increasing order of rarity, the ranks are 4 (limited distribution), 3 (more information needed), 2 (rare, threatened, or endangered in California but more common elsewhere), 1B (rare, threatened, or endangered in California), and 1A (presumed extinct).

Calyx. Outermost whorl of flower parts; hides the rest of the flower in the bud stage. Most often green, sometimes other colors, it may resemble the corolla. Its individual parts are called **sepals**.

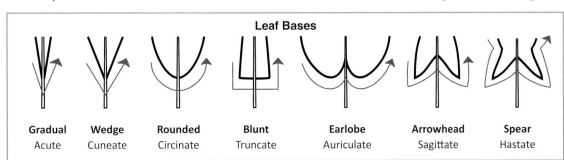

Carpel. Basic female structure of a flower, evolved from a fertile leaf.

Catkin. An inflorescence of small unisexual flowers spirally arranged around a central axis. It often droops (all catkins of silk-tassels, male catkins of alders and oaks) but some are held upright (willows).

Clade. A group of living things that share a common ancestor.

Cladogram. A diagram that represents the evolutionary history of a clade. On a cladogram, a clade includes an ancestral species and all (and only all) of its descendant species.

Classification. The process of placing living things into categories.

Cleistogamous, cleistogamy. See **in-bud fertilization**.

Composite head. Individual flowers arranged into complex groups. Found only in the aster family. Also called **flowerhead**.

Compound leaf. A leaf that is divided into distinct leaflets. Each leaflet may have its own stalk (**petiolule**) or may be directly attached to the main leaf stalk (**sessile**). Compound leaves have an axillary bud at the base of the petiole but no bud at the base of each leaflet. Sometimes called **divided**.

Compound pistil. Two or more fused or partially fused carpels.

Corm. A belowground structure composed of a short swollen stem topped with scale or foliage leaves (mariposa lilies, *Calochortus* spp.).

Corolla. The whorl of flower parts just inside the calyx. Its individual parts are called **petals**.

Cotyledon. The first leaf formed by an embryo within a seed. Sometimes called seed-leaf. It often functions in storage of food reserves or a conduit to those reserves. Magnoliids and eudicots have 2 cotyledons per seed, while monocots have only 1 cotyledon per seed. In magnoliids and eudicots, the cotyledons usually resemble leaves; they function primarily as storage organs and may grow to become photosynthesizing leaves or early support stems. In monocots, the cotyledon functions as an early stem, serves as a conduit to a food reserve, or remains within the seed (but is *never* a storage organ).

Cross-pollination. A form of plant fertilization whereby pollen from a flower on one plant is transferred to a flower on a different plant.

Cultivar (cultivated variety, abbreviated as "cult." or "c.v.") A form of a plant that differs from others, usually with attractive features and relatively easy to grow in a garden. When written out, the cultivar name appears after the species name and is placed within single quotes (*Eriogonum fasciculatum* 'Dana Point', *Ribes malvaceum* 'Dancing Tassels'). Some cultivars are hybrids between known or unknown parents; they are often written only with their genus and cultivar name (*Arctostaphylos* 'Howard McMinn', *Ceanothus* 'Ray Hartman').

Cyathium. The specialized inflorescence of some members of the spurge family. It consists of a cup-shaped involucre, often topped with red nectar glands that bear white petal-like appendages. Within it are a single female flower and several tiny male flowers.

Cypsela (plural **cypselae**). A dry one-seeded fruit that develops from a one-chambered inferior ovary (a true achene develops from a superior ovary). Found in the aster, teasel, and valerian families. Compare with **achene**.

Deciduous. Falling off at the end of a growing season (usually leaves, sometimes sepals or petals). Often used to describe a plant that drops its leaves in unison under particular conditions, such as **winter deciduous** (poison oak, California sycamore), **summer deciduous** (gooseberries and currants), or **drought deciduous** (California broom, California buckwheat).

Diapause. In arthropods such as insects, a period of slowed development and decreased metabolic rate.

Dioecious. A species or population of a species in which individuals have only one type of functional reproductive structures, either male or female, but never both (silk-tassels). A plant or individual flower that contains only one sex is said to be **unisexual**, **same-sex**, or **imperfect** (Greek, di- = two; oikos = household).

Diploid. Containing two sets of homologous chromosomes. In a diploid cell, one set of homologous chromosomes is inherited from each parent.

Disciform. A flowerhead of a plant that is a member of the aster family. A type of discoid flowerhead, it usually contains disk florets in the center surrounded by filiform florets (California sagebrush).

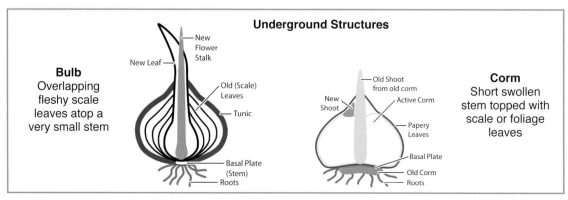

Underground Structures

Bulb — Overlapping fleshy scale leaves atop a very small stem

Corm — Short swollen stem topped with scale or foliage leaves

Discoid. A flowerhead of a plant that is a member of the aster family. The flowerhead contains disk florets only (thistles).

Disk floret. A floret in a flowerhead of a plant that is a member of the aster family. The corolla is radially symmetric (rarely bilateral), slender and tubular with 5 (rarely 4) small equal-sized lobes at the tip. Each floret includes both sexes (**bisexual disk floret**). In a few genera the pistil is non-functional (**staminate disk floret**).

Distal. Situated away from the center or from the point of attachment.

DNA. Deoxyribonucleic acid. A self-replicating material present in living organisms. Exists in discrete strands (chromosomes). The sequence of chemical groups in a chromosome constitutes genetic information (genes). Some gene sequences are identical or similar between species; others are unique to particular species, populations, and individuals. Comparing gene sequences can be used to identify living organisms.

Earlobe-shaped. Having two rounded lobes that project backward toward the stem, like a pair of earlobes. Technically called **auriculate**.

Edge. Outer margin of a leaf.

Elliptical. Widest in the middle, equally pointed or rounded at both ends, like a flattened circle or a football.

Elytra. The thickened front wings of some insects (beetles, earwigs, some true bugs). Singular **elytron**.

Endangered Species Act (ESA). A federal law to provide for the conservation of endangered and threatened species of fish, wildlife, and plants and for other purposes. Codified in Title 7 United States Code § 136 and Title 16 U.S.C. § 1531 et seq.

Endangered. In danger of extinction throughout all or a significant portion of its range (paraphrased from Section 2062, California Fish and Game Code).

Endemic. Restricted to a defined geographic area. For example, Laguna Beach live-forever is endemic to the San Joaquin Hills of Laguna Beach; Allen's daisy is endemic to Orange County. These plants occur naturally nowhere else.

Entire. Having edges (margins) that are smooth, neither toothed nor lobed.

Epaleate. Lacking **paleae** (bracts on the receptacle of the flowerhead).

Ephemeral perennial. A plant whose underground parts are perennial and whose aboveground parts are seasonal. Generally, such plants grow new aboveground parts, flower, set seed, then die off, while its belowground parts remain alive. They typically have tough roots or fibrous to fleshy tubers. They have neither bulbs nor corms. The term is sometimes modified with the name of the season during which aboveground parts start to grow or when those parts are in flower (ephemeral spring perennial, etc). Sometimes informally called root perennial. Compare with **geophyte**, **bulb**, **corm**.

Even-pinnate. Having an even number of leaflets.

Evergreen. Always having green leaves that remain on the plant for more than one growing season. The leaves mature and die naturally but do not drop off in unison; instead, leaves fall off individually all year long.

Evolution. Genetic change of populations over one or more generations. Also called descent with modification. A central tenet of the life sciences. Several ideas exist to explain *how* evolution takes place, such as natural selection initially proposed by both Charles Robert Darwin and Alfred Russell Wallace.

Extinct. Having no living members. For example, Los Angeles sunflower (*Helianthus nuttallii* ssp. *parishii*) was a huge sunflower that stood from 0.5 to 5 meters tall. It was known from very wet places in Huntington Beach (1924) and Newport Bay (1933), last seen in 1937 (in San Bernardino) and later declared extinct.

Extirpated. No longer surviving in regions that were once part of its range.

Family. A group to which closely related genera belong.

Female ♀. An organism or reproductive part that produces egg cells (ova). Often represented by the symbol of the Roman goddess Venus.

Filament. A male reproductive structure, the stalk that supports the anther.

Filiform floret. A floret in a flowerhead of a plant that is a member of the aster family. Resembles a disk floret in that the corolla is radially symmetric and tubular, but differs in the corolla's being cylindric, very narrow, and blunt-tipped (rarely with minute lobes) and in having no anthers (female [pistillate] florets of *Baccharis*).

Floret. In the aster family, an individual flower. In the grass family, an individual flower and the bracts immediately below it.

Flower. A structure in a flowering plant (Angiosperm) used for reproduction. It has a pistil and/or stamens, often surrounded by perianth parts (sepals and/or petals).

Flowerhead. Individual flowers arranged into complex groups. Found only in the sunflower family (Asteraceae). Also called **composite head**. Technically a type of **involucre**.

Flowering plant. An Angiosperm, characterized by production of a flower, a reproductive organ.

Free. Separate, not fused to any other part.

Fruit. A female reproductive structure that develops from one or more ovaries that contain maturing or mature seeds. A fruit results from the fertilization of a flower. Some fruits are fleshy (Pacific madrones, currants, toyons), others are dry (live-forevers, buttercups, popcorn flowers). Cultivated examples include peaches, avocados, and tomatoes. The word "fruit" does *not* imply that it is edible.

Fused. United with another part, not free. Leaves, sepals, petals, and stamens are commonly fused at the base.

Genus (plural **genera**). A group to which closely related species belong. Also the first word of a scientific name of a species. When referring to more than one species within a particular genus, it is often written as "Genus spp.," as in *Lupinus* spp. for lupines.

Geophyte. A plant with belowground storage organs that live for more than two years or growing seasons but aboveground parts that die off each year or growing season. Compare with ephemeral perennial, bulb, corm.

Glaucous. Covered with a whitish powder or wax.

Glochid. A short slender barbed spine that grows from the areoles of cactus plants.

Glucosinolate. A class of organic chemicals produced by members of the mustard clade, which includes mustard, caper, and spiderwort families.

Gradual or **tapered**. Forming a narrow angle on the petiole and gradually widening to the leaf blade. Technically called **acute**.

Guild. A group of living things that share certain characteristics. We use the term to represent certain plants and other living things that are tightly associated with them, as in the milkweed guild, the cactus guild, and the mistletoe guild.

Half-inferior ovary. An ovary that sits *below* the flower base; the base of the hypanthium is fused to the ovary. The top of the hypanthium only comes about halfway up the ovary (roses, water pimpernel, some saxifrages). Technically called **perigynous**.

Heart-shaped. Shaped like a non-scientific drawing of a heart, longer than wide or length and width about equal. Technically called **cordate**.

Hemiparasite. A parasitic plant that can make its own food via photosynthesis but also steals water and/or nutrients from a host plant (paintbrushes, mistletoes).

Herb. A plant with non-woody growth aboveground.

Herbaceous. Non-woody growth.

Holoparasite. A parasitic plant that does not (or cannot) make its own food via photosynthesis but steals all of its nutrients and water from a host plant (dodders, broomrapes).

Hypanthium. In some flowers, a tube-like structure formed from a fusion of the lower portion of sepals, petals, and stamens. Not found in all plants. In some plants, the hypanthium is free from (not attached to) the ovary.

Imperfect flower. A flower that possesses parts of only one sex. Also called **unisexual**.

In-bud fertilization. A form of plant self-fertilization in which the flower remains closed, bud-like, and never opens. Internally, it self-fertilizes and then goes directly into fruit and sets seed. Technically called **cleistogamy** (Venus' looking-glass, Shelton's violet).

Indigenous. See **native**.

Inferior ovary. An ovary that sits *below* the flower base. Sepals, petals, and stamens attach to the flower base *above* the ovary (aster, cucumber, evening primrose families). Technically called **epigynous**.

Inflorescence. An entire cluster of flowers.

Inflorescence axis. The main stalk *within* an inflorescence. Also called **rachis**. Compare with **peduncle**.

Internode. The region between two nodes.

Involucre. A group of bracts that form a unit, usually immediately below a flower, fruit, or inflorescence (sunflower flowerhead, buckwheat flowers, oak acorn cup).

Irregular flower. A flower of unusual shape. Often two-lipped, with unequally sized and shaped petals. Also called **bilateral** or **zygomorphic**.

Kidney-shaped. Shaped like a heart but wider than long. Technically called **reniform**.

Lance-shaped. Oval, longer than wide, the base wide and rounded, the tip long, narrow, and pointed. Technically called **lanceolate**.

Leaf node or **node**. A point on a stem or twig from which a leaf grows.

Leaflet. A section of a compound leaf that is divided into two or more parts.

Ligulate floret. A floret in a flowerhead of a plant that is a member of the aster family. The corolla is bilaterally symmetric; a short to long tube is topped on one side with a strap-like blade that is 5-lobed at its tip. Ligulate florets are always bisexual.

Liguliflorous. A flowerhead of a plant that is a member of the aster family. The flowerhead contains ligulate florets only (chichories, wreath plants).

Linear. Much longer than wide.

Linnaean System of Binomial Nomenclature. A system of two-part species names in Latin (or in Greek or with words changed to sound like Latin, called "Latinized"). In this system, closely related things share the same first word (the "genus") and have a unique second word (the "specific epithet"). The system currently in use was devised by the Swedish botanist Carolus Linnaeus and used in two landmark books, *Species Plantarum* (plant names, published in 1758) and *Systemae Naturae* (animal names, the tenth edition published in 1753).

Lobe. An expanded or divided region of a leaf or flower part. Leaves can be shallowly or deeply lobed. The calyx and corolla are often divided into lobes called **sepals** (singular: sepal or "calyx lobe") and **petals** (singular: petal or "corolla lobe"). The ovary and style may also be lobed.

Local concern. See **locally sensitive**.

Localized. Growing in a particular place, usually in a group with others of its species. Sometimes called a patchy distribution.

Locally sensitive. Few or dwindling in number, and/or in danger of extirpation within the area defined by this book. An informal rank, it is determined by local botanists familiar with the species and its population trends in the area. Given only to qualifying native plant species that have no other sensitive species designation but probably should have one. It is sometimes called **local concern**. Indicated by the symbol ✿

Locule. A chamber within an ovary.

Male ♂. An organism or reproductive part that produces sperm cells. Often represented by the symbol of the Roman god Mars.

Margin. The edge of a leaf, sepal, or petal.

Midrib. The leaf stalk that continues through the center of the leaf, surrounded by the blade.

Mima mound. Naturally occurring low mound of soil. In California, such mounds often surround vernal pools.

Monoecious. Having both male and female functional reproductive structures (milkweeds, sun cups). A flower that contains both sexes is said to be **bisexual** or **perfect**. Some monoecious species have both sexes on the same individual plant but in separate male-only or female-only flowers (California sycamore). (Greek, monos = single; oikos = household.)

Naked floret. A floret in a flowerhead of a plant that is a member of the aster family. Similar to filiform florets but has no corolla (female [pistillate] florets of ragweeds, bur-sages, and cockleburs).

Native. Occuring naturally in a defined geographic area, not as a consequence of human activity, either *evolving in that area* or evolving elsewhere and spreading to that area *on its own*. Also called **indigenous**. Compare with **non-native**. An organism can be native to one part of California and non-native to another. For example, elegant clarkia is native to northern and central California and a few places in southern California. It is non-native to Orange County but is found here occasionally, probably from seeds spread by gardeners and/or seeds spread by birds that obtained them from a garden.

Naturalized. Established, though non-native, in a wild (non-cultivated) area.

Node. Stem tissue from a which a leaf or a root grows.

Non-native or **not native**. Occurring in a defined geographical area without having evolved there or getting there naturally (on its own). Spread to that area by the direct or indirect activities of humans. Compare with **native**. Indicated by the symbol ✖

Oblong. Longer than wide, the sides parallel, both ends rounded.

Odd-pinnate. Having an odd number of leaflets; usually bearing a leaflet at the very tip of the pinnate leaf.

Opposite. Having two leaves per node, on either side of the stem (sages).

Orbicular leaf. A perfectly circular leaf with the petiole attached to it.

Outward. Away from its main stem.

Oval. Longer than wide, the base wide and rounded, the tip narrowed. Shaped about like a chicken egg. Technically called **ovate**. When the tip of an oval leaf is very long and narrow, it is said to be **lance-shaped**.

Ovary position. The location of the ovary relative to the place of attachment of the sepals (calyx) and petals (corolla). See **inferior**, **superior**, and **half-inferior**.

Ovary. A female reproductive structure; contains the ovules, female reproductive cells that mature into seeds after fertilization. The ovary develops into fruit as the seed(s) mature.

Ovule. A female reproductive structure that contains an egg and normally develops into a seed after fertilization.

Paleae. Bracts on the receptacle in the flowerhead of a plant in the aster family; subtend florets. Older names include pales, chaffy bracts, chaff scales, or receptacular bracts. Flowerheads that lack paleae are termed **epaleate**.

Palmate or **palmately compound**. Divided into leaflets that originate from a basal point, arranged like the palm of your hand (lupines).

Palmately lobed. Having lobes (not leaflets) that originate from a basal point, arranged like the palm of your hand (big-leaf maple, California sycamore, wild cucumber).

Pappus. In the aster family, a single part of the modified calyx.

Parallel. A form of leaf venation found in monocots (lilies, orchids, grasses, etc). where the primary or secondary veins run side by side, equidistant from each other.

Parthenogenesis. Development of an egg without fertilization by a sperm cell.

Pedicel. Stalk of an individual flower or fruit.

Peduncle. The stalk of an inflorescence *below the lowest flower*, or the stalk of a specialized inflorescence unit (such as the stalk below a flowerhead in the sunflower family). Also used to describe the stalk of an individual flower or fruit that is not in an inflorescence.

Peltate. Having the petiole attached at or near its center, like an umbrella.

Penni-parallel. A form of parallel leaf venation found in some monocots (palms, bananas) where parallel secondary veins branch from a single primary vein region. Also called pinnate-parallel.

Perennial. A plant that lives for more than two years or growing seasons.

Perfect. See **bisexual**.

Perianth. The calyx plus the corolla. See also tepals.

Petal. An individual element of the corolla. May be partly or wholly fused into a tube or exist as a free lobe.

Petiole. Stalk of a leaf, not surrounded by the blade.

Petiolule. Individual stalk of a leaflet, like a "miniature petiole" (western poison oak).

Phyllary. In the aster family, a bract of the involucre.

Phytophotodermatitis. An inflammation of the skin (rash) caused by contact with (a) chemical(s) that is activated by exposure to light. Known in bushrue and other plants in the rue family, and some plants in the carrot, pea, and fig families.

Pinnate-parallel. See penni-parallel.

Pinnate or **pinnately compound**. Divided into leaflets arranged on both sides of the midrib. See also 1-pinnate and 2-pinnate.

Pinnately lobed. Having lobes (not leaflets) arranged on both sides of the midrib (whispering bells).

Pistil. A female reproductive structure, composed of stigma, style, and ovary.

Plumose. Plume-like, with many fine filaments or branches that give a feathery appearance.

Poisonous. See **toxic.**

Pollen. A male reproductive structure; contains two reproductive cells; forms within the anther and is released from it when mature.

Pollen aperture. A region of weakness in the wall of a pollen grain though which the pollen tube exits.

Pollen tube. A root-like structure that grows from a pollen grain. It usually delivers 2 sperm cells to an ovule for fertilization.

Pollination. The transfer of pollen from the male flower parts to the female parts, eventually resulting in fertilization of ovules and production of seeds.

Pollinium (plural **pollinia**). A mass of pollen released as a single cohesive unit (chaparral yucca, milkweeds).

Polygamous. Bearing unisexual and bisexual flowers on the same plant (lemonade berry, sugar bush, Western nettle).

Population. A group of plants of the same species that live in a particular area. For example, we can refer to the population of California buckwheat that lives in the Santa Ana Mountains or only to the population of California buckwheat that lives in Trabuco Canyon within the Santa Ana Mountains.

Prickle. An outgrowth of epidermal and subepidermal tissue, not associated with a node and containing no vascular tissue. Most often found on stems but also on some branches, leaves, and flowers (California rose, California blackberry).

Pronotum. The dorsal (back) portion of an insect's thorax (the thorax is the second major body region behind the head).

Proximal. Situated nearer to the center of the body or the point of attachment.

Radial symmetry. Symmetry whereby an object, such as a flower or fruit, can be divided into halves if cut along 3 or more lengthwise planes that include the object's center. Also called **regular** or **actinomorphic.**

Radiant head. A type of discoid flowerhead that contains disk florets only but the outermost disk florets are large, often bilateral (yellow pincushion).

Radiate head. A type of flowerhead that contains both disk and ray florets (California bush sunflower).

Rare. Although not threatened with extinction, in such small numbers throughout its range that it may become endangered if its present environment worsens (paraphrased from Section 1901, California Fish and Game Code). Indicated herein by our sensitive symbol ★

Ray floret. A floret in which the corolla is bilaterally symmetric, a short to long tube topped on one side with a strap-like blade that is generally 3-lobed at tip (occasionally unlobed or with a different number of teeth). Ray florets lack stamens; each contains a fertile pistil (**pistillate ray floret**) or a small infertile pistil (**sterile ray floret**).

Receptacle. The structure to which a flower is attached.

Regular flower. A flower whose parts grow outward or upward away from its center. Often radiates outward like rays of sunshine, with equally sized and shaped petals. Technically called **radial** or **radial symmetry.**

Reverse-oval. Longer than wide, the base narrow and pointed, the tip wide and rounded. Shaped nearly like a chicken egg. Technically called **obovate.**

Rhizomatous. Growing from a rhizome.

Rhizome. A horizontal underground stem that forms roots and upright stems.

Root. An underground stem. Its smallest parts, root hairs, absorb water and nutrients from soil.

Samara. The winged dry fruit produced by plants such as big-leaf maple and ash.

Seed. A plant reproductive structure; develops from a fertilized, maturing or mature ovule. Made of 3 parts: embryo, nutritive tissue (endosperm), and a seed coat. The seed coat may be loose (peanuts) or bonded to the seed (poppies).

Self-fertilization. A form of plant fertilization whereby pollen is transferred within the same flower or from flower to flower on the same individual plant.

Senesce. Deteriorate with age.

Sepal. An individual element of the calyx. May be partly or wholly fused into a tube or exist as a free lobe.

Septum. A wall of tissue between chambers, often in fruit.

Sessile. Lacking a stalk, attached directly to something. Sessile leaves have no petiole. Sessile flowers have no peduncle or pedicel. Sessile stamens have no filaments and are attached directly to the ovary or to the petals.

Shrub. A woody plant, with multiple or single trunks, often 3 meters (about 9 feet) or shorter at maturity.

Silicle. A fruit produced by some members of the mustard family. A form of silique that is round or only slightly longer than wide and often flattened (lacepods, shepherd's purse, peppergrass).

Silique. A fruit produced by members of the mustard family. Typically long and narrow, it is a 2-chambered capsule that opens when the two sides fall away, exposing the seeds, which are attached to a paper-thin partition (false septum) within.

Simple leaf. A leaf that is not divided into leaflets but may be lobed or pinnately lobed. Has an axillary bud at the base of the petiole. Sometimes called **undivided.**

Simple pistil. A single, free carpel with a single ovary chamber and stigma.

Spear-shaped base. A leaf base that has two angled lobes that project backward toward the stem and outward from the petiole, like the base of a spear or an upside-down letter V. Technically called **hastate**.

Spear-shaped leaf. A leaf that is shaped like the head of a spear with basal lobes that point backward and outward, sometimes with multiple lobes. Technically called **hastate**.

Species name. A complete scientific name of a species; includes the genus, specific epithet, and authority. It may also include the year that the species name was published by the authority. Species may be abbreviated as "sp.," more than one species "spp."

Species. A group of living things that interbreed to produce fertile offspring. Put simply, living things that are of the same kind (California poppy, arroyo lupine, thick-leaved lilac). The word "species" is both singular and plural.

Specific epithet. The second word of a scientific name of a species.

Spine-tipped. Ending in a spine, often a continuation of the midrib or lateral vein.

Spine. A leaf or stipule modified into a hard sharp-tipped structure; contains vascular tissue (cacti, gooseberries).

Spoon-shaped. Narrow at the base, then abruptly widening into a much broader oval or rounded blade. Technically called **spatulate**.

Sprawling. Grows flat on the ground. Also called **prostrate**.

Spur. A projection from a calyx or corolla; often tubular, rounded, or pointed; nearly always contains nectar. Usually deepens away from the reproductive parts of the flower. Prevalent in larkspurs and orchids.

Stalk. A stick-like base of a leaf, flower, or fruit.

Stamen. A male reproductive structure, composed of a stalk (filament) topped with an anther.

Staminode. A sterile stamen. Usually composed of a filament (often modified) but no anther.

Stem leaves. Leaves on the stem and/or branches. Basically refers to all leaves other than basal leaves. Technically called **cauline leaves**.

Stems. The branches of a plant. Aboveground they bear leaves, buds, and flowers. Belowground they are called roots and are sometimes modified into structures called bulbs, corms, rhizomes, stolons, or tubers.

Stigma. A female reproductive structure; the surface and/or structure onto which pollen is deposited during pollination, often sticky, sometimes lobed.

Stipule. An appendage at the base of the petiole, often 2 (one at each side of the petiole). Stipules may look like tiny leaves, scales, corks, or spines.

Stolon. A creeping stem that lies flat on the ground and forms roots and upright stems. Also called **runner**.

Style. A female reproductive structure; the stalk that connects the stigma to the ovary. Usually 1 style per ovary, sometimes divided into 2 or more style lobes.

Subspecies (abbreviated as "ssp." or "subsp."). A group of plants within the same species, some of its members consistently different from others and recognizable as such. They are usually geographically separated from each other; they generally do not occur with other subspecies within the same population. Compare with **variety**.

Subtend. Lie immediately below another part.

Superior ovary. An ovary that sits *above* the flower base. Sepals, petals, and stamens attach to the flower base *below* the ovary (snapdragons, mints). Technically called **hypogynous**.

Symmetry. The presence, form, and number of mirror-image planes of an object. This is an important physical property used to describe all living things, especially flowers. See **bilateral**, **radial**.

Talus slope. A mass of rock fragments on a slope.

Tepals. A term for perianth parts, used when we cannot tell if they are sepals or petals. Applies to the laurel, cactus, and montia families and many of the monocot families.

Tetraploid. A cell that contains four sets of homologous chromosomes. In a tetraploid cell, two sets of homologous chromosomes are typically inherited from each parent.

Thorn. A stem modified into a hard sharp-tipped structure; contains vascular tissue (greenbark California lilac).

Thread-like. A linear leaf that is very slender, somewhat like thread.

Threatened. Likely to become endangered within the foreseeable future throughout all or a significant portion of its range (paraphrased from Section 2067 of the California Fish and Game Code, and U.S. Endangered Species Act of 1973).

Tip. The area of a leaf blade farthest from the petiole.

Toothed. Having teeth along its margin.

Toxic. Poisonous, relating to or caused by poison. The toxin can be acquired through contact, inhalation, and/or ingestion. Indicated by the symbol ☠.

Tree. A woody plant, generally with a single trunk, often taller than 3 meters (about 9 feet) at maturity.

Triangular. Shaped like a triangle or guitar flat pick. Technically called **deltoid**.

Truncate. Flattened perpendicular to the petiole.

Trunk. The main stalk of a plant that emerges from the ground, often thick and woody.

Unisexual. See **imperfect**.

Urticating. Causing a stinging or prickling sensation; generally describes the effects of nettles.

Urushiol. A class of oils produced by sumac plants such as Western poison oak.

Variety (abbreviated as "var.") A group of plants within the same species, some of its members consistently different from others and recognizable as such. They are usually not geographically separated from each other and may occur together within a single population. Compare with **subspecies**.

Vegetative. Not involved in reproduction, such as leaves, stems, and roots.

Vine. A plant with long stems that trail on the ground or climb onto others.

Wedge-shaped. Forming a wide angle on the petiole and widening to the leaf blade; shaped like a wedge used to split wood. Technically called **cuneate**.

Weed. A plant, native or non-native, that grows where a human does not want it to grow. A plant can be a weed (unwanted) in one situation but not a weed (wanted) in a different situation.

Whorled. Having 3 or more leaves per node (bedstraws, ocellated Humboldt lily).

Wildflower. Any flowering plant (Angiosperm) that lives in the wild.

Wort. An Old English word for plant, from wyrt. Used as a suffix in names of some plants, most often those employed for food and medicine.

Measurements

In this book, plant measurements are given in metric units, most often in meters, centimeters, and millimeters. For distances and elevations, we give both English and metric equivalents (miles, feet, and meters) because trip odometers in automobiles made for America measure in miles and topographic maps made by the US Geological Survey still measure elevation in feet.

The units of length used here are as follows:

Meter, m. The basic unit of the metric system. Made up of 10 decimeters, 100 centimeters, or 1000 millimeters (1 m = 10 dm; 1 m = 100 cm; 1 m = 1000 mm). Just a bit longer than a yard. A meter is equal to 39.4 inches or 3.28 feet.

Centimeter, cm. One-hundredth of a meter (100 cm = 1 m). Itself made up of 1000 millimeters. About equal to the width of an adult's little finger or fingernail. A centimeter is equal to 0.39 inches.

Millimeter, mm. One-thousandth of a meter (1000 mm = 1 m), one-tenth of a centimeter (10 mm = 1 cm). A millimeter is equal to 0.039 inches.

Coast Baby-star (*Leptosiphon parviflorus*, Polemoniaceae) below Elsinore Peak, April 27, 2008.

General Index

Primary entries for major plant groups and plant families are in boldface type, as are primary entries for plant and animal species.

Abronia 289-290
　umbellata 19, 147, **289**
　maritima **289**
　villosa var. *aurita* 19, **290**
Acalymma vittatum **225**
Acalypha californica **233**, 290
Acer macrophyllum 17, 333, **391**
Aceraceae: See Sapindaceae
Achillea millefolium **103**
Achyrachaena mollis **104**
Acmispon 242, 244-247
　americanus 242
　americanus var. *americanus* **244**
　argophyllus 242, **244**, 246
　brachycarpus 242, **244**
　glaber 25, 242
　glaber var. *brevialatus* 19, **245**
　glaber var. *glaber* 19, **245**, 356
　heermannii 242
　heermannii var. *heermannii* **246**
　heermannii var. *orbicularis* **246**
　maritimus 242
　maritimus var. *maritimus* **246**
　micranthus 242, 244, **246**
　parviflorus 242, **247**
　strigosus 242, **247**
　wrangelianus 242, **247**
Acourtia microcephala 63, **65**
Acrididae 105, 120, 448
Acroptilon 79-80
　repens **80**
Adela septentrionella **382**
Adelidae 382
Adenostoma 377-379
　fasciculatum 21, 305, 306
　fasciculatum var. *fasciculatum* **378**
　fasciculatum var. *obtusifolium* **378**
　sparsifolium 377
Adoxaceae 39-40, 202, 218, 307
Agavaceae 413-418, 424, 439
Agave americana 14
Agave Family 413-418
Agelaius phoeniceus 446, **448**
Ageratina adenophora **65**, 118
Agnorhiza ovata 120
Agoseris 64, 104, 131-133, 136
　grandiflora 131, **132**
　heterophylla var. *heterophylla* 131, **132**
　retrorsa 131, **133**
Aizoaceae 41, 91, 260
Alcea 286
Alchemilla occidentalis 380
Alder 410
　White 3, 17, 18, 34, 433
Alfalfa 93, 236, **254**
Algae, Green 13
Alisma 419
　plantago-aquatica 419
　triviale **419**
Alismataceae 419

Aliso 333
Alliaceae 420-423, 424, 440, 445
Allium 420-423, 424, 445
　haematochiton 420, **421**, 422
　lacunosum var. *lacunosum* 420, **421**, 422
　monticola 23, 420, **423**
　munzii 420-421, **422**
　peninsulare var. *peninsulare* 420, **423**
　praecox 420-421, **422**
Allophyllum glutinosum 25, **338**
Almonds 93
Alnus rhombifolia 17, 433
Aloë 424
　Candelabra 424
　Chabaud's 424
　Soap 424
Aloë 424
　arborescens **424**
　chabaudii **424**
　maculata **424**
　saponaria 424
Alpinegold, Red-rayed 124
Amaranths 34
Amaryllidaceae 420
Amaryllis Family 420, 440
Amblyopappus pusillus **66**
Ambrosia, San Diego 68; See also Ambrosia
Ambrosia 63, 66-68
　acanthicarpa **67**, 68
　chamissonis 19, 66, **67**, 147, 218
　confertiflora **67**, 68
　psilostachya 67, **68**
　pumila **68**
Ammannia 284
　coccinea **284**
　robusta 284
Amole 414
Amorpha 236-237, 243
　californica **237**
　fruticosa **237**
Amsinckia 144-145
　intermedia 19, **145**
　menziesii var. *intermedia* 145
　retrorsa **145**
Anacardiaceae 42-45, 218, 374, 386
Anagallis arvensis 270
Andrenidae 158, 159, 169
Anemone 38
Anemopsis californica **38**
Angel's Trumpet 395
Angiosperm 10, 13
Anise 50
　Wild 50
Ant 32, 59, 309, 347, 349, 355, 416
　California Harvester 32
　Red Imported Fire 32
　White Velvet 32

Velvety Tree 32
Antelope, American 351
Anthemis
　cotula **104**
　nobilis 94
Antilocapra americana 351
Antilocapridae 351
Antirrhinum 206, 320-321, 325
　coulterianum **320**
　kelloggii 320, **321**
　multiflorum 320, **321**
　nuttallianum 320, **321**
Aphanes occidentalis **380**
Aphanisma blitoides 19
Aphid
　Manzanita Leaf Gall 231
　Milkweed 60
　Oleander 60
Aphididae 60, 231
Aphis nerii **60**
Apiaceae 25, 39, **46-56**, 62, 162, 264, 362, 363, 380
Apiastrum angustifolium **46**
Apidae 93, 143, 196, 225, 236, 275, 338
Apioceridae 338
Apiognomonia veneta **334**
Apium graveolens **47**
Apis 32
　mellifera mellifera **32**
　mellifera scutellata 32
Apocynaceae 10, **57-61**, 411
　Vine-like Perennials 57
　Bushy Perennials 58
Apocynum cannabinum **58**
Apple 93, 377, 381
Aquifoliaceae 376, 381, 385
Arabis 173
　glabra 174
Arachnida 334
Aralia, California 11, **62**
Aralia californica **62**
Araliaceae 62
Arbutus 381
　menziesii 21, **227**
Arctostaphylos 15, 228-231, 260, 306
　glandulosa ssp. *glandulosa* 21, **228**, 305, 373, 401
　glauca **229**, 405
　peninsularis ssp. *keeleyi* 230
　pringlei ssp. *drupacea* **230**
　rainbowensis **230**
Arenaria douglasii **205**
Argemone 309
　munita **309**, 313
　munita ssp. *robusta* 309
Aristida 19
Arnica, Rayless 3, **69**
Arnica discoidea **69**
Arrow-grass 34
　Seaside 17

Arrowhead, Montevideo 419
Arrowweed, Desert 95
Artemisia 69-72
  californica 15, 19, 64, **70**, 120, 155
  douglasiana 17, 64, **71**
  dracunculus **71**
  tridentata ssp. *tridentata* 64, **72**
Arthrocnemum 210, 219, 304
  subterminale 17, **210**, 219
Artichoke 64, 86
  Cultivated 86
  Globe 86
Asclepiadaceae 57; See also Apocynaceae
Asclepias 58-59
  californica **59**, 411
  eriocarpa 58, **59**, 411
  fascicularis **58**
Ash 21, 40, 291, 410
  Arizona 291
  California 21, **291**, 450-451
Asphodelaceae 424
Asphodel, Hollow-stem 424
Asphodel Family 424
Asphodelus fistulosus 424
Aster
  California 107
  Cliff 135
  Cliff, Slender-leaved 135
  Lanceleaf, Western 106
  Roughleaf 105
  San Bernardino **105**, 454
  Sand 107, 306
  Sessileflower, Golden 25, **123**
  Slender 106
  True 105-107
Aster: See also Eurybia, Symphyotrichum
  bernardinus 105
  defoliatus 105
  exilis 106
  lanceolatus ssp. *hesperius* 106
  radulinus 105
  subulatus 106
Asteraceae 10-11, 13, 60, **63-142**, 147, 155, 165, 204, 218-219, 231, 237, 261, 274-275, 302, 306-307, 409, 411, 418
  Discoid, Radiant, and Disciform Heads 65-102
  Liguliflorous Heads 131-142
  Radiate Heads 103-130
Astragalus 25, 238-239
  brauntonii 23, **238**
  didymocarpus **238**
  gambelianus **238**
  pomonensis **239**
  trichopodus var. *lonchus* **239**
  trichopodus var. *trichopodus* **239**
Athysanus, Dwarf 182
Athysanus 171, 182
  pusillus **182**
Atlides halesus estesi **411**
Atriplex 19, 219
Avocado 13

Baby-star, Coast 342

Baby's
  Breath 205
  Tears 405
Baccharis 63-64, 72-76, 92
  Broom 73, **76**
  Encinitas 74
  Malibu 74
  Marsh 73
  Willow 73-75, **76**
Baccharis 63-64, 72-76, 92
  douglasii 73
  emoryi 73, 76
  glutinosa **73**, 75
  malibuensis **74**
  pilularis 19, 73, 75-76
  pilularis ssp. *consanguinea* **73**
  pilularis ssp. *pilularis* 73
  salicifolia 17, 72, **75**, 76, 218
  salicina 73, 75, **76**
  sarothroides 73, **76**
  vanessae **74**
  viminea **75**
Bacteria 32
Bahiopsis 116
  laciniata **119**
Barbarea 171, 172
  orthoceras **172**, 176
Barberry 3, 11, 13, 143
  California 143, 454, 457
  Shinyleaf 143
Barberry Family 143
Barley, Vernal Spring 16, 17
Basil 271
Basketbrush 43
Bather's Delight 67
Bay Laurel: See Laurel
Bean 233, 236
  Castor 233
Bear 386
  Black 386
  Grizzly 386
Beardtongue 3, 29, 324-328
  Buglar, Scarlet 326
  Bush, Blue-stemmed 326
  Bush, Heart-leaved 324, **325**
  Bush, Yellow 21, 29, 34, **325**, 457
  Foothill 327
  Foothill, Southern **327**
  Grinnell's 34, 324, **327**
  Royal 20, 324, **328**, 456
Beargrass
  Chaparral 3, 23, 187, 415, **438**, 439, 450
  Parry's 439
Bebbia 64
  juncea var. *aspera* 19, 63, **77**
Bedstraw 387-389
  Chaparral 387, **389**
  Climbing 387, **389**
  Common 387, **388**
  Goosegrass 388
  Graceful 389
  Lady's 387
  Narrow-leaved 389
  Parisian 387, **388**
  San Diego 387, **389**

  Three-petal, Pacific 387, **388**
Bee-plant 394
Bees 32, 59, 75, 77, 90, 96, 107, 115, 123-124, 154, 156, 162, 169, 199, 200, 205, 221-223, 227, 229, 231, 236, 241, 245, 261, 266, 271, 274, 275, 279, 282, 286, 291, 299, 301, 308, 309, 311, 314, 316, 320, 321, 327, 328, 330, 338, 347, 355, 358, 364, 374, 381, 385, 386, 394, 399, 402, 438, 441, 456
  Andrenid 158, 159, **169**
  Bumble 32, 169, **170**, 222-223, **236**, **238**, 241, 261, 271, 274-277, 279, 286, 289, 299, 301, 303, **304**, 305, 316, 327, **361**, 364, 399
  Bumble, Crotch's 236
  Cactus 196
  Carpenter 32, 143, 261, 271, 275, 276, 277, 316, 327
  Common 338
  Honey, Africanized 32
  Honey, European 32
  Leafcutter 32, 338
  Leafcutter, Slender 93
  Long-tongued 338
  Longhorn 93
  Mason 156
  Mason, Metallic 84
  Solitary Ground-nesting 355
  Squash 225
  Sweat 441
  Sweat, Metallic 174, 200, 392
Beet 210
Beetle 32, 59, 75, 90, 107, 309, 338, 355, 374, 384, 386, 395, 441, 446
  Blister, Black-winged 355
  Blister, Orange 355
  Blister, Squinting 92
  Borer, Milkweed 60
  Checkered 311
  Cucumber, Striped 225
  Darkling 32
  Darkling, Hairy 32
  Datura, Three-lined 396
  Flower, Cactus 195
  Flower, Soft-winged 124, 311
  Flower, Tumbling 156, **355**
  Goldenbush, Red-striped 92
  Goliath, African 93
  Green Fruit 93
  Japanese 93
  June 395
  June, Dark-headed 396
  Ladybird 195
  Ladybird, 20-spotted 334
  Ladybird, Northern 20-spotted 334
  Ladybird, Banded 195
  Ladybird, European Seven-spotted 195
  Ladybird, Three-forked 195
  Leaf, Milkweed 61
  Leaf, Southern Yerba Santa 155
  Leaf, Spotted 418
  Longhorn 311

Longhorn, California
  Elderberry 40
Longhorn, Ceanothus 374
Longhorn, Goldenbush 92
Longhorn, Six-spotted 355
Rain, Ceanothus 374
Scarab, Little Bear 145
Stink 32
Weevil, Yucca 418
Wood-boring, Metallic 124
Beggar-ticks, Common 87, 108
Belladonna 395
Bellflower Family 197
  Campanula-type 197-198
  Lobelia-type 197, 199-200
  Nemacladus-type 197, 201
Bells
  Canterbury, Desert 160-161, 166, 167
  Canterbury, Wild 160-161, 166, 167
  Mission 431
  School 19, 20 25, 26, **444**, 445
  Whispering 25, **154**
Berberidaceae 143
Berberis 143
  aquifolium var. *dictyota* **143**
  dictyota 143
  pinnata ssp. *pinnata* **143**
Berry
  Christmas 381
  Lemonade 18-19, 29, 42, **44**
Berula erecta **47**
Berytidae 279
Betula 380
Betulaceae 433
Bidens 108
  cernua 108
  frondosa **108**
  laevis **108**, 261
  pilosa 87, **108**
Bindweed 214-215
  Field 215
Birch 380
Bird 75, 271, 301, 355, 376, 381
  Blackbird 446
  Blackbird, Red-winged 446, **448**
  Bushtit 29
  Cuckoos 196
  Gnatcatcher, Blue-gray 70
  Gnatcatcher, Coastal California 70
  Goldfinch 29
  Goldfinch, Lesser **84**, 276
  Hummingbird 156, 204, 222, 223, 227, 229, 261, 266, 268, 273, 274, 276, 277, 279, 299, 273, 274, 276, 277, 299, 301, 304, 305, 314, 316, 320, 321, 326, 335, 362, 364, 424, 455-456
  Hummingbird, Allen's 314
  Hummingbird, Anna's 266, 268, **335**
  Phainopepla 410, **411**
  Roadrunner, Greater 196
  Sunbirds 424

Woodpeckers 196
Wren, Cactus 196
Wren, Marsh 446
Yellowthroat, Common 446
Bird's-beak 304
  Dark-tipped 304
  Salt Marsh 304
Bird's-foot Trefoil 243
Bittercress, Little 172
Blackberry 377, 386
  California 45, **386**
  Himalayan 386
Bladderpod 19, **213**
Blennosperma nanum **115**
Bloomeria crocea **440**
Blow-wives 104
Bluebells 166-167
  California **167**
  Desert 166
Bluecurls 282
  Parish's 3, **282**
  San Jacinto 282
  Vinegar 282
  Woolly **282**, 456
Blue-dicks: See School Bells
Blue-eyed Mary 323
  Parry's 323, 453
Blue-eyed-grass, Western 19, **425**
Blue-Eyes 157-158
  Baby, Meadow 158
  Baby, Menzies's 157
  Baby, Pale 158
Bluecup 25
  Southern, Blue-flowered 11, **197**
  Southern, White-flowered 197
Bobcat 33
Bombus 32, 275; *See also* Bumblebee
  crotchii **236**
Bombyliidae 355
Borage Family 11, 30, **144-170**, 231
  Classic Borages 144-154
  Lennoas 144, 170
  Waterleafs 144, 154-170
Boraginaceae 25, 30, 84, **144-170**, 231, 306
Bowlesia, American **48**, 380
Bowlesia incana **48**, 380
Box-thorn, California 19, **396**
Boykinia rotundifolia 23, **392**
Boykinia, Round-leaved 23, **392**, 451
Brass-buttons 88
  African 88
  Australian 88
Brassica 171, **177-179**
  campestris 177
  geniculata 178
  nigra 25, 177, **178**
  rapa 177
  tournefortii **179**
Brassicaceae **171-185**, 293, 294
Breath, Baby's 205
Brickellbush 78-79

California 78
Desert 78
Nevin's 78, **79**
Brickellia 78-79
  californica **78**
  desertorum **78**
  nevinii 78, **79**
Brittlebush
  California 116
  Desert **116**, 276
Brodiaea 20, 440-443
  Earth 441-442 **443**
  Mesa 443
  "Mystery" **443**
  Orcutt's 442-443
  Santa Rosa Basalt 441-442, 443
  Thread-leaved 441, **442**, 443
Brodiaea 20, 420, 440-443
  filifolia 441, **442**, 443
  jolonensis 443
  orcuttii 442-443
  santarosae 441-442, **443**
  terrestris 441-442, **443**
Brodiaea Family 420, **440-445**
Brookweed, Seaside 402
Broom 305
  Baccharis 73, **76**
  Butcher's 438
  California 25, **245**
  Chaparral 73
  Deer 19, 25, 242, **245**
  Deer, Coastal 245
  Deer, Desert 245
  Spanish 255
Broomrape 3, 305
  Goldenrod 3, 29, 130, **307**
  Chaparral **305**, 378
  Clustered 23, **306**, 357
  Parish's 306
  True 305
  Valley 40, **307**
Broomrape Family 301-307
Brownpuffs 136
  Grassland 136, **138**
Brush
  Coyote 19, **73**, 75-76
  Jim 21, 368, **371**, 372
  Southern Deer 21, 368, **372**, 373
  Whisker 340
Brushfoots: See Nymphalidae
Bryophytes 13
Buckthorn Family 367-376
Buckthorns 375-376
Buckwheat Family 10, 13, **347-360**
  Spineflowers 347-351
  Wild Buckwheats 352-359
  Assorted Buckwheats 360
Buckwheat 3, 29
  Ash-leaved 353
  Bluff 19, 352, 356, **357**
  California 17, 18, 19, 20, 21, 22, 26, 78, 217, 352, 353, 354, 355, 356
  California, Mojave Desert 352, **354**
  California, Coastal 352, **354**

California, Leafy 352, 353, **354**
Gray Coast 352, **353**
Davidson's 352, 358, 359
Long-stemmed 352, **353**, 357
Naked-stemmed 352, **357**
Rock 23, 24, 306, 352, **357**
Short-stemmed 352, 356, **358**
Slender 352, 356, **359**
Slender, Smooth-stemmed 352, **359**
Thurber's 352, **358**
Wild 352
Buffalo-bur 399, **400**
Bug
  Ambush, Pacific 101
  Assassin, Leafhopper 60
  Harlequin 213
  Lace, Western Sycamore 334
  Milkweed, Small 60
  Milkweed, Large 60
  Prickly-pear **194**
  Seed 279
  Stilt 279
Bugler, Scarlet 326
Buprestidae 124
Bur-marigold 108
  Nodding 108
  Smooth **108**, 261
Bur-sage 63, 66-67
  Annual 67
  Beach 18-19, 66, **67**, 147, 218
Burhead, Upright 419
Burweed, Weak-leaved 67, 68
Bush
  Button, San Diego 34
  Poodle-dog 84, **156**, 456
  Snowdrop, Southern 401
  Spice, Coastal 390
  Sugar 21, 42-43, **44**
  Wishbone, California 233, **290**
Bushrue 2, 118, **390**
Bushtits 29
Butcher's Broom Family 438
Buttercup 10
  California 11, **366**
  Aquatic 366
  Bermuda 308
  Delicate 366
  Desert 366
  Western 366
Buttercup Family 362
Butterflies, Butterfly 59, 75, 77, 90, 107, 123, 156, 227, 229, 231, 236, 245, 271, 274, 286, 291, 316, 338, 342, 347, 355, 374, 385
  Admiral, Red 231, **405**
  Anglewing, Satyr 405
  Blue, Acmon 356, **359**
  Blue, Bernardino 356
  Blue, Western Tailed 239
  Buckeye 154
  Checkerspot, Chalcedona 325
  Checkerspot, Quino 331
  Copper, Tailed 269
  Dogface, California 237, 316
  Fritillary, Comstock's 409

Fritillary, Semiramis 409
Hairstreak, Bramble Green 245
Hairstreak, Western Great Purple 411
Lady, American 77, **96**
Lady, Painted 231
Monarch 58, **61**, 433
Queen, Striated 61
Silverspots 409
Skipper 316
Swallowtail, Anise 50
Swallowtail, Pale 84, **156**, 375, 385, 433
Swallowtail, Western Carrot 50
Swallowtail, Western Tiger 335, 433
Butterweed
  California 128, **129**
  Sand-Wash 129, 457
Button-celery
  Pendleton 17, **49**
  San Diego 49

Cactaceae 10, 41, **186-196**
Cactus Family 10, **186-196**, 287
Cactus, Cacti 9, 11, 13, 31, 41; *See also* Cholla, Prickly-pears
Cakile 171, 182
  edentula 182
  maritima 19, **182**
Calabazilla 224
Calamagrostis koelerioides 23
Calandrinia 287
  ciliata **287**
  breweri **287**
  maritima 287
Calibrachoa parviflora 396
Calicoflower 17, 199
  Hoover's 17, **199**
  Toothed 17, **199**
California
  Largest Flower 313
  State Flower 311
  State Insect 237
California macrophylla **262**
Callimoxys sanguinicollis fuscipennis 374
Callophrys perplexa **245**
Calochortus 25, 426-430
  albus 426, **427**
  catalinae 19, 426, **427**
  davidsonianus 428
  invenustus 426, **428**
  plummerae 426, **429**, 430
  splendens 426, **428**
  weedii **429**
  weedii var. *intermedius* 426, 429, **430**
  weedii var. *peninsulare* 430
  weedii var. *weedii* 426, **430**
Calycadenia tenella **114**
Calypte anna 266, 268, **335**
Calyptridium monandrum **288**
Calystegia 214-215
  fulcrata 214

  macrostegia **214**
  occidentalis ssp. *fulcrata* 23, **214**
  soldanella 19, 147, **215**
Camas: *See* Death Camas
Camissonia 25, 292; *See also Camissoniopsis, Eulobus, Tetrapteron*
  bistorta 293
  californica 294
  cheiranthifolia 293
  graciliflora 294
  ignota 293
  micrantha 294
  strigulosa **292**
Camissoniopsis 25, 292-294
  bistorta **293**
  cheiranthifolia ssp. *suffruticosa* 19, 147, **293**
  ignota **293**
  micrantha **294**
Campanula 197
Campanulaceae 25, **197-201**
  Campanula-type 197-198
  Lobelia-type 197, 199-200
  Nemacladus-type 197, 201
Camphor 37
Campion 3, 206
  Lemmon's 208, 451, 457
  San Francisco 23-24, **208**
Campylorhynchus brunneicapillus **196**
Canchalagua 262, 449, 452-453
Canidae 224
Canis latrans 224
Cannabaceae 226
Cannabis sativa **226**
Caprifoliaceae 10, 39, **202-204**
Capsella 171, 183
  bursa-pastoris **183**
Caraway 46
Cardamine 171-172
  californica **172**
  oligosperma 172
Cardionema ramosissimum 23, **205**
Cardoon 86
Carduelis psaltria 84
Carduus 79-80
  pycnocephalus **80**, 85
Carex 17
Carnation 205
Carnation Family 10, **205-209**
Carpet-weed Family 41
Carpobrotus
  edulis **41**, 91
  chilensis **41**
Carpophilus pallipennis **195**
Carrot 46
Carrot Family 11, 39, **46-56**, 62, 162, 264, 362
Carthamus 79, 81
  baeticus 81
  creticus **81**
  tinctorius 81
Caryophyllaceae 25, **205-209**
Cashew 42
Castilleja 301-303
  affinis ssp. *affinis* 301, **302**
  applegatei ssp. *martinii* 301, **302**
  exserta 19, 301, **303**

foliolosa 301, **302**
minor ssp. *spiralis* 17, 301, **303**
Castor Bean 11, **233**
Catchflies, Catchfly 206-208
  Many-nerved **206**
  Snapdragon 25, **206**
Catnip 271
Cat's Ear **134**
  Hairy **134**
  Smooth **134**
Cat's-eyes 25, 146-148;
  Cleveland's 146, **147**, 148
  Coast 146, **147**
  Common 146, **147**, 148
  Guadalupe 146, **148**
  Jones's 146-147, **148**
  Tejon 146-147, **148**
  White-haired **148**
Cattail Family 446-448
Cattail 16, 17, 446-448
  Broad-leaved 446, **447**, 448
  Narrow-leaved 446-447
  Slender **447**
  Southern 446, **447**, 448
Caulanthus 171, 174-175
  heterophyllus **174**
  lasiophyllus **175**
Ceanothus **367**
Ceanothus 21, 25, 84, 218, 311, 367-374, 416
  crassifolius 21, 367-368
  crassifolius var. *crassifolius* 368, **369**
  crassifolius var. *planus* **369**
  integerrimus var. *macrothyrsus* 21, 368, **372**, 373
  integerrimus var. *puberulus* 372
  leucodermis 21, 368, **370**
  megacarpus ssp. *megacarpus* 21, 368, **369**
  oliganthus var. *oliganthus* 21, 368, **371**, 372
  oliganthus var. *orcuttii* 368, 371
  oliganthus var. *sorediatus* 21, 368, **371**, 372
  palmeri 368, 372, **373**
  papillosus 23, 368, **373**
  sorediatus 371
  spinosus 368, **370**, 373
  spinosus var. *spinosus* 370
  spinosus var. *palmeri* 373
  tomentosus 21, 368, 371, **372**
  tomentosus var. *olivaceous* 372
Cecidomyiidae 202, 440
Celery, Common **47**, 50
Centaurea 25, 64, 79, 81-82
  benedicta **81**
  melitensis **82**
  repens 80
  solstitialis **82**
Centarium venustum 262
Centaury, Charming **262**
Centranthus 326, 406
  ruber **406**
Centromadia 109-110
  parryi ssp. *australis* **109**
  pungens ssp. *laevis* 109, **110**
Centrostegia leptoceras **351**

Cerambycidae 40, 60, 92, 355, 374, 411
Ceratophyllales 13
Cercocarpus 380
  betuloides var. *betuloides* 21, **380**
  minutiflorus **380**
Cercopidae 92
Cervidae 413
Chaenactis 64, 87
  artemisiifolia **87**
  glabriuscula 19, 108
  glabriuscula var. *glabriuscula* **87**
  glabriuscula var. *orcuttii* 87
Chalice, Alkali **261**
Chamaesyce 235
  albomarginata 235
  polycarpa var. *polycarpa* 235
Chamise 3, 21, 23, 305-306, 377, 379, 411
  Common 20, 23, 377, **378-379**, 450, 453
  San Diego **378**
Chamomile, European 64, 94
Chamomilla 94
  occidentalis **94**
  suaveolens **94**
Charlock, Jointed **181**
Checkerbloom 286
  Coastal 19, **286**
  Salt Spring 286
Cheilanthes newberryi 23
Cheiranthus 293
Chenopodiaceae 119, **210-212**, 219, 304
Chenopodium 119
Cherries, Cherry 377
  Holly-leaved 21, 84, 376, **385**
Chia: *See* Sage
Chicalote **309**
Chickweed, Common **209**
Chicory 64, 139-140
  California **139**
Chili Peppers 395
Chinese Houses 322-323
  Purple 23, 25, **322**, 323, 452-453
  Southern **322**, **323**
Chive 420
Chloealtis gracilis 33
Chlorogalum 413-414, 424
  parviflorum **414**
  pomeridianum var. *pomeridianum* 14, **414**
Chloropyron 301, 304
  maritimum ssp. *maritimum* **304**
Cholla 186, 187
  Cane **187**
  Coast 18, **187**, 196
  Valley 19, **187**, 452
Chorizanthe 347-350
  coriacea 350
  fimbriata var. *fimbriata* 347, **348**
  leptoceras 351
  leptotheca 347, **349**
  parryi var. *fernandina* 347, **349**
  parryi var. *parryi* 347, **349**
  polygonoides var. *longispina* 347, **350**

procumbens 347, **350**
staticoides ssp. *chrysacantha* 19, 347, **348**
staticoides ssp. *staticoides* 23, 347, **348**
Chrysanthemum 115, 156
  coronarium 115
Chrysochus cobaltinus **62**
Chrysomelidae 61, 92, 155, 225, 396, 418
Chrysothamnus nauseosus ssp. *consimilis* 89
Cichorium intybus **140**
Cicuta 48, 162, 264
  douglasii **48**
Cinnamon 37
Cinquefoil
  Greene's **384**, 454
  Sticky 21, 383, **384**, 453
Cirsium 64, 79, 80, 83-85
  occidentale var. *californicum* **83**, 204, 237, 409
  occidentale var. *occidentale* **83**
  scariosum var. *citrinum* **85**
  vulgare 80, **85**
Cistaceae 25, **212**
Cistanthe 287-288
  maritima **287**
  monandra **288**
Cistothorus palustris 446
Citrus 10, 71, 390
Clarkia 25, 295-297
  Canyon **297**
  Diamond **297**
  Dudley's **295**
  Elegant **297**
  Four-spot **296**
  Punchbowl **295**
  Tongue **297**
  Waltham Creek **297**
  Willow-herb **296**
  Winecup 19, **296**
Clarkia 292, 295-297
  bottae **295**
  deflexa 295
  dudleyana **295**
  epilobioides **296**, 297
  modesta 297
  purpurea 19, **296**
  rhomboidea **297**
  similis **297**
  unguiculata **297**
Clay-cress, Hammitt's 3, **173**
Claytonia
  parviflora **288**
  perfoliata 23, 25, **288**
Clematis 45, 362-363
  lasiantha **362**
  ligusticifolia 362, 363
  pauciflora 19, 362, **363**
Cleomaceae **213**
Cleome isomeris 213
Cleridae 311
Clinopodium chandleri 271, **278**
Clotbur, Spiny **102**
Clover 236, 256, 303; *See also*

Owl's-clover, Sweetclover
Balloon 256
Clammy 257
Creek 257
Elk 62
Tomcat 257
Tree 256
Truncate Sack 17, **256**
Valley 257
Clusiaceae 270
Cluster Flower: *See* Phacelia
Cluster-lilies 441, 444; *See* also School Bells, Brodiaea
Cneoridium dumosum **390**
Cneorum 390
Cnicus benedictus 81
Coastweed, Dwarf 14, **66**
Coccinella septempunctata **195**
Coccinellidae 195, 334
Cochineal Scales 194, 195
Cocklebur 63, **102**, 261
Spiny 102
Cockscomb 210
Coffee 387
Coffeeberry 3, 374-375
California 21, 84, 232, **375**, 452
Chaparral 375
Desert 375
Colias eurydice **237**
Collinsia 322-323
concolor 322, **323**
heterophylla 23, 25, **322**, 323
parryi **323**
parviflora 323
Comarostaphylis diversifolia var. *diversifolia* **232**
Combseed 152-153
Bristly 153
Narrow-toothed 153
Northern 153
Round-nut 153
Slender 153
Stiff-stemmed 153
Winged 153
Conifers 13
Conium maculatum 47, **48**, 50
Convallariaceae 438
Convolvulaceae 67, 147, 210, **214-219**, 343
Convolvulus 214-215
arvensis **215**
simulans **215**, 343
Copestylum mexicanum **195**
Copperleaf, California 78, **233**, 290
Cordylanthus 301, 304
maritimus ssp. *maritimus* 304
rigidus ssp. *setigerus* **304**
Coreidae 194
Coreopsis 128
californica var. *californica* 128
gigantea 128
Corethrogyne 105, 107
filaginifolia **107**, 306
Corythucha confraterna **334**
Cotinis mutabilis **93**
Cotton-batting Plant 97
Cotton-thorn 101
Cottonweed, Summer 298
Cottonwood 16, 17, 410
Black 17
Fremont's 17
Cotula 88
australis **88**
coronopifolia **88**
Cotyledon viscida 14
Cougar 33
Coyote 224, 385
Crambidae 61
Cranesbills 265
Crassula 220
aquatica 17, **220**
connata 23, **220**
Crassulaceae 10, **220-223**
Cream Cups 25, **313**
Cress 171, 172
Bitter-, Little 172
Clay-, Hammitt's 3, **173**
Rock, Virginia 17, **173**
Water 17, 171, 176
Water, White 176
Winter, American 172, 176, 454
Yellow-, Marsh 172
Yellow-, Pacific 176
Yellow-, Western 172, **176**
Cressa truxillensis **216**, 219
Crete Weed 133
Crocus 425
Crossidius testaceus maculicollis **92**
Crotalus 33
mitchelli pyrrhus **33**
oreganus helleri **33**
ruber **33**
Croton, California 19, 25, **234**
Croton 234
californicus 19, **234**
setiger **234**
Crowfoot Family 362-366
Crowfoots 13, 362-366
Crownbeard 118-119
Big-leaved 118
Golden 119
Cryptantha 25, 144, 146-148, 152
clevelandii 146, **147**, 148
intermedia 146, **147**, 148
leiocarpa 146, **147**
maritima 146, **148**
microstachys 146-147, **148**
muricata var. *jonesii* 146-147, **148**
Ctenucha brunnea **381**
Ctenuchidae 381
Cuculidae 196
Cucumber 10, 224-226
Wild 11, 19, 48, 224, **225-226**
Cucumber Family 224-226
Cucurbita foetidissima **224**
Cucurbitaceae 10, **224-226**
Cudweed 96, 231; *See also* Everlasting
Bicolored 98
Bioletti's 98
Lowland 96, 97
Weedy 96, **97**
Cup, Cups
Cream 25, **313**
Sun 292-294; *See also* Evening-primrose
Cupressaceae 311, 410
Curculionidae 418
Currant 3, 13, 29, 43, 266-267
Chaparral 14, 267
Golden, Slender-flowered 266
Southern California 23, **267**, 455
White-flowered 267, 453
Currant Family 266-269
Currants 266-267
Gooseberries 268-269
Cuscuta 217-219
californica **217**, 218
campestris 67, 217, **218**
ceanothi 218
pacifica var. *pacifica* 124, 210, 217, **219**
pentagona 218
salina 216, 217, **219**
salina var. *major* 219
subinclusa 217, **218**
Cuscutaceae 217
Cycads 13
Cyclocephala melanocephala **396**
Cylindropuntia 186-187
californica var. *parkeri* 19, **187**
parryi 187
prolifera **187**
Cynara 79, 86
cardunculus **86**
Cynipidae 385
Cypress, Tecate 3, 22, 23, 410, 450
Cypresses 13

Dactylopiidae 194
Dactylopius 194
coccus 194
confusus 194
opuntiae 194
Daisy
Allen's 36, 126, **127**
Annual 125
False 115
Fleabane 107
Garland 115
Golden 14, **127**
Leafy 23, **107**
Danainae 61
Danaus
gilippus thersippus **61**
plexippus 58, **61**, 433
Dandelion
Cleveland's 134
Common 131
Mountain 3, 24, 104, 131, 136, 454
Mountain, Large-flowered 131, **132**
Mountain, Spear-leaved 131, **133**
Mountain, Woodland 131, **132**
Dasymutilla **32**
Datiscaceae 226

Datisca Family 226
Datisca glomerata 17, **226**
Datura 395
   meteloides 395
   stramonium **395**
   wrightii **395**
Daucus pusillus 25, **48**
Death Camas 434
   Fremont's 23, 25, **434**
   Meadow 434
Deer 374, 375
   California Mule 413
Deer Broom 19, 25, 242, 245, 356
   Coastal 18, **245**, 450
   Desert 245
Deer Brush, Southern 368, **372**
Deer's Ears, Parry's 3, 18, 26, **261**, 453-455
Deerweed 14, 245
   Coastal 245
   Desert 245
Deinandra 23, 109-111
   fasciculata 109, **110**
   kelloggii 109, **111**
   paniculata 109, **111**
Delphinium 363-365
   cardinale 25, 363, **364**
   parryi ssp. *parryi* 19, 363, **364**
   patens ssp. *hepaticoideum* 23, 363, **365**
Dendromecon rigida 25, 218, **309**
Dentaria californica 172
Dermacenter variabilis 31
Descurainia 171
   pinnata **175**
Desert-thorn 396
   Anderson's 396
   Santa Catalina Island 396
Desmocerus 40
   californicus californicus **40**
   californicus dimorphus 40
Dhatura 395
Diadasia **196**
Dicentra 310
   chrysantha 310
   ochroleuca 310
Dichelostemma
   capitatum ssp. *capitatum* 19, 25, **444**; See also front cover
   pulchellum var. *pulchellum* 444
Dichondra 216
   Asian 216
   Western 216
Dichondra 216
   occidentalis **216**
   micrantha 216
Dicots 13
Dill 46
Diplacus 19, 25, 314
   brevipes **318**
   clevelandii 36, **316**
   fremontii **317**, 318
   longiflorus **314**
   puniceus 26, **315**
   rattanii 317, **318**
   x *australis* **315**
Diplolepis polita **385**

Diplotaxis 171
Dipsacaceae 274
Disease
   Lyme 31
   Rocky Mountain Spotted Fever 31
Distichlis spicata 17
Dobie-pod, Slender 177
Dodder 3, 217-219
   California 18, **217**, 218
   Canyon 16, 217, **218**, 453
   Goldenthread 124, 210, 217, **219**
   Salt 216, 217, **219**
   Western Field 67, 217, **218**
Dodecahema 347
   leptoceras 347, **351**, 364
Dodecatheon clevelandii ssp. *clevelandii* 19, **361**
Dogbane 10, 57-58, 60-61
   Hemp 58
Dogbane Family 10, **57-61**, 411
   Vine-like Perennials 57
   Bushy Perennials 58-59
Douglas-fir, Bigcone 3, 15, 21-22, 391, 437, 451
Doveweed 234
Downingia 197, 199
   bella 17, **199**
   cuspidata 17, **199**
Dr. Seuss Plant 128
Draba, Wedge-leaf 175
Draba 171, 175
   cuneifolia **175**
Dragonfly, Cardinal Meadowhawk 448
Drymaria 360
Drymocallis 383-384
   glandulosa ssp. *glandulosa* 21, **384**
   glandulosa ssp. *reflexa* **384**
Duckweeds 17, 34
Dudleya, Dudleyas 221-223
Dudleya 14, 220-223
   blochmaniae ssp. *blochmaniae* **221**
   cymosa 23, **222**
   edulis **222**
   lanceolata 23, **223**
   multicaulis **221**
   pulverulenta **223**
   stolonifera **223**
   viscida 14, 23, **222**
Durango Root 17, **226**

Ear, Cat's 134
   Smooth 134
   Hairy 134
Eardrops 25, 310
   Golden 25, **310**, 364
   White 25, **310**, 450
Echinodorus berteroi **419**
Eclipta 115
   alba 115
   prostrata **115**
Eel-grasses 34
Eggplant 395
Ehrendorferia 310
   chrysantha 25, **310**, 364
   ochroleuca 25, **310**

Elderberry 39-40, 202
   Blue 2, 11, 16-19, 21-22, **39-40**, 218, 307
Eleocharis 17
Eleodes 32
   osculans **32**
Emmenanthe 144, 154
   penduliflora var. *penduliflora* 25, **154**
Encelia 64, 116, 119
   californica 19, **116**, 118
   farinosa 19, **116**, 155, 275
Epicauta straba **92**
Epilobium 292, 296, 298-299
   adenocaulon 298
   brachycarpum **298**
   campestre **298**
   canum 292
   canum ssp. *canum* **299**
   canum ssp. *latifolium* **299**
   ciliatum ssp. *ciliatum* 17, **298**
   densiflorum 298
   pygmaeum 298
Epipactis gigantea 17, **435**
Equisetophytes 13
Eremocarpus setigerus 234
Eriastrum 338-339
   densifolium ssp. *sanctorum* **338**
   sapphirinum **339**
Ericaceae 227-232, 260, 305-306, 373, 381, 405
Ericameria 89, 120-121
   cuneata var. *cuneata* **89**, 120
   nauseosa var. *oreophila* **89**, 121
   palmeri var. *pachylepis* 89-90, **120**, 121
   parishii var. *parishii* **90**
   pinifolia 89-90, 120, **121**
Erigeron 105, 107
   bonariensis 105
   canadensis 105
   foliosus 23, **107**
   sumatrensis 105
Eriodictyon 144, 155-156
   crassifolium var. *crassifolium* **155**, 306
   parryi 84, **156**
   trichocalyx ssp. *trichocalyx* **155**
Eriogonum 352-359
   cinereum 352, **353**
   davidsonii 352, **358**
   elongatum var. *elongatum* 352, **353**
   fasciculatum 17, 306, **353**, 354
   fasciculatum ssp. *fasciculatum* 352, **354**
   fasciculatum ssp. *foliolosum* 352, **354**
   fasciculatum ssp. *polifolium* 352, **354**
   fasciculatum var. *fasciculatum* 354
   fasciculatum var. *foliolosum* 354
   fasciculatum var. *polifolium* 354
   gracile var. *gracile* 352, **359**
   gracile var. *incultum* 352, **359**
   nudum 352
   nudum var. *pauciflorum* **357**
   parvifolium 19, 352, **357**

saxatile 23, 306, 352, **357**
thurberi 352, **358**
wrightii ssp. *subscaposum* 352, **358**
Eriophyllum 64
  confertiflorum var. *confertiflorum* 23, **103**, 302
Erodium 25, 262, 263-264
  botrys **264**
  brachycarpum 264
  cicutarium **264**
  moschatum **264**
Eryngium 49
  pendletonense 17, **49**
  aristulatum var. *parishii* **49**
Erysimum 171, 176, 293
  capitatum var. *capitatum* **176**
Erythranthe 25, 314, 316-319
  cardinalis 17, **316**
  diffusa **317**
  floribunda **319**
  guttata 17, **318**, 319
Eschscholzia 310-311
  caespitosa **310**
  californica 19, 25, **311**
Eucrypta, Common 156
Eucrypta 144, 156
  chrysanthemifolia var. *chrysanthemifolia* **156**
Eudicots 13, 34, **39-412**
  Basal 13
  Superrosids 13
  Superasterids 13
Eulobus 292, 294
  californicus 25, **294**
Eupatorium, Sticky 65, 118
Eupeodes volucris **435**
Euphilotes bernardino bernardino **356**
Euphorbia 233, 235
  albomarginata **235**
  misera 19, **235**
  polycarpa var. *polycarpa* **235**
Euphorbiaceae 233-235, 290
Euphydryas
  chalcedona **325**
  editha quino **331**
Eurybia 105
  radulina **105**
Eustoma exaltatum ssp. *exaltatum* **261**
Euthamia 130
  occidentalis **130**
Evening-primrose 3, 10, 25, 147, 292-294, 300; *See also* Sun Cup
  Beach 18, 19, 147, **293**
  California 300
  Hooker's 300
  Large 300
  Petioled 293
  Slender-flowered 294
  Small-flowered 294
Evening-primrose Family 283, **292-300**
Everes amyntula **239**
Everlasting 96-100, 231, 452; *See also* Cudweed
  California 65, **98**
  Fragrant 99, 100

  Pink 100
  Rabbit-tobacco, White 100
  Sonora 99, **100**, 452
  White 99
Exochomus fasciatus **195**

Fabaceae 10, 25, 218, **236-259**, 305, 346, 356, 418
Fagaceae 45, 89, 174, 203, 326, 335, 409, 411, 433
Fairy Lantern: *See* Lily
Fairyfans 295
False-hellebore Family 434
False Indigo 236-237, 243
  California 236, **237**
  Western 237
False-gilia, Blue 25, **338**
False-mustard, California 294
Farewell-to-spring 25, 295
Fat, Mule 16-17, **75**, 76, 217-18
Fennel, Sweet 48, **50**
Fern 13, 21, 23, 453
  Adder's-tongue 13
  Bracken 13
  Coffee 13, 23
  Cotton, California 23
  Chain 13
  Chain, Giant 451
  Goldenback 13
  King 13
  Leptosporangiate 13
  Lip 13
  Maidenhair 13, 451
  Marratioid 13
  Ophioglossoid 13
  Polypody 13
  Potato 13
  Whisk 13
Fiddleneck
  Common 19, **145**
  Gray 145
  Rigid 145
Fiesta Flower 159
  Blue 159
  San Diego 159
Fig, Hottentot 41, 91
Figwort 394
  California 394
  True 394
Figwort Family 301, 314, 320, **394**
Filaree 25, 262-264, 265
  Large-leaved 262
  Long-beaked 263, **264**
  Red-stemmed 263, **264**
  Short-fruited 263, 264
  White-stemmed 263, **264**
Fire Hearts 310
Fire-following Plants and Wildflowers 25-26
Firs 13
Fleshy Jaumea 17, **124**
Flax, Small-flowered Dwarf 283, 300
Flax Family 283, 300
Flies, Fly 59, 75, 77, 90, 96, 107, 115, 123, 162, 205, 221-222, 227, 229, 231, 271, 274, 291, 301, 309, 320, 327, 330, 341, 347, 355, 374, 381, 384-386, 402
  Bee 273, 338, **355**
  Biting 436
  Cactus, Mexican 195
  Deer 32, 436
  Flower 392, 435
  Flower, Swift 435
  Giant Flower-loving, Acton 338
  Horse 32, 436
  Midge 402
  Midge, Honeysuckle Bud Gall 202
  Midge, Lily Stem Gall 440
  Mosquitoes 32
  Horse 436
  Hover 392, 435
  Hover, Pied 435
  Snipe 32, 436
Flower
  California State 311
  Cluster 25, 160
  Fiesta 159
  Fiesta, Blue 159
  Fiesta, San Diego 159
  Popcorn 149-151
  Fried-Egg 313
Foeniculum vulgare 48, **50**
Four O'Clock Family 289-290
Fragaria 383
Frangula 375
  californica ssp. *californica* 21, 84, 232, **375**
  californica ssp. *cuspidata* 375
  californica ssp. *tomentella* **375**
Frankenia 260
  grandifolia 260
  salina 17, 42, 219, **260**
Frankenia Family 260
Frankeniaceae 42, 219, **260**
Frasera parryi 29, **261**
Fraxinus 40, 291
  dipetala 21, 40, **291**
  velutina 40, 291
Freesia 425
Fried-Egg Flowers 313
Fringepod 171, 182-183, 453
  Dwarf 182
  Hairy 26, **182**
  Southern 183
Fringillidae 84
Fritillaria biflora var. *biflora* 426, **431**
Frog-bits 34
Frog-fruit, Garden 407
Fuchsia, California 3, 29, **298-299**
  Broad-leaved 299
  Narrow-leaved 299
Funastrum cynanchoides ssp. *hartwegii* **57**
Fungus
  Mildew 334
  Mycorrhizal 177
  Sycamore Anthracnose 334

GENERAL INDEX • 483

Galium 387-389
  angustifolium var. *angustifolium* 387, **389**
  aparine 387, **388**
  nuttallii ssp. *nuttallii* 387, **389**
  parisiense 387, **388**
  porrigens var. *porrigens* 387, **389**
  trifidum var. *pacificum* 387, **388**
  verum 387
Garlic 420
Garryaceae 260
Garrya 260
  flavescens **260**
  veatchii 260
Gayophytum diffusum ssp. *parviflorum* 283, **300**
Gentianaceae 261-262
Gentian Family 261-262
Geococcyx californianus **196**
Geothlypis trichas 446
Geraniaceae 262-265
Geranium 262, 265
  Carolina 265
  Cutleaf 265
  Roundleaf 265
Geranium Family 262-265
Geranium 262, 265
  carolinianum **265**
  dissectum **265**
  rotundifolium **265**
Gilia 339-340
  Ball 339
  Blue False- 25, **338**
  Grassland 339
  Los Angeles 26, **339**
  Slender Desert 340
  Southern 340
Gilia 339-340
  angelensis **339**
  australis 340
  capitata ssp. *abrotanifolia* **339**
  ochroleuca ssp. *exilis* **340**
Ginkgo 13
Githopsis 197
  diffusa ssp. *candida* **197**
  diffusa ssp. *diffusa* **197**
Gladiolus 425
Glebionis coronaria **115**
Gnaphalium 96; See also *Pseudognaphalium*
  beneolens 99
  bicolor 98
  californicum 98
  canescens ssp. *beneolens* 99
  canescens ssp. *microcephalum* 99
  chilense 97
  leucocephalum 100
  luteo-album 97
  microcephalum 99
  palustre **96**, 97
  ramosissimum 100
  sanguineum 72
Gnatcatcher 70
  Blue-gray 70
  Coastal California 70
Goldenbush 3, 89, 120
  Coastal 90, **91**, 92-93

  Discoid-headed 89-91
  Grassland 89-90, **120**, 121
  Great Basin Rabbitbrush 89
  Parish's 90, 455
  Pine 89-90, 120, **121**
  Prostrate 19, **91**
  Radiate-headed 120-121
  Saw-toothed 90, 91, 121
  Wedge-leaved 89, 120
Goldenrod 3, 11, 130
  Broomrape 130, **307**
  California 130, 307
  Western 130
Goldenstar, Common 440
Goldfields 125-126
  Coastal 125, **126**
  Coulter's 125, 126
  Royal 125
  Southern 125, 126
Goliathus goliatus 93
Golondrina 235
Gooseberries, Gooseberry 10, 29, 266, 268-269
  Bitter 269
  Fuchsia-flowered 268
  Hillside, Southern California 21, **268**, 269, 455
  Sierra 269
Gooseberry Family 266-269
  Currants 266-267
  Gooseberries 268-269
Goosefoot Family 210-212
Goosefoots 34, 119, 210-212
Goosegrass 388
Gourd 10, 224-225
  Stinking 224
Gourd Family 224
Granny's Hairnet 23, 48, **360**
Grape 13, 412
  Desert Wild 3, 17, **412**
Grape Family 412
Grapefruits 390
Grappling Hook, Palmer's 152
Grass 13, 16, 19-21, 25, 34, 330
  Arrow- 34
  Arrow-, Seaside 17
  Barley, Vernal Spring 16
  Blue-eyed-, Western 19, **425**
  Eel- 34
  Muhly, Littleseed 23
  Needle, Foothill 19
  Needle, Giant 26
  Needle, Purple 19, 20
  Reed, San Diego 23
  Salt 16, 17, 18
  Three-awned 19
Grasshopper 105, 120, 355, 446, 448
  Cattail, Western Toothpick 448
  Rattlesnake 33
Greasewood 378
Greya politella **393**
Grindelia
  camporum 90, **121**
  hirsutula 121
  robusta 121

Grossulariaceae 10, 43, **266-269**
Ground-cherry 398
  Greene's 398
  Thick-leaved 368, **398**
Groundsmoke, Small-flowered 283, **300**
Guild
  Beargrass 439
  Bladderpod 213
  Cactus 194
  California Buckwheat 355
  California Lilac 374
  Carrot 50
  Cattail 448
  Common Thistle 84
  Currant 269
  Deer Broom 245
  Dogbane 60
  Elderberry 40
  False Indigo 237
  Fiddleneck 145
  Goldenbush 92
  Honeysuckle 202
  Jimsonweed 396
  Manzanita 231, 405
  Milkweed 60
  Mistletoe 411
  Nettle 405
  Oceanspray 382
  Rose 385
  Snowberry 204
  Squash 225
  Stream Orchid 435
  Sycamore 334
  Violet 409
  Woodland Star 393
  Woolly-star 338
  Yerba Santa 155
  Yucca 416
Guillenia lasiophylla 175
Gumplant 90, **121**
Gutierrezia californica **122**
Gymnosperms 13

Hairnet, Granny's 23, 48, **360**
Halictidae 174, 392, 441
Hareleaf, Common 114, 454
Harpagonella 144, 152
  palmeri **152**
Hasseanthus 221
Hazardia 89-91
  squarrosa 90-91, 121
  squarrosa var. *grindelioides* **90**
Hearts, Fire 310
Heath Family 227-232
Heath, Alkali 16, 17, 42, 217, 219, **260**
Hedera 72
  canariensis 72
  helix 62
Hedge-nettle 16, 279-281
  California 279, **280**
  Hillside 279, **281**
  Rigid 279, **281**
  Stebbins's 279, 281
  White-stem 279, **280**
Hedge-parsley

California 46
Field 51
Knotted 51
Hedypnois cretica **133**
Helenium puberulum **122**
Helianthemum scoparium 25, **212**
Helianthus 64, 116-117
   annuus **117**
   gracilentus **117**, 418
Heliotropaceae 154; See also Boraginaceae
Heliotrope
  Alkali 154
  Salt 154
  Wild 160
Heliotropium 144, **154**
   curassavicum ssp. *oculatum* **154**
Hemaris diffinis **204**
Hemileuca electra **356**
Hemizonia: See also *Deinandra, Centromadia*
   fasciculata 110
   kelloggii 111
   paniculata 111
   parryi ssp. *australis* 109
   pungens ssp. *laevis* 110
Hemlock
  Common Poison 46-47, **48**, 50
  Douglas's Water **48**
Hemp 58
  Dogbane 58
  Indian 58
Henbit, Common **272**
Heronsbill 263
Hesperocnide tenella **403**, 404
Hesperocyparis forbesii 23, **410**
Hesperolinon micranthum **283**, 300
Hesperoyucca 415
   whipplei 14, 21, **415**, 439
Heterocodon, Rareflower **198**
Heterocodon 197-198
   rariflorum **198**
Heteromeles arbutifolia 17, 333, **381**
Heterotheca **123**
   grandiflora 25, **123**
   sessiliflora ssp. *echioides* 25, **123**
Hierba del Pasmo **76**
Hirschfeldia 171, **177-178**
   incana 25, **178**
Hoita 240
   macrostachya **240**
   orbicularis **240**
Holly
  English 376, 381, 385
  Summer 2, **232**
Holocarpha 109, 112
   virgata var. *elongata* **112**
Holodiscus 382
   discolor var. *discolor* **382**
   discolor var. *microphyllus* **382**
   microphyllus 382
Honeysuckle 10, 202, 204
  Southern 21, **202**
Honeysuckle Family 39, **202-204**
Hookeria pulchella **444**
Hordeum intercedens 17

Horehound, Common **272**
Horkelia 21, **383**
  Mesa **383**, 453
  Ramona **383**
  Wedge-leaved **383**
Horkelia 21, **383**
   cuneata ssp. *cuneata* 383
   cuneata ssp. *puberula* 383
   truncata 383
Hornungia 171
Hornworts 13
  Aquatic 13
Horse-nettle, Silverleaf 399
Horsetails 13
Hosackia 242-243
   crassifolia var. *crassifolia* 242, **243**
Houses, Chinese 322-323
  Purple 23, 25, **322**, 323
  Southern **322**, **323**
Hulsea heterochroma **124**
Hummingbirds: See Birds
Hyacinth, Wild: See School Bells
Hyacinth Family 440, 444
Hyacinthaceae 440, 444
Hyalophora euryalus **374**
Hydrocotyle 62
   ranunculoides **62**
   umbellata **62**
   verticillata **62**
Hydrophyllaceae: 144; See also Boraginaceae
Hyperaspis trifurcata **195**
Hypericaceae **270**
Hypericum 270
   anagalloides **270**
   canariense **270**
   formosum ssp. *scoulteri* 270
   scouleri **270**
Hypochaeris 134
   glabra **134**
   radicata **134**
Hypochoeris: See *Hypochaeris*

Icaricia acmon acmon **356**
Iceplant 41
  Croceum 41
  Crystal 41
  Hottentot Fig 41
  Small-flowered 41
Icteridae 446, 448
Ilex aquifolium 376, 385
Incienso 116
Indigo: See False Indigo
Iridaceae **425**
Iris 425
Iris Family **425**
Isocoma 19, 89-91
   menziesii 90, **91**
   menziesii var. *sedoides* 19, **91**
   menziesii var. *vernonioides* **91**
Isomeris arborea 213
Ivy 62
  Canary Islands 62
  English 62
Ixodes pacificus **31**

Japanese Lacquer Tree 45

Jasmine 291
Jaumea, Fleshy 17, **124**, 217, 219
Jaumea carnosa 17, **124**, 219
Jepsonia, Coast **392**, 453-454
Jepsonia parryi **392**
Jewelflower 171, 174
  San Diego 26, **174**, 450
Jimsonweed **395**, 396
  Western **395**
Judolia sexspilota **355**
Juglandaceae 40
Juglans 21, 40
   californica 21
Jump-ups, Johnny 19, **408**
Juncus 17
   acutus ssp. *leopoldii* 17
Juniper 13
  California 72, 410, 453, 455
Juniperus californica 410
Junonia coenia **154**

Keckiella 324-326
   antirrhinoides ssp. *antirrhinoides* 21, **325**
   cordifolia **325**
   ternata ssp. *ternata* **326**
Knapweed 79-80
  Russian 80
Knotweed 350
Kotolo **59**

Lacepods 171, 182-183
  Hairy 26, **182**
  Southern **183**
Ladies's-fingers: See Live-forevers
Lady's Mantle, Western **380**
Laennicia 105
   coulteri 105
Lagophylla 109, 114
   ramosissima ssp. *ramosissima* **114**
Lamiaceae 25, 92, 261, **271-282**
Lamium amplexicaule **272**
Lantern, White Fairy **426**, **427**
Larkspur 3, 10-11, 363-365
  Liver-leaved 23-24, 363, **365**
  Parry's 19, 363, **364**, 455
  Scarlet 11, 25, 363, **364**, 451, 455
Lasioglossum **174**, **200**, **441**
Lasioptera 440
Lastarriaea 347, 350-351
   coriacea 347, **350**, 351
Lasthenia 125-126
   californica 126
   chrysostoma 126
   coronaria **125**, 126
   glabrata ssp. *coulteri* **125**, 126
   gracilis 125, **126**
Lathyrus 241, 258
   vestitus var. *alefeldii* **241**
   vestitus var. *vestitus* **241**, 416
Latrodectus hesperus **31**
Lauraceae **37**, 227, 437
Laurel 10, 13, 37: See also Sumac
  Bay 37
  Bay, California 3, 13, 17-18, 21, **37**, 227, 411, 437
Laurel Family **37**

Lavender 271
Layia 125-127
    platyglossa 63, **126**, 127
Leadwort Family 336-337, 348
Leek 420
Legenere 197
Legume Family 236-259
Lema 396
    daturaphila **396**
    trilineata 396
Lemna 17
Lemonade Berry 18-19, 29, **44**
Lemons 390
Lennoaceae: 144; See also Boraginaceae
Lennoas 144, 170
Lepechinia cardiophylla 21, 36, **271**
Lepidium 171, 183-185
    acutidens 183, **184**
    dictyotum var. acutidens 184
    lasiocarpum 183, **184**
    latifolium 183, 185
    nitidum 183, **184**
    pinnatifidum 183, **185**
    virginicum ssp. menziesii 185
    virginicum var. pubescens 185
    virginicum var. robinsonii 183, **185**
Lepidospartum squamatum 19, **77**, 411
Leptodactylon californicum 343
Leptosiphon 340-342
    ciliatus **340**
    floribundus ssp. floribundus **341**
    floribundus ssp. glaber 341
    liniflorus **341**, 342
    nuttallii ssp. floribundus 341
    parviflorus **342**
    pygmaeus ssp. continentalis **342**
Leptosyne 128-129
    californica **128**, 129
    gigantea **128**
Leptysma marginicollis hebardi **448**
Lessingia, Valley 101
Lessingia 64, 101, 107
    filaginifolia 107
    glandulifera **101**
Lettuce: See also Miner's Lettuce
    Wire 139
Libellulidae 448
Licorice-root 363
Ligusticum 363
Lilac, California 21, 25, 84, 217-218, 311, 367-374, 418, 455-457
    Bigpod 2, 21, 118, 368, **369**, 450
    Brush, Jim 21, 368, **371**, 373
    Brush, Southern Deer 368, **372**, 373
    Flat-leaved 369
    Greenbark 368, **370**, 373-374
    Hairy 20-21, 23, 368, **371**, 372
    Hairy, San Diego 368, 371
    Horned 368-369
    Mountain 368, 372-373
    Oval-leaf 368, 371-372
    Palmer 368, 372, **373**
    Thick-leaved 20, 367-368, **369**, 374, 377
    Thorny 368, 370
    True 291
    Wartleaf 22-23, 368, **373**, 451
    Whitethorn, Chaparral 21, 368, **370**, 451
    Woollyleaf 21, 29, 368, 371, **372**
Liliaceae 25, 84, 426, 440-444
Lilies, Lily 3, 10, 13, 330, 426-433
    Chocolate, California 426, **431**
    Fairy Lantern, White 426, **427**
    Humboldt, Ocellated 17, 84, 426, **432**, 433, 451
    Mariposa 20, 25, 426-430, 452, 454
    Mariposa, Catalina 19, 426-427, 452, 457
    Mariposa, Davidson's 428
    Mariposa, Intermediate 426, 429, **430**
    Mariposa, Peninsular 430
    Mariposa, Plummer's 426, **429**, 430
    Mariposa, Shy 426, **428**
    Mariposa, Splendid 426, **428**, 452
    Mariposa, Weed's 18, 26, 426, 429, **430**, 451-452, 454
Lilium humboldtii ssp. ocellatum 17, 84, 426, **432**, 433
Lily: See Lilies
Lily Family 426-433, 440-444
Lily-of-the-Valley, Pampas 398
Limonium 17, 336-337
    binervosum 337
    californicum 17, **336**
    duriusculum 337
    perezii **336**
    preauxii 336
    ramosissimum **337**
    sinuatum 337
    sventenii 336
Linaceae **283**, 300
Linanthus 341-342
    Flax-flowered 341, 342
    Many-flowered 341
    Many-flowered, Smooth 341
    Pygmy, Mainland 342
Linanthus 341-342
    californicus **343**
    ciliatus 340
    dianthiflorus iii, 19, 36, **343**
    floribundus 341, 342
    liniflorus 341
    nuttallii ssp. floribundus 341
    parviflorus 342
    pygmaeus 342
Linaria 328
    canadensis var. texana 328
    texana 328
Lindernia dubia **324**
Linderniaceae 324
Lion, Mountain 30, **33**
Lippia nodiflora 407
Lithophragma 23, 393
    affine 23, **393**
    heterophyllum 393
Live-forever 3, 10-11, 221-223
    Blochman's 221
    Canyon 23, **222**, 457
    Chalk 223
    Ladies's-fingers 222, 223, 452
    Laguna Beach 2, **223**
    Lance-leaved 23, **223**
    Many-stemmed 221
    Sticky 14, 23, **222**
Liverworts 13
Lizard's-Tail Family 13, **38**
Lizards 29, 355
    Fence 29
Loasaceae 25, **283**
Lobelia, Rothrock's 23, **200**, 449, 451
Lobelia 197, 200
    dunnii var. serrata 23, **200**
Lobeliaceae 197
Locoweed 11, 25, 238-239; See also Rattleweed, Milkvetch
    Gambel's 238
    Ocean 239
    Southern California 239
    White Dwarf 238
Lomatium 50, 52-54, 56
    dasycarpum ssp. dasycarpum **53**
    lucidum **53**, 56
    utriculatum **53**
Lonicera subspicata var. denudata 21, **202**
Looking-glass, Small Venus's 198, 452
Loosestrife 284
    California 17, 23, **284**
    Hyssop 284
    Three-bract 284
Loosestrife Family 284
Lopseed Family 314-319
Lotus 242-247
    Alkali 242, **246**
    American 242, **244**
    Bird's-foot Trefoil 242, **243**
    Buck 242, **243**
    California 242, **247**
    Coastal 246
    Deer Broom 242
    Deer Broom, Coastal 245
    Deer Broom, Desert 245
    Grab 246
    Hill 242, **244**
    Miniature 242, **247**
    San Diego 242, 244, **246**
    Silverleaf 242, **244**, 246
    Strigose 242, **247**
    Woolly 242
    Woolly, Northern 246
    Woolly, Southern 246
    Wrangel's 247
Lotus 242-247
    argophyllus 244, 246
    corniculatus 242, **243**
    crassifolius var. crassifolius 243
    hamatus 242
    heermannii var. heermannii 246
    heermannii var. orbicularis 246
    humistratus 244

micranthus 244, 247
purshianus 244
salsuginosus ssp. *salsuginosus* 246
scoparius var. *brevialatus* 245
scoparius var. *scoparius* 245
strigosus 247
subpinnatus 247
unifoliolatus 244
wrangelianus 247
Lotus Clade 242-247
Ludwigia 292, 299
hexapetala 299
peploides ssp. *peploides* **299**
repens 299
Lupine 3, 19, 25, 236, 248-253, 450, 452, 454, 456
 Agardh's 249
 Arroyo 249
 Bajada 249
 Chick 248
 Chick, Dense-flowered 248
 Collar 26, **250**
 Coulter's 250
 Dove 248
 Elegant 249
 Grape Soda 251, **252**
 Guard 252
 Miniature 248
 Parish's Stream 253
 Pauma 253
 Silver 251
 Southern Mountain 252
 Stinging 26, **250**
Lupinus 19, 25, 248-253
 agardhianus **249**
 albifrons var. *albifrons* **251**
 bicolor **248**
 concinnus **249**
 excubitus 251, 252
 excubitus var. *austromontanus* **252**
 excubitus var. *hallii* **252**
 hirsutissimus **250**
 latifolius 251
 latifolius var. *parishii* **253**
 longifolius 251, **253**
 microcarpus var. *microcarpus* **248**
 microcarpus var. *densiflorus* **248**
 sparsiflorus **250**
 succulentus **249**
 truncatus **250**
Lycaena arota **269**
Lycaenidae 239, 245, 269, 356, 358-359, 411
Lycium 19, 396
 andersoni 396
 brevipes var. *hassei* **396**
 californicum 19, **396**
Lycophytes 13
Lygaeidae 60, 279
Lygaeus kalmii **60**
Lynx rufus 33
Lythraceae 284
Lythrum 17, 23, 284
 californicum 17, 23, **284**
 hyssopifolia **284**
 tribracteatum 284

Madder Family 387-389
Madia 109, 112-113
 Elegant 112
 Slender 113
 Thread-stem 113
Madia 109, 112-113
 elegans ssp. *elegans* **112**
 exigua **113**
 gracilis **113**
Madrone, Pacific 3, 21, **227**, 381, 451
Madroño 227
Magnoliids 13, 34, **37-38**
Maianthemum racemosum **438**
Malacothamnus 25, 285
 densiflorus **285**
 fascicularis **285**
Malacothrix 64, 134-135
 clevelandii **134**
 saxatilis var. *saxatilis* **135**
 saxatilis var. *tenuifolia* **135**
Malephora crocea **41**
Mallow 262, 285-286
 Alkali 286
 Bush, Chaparral 25, **285**
 Bush, Many-flowered 285
Mallow Family 231, **285-286**
Malosma laurina 14, 19, **43**, 44, 218, 374
Malus 381
Malvaceae 25, 231, 262, **285-286**
Malvella leprosa **286**
Man-root 225
Mango 42
Mantle, Western Lady's 380
Manzanita 3, 11, 15, 21, 29, 93, 228-231, 260, 305-306, 411
 Bigberry 229, 231, 405, 455-456
 Eastwood 20-23, **228**, 230, 231, 373, 401, 454-456
 Mission 232
 Pink-bracted 230, 231
 Rainbow 230, 231
Maple 11, 391
 Big-leaf 3, 17-18, 21, 333, **391**, 451
Marah macrocarpus 19, 48, **225**
Marbles, Woolly 16, 17
Marijuana 226
Mariposa Lilies: *See* Lilies
Marrubium vulgare **272**
Marsh-fleabane 95
Marsh-rosemary, Western 336
Marsilea vestita **17**
Mary 323
 Blue-eyed 323
 Blue-eyed, Parry's 323
Matchweed, California 122
Matilija: *See* Poppy
Matricaria 25, 94
 discoidea 25, **94**
 globulifera 94
 matricarioides 94
 occidentalis **94**
Matthiola 171, 177, 179
 incana **179**
Mayweed, Dog 104
Meadow-rue, Common 365

Meconella denticulata 25, **312**
Medicago sativa **254**
Megachile angelarum **93**
Megachilidae 84, 93, 156, 338
Melanoplus 105, 120
Melanthiaceae 25, **434**
Melilotus 25, 218, 254
 albus **254**
 indicus **254**
Melissodes **93**
Meloidae 92, 355
Melon 224
 Coyote 224, 225
Melyridae 124, 311
Mentha 272
 aquatica 272
 spicata var. *spicata* **272**
 x *piperita* **272**
Mentzelia 25, 283
 micrantha 25, **283**
 veatchiana **283**
Mesembryanthemum 41
 crystallinum **41**
 nodiflorum **41**
Mesepiola specca **439**
Micranthes californica 23, **393**
Microrhopala rubrolineata **92**
Microseris 64, 131, 136-138
 douglasii ssp. *douglasii* 136, 137
 douglasii ssp. *platycarpha* 136, **137**
 elegans 136, **137**
 heterocarpa 136, **138**
 lindleyi 136, **138**
Microsteris gracilis 215, **343**
Midge: *See* Fly
Mignonette 34
Milkmaids, California 172
Milkvetch 238-239; *See also* Locoweed, Rattleweed
 Braunton's 23, **238**, 450
Milkweed 10, 20, **57-61**
 California 59, 61, 454
 Climbing 57
 Narrow-leaved 58, 61
 Woolly-fruited 58, **59**, 61, 411
Milkwort, Fish's 3, **346**, 454
Milkwort Family 346
Mimetanthe 25, 314, 319
 pilosa **319**
Mimulus 314-319; *See also Diplacus, Erythranthe, Mimetanthe*
 aurantiacus ssp. *australis* 315
 aurantiacus var. *pubescens* 314
 aurantiacus var. *puniceus* 315
 brevipes 318
 cardinalis 316
 clevelandii 316
 diffusus 317
 floribundus 319
 fremontii 317
 guttatus 318
 pilosus 319
 puniceus 315
 rattanii 318
Miner's Lettuce 23, 25, 288
 Common 23, 25, **288**
 Narrow-leaved 288

GENERAL INDEX • 487

Mint 271-273
  Mountain, California 271, 278
  Mustang 25, **273**, 456
  Pepper 271, **272**
  Spear 272
  Water 272
Mint Family 13, 14, **271-282**
Minuartia douglasii 23, **205**
Mirabilis 233, 290
  laevis var. *crassifolius* 233, **290**
  california 290
Mission
  Bells 431
  Manzanita 232
  Prickly-pear 188, **190**
Mistletoe 13, 410-411
  Dense 410, 411
  Hairy 411
  Large-leaved 410, 411
  Oak 411
Mistletoe Family 334, **410-411**
Mites 334
Monardella
  Hall's Large-flowered 3, **273**
  Intermediate Thick-leaved 3, **273**, 453, 456
  Mustang Mint 273
Monardella 25, 273
  breweri ssp. *lanceolata* **273**
  hypoleuca ssp. *intermedia* **273**
  lanceolata 25, 273
  macrantha ssp. *hallii* **273**
Monkeyflower 3, 16, 25-26, 314-319, 452, 456
  Bush 18-19, 314-316
  Bush, Cleveland's 36, **316**, 455-456
  Bush, Coastal 18, 26, **315**
  Bush, Hairy 314, 315
  Bush, Southern 315
  Downy 319
  Fremont's 317, 318
  Palomar 317, 453
  Rattan's 317, **318**
  Scarlet 17, 23, **316**
  Seep 17, 23, **318**, 319
  Slimy 319
  Wide-throated 318, 452, 457
Monocots 13, 34, 330, **413-448**
Montia Family 287-288
Montiaceae 25, **287-288**
Mordellidae 355
Morning-glory 214-215
  Beach 19, 147, **215**
  Field Bindweed 215
  Large-bracted 214
  Small-flowered 215, 343, 452
  Sonora 23, **214**
  True 214-215
Morning-glory Family 214-219
Mosquitoes: *See* Fly
Mosses, True 13
Moth, Moths 75, 77, 123, 274, 289, 338, 342, 355, 436, 446
  Beargrass 439
  Borer, Buckwheat Root 356
  Borer, Sycamore 335

Buck, Electric 356
Ctenucha, Brown 381
Dogbane, Western 61
Longhorn, Oceanspray 382
  Owlet 300
Silk, Ceanothus 374
  Sphinx 204, 292, 300, 314, 326, 395
Sphinx, Snowberry Clearwing 204
Woodland Star 393
  Yucca 416, 418
  Yucca, Bogus 416-417
Yucca, Bogus, Ash-colored 417
Yucca, Bogus, Bordered 417
Yucca, Southern California 416
  Yucca, True 416
Mountain, Mountains
  Dandelion 104, 131-133, 136
  Dandelion, Large-flowered 131, **132**
  Dandelion, Spear-leaved 131, **133**
  Dandelion, Woodland 131, **132**
  Lion 30, **33**
  Mint, California 271, 278
  Lupine, Southern 252
  Onion, San Bernardino 420, **423**
  Spiraea 382
  Violet, Yellow 21, 408
Mountain-mahogany 21, 380
  San Diego 380
  California 21, **380**
Mouse-tail, Little 17, **365**
Mud Nama 157
Mugwort, California 16-17, **71**
Muhlenbergia microsperma 23
Muilla, Common 445
Muilla maritima 445
Mule Fat 16-17, **75**, 76, 217-218
Mule's Ears, Southern 120, 454-455
Mullein 234, 394
  Turkey 234
  Wand 394
  Woolly 394
Murgantia histrionica 213
Muskroot Family 39-40
Mustard 3, 13, 171-185, 213, 294
  Black 24-25, 177, **178**
  California 175
  California False- 294
  Field 177
  Flat-pods 171, 180-185
  Fruits, Keeled 171, 177
  Hedge 179-181
  Hedge, Oriental 181
  Keeled-fruit 171, 177
  Linear-pods 171-181
  Oblong-pods 171, 182
  Sahara 179
  Shortpod 25, **178**
  Summer 178
  Tansy 171, **175**

Tower 174
Tumble 179, 180-181
Tumble, Oriental 179-180, **181**
Mustard Family **171-185**
Mycorrhizal Fungi 177
Myosurus minimus 17, **365**
Myrsinaceae 270
Myrsines 34

Nama, Mud 157
Nama 144, 157
  stenocarpum **157**
Narnia femorata 194
Nassella
  lepida: *See Stipa lepida*
  pulchra: *See Stipa pulchra*
Nasturtium 17, 171, 176, 402
  Garden 402
Nasturtium Family 402
Nasturtium officinale 176
Navarretia 16-17, 345
  Hooked 344
  Hooked, Southern 344
  Prostrate 16-17, **345**
  Spreading 345
Navarretia 17, 344-345
  atractyloides 344
  fossalis 345
  hamata ssp. *hamata* **344**
  hamata ssp. *leptantha* **344**
  prostrata 17, **345**
Nectarines 377
Nemacaulis denudata var. *denudata* **360**
Nemacladaceae 197
Nemacladus 197, 201
  longiflorus var. *longiflorus* **201**
  pinnatifidus **201**
  ramosissimus **201**
Nemognatha 355
  lurida 355
  nigripennis **355**
Nemophila 144, 157-159
  menziesii 158
  menziesii var. *integrifolia* **158**
  menziesii var. *menziesii* **157**
  pedunculata **158**
Nerium oleander 60
Nettle 226, 231, 233, 279, 403-404
  Dwarf 403, **404**, 405
  Hedge- 16, 279-281
  Hoary 403, **404**, 405
  Pellitory, California 403
  Pellitory, Western 403
  Western 403, 404
Nettle Family 30, **403-404**
Nicotiana 397
  attenuata 397
  bigelovii var. *wallacei* 397
  clevelandii 397
  glauca **397**
  quadrivalvis **397**
Nightshade 395, 399-400
  Black 399
  Blue Witch 399, **400**
  Buffalo-bur 399, **400**
  Chaparral 399-400

Douglas's 399, **400**
  Hairy 399
  Horse-nettle, Silverleaf 399
  Lance-leaf 399
  Parish's 399-400
  White 399
Nightshade Family 395-400
Nitidulidae 195
Nolina 438-439
  cismontana 23, 187, 415, **438**, 439
  parryi 439
Nolina Family 438
Nolinaceae 438
Nuttallanthus texanus 19, **328**
Nyctaginaceae 147, 233, **289-290**
Nymphaeaceae 242
Nymphaeales 13, 242
Nymphalidae 58, 61, 77, 96, 154, 231, 325, 331, 405, 409, 433

Oak 14, 34, 180, 203, 335, 381, 410-411; See also Poison Oak
  Engelmann 452
  Live, Canyon 3, 21-22, 89, 174, 391, 409, 452, 456
  Live, Coast 2-3, 16-17, 21-22, 26, 45, 89, 174, 203, 326, 335, 385, 411, 433, 453, 455
  Live, Interior 21
  Mistletoe 411
  Scrub 20
  Scrub, California 19-21, 411
Oceanspray 382
Odocoileus hemionus californicus 413
Odora 96
Oenanthe 50
Oenothera 292, 295, 300
  californica ssp. *californica* **300**
  elata ssp. *hirsutissima* **300**
Oleaceae 40, 291
Oleander 60
Olive Family **291**
Olive, European 291
Onagraceae 10, 25, 147, 283, **292-300**
Oncopeltis fasciatus 60
Oncosiphon piluliferum **94**
Onion 3, 13, 420-424, 440, 445
  Early 420-421, **422**
  Munz's 420-421, **422**
  Pitted 420, **421**, 422
  Red-flowered 420, **423**, 457
  Red-skinned 20, 420, **421**, 422
  San Bernardino Mountains 23-24, 420, **423**
  Weed 424
Onion Family 420-423, 440
Opuntia 186, 188-193
  basilaris var. *basilaris* 188, **189**
  engelmannii 193
  ficus-indica 188, **190**
  littoralis 19, 186, 188, **192**, 193
  oricola 188, **191**
  parryi 187
  phaeacantha 193
  prolifera **187**
  x *occidentalis* 188, **193**

x *vaseyi* 188, 192, **193**
Oranges 390
Orchid 3, 435-437
  Rein 436-437
  Rein, Cooper's 435, **436**, 437, 454
  Rein, Thin Wood 435-436, **437**
  Rein, Wood 435-436, **437**
  Stream 17, 29, **435**, 449, 455
  Vanilla 435
Orchid Family 10, 13, 64, 197, **435-437**
Orchidaceae 10, 13, **435-437**
Orobanchaceae 25, 29, 40, 130, **301-307**, 357, 378
Orobanche 29, 130, 301, 305-307
  bulbosa **305**, 378
  fasciculata 23, **306**, 357
  parishii ssp. *parishii* **306**
  sp. (new species) 29, 130, **307**
  vallicola 40, **307**
Orthocarpus purpurascens 303
Osmadenia 109, 114
  tenella 23, **114**
Osmia 84, **156**
Osmorrhiza brachypoda 21, **52**
Owl's-clover, Purple 19, 301, **303**, 455
Oxalidaceae **308**, 360
Oxalis 308, 360
  albicans ssp. *californica* 308
  californica **308**
  corniculata 308
  pes-caprae **308**
Oxalis Family **308**
Oxytheca trilobata 351

Paeonia 308
  californica **308**
Paeoniaceae 308
Paintbrush 3, 301-303
  Coastal 301, **302**
  Felt 18, 301, **302**, 450, 452, 456
  Martin's 301, **302**, 451, 457
  Owl's-clover, Purple 301, **303**
  Threadtorch, California 17, 301, **303**, 455
  Woolly 301, **302**
Papaver 312
  californicum 25, **312**
  heterophyllum **312**
Papaveraceae 10, 25, 212, 218, **309-313**, 364
Papilio
  eurymedon 84, **156**, 375, 385, 433
  rutulus rutulus **335**, 433
  zelicaon **50**
Papilionidae 50, 84, 156, 335, 375, 385, 433
Paracotalpa ursina **145**
Parietaria 403
  hespera var. *californica* 403
  hespera var. *hespera* **403**
Parsley: See also Hedge-parsley
  Garden 46, 50
  Mock 46
Parulidae 446

Pea
  Chaparral 3, **255**, 454
  Sweet 241, 258
  Sweet, Chaparral 241, 418
  Sweet, San Diego 241
Pea Family 10, 13, 217, 218, **236-259**, 346, 418
Peach, Domestic 84, 377
Peanut 236
Pears 377
Pectocarya 144, 152-153
  linearis ssp. *ferocula* 152, **153**
  penicillata 152, **153**
  setosa 152, **153**
Pedicularis 301, 305
  densiflora 25, **305**
Pellaea andromedifolia 23
Pellitory 403
  California 403
  Western 403
Pennyworts 62
  Marsh, Floating 62
  Marsh, Many-flowered 62
  Marsh, Whorled 62
Penstemon 324, 326-328
  centranthifolius **326**
  grinnellii 34
  grinnellii ssp. *grinnellii* 34, **327**
  heterophyllus var. *australis* **327**
  heterophyllus var. *heterophyllus* **327**
  spectabilis ssp. *spectabilis* **328**
Pentachaeta 125-127
  aurea 14, 64
  aurea ssp. *allenii* 14, 36, 126, **127**
  aurea ssp. *aurea* **127**
Pentatomidae 213
Peony Family 308
Peony, California 308, 453
Peplis portula 299
Peponapis 225
  angelica 225
  pruinosa **225**
Pepper, Bell 10
Peppergrass 171, 183-185
  Alkali 183, **184**
  Broad-leaved 184, **185**
  Hairy 17, 185
  Hairy-pod 184
  Menzies's 185
  Pinnate-leaved 184, **185**
  Robinson's 184, **185**
  Shining 184
Pepperwort, Hairy 17
Pepsis chrysothemis **32**
Peritoma arborea 19, **213**
Periwinkle 57
Persicaria 17, 360
  lapathifolia **360**
Petroselinum crispum 50
Petunia 395-396
  Wild, Small-flowered 396
Petunia parviflora **396**
Phacelia 11, 23, 160-170, 452
  Bluebells, California 167
  Bluebells, Desert 166
  Branching 160-161, **170**
  Canterbury Bells, Desert

160-161, **166**, 167
Canterbury Bells, Wild 160-161, 166, **167**
Caterpillar 160-161, **162**
Common 160-161, **165**
Davidson's 160-161, **164**
Great Valley 160-161, **163**
Hubby's 160-161, **163**
Imbricate 160
Imbricate, Northern 161, **169**
Imbricate, Southern 161, **169**
Large-flowered 160-161, **166**
Parry's 26, 160-161, 166, **167**
Santiago Peak 3, 23, 160-161, **168**
Short-lobed 160-161, **162**
Tansy 11, 160-161, **165**
Washoe 160-161, **164**
Phacelia 3, 23, 25, 144, 160-170
 brachyloba 160-161, **162**
 campanularia 160-161, **166**, 167
 cicutaria 160-161, 163
 cicutaria ssp. *hispida* **162**
 cicutaria var. *hispida* 162
 cicutaria var. *hubbyi* 162
 ciliata 160-161, **163**
 curvipes 160-161, **164**
 davidsonii 160-161, **164**
 distans 160-161, **165**
 grandiflora 160-161, **166**
 hubbyi 160-161, **163**
 imbricata 160, 169
 imbricata ssp. *bernardina* 169
 imbricata ssp. *imbricata* 161, **169**
 imbricata ssp. *patula* 161, **169**
 keckii 23, 160-161, 164, **168**
 minor 144, 160-161, 166, **167**
 parryi 144, 160-161, 166, **167**
 ramosissima 160-161, **170**
 suaveolens ssp. *keckii* 168
 tanacetifolia 160-161, **165**
Phainopepla 410, **411**
Phainopepla nitens **411**
Phlox 215, 343
 Prickly 343
 Slender 215, **343**
Phlox Family 338-345
Phlox gracilis 343
Pholisma arenarium 144, **170**
Pholistoma 144, 159
 auritum var. *auritum* **159**
 racemosum **159**
Phoradendron 410-411
 bolleanum **410**
 bolleanum ssp. *densum* 410
 densum 410
 macrophyllum 410
 serotinum ssp. *macrophyllum* **410**
 serotinum ssp. *tomentosum* **411**
 villosum 411
Phrymaceae 25, **314-319**
Phyla nodiflora **407**
Phymata pacifica **101**
Physalis 398
 crassifolia **398**
 greenei 398
Pickeringia montana var. *tomentosa*

255
Pickleweed 16-17, 210-211, 217, 219, 304
 Bigelow's 210, **211**
 Pacific **210**
 Parish's **210**, 219
Pieridae 237
Pillwort, American 17
Pilularia americana 17
Pimpernel
 Scarlet 270
 Water- 402
 Yellow False 324
Pinaceae 437
Pincushion 87
 Orcutt's 87
 Plant, European 274
 White 87
 Yellow 18-19, **87**, 108, 274, 454
Pine 3, 13
 Coulter 21-22, 454
 Knobcone 15, 22-23, 373
Pineapple Weed 94
 Common 25, **94**
 Valley 94
Pink
 Ground iii, 19, 36, **343**
 Mexican 207
 Southern 207, 457
 Windmill 207
Pink Family 10, **205-209**
Pinks 206-208
Pinus
 attenuata 15, 23
 coulteri 21
Piperia 436-437
 cooperi 435, **436**, 437
 elongata 435-436, **437**
 leptopetala 435-436, **437**
Pipestem 362
Pistachio 42
Pitcher-sage, Heart-leaved 3, 21, 23, 36, **271**, 450, 457
Plagiobothrys 25, 144, 146, 149-151
 acanthocarpus 17, **149**
 canescens var. *canescens* 149, **150**
 collinus var. *californicus* 149, **150**
 fulvus var. *campestris* 151
 nothofulvus 149, **151**
 tenellus 149, **151**
Plane Tree Family 333
Planodes 171-173
 virginicum 17, **173**
Plant
 Alkali 216, 217, 219
 Cotton-batting 97
 Dr. Seuss 128
 Pincushion, European 274
 Rattlesnake 25, **48**
 Sand, Scaly-stemmed 170
 Soap 3, 14, 413-414, 424, 450
 Soap, Small-flowered 414
 Soap, Wavy-leaved 413, **414**
Plant Groups
 Land 13
 Green 13
 Flowering 13

 Naked-seeded 13
 Non-vascular 13
 Seed 13
 Vascular 13
Plantaginaceae 25, 206, **320-332**, 365
Plantago 330-332, 365
 arenaria 330, **332**
 coronopus 330, 332
 elongata 330, **331**
 erecta ssp. *erecta* 23, 330, **331**
 indica 332
 lanceolata 330, **332**
 major 330, **332**
 ovata 330, **331**
 pusilla 330, 332
 rhodosperma 330, **332**
 subnuda 330, 332
 virginica 330, 332
Plantain 320, 330-332, 365
 Alkali 330, **331**
 California 23, 330, **331**
 Common 330, **332**
 Cutleaf 330, 332
 Dotseed 331
 Dwarf 330, 332
 English 330, **332**
 Mexican 330, 332
 Red-seeded 330, **332**
 Sand 330, **332**
 Virginia 330, 332
 Woolly 330, **331**
Platanaceae 227, **333**, 391, 410, 412
Platanus
 californica 14
 racemosa 17, 227, **333**, 391, 410, 412
Platystemon californicus 25, **313**
Plebejus acmon acmon **356**, **359**
Plectritis, Long-spurred 406, 454
Plectritis ciliosa **406**
Pleocoma puncticollis puncticollis **374**
Pleocomidae 374
Pluchea 64, 95
 odorata var. *odorata* **95**
 sericea **95**
Plum 377
Plumbaginaceae 336-337, 348
Pogonomyrmex californicus **32**
Poinsettia 233
Poison Ivy 42
Poison Oak, Western 14, 16-17, 19, 21, 30-31, 42-43, **45**, 71, 218, 227, 386; Dermatitis **31**, 45, 71; Lookalikes **45**
Polemoniaceae 25, 215, **338-345**
Polioptila
 caerulea **70**
 californica californica **70**
Polioptilidae 70
Polygalaceae 346
Polygala cornuta var. *fishiae* **346**
Polygonaceae 10, 306, **347-360**, 364
Polygonia satyrus **405**
Polygonum 350, 360

lapathifolium 360
Pondweeds 34
  Horned 34
Poodle-dog Bush 84, **156**, 456
Popcorn Flower 3, 25, 146, 149-151, 453
  Adobe 17, **149**
  California 149, **150**
  Field 151
  Pacific 149, **151**
  Rusty 20, 26, 149, **151**
  Slender 151
  Valley 149, **150**
Popillia japonica 93
Poppies, Poppy 3, 212, 309-313, 452
  Bush 9, 25, 218, **309**, 455
  California 11, 19-20, 25-26, 29, 310, **311**, 452
  Chicalote 309
  Fire 25, **312**, 454
  Matilija 309, 313
  Matilija, Coulter's 19-20, **313**, 450, 452
  Matilija, Hairy 313
  Prickly 309, 313
  Prickly, Robust 309
  Fairy, Small-flowered 25, **312**
  Tufted 310
  Western 310
  Wind 25, **312**
Poppy Family 10, 13, **309-313**
Populus 17
  fremontii 17
  trichocarpa 17
Poreleaf 96
Porophyllum gracile **96**
Porterella 197
Portulacaceae 287; *Also see* Montiaceae
Potato 395
Potentilla 383-384
  anserina ssp. *pacifica* **384**
  glandulosa ssp. *glandulosa* 384
  glandulosa ssp. *reflexa* 384
Prickly-pear 186, 188-193
  Beavertail 188, **189**
  Brown-spined 193
  Chaparral 188, **191**, 196
  Coast 18-19, 26, 188, **192**, 193
  Engelmann's 193
  Mesa 188, 192, **193**
  Mission 188, **190**
  Western 188, **193**
Primrose, European 215
Primrose Family 361, 402
Primulaceae 361, 402
Privet 291
Prodoxidae 393, 416, 439
Prodoxus 416-417
  cinereus **417**
  marginatus **417**
  pulverulentus 417
Protozoa 32
Prunus 385
  ilicifolia ssp. *ilicifolia* 21, 84, 376, **385**

persica 84
Pseudognaphalium 64, 96-100; *See also* Gnaphalium
  beneolens **99**, 100
  biolettii **98**
  californicum 65, **98**, 312
  leucocephalum 99, **100**
  luteoalbum 96, **97**
  microcephalum **99**
  ramosissimum **100**
  stramineum **97**
Pseudoluperus maculicollis **418**
Pseudoscorpiones, Pseudoscorpions 334
Pseudotsuga macrocarpa 15, 21, 437
Psilocarphus brevissimus 17
Psoralea 240
  macrostachya 240
  orbicularis 240
  physodes 240
Psyllobora 334
  borealis 334
  vigintimaculata **334**
Pterostegia drymarioides 23, 48, **360**
Ptilogonatidae 411
Pulicaria 128
  hispanica 128
  paludosa **128**
Puma concolor **33**
Pumpkin 224
Puncturebract: *See* Spineflower
Purslane Family 287
Pussypaws 287-288
  Common 288
  Seaside 287
Pycnanthemum californicum **271**, 278
Pygmy-stonecrop 220
  Sand 23, **220**
  Water 17, **220**

Quercus
  agrifolia 17, 45, 174, 203, 326, 335, 411, 433
  berberidifolia 19, 411
  chrysolepis 21, 89, 174, 409
  wislizenii 21
Quillworts 13

Rabbitbrush, Great Basin 89, 120-121
Rabbit-tobacco: *See* Everlasting
Radish 181
  Charlock, Jointed 181
  Wild 181
Rafinesquia californica **139**
Ragweed 66-68
  Western 67, **68**
Ranunculaceae 10, 25, 45, **362-366**
Ranunculus 366
  aquatilis var. *diffusus* **366**
  californicus **366**
  cymbalaria 366
  hebecarpus **366**
  occidentalis **366**
Raphanus 171, 177, 181
  raphanistrum **181**

sativus **181**
Rattlesnake 33
  Red Diamond 33
  Southern Pacific 33
  Southwestern Speckled 33
Rattleweed 238-239; *See also* Locoweed, Milkvetch
  Pomona 239
Red Maids 287
  Brewer's 287
  Common 287
Redberry 3, 375-376
  Holly-leaved 18, 21, **376**, 385
  Spiny 84, 118, **376**
Reduviidae 60, 101
Red-stem 284
  Robust 284
  Valley 284
Redwood 311
  Coast 311
  Giant 311
Rhamnaceae 25, 84, 218, 232, 311, **367-376**, 385, 418
Rhamnus 375-376
  californica ssp. *californica* 21, 375
  crocea 84, **376**
  crocea ssp. *ilicifolia* 376
  ilicifolia 21, **376**, 385
  tomentella ssp. *cuspidata* 375
  tomentella ssp. *tomentella* 375
Rhaphiomidas acton **338**
Rhopalomyia lonicera **202**
Rhus 43-44
  aromatica **43**, 45
  integrifolia 19, **44**
  laurina 14
  ovata 21, 43, **44**
  trilobata 43
Ribbonshanks **377**, 453
Ribes 43, 266-269
  amarum **269**
  aureum var. *gracillimum* **266**
  californicum var. *hesperium* 21, **268**
  indecorum **267**
  malvaceum 14
  malvaceum var. *malvaceum* 23, 267
  malvaceum var. *viridifolium* **267**
  roezlii var. *roezlii* **269**
  speciosum **268**
Ricinus communis **233**
Rocket 179
  Horned Sea 19, 171, **182**
  London 179, **180**, 181
Rock-rose Family 212
Romneya 313
  coulteri 19, 309, **313**
  trichocalyx 309, **313**
Root, Durango 17, **226**
Ropevine 363
Rorippa 17, 171, 176
  curvisiliqua 172, **176**
  nasturtium-aquaticum 176
  palustris 172, 176
Rosa californica 21, **385**
Rosaceae 84, 305-306, 333, 376, **377-386**
Rose 13, 21, 377, 385

California 21, **385**, 452
Rose Family 377-386
Rosemary 271
Rosewood 37
Rosilla 122
Rosinweed, False 23, **114**, 452
Rubiaceae 387-389
Rubus 386
  armeniacus **386**
  discolor 386
  ursinus 45, **386**
Rue 50, 390
  Common 390
  Fringed 390
Rue Family 390
Rupertia 240
  physodes **240**
Ruscaceae 187, 415, **438-439**
Rush 16-17, 34, 330
  Southwestern Spiny 16-17
Rush-rose, California 20, 25, **212**
Rutaceae 50, **390**
Ruta 50, 390
  chalepensis **390**
  graveolens **390**

Sacapellote 63, **65**
Safflower 81, 64
Sage 3, 14, 29, 64, 92, 271, 274-277
  Black 14-15, 19, 26, 29, 217, 275, **276**
  Chia 25-26, **274**, 454
  Cleveland 276, 277
  Heart-leaved Pitcher- 3, 21, 23, 36, **271**, 450, 457
  Hummingbird 29, **277**
  Purple 14, **277**
  Thistle 274
  White 15, 18-21, 217, 261, **275**, 276, 455; Pollination **275**; In Peril? 276
Sagebrush 14, 69-72
  An Imperiled Habitat 70
  Big 72
  California 14-15, 18-19, 69, **70**, 120, 155
  Great Basin 72
Sagewort, Dragon 71
Sagittaria 419
  latifolia 419
  montevidensis ssp. *calycina* 419
Salicaceae 75
Salicornia 210-211, 219, 304
  bigelovii 17, **211**
  pacifica 17, **210**
  subterminalis 210
  virginica 210
Salix 17, 75
  exigua 17
  lasiolepis 17
Salpichroa origanifolia **398**
Salsify 142
  Purple 142
  Yellow 142
Salsola tragus 25, 210, **212**
Saltbush 19, 217, 219

Saltugilia australis **340**
Saltwort 34
Salvia 92, 274-277
  apiana 15, 19, 261, **275**
  carduacea **274**
  clevelandii 25, **276**
  columbariae **274**
  leucophylla **277**
  mellifera 15, 19, **276**
  spathacea **277**
Sambucus 39-40
  caerulea 39
  mexicana 39-40
  nigra ssp. *caerulea* 17, **39**, 218, 307
Samolaceae 402
Samolus parviflorus **402**
Sand Plant, Scaly-stemmed 170
Sand-bur 67, 68
Sand-verbena 289-290
  Beach 18-19, 147, **289**, 290
  Chaparral 19, **290**
  Red 289
Sand-spurrey 209
  Salt Marsh 209
  Sticky 209
Sandmat 23, **205**
  Small-seed 235
Sanicle 52, 54-55
  Pacific 21, **55**
  Poison 54
  Purple 11, **54**
  Sharp-toothed 54
  Sierra 55, 56
  Tuberous 23-24, **55**
Sanicula 52, 54-55
  arguta **54**
  bipinnata **54**
  bipinnatifida **54**
  crassicaulis var. *crassicaulis* 21, **55**
  graveolens **55**, 56
  tuberosa 23, **55**
Sapindaceae 333, **391**
Sarcocornia pacifica 210
Sarcostemma cynanchoides ssp. *hartwegii* 57
Sassafras 37
Satureja chandleri 278
Saturniidae 356, 374
Saucrobotys futilalis inconcinnalis **61**
Saururaceae 38
Savory, San Miguel 3, 271, **278**, 454
Saxifraga californica 393
Saxifragaceae 266, **392-393**
Saxifrage, California 23, **393**, 454
Saxifrage Family 266, **392-393**
Scabiosa columbaria 274
Scaeva pyrastri **435**
Scale-broom 18-19, **77**, 411, 452
Scale Insect
  Cochineal 190, **194**, 195
  Sycamore 334
Scarabaeidae 93, 145, 396
School Bells 19, 20, 25-26, **444**, 445, 457

Scorpion 31
  Burrowing 31
Scorpionweed: *See* Phacelia
Scrophularia californica **394**
Scrophulariaceae 301, 314, 320, **394**
Scutellaria tuberosa **278**
Scyphophorus yuccae **418**
Sea Rocket 19, 171, 182
  Horned 182
  Oval 182
Sea-blite 19, 211
  Estuary 211
  Horned 211
  Woolly 211
Sea-fig 41
Sea-lavender 336-337
  Algerian 337
  California 16-17, **336**
  Dèspréaux's 336
  European 337
  Perez's 336
  Rock 337
  Sventenius's 336
  Winged 337
Sea-purslane, Western 42, 260
Sedges 16-17, 34
Selaginella bigelovii 23
Selasphorus sasin 314
Senecio 129
  californicus 128, **129**
  flaccidus var. *douglasii* **129**
Sequoia sempervirens 311
Sequoiadendron gigantea 311
Sesiidae 335, 356
Sesuvium verrucosum **42**, 260
Shanks, Red 377
Shepherd's-purse 171, **183**
Shooting Star, Padre's 19-20, **361**, 452, 457
Sibara 173
  virginica 173
Sibaropsis 171-173
  hammittii **173**
Sida 286
Sidalcea 286
  malviflora 19, **286**
  neomexicana 286
Sidotheca 347, 351
  trilobata 347, **351**
Silene 206-208
  antirrhina 25, **206**
  coniflora **206**
  gallica **207**
  laciniata ssp. *laciniata* **207**
  laciniata ssp. *major* 207
  lemmonii **208**
  multinervia 206
  verecunda 23, **208**
Silk-tassel 260
  Pale 260, 455
  Southern 260
Silk-tassel Family 260
Silverpuffs 136-138
  Douglas's 136, 137
  Elegant 136, **137**
  Lindley's 136, **138**

San Diego 16, 136, **137**
Silverweed, Pacific 384
Silybum 79, 86
  marianum **86**
Sisymbrium 171, 177, 179-181
  altissimum **179**, 180-181
  irio 179, **180**, 181
  officinale 179, **180**, 181
  orientale 179-180, **181**
Sisyrinchium bellum 19, **425**
Skullcap, Danny's 25, **278**
Skunkbrush 43, 45
Skunkweed, Holly-leaved 344
Smartweed 16-17, 360
  Pale 360
Smilacina racemosa 438
Snakeroots 52, 54-55
Snakes 33, 355
  Rattlesnake, Red Diamond 33
  Rattlesnake, Southern Pacific 33
  Rattlesnake, Southwestern Speckled 33
Snapdragon 3, 13, 25, 206, 320-321, 325, 328, 330
  Catchfly 25, **206**
  Chaparral 320, **321**
  Coulter's 320, 321, 457
  Kellogg's 320, **321**
  Nuttall's 320, **321**, 454
Sneezeweed 122
Snowberry 203-204
  Common 203
  Parish's 204
  Spreading 203
Snowdrop Bush, Southern 401
Soap Plant 3, 14, 413-414, 424, 450
  Small-flowered 414
  Wavy-leaved 413, **414**
Soapberry Family 391
Sock-destroyer 51
  Short 51
  Tall 51
Solanaceae **395**
Solanum 399-400
  americanum 399
  douglasii 399, **400**
  eleagnifolium 399
  lanceolatum 399
  nigrum 399
  parishii 399-400
  physalifolium 399
  rostratum 399, **400**
  umbelliferum 399, **400**
  xanti 399-400
Soleirolia soleirolii 405
Solenopsis invicta 32
Solidago 130
  californica 130
  occidentalis 130
  velutina ssp. *californica* **130**, 307
Solomon's Seal, Western False 438
Sourgrass 308
Soybeans 236

Spartium junceum **255**
Speedwell 329
  Great Water 329
  Mexican 17, **329**
  Persian 329
Spergularia **209**
  macrotheca var. *macrotheca* **209**
  marina **209**
  salina 209
Speyeria 409
  callippe comstocki **409**
  coronis semiramis **409**
Sphingidae 204, 292
Spice Bush, Coastal 390
Spiderflower Family 213
Spider 31
  Black Widow, Western 31
Spike-moss 13
  Bigelow's 23
Spike-rushes 16-17
Spikeweed, Southern 109
Spinach 210
  New Zealand 42
Spineflower 3, 347-351
  Fringed 26, 347, **348**
  Leather 347, **350**, 351
  Long-spined 347, **350**
  Parry's 347, **349**
  Peninsular 347, **349**
  Prostrate 347, **350**
  Puncturebract 347
  Puncturebract, Three-lobed Starry 351
  San Fernando Valley 347, **349**
  Slender--horned 347, **351**, 364
  Turkish-rugging 347, **348**
  Turkish-rugging, Orange County 347, **348**
Spinus psaltria **84**, 276
Spiraea, Mountain 382
Spittlebugs 92
Spurge 235
  Cliff 19, **235**
  Golondrina 235
  Rattlesnake 235
Spurge Family 233
Squash 224-225
St. John's Wort 270
  Canary Islands 270
  Scouler's 270, 454
  Tinker's Penny 270
St. John's Wort Family 270
Stachys 279-281
  albens 279, **280**
  bullata 279, **280**
  rigida var. *rigida* 279, **281**
  rigida var. *quercetorum* 279, **281**
  stebbinsii 279, 281
Star
  Coast Baby- 342
  Hill 393
  Shooting, Padre's 19-20, **361**
  Woodland 23, **393**
Star-thistle 25, 79, 81-82
  Malta 82
  Maltese 82
  Yellow 82

State
  Flower of California 311
  Insect of California 237
  Largest Flower 313
Statice 348
Stebbinsoseris 131, 136, 138
  heterocarpa 136, 138
Steingeliidae 334
Stellaria media **209**
Stephanomeria 63-64, 139-141, 218
  cichoriacea 139, **140**
  diegensis 139, **141**
  exigua ssp. *deanei* 139, **140**
  virgata ssp. *virgata* 139, **141**
Stick-leaf 283
  Small-flowered 25, **283**
  Veatch's 283
Stick-leaf Family 283
Sticktight 108
Stillingia, Linear-leaved 234
Stillingia linearifolia **234**
Stinknet 94
Stinkweed 104
Stipa 19
  lepida 19
  pulchra 19
Stitchwort, Douglas's 23, **205**, 454
Stock 179
Stomaccocus platani 334
Stonecrop 10, 220
  Pygmy- 220
  Pygmy-, Sand 23, **220**
  Pygmy-, Water 23, **220**
Stonecrop Family 220-223
Storax Family 401
Storax, Southern California 3, **401**
Storksbills 263-264
Strawberries 377, 383
Strawberry Tree 381
Streptanthus heterophyllus 174
Stylomecon heterophylla 312
Stylophyllum 221-222
Styracaceae **401**
Styrax redivivus **401**
Suaeda 211
  calceoliformis 211
  californica var. *pubescens* 211
  esteroa 211
  taxifolia **211**
Sugar Bush 21, 42-43, **44**, 411, 453
Sumac 3, 13-14, 42-45
  Laurel 14, 18-19, **43**, 44, 118, 217-218, 374
Sumac Family 31, **42-45**
Sun Cup 292-294; *See also* Evening-primrose
  California 293, 294
  Sandysoil 292
  Southern 293, 294
Sun Cups Clade 292-294
Sunflower 10-11, 14, 116
  Bush, California 18-19, 29, **116**, 118
  Bush, Desert 19, **116**, 155
  Canyon 2, 65, **118**

San Diego 119
Slender 117, 418, 453, 456
Spanish 128
Western 64, 117
Sunflower Family 13, **63-142**, 197, 217-218, 449
   Discoid, Radiant, and Disciform Heads 65-102
   Liguliflorous Heads 131-142
   Radiate Heads 103-130
Superasterids 13
Superrosids 13
Sweet Pea: See Pea
Sweetbush, Scabrid 19, 63, **77**
Sweetclover 25, 218, 254
   White 254
   Yellow 254
Sweet-cicely, California 21, **52**
Swertia parryi 261
Sycamore 11, 13, 334-335
   California 3, 14, 16-17, 19, 187, 227, **333**, 385, 391, 410-412, 452
Sycamore Family 333-335
Sympetrum illotum **448**
Symphoricarpos 203-204
   albus var. *laevigatus* **203**
   mollis **203**
   rotundifolius var. *parishii* **204**
Symphyotrichum 105-106
   defoliatum **105**
   lanceolatum ssp. *hesperium* **106**
   subulatum **106**
Synanthedon
   polygoni **356**
   resplendens **335**
Syrphidae 195, 392, 435

Tamalia coweni **231**
Tanacetum vulgare **165**
Taraxacum officinale 131
Tarplant 20, 23, 25, 34, 109-114
   Fascicled 20, 109, **110**
   Graceful 109, **112**
   Kellogg's 109, **111**
   Paniculate 109, **111**, 112
   San Diego 109, **111**
   Slender **113**
   Smooth 109, **110**
   Southern **109**
Tarragon 64, **71**
Tarweed 14, 34; *See also* Tarplant
Tauschia 50, 52, 56
   Parish's 11, 23, 55, **56**
   Southern 21, 53, **56**
Tauschia 52, 56
   arguta 21, 53, **56**
   parishii 23, 55, **56**
Taxaceae 211
Taxus 211
Tea 240
   California **240**, 453
   Leather-root **240**
   Round-leaved **240**
Tegeticula 416-417
   maculata extranea **416**
Telegraph-weed **123**
Tetradymia comosa **101**

Tetragonia tetragonioides **42**
Tetraopes basalis **60**
Tetrapteron 292, 294
   graciliflorum **294**
Thalictrum 365
   fendleri var. *polycarpum* **365**
   polycarpum 365
Thelypodium lasiophyllum 175
Themidaceae 25, 420, **440-445**
Theophrastaceae 402
Theophrasta Family 402
Thistle 64, 79-86, 231, 274
   Artichoke 79, **86**
   Blessed **81**
   Bull 80, **85**
   California **83**, 84, 204, 237, 409
   Cobwebby **83**
   Common 79, 83-85
   Distaff 79
   Distaff, Smooth **81**
   Italian 79, **80**, 85
   Knapweed, Russian 79, **80**
   Milk 79, **86**
   Morocco 79
   Russian 25, 210, **212**
   Sage **274**
   Smooth Distaff **81**
   Southern Meadow **85**
   Star- 25, 79, 81-82
   Star-, Malta **82**
   Star-, Maltese **82**
   Star-, Yellow **82**
Thorn-apples 395
Threadplant **201**
   Comb-leaved **201**
   Long-flowered **201**
   Nuttall's **201**
Threadstem, Woodland **360**
Threadtorch: See Paintbrush
Three-spot **114**
Thrips 75
Thysanocarpus 171, 182-183
   curvipes **182**
   laciniatus **183**
Tick 31
   Western Black-legged **31**
   Western Dog **31**
Tickseed 128-129
   California **128**, 129
   Giant **128**
Tidy-tips, Common 63, **126**, 127, 331, 457
Tillaea erecta 220
Tingidae 334
Tinker's Penny **270**
Toadflax, Blue 19, **328**
Tobacco 395, 397
   Cleveland 397
   Coyote 397
   Tree **397**
   Wallace's **397**
   White Rabbit- **100**
Tocalote **82**
Tomato 395
Toothwort, California **172**
Torilis 51
   arvensis **51**

nodosa **51**
Toxicodendron 45
   diversilobum 17, 31, 43, **45**, 218, 386
   vernicifluum 45
Toxicoscordion 434
   fremontii 23, 25, **434**
   venenosum var. *venenosum* **434**
Toyon 3, 17-19, 21-22, 333, **381**, 451
Tragopogon 64, 142
   dubius **142**
   porrifolius **142**
Trichostema 282
   austromontanum **282**
   lanatum **282**
   lanceolatum **282**
   parishii **282**
Trifolium 256-257
   ciliolatum **256**
   depauperatum var. *truncatum* 17, **256**
   obtusiflorum **257**
   tridentatum 257
   truncatum 256
   willdenovii **257**
Triglochin concinna 17
Triodanis 197-198
   biflora 25, **198**
Trirhabda 155
   eriodictyonis **155**
Trochilidae 266, 314, 335
Troglodytidae 196, 446
Tropaeolaceae 402
Tropaeolum majus **402**
Tropidocarpum 171, 177
   gracile **177**
Tule-potato 419
Tumbleweed 210, **212**
Tulips, Butterfly: See Lily, Mariposa
Turkish Rugging 23, 347, **348**
   Orange County 19, 347, **348**
Turnips 178
Turricula parryi 156
Turritis 171, 172, 174
   glabra **174**
Typha 17, 446-447
   angustifolia 446, **447**
   domingensis 446, **447**
   latifolia 446, **447**
Typhaceae 446

Umbellularia californica 17, **37**, 227, 437
Uropappus 131, 136, 138
   lindleyi 136, 138
Urticaceae 30, 226, 231, 279, **403-405**
Urtica 279, 404
   dioica ssp. *holosericea* **404**
   urens 403, **404**
Urushi 45

Valerian 326, 406
   Red **406**
Valerian Family 406
Valerianaceae 326, **406**
Valsaceae 334
Vanessa

atalanta rubria **231**, **405**
cardui **231**
virginiensis **77**, 96
Venegasia 116, 118
carpesioides 65, **118**
Venus's Looking-glass, Small 25, **198**
Verbascum 394
thapsus **394**
virgatum **394**
Verbena, Western 407
Verbena 29, 407
bracteata 17, **407**
lasiostachys **407**
Verbenaceae 407
Verbesina 116, 118-119
dissita **118**
encelioides **119**
Veronica 329
anagallis-aquatica **329**
arvensis 329
peregrina ssp. *xalapensis* 17, **329**
persica **329**
Vervain, Bracted 17, **407**
Vervain Family 407
Vespula pensylvanica 32
Vetch 258-259, 305
American 258
Bengal 259
Common 259
Hairy 259
Hasse's 258
Purple 259
Slender, Southern 258
Spring 259
Winter 259
Winter, Variable 259
Viburnum 39
Vicia 258-259, 305
americana var. *americana* **258**
benghalensis **259**
exigua 258
hassei **258**
ludoviciana var. *ludoviciana* **258**
sativa **259**
villosa ssp. *varia* 259
villosa ssp. *villosa* **259**
Viguiera laciniata 119
Vinca major **57**
Violaceae 10, **408**
Viola 408-409
pedunculata 19, **408**, 409
purpurea 409
purpurea ssp. *purpurea* 21, **408**
purpurea ssp. *quercetorum* **408**
sheltonii **409**
Violet 408-409
Johnny Jump-ups 19, **408**
Yellow, Mountain 21, 408
Yellow, Oak 408
Shelton's 11, **409**
Violet Family 3, 10, **408**-**409**
Virgin's-bower 16, 20, 45, 362-363
Chaparral 362
Pipestem 362
Ropevine 363
Southern California 362, **363**

Western 362, **363**
Virus 32
West Nile 32
Viscaceae 334, **410**-**411**
Vitaceae 412
Vitis girdiana 17, **412**
Volutaria 79

Wallflower 171, 176
Mediterranean 293
Western 176
Walnut 34, 40, 410
Black, Southern California 2, 21-22
Warrior's Plume 11, 25, **305**
Wasp, Wasps 32, 59, 90, 96, 107, 271, 301, 328, 355, 394, 416, 438
Gall, Spiny Leaf 385
Paper 335
Parasitoid 60
Sawfly, Stem 384
Tarantula Hawk 32
Ant, Velvet 32
Yellow Jacket 32
Water-milfoil 34
Water-parsnip, Cutleaf 47
Water-pimpernel 402
Water-plantain, Northern 419
Water-plantain Family 419
Water-primrose 299
Red 299
Uruguayan 299
Yellow 299
Waterleaf Family: *See* Borage Family
Waterleafs 144, 154-170
Waterlilies 13
Waterlotus 13, 242
Watermelon 400
Watershields 13
Waterworts 34
Weed 14
Crete 133
Crofton 65, 118
Onion 424
Pineapple, Common 25, **94**
Pineapple, Valley 94
Telegraph- 25, **123**
Vinegar- 282
Weevil: *See* Beetle
Whitethorn: *See* Lilac
Whitlow, Desert 171, **175**
William, Sweet 205
Willow 16-17, 26, 34, 75, 335, 410
Arroyo 17
Baccharis 73-75, **76**
Sandbar 17
Willow-herb 296, 298
Clarkia 296, 297
Dense-flowered 298
Green 17, **298**
Smooth 298
Wing-fruit 52-53
Foothill 53
Shiny 18, **53**, 56
Woolly-fruited 53, 454
Winter cress, American 172, 176

Wire Lettuce 139-141
Wishbone Bush, California 18, 233, **290**
Witch, Blue 399, **400**
Witch's Hair 217-219
Wood-sorrel 308, 360
California 308
Creeping 308
Wood-sorrel Family 308
Woollyheads, Coastal 360
Woolly-star 338-339
Santa Ana River 338
Sapphire 14, **339**, 454
Wreath-plant 63, 139-141, 218
Chicory-leaved 139, **140**
Deane's 139, **140**
San Diego 139, **141**
Tall 139, **141**
Wyethia 116, 120
ovata **120**

Xanthium 63, 102
spinosum **102**
strumarium **102**, 261
Xylococcus bicolor **232**
Xylocopa 32, **143**, 275

Yabea microcarpa **46**
Yarrow 60, 103
Golden 20, 23, 29, **103**, 302
White 103
Yellow Carpet, Common 11, **115**
Yellow-cress
Marsh 172
Pacific 176
Western 172, **176**
Yerba
de la Vibora 48
Mansa 13, 16, **38**
Yerba Santa 155-156
Hairy 155
Poodle-dog Bush 156
Thick-leaved 20, **155**, 306, 456
Yew Tree 211
Yucca 13, 415-416
Chaparral 3, 18, 21, **415**, 416-418, 439
Western 415
Yucca whipplei 415

Zauschneria 299
californica ssp. *californicum* 299
californica ssp. *latifolia* 299
Zeltnera venusta **262**
Zelus renardii **60**
Zerene eurydice **237**
Zigadenus 434
fremontii 434
venenosus var. *venenosus* 434

Quick Index by Flower Color

Listed by common name. Includes major genera often used in common names.

Whites: White or Whitish

- [ ] Aralia 62
- [ ] Ash 291
- [ ] Asphodel 424
- [ ] Asters 105, 135
- [ ] Baccharis 72
- [ ] Beargrass 438
- [ ] Bedstraw 388
- [ ] Bird's-beaks 304
- [ ] Blackberries 386
- [ ] Bluecup 197
- [ ] Box-thorn 396
- [ ] Boykinia 391
- [ ] Brickellbushes 78
- [ ] Buckwheats 353-354, 358-359
- [ ] Bushrue 390
- [ ] Buttercup 366
- [ ] Button-celery 49
- [ ] Calif. Lilacs 367, 369-370, 372-373
- [ ] Calochortus 427-428
- [ ] Canchalagua 262
- [ ] Carrots 46-48, 51
- [ ] Catchflies 206-207
- [ ] Ceanothus 367, 369, 370, 272, 373
- [ ] Celery 47
- [ ] Chamise 378-379
- [ ] Cherry 385
- [ ] Chickweed 209
- [ ] Chicory 139
- [ ] Cinquefoil 384
- [ ] Clarkia 296
- [ ] Coastweed 66
- [ ] Cockleburs 102
- [ ] Combseeds 152-153
- [ ] Croton 234
- [ ] Cryptantha 146-148
- [ ] Cucumber 225
- [ ] Cudweeds 96-100
- [ ] Currant 267
- [ ] Daisy 115
- [ ] Death Camas 434
- [ ] Deer's-ears 261
- [ ] Desert-thorns 396
- [ ] Dichondra 216
- [ ] Dodder 217-219
- [ ] Dogbane 58
- [ ] Doveweed 234
- [ ] Dudleya 221, 222
- [ ] Eardrops 310
- [ ] Eucrypta 156
- [ ] Evening-primrose 300
- [ ] Everlastings 96-100
- [ ] False Pimpernel 324
- [ ] Fiesta Flower 159
- [ ] Filaree 262
- [ ] Flax 283
- [ ] Fringepods 182-183
- [ ] Frog-fruit 407
- [ ] Forget-me-nots 146-148
- [ ] Gilia 340
- [ ] Gooseberries 268-269
- [ ] Grappling Hook 152
- [ ] Groundsmoke 300
- [ ] Hedge-nettles 279-281
- [ ] Heliotrope 154
- [ ] Hemlock 48
- [ ] Hemp 58
- [ ] Honeysuckle 202
- [ ] Horehound 272
- [ ] Horkelia 383
- [ ] Iceplants 41
- [ ] Iris 425
- [ ] Jepsonia 391
- [ ] Jimsonweeds 395
- [ ] Lacepods 182-183
- [ ] Lilies 427-428
- [ ] Lily-of-the-Valley 398
- [ ] Linanthus 341-342
- [ ] Live-forevers 221, 222
- [ ] Lobelia 200
- [ ] Locoweeds 238-239
- [ ] Lotus 244, 247
- [ ] Lupine 248
- [ ] Madrone 227
- [ ] Mallow 286
- [ ] Manroot 225
- [ ] Matilija Poppy
- [ ] Manzanita 228-230, 232
- [ ] Milkweeds 58, 59
- [ ] Milkwort 172
- [ ] Miner's Lettuce 288
- [ ] Mints 271-272, 273
- [ ] Monardella 273
- [ ] Morning-glories 214-216
- [ ] Mountain-mahogany 380
- [ ] Mouse-tail 365
- [ ] Muilla 445
- [ ] Mustards 172-173, 175, 181-185
- [ ] Nama 157
- [ ] Navarretia 345
- [ ] Nightshades 399-400
- [ ] Oceanspray 382
- [ ] Onions 420-422
- [ ] Orchid 435
- [ ] Owl's-clover 303
- [ ] Pennyworts 62
- [ ] Peppergrass 184-185
- [ ] Petunia 396
- [ ] Phacelia 161-164, 169
- [ ] Phlox 343
- [ ] Pincushion 87
- [ ] Pinks 206-207
- [ ] Plagiobothrys 149-152
- [ ] Plantains 331-332
- [ ] Popcorn Flowers 149-151
- [ ] Poppies 309, 312, 313
- [ ] Poreleaf 96
- [ ] Pygmy-stonecrops 220
- [ ] Radishes 181
- [ ] Rattlesnake Weed 48
- [ ] Ribbonshanks 377
- [ ] Rosinweed 114
- [ ] Sages 271, 275-276
- [ ] Salvia 271, 275-276
- [ ] Sand-spurreys 209
- [ ] Sandmat 205
- [ ] Savory 278
- [ ] Saxifrage 393
- [ ] Sea-lavenders 336-337
- [ ] Shooting Star 361
- [ ] Smartweed 360
- [ ] Snapdragon 320
- [ ] Snowberry 204
- [ ] Soap Plants 413-414
- [ ] Solomon's Seal 438
- [ ] Speedwell 329
- [ ] Spineflowers 349-351
- [ ] Spiraea 382
- [ ] Spurges 235
- [ ] Stitchwort 205
- [ ] Storax 401
- [ ] Summer Holly 232
- [ ] Sweetclover 254
- [ ] Thistles 80, 83, 85
- [ ] Threadplants 201
- [ ] Tidy-tips 126
- [ ] Tobacco 397
- [ ] Toyon 381
- [ ] Vervain 407
- [ ] Virgin's-bowers 362-363
- [ ] Water-pimpernel 402
- [ ] Water-plantains 419
- [ ] Willow-herb 298
- [ ] Woodland Star 393
- [ ] Wreath-plants 139-141
- [ ] Yarrow 103
- [ ] Yerba Mansa 38
- [ ] Yerba Santa 155
- [ ] Yucca 415

Yellows: Yellow, Lemon, Golden, Mustard, Cream

- Alpinegold 124
- Ambrosia 66
- Anise 50
- Arnica 69
- Artemisia 69-72
- Asters 105-108
- Barberries 143
- Bay Laurel 37
- Beardtongue 325
- Beargrass 438
- Bedstraws 389
- Bird's-beaks 304
- Bladderpod 213
- Blow-wives 104
- Bowlesia 48
- Brass Buttons 88
- Broomrapes 306-307
- Buckwheats 357, 359
- Bur-marigolds 108
- Bur-sages 69-72
- Buttercups 366
- Butterweeds 129
- Cactus 189
- Calochortus 429-430
- Cat's Ears 134
- Catchflies 208
- Charlock 181

- Cholla 187
- Cinquefoil 384
- Clovers 265-257
- Coffeeberries 375
- Cotton-thorn 101
- Cresses 172, 176
- Cucumber 225
- Cudweeds 96-100
- Currant 266
- Daisies 115, 125-127
- Dandelions 131-134
- Death Camas 434
- Deer Broom 245
- Deer's Ears 261
- Desert-thorns 396
- Dudleya 221-222
- Eardrops 310
- Elderberry 39
- Evening-primrose 292-294, 300
- Fennel 50
- Frog-fruit 407
- Goldenstar 440
- Goldenbushes 89-91, 120-121
- Goldenrods 130
- Goldfields 125-126
- Gourd 224
- Granny's Hairnet 360
- Grape 412
- Ground-cherry 398
- Gumplant 121
- Hemlock 48
- Jaumea 124
- Jewelflower 174
- Keckiella 325
- Lessingia 101
- Lilies 429-430
- Linanthus 342
- Live-forevers 221-223
- Lotus 243-247
- Lupine 248
- Madrone 227
- Manroot 225
- Maple 391
- Matchweed 122
- Meadow-rue 365
- Melon 224
- Monkeyflowers 314, 316, 318, 319
- Mountain-mahogany 380
- Mulleins 394
- Mustards 172, 174-181
- Nasturtium 402
- Orchids 436-437
- Paintbrushes 302-303
- Pincushion 87
- Pineapple Weeds 94
- Poison Oak 45
- Poppies 309, 310, 313
- Prickly-pears 190-193
- Radishes 181
- Redberries 376
- Rosilla 122
- Rues 390
- Rush-rose 212
- Sagebrushes 69-72
- Salsify 142
- Sanicles 54-55
- Scale-broom 77

- Silverpuffs 136-138
- Silverweed 384
- Snakeroot 54-55
- Spanish Broom 255
- Spinach 42
- Spineflower 350
- Stick-leaf 283
- Stinknet 94
- St. John's Worts 270
- Sumacs 43
- Sun Cups 292-294
- Sunflowers 116-128, 128
- Sweetbush 77
- Sweetclover 254
- Tarplants 109-114
- Tauschia 56
- Tea 240
- Telegraph-weeds 123
- Thistle 81-82
- Tickseeds 128
- Tidy Tips 126
- Tobacco 397
- Violets 408-409
- Virgin's-bower 361
- Wallflower 176
- Waterweeds 299
- Whispering Bells 154
- Wing-fruits 53
- Woollyheads 360
- Wood-Sorrels 308
- Yarrow 103
- Yellow Carpet 115
- Yucca 415

Oranges: Orange, Salmon, Yellow-Orange

- ■ Aloë 424
- ■ Coyote Melon 224
- ■ Fiddlenecks 145
- ■ Lilies 432-433
- ■ Live-forever 223
- ■ Lotus 243-247
- ■ Monkeyflower 315
- ■ Nasturtium 402
- ■ Orchid 435
- ■ Poppies 311, 312
- ■ Prickly-pears 190, 193
- ■ Radish 181
- ■ Threadtorch 303
- ■ Wallflower 176

Reds: Red-Orange, Red, Pink, Lavender, Magenta, Maroon, Purple, Red-Violet, Violet

- ■ Alkali Heath 260
- ■ Alpinegold 124
- ■ Arrowweed 95
- ■ Asters 106
- ■ Beardtongues 325-327
- ■ Blue-eyed Mary 323
- ■ Blow-wives 104
- ■ Broomrapes 306-307
- ■ Buckwheats 353-354, 357-359
- ■ Cactus 187, 189, 192-193
- ■ Calochortus 427-430

- ■ Campions 206-208
- ■ Canchalagua 262
- ■ Castor Bean 233
- ■ Catchflies 206-208
- ■ Checkerbloom 286
- ■ Chinese Houses 322-323
- ■ Cholla 187
- ■ Clarkia 295-297
- ■ Clovers 256-257
- ■ Copperleaf 233
- ■ Cranesbills 265
- ■ Currant 267
- ■ Desert-thorns 396
- ■ Dudleya 222-223
- ■ False Indigo 236-237
- ■ False Pimpernel 324
- ■ Fiesta Flower 159
- ■ Figwort 394
- ■ Filaree 263-264
- ■ Frog-fruit 407
- ■ Fuchsias 299
- ■ Geraniums 265
- ■ Gilia 339-340
- ■ Gooseberries 268-269
- ■ Granny's Hairnet 360
- ■ Groundsmoke 300
- ■ Hedge-nettle 280-281
- ■ Henbit 272
- ■ Iceplants 41
- ■ Keckiella 325-326
- ■ Larkspur 364
- ■ Leather Root 240
- ■ Lilies 428-430
- ■ Linanthus 341-343
- ■ Live-Forevers 221-223
- ■ Locoweed 238
- ■ Loosestrife 284
- ■ Lotus 244, 247
- ■ Lupines 248, 250
- ■ Madrone 227, 230
- ■ Mallows 285-286
- ■ Manzanita 228-230, 232
- ■ Marsh-fleabane 95
- ■ Meadow-rue 365
- ■ Milkvetch 238
- ■ Milkweeds 57, 59
- ■ Milkwort 346
- ■ Miner's Lettuce 288
- ■ Mints 272-273
- ■ Monardella 273
- ■ Monkeyflowers 315-318
- ■ Morning-glories 214-215
- ■ Mustang Mint 273
- ■ Mustards 173
- ■ Nama 157
- ■ Oceanspray 382
- ■ Onions 421-423
- ■ Orchid 435
- ■ Owl's-clover 303
- ■ Paintbrushes 302-303
- ■ Peas 241, 255
- ■ Penstemon 326-328
- ■ Peony 308
- ■ Petunia 396
- ■ Phacelia 161-162, 168
- ■ Phlox 343
- ■ Pinks 206-208

- Pitcher-sage 271
- Plectritis 406
- Poppies 312
- Prickly-pears 189, 192-193
- Pussypaws 287-288
- Radishes 181
- Red Maids 287
- Rose 385
- Sacapellote 65
- Sages 274, 275-277
- Salvia 274, 275-277
- Sand Plant 170
- Sand-spurrey 209
- Sand-verbenas 289-290
- Sanicle 54
- School Bells 444-445
- Sea Rockets 182
- Sea-lavenders 336-337
- Sea-purslane 42
- Shooting Star 361
- Skunkweeds 344
- Snakeroot 54
- Snapdragons 320-321
- Snowberries 203
- Speedwell 329
- Spineflowers 348-351
- Spiraea 382
- Stillingia 234
- Stock 179
- Storksbills 263-264
- Sumacs 44
- Sycamore 333
- Teas 240
- Thistles 80, 83, 85
- Tobacco 397
- Valerian 406
- Vervain 407
- Vetch 259
- Warrior's Plume 305
- Willow-herbs 298
- Wishbone Bush 290
- Woollyheads 360
- Yucca 415

Blues: Blue-Violet, Blue, Indigo, Blue-Green

- Alfalfa 254
- Alkali Chalice 261
- Alkali Heath 260
- Artichoke 86
- Asters 105-107
- Beardtongues 327-328
- Bellflowers 198-199
- Blue-eyes 157-158
- Bluecup 197
- Bluecurls 282
- Brodiaeas 441-443
- Broomrape 305
- California Lilacs 370-373
- Canterbury Bell
- Cardoon 86
- Ceanothus 370-373
- Chia 274
- Chinese Houses 322-323
- False Pimpernel 324
- Fiesta Flowers 159
- Gilias 338-339
- Iris 425
- Larkspurs 364-365
- Linanthus 342
- Lobelia 200
- Locoweed 238
- Lupines 248-253
- Morning-glory 215
- Navarretia 345
- Nightshades 399-400
- Penstemon 327-328
- Periwinkle 57
- Phacelia 161, 163-167, 170
- Pholistoma 159
- Sages 274, 276-277
- Salsify 142
- Salvia 274, 276-277
- Sand Plant 170
- School Bells 444-445
- Sea-lavenders 336-337
- Skullcap 278
- Snapdragons 321
- Speedwell 329
- Thistle 86
- Toadflax 328
- Yerba Santas 155-156
- Verbena 407
- Woolly Stars 338-339

Greens: Green, Greenish, Olive, Yellow-Green

- Ash 291
- Calochortus 428
- Castor Bean 233
- Chocolate Lily 431
- Cholla 187
- Cockleburs 102
- Deer's Ears 261
- Durango Root 226
- Lady's Mantle 380
- Lotus 243
- Maple 391
- Meadow-rue 365
- Mistletoes 410-411
- Mouse-tail 365
- Muilla 445
- Nettles 403
- Orchids 435-437
- Peppergrass 184
- Pickleweeds 210-211
- Pineapple Weeds 94
- Plantains 331-332
- Redberries 376
- Sandmat 205
- Sea-blite 211
- Spineflowers 350
- Sycamore 333
- Tobacco 397
- Tumbleweed 212

Browns: Brown, Deep Rust

- Cattails 446-447
- Chocolate Lily 431
- Cockleburs 102
- Plantains 331-332

There is nothing so beautiful as a plant or animal in its native habitat, a product of ongoing evolution, competing to stay alive and reproduce, yet taking the time and energy to put forth flowers or plumage that dazzles the senses and makes one's heart soar.
– Robert Lee Allen (1959-)

Documenting the rediscovery of Santiago Peak phacelia (*Phacelia keckii*) on Modjeska Peak, June 8, 2008. Santiago Peak is in the background, nearly 1 mile away.

About the Authors...

Fred M. Roberts, Jr., is a consulting botanist, artist, and author. He grew up in Dana Point, where thousands of acres of undeveloped grasslands and hiking opportunities contributed to a strong interest in natural history. While attending Dana Hills High School he became interested in reptiles, amphibians, birds, and plants. At Saddleback College, botany became his primary interest and he put together his first checklist of Orange County plants which he later published and is now in its third edition. In 1982, he received his Bachelor of Science in Geography with an emphasis on Botany from the University of California, Santa Barbara. From 1982 to 1991 he worked as an assistant curator in the Museum of Systematic Biology, University of California, Irvine. From 1991 to 1999 he worked as a botanist for the U.S. Fish and Wildlife Service before starting his own botanical consulting business. Fred is a Research Associate at the Rancho Santa Ana Botanic Garden. He is currently working on several book projects, including *A Flora of Orange County, California*. He lives in Oceanside with his wife, Carol, and their two personable cats.

Robert L. "Bob" "BugBob" Allen is an entomologist, botanist, instructor, and photographer. Raised in San Juan Capistrano, he studied insects from a very early age. He and Fred attended Marco Forster Junior High School in San Juan Capistrano but never met until they entered Dana Hills High School, where both learned native plants in a natural history class. He attended Saddleback College in Mission Viejo, then California Polytechnic State University, San Luis Obispo. In 1982, Cal Poly awarded him a Bachelor of Science degree in Environmental and Systematic Biology. In 2006, he received his Master of Science degree in Environmental Studies from California State University, Fullerton. At CSUF, he taught entomology and biological illustration. He teaches nature photography, entomology, botany, and pollination classes in classroom and in the field. Bob is a Research Associate at the Rancho Santa Ana Botanic Garden and a Research Associate in Entomology at the Natural History Museum of Los Angeles County. He is now working on a book about California's pollinating insects. He lives in Mission Viejo with his wife, Linnell, and son, Charles.

It is not gold or gems he seeks, But other sort of treasure. Sweat and fatigue and rocks and mire
And gnats and ticks and tangled briar, All these he deems a pleasure,

If only he can find some tree, Some shrub or flower or grass or weed
That's rare or new or strange, Or growing somewhat out of range, He has reward indeed.

– Ernest Jesse Palmer (1875-1962)
"The Botanist"

This book is a treasure trove of information on a surprisingly varied range of subjects—part field guide, part encyclopedia, and part love letter to Mother Nature. Beyond identifying the plants, it offers structural characterizations and informative descriptions of the clustering of certain plant species together in families and how the families are related. The concept of "guilds" is wonderfully used to present invertebrates, birds, or mammals that make use of, depend upon, or play roles in dispersing a particular plant. A lifetime of fieldwork is condensed into a volume that is instantly an irreplaceable resource for biologists, conservationists, and naturalists in Orange County and beyond.

—Bruce A. Aird
President, Sea and Sage Chapter
National Audubon Society

This extraordinary book will captivate aspiring and professional botanists alike with its wealth of gorgeous, informative images and compelling descriptions of cismontane Southern California's wild plants and associated fauna. Anyone interested in the flora of California in general will find much of value here, including aspects of natural history that are rarely if ever encountered in plant guides and that speak to the authors' vast field experience and observational skills.

—Bruce G. Baldwin
Professor of Integrative Biology
Curator of the Jepson Herbarium
University of California, Berkeley

This is the definitive Orange County flower guide we have all been waiting for! Not only have Allen and Roberts spent a lifetime studying the region's botany but they have educated the public on the region's natural resources. This book is the culmination of those efforts, and Orange County nature lovers are thrilled!

—Max Borella
Executive Director
Laguna Canyon Foundation

This book has something for all, from outdoor enthusiasts to experienced botanists—amazing illustrations, descriptions, photographs, and fun-filled facts. It meets a long-standing need for a well-illustrated text on the flora of Orange County and surrounding areas. Packed with pearls of botanic wisdom, it is an excellent resource for all who enjoy the outdoors.

—John Gannaway
Parks Division Manager
Orange County Parks

In the past, we in Orange County compiled our wildflower information from materials that were disparate, incomplete, or from other regions. Now we have all the carry-able information we could ask for in one handy book—spectacular photos that aid in identifying species, plants grouped by family, flowers indexed by color, clearly written descriptions, locations, botanical terms simply explained, information on plant communities and habitats and on associated insects and animals, natural history, and the history of plant names. Experienced naturalists and native plant enthusiasts will quickly scan the book and find multiple reasons why they need it. People new to wildflowers will find that its usefulness grows with their interests, offering the answers to their questions every step of the way. Allen and Roberts are two of our region's top field botanists/biologists. Their list of people acknowledged for some type of assistance is essentially a *Who's Who* of Orange County and Southern California nature experts. With so much carefully prepared content, this book will become a well-used reference for novice and expert.

—Brad Jenkins
Past President, Orange County Chapter
California Native Plant Society

This remarkable publication represents a love of nature and a dedication to sharing it. The authors know Orange County and the Santa Ana Mountains region, and they lovingly describe its special places, its plant life, and some of its wildlife. Plant descriptions are provided at various taxonomic levels along with habitat and locality information and natural history notes. A section on watching wildflowers includes practical tips for observing and documenting wildflowers. An unexpected, unadvertised, and most informative feature is the description of guilds of animals, especially insects, that are associated with particular species or groups of plants. A discussion of where to go wildflower watching includes maps, directions, and highlights of what may be observed. In such an urbanized area, it is very useful to know that there are so many special places to see wildflowers. The authors are to be congratulated on an outstanding achievement. I look forward to using this book and highly recommend it.

—David J. Keil
Professor Emeritus of Biological Sciences
California Polytechnic State University
San Luis Obispo

This amazing new field guide is destined to become the new, lofty standard for regional natural history guides in the future. It not only covers plants of Southern California in a detailed and interesting way that will appeal to both professionals and beginners but also provides information on local nature viewing, biogeography, animal associations, and other ecological relationships. It will be a "must have" for anyone interested in natural history in California.

—Jon P. Rebman
Curator of Botany
San Diego Natural History Museum

Wow! *Wildflowers of Orange County and the Santa Ana Mountains* is a gem! The photographs are among the best I've seen. The entries are taxonomically complete, with useful information on diagnostic features for field identification, habitat, range, economic uses, and sometimes even chemistry. This is a "must" field book for the professional and amateur botanists, naturalists, and plant hobbyists of Southern California.

—Michael G. Simpson
Professor of Botany
San Diego State University

Quick Index to Orange County Wildflowers

Major plant groups and family names are in bold type.

Agave 413-418
Alkali Heath 260
Aralia 62
Artemisia 69-72
Ashes 291
Asphodel 424
Asters 105-107
Baby Blue-eyes 157-158
Baccharis 72-76
Barberry 143
Beardtongues 324-328
Beargrass 438-439
Bedstraws 387-389
Bellflower 197-201
Blackberries 386
Bladderpod 213
Bluecurls 282
Borage 144-170
Brodiaea 440-445
Broomrape 301-307
Buckthorn 367
Buckwheat 347-360
Buckwheats 352-360
Butcher's-broom 438
Buttercups 366
Cactus 186-196
California Lilacs 367-373
Calochortus 426-430
Carpet-weed 41
Carrot 46-56
Cat's-eyes 146-148
Cattail 446-448
Ceanothus 367-373
Chamise 378-379
Chinese Houses 322-323
Cinquefoils 383
Clarkia 295-297
Cleome 213
Clovers 256-257
Coffeeberries 375
Combseeds 152-153
Coyote Melon 224
Crowfoot 362-366
Cryptantha 146-148
Cucumber 224-226
Cudweeds 96-100
Currant 266-269
Daisies 127
Dandelions 131-133
Datisca 226
Death Camas 434

Dodder 217-219
Dogbane 57-61
Dudleya 221-223
Durango Root 226
Elderberry 39-40
Eriogonum 347-360
Eudicots 39-412
Evening-primrose 292-300
False-hellebore 434
Figwort 394
Filarees 262-264
Flax 283
Four O'Clock 289-290
Frankenia 260
Fuchsia 299
Gentian 261-262
Geranium 262-265
Ginseng 62
Goldenbushes: *Discoid Heads* 89-91
Goldenbushes: *Radiate Heads* 120-121
Goldenrods 130
Goldenstar 440
Goldfields 125-126
Gooseberry 266-269
Goosefoot 210-212
Grape 412
Heath 227-232
Hedge-nettles 279-281
Honeysuckle 202-204
Horkelia 383
Iceplants 41-42
Iris 425
Jimsonweeds 395
Keckiella 324-326
Larkspurs 363-365
Laurel 37
Leadwort 336
Lily 426-433
Live-forevers 221-223
Lizard's-tail 38
Locoweeds 238-239
Loosestrife 284
Lopseed 314-319
Lotus 242-247
Lupines 248-253
Madder 387-389
Madrone 227
Magnoliids 37-38
Mallow 285-286
Manzanitas 228-232
Maple 391

Milkweeds 57-61
Milkwort 346
Mint 271-282
Mistletoe 410-412
Monardella 273
Monkeyflowers 314-319
Monocots 413-448
Montia 287-288
Morning-glory 214-216
Muilla 445
Muskroot 39-40
Mustard 171-185
Nettle 403-405
Nightshade 399-400
Olive 291
Onion 420-423
Orchid 435-437
Paintbrushes 301-303
Pea 236-259
Peas 241, 255
Pennyworts 62
Penstemon 324, 326-328
Peony 308
Phacelia 160-170
Phlox 338-345
Pink 205-209
Plagiobothrys 149-151
Plantain 320-332
Plantains 330-332
Poison Oak 45
Popcorn Flowers 149-151
Poppy 309-313
Primrose 361
Pygmy-stonecrops 220
Redberries 376
Ribbonshanks 377
Rock-rose 212
Rose 377-386
Rose 385
Rue 390
Rush-rose 212
Sagebrushes 69-72
Sages 271, 274-277
Salvia 274-277
Sand Plant 170
Sand-verbenas 289-290
Saxifrage 392-393
School Bells 444-445
Sea-lavenders 336-337

Shooting Star 361
Silk-tassel 260
Silverpuffs 136-138
Snapdragons 320-321
Snowberries 203-204
Soap Plants 413-414
Soapberry 391
Speedwells 329
Spineflowers 347-351
Spiderflower 213
Spurge 233-235
St. John's Wort 270
Stick-leaf 283
Stonecrop 220-223
Storax 401
Sumac 42-45
Summer Holly 232
Sun Cups 292-294
Sunflower 63-142
Sunflowers 116-120
Sunflowers: *Discoid, Radiant, Disciform Heads* 65-102
Sunflowers: *Liguliflorous Heads* 131-142
Sunflowers: *Radiate Heads* 103-130
Sycamore 333
Theophrasta 402
Tobaccos 397
Toyon 381
Valerian 406
Verbenas 407
Vervain 407
Violet 408-409
Virgin's-bowers 362-363
Water-pimpernel 402
Water-plantain 419
Waterleafs 154-170
Wishbone Bush 290
Wood-sorrel 308
Yerba Mansa 38
Yucca 415-418